INTRODUCTION TO THE

Theory of Fuzzy Subsets

VOLUME I

Fundamental Theoretical Elements

INTRODUCTION TO THE

Theory of Fuzzy Subsets

VOLUME I

Fundamental Theoretical Elements

A. Kaufmann

Universite de Louvain
Louvain, Belgium

FOREWORD BY

L. A. Zadeh

TRANSLATED BY

D. L. Swanson

Academic Press *New York San Francisco London 1975*

A Subsidiary of Harcourt Brace Jovanovich, Publishers

Original edition, Introduction a la Theorie des Sous–Ensembles Flous, copyright Masson et Cie Editeurs, Paris, 1973.

ACADEMIC PRESS, INC.
111 Fifth Avenue, New York, New York 10003

United Kingdom Edition published by
ACADEMIC PRESS, INC. (LONDON) LTD.
24/28 Oval Road, London NW1

Library of Congress Cataloging in Publication Data

Kaufmann, Arnold.
Introduction to the theory of fuzzy subsets.

Translation of Introduction à la théorie des sous-ensembles flous à l'usage des ingenieurs.
Includes index.
Bibliography: v. 1, p.
CONTENTS: v. 1. Fundamental theoretical elements.
1. Set theory. I. Title.
QA248.K3813 511'.3 75-19296
ISBN 0–12–402301–0

PRINTED IN THE UNITED STATES OF AMERICA

Contents

FOREWORD ix
PREFACE xi
LIST OF PRINCIPAL SYMBOLS xv

CHAPTER I FUNDAMENTAL NOTIONS

1. Introduction 1
2. Review of the notion of membership 1
3. The concept of a fuzzy subset 4
4. Dominance relations 7
5. Simple operations on fuzzy subsets 8
6. Set of fuzzy subsets for **E** and **M** finite 30
7. Properties of the set of fuzzy subsets 33
8. Product and algebraic sum of two fuzzy subsets 34
9. Exercises 38

CHAPTER II FUZZY GRAPHS AND FUZZY RELATIONS

10. Introduction 41
11. Fuzzy graphs 41
12. Fuzzy relations 46
13. Composition of fuzzy relations 60
14. Fuzzy subsets induced by a mapping 69
15. Conditioned fuzzy subsets 71
16. Properties of fuzzy binary relations 78
17. Transitive closure of a fuzzy binary relation 87
18. Paths in a finite fuzzy graph 94
19. Fuzzy preorder relations 98
20. Similitude relations 101
21. Similitude subrelations in a fuzzy preorder 104
22. Antisymmetry 105
23. Fuzzy order relations 110
24. Antisymmetric relations without loops. Ordinal relations. Ordinal functions in a fuzzy order relation 118
25. Dissimilitude relations 127
26. Resemblance relations 131
27. Various properties of similitude and resemblance 142
28. Various properties of fuzzy perfect order relations 160
29. Ordinary membership functions 167
30. Exercises 180

CHAPTER III FUZZY LOGIC

31. Introduction 191
32. Characteristic function of a fuzzy subset. Fuzzy variables 192
33. Polynomial forms 201
34. Analysis of a function of fuzzy variables. Method of Marinos 208
35. Logical structure of a function of fuzzy variables 214
36. Composition of intervals 218
37. Synthesis of a function of fuzzy variables 225
38. Networks of fuzzy elements 235
39. Fuzzy propositions and their functional representations 243
40. The theory of fuzzy subsets and the theory of probability 250
41. The theory of fuzzy subsets and the theory of functions of structure 253
42. Exercises 264

CHAPTER IV THE LAWS OF FUZZY COMPOSITION

43. Introduction 267
44. Review of the notion of a law of composition 267
45. Laws of fuzzy internal composition. Fuzzy groupoids 269
46. Principal properties of fuzzy groupoids 273
47. Fuzzy monoids 278
48. Fuzzy external composition 284
49. Operations on fuzzy numbers 290
50. Exercises 295

CHAPTER V GENERALIZATION OF THE NOTION OF FUZZY SUBSET

51. Introduction 297
52. Operations on ordinary sets 298
53. Fundamental properties of the set of mappings from one set to another 301
54. Review of several fundamental structures 304
55. Generalization of the notion of fuzzy subsets 315
56. Operations on fuzzy subsets when **L** is a lattice 334
57. Review of various notions with a view toward explaining the concept of category 337
58. The concept of a category 358
59. Fuzzy **C**-Morphisms 370
60. Exercises 378

Appendix A General proof procedure for operations concerning maxima and minima 383

Appendix B Decomposition into maximal similitude subrelations 387

Conclusion 401

Bibliography 403

References 411

Index 413

Foreword

The theory of fuzzy subsets is, in effect, a step toward a rapprochement between the precision of classical mathematics and the pervasive imprecision of the real world —a rapprochement born of the incessant human quest for a better understanding of mental processes and cognition.

At present, we are unable to design machines that can compete with humans in the performance of such tasks as recognition of speech, translation of languages, comprehension of meaning, abstraction and generalization, decisionmaking under uncertainty and, above all, summarization of information.

In large measure, our inability to design such machines stems from a fundamental difference between human intelligence, on the one hand, and machine intelligence, on the other. The difference in question lies in the ability of the human brain —an ability which present-day digital computers do not possess— to think and reason in imprecise, nonquantitative, fuzzy terms. It is this ability that makes it possible for humans to decipher sloppy handwriting, understand distorted speech, and focus on that information which is relevant to a decision. And it is the lack of this ability that makes even the most sophisticated large scale computers incapable of communicating with humans in natural —rather than artificially constructed— languages.

The fundamental concept in mathematics is that of a set – a collection of objects. We have been slow in coming to the realization that much perhaps most, of human cognition and interaction with the outside world involves constructs which are not sets in the classical sense, but rather "fuzzy sets" (or subsets), that is, classes with unsharp boundaries in which the transition from membership to nonmembership is gradual rather than abrupt. Indeed, it may be argued that much of the logic of human reasoning is not the classical two-valued or even multivalued logic but a logic with fuzzy truths, fuzzy connectives, and fuzzy rules of inference.

In our quest for precision, we have attempted to fit the real world to mathematical models that make no provision for fuzziness. We have tried to describe the laws governing the behavior of humans, both singly and in groups, in mathematical terms similar to those employed in the analysis of inanimate systems. This, in my view, has been and will continue to be a misdirected effort, comparable to our long-forgotten searches for the perpetuum mobile and the philosopher's stone.

What we need is a new point of view, a new body of concepts and techniques in which fuzziness is accepted as an all pervasive reality of human existence. Clearly, we need an understanding of how to deal with fuzzy sets within the framework of classical mathematics. More important, we have to develop novel methods of treating fuzziness in a systematic—but not necessarily quantitative— manner. Such methods could open many new frontiers in psychology, sociology, political science, philosophy, physiology, economics, linguistics, operations research, management science, and other fields, and provide a basis for the design of systems far superior in artificial intelligence to those we can conceive today.

Professor Kaufmann's work is a highly important contribution to the attainment of these objectives. With his characteristic thoroughness and lucidity, he has given us a first systematic exposition of a subject area that, in the years to come is likely to have a significant impact on the orientation of science and engineering. Reflecting his exceptionally broad expertise in a wide variety of areas in applied mathematics, system theory, and engineering, Professor Kaufmann's treatment of fuzzy sets casts much light on their theory and enables him to extend it in many new and important directions.

The present volume is concerned, in the main, with the mathematical aspects of the theory of fuzzy sets. The volume to follow will cover the applications to problem areas centering on information and decision processes in humans and machines. In such applications, the concept of a fuzzy algorithm plays a central role and the existence of fuzzy feedback makes it possible to deal with fuzzy sets in a qualitative manner closely approximating the nonquantitative thought processes of the human brain.

Professor Kaufmann's treatise is clearly a very important accomplishment. It may well exert a significant influence on scientific thinking in the years ahead and stimulate much further research on the theory of fuzzy sets and their applications in various fields of science and engineering.

L.A. ZADEH
Berkeley, California

Preface

In an epoch in which the elite make science their guide while the masses return to magic, may one say that science is objective, without weaknesses for inexactitude? One is, in many scientific disciplines, more careful to make distinctions. Certainly, objective knowledge must be the purpose of human evolution, but a permanent modesty must be present in the researcher and in the engineer; we know nothing of reality other than through our models, our representations, our more or less true laws, our acceptable approximations in the state of our knowledge. And the model of something for one is not exactly the same model of this thing for another; the formula may remain the same, but the interpretation may be different. The universe is perceived with the aid of models that are indeed perfecting themselves through embodying one in another, at least until some revolution in ideas appears, no longer permitting a correct embodyment.

Our models are fuzzy; our thoughts formed from more or less independent models are fuzzy; in such a manner are we different from a computer! A computer is a logical sequential machine that may not have, by the nature of its definition, any theoretical error; it is a nonfuzzy machine by definition. But man possesses, in addition to the faculty of reckoning and thinking logically, that of taking things into account globally or in parallel, as all living beings. This global or parallel reasoning, as opposed to logical reckoning, is fuzzy and must be fuzzy. A living being, having the possibility of initiative, perceives and treats a piece of more or less fuzzy information and adapts itself. When the living being has almost no more initiative, when its entropy function is almost zero, the fuzziness may then disappear; the being is programmed. A cell, in biology, functions as a small computer commanding a small factory (the word *small* applies here to size but not complexity); there is almost no entropy in this system. A man, on the contrary, has an immense entropy function; he may choose, decide, evolve, err, set right, begin again, understand a little, and build his knowledge in the adventure of science, with a formal program.

How does one join conceptually global reasoning and logical reasoning; how does one associate that which is physically true and that which is an interpretation of human thought in order to be close to both at the same time? How does one introduce fuzziness into mathematics, since it is finally in this most clear form that it will be necessary to express this at first strange association.

For a mathematician, what does the word *fuzzy* signify (or synonymous words)? This will mean that an element is a member of a subset only in an uncertain fashion; while, on the other hand, in mathematics we understand that there are only two acceptable situations for an element: being a member of or not being a member of a subset. Any formal logic, boolean logic, rests on this base: membership or nonmembership in a subset of a reference set.

The merit of L. A. Zadeh has been to attempt to leave this impasse by introducing the notion of weighted membership. An element may then belong more or less to a subset, and, from there, introducing a fundamental concept, that of a fuzzy subset.

In an obviously different presentation, constructed on n-ary logic, Post (1921), Lukasiewicz (1937), and Moisil (1940) have given general theories in which the theory of fuzzy subsets may be placed in a number of its aspects. It may be considered that the two schools have converged as expressed by the American author Zadeh and the Romanian author Moisil.

Speaking at various conferences on the theory of fuzzy subsets, I have frequently met the same proposition: what may be done with this theory may just as well be done without it. But this is true for any theory. I recall meeting such a proposition twenty years ago when I wrote with one of my friends one of the first works on matrix calculus; it was criticized as another book on tensor calculus. This criticism has been offered for graph theory, for applied modern mathematics, and for many other things. Perhaps those who say this have not understood to what uses mathematics is put, more or less independently of the fact that it constitutes a science in a pure state, that is, the science of sciences. Mathematics is our means of access to knowledge through logical models; their providing practical application is the explanation; they give us structure, and numbers are only totally ordered structures, a small particular, very convenient case of an infinite set of structures.

The theory of fuzzy subsets at least allows the structuring of all that which is separated by frontiers only a little precise, as thought, language, and perception among men. The humanistic sciences are replete with all sorts of abstract or concrete forms; and the sciences said to be exact also may be concerned with situations where uncertainty is in the nature of things. This relatively new theory is useful and important; scientists of all disciplines ought to be interested, but also the literati and artists, those who construct truth and beauty with fuzziness, with an indispensable fuzziness, which allows all nuances and which is the stimulant of the imagination, with this fuzziness that one may call mental entropy.

I have done my best to make this work accessible to the usual readers of my books: engineers, designers, professors, students, executives, decision makers. This theory is not always easy, it falls far short of it. The pedagogical effort will have its reward if this work compiling various personal works excites the reader to publish on the subject of this work applications and new ideas. I have asked Professor Zadeh to offer a foreword for this book; his support and his encouragement have been extremely important for the mathematician-engineer and author who has carried out this modest work.

As all my books, this one is didactic; and I have introduced examples throughout; this has lengthened the text, the number of pages, and as a consequence, increased the price of the work. This didactic aspect seems to me to be indispensable for the readership imagined for this book: the tens of thousands of engineers in the world. These people have a good background and good training for understanding and using mathematics; they appreciate that it facilitates their tasks. Of course, this makes the text, in its setting, somewhat heavy; and any professional mathematician who reads my book will often find it drawn and long. The mathematician likes explicit aesthetics; but the engineer, if he is sensitive to elegance, is not concerned only with the mathematics, he has other fish to fry. My experience as a professor, in more than forty countries where international organizations and the university have allowed me to go, and my experience as an author encourage me to continue on this didactic path where the objective is to attain in others knowledge of his own research efforts. Thus various flaws that are often consequences of these intentions ought to be pardonned.

Of course, the notion of fuzziness may be considered from various points of view, in the specification of variables and of their values, in configurations, and more generally at the conceptual level. The present theory is limited to variables and configurations, but one may already envisage how to attack the conceptual aspect, at the cost of some difficulties as one may guess. The humane sciences, and biology in particular, essentially pose fundamental questions concerning conceptual fuzziness. How does one describe mechanisms where almost everything depends on almost everything else, where the smallest element has its role, where all is based on information, messages, and their treatment? The theory of fuzzy subsets is a limited method for seeing these things; the future will reveal whether it is a rather large and interesting base or whether it is only a provisionary way of treating uncertainty or an aspect of uncertainty. This theory is surely only a subtheory of something much larger, as we shall see. Perhaps it would be better to attend to and propose more ambititious theories. We begin by attacking an elementary aspect of fuzziness, that concerning variables and configurations; this is a step forward. So many disciplines, economists, linguists, information theorists, biologists, psychologists, sociologists, etc., are involved with conceptual fuzziness that we will advance boldly. A multitude of work is to be found here and in the worlds between these disciplines. I am optimistic on this subject, but as all those who take a deep interest in these researches, this optimism is tempered with an attitude of prudence and patience.

An important point deserves to be discussed with emphasis in this preface. Will it be possible to treat these fuzzy problems with computers that are sequential machines using binary logic? The answer, as we shall see, is affirmative given a new hardware, a new software, or what is the same a new mixture of kinds of firmware. Thus this will change several habits among analysts and programmers, while very probably and soon the designers of technologies will include fuzzy logic circuits (how poorly the two words go together—we shall explain later), giving binary logical elements, semi *n*-ary logics or degree logics in new classes of computers. It is more and more evident and necessary that the man–machine dialogue must be put within reach of all, and no longer only by passing through formal languages and very simplified programming. And the language of man as his thoughts is fuzzy or/and logical; between him and the computer, intermediate stages, finite automata of all kinds, must permit better communication. Thus, we may predict without great risk of error that technologies will evolve toward an incorporation of these concepts closer to the usual thought of man, this while waiting for the day when machines other than computers will treat global information, treating it in parallel intrinsically without necessarily passing through a sequential treatment. Then these machines, which may be called combiners or parallel processors, will allow treating fuzziness through fuzziness and no longer treating fuzziness with the same microminature binary logic. Utopia today, reality in the next ten years. Artificial intelligence such as one has been able to conceive of until now with the aid of autoadaptive programs on computers will more closely approximate the intelligence of men (but is this then human intelligence?).

To comment on another point of detail, why do we use the term fuzzy subset instead of fuzzy set? This is because a fuzzy set will never be a concept proper to the present theory; the reference set will always be an ordinary set, that is, such as one defined intuitively in modern mathematics, that is again, a collection of well-specified and distinct objects. It is the subsets that will be fuzzy, as we shall see. I am not a stubborn and rigorous bourbakiste, those who have read one of my books will know better, but there are some definitions that must not be fuzzy. There are times when it is convenient to have a touch of the Bourbaki!

Volume I constitutes the first part of this work and contains the theoretical bases. Volume II contains applications with principal titles: fuzzy languages, fuzzy systems, fuzzy automata, fuzzy algorithms, machines and control, decision problems in a fuzzy universe, recognition of forms, problems of classification and selection, documentary research, etc. Volume II will be presented in the same didactic spirit with very numerous examples and in addition a detailed review of basic knowledge on each subject treated. While waiting for the second volume, the reader is referred to the bibliography where the cited articles specify well the first applications under consideration.

Several persons have given me their cooperation in reading the manuscript and offering constructive criticisms: Mme. Monique Peteau, Docteur en Sciences and MM. Michel Cools and Thierry Dubois, Ingenieurs de Recherche au Centre I.M.A.G.O. de l'Universite de Louvain; M. Etienne Pichat, Professeur a l'Institut d'Informatique d'Entreprise, Conservatoire National des Arts et Metiers; M. Jules Kun, Conseiller Scientifique a la Cie. Honeywell Bull; M. Combe, Ingenieur a la Cie. Honeywell Bull; M. Arnaud Henri-Labordere, Ingenieur a la Societe d'Economie et de Mathematiques Appliquees.

My son Alain, as usual, has borne the burden of corrections. He is a student of medicine who thinks, as do others, that medicine is at the same time an art and a science, and that mathematics and thus information are precise instruments that it will also be necessary for him to be able to use tomorrow.

It is intuitive but pretentious to affirm that the human mind is not a simple mechanism, reducible to more or less complicated programs; Kurt Gödel has demonstrated this formally; then we reconcile our desires for rigor and for imagination. Since it is impossible to construct a program of all programs, today or tomorrow, we remain fuzzy and creative.

For this second edition, I have rectified several errors and omissions that have been indicated to me by several readers. In several places I have modified the terminology.

A little before publication of the second edition of Volume I, Volumes II and III concerning the applications have appeared, several months in advance of the predicted dates. This very great effort of reading, compilation, of imagination, and of publication has already found its reward in the reception given to Volume I by my readers and friends. In a soil if fertile, ideas germinate, and indeed all the humanistic sciences will be better modeled with this mathematics of nuances and of subjectivity.

A. Kaufmann

List of Principal Symbols

$\mathbf{A}$	ordinary set or subset
$\underset{\sim}{\mathbf{A}}$	fuzzy subset
$\in$	symbol of membership
$\notin$	symbol of nonmembership
$\lvert\mathbf{A}\rvert$, card $\mathbf{A}$	number of elements or cardinality of $\mathbf{A}$
$\subset$	is a subset of (inclusion)
$\subset\subset$	is a strict subset of (one also says, a true subset of)
$\not\subset$	noninclusion
$\cup$	union
$\cap$	intersection
$\overline{\mathbf{A}}$	complement of $\mathbf{A}$
$\overline{\underset{\sim}{\mathbf{A}}}$	pseudo-complement of $\underset{\sim}{\mathbf{A}}$
$\mathscr{P}(\mathbf{E})$	set of ordinary subsets of $\mathbf{E}$, the power set of $\mathbf{E}$
ϕ	empty subset
$\mathbf{L}^{\mathbf{E}}$	set of fuzzy subsets of $\mathbf{E}$ when the membership function takes its values in $\mathbf{L}$. In certain cases this set is also denoted $\underset{\sim}{\mathscr{P}}(\mathbf{E})$
$\mathbf{E}_1 \times \mathbf{E}_2$	cartesian product or product of $\mathbf{E}_1$ and $\mathbf{E}_2$
$\Rightarrow$	metaimplication (one also says, usually but improperly, implication)
$\Leftrightarrow$	logical equivalence
$\exists x$	existential quantifier (there exists an x)
$\exists ! x$	unique existential quantifier (there exists one and only one x)
$\forall x$	universal quantifier (for all x)
$\mathbf{E}_1 \overset{\Gamma}{\rightsquigarrow} \mathbf{E}_2$	a mapping Γ of $\mathbf{E}_1$ into $\mathbf{E}_2$
$\mathbf{E}_2 \overset{\Gamma^{-1}}{\rightsquigarrow} \mathbf{E}_1$	the mapping Γ^{-1} of $\mathbf{E}_2$ into $\mathbf{E}_1$ (inverse mapping of Γ)
$\Gamma_2 \circ \Gamma_1$	composition of the two mappings Γ_1 and Γ_2
iff	if and only if
$\oplus$	disjunctive sum
$\mathrm{X} \leqslant \mathrm{Y}$	order relation
$\mathrm{X} < \mathrm{Y}$	strict order relation
$\mathbf{N}$	set of natural numbers, $\mathbf{N} = \{0, 1, 2, 3, \ldots\}$
$\mathbf{N}_0$	the set $\mathbf{N}$ but excluding 0
$\mathbf{Z}$	set of integers, $\mathbf{Z} = \{0, +1, -1, +2, -2, +3, -3, \ldots\}$
$\mathbf{Z}_0$	the set $\mathbf{Z}$ excluding 0
$\mathbf{R}$	set of real numbers
$\mathbf{R}_0$	the set $\mathbf{R}$ excluding 0
$\mathbf{R}^+$	set of nonnegative real numbers

$\mathbf{R}_0^+$	set of positive real numbers
$\mathbf{R}^n$	the set product $\mathbf{R} \times \mathbf{R} \times \cdots \times \mathbf{R}$ (n factors) or the real space with dimension n
$]a, b[$	interval of **R** "open on the left and on the right," thus, $\{x \mid a < x < b\}$
$]a, b]$	interval of **R** "open on the left and closed on the right," thus, $\{x \mid a < x \leqslant b\}$
$[a, b[$	interval of **R** "closed on the left and open on the right," thus, $\{x \mid a \leqslant x < b\}$
$[a, b]$	interval of **R** "closed on the left and on the right," thus, $\{x \mid a \leqslant x \leqslant b\}$; one also says *segment*
$\mu_{\underset{\sim}{\mathbf{A}}}(x)$	membership function for the element x with respect to the fuzzy subset $\underset{\sim}{\mathbf{A}}$
$d(\underset{\sim}{\mathbf{A}}, \underset{\sim}{\mathbf{B}})$	generalized Hamming distance between two fuzzy subsets $\underset{\sim}{\mathbf{A}}$ and $\underset{\sim}{\mathbf{B}}$
$\delta(\underset{\sim}{\mathbf{A}}, \underset{\sim}{\mathbf{B}})$	generalized relative Hamming distance between two fuzzy subsets $\underset{\sim}{\mathbf{A}}$ and $\underset{\sim}{\mathbf{B}}$
$\underset{\approx}{\mathbf{A}}$	the ordinary subset nearest to the fuzzy subset $\underset{\sim}{\mathbf{A}}$
$\nu(\underset{\sim}{\mathbf{A}})$	index of fuzziness of the fuzzy subset $\underset{\sim}{\mathbf{A}}$
$\mathbf{A}_\alpha$	ordinary subset of level α of a fuzzy subset $\underset{\sim}{\mathbf{A}}$
max(X, Y) or X ∨ Y	maximum of X and Y
min(X, Y) or X ∧ Y	minimum of X and Y
sup(X, Y) or X ∇ Y	limit superior of X and Y
inf(X, Y) or X Δ Y	limit inferior of X and Y
$\hat{+}$	symbol for an algebraic sum, $a \hat{+} b = a + b - a \,.\, b$
$\underset{\sim}{\mathbf{G}} \subset \mathbf{E}_1 \times \mathbf{E}_2$	fuzzy graph
$\underset{\sim}{\mathcal{R}}_2 \circ \underset{\sim}{\mathcal{R}}_1$	composition of two fuzzy relations
$\underset{\approx}{\mathcal{R}}$	ordinary relation nearest to a fuzzy relation $\underset{\sim}{\mathcal{R}}$
$\mu_{\underset{\sim}{\mathbf{B}}}(y \parallel x)$	membership function of a conditioned fuzzy subset
$\mu_{\underset{\sim}{\mathcal{R}}}(x, y)$	membership function of the ordered pair (x, y) for the fuzzy relation $x \underset{\sim}{\mathcal{R}} y$
$\underset{\sim}{\mathcal{R}}^n$	represents $\mathcal{R} \circ \mathcal{R} \circ \cdots \circ \mathcal{R}$ (n times)
$\overline{\underset{\sim}{\mathcal{R}}}$	complementary relation of $\underset{\sim}{\mathcal{R}}$ such that $\mu_{\overline{\underset{\sim}{\mathcal{R}}}}(x, y) = 1 - \mu_{\underset{\sim}{\mathcal{R}}}(x, y)$
$\hat{\underset{\sim}{\mathcal{R}}}$	transitive closure of $\underset{\sim}{\mathcal{R}}$
$l(x_{i_1}, x_{i_2}, \ldots, x_{i_r})$	value of a path from x_{i_1} to x_{i_r}
$l^*(x_i, x_j)$	strongest path from x_i to x_j
$\underset{\sim}{a}, \underset{\sim}{b}, \ldots$	fuzzy variables
$\underset{\sim}{f}(\underset{\sim}{a}, \underset{\sim}{b}, \ldots)$	function of fuzzy variables
$(\mathbf{E}, *)$	groupoid
$\mathscr{D}(\mathrm{X}_i, \mathrm{X}_j)$	distance between two elements X_i and X_j
MOR(X, Y)	set of morphisms of a category
$\underset{\sim}{\Gamma}$	fuzzy mapping

CHAPTER I

FUNDAMENTAL NOTIONS

1. INTRODUCTION

In this first chapter we review the principal definitions and concepts of the theory of ordinary sets, that is, those that are at the foundation of present-day mathematics; but these definitions and concepts will be reexamined and extended to notions that pertain to fuzzy subsets.

We shall progress rather slowly so that the reader who is not a mathematician but rather a user of mathematics will be able to follow without difficulty.

The examples will allow the reader to verify, step by step, whether the new notions have been well understood. But all that is presented in this first chapter is very simple; the difficulties will appear later.

The theory of ordinary sets is a particular case of the theory of fuzzy subsets (we shall see presently why it is necessary to say *fuzzy subset* and not *fuzzy set*—the reference set will not be fuzzy). We have here a new and very useful extension; but, as we shall note several times, what may be described or explained with the theory of fuzzy subsets may also be considered without this theory, using other concepts. One may always replace one mathematical concept with another. But will it be so clear or generative of properties that are easier to discover and prove, or to use?

2. REVIEW OF THE NOTION OF MEMBERSHIP

Let $\mathbf{E}$ be a set and $\mathbf{A}$ a subset of $\mathbf{E}$:

$$\mathbf{A} \subset \mathbf{E}. \tag{2.1}$$

One usually indicates that an element x of $\mathbf{E}$ is a member of $\mathbf{A}$ using the symbol $\in$:

$$x \in \mathbf{A}. \tag{2.2}$$

In order to indicate this membership one may also use another concept, a characteristic function $\mu_{\mathbf{A}}(x)$, whose value indicates (yes or no) whether x is a member of $\mathbf{A}$:

$$\begin{aligned} \mu_{\mathbf{A}}(x) &= 1 \quad \text{if} \quad x \in \mathbf{A} \\ &= 0 \quad \text{if} \quad x \notin \mathbf{A}. \end{aligned} \tag{2.3}$$

Example. Consider a finite set with five elements:

(2.4) $$\mathbf{E} = \{x_1, x_2, x_3, x_4, x_5\}$$

and let

(2.5) $$\mathbf{A} = \{x_2, x_3, x_5\}.$$

And one writes

(2.6) $$\mu_{\mathbf{A}}(x_1) = 0, \mu_{\mathbf{A}}(x_2) = 1, \mu_{\mathbf{A}}(x_3) = 1, \mu_{\mathbf{A}}(x_4) = 0, \mu_{\mathbf{A}}(x_5) = 1.$$

This allows us to represent **A** by accompanying the elements of **E** with their characteristic-function values:

(2.7) $$\mathbf{A} = \{(x_1, 0), (x_2, 1), (x_3, 1), (x_4, 0), (x_5, 1)\}.$$

Recall the well-known properties of a boolean binary algebra:

Let $\overline{\mathbf{A}}$ be the complement of **A** with respect to **E**:

(2.8) $$\mathbf{A} \cap \overline{\mathbf{A}} = \phi,$$

(2.9) $$\mathbf{A} \cup \overline{\mathbf{A}} = \mathbf{E}.$$

(2.10) If $x \in \mathbf{A}$, $x \notin \overline{\mathbf{A}}$, and one writes

(2.11) $$\mu_{\mathbf{A}}(x) = 1 \text{ and } \mu_{\overline{\mathbf{A}}}(x) = 0.$$

Considering the example in (2.6) and (2.7), one sees:

(2.12) $$\mu_{\overline{\mathbf{A}}}(x_1) = 1, \mu_{\overline{\mathbf{A}}}(x_2) = 0, \mu_{\overline{\mathbf{A}}}(x_3) = 0, \mu_{\overline{\mathbf{A}}}(x_4) = 1, \mu_{\overline{\mathbf{A}}}(x_5) = 0,$$

and one writes

(2.13) $$\overline{\mathbf{A}} = \{(x_1, 1), (x_2, 0), (x_3, 0), (x_4, 1), (x_5, 0)\}.$$

Given now two subsets **A** and **B**, one may consider the intersection

(2.14) $$\mathbf{A} \cap \mathbf{B}.$$

One has

(2.15) $$\begin{aligned} \mu_{\mathbf{A}}(x) &= 1 \quad \text{if} \quad x \in \mathbf{A} \\ &= 0 \quad \text{if} \quad x \notin \mathbf{A}, \end{aligned}$$

(2.16) $$\begin{aligned} \mu_{\mathbf{B}}(x) &= 1 \quad \text{if} \quad x \in \mathbf{B} \\ &= 0 \quad \text{if} \quad x \notin \mathbf{B}, \end{aligned}$$

(2.17) $$\begin{aligned} \mu_{\mathbf{A} \cap \mathbf{B}}(x) &= 1 \quad \text{if} \quad x \in \mathbf{A} \cap \mathbf{B} \\ &= 0 \quad \text{if} \quad x \notin \mathbf{A} \cap \mathbf{B}. \end{aligned}$$

This allows us to write

(2.18) $$\mu_{\mathbf{A} \cap \mathbf{B}}(x) = \mu_{\mathbf{A}}(x) \cdot \mu_{\mathbf{B}}(x),$$

where the operation . corresponds to the table in Figure 2.1 and is called the boolean product.

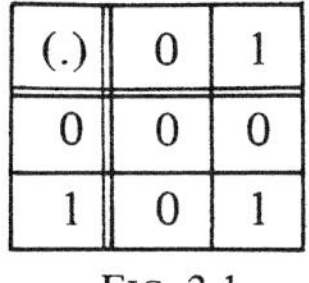

(.)	0	1
0	0	0
1	0	1

FIG. 2.1

In the same fashion for the two subsets **A** and **B**, one defines the union or join:

$$\mu_{\mathbf{A}\cup\mathbf{B}}(x) = 1 \quad \text{if} \quad x \in \mathbf{A} \cup \mathbf{B} \tag{2.19}$$
$$= 0 \quad \text{if} \quad x \notin \mathbf{A} \cup \mathbf{B}\ ;$$

with the property

$$\mu_{\mathbf{A}\cup\mathbf{B}}(x) = \mu_{\mathbf{A}}(x) \dot{+} \mu_{\mathbf{B}}(x), \tag{2.20}$$

where the operation $\dot{+}$, the boolean sum, is defined by the table in Figure 2.2.

$(\dot{+})$	0	1
0	0	1
1	1	1

FIG. 2.2

Example. Consider the reference set (2.4) and the two subsets

$$\mathbf{A} = \{(x_1, 0), (x_2, 1), (x_3, 1), (x_4, 0), (x_5, 1)\}, \tag{2.21}$$

$$\mathbf{B} = \{(x_1, 1), (x_2, 0), (x_3, 1), (x_4, 0), (x_5, 1)\}. \tag{2.22}$$

One sees

$$\mathbf{A} \cap \mathbf{B} = \{(x_1, 0.1), (x_2, 1.0), (x_3, 1.1), (x_4, 0.0), (x_5, 1.1)\} \tag{2.23}$$
$$= \{(x_1, 0), (x_2, 0), (x_3, 1), (x_4, 0), (x_5, 1)\}.$$

$$\mathbf{A} \cup \mathbf{B} = \{(x_1, 0 \dot{+} 1), (x_2, 1 \dot{+} 0), (x_3, 1 \dot{+} 1), (x_4, 0 \dot{+} 0), (x_5, 1 \dot{+} 1)\} \tag{2.24}$$
$$= \{(x_1, 1), (x_2, 1), (x_3, 1), (x_4, 0), (x_5, 1)\}.$$

To continue, emanating from these two subsets one has

$$\overline{\mathbf{A} \cap \mathbf{B}} = \{(x_1, 1), (x_2, 1), (x_3, 0), (x_4, 1), (x_5, 0)\}, \tag{2.25}$$

$$\overline{\mathbf{A} \cup \mathbf{B}} = \{(x_1, 0), (x_2, 0), (x_3, 0), (x_4, 1), (x_5, 0)\}. \tag{2.26}$$

These few exercises constitute only a didactic preamble to an understanding of fuzzy subsets.

3. THE CONCEPT OF A FUZZY SUBSET

We shall begin with an example. Consider the subset **A** of **E** defined by (2.7). The five elements of **E** belong or do not belong to **A**, one or the other. The characteristic function takes only the values 0 or 1.

Imagine now that this characteristic function may take any value whatsoever in the interval [0, 1]. Thus, an element x_i of **E** may not be a member of **A** ($\mu_{\mathbf{A}} = 0$), could be a member of **A** a little ($\mu_{\mathbf{A}}$ near 0), may more or less be a member of **A** ($\mu_{\mathbf{A}}$ neither too near 0 nor too near 1), could be strongly a member of **A** ($\mu_{\mathbf{A}}$ near 1), or finally might be a member of **A** ($\mu_{\mathbf{A}} = 1$). In this manner the notion of membership takes on an interesting extension and leads, as we shall see, to very useful developments.

The mathematical concept is defined by the expression

$$\underset{\sim}{\mathbf{A}} = \{(x_1 | 0{,}2), (x_2 | 0), (x_3 | 0{,}3), (x_4 | 1), (x_5 | 0{,}8)\}, \tag{3.1}$$

where x_i is an element of the reference set **E** and where the number placed after the bar† is the value of the characteristic function for the element; this mathematical concept will be called a *fuzzy subset* of **E** and be denoted

$$\underset{\sim}{\mathbf{A}} \subset \mathbf{E} \qquad \text{or} \qquad \mathbf{A} \underset{\sim}{\subseteq} \mathbf{E} \tag{3.2}$$

One may denote membership in a fuzzy subset by

$$x \underset{0,2}{\in} \underset{\sim}{\mathbf{A}}, \quad y \underset{1}{\in} \underset{\sim}{\mathbf{A}}, \quad z \underset{0}{\in} \underset{\sim}{\mathbf{A}}. \tag{3.3}$$

The symbol $\underset{1}{\in}$ may be taken to be equivalent to $\in$, and $\underset{0}{\in}$ to $\notin$. In order to avoid encumbering the notation, one uses simply $\in$ to indicate membership and $\notin$, nonmembership.

Thus, the fuzzy subset defined by (3.1) contains a little x_1, does not contain x_2, contains a little more x_3, contains x_4 completely, and a large part of x_5. This will allow us to construct a mathematical structure with which one may be able to manipulate concepts that are rather poorly defined but for which membership in a subset is somewhat hierarchical. Thus, one may consider: in the set of men, the fuzzy subset of very tall men; in the set of basic colors, the fuzzy subset of deep green colors; in the set of decisions, a fuzzy subset of good decisions; and so forth. We shall go on to see how to manipulate these concepts that seem particularly well adapted to the imprecision prevalent in the social sciences.

We shall now give a rigorous definition‡ of the concept due to Zadeh [Z1].

†A vertical bar has been used in place of a comma, as in (2.7), in order to avoid confusion. When one is using the American decimal point, a comma may, of course, be used in place of the bar.

‡We have, however, adapted this definition to the terminology and presentation of the present work.

Let $\mathbf{E}$ be a set, denumerable or not, and let x be an element of $\mathbf{E}$. Then a *fuzzy subset* $\underset{\sim}{\mathbf{A}}$ of $\mathbf{E}$ is a set of ordered pairs

$$\{(x \mid \mu_{\underset{\sim}{\mathbf{A}}}(x))\}, \ \forall x \in \mathbf{E} \ ; \tag{3.4}$$

where $\mu_{\underset{\sim}{\mathbf{A}}}(x)$ is the grade or degree of membership of x in $\underset{\sim}{\mathbf{A}}$. Thus, if $\mu_{\underset{\sim}{\mathbf{A}}}(x)$ takes its values in a set $\mathbf{M}$, called the *membership set*, one may say that x takes its values in $\mathbf{M}$ through the function $\mu_{\underset{\sim}{\mathbf{A}}}(x)$. Let us write

$$x \underset{\mu_{\underset{\sim}{\mathbf{A}}}}{\rightsquigarrow} \mathbf{M}. \tag{3.5}$$

This function will likewise be called the *membership function.*

With a view toward considering binary boolean functions as a particular case of these membership functions, in the present work we have replaced the above definition by the following.

Let $\mathbf{E}$ be a set, denumerable or not, and let x be an element of $\mathbf{E}$. Then a *fuzzy subset* $\underset{\sim}{\mathbf{A}}$ of $\mathbf{E}$ is a set of ordered pairs

$$\{(x, \mu_{\underset{\sim}{\mathbf{A}}}(x))\}, \forall x \in \mathbf{E}, \tag{3.6}$$

where $\mu_{\underset{\sim}{\mathbf{A}}}(x)$ is a *membership characteristic function* that takes its values in a totally ordered set† $\mathbf{M}$, and which indicates the *degree* or *level* or membership. $\mathbf{M}$ will be called a *membership set.*

If $\mathbf{M} = \{0, \ 1\}$, the "fuzzy subset" $\underset{\sim}{\mathbf{A}}$ will be understood as a "nonfuzzy subset" or "ordinary subset." The functions $\mu_{\mathbf{A}}(x)$ are then binary boolean functions.

Thus, the notion of a fuzzy subset is linked with the notion of a set and allows one to study, using mathematical structures, imprecise concepts.

We shall consider some examples:

the fuzzy subset of numbers x approximately equal to a given real number n, where $n \in \mathbf{R}$ ($\mathbf{R}$ being the set of reals);

the fuzzy subset of integers very near 0;

let a be a real number and let x be a small positive increment given to a; then the numbers $a + x$ form a fuzzy subset in the set of reals;

let H be an element of a lattice‡; the elements most near to H in the order relation form a fuzzy subset in the set of elements of the lattice.

A set or a subset is denoted in the present work by a boldface letter: $\mathbf{A}, \mathbf{X}, \mathbf{a}, \mathbf{p}, \ldots$. A fuzzy subset will be designated by a boldface letter under which is placed the symbol $\sim$. Thus

$$\underset{\sim}{\mathbf{A}}, \underset{\sim}{\mathbf{X}}, \underset{\sim}{\mathbf{a}}, \underset{\sim}{\mathbf{p}}, \tag{3.7}$$

†Eventually in a set structured considerably more generally, as we will see in Chapter V.

‡The reader who is unfamiliar with the concept of a lattice is referred to [1F, 1K, 3K]. Also a review of this concept will be given in Section 54.

represent fuzzy subsets. When all the subsets turn out to be ordinary, one may, if it is useful, suppress the small supplementary symbol $\sim$.

Membership and nonmembership will be indicated by

(3.8) $$\in \quad \text{and} \quad \notin;$$

fuzzy membership and fuzzy nonmembership will be represented by

(3.9) $$\underset{\sim}{\in} \quad \text{and} \quad \underset{\sim}{\notin}, \text{ if this is necessary.}$$

In certain cases where the totally ordered set **M**, in which $\mu_{\mathbf{A}}(x)$ takes its values, is the doubly closed interval $[0, 1]$, it may be convenient to accompany the symbol $\in$ by a number from $[0, 1]$ placed beneath it. Thus

(3.10)
$$x \underset{1}{\in} \underset{\sim}{\mathbf{A}} \text{ indicates } x \in \underset{\sim}{\mathbf{A}}, \text{ that is, "}x\text{ is a member of } \underset{\sim}{\mathbf{A}}\text{,"}$$
$$x \underset{0}{\in} \underset{\sim}{\mathbf{A}} \text{ indicates } x \notin \underset{\sim}{\mathbf{A}}, \text{ that is, "}x\text{ is not a member of } \underset{\sim}{\mathbf{A}}\text{,"}$$
$$x \underset{0,8}{\in} \underset{\sim}{\mathbf{A}} \text{ indicates that } x \text{ is a member of } \underset{\sim}{\mathbf{A}} \text{ with degree 0,8,}$$

and so forth. The next examples will be very useful.

Example 1. Consider a finite set:

(3.11) $$\mathbf{E} = \{a, b, c, d, e, f\}$$

and the finite ordered set

(3.12) $$\mathbf{M} = \{0, 1/2, 1\}.$$

Then

(3.13) $$\underset{\sim}{\mathbf{A}} = \{(a \mid 0), (b \mid 1), (c \mid \tfrac{1}{2}), (d \mid 0), (e \mid \tfrac{1}{2}), (f \mid 0)\}$$

is a fuzzy subset of **E** and one may write

$$a \underset{0}{\in} \underset{\sim}{\mathbf{A}}, \quad b \underset{1}{\in} \underset{\sim}{\mathbf{A}}, \quad c \underset{\frac{1}{2}}{\in} \underset{\sim}{\mathbf{A}}, \quad d \underset{0}{\in} \underset{\sim}{\mathbf{A}}, \text{ etc.}$$

Example 2. Let **N** be the set of natural numbers:

(3.14) $$\mathbf{N} = \{0, 1, 2, 3, 4, 5, 6, \ldots\}$$

and consider the fuzzy subset $\underset{\sim}{\mathbf{A}}$ of "small" natural numbers:

(3.15) $$\underset{\sim}{\mathbf{A}} = \{(0 \mid 1), (1 \mid 0{,}8), (2 \mid 0{,}6), (3 \mid 0{,}4), (4 \mid 0{,}2), (5 \mid 0), (6 \mid 0), \ldots\}.$$

Here, of course, the functional values $\mu_{\underset{\sim}{\mathbf{A}}}(x)$, where $x = 0, 1, 2, 3, \ldots$, have been given subjectively. One may write

(3.16) $$0 \underset{1}{\in} \underset{\sim}{\mathbf{A}}, \quad 1 \underset{0,8}{\in} \underset{\sim}{\mathbf{A}}, \quad 2 \underset{0,6}{\in} \underset{\sim}{\mathbf{A}}, \quad 3 \underset{0,4}{\in} \underset{\sim}{\mathbf{A}}, \ldots.$$

Example 3. Let **E** be the finite set of the first ten integers:

$$\mathbf{E} = \{0, 1, 2, 3, 4, 5, 6, 7, 8, 9\} \tag{3.17}$$

and consider the fuzzy subset $\underset{\sim}{\mathbf{A}}$ containing the numbers of **E** in the following fashion:

$$\underset{\sim}{\mathbf{A}} = \{(0 \mid 0), (1 \mid 0{,}2), (2 \mid 0{,}3), (3 \mid 0), (4 \mid 1), (5 \mid 1), (6 \mid 0{,}8), (7 \mid 0{,}5), (8 \mid 0), (9 \mid 0)\}, \tag{3.18}$$

where again the $\mu_{\underset{\sim}{\mathbf{A}}}(x)$ are subjective.

One may write

$$0 \underset{0}{\in} \underset{\sim}{\mathbf{A}}, \quad 1 \underset{0,2}{\in} \underset{\sim}{\mathbf{A}}, \quad 2 \underset{0,3}{\in} \underset{\sim}{\mathbf{A}}, \quad 3 \underset{0}{\in} \underset{\sim}{\mathbf{A}}, \ldots . \tag{3.19}$$

The reader will note that this symbol of generalized membership may be employed in the opposite sense. Thus, for (3.13) one may write

$$\underset{\sim}{\mathbf{A}} \underset{0}{\ni} a, \quad \underset{\sim}{\mathbf{A}} \underset{1}{\ni} b, \quad \underset{\sim}{\mathbf{A}} \underset{0,5}{\ni} c. \tag{3.20}$$

and for (3.19),

$$\underset{\sim}{\mathbf{A}} \underset{0}{\ni} 0, \quad \underset{\sim}{\mathbf{A}} \underset{0,2}{\ni} 1, \quad \underset{\sim}{\mathbf{A}} \underset{0,3}{\ni} 2. \tag{3.21}$$

4. DOMINANCE RELATIONS

Recall first the nature of a dominance relation existing between two ordered n-tuples. Consider the two ordered n-tuples.

$$\nu = (k_1, k_2, \ldots, k_n) \tag{4.1}$$

and

$$\nu' = (k'_1, k'_2, \ldots, k'_n), \tag{4.2}$$

in which the k_i and the k'_i, $i = 1, 2, \ldots, n$, belong to the same totally ordered set **K**, in which the relation of order will be represented by the symbol $\geqslant$.

We shall say that ν' *dominates* ν, which is written

$$\nu' \succcurlyeq \nu, \tag{4.3}$$

if and only if

$$k'_1 \geqslant k_1, \quad k'_2 \geqslant k_2, \ldots, \quad k'_n \geqslant k_n. \tag{4.4}$$

The symbols $\geqslant$ and $\succcurlyeq$ for the order relation correspond to a nonstrict order relation. If we then use the symbols $>$ and $\succ$ corresponding to a strict order relation, we say that ν' *strictly dominates* ν. One may then see that

$$\nu' \succ \nu \tag{4.5}$$

if

$$k_1' \geqslant k_1,\ k_2' \geqslant k_2, \ldots,\ k_n' \geqslant k_n, \tag{4.6}$$

with at least one k_i' and one k_i between which there exists a strict relation.

Granting what has been developed here, one may then say that all dominance relations introduce an order relationship (total or partial) between n-tuples such as v and v'.

Example 1. Consider the four-tuples

$$u = (7, 3, 0, 5), \tag{4.7}$$

$$v = (2, 2, 0, 4), \tag{4.8}$$

$$w = (3, 4, 1, 4), \tag{4.9}$$

One sees that

$$u \succcurlyeq v \quad \text{since } 7 > 2,\ 3 > 2,\ 0 = 0,\ 5 > 4. \tag{4.10}$$

Since one of the terms of u, at least, is greater than the corresponding term of v, one may likewise write $u \succ v$. In the same manner one may verify that $w \succcurlyeq v$. But u and w are not comparable. In fact,

$$7 > 3,\quad 3 < 4,\quad 0 < 1,\quad 5 > 4. \tag{4.11}$$

Example 2. Consider the set **P** of points (x_1, x_2) in the plane indicated in Figure 4.1 and defined by $x_1 \geqslant 0$ and $x_2 \geqslant 0$. All points of the shaded domain II, that is, those with $x_1 \geqslant a$, $x_2 \geqslant b$, dominate and in fact strictly dominate all points of domain I, $0 \leqslant x_2 < a, 0 \leqslant x_2 < b$. All points of domain III are not necessarily comparable with all points of I or of II; and the same holds for IV on the one hand with I and II on the other. Finally, each point of III is not comparable to a point of IV and vice versa, evidently, except those points such that $x_1 = a$ or $x_2 = b$.

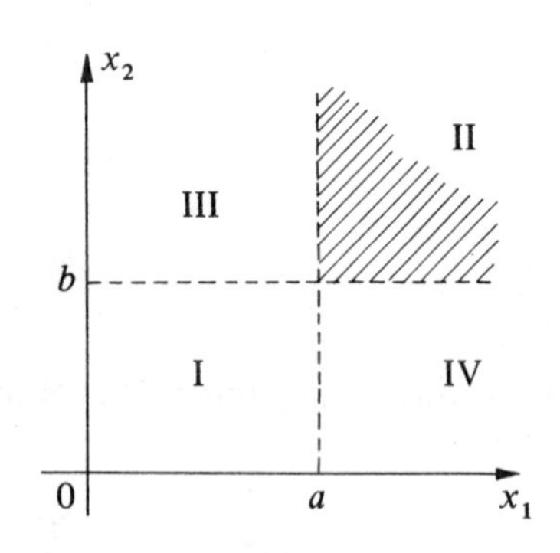

Fig. 4.1

5. SIMPLE OPERATIONS ON FUZZY SUBSETS

Inclusion. Let **E** be a set and **M** its associated membership set, and let $\underset{\sim}{\mathbf{A}}$ and $\underset{\sim}{\mathbf{B}}$ be two fuzzy subsets of **E**; we say that $\underset{\sim}{\mathbf{A}}$ is included in $\underset{\sim}{\mathbf{B}}$ if

$$\forall x \in \mathbf{E}: \quad \mu_{\underset{\sim}{\mathbf{A}}}(x) \leqslant \mu_{\underset{\sim}{\mathbf{B}}}(x). \tag{5.1}$$

This will be denoted by

(5.2) $$\underset{\sim}{A} \subset \underset{\sim}{B}$$

and, if necessary to avoid confusion,

(5.3) $$\underset{\sim}{A} \underset{\sim}{\subset} \underset{\sim}{B} \,,$$

which says very precisely that it is a case of inclusion in the sense of the theory of fuzzy subsets.

Strict inclusion, corresponding to the case where at least one relation in (5.1) is strict, will be denoted

(5.4) $$\underset{\sim}{A} \subset\subset \underset{\sim}{B} \quad \text{or} \quad \underset{\sim}{A} \underset{\sim}{\subset\subset} \underset{\sim}{B}$$

We will consider three examples.

(1) Let

(5.5) $$E = \{x_1, x_2, x_3, x_4\}, \quad M = [0,1].$$

(5.6) $$\underset{\sim}{A} = \{(x_1 | 0{,}4), (x_2 | 0{,}2), (x_3 | 0), (x_4 | 1)\}.$$

(5.7) $$\underset{\sim}{B} = \{(x_1 | 0{,}3), (x_2 | 0), (x_3 | 0), (x_4 | 0)\}.$$

One has

(5.8) $$\underset{\sim}{B} \subset \underset{\sim}{A} \quad \text{since} \quad 0{,}3 < 0{,}4, \quad 0 < 0{,}2, \quad 0 = 0, \quad 0 < 1.$$

(2) Let

(5.9) $$\underset{\sim}{A} \subset E, \quad \underset{\sim}{B} \subset E, \quad M = [0,1].$$

If

(5.10) $$\forall x \in E: \quad \mu^2_{\underset{\sim}{A}}(x) = \mu_{\underset{\sim}{B}}(x),$$

then

(5.11) $$\underset{\sim}{B} \subset \underset{\sim}{A}.$$

(3) Let

$$E = \{x_1, x_2, x_3, x_4, x_5\}, \quad M = [0,1].$$

One can write

(5.12) $$E = \{(x_1, 1), (x_2, 1), (x_3, 1), (x_4, 1), (x_5, 1)\}$$

Thus **E** is also included in itself in the sense of the theory of fuzzy subsets:

(5.13) $$E \underset{\sim}{\subset} E.$$

And this property remains true whatever the set **E** may be.

Equality. Let **E** be a set and **M** its associated membership set, and let $\underset{\sim}{\mathbf{A}}$ and $\underset{\sim}{\mathbf{B}}$ be two fuzzy subsets of **E**; we say that $\underset{\sim}{\mathbf{A}}$ and $\underset{\sim}{\mathbf{B}}$ are equal if and only if

(5.14) $$\forall x \in \mathbf{E}: \quad \mu_{\underset{\sim}{\mathbf{A}}}(x) = \mu_{\underset{\sim}{\mathbf{B}}}(x).$$

This will be denoted by

(5.15) $$\underset{\sim}{\mathbf{A}} = \underset{\sim}{\mathbf{B}}.$$

If at least one x of **E** is such that the equality $\mu_{\underset{\sim}{\mathbf{A}}}(x) = \mu_{\underset{\sim}{\mathbf{B}}}(x)$ is not satisfied, we say that $\underset{\sim}{\mathbf{A}}$ and $\underset{\sim}{\mathbf{B}}$ are not equal, and this will be denoted

(5.16) $$\underset{\sim}{\mathbf{A}} \neq \underset{\sim}{\mathbf{B}}.$$

Complementation. Let **E** be a set and **M** = [0, 1] its associated membership set, and let $\underset{\sim}{\mathbf{A}}$ and $\underset{\sim}{\mathbf{B}}$ be two fuzzy subsets of **E**; we say that $\underset{\sim}{\mathbf{A}}$ and $\underset{\sim}{\mathbf{B}}$ are complementary if

(5.17) $$\forall x \in \mathbf{E}: \mu_{\underset{\sim}{\mathbf{B}}}(x) = 1 - \mu_{\underset{\sim}{\mathbf{A}}}(x).$$

This will be denoted

(5.18) $$\underset{\sim}{\mathbf{B}} = \overline{\underset{\sim}{\mathbf{A}}} \quad \text{or} \quad \overline{\underset{\sim}{\mathbf{A}}} = \underset{\sim}{\mathbf{B}}.$$

One obviously always has

(5.19) $$\overline{(\overline{\underset{\sim}{\mathbf{A}}})} = \underset{\sim}{\mathbf{A}}.$$

We note that here complementation is defined for **M** = [0, 1], but one may extend this to other ordered membership sets **M** using other appropriate definitions.†

We consider an example:

(5.20) $$\mathbf{E} = \{x_1, x_2, x_3, x_4, x_5, x_6\}, \quad \mathbf{M} = [0,1].$$

(5.21) $$\underset{\sim}{\mathbf{A}} = \{(x_1 | 0{,}13), (x_2 | 0{,}61), (x_3 | 0), (x_4 | 0), (x_5 | 1), (x_6 | 0{,}03)\},$$
$$\underset{\sim}{\mathbf{B}} = \{(x_1 | 0{,}87), (x_2 | 0{,}39), (x_3 | 1), (x_4 | 1), (x_5 | 0), (x_6 | 0{,}97)\}.$$

Then certainly

(5.22) $$\overline{\underset{\sim}{\mathbf{A}}} = \underset{\sim}{\mathbf{B}}.$$

Intersection. Let **E** be a set and **M** = [0, 1] its associated membership set, and let $\underset{\sim}{\mathbf{A}}$ and $\underset{\sim}{\mathbf{B}}$ be two fuzzy subsets of **E**; one defines the intersection

(5.23) $$\underset{\sim}{\mathbf{A}} \cap \underset{\sim}{\mathbf{B}},$$

†As will be made precise in Section 56, that expressed by (5.17) must be called *pseudo-complementation.*

as the largest fuzzy subset contained at the same time in $\underset{\sim}{\mathbf{A}}$ and $\underset{\sim}{\mathbf{B}}$. That is,

(5.24) $$\forall x \in \mathbf{E} \ : \ \mu_{\underset{\sim}{\mathbf{A}} \cap \underset{\sim}{\mathbf{B}}}(x) = \mathrm{MIN}(\mu_{\underset{\sim}{\mathbf{A}}}(x), \mu_{\underset{\sim}{\mathbf{B}}}(x)).$$

Example

(5.25) $$\mathbf{E} = \{x_1, x_2, x_3, x_4, x_4, x_5\}, \quad \mathbf{M} = [0,1].$$

(5.26) $$\underset{\sim}{\mathbf{A}} = \{(x_1 | 0,2), (x_2 | 0,7), (x_3 | 1), (x_4 | 0), (x_5 | 0,5)\}.$$

(5.27) $$\underset{\sim}{\mathbf{B}} = \{(x_1 | 0,5), (x_2 | 0,3), (x_3 | 1), (x_4 | 0,1), (x_5 | 0,5)\}.$$

(5.28) $$\underset{\sim}{\mathbf{A}} \cap \underset{\sim}{\mathbf{B}} = \{(x_1 | 0,2), (x_2 | 0,3), (x_3 | 1), (x_4 | 0), (x_5 | 0,5)\}.$$

Referring to the general definition (5.23) and (5.24), one may, moreover, write

(5.29) $$\forall x \in \mathbf{E} \ : \ x \underset{\mu_{\underset{\sim}{\mathbf{A}}}}{\in} \underset{\sim}{\mathbf{A}} \ \text{and} \ x \underset{\mu_{\underset{\sim}{\mathbf{B}}}}{\in} \underset{\sim}{\mathbf{B}} \Rightarrow x \underset{\mu_{\underset{\sim}{\mathbf{A}} \cap \underset{\sim}{\mathbf{B}}}}{\in} \underset{\sim}{\mathbf{A}} \cap \underset{\sim}{\mathbf{B}}.$$

This permits us to introduce a fuzzy *and* to be symbolized $\underset{\sim}{\text{and}}$.

Thus one may say: If $\underset{\sim}{\mathbf{A}}$ is the fuzzy subset of real numbers very near 5 and $\underset{\sim}{\mathbf{B}}$ the fuzzy subset of real numbers very near 10, then $\underset{\sim}{\mathbf{A}} \cap \underset{\sim}{\mathbf{B}}$ is the fuzzy subset of real numbers very near to 5 $\underset{\sim}{\text{and}}$ to 10. The fuzzy conjunction $\underset{\sim}{\text{and}}$ is pronounced as *and*, but except where necessary, one may omit placing $\sim$ beneath it.

Union. Let **E** be a set and **M** = [0, 1] its associated membership set, and let $\underset{\sim}{\mathbf{A}}$ and $\underset{\sim}{\mathbf{B}}$ be two fuzzy subsets of **E**; we define the union

(5.30) $$\underset{\sim}{\mathbf{A}} \cup \underset{\sim}{\mathbf{B}}$$

as the smallest fuzzy subset that contains both $\underset{\sim}{\mathbf{A}}$ and $\underset{\sim}{\mathbf{B}}$. That is,

(5.31) $$\forall x \in \mathbf{E} \ : \ \mu_{\underset{\sim}{\mathbf{A}} \cup \underset{\sim}{\mathbf{B}}}(x) = \mathrm{MAX}(\mu_{\underset{\sim}{\mathbf{A}}}(x), \mu_{\underset{\sim}{\mathbf{B}}}(x)).$$

Recalling the example presented in (5.25)–(5.27), one sees

(5.32) $$\underset{\sim}{\mathbf{A}} \cup \underset{\sim}{\mathbf{B}} = \{(x_1 | 0,5), (x_2 | 0,7), (x_3 | 1), (x_4 | 0,1), (x_5 | 0,5)\}.$$

And recalling the general definitions (5.30), (5.31), one may, moreover, write

(5.33) $$\forall x \in \mathbf{E} \ : \ x \underset{\mu_{\underset{\sim}{\mathbf{A}}}}{\in} \underset{\sim}{\mathbf{A}} \ \text{or/and} \ x \underset{\mu_{\underset{\sim}{\mathbf{B}}}}{\in} \underset{\sim}{\mathbf{B}} \Rightarrow x \underset{\mu_{\underset{\sim}{\mathbf{A}} \cup \underset{\sim}{\mathbf{B}}}}{\in} \underset{\sim}{\mathbf{A}} \cup \underset{\sim}{\mathbf{B}}.$$

This allows us to introduce a fuzzy *or/and*, to be symbolized $\underset{\sim}{\text{or/and}}$.[†] Except where necessary, one omits the symbol $\sim$.

Thus one may say: If $\underset{\sim}{\mathbf{A}}$ is the fuzzy subset of real numbers very near 5 and $\underset{\sim}{\mathbf{B}}$ the fuzzy subset of real numbers very near 10, then $\underset{\sim}{\mathbf{A}} \cup \underset{\sim}{\mathbf{B}}$ is the fuzzy subset of real numbers very near to 5 $\underset{\sim}{\text{or/and}}$ to 10. The conjunction $\underset{\sim}{\text{or/and}}$ is pronounced as *or/and*.

[†] *Or/and* is the author's preferred translation of *et/ou*.

Remark. When there is no possibility of error in interpretation, one will write "and" for "$\underset{\sim}{\text{and}}$," and in the same manner "or/and" for "$\text{or/}\underset{\sim}{\text{and}}$".

Disjunctive sum. The disjunctive sum of two fuzzy subsets is defined in terms of unions and intersections in the following fashion:

(5.34)
$$\underset{\sim}{A} \oplus \underset{\sim}{B} = (\underset{\sim}{A} \cap \overline{\underset{\sim}{B}}) \cup (\overline{\underset{\sim}{A}} \cap \underset{\sim}{B}) .$$

This operation corresponds to "fuzzy disjunctive or," where "$\underset{\sim}{\text{or}}$" is read "or" and will be written "or" when there is no risk of error.

We consider an example (the example that has served for union and intersection):

(5.35)
$$\underset{\sim}{A} = \{(x_1 | 0{,}2) , (x_2 | 0{,}7) , (x_3 | 1) , (x_4 | 0) , (x_5 | 0{,}5)\},$$

(5.36)
$$\underset{\sim}{B} = \{(x_1 | 0{,}5) , (x_2 | 0{,}3) , (x_3 | 1) , (x_4 | 0{,}1) , (x_5 | 0{,}5)\},$$

(5.37)
$$\overline{\underset{\sim}{A}} = \{(x_1 | 0{,}8) , (x_2 | 0{,}3) , (x_3 | 0) , (x_4 | 1) , (x_5 | 0{,}5)\},$$

(5.38)
$$\overline{\underset{\sim}{B}} = \{(x_1 | 0{,}5) , (x_2 | 0{,}7) , (x_3 | 0) , (x_4 | 0{,}9) , (x_5 | 0{,}5)\},$$

(5.39)
$$\underset{\sim}{A} \cap \overline{\underset{\sim}{B}} = \{(x_1 | 0{,}2) , (x_2 | 0{,}7) , (x_3 | 0) , (x_4 | 0) , (x_5 | 0{,}5)\},$$

(5.40)
$$\overline{\underset{\sim}{A}} \cap \underset{\sim}{B} = \{(x_1 | 0{,}5) , (x_2 | 0{,}3) , (x_3 | 0) , (x_4 | 0{,}1) , (x_5 | 0{,}5)\},$$

(5.41)
$$\underset{\sim}{A} \oplus \underset{\sim}{B} = \{(x_1 | 0{,}5) , (x_2 | 0{,}7) , (x_3 | 0) , (x_4 | 0{,}1) , (x_5 | 0{,}5)\}.$$

Difference. The difference is defined by the relation

(5.42)
$$\underset{\sim}{A} - \underset{\sim}{B} = \underset{\sim}{A} \cap \overline{\underset{\sim}{B}} .$$

Considering again the example (5.26) and (5.27), and using (5.38) and (5.39),

(5.43)
$$\underset{\sim}{A} \cap \overline{\underset{\sim}{B}} = \{(x_1 | 0{,}2) , (x_2 | 0{,}7) , (x_3 | 0) , (x_4 | 0) , (x_5 | 0{,}5)\}.$$

Of course, except in particular cases,

(5.44)
$$\underset{\sim}{A} - \underset{\sim}{B} \neq \underset{\sim}{B} - \underset{\sim}{A} .$$

Visual representation of simple operations on fuzzy subsets. For fuzzy subsets, one may construct a visual representation allied with that for ordinary subsets (Venn–Euler diagrams).

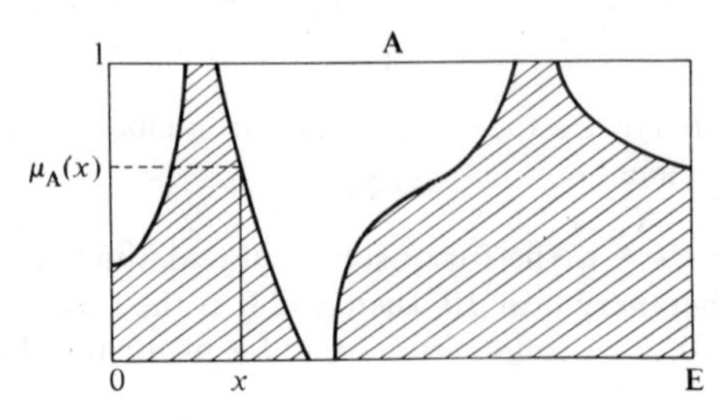

Fig. 5.1

Consider a rectangle (Figure 5.1) with the values of $\mu_{\mathbf{A}}(x)$ as ordinate and as an abscissa the elements of **E** in an arbitrary order (if there is in the nature of **E** a total order, that order will be taken). In Figure 5.1 the membership of each element is represented by its ordinate. The shaded part conveniently represents† the fuzzy subset $\underset{\sim}{\mathbf{A}} \subset \mathbf{E}$.

With this representation we see how to visualize the various simple operations on fuzzy subsets. A series of figures will show how to use this representation.

In Figures 5.2a–c the property of inclusion is presented. Figures 5.3a–c illustrate complementation. The properties of union and intersection are shown in Figures 5.4a–d,

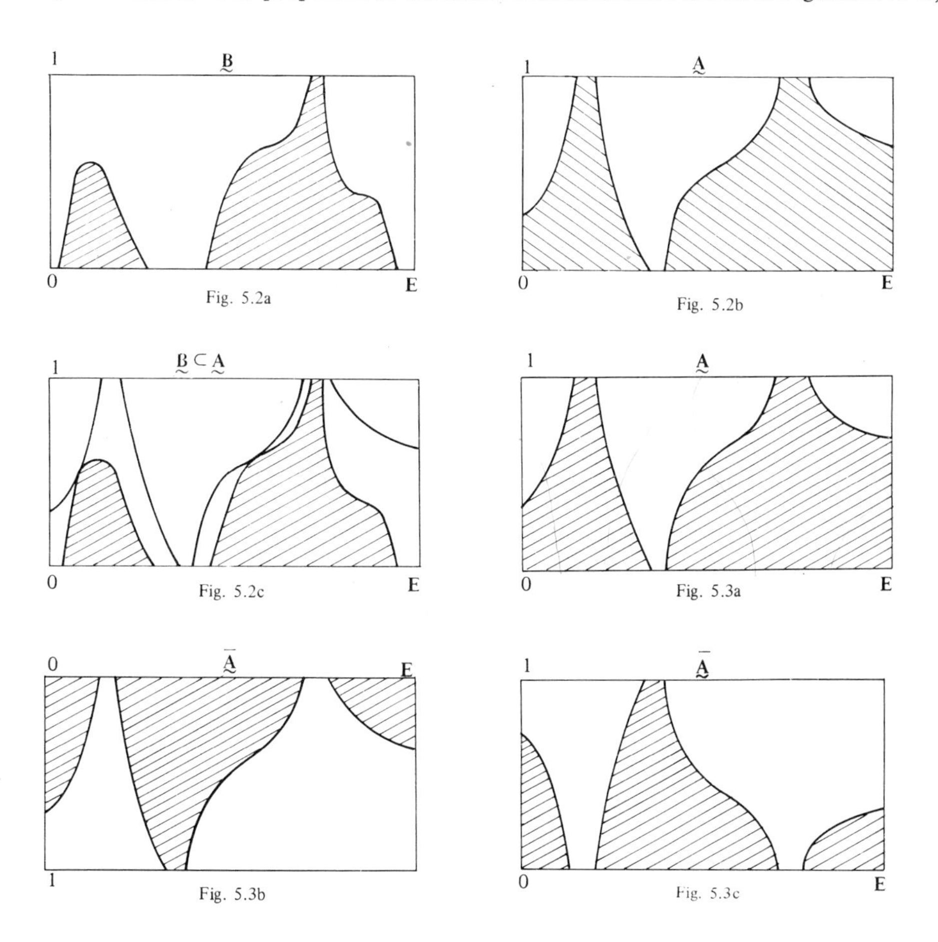

Fig. 5.2a

Fig. 5.2b

Fig. 5.2c

Fig. 5.3a

Fig. 5.3b

Fig. 5.3c

†This represents the fuzzy subset **A** and contains all the fuzzy subsets that are included in $\underset{\sim}{\mathbf{A}}$. These shadings are convenient for distinguishing one fuzzy subset from another.

In Figures 5.5a–g are represented the properties of the difference $\underset{\sim}{\mathbf{A}} - \underset{\sim}{\mathbf{B}} = \underset{\sim}{\mathbf{A}} \cap \overline{\underset{\sim}{\mathbf{B}}}$ and the disjunctive sum $\underset{\sim}{\mathbf{A}} \oplus \underset{\sim}{\mathbf{B}} = (\underset{\sim}{\mathbf{A}} \cap \overline{\underset{\sim}{\mathbf{B}}}) \cup (\overline{\underset{\sim}{\mathbf{A}}} \cap \underset{\sim}{\mathbf{B}})$.

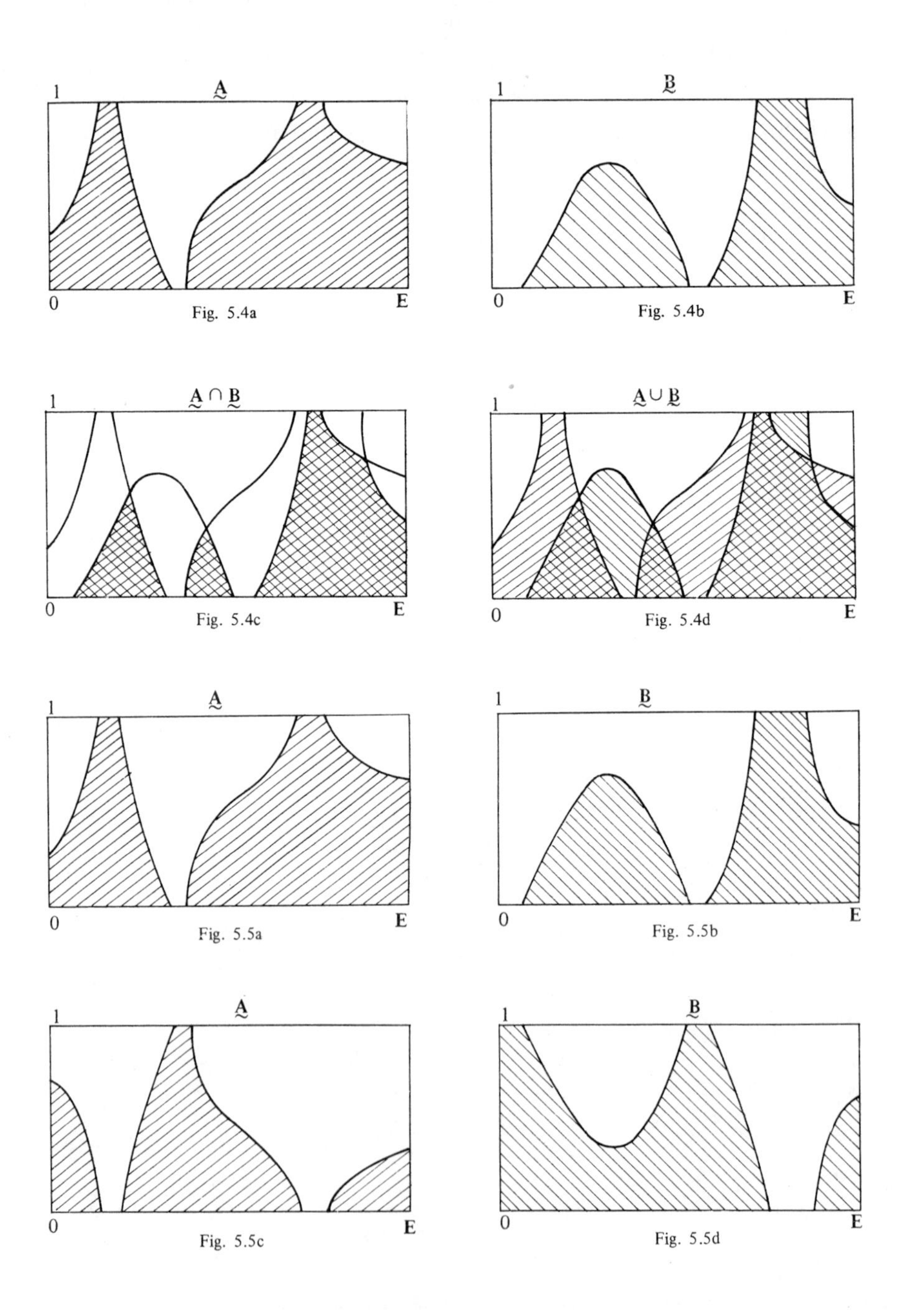

Fig. 5.4a

Fig. 5.4b

Fig. 5.4c

Fig. 5.4d

Fig. 5.5a

Fig. 5.5b

Fig. 5.5c

Fig. 5.5d

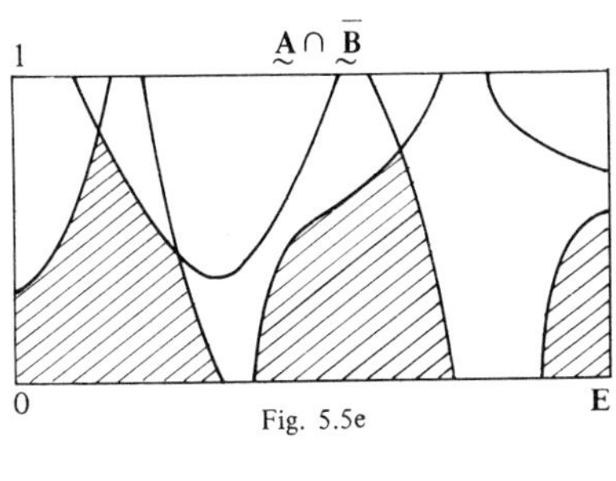

Fig. 5.5e

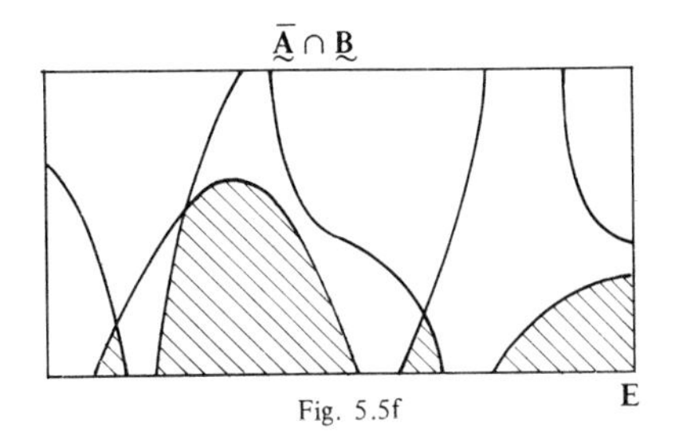

Fig. 5.5f

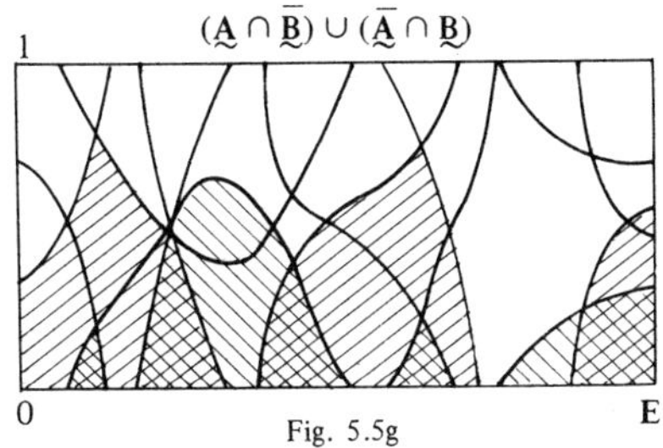

Fig. 5.5g

Hamming distance. We recall first what is meant by *Hamming distance* in the theory of ordinary subsets. Consider two ordinary subsets $\mathbf{A} \subset \mathbf{E}$, $\mathbf{B} \subset \mathbf{E}$, $\mathbf{E}$ finite,

(5.45)

	x_1	x_2	x_3	x_4	x_5	x_6	x_7
A =	1	0	0	1	0	1	0

,

(5.46)

	x_1	x_2	x_3	x_4	x_5	x_6	x_7
B =	0	1	0	0	0	1	1

.

The *Hamming distance* between **A** and **B** is the quantity

$$d(\mathbf{A}, \mathbf{B}) = \sum_{i=1}^{n} |\mu_{\mathbf{A}}(x_i) - \mu_{\mathbf{B}}(x_i)|. \tag{5.47}$$

For the example in (5.45) and (5.46), one has

$$\begin{aligned} d(\mathbf{A}, \mathbf{B}) &= |1-0| + |0-1| + |0-0| + |1-0| + |0-0| + |1-1| + |0-1| \\ &= 1 + 1 + 0 + 1 + 0 + 0 + 1 = 4. \end{aligned} \tag{5.48}$$

The reader knows that the word *distance* may not be used arbitrarily in mathematics. If X and Y are two elements between which one wishes to define a distance, it is necessary, you will recall, that one have, for some operation∗:

$\forall\ \mathrm{X}, \mathrm{Y}, \mathrm{Z} \in \mathbf{E}$:

(5.49) 1) $d(\mathrm{X}, \mathrm{Y}) \geqslant 0$, nonnegativity

(5.50) 2) $d(\mathrm{X}, \mathrm{Y}) = d(\mathrm{Y}, \mathrm{X})$, symmetry

(5.51) 3) $d(\mathrm{X}, \mathrm{Z}) \leqslant d(\mathrm{X}, \mathrm{Y}) * d(\mathrm{Y}, \mathrm{Z})$.

transitivity for the operation $*$ associated with the notion of distance.

To these three conditions, one may add a fourth:

(5.52) 4) $d(\mathrm{X}, \mathrm{X}) = 0$.

One may easily verify that a Hamming distance is indeed a distance in the sense given by (5.49)−(5.52) with the operation $* = +$ (ordinary sum).

We define also, for a finite **E** with $n = \text{card } \mathbf{E}$ (the number of elements in **E**), a relative Hamming distance:

(5.53)
$$\delta(\mathbf{A}, \mathbf{B}) = \frac{1}{n} d(\mathbf{A}, \mathbf{B}).$$

For the example in (5.45) and (5.46), one has

$$\delta(\mathbf{A}, \mathbf{B}) = \frac{d(\mathbf{A}, \mathbf{B})}{7} = \frac{4}{7}.$$

One has always

(5.54)
$$0 \leqslant \delta(\mathbf{A}, \mathbf{B}) \leqslant 1.$$

With a view toward generalizing the notion of Hamming distance to the case where one considers fuzzy subsets and not only ordinary subsets, we state two theorems.

Theorem I. Let $p_i, m_i, n_i \in \mathbf{R}^+$, $i = 1, 2, \ldots, k$; then

(5.55)
$$(p_i \leqslant m_i + n_i, \quad i = 1, 2, \ldots, k) \quad \Rightarrow \quad \sum_{i=1}^{k} p_i \leqslant \sum_{i=1}^{k} m_i + \sum_{i=1}^{k} n_i.$$

Proof. This result is immediate upon forming the sums from $i = 1$ to k on the left and right sides of the inequality.

Theorem II. Let $p_i, m_i, n_i \in \mathbf{R}^+$, $i = 1, 2, \ldots, k$; then

(5.56)
$$(p_i \leqslant m_i + n_i, \quad i = 1, 2, \ldots, k) \Rightarrow \sqrt{\sum_{i=1}^{k} p_i^2} \leqslant \sqrt{\sum_{i=1}^{k} m_i^2} + \sqrt{\sum_{i=1}^{k} n_i^2}.$$

Proof.† This result is less immediate.

†One may give another proof based on the theory of complex numbers; we have preferred a direct presentation.

We treat the evident inequality

$$\sum_{\substack{i,j=1\\i\neq j}}^{k} (m_i\, n_j - m_j\, n_i)^2 \geqslant 0\,. \tag{5.57}$$

Developing this sum of squares, we have

$$\sum_{\substack{i,j=1\\i\neq j}}^{k} m_i^2\, n_j^2 - 2 \sum_{\substack{i,j=1\\i\neq j}}^{k} m_i\, n_i\, m_j\, n_j \geqslant 0\,. \tag{5.58}$$

That is,

$$\sum_{\substack{i,j=1\\i\neq j}}^{k} m_i^2\, n_j^2 \geqslant 2 \sum_{\substack{i,j=1\\i\neq j}}^{k} m_i\, n_i\, m_j\, n_j. \tag{5.59}$$

Adding $\Sigma_{i=1}^{k}\, m_i^2\, n_i^2$ to the two members of this inequality,

$$\sum_{\substack{i,j=1\\i\neq j}}^{k} m_i^2\, n_j^2 + \sum_{i=1}^{k} m_i^2\, n_i^2 \geqslant \sum_{i=1}^{k} m_i^2\, n_i^2 + \sum_{\substack{i,j=1\\i\neq j}}^{k} 2 m_i\, n_i\, m_j\, n_j\;; \tag{5.60}$$

which may be rewritten

$$\Big(\sum_{i=1}^{k} m_i^2\Big) . \Big(\sum_{i=1}^{k} n_i^2\Big) \geqslant \Big(\sum_{i=1}^{k} m_i\, n_i\Big)^2, \tag{5.61}$$

$$\sqrt{\sum_{i=1}^{k} m_i^2} \,.\, \sqrt{\sum_{i=1}^{k} n_i^2} \geqslant \sum_{i=1}^{k} m_i\, n_i\,, \tag{5.62}$$

$$2\sqrt{\sum_{i=1}^{k} m_i^2} \,.\, \sqrt{\sum_{i=1}^{k} n_i^2} \geqslant 2 \sum_{i=1}^{k} m_i\, n_i\,. \tag{5.63}$$

Adding $\Sigma_{i=1}^{k}\, m_i^2 + \Sigma_{i=1}^{k}\, n_i^2$, one has

$$\sum_{i=1}^{k} m_i^2 + \sum_{i=1}^{k} n_i^2 + 2\sqrt{\sum_{i=1}^{k} m_i^2} \,.\, \sqrt{\sum_{i=1}^{k} n_i^2} \geqslant \sum_{i=1}^{k} m_i^2 + \sum_{i=1}^{k} n_i^2 + 2 \sum_{i=1}^{k} m_i\, n_i\;; \tag{5.64}$$

which may be rewritten

$$\left(\sqrt{\sum_{i=1}^{k} m_i^2} + \sqrt{\sum_{i=1}^{k} n_i^2}\right)^2 \geqslant \sum_{i=1}^{k} (m_i + n_i)^2 \tag{5.65}$$

or

$$\sqrt{\sum_{i=1}^{k} m_i^2} + \sqrt{\sum_{i=1}^{k} n_i^2} \geqslant \sqrt{\sum_{i=1}^{k} (m_i + n_i)^2}\,. \tag{5.66}$$

But, by hypothesis,

$$\forall_i = 1\,, 2\,, \ldots\,, k: \quad m_i + n_i \geqslant p_i\,, \tag{5.67}$$

and then

$$\sqrt{\sum_{i=1}^{k} m_i^2} + \sqrt{\sum_{i=1}^{k} n_i^2} \geqslant \sqrt{\sum_{i=1}^{k} p_i^2}. \tag{5.68}$$

Q.E.D.

Generalization of the notion of Hamming distance. Consider now three fuzzy subsets $\underset{\sim}{\mathbf{A}}, \underset{\sim}{\mathbf{B}}, \underset{\sim}{\mathbf{C}}, \subset \mathbf{E}$, $\mathbf{E}$ finite, card $\mathbf{E} = n$:

(5.69)

	x_1	x_2	x_3	...	x_n
$\underset{\sim}{\mathbf{A}} =$	a_1	a_2	a_3	...	a_n

,

(5.70)

	x_1	x_2	x_3	...	x_n
$\underset{\sim}{\mathbf{B}} =$	b_1	b_2	b_3	...	b_n

.

(5.71)

	x_1	x_2	x_3	...	x_n
$\underset{\sim}{\mathbf{C}} =$	c_1	c_2	c_3	...	c_n

.

Suppose that one has defined a distance, denoted $\mathcal{D}(a_i, b_i)$, between a_i and b_i for all $i = 1, 2, \ldots, n$, and that the same holds for (b_i, c_i) and for (a_i, c_i). One must then have, since it is a distance according to (5.49)–(5.52),

$$\forall_i = 1,2,\ldots, n: \quad \mathcal{D}(a_i, c_i) \leqslant \mathcal{D}(a_i, b_i) * \mathcal{D}(b_i, c_i). \tag{5.72}$$

And, according to Theorems I (5.55) and II (5.56), one may write

$$\sum_{i=1}^{n} \mathcal{D}(a_i, c_i) \leqslant \sum_{i=1}^{n} \mathcal{D}(a_i, b_i) + \sum_{i=1}^{n} \mathcal{D}(b_i, c_i), \tag{5.73}$$

$$\sqrt{\sum_{i=1}^{n} \mathcal{D}^2(a_i, c_i)} \leqslant \sqrt{\sum_{i=1}^{n} \mathcal{D}^2(a_i, b_i)} + \sqrt{\sum_{i=1}^{n} \mathcal{D}^2(b_i, c_i)}. \tag{5.74}$$

These two formulas give two evaluations of the distance between fuzzy subsets, one linear and the other quadratic.†

Now we consider the case where, in fuzzy subsets, the membership function takes its values in $\mathbf{M} = [0, 1]$, that is, where one has in (5.69)–(5.71), $a_i, b_i, c_i \in [0,1]$, $i = 1, 2, \ldots, n$.

Now take

$$\mathcal{D}(a_i, b_i) = |a_i - b_i|, \quad \mathcal{D}(b_i, c_i) = |b_i - c_i|, \quad \mathcal{D}(a_i, c_i) = |a_i - c_i| \tag{5.75}$$

†The notion of distance has been the subject of a number of works. We present here two notions among those used most often. Of course, one may define other notions of distance for fuzzy subsets.

and we define two types of distance corresponding to (5.73) and (5.74):

Generalized Hamming distance or linear distance. This will be defined by†

(5.76) $$d(\underset{\sim}{A}, \underset{\sim}{B}) = \sum_{i=1}^{n} |\mu_{\underset{\sim}{A}}(x_i) - \mu_{\underset{\sim}{B}}(x_i)| .$$

This generalizes (5.47) to the case where

(5.76a) $$\mu_{\underset{\sim}{A}}(x_i),\ \mu_{\underset{\sim}{B}}(x_i) \in [0,1],\ i = 1, 2, \ldots, n$$

And one has

(5.77) $$0 \leqslant d(\underset{\sim}{A}, \underset{\sim}{B}) \leqslant n .$$

Euclidean distance or quadratic distance. This will be defined by

(5.78) $$e(\underset{\sim}{A}, \underset{\sim}{B}) = \sqrt{\sum_{i=1}^{n} (\mu_{\underset{\sim}{A}}(x_i) - \mu_{\underset{\sim}{B}}(x_i))^2}.$$

One has

(5.79) $$0 \leqslant e(\underset{\sim}{A}, \underset{\sim}{B}) \leqslant \sqrt{n}.$$

The quantity $e^2(\underset{\sim}{A}, \underset{\sim}{B})$ is called the euclidean norm:

(5.80) $$e^2(\underset{\sim}{A}, \underset{\sim}{B}) = \sum_{i=1}^{n} (\mu_{\underset{\sim}{A}}(x_i) - \mu_{\underset{\sim}{B}}(x_i))^2 .$$

We now define some relative distances.

Generalized relative Hamming distance

(5.81) $$\delta(\underset{\sim}{A}, \underset{\sim}{B}) = \frac{d(\underset{\sim}{A}, \underset{\sim}{B})}{n} = \frac{1}{n} \sum_{i=1}^{n} |\mu_{\underset{\sim}{A}}(x_i) - \mu_{\underset{\sim}{B}}(x_i)| .$$

One may verify that this is indeed a distance according to (5.49)–(5.52), and with reference to (5.73), where the property has not been altered by dividing the two members by n, one has

(5.82) $$0 \leqslant \delta(\underset{\sim}{A}, \underset{\sim}{B}) \leqslant 1 ,$$

and (5.81) generalizes (5.53) for the case where $\mu_{\underset{\sim}{A}}(x_i), \mu_{\underset{\sim}{B}}(x_i) \in [0, 1]$.

Relative euclidean distance

(5.83) $$\epsilon(\underset{\sim}{A}, \underset{\sim}{B}) = \frac{e(\underset{\sim}{A}, \underset{\sim}{B})}{\sqrt{n}} = \sqrt{\frac{1}{n} \sum_{i=1}^{n} (\mu_{\underset{\sim}{A}}(x_i) - \mu_{\underset{\sim}{B}}(x_i))^2} .$$

One may verify that this is indeed a distance according to (5.49)–(5.52), and with reference to (5.74), where the property is not altered by dividing the two members by $\sqrt{n}$, one has

†Note that we have $|\mu_{\underset{\sim}{A}}(x_i) - \mu_{\underset{\sim}{B}}(x_i)| = \mathrm{MAX}[\mu_{\underset{\sim}{A}}(x_i), \mu_{\underset{\sim}{B}}(x_i)] - \mathrm{MIN}[\mu_{\underset{\sim}{A}}(x_i), \mu_{\underset{\sim}{B}}(x_i)]$.

(5.84) $$0 \leqslant \epsilon(\underset{\sim}{A}, \underset{\sim}{B}) \leqslant 1 .$$

$\epsilon^2(\underset{\sim}{A}, \underset{\sim}{B})$ is called the relative euclidean norm:

(5.85) $$\epsilon^2(\underset{\sim}{A}, \underset{\sim}{B}) = \frac{e^2(\underset{\sim}{A}, \underset{\sim}{B})}{n} = \frac{1}{n}\sum_{i=1}^{n}(\mu_{\underset{\sim}{A}}(x_i) - \mu_{\underset{\sim}{B}}(x_i))^2.$$

One will not be astonished that, in the particular case where $\mu_{\underset{\sim}{A}}(x_i), \mu_{\underset{\sim}{B}}(x_i) \in \{0, 1\}$,

(5.86) $$e^2(\underset{\sim}{A}, \underset{\sim}{B}) = d(\underset{\sim}{A}, \underset{\sim}{B}) ,$$

(5.87) $$\epsilon^2(\underset{\sim}{A}, \underset{\sim}{B}) = \delta(\underset{\sim}{A}, \underset{\sim}{B}) .$$

These correspond to the boolean property

(5.88) $$a^2 = a , \quad a \in \{0,1\} .$$

Thus one may say that (5.76) and (5.81) generalize the notions of Hamming distance, absolute or relative, (5.47), and (5.53); one may not classify the euclidean norm as a distance since this norm does not satisfy the inequality (5.51) of the notion of distance.

The choice of a notion of distance, whether generalized (absolute or relative) Hamming or euclidean (absolute or relative), depends on the nature of the problem to be treated. These possess, respectively, advantages and inconveniences, which become evident in applications; we shall occupy ourselves with this in Volume III. One may, obviously, imagine and define other notions of distance.

Example. Let

(5.89) $\underset{\sim}{A} =$

x_1	x_2	x_3	x_4	x_5	x_6	x_7
0,7	0,2	0	0,6	0,5	1	0

and

(5.90) $\underset{\sim}{B} =$

x_1	x_2	x_3	x_4	x_5	x_6	x_7
0,2	0	0	0,6	0,8	0,4	1

.

One has

(5.91) $$d(\underset{\sim}{A}, \underset{\sim}{B}) = |0,7 - 0,2| + |0,2 - 0| + |0 - 0| + |0,6 - 0,6| + |0,5 - 0,8| + |1 - 0,4| + |0 - 1| .$$
$$= 0,5 + 0,2 + 0 + 0 + 0,3 + 0,6 + 1 = 2,6 ,$$

(5.92) $$\delta(\underset{\sim}{A}, \underset{\sim}{B}) = \frac{1}{7} d(\underset{\sim}{A}, \underset{\sim}{B}) = \frac{2,6}{7} = 0,37 ,$$

$$e^2(\underset{\sim}{\mathbf{A}}, \underset{\sim}{\mathbf{B}}) = (0{,}7 - 0{,}2)^2 + (0{,}2 - 0)^2 + (0 - 0)^2 + (0{,}6 - 0{,}6)^2 + (0{,}5 - 0{,}8)^2 + (1 - 0{,}4)^2 + (0 - 1)^2 \tag{5.93}$$

$$= (0{,}5)^2 + (0{,}2)^2 + (0)^2 + (0)^2 + (0{,}3)^2 + (0{,}6)^2 + (1)^2$$

$$= 1{,}74\,,$$

$$e(\underset{\sim}{\mathbf{A}}, \underset{\sim}{\mathbf{B}}) = \sqrt{1{,}74} = 1{,}32\,, \tag{5.94}$$

$$\epsilon(\underset{\sim}{\mathbf{A}}, \underset{\sim}{\mathbf{B}}) = \frac{e(\underset{\sim}{\mathbf{A}}, \underset{\sim}{\mathbf{B}})}{\sqrt{7}} = \frac{1{,}32}{\sqrt{7}} = 0{,}49\,. \tag{5.95}$$

Case of a nonfinite reference set. The distances $d(\underset{\sim}{\mathbf{A}}, \underset{\sim}{\mathbf{B}})$ and $e(\underset{\sim}{\mathbf{A}}, \underset{\sim}{\mathbf{B}})$, and thus evidently the norm $e^2(\underset{\sim}{\mathbf{A}}, \underset{\sim}{\mathbf{B}})$, may be extended to the case where the reference set is not finite (denumerable or not), with the reservation, of course, that the corresponding summations be convergent.

If **E** is denumerable, one writes

$$d(\underset{\sim}{\mathbf{A}}, \underset{\sim}{\mathbf{B}}) = \sum_{i=1}^{\infty} |\mu_{\underset{\sim}{\mathbf{A}}}(x_i) - \mu_{\underset{\sim}{\mathbf{B}}}(x_i)| \tag{5.96}$$

if this series is convergent.

If **E** = **R**, one writes

$$d(\underset{\sim}{\mathbf{A}}, \underset{\sim}{\mathbf{B}}) = \int_{-\infty}^{\infty} |\mu_{\underset{\sim}{\mathbf{A}}}(x) - \mu_{\underset{\sim}{\mathbf{B}}}(x)| \cdot dx \tag{5.97}$$

if this integral is convergent.

And similarly (see Figure 5.6),

$$e(\underset{\sim}{\mathbf{A}}, \underset{\sim}{\mathbf{B}}) = \sqrt{\sum_{i=1}^{\infty} (\mu_{\underset{\sim}{\mathbf{A}}}(x_i) - \mu_{\underset{\sim}{\mathbf{B}}}(x_i))^2} \tag{5.98}$$

if this series is convergent.

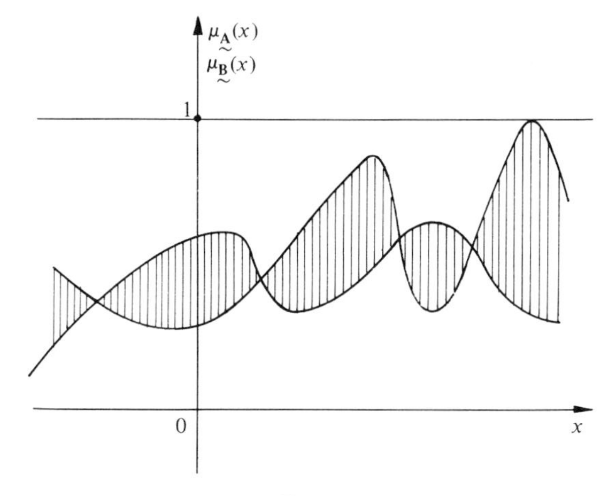

Fig. 5.6

And

(5.99) $$e(\underset{\sim}{\mathbf{A}}, \underset{\sim}{\mathbf{B}}) = \sqrt{\int_{-\infty}^{\infty} (\mu_{\underset{\sim}{\mathbf{A}}}(x) - \mu_{\underset{\sim}{\mathbf{B}}}(x))^2 \, . \, dx}$$

if this integral is convergent.

Generally, $\delta(\underset{\sim}{\mathbf{A}}, \underset{\sim}{\mathbf{B}})$ and $\epsilon(\underset{\sim}{\mathbf{A}}, \underset{\sim}{\mathbf{B}})$ are not used in the case of a nonfinite reference set, but one may, if necessary, at the cost of using a different definition or interposing other notions of convergence.

If one considers the case where $\mathbf{E} \subset \mathbf{R}$ is bounded above and below, then the integral (5.97) is convergent and likewise (5.98); then $d(\underset{\sim}{\mathbf{A}}, \underset{\sim}{\mathbf{B}})$ and $e(\underset{\sim}{\mathbf{A}}, \underset{\sim}{\mathbf{B}})$ are always finite.

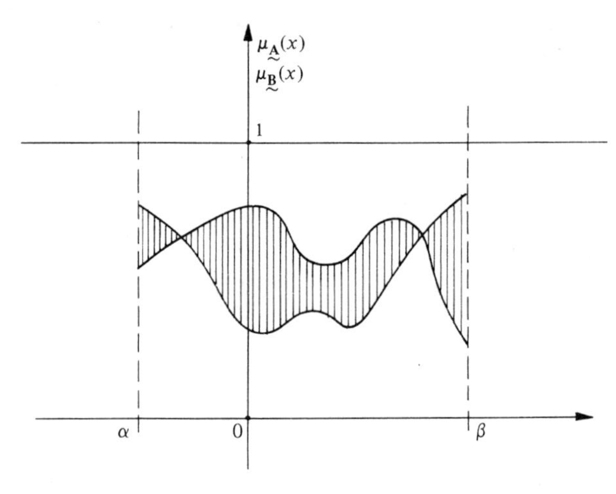

Fig. 5.7

In this case one will always be able to define $\delta(\underset{\sim}{\mathbf{A}}, \underset{\sim}{\mathbf{B}})$ and $\epsilon(\underset{\sim}{\mathbf{A}}, \underset{\sim}{\mathbf{B}})$ (Figure 5.7):

(5.100) $$\delta(\underset{\sim}{\mathbf{A}}, \underset{\sim}{\mathbf{B}}) = \frac{d(\underset{\sim}{\mathbf{A}}, \underset{\sim}{\mathbf{B}})}{\beta - \alpha},$$

(5.101) $$\epsilon(\underset{\sim}{\mathbf{A}}, \underset{\sim}{\mathbf{B}}) = \frac{e(\underset{\sim}{\mathbf{A}}, \underset{\sim}{\mathbf{B}})}{\sqrt{\beta - \alpha}}.$$

Ordinary subset nearest to a fuzzy subset. We pose the following question: Which is the ordinary subset (or subsets) **A** that has, with respect to a given fuzzy subset $\underset{\sim}{\mathbf{A}}$, the smallest euclidean distance (or, if one wishes, the smallest norm)?

It is trivial to prove that this will be the ordinary subset, denoted $\underset{\approx}{\mathbf{A}}$, such that

(5.102) $$\begin{aligned} \mu_{\underset{\approx}{\mathbf{A}}}(x_i) &= 0 && \text{if } \mu_{\underset{\sim}{\mathbf{A}}}(x_i) < 0{,}5\,, \\ &= 1 && \text{if } \mu_{\underset{\sim}{\mathbf{A}}}(x_i) > 0{,}5\,, \\ &= 0 \text{ or } 1 && \text{if } \mu_{\underset{\sim}{\mathbf{A}}}(x_i) = 0{,}5\,. \end{aligned}$$

Where, by convention, we take $\mu_{\underset{\approx}{A}}(x_i) = 0$ if $\mu_{\underset{\sim}{A}}(x_i) = 0{,}5$.

Example. Let

(5.103) $\underset{\sim}{A} =$

x_1	x_2	x_3	x_4	x_5	x_6	x_7	x_8
0,2	0,8	0,5	0,3	1	0	0,9	0,4

Then one has

(5.104) $\underset{\approx}{A} =$

x_1	x_2	x_3	x_4	x_5	x_6	x_7	x_8
0	1	0	0	1	0	1	0

Index of fuzziness. One may consider, among others, two indexes of fuzziness: the linear index of fuzziness, defined with respect to the generalized relative Hamming distance, and the quadratic index of fuzziness, defined with respect to the relative euclidean distance. One designates these respectively by $\nu(\underset{\sim}{A})$ and $\eta(\underset{\sim}{A})$:

$$\nu(\underset{\sim}{A}) = \frac{2}{n} \cdot d(\underset{\sim}{A}, \underset{\approx}{A}), \tag{5.105}$$

$$\eta(\underset{\sim}{A}) = \frac{2}{\sqrt{n}} \, e(\underset{\sim}{A}, \underset{\approx}{A}). \tag{5.106}$$

The number 2 appears in the numerator in order to obtain

(5.107)
(5.108)

$$0 \leqslant \nu(\underset{\sim}{A}) \leqslant 1 \qquad \text{and} \qquad 0 \leqslant \eta(\underset{\sim}{A}) \leqslant 1$$

because

(5.109)
(5.110)

$$0 \leqslant \delta(\underset{\sim}{A}, \underset{\approx}{A}) \leqslant \frac{1}{2} \qquad \text{and} \qquad 0 \leqslant \epsilon(\underset{\sim}{A}, \underset{\approx}{A}) \leqslant \frac{1}{2}.$$

The notion of the subset closest to a given fuzzy subset and the notion of the index of fuzziness may be extended to the case of a nonfinite reference set, with reservations–for example, concerning the index of fuzziness, that the summation be convergent. We shall consider the case of reference set $\mathbf{E} = [a, b] \in \mathbf{R}$.

Figure 5.8 indicates how to evaluate the ordinary subset nearest and, from there, the index of fuzziness. For example, formula (5.105) gives

$$\nu_{(\underset{\sim}{A})} = \frac{2}{b-a} \int_a^b |\mu_{\underset{\sim}{A}}(x) - \mu_{\underset{\approx}{A}}(x)| \, dx. \tag{5.111}$$

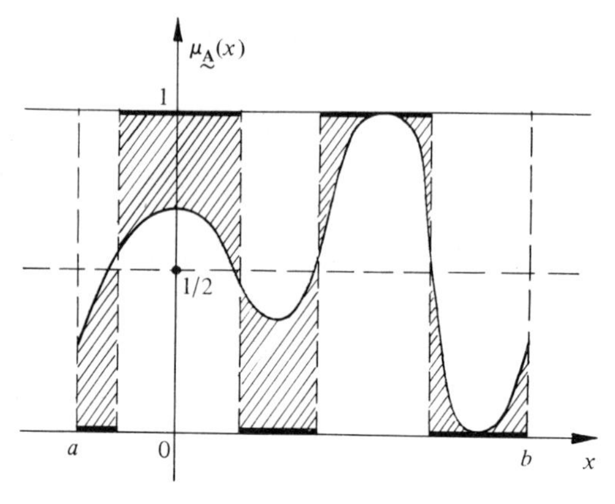

Fig. 5.8

Principal properties concerning the nearest ordinary subset. The following properties may be easily verified:

(5.112) $$\underline{\underset{\sim}{A} \cap \underset{\sim}{B}} = \underline{\underset{\sim}{A}} \cap \underline{\underset{\sim}{B}}\,,$$

(5.113)† $$\underline{\underset{\sim}{A} \cup \underset{\sim}{B}} = \underline{\underset{\sim}{A}} \cup \underline{\underset{\sim}{B}}\,,$$

Another interesting property is

(5.115) $$\forall\, x_i \in E: \quad |\mu_{\underset{\sim}{A}}(x_i) - \mu_{\underline{\underset{\sim}{A}}}(x_i)| = \mu_{\underset{\sim}{A} \cap \bar{\underset{\sim}{A}}}(x_i)\,,$$

which is proved with reference to properties (5.112) and (5.114)

We shall see an example by reconsidering (5.103) and (5.104):

(5.116) $\bar{\underset{\sim}{A}} =$

x_1	x_2	x_3	x_4	x_5	x_6	x_7	x_8
0,8	0,2	0,5	0,7	0	1	0,1	0,6

,

(5.117) $\underset{\sim}{A} \cap \bar{\underset{\sim}{A}} =$

x_1	x_2	x_3	x_4	x_5	x_6	x_7	x_8
0,2	0,2	0,5	0,3	0	0	0,1	0,4

.

One sometimes calls the fuzzy subset whose membership function is $2\mu_{\underset{\sim}{A} \cap \bar{\underset{\sim}{A}}}(x)$ the vectorial indicator of fuzziness.‡ Thus for (5.103) one has

†Equation (5.114) has been omitted.
‡Proposed by M. Nadler, research engineer at Honeywell Bull Cie.

(5.118)

x_1	x_2	x_3	x_4	x_5	x_6	x_7	x_8
0,4	0,4	1	0,6	0	0	0,2	0,8

Formula (5.105) may be written more conveniently as

(5.119) $$\nu(\underset{\sim}{\mathrm{A}}) = \frac{2}{n} \sum_{i=1}^{n} \mu_{\underset{\sim}{\mathrm{A}} \cap \overline{\underset{\sim}{\mathrm{A}}}}(x_i) .$$

One again has

(5.120) $$\nu(\overline{\underset{\sim}{\mathrm{A}}}) = \nu(\underset{\sim}{\mathrm{A}}) .$$

One may ask the following interesting question: Suppose $\underset{\sim}{\mathrm{A}}$ and $\underset{\sim}{\mathrm{B}}$ are two fuzzy subsets of the same reference set **E**; then do $\underset{\sim}{\mathrm{A}} \cap \underset{\sim}{\mathrm{B}}$ or $\underset{\sim}{\mathrm{A}} \cup \underset{\sim}{\mathrm{B}}$ have indexes of fuzziness larger (or smaller) than $\underset{\sim}{\mathrm{A}}$ or/and $\underset{\sim}{\mathrm{B}}$? The following counterexamples show that, unfortunately, one may not say anything on this subject:

(5.121)

$\underset{\sim}{\mathrm{A}}$ =

x_1	x_2	x_3
0,2	0,6	0,1

, $\nu(\underset{\sim}{\mathrm{A}}) = 0{,}46$

$\underset{\sim}{\mathrm{B}}$ =

x_1	x_2	x_3
0,6	0,3	0,8

, $\nu(\underset{\sim}{\mathrm{B}}) = 0{,}60$

$\underset{\sim}{\mathrm{A}} \cap \underset{\sim}{\mathrm{B}}$ =

x_1	x_2	x_3
0,2	0,3	0,1

, $\nu(\underset{\sim}{\mathrm{A}} \cap \underset{\sim}{\mathrm{B}}) = 0{,}40$

(5.122)

$\underset{\sim}{\mathrm{A}}'$ =

x_1	x_2	x_3
0,8	0,6	0,8

, $\nu(\underset{\sim}{\mathrm{A}}') = 0{,}53$

$\underset{\sim}{\mathrm{B}}'$ =

x_1	x_2	x_3
0,4	0,7	0,2

, $\nu(\underset{\sim}{\mathrm{B}}') = 0{,}60$

$\underset{\sim}{\mathrm{A}}' \cap \underset{\sim}{\mathrm{B}}'$ =

x_1	x_2	x_3
0,4	0,6	0,2

. $\nu(\underset{\sim}{\mathrm{A}}' \cap \underset{\sim}{\mathrm{B}}') = 0{,}66$

The same holds concerning $\underset{\sim}{\mathrm{A}} \cup \underset{\sim}{\mathrm{B}}$, also unfortunately, and similarly also using $\eta(\mathrm{A})$.

Since we have seen that a fuzzy subset and its complement have the same index of fuzziness, one then sees that each operation ($\cap$, $\cup$, $\overline{\ }$) does not ensure any systematic effect of increasing or decreasing fuzziness.

Evaluation of fuzziness through entropy. We here restrict ourselves to the case of a finite reference set. We know that the entropy of a system measures the degree of disorder of the components of the system with respect to the probabilities of state.

Consider N states $\mathcal{E}_1, \mathcal{E}_2, \ldots, \mathcal{E}_N$ of a system with which are associated the probabilities $p_1, p_2, \ldots, p_N$; then the *entropy* of the system is defined by†

†Taken to a multiplicative coefficient K, to be placed before the summation sign Σ. ln indicates the naperian logarithm, using the base e.

(5.123) $$H(p_1, p_2, \ldots, p_N) = -\sum_{i=1}^{N} p_i \ln p_i \,.$$

It is easy to show that

(5.124) $H = 0$ (H minimal) for $p_r = 1$, $r \in \{1, 2, \ldots, N\}$

$p_i = 0 \quad i \neq r$.

(5.125) $H = \ln N$ (H maximal) for $p_1 = p_2 = \ldots = p_N = p = \frac{1}{N}$.

If we take the formula

(5.126) $$H(p_1, p_2, \ldots, p_N) = -\frac{1}{\ln N}\sum_{i=1}^{N} p_i \ln p_i \,,$$

the entropy is then a quantity that varies between 0 and 1:

(5.127) $H_{min} = 0$ and $H_{max} = 1$.

We shall see how to use this notion to evaluate the fuzziness of a subset. Consider a fuzzy subset $\underset{\sim}{A}$:

(5.128) $$\mu_{\underset{\sim}{A}}(x_1) = 0{,}7\,,\ \mu_{\underset{\sim}{A}}(x_2) = 0{,}9\,,\ \mu_{\underset{\sim}{A}}(x_3) = 0\,,\ \mu_{\underset{\sim}{A}}(x_4) = 0{,}6\,,$$
$$\mu_{\underset{\sim}{A}}(x_5) = 0{,}5\,,\ \mu_{\underset{\sim}{A}}(x_6) = 1\,.$$

Putting

(5.129) $$\pi_{\underset{\sim}{A}}(x_i) = \frac{\mu_{\underset{\sim}{A}}(x_i)}{\sum_{i=1}^{6} \mu_{\underset{\sim}{A}}(x_i)}\,,$$

one obtains

(5.130)
$$\pi_{\underset{\sim}{A}}(x_1) = \frac{7}{37},\quad \pi_{\underset{\sim}{A}}(x_2) = \frac{9}{37},\quad \pi_{\underset{\sim}{A}}(x_3) = 0\,,\quad \pi_{\underset{\sim}{A}}(x_4) = \frac{6}{37},\quad \pi_{\underset{\sim}{A}}(x_5) = \frac{5}{37},$$
$$\pi_{\underset{\sim}{A}}(x_6) = \frac{10}{37}.$$

Then

(5.131) $$H(\pi_1, \pi_2, \ldots, \pi_6) = -\frac{1}{\ln 6}\sum_{i=1}^{6} \pi_{\underset{\sim}{A}}(x_i)\,.\,\ln \pi_{\underset{\sim}{A}}(x_i)$$
$$= -\frac{1}{\ln 6}\Big(\frac{7}{37}\ln\frac{7}{37} + \frac{9}{37}\ln\frac{9}{37}$$
$$+ \frac{6}{37}\ln\frac{6}{37} + \frac{5}{37}\ln\frac{5}{37} + \frac{10}{37}\ln\frac{10}{37}\Big) \doteq 0{,}89\,.$$

The general formula permitting the calculation of the entropy from the fuzziness may be rewritten as

(5.132) $$H(\pi_{\underset{\sim}{A}}(x_1), \pi_{\underset{\sim}{A}}(x_2), \ldots, \pi_{\underset{\sim}{A}}(x_n)) = -\frac{1}{\ln N} \sum_{i=1}^{N} \pi_{\underset{\sim}{A}}(x_i) \,.\, \ln \pi_{\underset{\sim}{A}}(x_i)$$

$$= \frac{1}{\ln N \,.\, \sum_{i=1}^{N} \mu_{\underset{\sim}{A}}(x_i)} \left[\left(\sum_{i=1}^{N} \mu_{\underset{\sim}{A}}(x_i)\right) . \left(\ln \sum_{i=1}^{N} \mu_{\underset{\sim}{A}}(x_i)\right) - \sum_{i=1}^{N} \mu_{\underset{\sim}{A}}(x_i) \,.\, \ln \mu_{\underset{\sim}{A}}(x_i)\right] .$$

We remark that this method of calculating fuzziness through entropy does not depend on accounting the effective values of μ but their relative values. Thus, the two fuzzy subsets below

(5.133)
(5.134)

$\underset{\sim}{A}$ =

x_1	x_2	x_3	x_4	x_5	x_6
0,1	0,1	0,1	0,1	0,1	0,1

$\underset{\sim}{B}$ =

x_1	x_2	x_3	x_4	x_5	x_6
0,8	0,8	0,8	0,8	0,8	0,8

have the same entropy, H = 1. The same holds for the ordinary subset

(5.135) C =

x_1	x_2	x_3	x_4	x_5	x_6
1	1	1	1	1	1

All ordinary subsets having a single nonzero element have entropy 0. Finally, the empty subset will always have an entropy equal to 1.

Entropy may be used in the theory of fuzzy subsets, but it is not a good indicator†; it relates to the theory of probabilities in systems, a different theory, as we will see in Section 40, than that we shall examine here. Sometimes it is possible to have some rapprochement, but only sometimes.

Ordinary subset of level α. Let $\alpha \in [0, 1]$; one will call the *ordinary subset of level* α of a fuzzy subset $\underset{\sim}{A}$, the ordinary subset

(5.136) $$A_\alpha = \{x \mid \mu_{\underset{\sim}{A}}(x) \geqslant \alpha\}.$$

Example 1. Let

(5.137) $\underset{\sim}{A}$ =

x_1	x_2	x_3	x_4	x_5	x_6	x_7
0,8	0,1	1	0,3	0,6	0,2	0,5

.

†Since publication of the present work (first French edition), A. De Luca and S. Termini [D2] have defined a new and interesting extension of the concept of entropy for fuzzy subsets.

One has

(5.138)

	x_1	x_2	x_3	x_4	x_5	x_6	x_7
$\mathbf{A}_{0,3} =$	1	0	1	1	1	0	1

,

(5.139)

	x_1	x_2	x_3	x_4	x_5	x_6	x_7
$\mathbf{A}_{0,55} =$	1	0	1	0	1	0	0

.

Example 2. In Figure 5.9 we present an example where the reference set is $\mathbf{R}^+$.

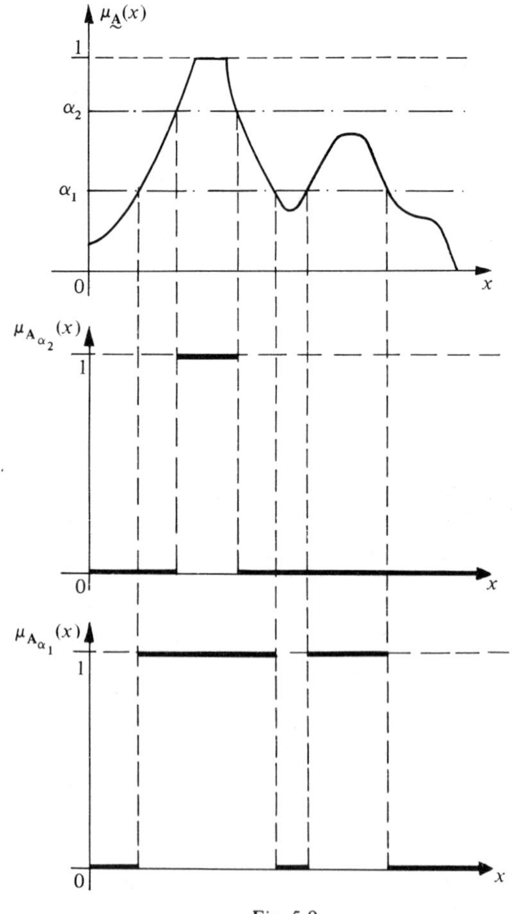

Fig. 5.9

Important property. At once we have the evident property

$$\alpha_2 \geqslant \alpha_1 \quad \Rightarrow \quad \mathbf{A}_{\alpha_2} \subset \mathbf{A}_{\alpha_1} . \tag{5.140}$$

We now consider an important theorem:

Decomposition theorem. Any fuzzy subset $\underset{\sim}{A}$ may be decomposed in the following form, clearly as products of ordinary subsets by the coefficients α_i:

(5.141)

$$\underset{\sim}{A} = \underset{\alpha_i}{\text{MAX}} [\alpha_1 . A_{\alpha_1}, \alpha_2 . A_{\alpha_2}, \ldots, \alpha_n . A_{\alpha_n}], \quad 0 < \alpha_i \leqslant 1, \quad i = 1, 2, \ldots, n.$$

The proof is immediate:

(5.142)

$$\begin{aligned} \mu_{A_{\alpha_i}}(x) &= 1 \quad \text{if} \quad \mu_{\underset{\sim}{A}}(x) \geqslant \alpha_i \\ &= 0 \quad \text{if} \quad \mu_{\underset{\sim}{A}}(x) < \alpha_i \end{aligned} .$$

Thus, the membership function of $\underset{\sim}{A}$ may be written

(5.143)

$$\begin{aligned} \mu_{\underset{\alpha_i}{\text{MAX}} [\alpha_i . A_{\alpha_i}]}(x) &= \underset{\alpha_i}{\text{MAX}} [\alpha_i \, A_{\alpha_i}] \\ &= \underset{\alpha_i \leqslant \mu_{\underset{\sim}{A}}(x)}{\text{MAX}} [\alpha_i] \\ &= \mu_{\underset{\sim}{A}}(x) . \end{aligned}$$

Example 1

(5.144)

x_1	x_2	x_3	x_4	x_5
0,2	0	0,5	1	0,7

= MAX((0,2) .

x_1	x_2	x_3	x_4	x_5
1	0	1	1	1

, (0,5) .

x_1	x_2	x_3	x_4	x_5
0	0	1	1	1

, (0,7) .

x_1	x_2	x_3	x_4	x_5
0.	0	0	1	1

, (1) .

x_1	x_2	x_3	x_4	x_5
0	0	0	1	0

).

Example 2. The decomposition formula (5.142) is still valid when the reference set has the power of the continuum. Let, for example,

(5.145)

$$\mu_{\underset{\sim}{A}}(x) = 1 - \frac{1}{1 + x^2}, \quad x \in \mathbf{R}^+ .$$

Considering the interval $[\alpha, 1]$, where $0 < \alpha \leqslant 1$, we may write

(5.146) $$\mu_{A_\alpha}(x) = 1 \quad \text{if} \quad \mu_{\underset{\sim}{A}}(x) \in [\alpha, 1]$$
$$= 0 \quad \text{if} \quad \mu_{\underset{\sim}{A}}(x) \notin [\alpha, 1] .$$

Thus, in the given example

(5.147) $$\mu_{A_\alpha}(x) = 1 \quad \text{if} \quad x \geqslant \sqrt{\frac{\alpha}{1-\alpha}}$$
$$= 0 \quad \text{if} \quad x < \sqrt{\frac{\alpha}{1-\alpha}} .$$

And for all arbitrary sets of value α, $0 < \alpha \leqslant 1$, one may decompose (5.145).

Synthesis of a fuzzy subset by joining ordinary subsets. The decomposition theorem may be applied not only for analysis but also for synthesis. If one then considers a sequence of ordinary subsets

(5.148) $$A_1 \subset\subset A_2 \subset\subset \ldots \subset\subset A_n$$

and attributes α_1 to A_1, α_2 to $A_2, \ldots, \alpha_n$ to A_n, with

(5.149) $$\alpha_1 > \alpha_2 > \ldots > \alpha_n .$$

then one obtains a fuzzy subset with the aid of (5.140).

6. SET OF FUZZY SUBSETS FOR E AND M FINITE

We restrict ourselves to the case where **E** and **M** are finite. Recall the definition of the *set of subsets* (or *power set*) of a set by considering a simple example. Let

(6.1) $$E = \{x_1, x_2, x_3\} \cdot$$

Then

(6.2) $$\mathcal{P}(E) = \{\phi, \{x_1\}, \{x_2\} \ \{x_3\}, \{x_1, x_2\}, \{x_1, x_3\}, \{x_2, x_3\}, E\} .$$

There are $2^3 = 8$ elements in this set. More generally, for a set

(6.3) $$E = \{x_1, x_2, \ldots, x_n\},$$

one may define 2^n elements in the same manner.

For fuzzy subsets, the power set or "set of fuzzy subsets" is presented in a different manner. First we consider an example. Let

(6.4)
(6.5) $$E = \{x_1, x_2\} \quad \text{and} \quad M = \{0, \tfrac{1}{2}, 1\} .$$

The set of fuzzy subsets† $\underset{\sim}{\mathcal{P}}(\mathbf{E})$ will be

$$(6.6)\quad \underset{\sim}{\mathcal{P}}(\mathbf{E}) = \{\{(x_1 \mid 0), (x_2 \mid 0)\}, \{(x_1 \mid 0), (x_2 \mid 0{,}5)\}, \{(x_1 \mid 0{,}5), (x_2 \mid 0)\}, \{(x_1 \mid 0{,}5), (x_2 \mid 0{,}5)\}, \{(x_1 \mid 0), (x_2 \mid 1)\}, \{(x_1 \mid 1), (x_2 \mid 0)\}, \{(x_1 \mid 1), (x_2 \mid 0{,}5)\}, \{(x_1 \mid 0{,}5), (x_2 \mid 1)\}, \{(x_1 \mid 1), (x_2 \mid 1)\}\} .$$

More generally, if

$$(6.6a)\qquad \text{card } \mathbf{E} = n \quad \text{and} \quad \text{card } \mathbf{M} = m ,$$

where card means "cardinality of," that is, gives the number of elements of the set, then

$$(6.7)\qquad \text{card } \underset{\sim}{\mathcal{P}}(\mathbf{E}) = m^n .$$

It follows that card $\underset{\sim}{\mathcal{P}}(\mathbf{E})$ is finite if and only if m and n are finite. The set $\underset{\sim}{\mathcal{P}}(\mathbf{E})$ contains 2^n ordinary subsets.

Consider another example for better comparison with (6.2):

$$(6.8)\qquad \mathbf{E} = \{x_1, x_2, x_3\} \quad \text{and} \quad \mathbf{M} = \{0, \tfrac{1}{2}, 1\} .$$

$$(6.9)\quad \underset{\sim}{\mathcal{P}}(\mathbf{E}) = \Big\{ \{(x_1 \mid 0), (x_2 \mid 0), (x_3 \mid 0)\}, \{(x_1 \mid 0), (x_2 \mid 0), (x_3 \mid \tfrac{1}{2})\}, \{(x_1 \mid 0), (x_2 \mid \tfrac{1}{2}), (x_3 \mid 0)\}, \{(x_1 \mid \tfrac{1}{2}), (x_2 \mid 0), (x_3 \mid 0)\}, \{(x_1 \mid 0), (x_2 \mid 0), (x_3 \mid 1)\}, \{x_1 \mid 0), (x_2 \mid \tfrac{1}{2}), (x_3 \mid \tfrac{1}{2})\}, \{(x_1 \mid 0), (x_2 \mid 1), (x_3 \mid 0)\}, \ldots, \{(x_1 \mid 1), (x_2 \mid \tfrac{1}{2}), (x_3 \mid 1)\}, \{(x_1 \mid 1), (x_2 \mid 1), (x_3 \mid \tfrac{1}{2})\}, \{(x_1 \mid 1), (x_2 \mid 1), (x_3 \mid 1)\} \Big\} .$$

It is well known that the structure of a power set $\mathcal{P}(\mathbf{E})$ of a set is a distributive and complementary lattice, that is, a boolean lattice.‡ The set of fuzzy subsets $\underset{\sim}{\mathcal{P}}(\mathbf{E})$, however, has the structure of a vectorial lattice that is distributive but not complementary.

†Note that one always has

$$\mathcal{P}(\mathbf{E}) \subset \underset{\sim}{\mathcal{P}}(\mathbf{E}) .$$

Thus, in considering (6.4) and (6.5), one may write

$$\mathcal{P}(\mathbf{E}) = \{\phi, \{x_1\}, \{x_2\}\ \{x_1, x_2\}\} = \{\{(x_1 \mid 0), (x_2 \mid 0)\}, \{(x_1 \mid 1), (x_2 \mid 0)\}, \{(x_1 \mid 0), (x_2 \mid 1)\}, \{(x_1 \mid 1), (x_2 \mid 1)\}\};$$

this is of course a subset of (6.6). One will see this more clearly later in Figures 6.1 and 6.2, 6.3 and 6.4, and 6.5 and 6.6.

‡Readers who have not studied the theory of a lattice are referred to [1F, 1K, 3K]. A review of lattice theory will be given in Section 54.

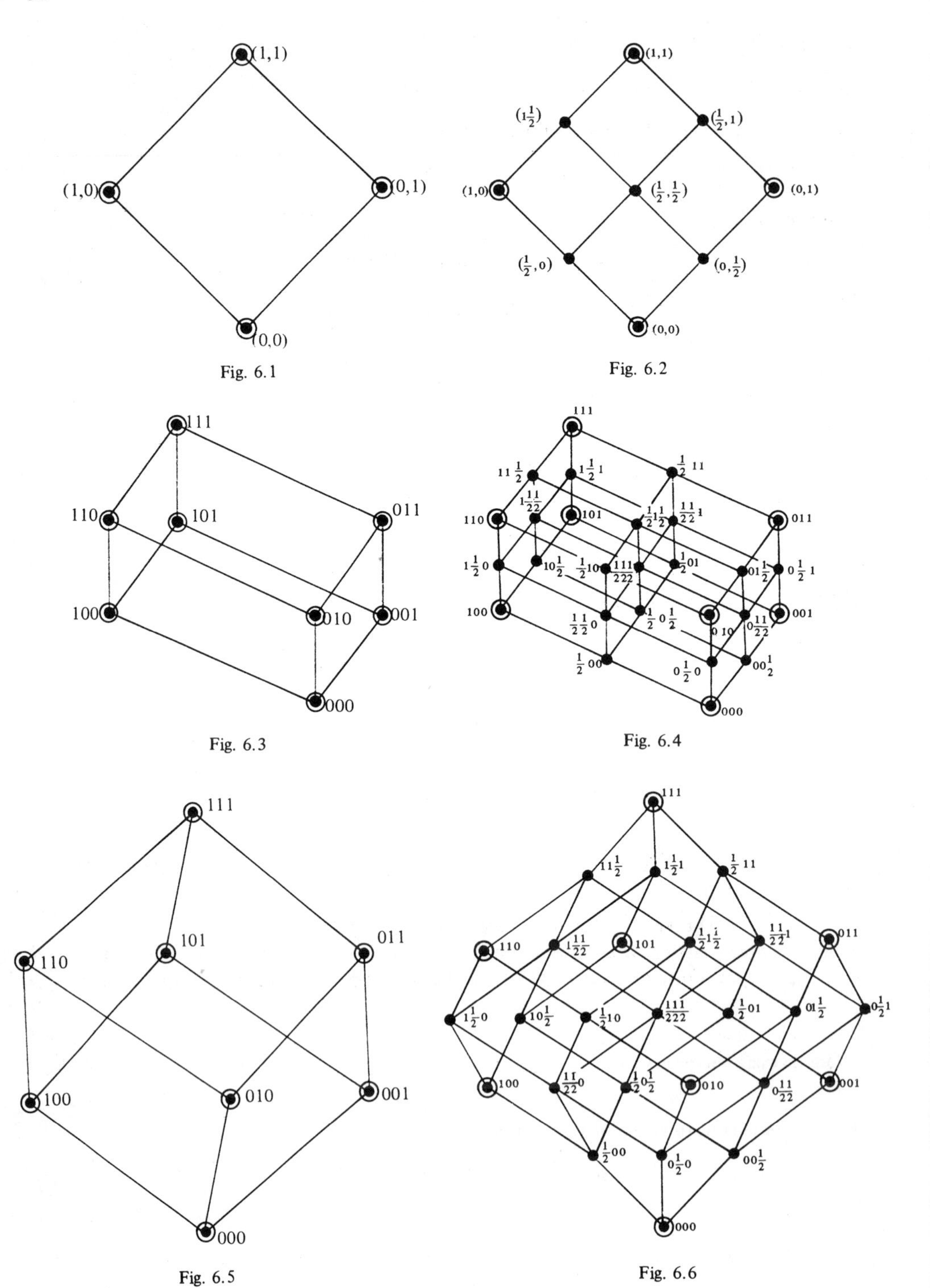

Fig. 6.1

Fig. 6.2

Fig. 6.3

Fig. 6.4

Fig. 6.5

Fig. 6.6

Recall that in a distributive lattice if the complement of an element exists, it is unique; and that this is the case for a vectorial lattice. The complementation considered here has a different sense from that given in (5.17).

This complementation does not necessarily give, as one says *complement* in a lattice, $\mathbf{A} \cap \overline{\mathbf{A}} = \phi$ and $\mathbf{A} \cup \overline{\mathbf{A}} = \mathbf{E}$. This is all the difference, but it is primary.

In Figures 6.1–6.6 we present several simple examples where, in order to simplify the labels, the fuzzy subsets are represented by their respective membership functions.

Figure 6.2: $\mathbf{E} = \{x_1, x_2\}$, $\mathbf{M} = \{0, 1/2, 1\}$. This figure represents a vector lattice of fuzzy subsets, and Figure 6.1 a boolean lattice of ordinary sets.

Figure 6.4: $\mathbf{E} = \{x_1, x_2, x_3\}$, $\mathbf{M} = \{0, 1/2, 1\}$. This figure represents a vector lattice of fuzzy subsets, and Figure 6.3, a boolean lattice of ordinary sets.

Figure 6.6: This is another representation of the vector lattice of Figure 6.4, to the left of which has been placed a boolean lattice of ordinary sets (Figure 6.5).

7. PROPERTIES OF THE SET OF FUZZY SUBSETS

Recall that the principal properties of the power set of an ordinary set **E** are as follows. Given $\mathbf{A} \subset \mathbf{E}$, $\mathbf{B} \subset \mathbf{E}$, $\mathbf{C} \subset \mathbf{E}$, one has:

(7.1) $\quad \mathrm{A} \cap \mathrm{B} = \mathrm{B} \cap \mathrm{A}\,,$

(7.2) $\quad \mathrm{A} \cup \mathrm{B} = \mathrm{B} \cup \mathrm{A}\,,$ — commutativity properties

(7.3) $\quad (\mathrm{A} \cap \mathrm{B}) \cap \mathrm{C} = \mathrm{A} \cap (\mathrm{B} \cap \mathrm{C})\,,$

(7.4) $\quad (\mathrm{A} \cup \mathrm{B}) \cup \mathrm{C} = \mathrm{A} \cup (\mathrm{B} \cup \mathrm{C})\,,$ — associativity properties

(7.5) $\quad \mathrm{A} \cap \mathrm{A} = \mathrm{A}\,,$

(7.6) $\quad \mathrm{A} \cup \mathrm{A} = \mathrm{A}\,.$ — idempotence

(7.7) $\quad \mathrm{A} \cap (\mathrm{B} \cup \mathrm{C}) = (\mathrm{A} \cap \mathrm{B}) \cup (\mathrm{A} \cap \mathrm{C})\,,$

(7.8) $\quad \mathrm{A} \cup (\mathrm{B} \cap \mathrm{C}) = (\mathrm{A} \cup \mathrm{B}) \cap (\mathrm{A} \cup \mathrm{C})\,,$ — distributivity of intersection with respect to union and of union with respect to intersection

(7.9) $\quad \mathrm{A} \cap \overline{\mathrm{A}} = \phi\,,$

(7.10) $\quad \mathrm{A} \cup \overline{\mathrm{A}} = \mathrm{E}\,,$

(7.11) $\quad \mathrm{A} \cap \phi = \phi\,,$

(7.12) $\quad \mathrm{A} \cup \phi = \mathrm{A}\,,$

(7.13) $\quad \mathrm{A} \cap \mathrm{E} = \mathrm{A}\,,$

(7.14) $\quad \mathrm{A} \cup \mathrm{E} = \mathrm{E}\,,$

(7.15) $\quad \overline{(\overline{\mathrm{A}})} = \mathrm{A}\,,$ — involution

(7.16) $$\overline{A \cap B} = \bar{A} \cup \bar{B},$$

(7.17) $$\overline{A \cup B} = \bar{A} \cap \bar{B}.$$

De Morgan's theorems

If $\underset{\sim}{A}$, $\underset{\sim}{B}$, and $\underset{\sim}{C}$ are fuzzy subsets of **E**, all the properties (7.1)–(7.17) are satisfied except (7.9) and (7.10). One may define a unique complement, but the properties (7.9) and (7.10) hold only for ordinary subsets.

(7.18) $$\underset{\sim}{A} \cap \underset{\sim}{B} = \underset{\sim}{B} \cap \underset{\sim}{A},$$

(7.19) $$\underset{\sim}{A} \cup \underset{\sim}{B} = \underset{\sim}{B} \cup \underset{\sim}{A},$$

commutativity

(7.20) $$(\underset{\sim}{A} \cap \underset{\sim}{B}) \cap \underset{\sim}{C} = \underset{\sim}{A} \cap (\underset{\sim}{B} \cap \underset{\sim}{C}),$$

(7.21) $$(\underset{\sim}{A} \cup \underset{\sim}{B}) \cup \underset{\sim}{C} = \underset{\sim}{A} \cup (\underset{\sim}{B} \cup \underset{\sim}{C}),$$

associativity

(7.22) $$\underset{\sim}{A} \cap \underset{\sim}{A} = \underset{\sim}{A},$$

(7.23) $$\underset{\sim}{A} \cup \underset{\sim}{A} = \underset{\sim}{A},$$

idempotence

(7.24) $$\underset{\sim}{A} \cap (\underset{\sim}{B} \cup \underset{\sim}{C}) = (\underset{\sim}{A} \cap \underset{\sim}{B}) \cup (\underset{\sim}{A} \cap \underset{\sim}{C}),$$

(7.25) $$\underset{\sim}{A} \cup (\underset{\sim}{B} \cap \underset{\sim}{C}) = (\underset{\sim}{A} \cup \underset{\sim}{B}) \cap (\underset{\sim}{A} \cup \underset{\sim}{C}),$$

distributivity

(7.26) $$\underset{\sim}{A} \cap \phi = \phi,$$ where ϕ is the ordinary set such that $\forall x_i \in \mathbf{E}: \ \mu_\phi(x_i) = 0,$

(7.27) $$\underset{\sim}{A} \cup \phi = \underset{\sim}{A},$$

(7.28) $$\underset{\sim}{A} \cap \mathbf{E} = \underset{\sim}{A},$$ where **E** is the ordinary set such that $\forall x_i \in \mathbf{E} : \mu_{\mathbf{E}}(x_i) = 1$, that is, the reference set.

(7.29) $$\underset{\sim}{A} \cup \mathbf{E} = \mathbf{E},$$

(7.30) $$\overline{(\bar{\underset{\sim}{A}})} = \underset{\sim}{A},$$ involution

(7.31) $$\overline{\underset{\sim}{A} \cap \underset{\sim}{B}} = \bar{\underset{\sim}{A}} \cup \bar{\underset{\sim}{B}}$$

(7.32) $$\overline{\underset{\sim}{A} \cup \underset{\sim}{B}} = \bar{\underset{\sim}{A}} \cap \bar{\underset{\sim}{B}}$$

De Morgan's theorems for the case of fuzzy subsets

Thus, we stress: All the properties of an ordinary power set are found again in a power set of fuzzy subsets, except (7.9) and (7.10). Thus, we no longer have an *algebra* in the sense of the theory of ordinary sets; the structure is that of a vector lattice.

8. ALGEBRAIC PRODUCT AND SUM OF TWO FUZZY SUBSETS

Let **E** be a set and **M** = [0, 1] its associated membership set, let $\underset{\sim}{A}$ and $\underset{\sim}{B}$ be two fuzzy subsets of **E**; one defines the *algebraic product* of $\underset{\sim}{A}$ and $\underset{\sim}{B}$, denoted

(8.1) $$\underset{\sim}{A} \,.\, \underset{\sim}{B} \,,$$

in the following manner:

(8.2) $$\forall x \in E: \quad \mu_{\underset{\sim}{A}.\underset{\sim}{B}}(x) = \mu_{\underset{\sim}{A}}(x) \,.\, \mu_{\underset{\sim}{B}}(x)\,.$$

Likewise one defines the *algebraic sum* of these two subsets, denoted

(8.3) $$\underset{\sim}{A} \hat{+} \underset{\sim}{B}\,,$$

in the following manner:

(8.4) $$\forall x \in E: \quad \mu_{\underset{\sim}{A}\hat{+}\underset{\sim}{B}}(x) = \mu_{\underset{\sim}{A}}(x) + \mu_{\underset{\sim}{B}}(x) - \mu_{\underset{\sim}{A}}(x) \,.\, \mu_{\underset{\sim}{B}}(x)\,.$$

Consider again the example in (5.25)−(5.27):

(8.5) $$\underset{\sim}{A} = \{(x_1 \mid 0{,}2), (x_2 \mid 0{,}7), (x_3 \mid 1), (x_4 \mid 0), (x_5 \mid 0{,}5)\},$$

(8.6) $$\underset{\sim}{B} = \{(x_1 \mid 0{,}5), (x_2 \mid 0{,}3), (x_3 \mid 1), (x_4 \mid 0{,}1), (x_5 \mid 0{,}5)\}.$$

(8.7) $$\underset{\sim}{A} \,.\, \underset{\sim}{B} = \{(x_1 \mid 0{,}10), (x_2 \mid 0{,}21), (x_3 \mid 1), (x_4 \mid 0), (x_5 \mid 0{,}25)\},$$

(8.8) $$\underset{\sim}{A} \hat{+} \underset{\sim}{B} = \{(x_1 \mid 0{,}60), (x_2 \mid 0{,}79), (x_3 \mid 1), (x_4 \mid 0{,}1), (x_5 \mid 0{,}75)\}.$$

We now make the following important remark: If $\mathbf{M} = \{0, 1\}$, that is, if we are in the case of ordinary subsets, then

(8.9) $$A \cap B = A \,.\, B\,,$$

(8.10) $$A \cup B = A \hat{+} B\,.$$

In fact, if $\mu_{A}(x) \in \{0, 1\}$ and $\mu_{B}(x) \in \{0, 1\}$, the following tables are equivalent, but this is not true for $\mathbf{M} \neq \{0, 1\}$, except in a few trivial cases.

(8.11)

MIN	0	1
0	0	0
1	0	1

is equivalent to

(.)	0	1
0	0	0
1	0	1

,

(8.12)

MAX	0	1
0	0	1
1	1	1

is equivalent to

$(\hat{+})$	0	1
0	0	1
1	1	1

.

In the present work we use the algebraic product and sum operations rather infrequently, but these constitute an interesting direction for other research.

If one considers the two operations $\cdot$ and $\hat{+}$ on the power set of fuzzy subsets, only the following properties may be verified; these are obviously more restricted than those for $\cap$ and $\cup$ for the power set of fuzzy subsets, and a fortiori those concerning $\cap$ and $\cup$ for ordinary power sets. One may easily verify

(8.13) $\underset{\sim}{A} . \underset{\sim}{B} = \underset{\sim}{B} . \underset{\sim}{A}$ } commutativity

(8.14) $\underset{\sim}{A} \hat{+} \underset{\sim}{B} = \underset{\sim}{B} \hat{+} \underset{\sim}{A}$

(8.15) $(\underset{\sim}{A} . \underset{\sim}{B}) . \underset{\sim}{C} = \underset{\sim}{A} . (\underset{\sim}{B} . \underset{\sim}{C})$ } associativity

(8.16) $(\underset{\sim}{A} \hat{+} \underset{\sim}{B}) \hat{+} \underset{\sim}{C} = \underset{\sim}{A} \hat{+} (\underset{\sim}{B} \hat{+} \underset{\sim}{C})$

(8.17) $\underset{\sim}{A} . \phi = \phi ,$

(8.18) $\underset{\sim}{A} \hat{+} \phi = \underset{\sim}{A} ,$

(8.19) $\underset{\sim}{A} . E = \underset{\sim}{A} ,$

(8.20) $\underset{\sim}{A} \hat{+} E = E ,$

(8.21) $\overline{(\overline{\underset{\sim}{A}})} = \underset{\sim}{A} ,$ involution

(8.22) $\overline{\underset{\sim}{A} . \underset{\sim}{B}} = \overline{\underset{\sim}{A}} \hat{+} \overline{\underset{\sim}{B}}$ } De Morgan's theorems for the operations $\cdot$ and $\hat{+}$ on fuzzy subsets

(8.23) $\overline{\underset{\sim}{A} \hat{+} \underset{\sim}{B}} = \overline{\underset{\sim}{A}} . \overline{\underset{\sim}{B}}$

Thus properties, (7.5) and (7.6) (idempotence) are not satisfied, nor are (7.7) and (7.8) (distributivity), nor likewise (7.9) and (7.10). This gives a noticeably poorer structure, especially because of the absence of distributivity. We shall show through several examples how to prove properties (8.13)–(8.23).

We prove (8.16), for example, by putting

(8.24) $$a = \mu_{\underset{\sim}{A}}(x) , b = \mu_{\underset{\sim}{B}}(x) , c = \mu_{\underset{\sim}{C}}(x) :$$

(8.25) $(\underset{\sim}{A} \hat{+} \underset{\sim}{B}) \hat{+} \underset{\sim}{C} = \underset{\sim}{A} \hat{+} (\underset{\sim}{B} \hat{+} \underset{\sim}{C})$ is verified if

(8.26) $$(a + b - ab) + c - (a + b - ab)\, c = a + (b + c - bc) - a\,(b + c - bc)$$

is verified. By expanding the two members one has

(8.27)

$$a + b - ab + c - ac - bc + abc = a + b + c - bc - ab - ac + abc .$$

The two sides are indeed identical. Thus, (8.25) is a correct formula.

We prove (8.22). The equation

(8.28) $$\overline{\underset{\sim}{A} \,.\, \underset{\sim}{B}} = \overline{\underset{\sim}{A}} \,\hat{+}\, \overline{\underset{\sim}{B}} \quad \text{is verified if}$$

(8.29)
$$\begin{aligned} 1 - ab &= (1 - a) + (1 - b) - (1 - a)(1 - b) \\ &= 1 - a + 1 - b - 1 + a + b - ab \\ &= 1 - ab \,. \end{aligned}$$

We now prove that distributivity does not hold; for example,

(8.30) $$\underset{\sim}{A} \,.\, (\underset{\sim}{B} \,\hat{+}\, \underset{\sim}{C}) \neq (\underset{\sim}{A} \,.\, \underset{\sim}{B}) \,\hat{+}\, (\underset{\sim}{A} \,.\, \underset{\sim}{C}) \,.$$

For the left-hand side of this equation, one must have

(8.31) $$a \,.\, (b + c - bc) = ab + ac - abc \,.$$

For the member on the right.

(8.32) $$ab + ac - (ab)(ac) = ab + ac - a^2bc \,.$$

These then prove nondistributivity.

We note that $\cup$ is not distributive with respect to $\cdot$ or $\hat{+}$, and likewise $\cap$; but on the other hand one has

(8.33) $$\underset{\sim}{A} \,.\, (\underset{\sim}{B} \cap \underset{\sim}{C}) = (\underset{\sim}{A} \,.\, \underset{\sim}{B}) \cap (\underset{\sim}{A} \,.\, \underset{\sim}{C}) \,,$$

(8.34) $$\underset{\sim}{A} \,.\, (\underset{\sim}{B} \cup \underset{\sim}{C}) = (\underset{\sim}{A} \,.\, \underset{\sim}{B}) \cup (\underset{\sim}{A} \,.\, \underset{\sim}{C}) \,,$$

(8.35) $$\underset{\sim}{A} \,\hat{+}\, (\underset{\sim}{B} \cap \underset{\sim}{C}) = (\underset{\sim}{A} \,\hat{+}\, \underset{\sim}{B}) \cap (\underset{\sim}{A} \,\hat{+}\, \underset{\sim}{C}) \,,$$

(8.36) $$\underset{\sim}{A} \,\hat{+}\, (\underset{\sim}{B} \cup \underset{\sim}{C}) = (\underset{\sim}{A} \,\hat{+}\, \underset{\sim}{B}) \cup (\underset{\sim}{A} \,\hat{+}\, \underset{\sim}{C}) \,.$$

Index of fuzziness for a product. It is possible to define an index of fuzziness for a product similar to (5.108); one puts

(8.37) $$\eta(\underset{\sim}{A}) = \frac{4}{N} \sum_{i=1}^{N} \mu_{\underset{\sim}{A} \,.\, \overline{\underset{\sim}{A}}}(x_i).$$

Example. Let

(8.38) $\underset{\sim}{A}$ =

x_1	x_2	x_3	x_4	x_5	x_6	x_7
0,7	0,2	0,9	1	0	0,4	1

,

(8.39) $\overline{\underset{\sim}{A}}$ =

x_1	x_2	x_3	x_4	x_5	x_6	x_7
0,3	0,8	0,1	0	1	0,6	0

,

(8.40)

	x_1	x_2	x_3	x_4	x_5	x_6	x_7
$\underset{\sim}{A} \cdot \overline{\underset{\sim}{A}} =$	0,21	0,16	0,09	0	0	0,24	0

(8.41) $$\eta(\underset{\sim}{A}) = \frac{4}{7}(0{,}21 + 0{,}16 + 0{,}09 + 0 + 0 + 0{,}24 + 0) = 0{,}40 .$$

One may then, on the other hand, raise the following question, as we have in Section 5 for $\cap$ and $\cup$: Are the indexes of fuzziness for $\underset{\sim}{A} \cdot \underset{\sim}{B}$ or $\underset{\sim}{A} \hat{+} \underset{\sim}{B}$ greater than or less than those of $\underset{\sim}{A}$ or/and $\underset{\sim}{B}$? Unfortunately, the operations $\cdot$ and $\hat{+}$ do not always modify the fuzziness in the same sense, as may be seen with examples.

General remark on the subject of fuzziness. We have seen that each of the operations $\cap$, $\cup$, $\cdot$, $\hat{+}$ does not systematically increase or decrease the fuzziness of a subset $\underset{\sim}{A}$ in applying these operations with other subsets of the same reference set. It should be borne in mind that the membership function is supposed known in order to treat fuzzy subsets adequately.

If $*$ is one of the four operations considered above, one may not say, a priori, that for $\underset{\sim}{A} \subset E$, $\underset{\sim}{B} \subset E$, $\underset{\sim}{A}$ and $\underset{\sim}{B}$ arbitrary, whether $\nu(\underset{\sim}{A} * \underset{\sim}{B})$ is greater than or less than $\nu(\underset{\sim}{A})$ or $\nu(\underset{\sim}{B})$.

One has the same situation in considering entropy. It recurs there that if one wishes to increase or decrease the entropy H, it is necessary that one have knowledge of $\underset{\sim}{A}$: knowledge of H is not sufficient, one may surmise.

9. EXERCISES†

I1. Consider a reference set

(9.1) $$E = \{A, B, C, D, E, F, G\}$$

and the fuzzy subsets

(9.2) $$\underset{\sim}{A} = \{(A \mid 0), (B \mid 0{,}3), (C \mid 0{,}7), (D \mid 1), (E \mid 0), (F \mid 0{,}2), (G \mid 0{,}6)\},$$

$$\underset{\sim}{B} = \{(A \mid 0{,}3), (B \mid 1), (C \mid 0{,}5), (D \mid 0{,}8), (E \mid 1), (F \mid 0{,}5), (G \mid 0{,}6)\},$$

$$\underset{\sim}{C} = \{(A \mid 1), (B \mid 0{,}5), (C \mid 0{,}5), (D \mid 0{,}2), (E \mid 0), (F \mid 0{,}2), (G \mid 0{,}9)\}.$$

Calculate:

a) $\underset{\sim}{A} \cap \underset{\sim}{B}$, b) $\underset{\sim}{A} \cup \underset{\sim}{B}$, c) $\underset{\sim}{A} \cap \overline{\underset{\sim}{B}}$, d) $(\underset{\sim}{A} \cup \overline{\underset{\sim}{B}}) \cap \underset{\sim}{C}$, e) $\overline{(\underset{\sim}{A} \cap \underset{\sim}{B})} \cup \overline{\underset{\sim}{C}}$, f) $\underset{\sim}{A} \oplus \underset{\sim}{B}$,

g) $\overline{\underset{\sim}{A} \oplus \underset{\sim}{B}}$, h) $(\underset{\sim}{A} \cap \overline{\underset{\sim}{A}}) \cup \underset{\sim}{A}$.

I2. Considering the fuzzy subsets of Exercise I1, give:

†These exercises are all very simple; they are intended to check quickly that one has understood.

a) $\delta(\underset{\sim}{A} . \underset{\sim}{B})$, $\delta(\underset{\sim}{B} . \underset{\sim}{C})$, $\delta(\underset{\sim}{A} . \underset{\sim}{C})$,

b) $\epsilon(\underset{\sim}{A} . \underset{\sim}{B})$, $\epsilon(\underset{\sim}{B} . \underset{\sim}{C})$, $\epsilon(\underset{\sim}{A} , \underset{\sim}{C})$,

c) $\nu(\underset{\sim}{A})$, $\nu(\underset{\sim}{B})$, $\nu(\underset{\sim}{A} \cap \underset{\sim}{B})$, $\nu(\underset{\sim}{A} \cup \underset{\sim}{B})$, $\nu(\overline{\underset{\sim}{A}})$,

d) $\eta(\underset{\sim}{A})$, $\eta(\underset{\sim}{B})$, $\eta(\underset{\sim}{A} \cap \underset{\sim}{B})$, $\eta(\underset{\sim}{A} \cup \underset{\sim}{B})$, $\eta(\overline{\underset{\sim}{A}})$.

I3. Consider the reference set

$$\mathrm{E} = [0\,, a] \subset \mathrm{R}\,, \tag{9.3}$$

If $\underset{\sim}{A}$ is the fuzzy subset defined by $\mu_{\underset{\sim}{A}}(x)$, give the index ν of fuzziness of $\underset{\sim}{A}$.

a) $\mu_{\underset{\sim}{A}}(x) = \dfrac{x^2}{a^2}\,, \quad x \in [0\,, a]\,.$

b) $\mu_{\underset{\sim}{A}}(x) = \dfrac{(x-a)^2}{a^2}\,, \quad x \in [0\,, a]\,.$

c) $$\begin{aligned} \mu_{\underset{\sim}{A}}(x) &= \frac{4\,x^2}{a^2}\,, & 0 \leqslant x \leqslant \frac{a}{2} \\ &= \frac{4\,(x-a)^2}{a^2}\,, & \frac{a}{2} \leqslant x \leqslant a\,. \end{aligned}$$

I4. Give the ordinary subset of level α for the fuzzy subset

$$\underset{\sim}{A} = \{(A \mid 0{,}7)\,, (B \mid 0{,}5)\,, (C \mid 1)\,, (D \mid 0{,}2)\,, (E \mid 0{,}6)\} \tag{9.4}$$

a) $\alpha = 0{,}1$, b) $\alpha = 0{,}6$, c) $\alpha = 0{,}8$, d) $\alpha = 0{,}9$.

Give the decomposition of the fuzzy subset in the form (5.140).

I5. Give the power set of fuzzy subsets for the following cases:

a) $\mathrm{E} = \{x_1\,, x_2\}\,, \quad \mathrm{M} = \left\{0\,, \frac{1}{3}, \frac{2}{3}, 1\right\}.$

b) $\mathrm{E} = \{x_1\,, x_2\,, x_3\}\,, \quad \mathrm{M} = \{a\,, b\,, c\}, \quad a < b < c\,.$

I6. Prove the following properties:

a) $\underset{\sim}{A} \cap (\underset{\sim}{A} \cup \underset{\sim}{B}) = \underset{\sim}{A}$ and $\underset{\sim}{A} \cup (\underset{\sim}{A} \cap \underset{\sim}{B}) = \underset{\sim}{A}$.

b) $\phi \subset \underset{\sim}{A} \cap \overline{\underset{\sim}{A}} \subset \underset{\sim}{A} \cup \overline{\underset{\sim}{A}} \subset \mathrm{E}$.

c) $(\underset{\sim}{A} \cap \underset{\sim}{B}) \cup (\underset{\sim}{B} \cap \underset{\sim}{C}) \cup (\underset{\sim}{C} \cap \underset{\sim}{A}) = (\underset{\sim}{A} \cup \underset{\sim}{B}) \cap (\underset{\sim}{B} \cup \underset{\sim}{C}) \cap (\underset{\sim}{C} \cup \underset{\sim}{A})$.

I7. Considering the three fuzzy subsets of Exercise I1, calculate:

a) $\underset{\sim}{A} \mathbin{\hat{+}} \underset{\sim}{B} \mathbin{\hat{+}} \underset{\sim}{C}$.

b) $\underset{\sim}{\mathbf{A}} \,.\, (\underset{\sim}{\mathbf{B}} \mathbin{\hat{+}} \underset{\sim}{\mathbf{C}})\,.$

and prove:

c) $\underset{\sim}{\mathbf{A}} \,.\, \underset{\sim}{\mathbf{A}} \subset \underset{\sim}{\mathbf{A}}$ and $\underset{\sim}{\mathbf{A}} \mathbin{\hat{+}} \underset{\sim}{\mathbf{A}} \supset \underset{\sim}{\mathbf{A}}\,.$

d) $\underset{\sim}{\mathbf{A}} \,.\, \underset{\sim}{\mathbf{B}} \mathbin{\hat{+}} \underset{\sim}{\mathbf{A}} \,.\, \underset{\sim}{\mathbf{C}} \supset \underset{\sim}{\mathbf{A}} \,.\, (\underset{\sim}{\mathbf{B}} \mathbin{\hat{+}} \underset{\sim}{\mathbf{C}})\,.$

18. Simplify the expression

$$[\underset{\sim}{\mathbf{A}} \cap [(\underset{\sim}{\mathbf{B}} \cap \underset{\sim}{\mathbf{C}}) \cup (\underset{\sim}{\bar{\mathbf{A}}} \cap \underset{\sim}{\bar{\mathbf{C}}})]] \cup \underset{\sim}{\bar{\mathbf{C}}}\,. \tag{9.5}$$

CHAPTER II

FUZZY GRAPHS AND FUZZY RELATIONS

10. INTRODUCTION

The notions of graph, correspondence, and relation play a fundamental role in applications of mathematics. They may be generalized with respect to the notion of fuzzy subsets. One will then discover some new and very interesting properties. For example, the notion of an equivalence class will be found to be replaced by that of similitude, stronger and more apt for representing some less-precise but more often encountered situations. Preorder and order are likewise generalized; whereas some other relations, such as resemblance and dissemblance, are defined. This, then, is a new theory that may be formed with fuzzy relations. It is only a beginning. It is likely that research on fuzzy concepts will develop progressively in importance and will permit at least good descriptions of complex phenomena, constrained until now to the specifications all or nothing.

11. FUZZY GRAPHS

Consider two sets $\mathbf{E}_1$ and $\mathbf{E}_2$; let x designate an element of $\mathbf{E}_1$ and y an element of $\mathbf{E}_2$. The set of ordered pairs (x, y) defines the product set $\mathbf{E}_1 \times \mathbf{E}_2$.

The fuzzy subset $\underset{\sim}{G}$ such that

(11.1) $$\forall (x, y) \in \mathbf{E}_1 \times \mathbf{E}_2 : \mu_{\underset{\sim}{G}}(x, y) \in \mathbf{M}$$

where $\mathbf{M}$ is the membership set of $\mathbf{E}_1 \times \mathbf{E}_2$, is called a *fuzzy graph*.

Example 1. Let

(11.2) $$\mathbf{E}_1 = \{x_1, x_2, x_3\}$$

and

(11.3) $$\mathbf{E}_2 = \{y_1, y_2\}.$$

(11.4) $$\mathbf{E}_1 \times \mathbf{E}_2 = \{(x_1, y_1), (x_1, y_2), (x_2, y_1), (x_2, y_2), (x_3, y_1), (x_3, y_2)\}.$$

Set, in order to simplify notation,

(11.5) $$\mu(x_i, y_j) = \mu_{\underset{\sim}{G}}(x_i, y_j) \quad , i = 1, 2, 3 \; , j = 1, 2 \; ,$$

which will be called the *value* of the ordered pair (x_i, y_i).

Consider for example:

$$\mu(x_1, y_1) = 0{,}3 \quad , \quad \mu(x_1, y_2) = 0{,}7 \quad , \quad \mu(x_2, y_1) = 1 \; ,$$
$$\mu(x_2, y_2) = 0 \quad , \quad \mu(x_3, y_1) = 0{,}5 \quad , \quad \mu(x_3, y_2) = 0{,}2 \; . \tag{11.6}$$

This function defines the fuzzy subset

$$\underset{\sim}{G} = \{((x_1, y_1) \mid 0{,}3), ((x_1, y_2) \mid 0{,}7), ((x_2, y_1) \mid 1), ((x_2, y_2) \mid 0), ((x_3, y_1) \mid 0{,}5), ((x_3, y_2) \mid 0{,}2)\} \; . \tag{11.7}$$

This fuzzy subset may be represented by a matrix such as that shown in Figure 11.1.

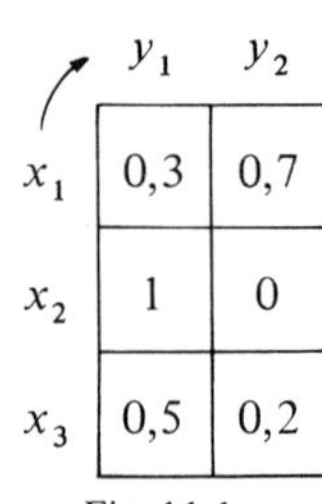

	y_1	y_2
x_1	0,3	0,7
x_2	1	0
x_3	0,5	0,2

Fig. 11.1

The graph

$$\underset{\sim}{G} \subset \mathbf{E}_1 \times \mathbf{E}_2 \tag{11.8}$$

is a *fuzzy graph.*

The graph

$$\mathbf{G} = \{((x_1, y_1) \mid 0), ((x_1, y_2) \mid 1), ((x_2, y_1) \mid 1), ((x_2, y_2) \mid 1), ((x_3, y_1) \mid 1), ((x_3, y_2) \mid 0)\} \; , \tag{11.9}$$

is an ordinary graph in set theory (Figure 11.2).

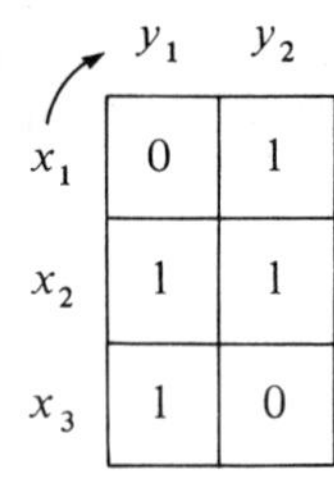

	y_1	y_2
x_1	0	1
x_2	1	1
x_3	1	0

Fig. 11.2

Example 2. Let $\mathbf{E}_1 = \mathbf{E}_2 = \mathbf{R}^+$, where $\mathbf{R}^+$ is the set of nonnegative real numbers. Let $x \in \mathbf{R}^+$ and $y \in \mathbf{R}^+$ and consider the product set $\mathbf{R}^+ \times \mathbf{R}^+$. Then the relation $y \geqslant x$ defines a fuzzy graph in $\mathbf{R}^{+2}$.

Suppose that one has a use for the function

$$\begin{aligned} &\mu(x, y \mid y = x) = e^0 = 1 \; , \\ &\mu(x, y \mid y = 2x) = e^{-1} \; , \\ &\ldots . \\ &\mu(x, y \mid y = kx) = e^{-(k-1)} \\ &\ldots\ldots . \qquad k = 1, 2, 3, 4, \ldots , \end{aligned} \tag{11.10}$$

with $\mathbf{M} = \{1, e^{-1}, e^{-2}, \ldots, e^{-k-1}, \ldots, 0\}$.

Figure 11.3 gives a visual representation of this fuzzy subset for the points $y = kx$, $k \geqslant 1$.

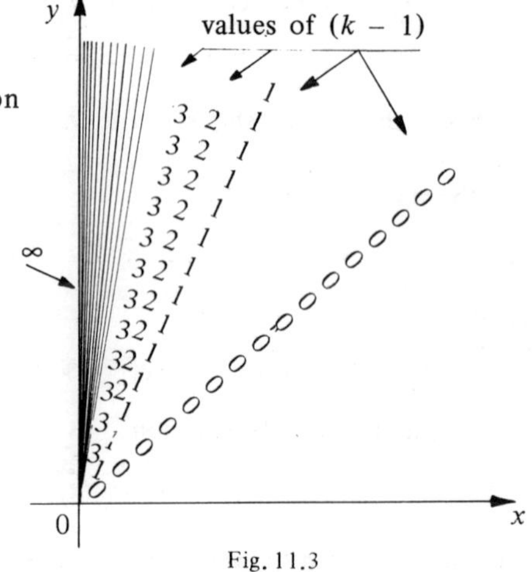

Fig. 11.3

Example 3 (Berge graphs). A graph in the sense of Berge† is one such that

(11.11) $$\mathbf{E}_1 = \mathbf{E}_2 = \mathbf{E} \quad \text{countable}$$

and is formed by the subset of ordered pairs

$$(x\,,\,y) \in \mathbf{G} \subset \mathbf{E} \times \mathbf{E}\,,$$

(11.12) such that $$\mathbf{G} \cap \overline{\mathbf{G}} = \phi\,,$$

(11.13) and $$\mathbf{G} \cup \overline{\mathbf{G}} = \mathbf{E} \times \mathbf{E}\,.$$

For such graphs, which evidently are only a particular case of the notion of graphs in set theory, one may define a generalization to fuzzy graphs. Thus, Figures 11.4, 11.6, 11.8, and 11.10 represent the same Berge fuzzy graph, whereas Figures 11.5, 11.7, 11.9, and 11.11 show the same ordinary Berge graph.‡

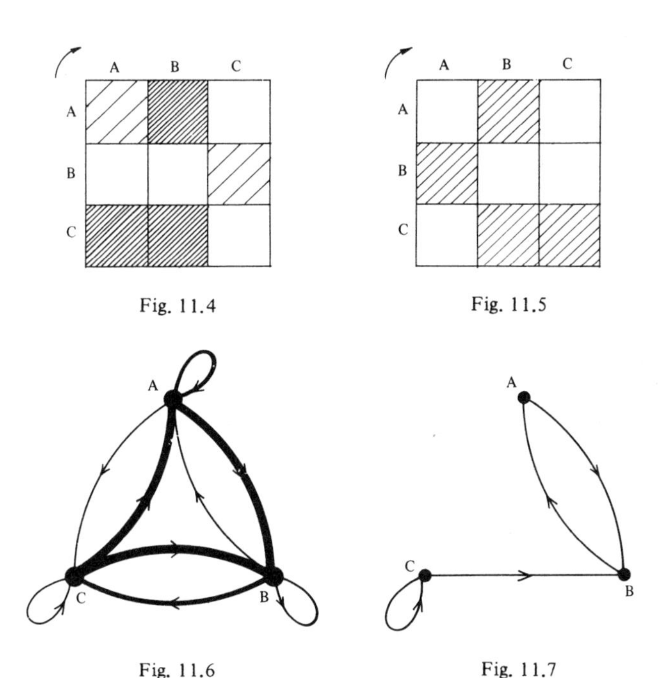

Fig. 11.4 Fig. 11.5

Fig. 11.6 Fig. 11.7

† One also says in the sense of Berge and Konig. See references [1B, 1K, 2K, 5K].

‡ I hope that my good friend Professor Claude Berge will not be offended by such an adjective given a most useful concept in modern mathematics, for one who has brought so much to mathematical science.

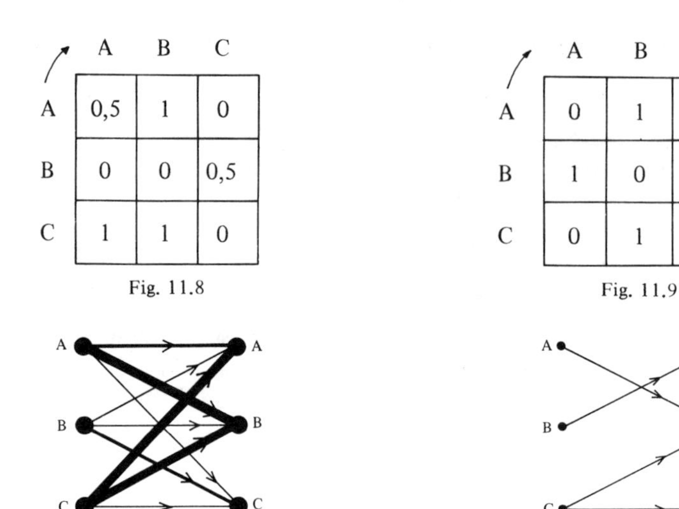

Fig. 11.8

Fig. 11.9

Fig. 11.10

Fig. 11.11

Using Berge's notation, for the ordinary graph in Figures 11.5, 11.7, 11.9, and 11.11, one puts

$$\Gamma\{A\} = \{B\} \, ,$$
$$(11.14) \qquad \Gamma\{B\} = \{A\} \, ,$$
$$\Gamma\{C\} = \{B \, , C\} \, ,$$

where $\Gamma\{\mathbf{X}\}$ is called a *multivalued mapping of* $\{\mathbf{X}\}$ *in its reference set* **E**.

In the spirit of this notation, one will write for the fuzzy graph represented in various fashions in Figures 11.4, 11.6, 11.8, and 11.10

$$\Gamma\{A\} = \{(A \mid 0{,}5) \, , (B \mid 1) \, , (C \mid 0)\} \, ,$$
$$(11.15) \qquad \Gamma\{B\} = \{(A \mid 0) \, , (B \mid 0) \, , (C \mid 0{,}5)\} \, ,$$
$$\Gamma\{C\} = \{(A \mid 1) \, , (B \mid 1) \, , (C \mid 0)\} \, .$$

Example 4. Figure 11.12 represents a fuzzy graph and Figure 11.13 an ordinary graph.

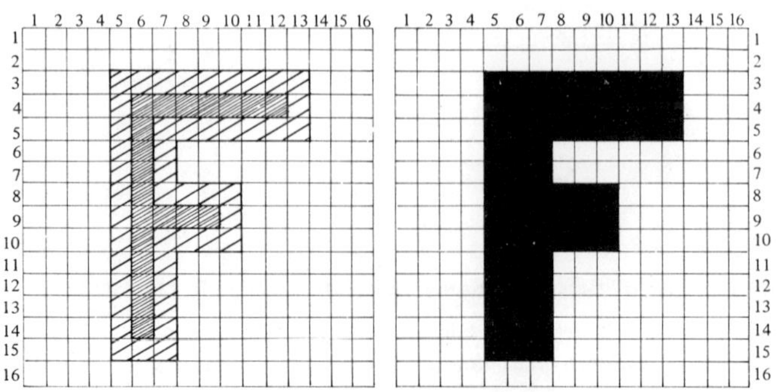

Fig. 11.12

Fig. 11.13

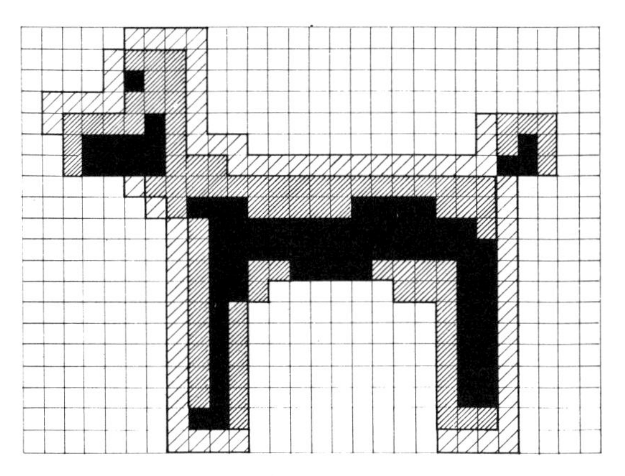

Fig. 11.14

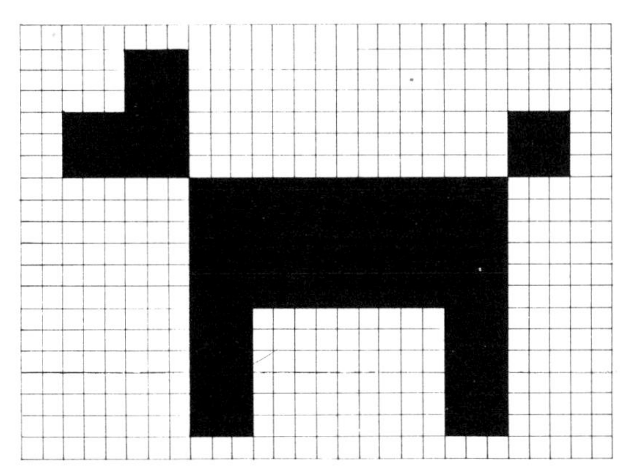

Fig. 11.15

Also, Figure 11.14 represents a fuzzy graph and Figure 11.15 an ordinary graph.

Example 5. The shaded parts of Figure 11.16, where we attribute a value $\mu(x, y)$, to each point (x, y), represents a fuzzy graph.

Generalization. What has been presented for a product set $\mathbf{E}_1 \times \mathbf{E}_2$ may be generalized for a product set

$$\mathbf{E}_1 \times \mathbf{E}_2 \times \cdots \times \mathbf{E}_n .$$

The fuzzy subset such that $x^{(i)} \in \mathbf{E}_i$, $i = 1, 2, \ldots, n$,

(11.16) $\forall (x^{(1)}, x^{(2)}, \ldots, x^{(n)}) \in \mathbf{E}_1 \times \mathbf{E}_2 \times \cdots \times \mathbf{E}_n$:

$\mu(x^{(1)}, x^{(2)}, \ldots, x^{(n)}) \in \mathbf{M}$,

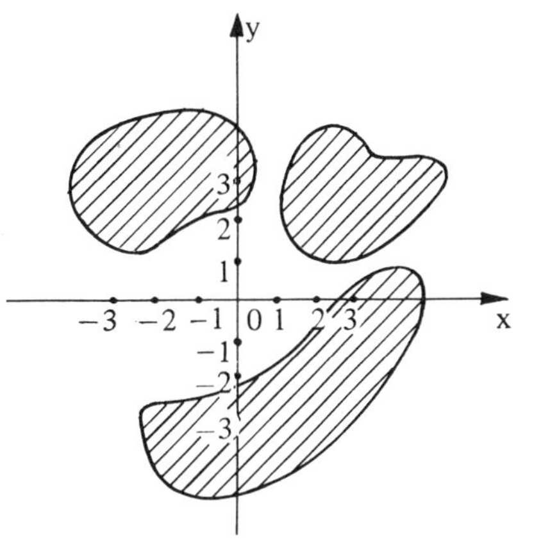

Fig. 11.16

where $\mathbf{M}$ is the membership set of $\mathbf{E}_1 \times \mathbf{E}_2 \times \ldots \times \mathbf{E}_n$, is called a *fuzzy graph*.

Example. Let

(11.17) $$\mathbf{E}_1 = \{x_1, x_2\}, \ \mathbf{E}_2 = \{y_1, y_2\}, \ \mathbf{E}_3 = \{z_1, z_2\}, \quad \mathbf{M} = [0, 1].$$

(11.18) $$\underset{\sim}{G} = \{((x_1, y_1, z_1) | 0{,}3), ((x_1, y_1, z_2) | 0{,}2), ((x_1, y_2, z_1) | 1), ((x_1, y_2, z_2) | 0), ((x_2, y_1, z_1) | 0), ((x_2, y_1, z_2) | 0{,}1), ((x_2, y_2, z_1) | 0.9), ((x_2, y_2, z_2) | 0{,}7)\}.$$

is a fuzzy graph in $\mathbf{E}_1 \times \mathbf{E}_2 \times \mathbf{E}_3$.

12. FUZZY RELATION

As is done in the theory of ordinary sets,† the notion of a fuzzy graph may be explained in terms of the notion of a *fuzzy relation.* Let **P** be a product set of n sets and **M** its membership function; a fuzzy n-ary relation is a fuzzy subset of **P** taking its values in **M.**

Example 1. Let

(12.1) $$\mathbf{E}_1 = \{x_1, x_2, x_3\},$$

(12.2) $$\mathbf{E}_2 = \{y_1, y_2, y_3, y_4, y_5\}$$

(12.3) $$\mathbf{M} = [0, 1].$$

The table presented in Figure 12.1 expresses a fuzzy 2-ary relation (which we may describe as binary if no confusion with other interpretations of the word *binary* is possible).

	y_1	y_2	y_3	y_4	y_5
x_1	0	0	0,1	0,3	1
x_2	0	0,8	0	0	1
x_3	0,4	0,4	0,5	0	0,2

Fig. 12.1

Example 2. Let

(12.4) $$\mathbf{E}_1 = \mathbf{E}_2 = \mathbf{R},$$

where $\mathbf{R} = (-\infty, \infty)$, that is, the set of real numbers. Then the relation $y \ll x$, where $x \in \mathbf{R}$, $y \in \mathbf{R}$, is a fuzzy relation in $\mathbf{R}^2$.

For example, a subjective expression (that is, a valuation that may depend on a subjective estimate) of $y \ll x$ may be given by

(12.5) $$\mu_{\mathbf{R}^2}(x, y) = 0 \quad \text{if} \quad y \geqslant x,$$
$$= \frac{1}{1 + \dfrac{1}{(x-y)^2}} \quad \text{if} \quad y < x.$$

†That is, in the theory of sets.

Notation. A fuzzy relation in $\mathbf{E}_1 \times \mathbf{E}_2$ will be written as

(12.6) $$x \in \mathbf{E}_1 \ , \ y \in \mathbf{E}_2 \quad : \quad x \underset{\sim}{\mathcal{R}} y \ .$$

Symbols used for extrema. In what follows we will use the symbols

$\bigvee_x$ to represent the maximum with respect to an element or variable x,

$\bigwedge_x$ to represent the minimum with respect to an element or variable x,

Thus, writing

(12.7) $$\mu_1(x) = \bigvee_y \mu(x, y) \ ,$$

will be equivalent to

(12.8) $$\mu_1(x) = \underset{y}{\mathrm{MAX}} \ \mu(x, y) \ .$$

Likewise, writing

(12.9) $$\mu_2(x) = \bigwedge_y \mu(x, y) \ ,$$

will be equivalent to

(12.10) $$\mu_2(x) = \underset{y}{\mathrm{MIN}} \ \mu(x, y) \ .$$

Projection of a fuzzy relation. The membership function

(12.11) $$\mu^{(1)}_{\underset{\sim}{\mathcal{R}}}(x) = \bigvee_y \mu_{\underset{\sim}{\mathcal{R}}}(x, y)$$

defines the *first projection* of $\underset{\sim}{\mathcal{R}}$.

In the same fashion, the membership function

(12.12) $$\mu^{(2)}_{\underset{\sim}{\mathcal{R}}}(y) = \bigvee_x \mu_{\underset{\sim}{\mathcal{R}}}(x, y)$$

defines the *second projection* of $\underset{\sim}{\mathcal{R}}$.

The second projection of the first projection (or vice versa) will be called the *global projection* of the fuzzy relation and will be denoted $h(\underset{\sim}{\mathcal{R}})$ Thus

(12.13) $$h(\underset{\sim}{\mathcal{R}}) = \bigvee_x \bigvee_y \mu_{\underset{\sim}{\mathcal{R}}}(x, y)$$
$$= \bigvee_y \bigvee_x \mu_{\underset{\sim}{\mathcal{R}}}(x, y) \ .$$

If $h(\underset{\sim}{\mathcal{R}}) = 1$, the relation is said to be *normal.* If $h(\underset{\sim}{\mathcal{R}}) < 1$, the relation is called *subnormal.*

$\underset{\sim}{\mathcal{R}}$	y_1	y_2	y_3	y_4	1st proj.
x_1	0,1	0,2	1	0,3	1
x_2	0,6	0,8	0	0,1	0,8
x_3	0	1	0,3	0,6	1
x_4	0,8	0,1	1	0	1
x_5	0,9	0,7	0	0,5	0,9
x_6	0,9	0	0,3	0,7	0,9
2nd proj.	0,9	1	1	0,7	1 (global projection)

Fig. 12.2

Example 1. (Figure 12.2). We calculate the first projection:

(12.14)
$$\mu^{(1)}_{\underset{\sim}{\mathcal{R}}}(x) = \bigvee_y \mu_{\underset{\sim}{\mathcal{R}}}(x, y),$$
$$\mu^{(1)}_{\underset{\sim}{\mathcal{R}}}(x_1) = \bigvee_y \mu_{\underset{\sim}{\mathcal{R}}}(x_1, y) = \text{MAX}[0{,}1\,,\,0{,}2\,,\,1\,,\,0{,}3] = 1\,,$$
$$\mu^{(1)}_{\underset{\sim}{\mathcal{R}}}(x_2) = \bigvee_y \mu_{\underset{\sim}{\mathcal{R}}}(x_2, y) = \text{MAX}[0{,}6\,,\,0{,}8\,,\,0\,,\,0{,}1] = 0{,}8\,,$$
....
$$\mu^{(1)}_{\underset{\sim}{\mathcal{R}}}(x_6) = \bigvee_y \mu_{\underset{\sim}{\mathcal{R}}}(x_6, y) = \text{MAX}[0{,}9\,,\,0\,,\,0{,}3\,,\,0{,}7] = 0{,}9\,.$$

One may similarly calculate the second projection. The results are given in Figure 12.2. We see that this relation $\underset{\sim}{\mathcal{R}}$ is normal.

Example 2. We consider the case of a relation $x \underset{\sim}{\mathcal{R}} y$ where $x \in \mathbf{R}^+$ and $y \in \mathbf{R}^+$ with

(12.15)
$$\mu_{\underset{\sim}{\mathcal{R}}}(x, y) = e^{-k(y-x)^2}, \quad k > 1$$

(Figure 12.3), which one may take to be defined by the fuzzy phrase: x and y are very near to one another (for a sufficient value of k).

In this case we see, for a fixed value x_0,

(12.16)
$$\mu^{(1)}_{\underset{\sim}{\mathcal{R}}}(x_0) = \bigvee_y \mu_{\underset{\sim}{\mathcal{R}}}(x_0, y)$$
$$= \bigvee_y e^{-k(y-x_0)^2}$$
$$= e^{-k(y-x_0)^2} \quad \text{for } y = x_0$$
$$= 1\,.$$

One will find the same value for $\mu^{(2)}_{\underset{\sim}{\mathcal{R}}}(y_0)$ and therefore $h(\underset{\sim}{\mathcal{R}}) = 1$.

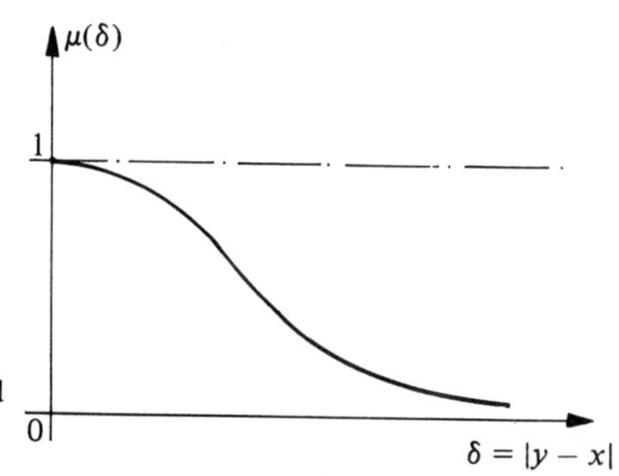

Fig. 12.3

Support of a fuzzy relation. One will call the *support of* $\underset{\sim}{\mathcal{R}}$ the ordinary subset of ordered pairs (x, y) for which the membership function is nonzero. Thus

(12.17)
$$\mathbf{S}(\underset{\sim}{\mathcal{R}}) = \{(x, y) \mid \mu_{\underset{\sim}{\mathcal{R}}}(x, y) > 0\}\,.$$

Example 1. (Figure 12.4)

(12.18)
$$\mathbf{S}(\underset{\sim}{\mathcal{R}}) = \{(x_1, y_1), (x_1, y_3), (x_2, y_1), (x_2, y_4), (x_3, y_1), (x_3, y_2), (x_3, y_3), (x_3, y_4)\}\,.$$

$\underset{\sim}{\mathcal{R}}$	y_1	y_2	y_3	y_4
x_1	0,1	0	0,2	0
x_2	0,3	0	0	0,9
x_3	0,4	0,7	1	1

Fig. 12.4

Example 2. Consider a relation $x \underset{\sim}{\mathcal{R}} y$ where $x \in \mathbf{R}^+$ and $y \in \mathbf{R}^+$ with

(12.19)

$$\mu_{\underset{\sim}{\mathcal{R}}}(x, y) = e^{-(y-x)^2}, \quad |y - x| \leqslant 0{,}46,$$

$$= 0 \quad , \quad |y - x| > 0{,}46 .$$

(fig. 12.5).

One then has

(12.20) $\mathbf{S}(\underset{\sim}{\mathcal{R}}) = \{(x, y) \mid 0 \leqslant |y - x| \leqslant 0{,}46\}$.

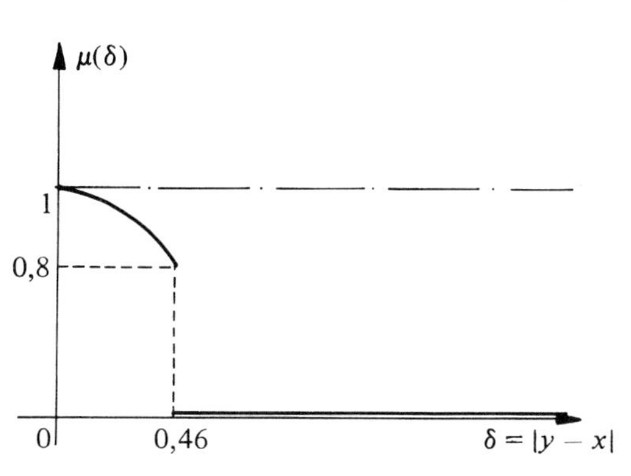

Fig. 12.5

Envelope of a fuzzy relation. Let $\underset{\sim}{\mathcal{R}}$ and $\underset{\sim}{\mathcal{Q}}$ be two fuzzy relations such that

$$\forall (x, y) \in \mathbf{E}_1 \times \mathbf{E}_2 \; : \; \mu_{\underset{\sim}{\mathcal{R}}}(x, y) \leqslant \mu_{\underset{\sim}{\mathcal{Q}}}(x, y) \; ; \tag{12.21}$$

one then says that $\underset{\sim}{\mathcal{Q}}$ is *an envelope* of $\underset{\sim}{\mathcal{R}}$ or that $\underset{\sim}{\mathcal{R}}$ is an *enclosure* of $\underset{\sim}{\mathcal{Q}}$.

We note

$$\underset{\sim}{\mathcal{R}} \subset \underset{\sim}{\mathcal{Q}} , \tag{12.22}$$

If $\underset{\sim}{\mathcal{Q}}$ is *an envelope* of $\underset{\sim}{\mathcal{R}}$.

Example 1. (Figure 12.6). One may verify that $\underset{\sim}{\mathcal{Q}}$ is *an envelope* of $\underset{\sim}{\mathcal{R}}$.

$\underset{\sim}{\mathcal{R}}$	y_1	y_2	y_3	y_4
x_1	0,3	0,4	0,2	0
x_2	0,5	0	1	0,9
x_3	0,4	0	0,1	0,8

(1)

$\underset{\sim}{\mathcal{Q}}$	y_1	y_2	y_3	y_4
x_1	0,4	0,4	0,2	0,1
x_2	0,5	0	1	1
x_3	0,5	0,1	0,2	0,9

(2)

Fig. 12.6

Example 2. Consider the fuzzy relation $x \underset{\sim}{\mathcal{R}}_1 y$ with $x \in \mathbf{R}^+$ and $y \in \mathbf{R}^+$ such that $y \gg x$, that is, "y is much larger than x," expressed by

(12.23) $\mu_{\underset{\sim}{\mathcal{R}}_1}(x, y) = 0 \quad y - x < 0$

$= 1 - e^{-k_1(y-x)^2}$

$y - x \geqslant 0 .$

Let now $k_2 > k_1$; then

(12.24) $\mu_{\underset{\sim}{\mathcal{R}}_2}(x, y) = 0 \quad y - x < 0$

$= 1 - e^{-k_2(y-x)^2}$

$y - x \geqslant 0 .$

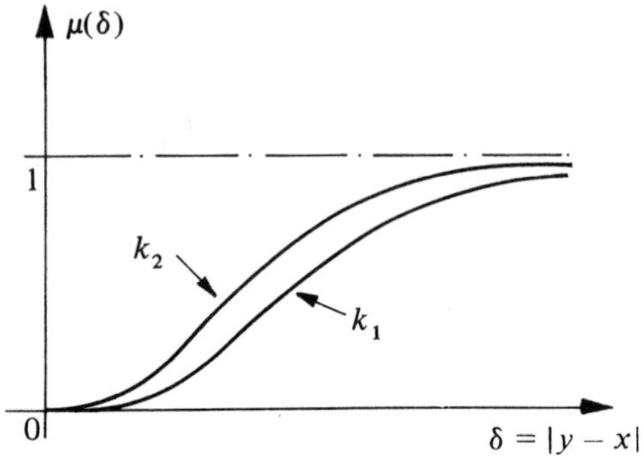

Fig. 12.7

is an envelope of (12.33) (see Figure 12.7).

Union of two relations. The *union* of two relations $\underset{\sim}{\mathcal{R}}$ and $\underset{\sim}{\mathcal{Q}}$, denoted $\underset{\sim}{\mathcal{R}} \cup \underset{\sim}{\mathcal{Q}}$, or $\underset{\sim}{\mathcal{R}} + \underset{\sim}{\mathcal{Q}}$, is a relation such that

$$\mu_{\underset{\sim}{\mathcal{R}} \cup \underset{\sim}{\mathcal{Q}}}(x, y) = \mu_{\underset{\sim}{\mathcal{R}}}(x, y) \vee \mu_{\underset{\sim}{\mathcal{Q}}}(x, y) \tag{12.25}$$

$$= \text{MAX}[\mu_{\underset{\sim}{\mathcal{R}}}(x, y), \mu_{\underset{\sim}{\mathcal{Q}}}(x, y)] .$$

If $\underset{\sim}{\mathcal{R}}_1, \underset{\sim}{\mathcal{R}}_2, \ldots, \underset{\sim}{\mathcal{R}}_n$ are relations,

$$\mu_{\underset{\sim}{\mathcal{R}}_1 \cup \underset{\sim}{\mathcal{R}}_2 \cup \cdots \cup \underset{\sim}{\mathcal{R}}_n}(x, y) = \bigvee_{\underset{\sim}{\mathcal{R}}_i} \mu_{\underset{\sim}{\mathcal{R}}_i}(x, y) . \tag{12.26}$$

We note the result

$$\underset{\sim}{\mathcal{R}} = \bigcup_i \underset{\sim}{\mathcal{R}}_i \quad \text{or} \quad \sum_i \underset{\sim}{\mathcal{R}}_i . \tag{12.27}$$

Example 1. (Figure 12.8)

(1)

$\underset{\sim}{\mathcal{R}}$	y_1	y_2	y_3	y_4
x_1	0,3	0,2	1	0
x_2	0,8	1	0	0,2
x_3	0,5	0	0,4	0

(2)

$\underset{\sim}{\mathcal{Q}}$	y_1	y_2	y_3	y_4
x_1	0,3	0	0,7	0
x_2	0,1	0,8	1	1
x_3	0,6	0,9	0,3	0,2

(3)

$\underset{\sim}{\mathcal{R}} \cup \underset{\sim}{\mathcal{Q}}$	y_1	y_2	y_3	y_4
x_1	0,3	0,2	1	0
x_2	0,8	1	1	1
x_3	0,6	0,9	0,4	0,2

Fig. 12.8

Example 2. In Figure 12.9a we have expressed a fuzzy relation $x \underset{\sim}{\mathcal{R}}_1 y$, $x \in \mathbf{R}^+$, and $y \in \mathbf{R}^+$, such that "x and y are very near." In Figure 12.9b one sees a relation $x \underset{\sim}{\mathcal{R}}_2 y$, $x \in \mathbf{R}^+$, and $\underset{\sim}{\mathbf{Y}} \in \mathbf{R}^+$, such that "$x$ and y are very different."

The relation $x \underset{\sim}{\mathcal{R}}_3 y$ such that "x and y are very near or/and very different" is defined by the curve $\mu_3(x, y)$ such that

(12.28)

$$\mu_{\underset{\sim}{\mathcal{R}}_3}(x, y) = 0 \quad , \quad |y - x| < 0$$
$$= \mu_{\underset{\sim}{\mathcal{R}}_1}(x, y) \quad , \quad 0 \leqslant |y - x| \leqslant \alpha$$
$$= \mu_{\underset{\sim}{\mathcal{R}}_2}(x, y) \quad , \quad \alpha \leqslant |y - x| \; ,$$

with

$|y-x| = \alpha$ such that

(12.29) $$\mu_{\underset{\sim}{\mathcal{R}}_1}(x, y) = \mu_{\underset{\sim}{\mathcal{R}}_2}(x, y) \; .$$

In a logic constructed on the theory of ordinary sets, a proposition like "x and y are very near or/and very different" must be reduced to "x and y are very near or very different." But with respect to the theory of fuzzy subsets, the first proposition is coherent; it expresses that the "and" case is conceivable with a very weak weight, corresponding to the case where x and y are neither very near nor very different.

This example illustrates well the propositional flexibility that one finds in the present theory.

Intersection of two relations. The intersection of two relations $\underset{\sim}{\mathcal{R}}$ and $\underset{\sim}{\mathcal{S}}$, denoted $\underset{\sim}{\mathcal{R}} \cap \underset{\sim}{\mathcal{S}}$, is defined by the expression

(12.30) $$\mu_{\underset{\sim}{\mathcal{R}} \cap \underset{\sim}{\mathcal{S}}}(x, y) = \mu_{\underset{\sim}{\mathcal{R}}}(x, y) \wedge \mu_{\underset{\sim}{\mathcal{S}}}(x, y) = \text{MIN}\,[\mu_{\underset{\sim}{\mathcal{R}}}(x, y), \mu_{\underset{\sim}{\mathcal{S}}}(x, y)] \, .$$

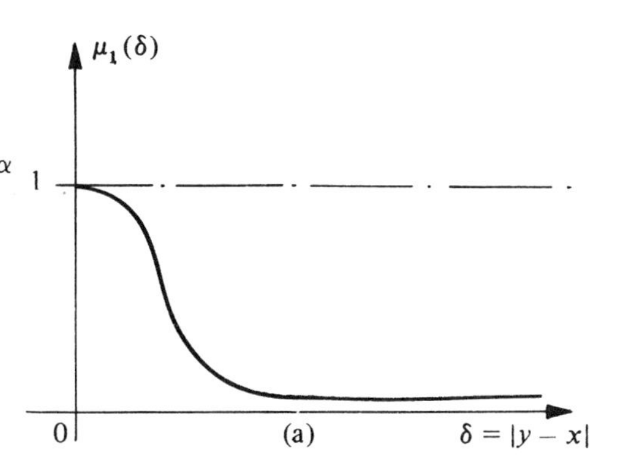

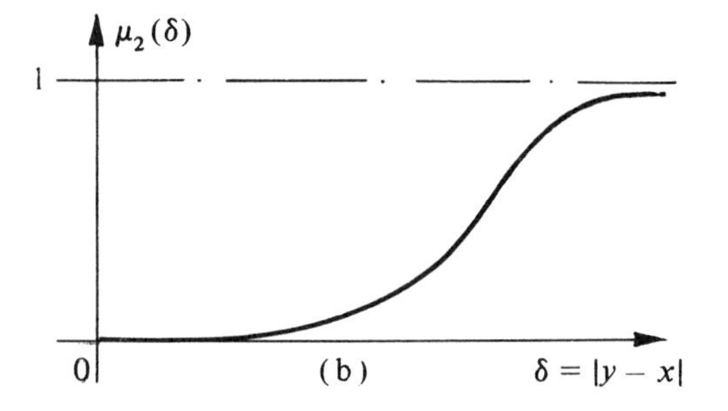

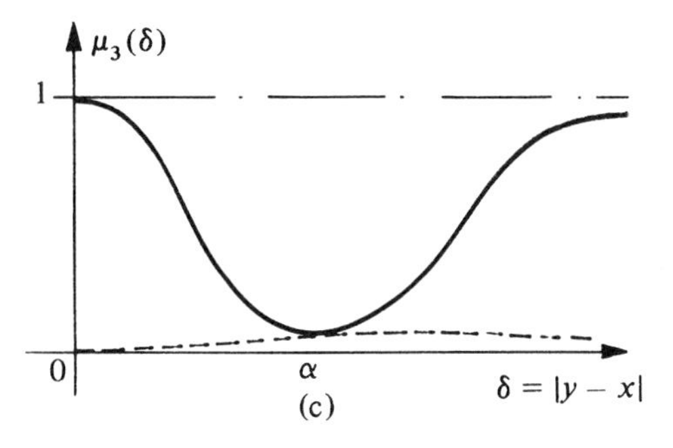

Fig. 12.9

If $\underset{\sim}{\mathcal{R}}_1, \underset{\sim}{\mathcal{R}}_2, \ldots, \underset{\sim}{\mathcal{R}}_n$ are relations,

(12.31) $$\mu_{\underset{\sim}{\mathcal{R}}_1 \cap \underset{\sim}{\mathcal{R}}_2 \cap \cdots \cap \underset{\sim}{\mathcal{R}}_n}(x, y) = \bigwedge_{\underset{\sim}{\mathcal{R}}_i} \mu_{\underset{\sim}{\mathcal{R}}_i}(x, y) \; .$$

We note the result

(12.32) $$\underset{\sim}{\mathcal{R}} = \bigcap_i \underset{\sim}{\mathcal{R}}_i \; .$$

$\underset{\sim}{\mathcal{R}} \cap \underset{\sim}{\mathcal{S}}$	y_1	y_2	y_3	y_4
x_1	0,3	0	0,7	0
x_2	0,1	0,8	0	0,2
x_3	0,5	0	0,3	0

Fig. 12.10

Example 1. (Figure 12.10). We have reconsidered the example given in Figure 12.8.

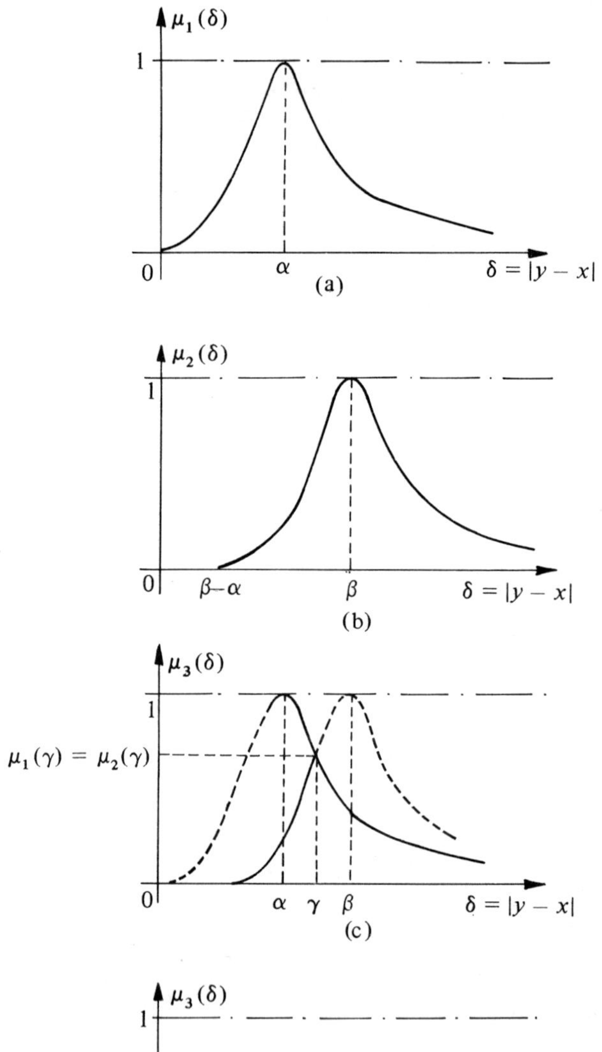
μ1 (δ)
1
0
α
δ = |y − x|
(a)
μ2 (δ)
1
0
β−α
β
δ = |y − x|
(b)
μ3 (δ)
1
μ1 (γ) = μ2(γ)
0
α γ β
δ = |y − x|
(c)

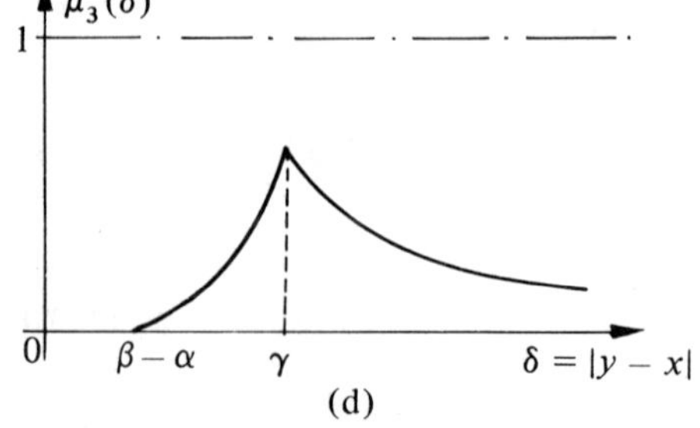
μ3 (δ)
1
0
β − α
γ
δ = |y − x|
(d)

Fig. 12.11

Example 2. In Figure 12.11a is expressed the fuzzy relation $x \underset{\sim}{\mathcal{R}}_1 y$, $x \in \mathbf{R}^+$, $y \in \mathbf{R}^+$, that "$|y-x|$ is very near α." In Figure 12.11b is represented the similar relation "$|y-x|$ is very near β" (with $\beta > \alpha$).

Figure 12.11c shows how to obtain

$$\underset{\sim}{\mathcal{R}}_3 = \underset{\sim}{\mathcal{R}}_1 \cap \underset{\sim}{\mathcal{R}}_2 . \tag{12.33}$$

One has

$$\begin{aligned} \mu_{\underset{\sim}{\mathcal{R}}_3}(x, y) &= 0 && , \ |y-x| < \beta - \alpha , \\ &= \mu_{\underset{\sim}{\mathcal{R}}_2}(x, y) && , \ \beta - \alpha \leqslant |y-x| \leqslant \gamma , \\ &= \mu_{\underset{\sim}{\mathcal{R}}_1}(x, y) && , \ \gamma \leqslant |y-x| , \end{aligned} \tag{12.34}$$

where γ is the value of $|y-x|$ such that $\mu_{\underset{\sim}{\mathcal{R}}_1}(x, y) = \mu_{\underset{\sim}{\mathcal{R}}_2}(x, y)$.

The result appears in Figure 12.11d.

Algebraic product of two relations. One defines the algebraic product of two relations $\underset{\sim}{\mathcal{R}}$ and $\underset{\sim}{\mathcal{Q}}$, denoted $\underset{\sim}{\mathcal{R}} . \underset{\sim}{\mathcal{Q}}$, by the expression

$$\mu_{\underset{\sim}{\mathcal{R}} . \underset{\sim}{\mathcal{Q}}}(x, y) = \mu_{\underset{\sim}{\mathcal{R}}}(x, y) \cdot \mu_{\underset{\sim}{\mathcal{Q}}}(x, y) . \tag{12.35}$$

In the right-hand side of this expression, the . indicates a numerical product (ordinary multiplication).

Example 1. (Figure 12.12). Again we reconsider the example of Figure 12.8.

$\underset{\sim}{\mathcal{R}} \cdot \underset{\sim}{\mathcal{Q}}$	y_1	y_2	y_3	y_4
x_1	0,09	0	0,7	0
x_2	0,08	0,8	0	0,2
x_3	0,3	0	0,12	0

Fig. 12.12

Example 2. Taking again the example considered in Figures 12.11a and 12.11b, let

$$\underset{\sim}{\mathcal{R}}_3 = \underset{\sim}{\mathcal{R}}_1 \cdot \underset{\sim}{\mathcal{R}}_2 ; \tag{12.36}$$

then one has

$$\begin{aligned} \mu_{\underset{\sim}{\mathcal{R}}_3}(x, y) &= 0 , \ |y-x| < \beta - \alpha , \\ &= \mu_{\underset{\sim}{\mathcal{R}}_1}(x, y) \cdot \mu_{\underset{\sim}{\mathcal{R}}_2}(x, y) , \ \beta - \alpha \leqslant |y-x| . \end{aligned} \tag{12.37}$$

See Figures 12.13a–c.

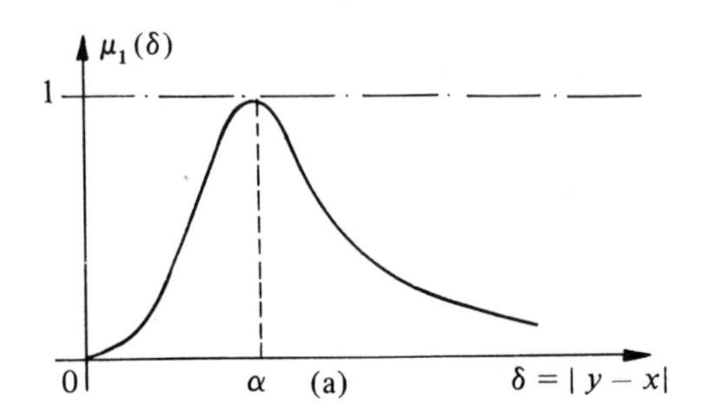

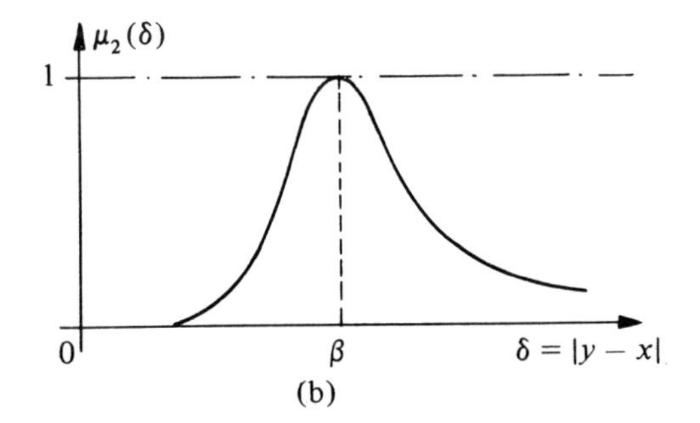

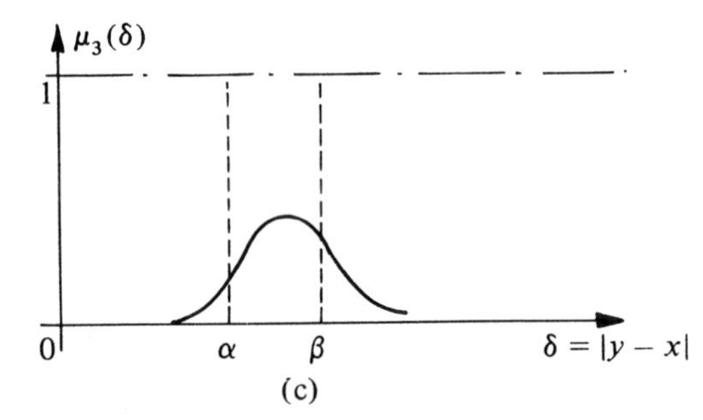

Fig. 12.13

Distributivity. We note the properties of distributivity:

(12.38) $$\underset{\sim}{\mathcal{R}} \cap (\underset{\sim}{\mathcal{Q}} \cup \underset{\sim}{\mathcal{C}}) = (\underset{\sim}{\mathcal{R}} \cap \underset{\sim}{\mathcal{Q}}) \cup (\underset{\sim}{\mathcal{R}} \cap \underset{\sim}{\mathcal{C}}) ,$$

(12.39) $$\underset{\sim}{\mathcal{R}} \cup (\underset{\sim}{\mathcal{Q}} \cap \underset{\sim}{\mathcal{C}}) = (\underset{\sim}{\mathcal{R}} \cup \underset{\sim}{\mathcal{Q}}) \cap (\underset{\sim}{\mathcal{R}} \cup \underset{\sim}{\mathcal{C}}) ,$$

(12.40) $$\underset{\sim}{\mathcal{R}} \cdot (\underset{\sim}{\mathcal{Q}} \cup \underset{\sim}{\mathcal{C}}) = (\underset{\sim}{\mathcal{R}} \cdot \underset{\sim}{\mathcal{Q}}) \cup (\underset{\sim}{\mathcal{R}} \cdot \underset{\sim}{\mathcal{C}}) .$$

(12.41) $$\underset{\sim}{\mathcal{R}} \cdot (\underset{\sim}{\mathcal{Q}} \cap \underset{\sim}{\mathcal{C}}) = (\underset{\sim}{\mathcal{R}} \cdot \underset{\sim}{\mathcal{Q}}) \cap (\underset{\sim}{\mathcal{R}} \cdot \underset{\sim}{\mathcal{C}}) .$$

Algebraic sum of two relations. One defines the algebraic sum of two relations $\underset{\sim}{\mathcal{R}}$ and $\underset{\sim}{\mathcal{Q}}$, denoted $\underset{\sim}{\mathcal{R}} \hat{+} \underset{\sim}{\mathcal{Q}}$, by the expression

(12.42) $$\mu_{\underset{\sim}{\mathcal{R}} \hat{+} \underset{\sim}{\mathcal{Q}}}(x, y) = \mu_{\underset{\sim}{\mathcal{R}}}(x, y) + \mu_{\underset{\sim}{\mathcal{Q}}}(x, y) - \mu_{\underset{\sim}{\mathcal{R}}}(x, y) \cdot \mu_{\underset{\sim}{\mathcal{Q}}}(x, y)$$

The $\cdot$ indicates ordinary multiplication and the sign +, ordinary addition.

Example. (Figure 12.14). The example of Figure 12.8 is considered once again.

$\underset{\sim}{\mathcal{R}} \hat{+} \underset{\sim}{\mathcal{Q}}$	y_1	y_2	y_3	y_4
x_1	0,51	0,20	1	0
x_2	0,82	1	1	1
x_3	0,80	0,90	0,58	0,20

Fig. 12.14

We note two properties of distributivity:

$$\underset{\sim}{\mathcal{R}} \hat{+} (\underset{\sim}{\mathcal{Q}} \cup \underset{\sim}{\mathcal{C}}) = (\underset{\sim}{\mathcal{R}} \hat{+} \underset{\sim}{\mathcal{Q}}) \cup (\underset{\sim}{\mathcal{R}} \hat{+} \underset{\sim}{\mathcal{C}}) , \tag{12.43}$$

$$\underset{\sim}{\mathcal{R}} \hat{+} (\underset{\sim}{\mathcal{Q}} \cap \underset{\sim}{\mathcal{C}}) = (\underset{\sim}{\mathcal{R}} \hat{+} \underset{\sim}{\mathcal{Q}}) \cap (\underset{\sim}{\mathcal{R}} \hat{+} \underset{\sim}{\mathcal{C}}) . \tag{12.44}$$

Complement of a relation. The complement of $\underset{\sim}{\mathcal{R}}$, denoted $\overline{\underset{\sim}{\mathcal{R}}}$, is the relation such that

$$\forall (x , y) \in \mathbf{E}_1 \times \mathbf{E}_2 \ : \quad \mu_{\overline{\underset{\sim}{\mathcal{R}}}}(x , y) = 1 - \mu_{\underset{\sim}{\mathcal{R}}}(x , y) . \tag{12.45}$$

Example 1. (Figure 12.15).

(1)

$\underset{\sim}{\mathcal{R}}$	y_1	y_2	y_3	y_4
x_1	0,3	0,4	0,2	0
x_2	0,5	0	1	0,9
x_3	0,4	0	0,1	0,8

(2)

$\overline{\underset{\sim}{\mathcal{R}}}$	y_1	y_2	y_3	y_4
x_1	0,7	0,6	0,8	1
x_2	0,5	1	0	0,1
x_3	0,6	1	0,9	0,2

Fig. 12.15

Example 2. In Figure 12.16a is represented the membership function $\mu_{\mathcal{R}_1}(x, y)$, corresponding to the relation $x \underset{\sim}{\mathcal{R}}_1 y$ signifying "x and y are very near to one another," $x \in \mathbf{R}^+$ and $y \in \mathbf{R}^+$.

Figure 12.16b then represents the membership function

$$\mu_{\underset{\sim}{\mathcal{R}}_2}(x , y) = 1 - \mu_{\underset{\sim}{\mathcal{R}}_1}(x , y) , \tag{12.46}$$

which may be associated with the relation "x and y are not very near."

Figure 12.16c may be taken to represent a membership function $\mu_{\mathcal{R}_3}(x, y)$ relative to the relation "x and y are very different from one another."

We note that the two propositions "x and y are not very near" and "x and y are very different" are not generally identical, unless one chooses membership functions that represent both propositions rather poorly.

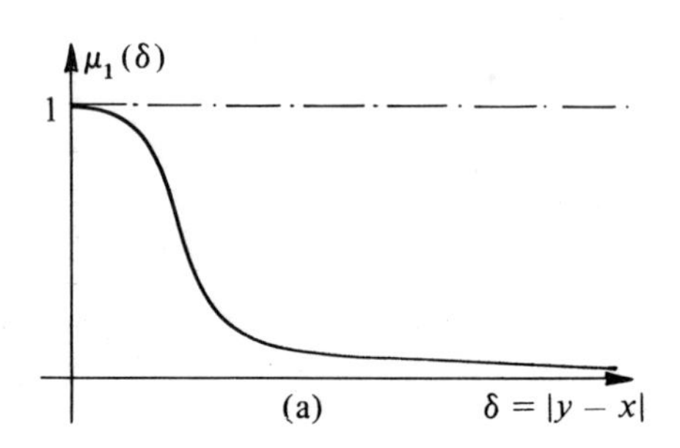

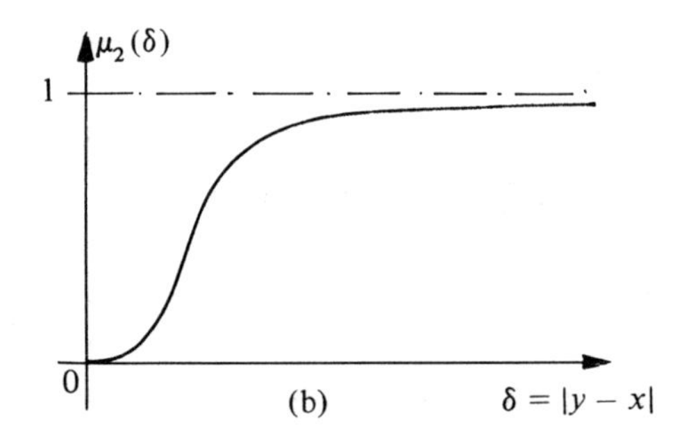

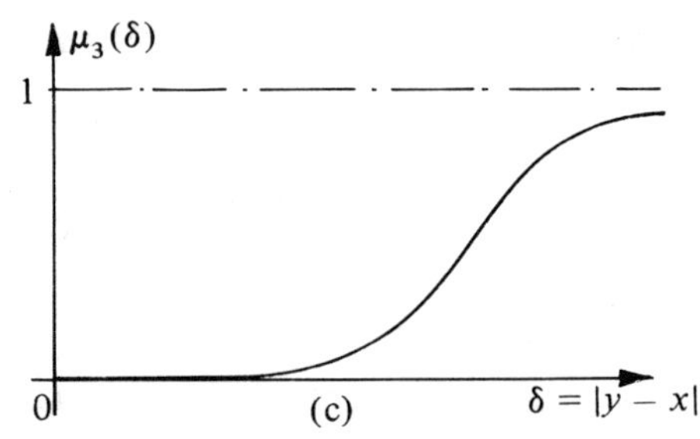

Fig. 12.16

Disjunctive sum of two relations. The disjunctive sum, denoted $\underset{\sim}{\mathcal{R}} \oplus \underset{\sim}{\mathcal{Q}}$, is defined by the expression

$$\underset{\sim}{\mathcal{R}} \oplus \underset{\sim}{\mathcal{Q}} = (\underset{\sim}{\mathcal{R}} \cap \overline{\underset{\sim}{\mathcal{Q}}}) \cup (\overline{\underset{\sim}{\mathcal{R}}} \cap \underset{\sim}{\mathcal{Q}}) \,. \tag{12.47}$$

Example 1. (Figure 12.17).

$\underset{\sim}{\mathcal{R}}$	y_1	y_2	y_3
x_1	1	0,3	0
x_2	0,1	1	0,8

$\underset{\sim}{\mathcal{Q}}$	y_1	y_2	y_3
x_1	0,2	0,6	1
x_2	0,8	0	0,2

$\overline{\underset{\sim}{\mathcal{R}}}$	y_1	y_2	y_3
x_1	0	0,7	1
x_2	0,9	0	0,2

$\overline{\underset{\sim}{\mathcal{Q}}}$	y_1	y_2	y_3
x_1	0,8	0,4	0
x_2	0,2	1	0,8

$\underset{\sim}{\mathcal{R}} \cap \overline{\underset{\sim}{\mathcal{Q}}}$	y_1	y_2	y_3
x_1	0,8	0,3	0
x_2	0,1	1	0,8

$\overline{\underset{\sim}{\mathcal{R}}} \cap \underset{\sim}{\mathcal{Q}}$	y_1	y_2	y_3
x_1	0	0,6	1
x_2	0,8	0	0,2

$\underset{\sim}{\mathcal{R}} \oplus \underset{\sim}{\mathcal{Q}}$	y_1	y_2	y_3
x_1	0,8	0,6	1
x_2	0,8	1	0,8

Fig. 12.17

Example 2. Consider again the example given in Figures 12.11a and 12.11b; let $\underset{\sim}{\mathcal{R}}$ be the relation induced by the membership function in Figure 12.11a and $\underset{\sim}{\mathcal{Q}}$ that pertaining to Figure 12.11b. By following Figures 12.18a–i, the reader may see how to obtain the membership function relative to the relation $\underset{\sim}{\mathcal{R}} \oplus \underset{\sim}{\mathcal{Q}}$.

Compare Figures 12.11d and 12.18i; as may be seen, the *disjunctive or* (Figure 12.18i) gives a considerably different result than *and*, and also rather different than that of *or/and* (Figure 12.18j).

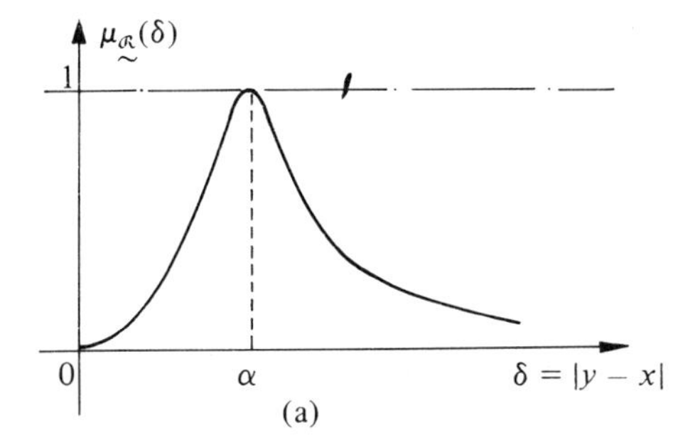

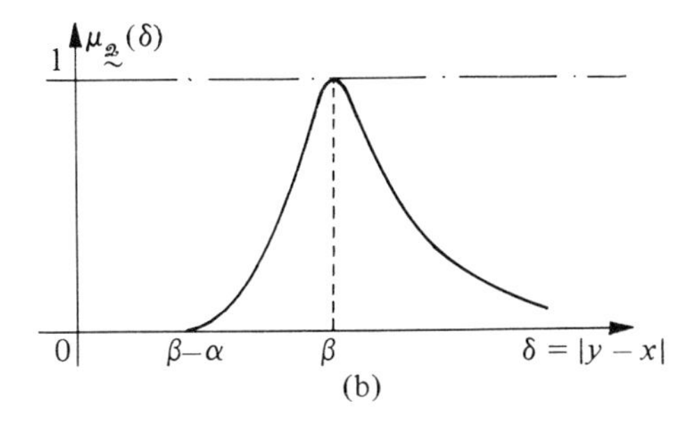

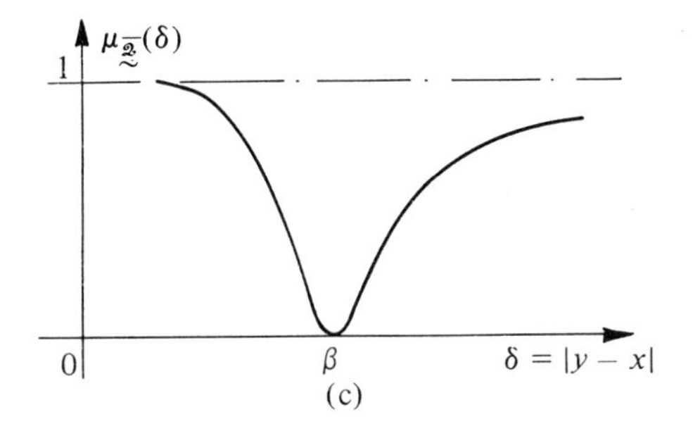

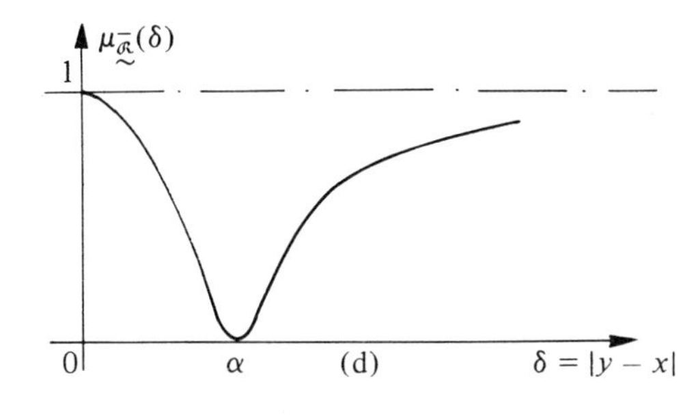

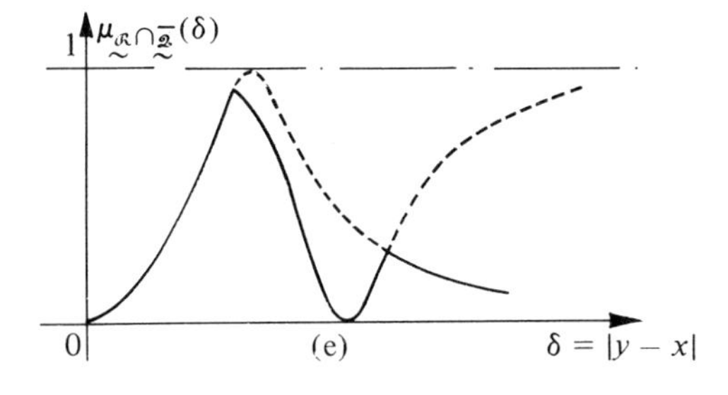

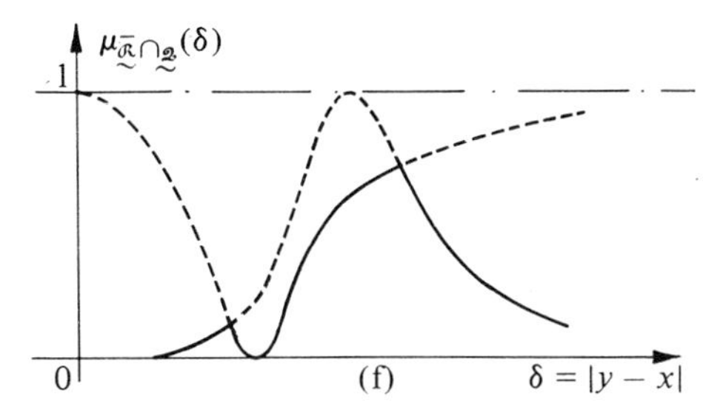

Fig. 12.18

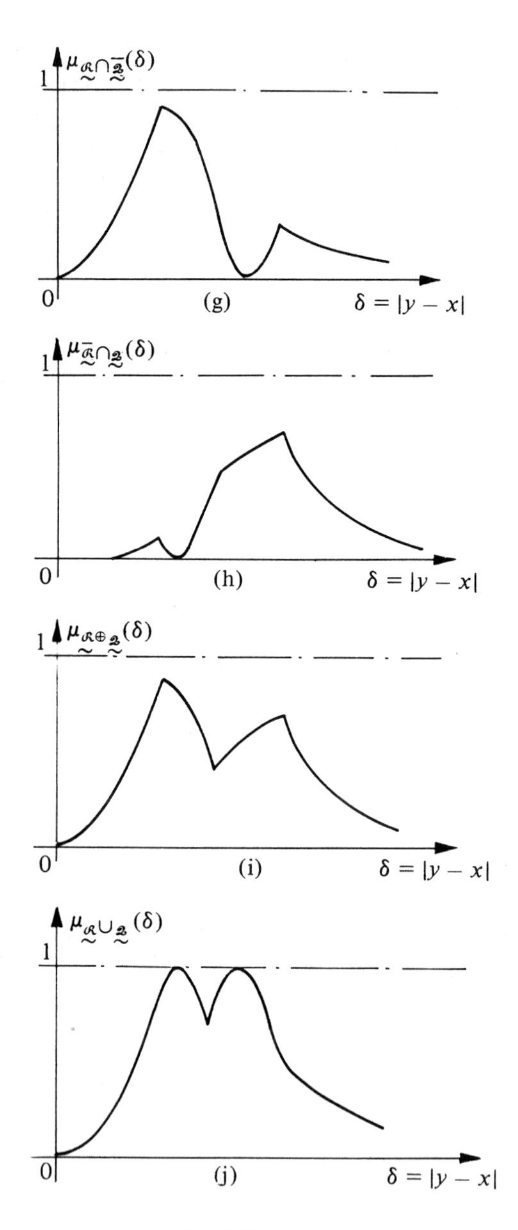

Fig. 12.18 (suite)

One likewise defines the operation of complementation:

(12.48)
$$\underset{\sim}{\mathcal{R}} \mathbin{\overline{\oplus}} \underset{\sim}{\mathcal{Q}} = \overline{\underset{\sim}{\mathcal{R}} \oplus \underset{\sim}{\mathcal{Q}}}$$
$$= (\underset{\sim}{\mathcal{R}} \cup \overline{\underset{\sim}{\mathcal{Q}}}) \cap (\overline{\underset{\sim}{\mathcal{R}}} \cup \underset{\sim}{\mathcal{Q}}) .$$

We reconsider the preceding examples in Figures 12.19 and 12.20. Figure 12.20 has been obtained with reference to Figure 12.18i.

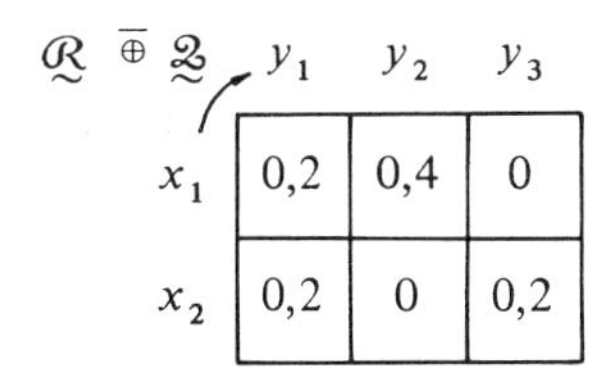

$\underset{\sim}{\mathcal{R}} \,\bar{\oplus}\, \underset{\sim}{\mathcal{Q}}$	y_1	y_2	y_3
x_1	0,2	0,4	0
x_2	0,2	0	0,2

Fig. 12.19

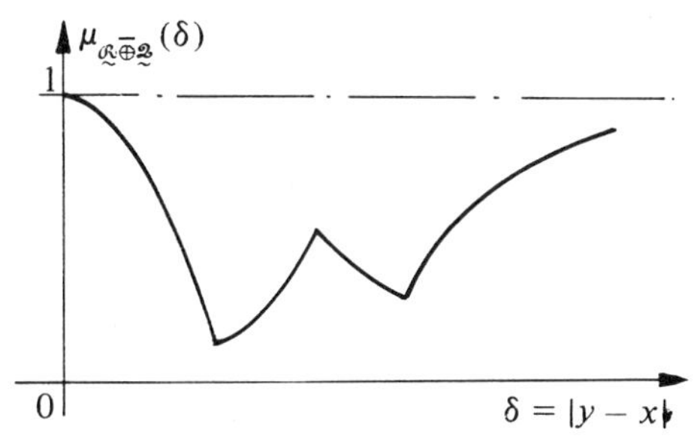

Fig. 12.20

Ordinary relation closest to a fuzzy relation. We have seen in Section 5 how to obtain an ordinary subset closest to a given fuzzy subset [see (5.102)]. In the same manner, let $\underset{\sim}{\mathcal{R}}$ be a fuzzy relation; an ordinary relation closest to $\underset{\sim}{\mathcal{R}}$ will be given by

(12.49)
$$\begin{aligned} \mu_{\underset{\approx}{\mathcal{R}}}(x\,,\,y) &= 0 \qquad \text{if} \quad \mu_{\underset{\sim}{\mathcal{R}}}(x\,,\,y) < 0{,}5 \\ &= 1 \qquad \text{if} \quad \mu_{\underset{\sim}{\mathcal{R}}}(x\,,\,y) > 0{,}5 \\ &= 0 \text{ or } 1 \quad \text{if} \quad \mu_{\underset{\sim}{\mathcal{R}}}(x\,,\,y) = 0{,}5\,. \end{aligned}$$

This definition is usable whatever may be the reference sets $\mathbf{E}_1$ and $\mathbf{E}_2$ forming $\mathbf{E}_1 \times \mathbf{E}_2$, where $x \in \mathbf{E}_1$, $y \in \mathbf{E}_2$, and regardless of whether these reference sets are finite or not.

By convention, we take

(12.50)
$$\mu_{\underset{\sim}{\mathcal{R}}}(x\,,\,y) = 0{,}5 \;\Rightarrow\; \mu_{\underset{\approx}{\mathcal{R}}}(x\,,\,y) = 0\,.$$

Example. Figures 12.21 and 12.22 show easily how to pass from $\underset{\sim}{\mathcal{R}}$ to $\underset{\approx}{\mathcal{R}}$.

$\underset{\sim}{\mathcal{R}}$	y_1	y_2	y_3	y_4	y_5	y_6
x_1	0,7	0,3	0,2	1	0	0,8
x_2	0,5	0,4	0	0,6	0,9	0,1
x_3	0,6	1	0,8	0	0	0,7

Fig. 12.21

$\underset{\approx}{\mathcal{R}}$	y_1	y_2	y_3	y_4	y_5	y_6
x_1	1	0	0	1	0	1
x_2	0	0	0	1	1	0
x_3	1	1	1	0	0	1

Fig. 12.22

The presence of an element equal to 1/2, that corresponding to (x_2, y_1), shows that $\underset{\sim}{\mathcal{R}}$ is not unique. There are two ordinary relations closest to $\underset{\sim}{\mathcal{R}}$, one with $\mu_{\underset{\approx}{\mathcal{R}}}(x_2, y_1) = 1$ and the other with $\mu_{\underset{\approx}{\mathcal{R}}}(x_2, y_1) = 0$. According to the chosen convention, we will take $\mu_{\underset{\approx}{\mathcal{R}}}(x_2, y_1) = 0$.

13. COMPOSITION OF TWO FUZZY RELATIONS

We mention now that sometimes we will use the notation

$$\underset{\sim}{\mathcal{R}} \subset \mathbf{X} \times \mathbf{Y} \quad \text{signifying} \tag{13.1}$$

$$\underset{\sim}{\mathbf{G}} \subset \mathbf{X} \times \mathbf{Y} ,$$

where $\underset{\sim}{\mathcal{R}}$ is the fuzzy relation corresponding to the fuzzy graph $\underset{\sim}{\mathbf{G}}$.

Max–min composition. Let $\underset{\sim}{\mathcal{R}}_1 \subset \mathbf{X} \times \mathbf{Y}$ and $\underset{\sim}{\mathcal{R}}_2 \subset \mathbf{Y} \times \mathbf{Z}$. We define the *min–max-composition* of $\underset{\sim}{\mathcal{R}}_1$ and $\underset{\sim}{\mathcal{R}}_2$, denoted $\underset{\sim}{\mathcal{R}}_2 \circ \underset{\sim}{\mathcal{R}}_1$, by the expression

$$\begin{aligned} \mu_{\underset{\sim}{\mathcal{R}}_2 \circ \underset{\sim}{\mathcal{R}}_1}(x, z) &= \bigvee_y [\mu_{\underset{\sim}{\mathcal{R}}_1}(x, y) \wedge \mu_{\underset{\sim}{\mathcal{R}}_2}(y, z)] \\ &= \underset{y}{\mathrm{MAX}}[\mathrm{MIN}(\mu_{\underset{\sim}{\mathcal{R}}_1}(x, y), \mu_{\underset{\sim}{\mathcal{R}}_2}(y, z))] , \end{aligned} \tag{13.2}$$

where $x \in \mathbf{X}$, $y \in \mathbf{Y}$, and $z \in \mathbf{Z}$.

Example 1. Consider two fuzzy relations $x \underset{\sim}{\mathcal{R}}_1 y$ and $y \underset{\sim}{\mathcal{R}}_2 z$, where x, y, and $z \in \mathbf{R}^+$. We suppose

$$\mu_{\underset{\sim}{\mathcal{R}}_1}(x, y) = e^{-k(x-y)^2}, \quad k \geqslant 1 , \tag{13.3}$$

$$\mu_{\underset{\sim}{\mathcal{R}}_2}(y, z) = e^{-k(y-z)^2}, \quad k \geqslant 1 . \tag{13.4}$$

We now propose to determine $\mu_{\underset{\sim}{\mathcal{R}}_2 \circ \underset{\sim}{\mathcal{R}}_1}(x, z)$.

Consider two values $x = a$ and $z = b$ of the variables x and z. Here the membership functions (13.3) and (13.4) are continuous on the interval $[0, \infty)$; we may write

$$\begin{aligned} \mu_{\underset{\sim}{\mathcal{R}}_2 \circ \underset{\sim}{\mathcal{R}}_1}(a, b) &= \bigvee_y [\mu_{\underset{\sim}{\mathcal{R}}_1}(a, y) \wedge \mu_{\underset{\sim}{\mathcal{R}}_2}(y, b)] \\ &= \bigvee_y [e^{-k(a-y)^2} \wedge e^{-k(y-b)^2}]. \end{aligned} \tag{13.5}$$

The composition of $\underset{\sim}{\mathcal{R}}_1$ and $\underset{\sim}{\mathcal{R}}_2$ with respect to the max–min operation is represented in Figure 13.1. As may be readily seen, one has

$$\begin{aligned} \mu_{\underset{\sim}{\mathcal{R}}_2 \circ \underset{\sim}{\mathcal{R}}_1}(a, b) &= e^{-k\left(a - \frac{a+b}{2}\right)^2} \\ &= e^{-k\left(\frac{a-b}{2}\right)^2} . \end{aligned}$$

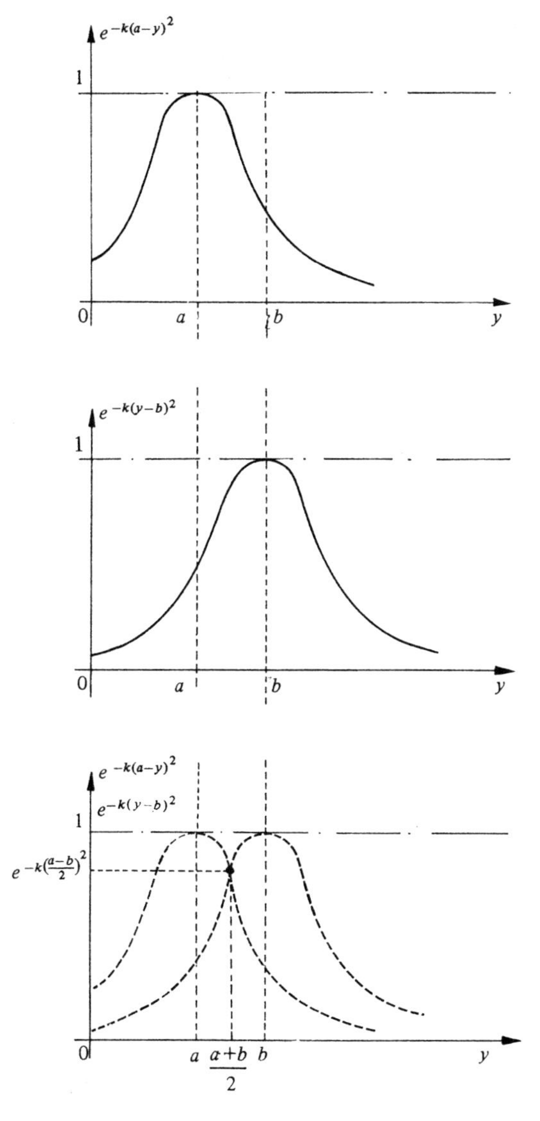

Fig. 13.1

and, for all values of x and z,

(13.6)
$$\mu_{\underset{\sim}{\mathcal{R}}_2 \circ \underset{\sim}{\mathcal{R}}_1}(x, z) = e^{-k \frac{(x-z)^2}{4}}$$

For simplicity, we have here considered two identical functions $\mu_{\underset{\sim}{\mathcal{R}}_1}(x, y)$ and $\mu_{\underset{\sim}{\mathcal{R}}_2}(y, z)$, but the reasoning remains the same for two distinct functions. Superimpose $\mu_{\underset{\sim}{\mathcal{R}}_1}(a, y)$ and $\mu_{\underset{\sim}{\mathcal{R}}_2}(y, b)$ and determine the curve $\mu_{\underset{\sim}{\mathcal{R}}_1}(a, y) \wedge \mu_{\underset{\sim}{\mathcal{R}}_2}(y, b)$ as a function of y. Take the maximum value of this curve, that which gives the abscissa and the ordinate of this maximum as a function of y.

The problem becomes more complicated if the abscissa of the maximum is not unique. It is then necessary to impose some more complicated considerations.

We consider another example with membership functions having a finite reference set.

Example 2. (Figure 13.2). Let

$$
\begin{aligned}
&(x\,,\,z) = (x_1\,,\,z_1)\\
&\mathrm{MIN}\,(\mu_{\underset{\sim}{\mathcal{R}}_1}(x_1\,,\,y_1)\,,\,\mu_{\underset{\sim}{\mathcal{R}}_2}(y_1\,,\,z_1))\\
&\qquad = \mathrm{MIN}\,(0{,}1\,,\,0{,}9)\\
&\qquad = 0{,}1\,,\\
&\mathrm{MIN}\,(\mu_{\underset{\sim}{\mathcal{R}}_1}(x_1\,,\,y_2)\,,\,\mu_{\underset{\sim}{\mathcal{R}}_2}(y_2\,,\,z_1))\\
&\qquad = \mathrm{MIN}\,(0{,}2\,,\,0{,}2)\\
&\qquad = 0{,}2\,,\\
&\mathrm{MIN}\,(\mu_{\underset{\sim}{\mathcal{R}}_1}(x_1\,,\,y_3)\,,\,\mu_{\underset{\sim}{\mathcal{R}}_2}(y_3\,,\,z_1))\\
&\qquad = \mathrm{MIN}\,(0\,,\,0{,}8)\\
&\qquad = 0\,,\\
&\mathrm{MIN}\,(\mu_{\underset{\sim}{\mathcal{R}}_1}(x_1\,,\,y_4)\,,\,\mu_{\underset{\sim}{\mathcal{R}}_2}(y_4\,,\,z_1))\\
&\qquad = \mathrm{MIN}\,(1\,,\,0{,}4)\\
&\qquad = 0{,}4\,,\\
&\mathrm{MIN}\,(\mu_{\underset{\sim}{\mathcal{R}}_1}(x_1\,,\,y_5)\,,\,\mu_{\underset{\sim}{\mathcal{R}}_2}(y_5\,,\,z_1)\\
&\qquad = \mathrm{MIN}\,(0{,}7\,,\,0)\\
&\qquad = 0
\end{aligned}
\tag{13.7}
$$

$\underset{\sim}{\mathcal{R}}_1$	y_1	y_2	y_3	y_4	y_5
x_1	0,1	0,2	0	1	0,7
x_2	0,3	0,5	0	0,2	1
x_3	0,8	0	1	0,4	0,3

(a)

$\underset{\sim}{\mathcal{R}}_2$	z_1	z_2	z_3	z_4
y_1	0,9	0	0,3	0,4
y_2	0,2	1	0,8	0
y_3	0,8	0	0,7	1
y_4	0,4	0,2	0,3	0
y_5	0	1	0	0,8

(b)

$\underset{\sim}{\mathcal{R}}_2 \circ \underset{\sim}{\mathcal{R}}_1$	z_1	z_2	z_3	z_4
x_1	0,4	0,7	0,3	0,7
x_2	0,3	1	0,5	0,8
x_3	0,8	0,3	0,7	1

(c)

Fig. 13.2

$$
\begin{aligned}
&\underset{y_i}{\mathrm{MAX}}\,[\mathrm{MIN}((\mu_{\underset{\sim}{\mathcal{R}}_1}(x_1\,,\,y_i)\,,\,\mu_{\underset{\sim}{\mathcal{R}}_2}(y_i\,,\,z_1))]\\
&\qquad = \mathrm{MAX}\,(0{,}1\,,\,0{,}2\,,\,0\,,\,0{,}4\,,\,0)\\
&\qquad = 0{,}4\,.
\end{aligned}
$$

Now let

$$(x, z) = (x_1, z_2),$$

$$\text{MIN}\,(\mu_{\underset{\sim}{\mathcal{R}}_1}(x_1, y_1), \mu_{\underset{\sim}{\mathcal{R}}_2}(y_1, z_2))$$
$$= \text{MIN}\,(0{,}1, 0)$$
$$= 0,$$

(13.8)
$$\text{MIN}\,(\mu_{\underset{\sim}{\mathcal{R}}_1}(x_1, y_2), \mu_{\underset{\sim}{\mathcal{R}}_2}(y_2, z_2))$$
$$= \text{MIN}\,(0{,}2, 1)$$
$$= 0{,}2, \ldots .$$

and so on. The results are given in Figure 13.2.

It is interesting to compare this composition of fuzzy relations with the composition of ordinary relations.

In the composition of ordinary relations one has

$$\mu_{\mathcal{R}_2 \circ \mathcal{R}_1}(x, z) = \underset{y}{\text{MAX}} \cdot [\text{MIN}\,(\mu_{\mathcal{R}_1}(x, y), \mu_{\mathcal{R}_2}(y, z)] \tag{13.9}$$

where $\mu_{\mathcal{R}_1}(x, y) \in \{0, 1\}$, $\mu_{\mathcal{R}_2}(y, z) \in \{0, 1\}$. Then expression (13.9) may be written

$$\mu_{\mathcal{R}_2 \circ \mathcal{R}_1}(x, z) = \dot{\sum_{y}} \mu_{\mathcal{R}_1}(x, y) \cdot \mu_{\mathcal{R}_2}(y, z) \tag{13.10}$$

where $\cdot$ represents boolean multiplication, and $\dot{\sum\limits_{y}}$ represents the boolean sum of the products obtained.

$\mathcal{R}_1$	y_1	y_2	y_3	y_4	y_5
x_1	0	1	1	0	0
x_2	1	0	0	1	0
x_3	0	1	0	1	1

(a)

$\mathcal{R}_2$	z_1	z_2	z_3	z_4
y_1	1	0	1	0
y_2	0	0	0	0
y_3	1	0	0	0
y_4	0	1	1	0
y_5	0	1	0	1

(b)

$\mathcal{R}_2 \circ \mathcal{R}_1$	z_1	z_2	z_3	z_4
x_1	1	0	0	0
x_2	1	1	1	0
x_3	0	1	1	1

(c)

Fig. 13.3

In Figure 13.3 is an example using (13.9), or, what is the same thing, (13.10).

Example 3. In Figure 13.4 an example of the composition of three relations is presented.

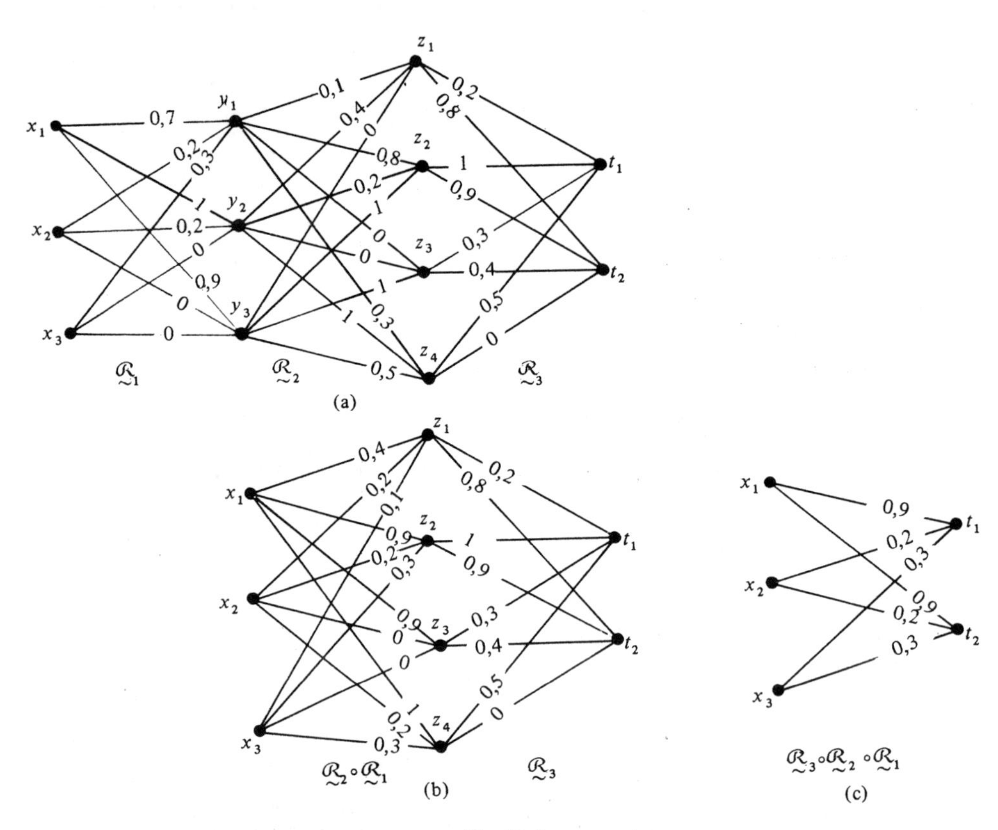

Fig. 13.4

The max–min composition operation is associative, that is,

$$(\underset{\sim}{\mathcal{R}}_3 \circ \underset{\sim}{\mathcal{R}}_2) \circ \underset{\sim}{\mathcal{R}}_1 = \underset{\sim}{\mathcal{R}}_3 \circ (\underset{\sim}{\mathcal{R}}_2 \circ \underset{\sim}{\mathcal{R}}_1)\,. \tag{13.11}$$

On the other hand, if $\underset{\sim}{\mathcal{R}}$ is a relation defined on $\mathbf{E} \times \mathbf{E}$, then $\underset{\sim}{\mathcal{R}} \subset \mathbf{E} \times \mathbf{E}$; one may write

$$\underset{\sim}{\mathcal{R}} \circ \underset{\sim}{\mathcal{R}} = \underset{\sim}{\mathcal{R}}^2\,. \tag{13.12}$$

and from this

$$\underset{\sim}{\mathcal{R}} \circ \underset{\sim}{\mathcal{R}}^2 = \underset{\sim}{\mathcal{R}}^2 \circ \underset{\sim}{\mathcal{R}} = \underset{\sim}{\mathcal{R}}^3\,; \tag{13.13}$$

and more generally

$$\underbrace{\underset{\sim}{\mathcal{R}} \circ \underset{\sim}{\mathcal{R}} \circ \cdots \circ \underset{\sim}{\mathcal{R}}}_{k \text{ times}} = \underset{\sim}{\mathcal{R}}^k\,. \tag{13.14}$$

We note the distributivity of max–min composition with respect to union but its non-distributivity with respect to intersection:

(13.15) $$\underset{\sim}{\mathcal{R}} \circ (\underset{\sim}{\mathcal{Q}}_1 \cup \underset{\sim}{\mathcal{Q}}_2) = (\underset{\sim}{\mathcal{R}} \circ \underset{\sim}{\mathcal{Q}}_1) \cup (\underset{\sim}{\mathcal{R}} \circ \underset{\sim}{\mathcal{Q}}_2) \,,$$

(13.16) $$\underset{\sim}{\mathcal{R}} \circ (\underset{\sim}{\mathcal{Q}}_1 \cap \underset{\sim}{\mathcal{Q}}_2) \neq (\underset{\sim}{\mathcal{R}} \circ \underset{\sim}{\mathcal{Q}}_1) \cap (\underset{\sim}{\mathcal{R}} \circ \underset{\sim}{\mathcal{Q}}_2) \,.$$

The proof of (13.15) and (13.16) is given in the Appendix, pages 384 and 385. A further important property is the following:

(13.17) $$\underset{\sim}{\mathcal{Q}} \subset \underset{\sim}{\mathcal{T}} \Rightarrow \underset{\sim}{\mathcal{R}} \circ \underset{\sim}{\mathcal{Q}} \subset \underset{\sim}{\mathcal{R}} \circ \underset{\sim}{\mathcal{T}} \,,$$

which is easy to prove, and we leave it to the reader to do so.

Max–star composition. One may replace the operation $\wedge$ in (13.2) arbitrarily with another, under the restriction that one uses an operation, like $\wedge$, that is associative and monotone nondecreasing in each argument. One may then write

(13.18) $$\mu_{\underset{\sim}{\mathcal{Q}} * \underset{\sim}{\mathcal{R}}}(x\,,\,z) = \bigvee_{y} \, [\mu_{\underset{\sim}{\mathcal{R}}}(x\,,\,y) * \mu_{\underset{\sim}{\mathcal{Q}}}(y\,,\,z)] \,.$$

Max–product composition. Among the max–star compositions that may be imagined, the max–product composition deserves our particular attention. In this case, the operation $*$ will be the product designated by $\cdot$, formula (13.18) then becomes

(13.19) $$\mu_{\underset{\sim}{\mathcal{Q}} \cdot \underset{\sim}{\mathcal{R}}}(x\,,\,z) = \bigvee_{y} \, [\mu_{\underset{\sim}{\mathcal{R}}}(x\,,\,y) \cdot \mu_{\underset{\sim}{\mathcal{Q}}}(y\,,\,z)] \,.$$

We shall have occasion later to speak of the max–product composition and at the same time of justifications for some particular uses.

Ordinary subset of level α in a fuzzy relation. Let $\alpha \in [0,\,1]$; we shall call the *ordinary subset of level* α of a fuzzy relation $\underset{\sim}{\mathcal{R}} \subset \mathbf{X} \times \mathbf{Y}$, the ordinary subset

(13.20) $$\mathbf{G}_\alpha = \{(x\,,\,y) \mid \mu_{\underset{\sim}{\mathcal{R}}}(x\,,\,y) \geqslant \alpha\} \,.$$

Example 1. (Figure 13.5)

(13.21) $$\mathbf{G}_{0,8} = \{(x_1\,,\,y_2)\,,\,(x_1\,,\,y_3)\,,\,(x_2\,,\,y_2)\,,\,(x_2\,,\,y_4)\,,\,(x_3\,,\,y_1)\} \,.$$

$\underset{\sim}{\mathcal{R}}$	y_1	y_2	y_3	y_4
x_1	0,3	0,8	1	0
x_2	0,5	1	0,3	0,9
x_3	1	0,2	0,6	0,7

Fig. 13.5

Example 2. Consider the fuzzy relation defined in $\mathbf{R}^2$ by

(13.22) $$\mu_{\underset{\sim}{\mathcal{R}}}(x, y) = 1 - \frac{1}{1 + x^2 + y^2} .$$

The subset of level 0,3 will be defined by

(13.23) $$1 - \frac{1}{1 + x^2 + y^2} \geqslant 0{,}3 .$$

Thus

(13.24) $$x^2 + y^2 \geqslant \frac{3}{7} .$$

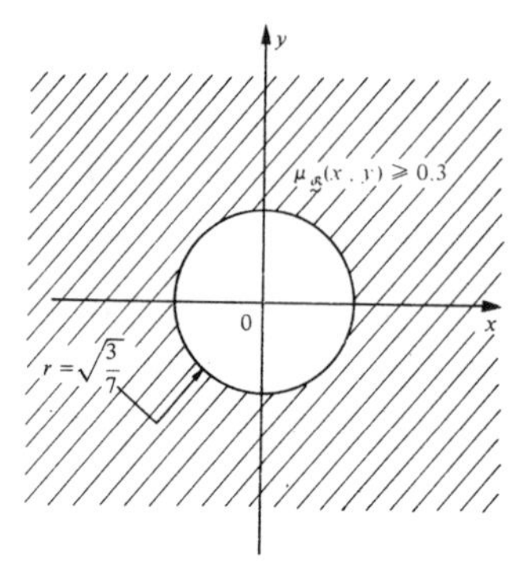

Fig. 13.6

This subset is the exterior and circumference of the circle with radius $r = \sqrt{3/7}$. (See Figure 13.6.)

One may define the ordinary subset $\mathbf{G}_\alpha$ in another fashion, with the aid of an ordinary relation $\mathcal{R}_\alpha$ such that

(13.25) $$\mu_{\mathcal{R}_\alpha}(x, y) = 1 \quad \text{if} \quad \mu_{\underset{\sim}{\mathcal{R}}}(x, y) \geqslant \alpha$$
$$= 0 \quad \text{if} \quad \text{"} \quad < \alpha .$$

Reconsidering the preceding examples, one may write, referring now to Figure 13.5:

(13.26) $\mathcal{R}_{0,4} =$

	y_1	y_2	y_3	y_4
x_1	0	1	1	0
x_2	1	1	0	1
x_3	1	0	1	1

(13.27) $\mathcal{R}_{0,7} =$

	y_1	y_2	y_3	y_4
x_1	0	1	1	0
x_2	0	1	0	1
x_3	1	0	0	1

For the example of Figure 13.6, one sees

(13.28) $$\mu_{\mathcal{R}_{0,3}}(x, y) = 0 \quad \text{for} \quad x^2 + y^2 < \frac{3}{7} ,$$
$$= 1 \quad \text{for} \quad x^2 + y^2 \geqslant \frac{3}{7} ,$$

which is defined by the ordinary relation $\mathcal{R}_{0,3}$.

Important property. One now has the evident property

(13.29) $$\alpha_1 \geqslant \alpha_2 \Rightarrow \mathbf{G}_{\alpha_1} \subset \mathbf{G}_{\alpha_2} ,$$

or, what is the same thing,

(13.30) $$\mathcal{R}_{\alpha_1} \subset \mathcal{R}_{\alpha_2} .$$

We now present an important theorem:

Decomposition Theorem †. Any fuzzy relation $\underset{\sim}{\mathcal{R}}$ may be decomposed in the form

$$\underset{\sim}{\mathcal{R}} = \bigvee_{\alpha} \alpha \cdot \mathcal{R}_{\alpha}\,, \quad 0 < \alpha \leqslant 1 \tag{13.31}$$

where

$$\begin{aligned} \mu_{\mathcal{R}_{\alpha}}(x\,,y) &= 1 \quad \text{if} \quad \mu_{\underset{\sim}{\mathcal{R}}}(x\,,y) \geqslant \alpha \\ &= 0 \quad \text{if} \quad \mu_{\underset{\sim}{\mathcal{R}}}(x\,,y) < \alpha\,. \end{aligned} \tag{13.32}$$

Here $\alpha\,\mathcal{R}_{\alpha}$ indicates that all the elements of the ordinary relation $\mathcal{R}_{\alpha}$ are multiplied by α.

Proof. The membership function of (13.31) may be written

$$\begin{aligned} \mu_{\bigvee_{\alpha} \alpha \mathcal{R}_{\alpha}}(x\,,y) &= \bigvee_{\alpha} \alpha\, \mu_{\mathcal{R}_{\alpha}}(x\,,y) \\ &= \bigvee_{\alpha \leqslant \mu_{\underset{\sim}{\mathcal{R}}}(x,y)} \alpha \\ &= \mu_{\underset{\sim}{\mathcal{R}}}(x\,,y)\,. \end{aligned} \tag{13.33}$$

Example 1.

$$\begin{bmatrix} 0{,}3 & 0{,}8 & 1 & 0 \\ 0{,}5 & 1 & 0{,}3 & 0{,}9 \\ 1 & 0{,}2 & 0{,}6 & 0{,}7 \end{bmatrix} = \bigvee \left(0{,}2 \cdot \begin{bmatrix} 1 & 1 & 1 & 0 \\ 1 & 1 & 1 & 1 \\ 1 & 1 & 1 & 1 \end{bmatrix},\ 0{,}3 \cdot \begin{bmatrix} 1 & 1 & 1 & 0 \\ 1 & 1 & 1 & 1 \\ 1 & 0 & 1 & 1 \end{bmatrix},\right.$$

$$0{,}5 \cdot \begin{bmatrix} 0 & 1 & 1 & 0 \\ 1 & 1 & 0 & 1 \\ 1 & 0 & 1 & 1 \end{bmatrix},\ 0{,}6 \cdot \begin{bmatrix} 0 & 1 & 1 & 0 \\ 0 & 1 & 0 & 1 \\ 1 & 0 & 1 & 1 \end{bmatrix},$$

$$0{,}7 \cdot \begin{bmatrix} 0 & 1 & 1 & 0 \\ 0 & 1 & 0 & 1 \\ 1 & 0 & 0 & 1 \end{bmatrix},\ 0{,}8 \cdot \begin{bmatrix} 0 & 1 & 1 & 0 \\ 0 & 1 & 0 & 1 \\ 1 & 0 & 0 & 0 \end{bmatrix},$$

$$\left. 0{,}9 \begin{bmatrix} 0 & 0 & 1 & 0 \\ 0 & 1 & 0 & 1 \\ 1 & 0 & 0 & 0 \end{bmatrix},\ 1 \cdot \begin{bmatrix} 0 & 0 & 1 & 0 \\ 0 & 1 & 0 & 0 \\ 1 & 0 & 0 & 0 \end{bmatrix} \right) \tag{13.34}$$

†The word *decomposition* employed here is not connected to the word *composition* employed in the sense of max–min composition or other compositions of relations.

Example 2. The decomposition according to (13.25) is still valid when **X** or/and **Y** have the power of the continuum. But the operation V (max) must then be considered as being carried out (if necessary) for the interval considered with continuous values.

Thus, in considering the example (13.22) (Figure 13.6), one may write

$$\mu_{\underset{\sim}{\mathcal{R}}}(x\,,\,y) = \bigvee_{\alpha} \alpha \mathcal{R}_{\alpha}\,, \tag{13.35}$$

(13.36) where
$$\mu_{\mathcal{R}_{\alpha}}(x\,,\,y) = 1 \quad \text{if} \quad (x\,,\,y) \in \mathbf{D}(\alpha)$$
$$= 0 \quad \text{if} \quad (x, y) \notin \mathbf{D}\,(\alpha),$$

in which $\mathbf{D}(\alpha) \subset \mathbf{X} \times \mathbf{Y}$ is the domain such that

$$1 - \frac{1}{1 + x^2 + y^2} \geqslant \alpha\,. \tag{13.37}$$

Composition of nearest ordinary relations. Recall that $\underset{\approx}{\mathcal{R}}$ is an ordinary relation nearest to a fuzzy relation $\underset{\sim}{\mathcal{R}}$. It is then easy to prove

$$\underset{\sim}{\mathcal{R}}_2 \circ \underset{\sim}{\mathcal{R}}_1 = \underset{\sim}{\mathcal{R}} \;\Rightarrow\; \underset{\approx}{\mathcal{R}}_2 \circ \underset{\approx}{\mathcal{R}}_1 = \underset{\approx}{\mathcal{R}}\,. \tag{13.38}$$

where $\circ$ represents the max–min composition.

Example.

(13.39)

$\underset{\sim}{\mathcal{R}}_1$	y_1	y_2	y_3	y_4	y_5
x_1	0,1	0,2	0	1	0,7
x_2	0,3	0,5	0	0,2	1
x_3	0,8	0	1	0,4	0,3

$\circ$

$\underset{\sim}{\mathcal{R}}_2$	z_1	z_2	z_3	z_4
y_1	0,9	0	0,3	0,4
y_2	0,2	1	0,8	0
y_3	0,8	0	0,7	1
y_4	0,4	0,2	0,3	0
y_5	0	1	0	0,8

$=$

$\underset{\sim}{\mathcal{R}}$	z_1	z_2	z_3	z_4
x_1	0,4	0,7	0,3	0,7
x_2	0,3	1	0,5	0,8
x_3	0,8	0,3	0,7	1

(13.40)

$\underset{\approx}{\mathcal{R}}_1$	y_1	y_2	y_3	y_4	y_5
x_1	0	0	0	1	1
x_2	0	0	0	0	1
x_3	1	0	1	0	0

$\circ$

$\underset{\approx}{\mathcal{R}}_2$	z_1	z_2	z_3	z_4
y_1	1	0	0	0
y_2	0	1	1	0
y_3	1	0	1	1
y_4	0	0	0	0
y_5	0	1	0	1

$=$

$\underset{\approx}{\mathcal{R}}$	z_1	z_2	z_3	z_4
x_1	0	1	0	1
x_2	0	1	0	1
x_3	1	0	1	1

14. FUZZY SUBSET INDUCED BY A MAPPING

Consider a mapping† of set $\mathbf{E}_1$ into a set $\mathbf{E}_2$, denoted

(14.1) $$\mathbf{E}_1 \underset{\Gamma}{\rightsquigarrow} \mathbf{E}_2$$

where, if $x \in \mathbf{E}_1$ and $y \in \mathbf{E}_2$,

(14.2) $$y \in \Gamma\,\{x\}$$

Let $\mu_{\underset{\sim}{A}}(x)$ be the membership function of a fuzzy subset $\underset{\sim}{A} \subset \mathbf{E}_1$; then the mapping Γ induces in $\mathbf{E}_2$ a fuzzy subset $\underset{\sim}{B} \subset \mathbf{E}_2$ whose membership function is

(14.3) $$\mu_{\underset{\sim}{B}}(y) = \underset{x \in \Gamma^{-1}\{y\}}{\mathrm{MAX}} [\mu_{\underset{\sim}{A}}(x)]\,, \quad \text{if } \Gamma^{-1}\{y\} \neq \phi$$
$$= 0 \quad\quad , \text{ if } \Gamma^{-1}\{y\} = \phi$$

Example 1. (Figure 14.1). Let

(14.4) $\mathbf{E}_1 = \{x_1, x_2, x_3, x_4, x_5, x_6, x_7\}$

(14.5) $\mathbf{E}_2 = \{y_1, y_2, y_3, y_4\}$.

Consider a mapping such that

(14.6)
$$\Gamma\{x_1\} = \{y_2\}\ ,\quad \Gamma\{x_2\} = \{y_1, y_4\}$$
$$\Gamma\{x_3\} = \{y_1\}\ ,\quad \Gamma\{x_4\} = \{y_3\}\ ,$$
$$\Gamma\{x_5\} = \{y_1\}\ ,\quad \Gamma\{x_6\} = \{y_2\}\ .$$
$$\Gamma\{x_7\} = \{y_4\}\ .$$

Also consider the inverse mapping Γ^{-1}:

(14.7)
$$\Gamma^{-1}\{y_1\} = \{x_2, x_3; x_5\},$$
$$\Gamma^{-1}\{y_2\} = \{x_1, x_6\}\ ,$$
$$\Gamma^{-1}\{y_3\} = \{x_4\},\ \Gamma^{-1}\{y_4\} = \{x_2, x_7\}\ .$$

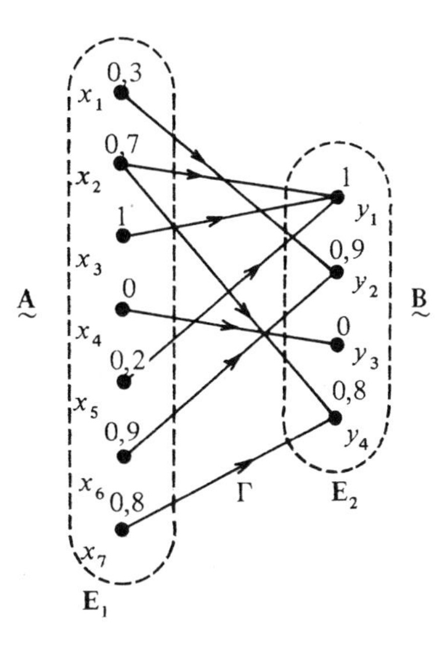

Fig. 14.1

And finally consider the fuzzy subset $\underset{\sim}{A} \subset \mathbf{E}_1$:

(14.8) $\underset{\sim}{A} = \{(x_1\,|\,0{,}3), (x_2\,|\,0{,}7), (x_3\,|\,1), (x_4\,|\,0), (x_5\,|\,0{,}2), (x_6\,|\,0{,}9), (x_7\,|\,0{,}8)\}$.

One then has

(14.9)
$$\mu_{\underset{\sim}{B}}(y_1) = \underset{\{x_2, x_3, x_5\}}{\mathrm{MAX}} (0{,}7\ ;1\ ;0{,}2) = 1\ ,$$
$$\mu_{\underset{\sim}{B}}(y_2) = \underset{\{x_1, x_6\}}{\mathrm{MAX}} (0{,}3\ ;0{,}9) = 0{,}9\ ,$$
$$\mu_{\underset{\sim}{B}}(y_3) = \underset{\{x_4\}}{\mathrm{MAX}} (0) = 0\ ,$$
$$\mu_{\mathbf{B}}(y_4) = \underset{\{x_2, x_7\}}{\mathrm{MAX}} (0{,}7\ ;0{,}8) = 0{,}8\ .$$

†We are concerned here with a mapping that is not necessarily single valued.

These results have been portrayed in Figure 14.1.

It is interesting to compare this notion with the corresponding one for ordinary subsets. Consider Figures 14.2.

Let

(14.10) $E_1 = \{x_1, x_2, x_3, x_4, x_5, x_6, x_7, x_8\}$,

(14.11) $E_2 = \{y_1, y_2, y_3, y_4, y_5\}$.

One has

(14.12) $\Gamma\{x_4, x_5, x_6, x_8\} = \{y_1, y_2, y_3\}$.

To the subset $\mathbf{A} = \{x_4, x_5, x_6, x_8\}$, the mapping Γ associates the subset $\mathbf{B} = \{y_1, y_2, y_3\}$.

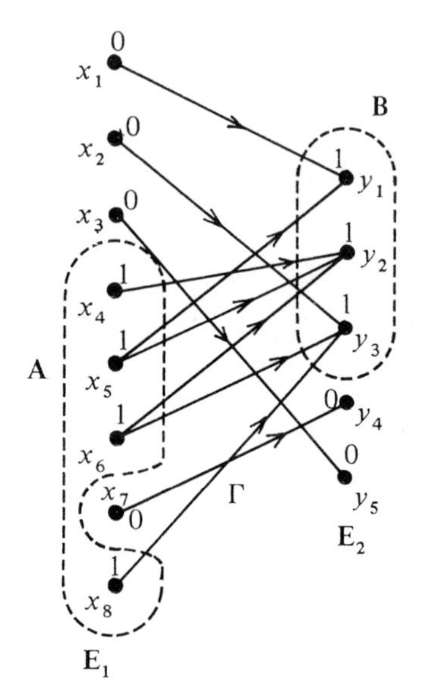

Fig. 14.2

Example 2 Let $x \in \mathbf{R}$, $y \in \mathbf{R}$, where $\mathbf{R}$ is the set of real numbers. We consider the fuzzy subset $\underset{\sim}{\mathbf{A}}$ defined by "x near $(4k + 1)\pi/2$, $k = \ldots, -2, -1, 0, 1, 2, \ldots$." We consider also the function

(14.13) $$y = f(x) = \sin x \ ;$$

then the fuzzy subset $\underset{\sim}{\mathbf{B}}$ induced by $f(x)$ will be

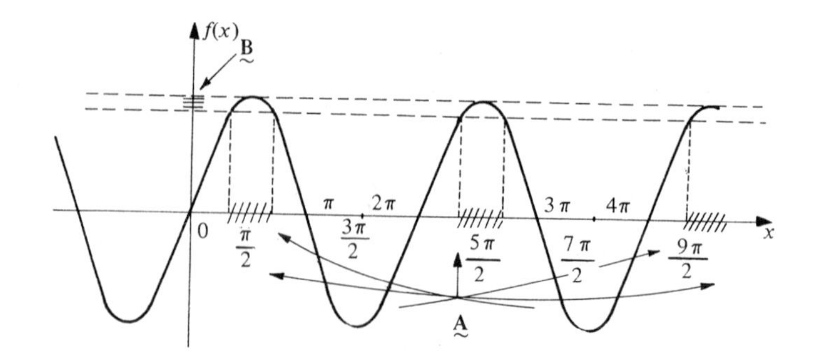

Fig. 14.3

(14.14) $$\underset{\sim}{\mathbf{B}} = \{y \mid y \leqslant 1 \text{ and } y \text{ is near } 1\}.$$

(See Figure 14.3.)

15. CONDITIONED FUZZY SUBSETS

A fuzzy subset $\underset{\sim}{B}(x) \subset E_2$ will be said to be *conditioned on* E_1 if its membership function depends on $x \in E_1$ as a parameter.

The conditional membership function will then be written

(15.1) $$\mu_{\underset{\sim}{B}}(y \| x), \quad \text{where } x \in E_1 \quad \text{and} \quad y \in E_2 .$$

This function defines a mapping of E_1 into the set of fuzzy subsets defined on E_2.

Thus, a fuzzy subset $\underset{\sim}{A} \subset E_1$ will induce a fuzzy subset $\underset{\sim}{B} \subset E_2$ whose membership function will be

(15.2) $$\mu_{\underset{\sim}{B}}(y) = \underset{x \in E_1}{\text{MAX}}\Big(\text{MIN}\,[\mu_{\underset{\sim}{B}}(y \| x), \mu_{\underset{\sim}{A}}(x)]\Big)$$

We immediately see an example.

Example. Consider a fuzzy relation existing between

(15.3)
(15.4) $$E_1 = \{x_1, x_2, x_3, x_4, x_5, x_6\} \quad , \quad E_2 = \{y_1, y_2, y_3\} ,$$

and defined by

(15.5)

$\underset{\sim}{\mathcal{R}}$	y_1	y_2	y_3
x_1	0,3	0,7	0
x_2	0,2	0,5	0
x_3	1	0	0,8
x_4	0	1	0,5
x_5	0,3	1	0,4
x_6	0,8	0	0

This relation $\underset{\sim}{\mathcal{R}}$ expresses a conditional membership function

(15.6) $$\mu_{\underset{\sim}{B}}(y \| x) .$$

Thus,

(15.7) $$\mu_{\underset{\sim}{B}}(y_3 \| x_5) = 0{,}4 .$$

Suppose that one has a fuzzy subset $\underset{\sim}{A}$ of E_1 defined by

(15.8) $$\underset{\sim}{A} = \{(x_1 | 0{,}5), (x_2 | 0{,}2), (x_3 | 0{,}8), (x_4 | 1), (x_5 | 0{,}7), (x_6 | 0)\} .$$

To this fuzzy subset $\underset{\sim}{A} \subset E_1$ corresponds a fuzzy subset in E_2, say $\underset{\sim}{B} \subset E_2$, which will be given by (15.2). We carry out the calculations.

First we calculate $\mu_{\underset{\sim}{B}}(y_1)$. One has

(15.9)
$$\begin{aligned}
&\text{MIN}\,[\mu_{\underset{\sim}{B}}(y_1 \,\|\, x_1)\,,\,\mu_{\underset{\sim}{A}}(x_1)] = \text{MIN}\,[0{,}3\,,\,0{,}5] = 0{,}3\,,\\
&\text{MIN}\,[\mu_{\underset{\sim}{B}}(y_1 \,\|\, x_2)\,,\,\mu_{\underset{\sim}{A}}(x_2)] = \text{MIN}\,[0{,}2\,,\,0{,}2] = 0{,}2\,,\\
&\text{MIN}\,[\mu_{\underset{\sim}{B}}(y_1 \,\|\, x_3)\,,\,\mu_{\underset{\sim}{A}}(x_3)] = \text{MIN}\,[1\,,\,0{,}8] = 0{,}8\,,\\
&\text{MIN}\,[\mu_{\underset{\sim}{B}}(y_1 \,\|\, x_4)\,,\,\mu_{\underset{\sim}{A}}(x_4)] = \text{MIN}\,[0\,,\,1] = 0\,,\\
&\text{MIN}\,[\mu_{\underset{\sim}{B}}(y_1 \,\|\, x_5)\,,\,\mu_{\underset{\sim}{A}}(x_5)] = \text{MIN}\,[0{,}3\,,\,0{,}7] = 0{,}3\,,\\
&\text{MIN}\,[\mu_{\underset{\sim}{B}}(y_1 \,\|\, x_6)\,,\,\mu_{\underset{\sim}{A}}(x_6)] = \text{MIN}\,[0{,}8\,,\,0] = 0\,,
\end{aligned}$$

Then

(15.10)
$$\begin{aligned}
\underset{x_i}{\text{MAX}}\;\text{MIN}\,[\mu_{\underset{\sim}{B}}(y_1 \,\|\, x_i)\,,\,\mu_{\underset{\sim}{A}}(x_i)]
&= \text{MAX}\,[0{,}3\,,\,0{,}2\,,\,0{,}8\,,\,0\,,\,0{,}3\,,\,0]\\
&= 0{,}8\,.
\end{aligned}$$

One should then do the same for y_2, then y_3. One will finally obtain

$$\mu_{\underset{\sim}{B}}(y_1) = 0{,}8 \quad , \quad \mu_{\underset{\sim}{B}}(y_2) = 1 \quad , \quad \mu_{\underset{\sim}{B}}(y_3) = 0{,}8\,.$$

Thus,

(15.11)
$$\underset{\sim}{B} = \{(y_1 \,|\, 0{,}8)\,,\,(y_2 \,|\, 1)\,,\,(y_3 \,|\, 0{,}8)\}.$$

Another presentation of this concept. The expression (15.2), as we shall see later, plays for fuzzy subsets the same role as the notion of function for the elements of formal sets. The notion of function for these elements may be expressed by the phrase: "if $x = a$, then $y = b$ by the function f," which may be written

(15.12)
$$x \underset{f}{\rightsquigarrow} y$$

or likewise

(15.13)
$$y = f(x)\,.$$

The notion of conditioned fuzzy subsets plays exactly the same role, but instead of considering elements $x \in \mathbf{E}_1$, $y \in \mathbf{E}_2$, and the relation f, which is a function, one will make the following definition.

Let $\underset{\sim}{X} \subset \mathbf{E}_1$ and $\underset{\sim}{Y} \subset \mathbf{E}_2$; consider the fuzzy relation $\underset{\sim}{\mathcal{R}}$ existing between $\mathbf{E}_1$ and $\mathbf{E}_2$. One then defines: If $\underset{\sim}{X} = \underset{\sim}{A}$, then $\underset{\sim}{Y} = \underset{\sim}{B}$ through the relation $\underset{\sim}{\mathcal{R}}$; this may be written:

(15.14)
$$\underset{\sim}{A} \underset{\underset{\sim}{\mathcal{R}}}{\rightsquigarrow} \underset{\sim}{B}\,.$$

If $\mu_{\underset{\sim}{\mathcal{R}}}(x, y)$ is the membership function of the fuzzy relation $\underset{\sim}{\mathcal{R}}$, $\mu_{\underset{\sim}{\mathbf{A}}}(x)$ that of $\underset{\sim}{\mathbf{A}}$, and $\mu_{\underset{\sim}{\mathbf{B}}}(x)$ that of $\underset{\sim}{\mathbf{B}}$, one sees that then

$$\mu_{\underset{\sim}{\mathbf{B}}}(y) = \underset{x \in \mathbf{E}_1}{\text{MAX}}\ \text{MIN}\ [\mu_{\underset{\sim}{\mathbf{A}}}(x), \mu_{\underset{\sim}{\mathcal{R}}}(x, y)] \tag{15.15}$$

$$= \bigvee_x [\mu_{\underset{\sim}{\mathbf{A}}}(x) \wedge \mu_{\underset{\sim}{\mathcal{R}}}(x, y)] .$$

This constitutes another presentation of conditioned fuzzy subsets. Moreover, in Section 39 we shall see the importance of this notion.

We consider an example using this presentation.

Example 1.

$$\mathbf{E}_1 = \{x_1, x_2, x_3\} \tag{15.16}$$

$$\underset{\sim}{\mathbf{A}} = \{(x_1 | 0{,}3), (x_2 | 0{,}7), (x_3 | 1)\} \tag{15.17}$$

$$\mathbf{E}_2 = \{y_1, y_2, y_3, y_4, y_5\}, \tag{15.18}$$

(15.19)

$\underset{\sim}{\mathcal{R}}$	y_1	y_2	y_3	y_4	y_5
x_1	0,8	1	0	0,3	0,7
x_2	0,8	0,3	0,8	0,4	0,7
x_3	0,2	0,3	0	0,2	1

.

We present (15.17) as follows:

(15.20)

$\underset{\sim}{\mathbf{A}} =$

x_1	x_2	x_3
0,3	0,7	1

.

We now carry out the operation min for all the elements of the row (15.20) with the column y_1 of (15.19); this gives

(15.21)

x_1	x_2	x_3
0,3	0,7	1

$\wedge$

y_1
0,8
0,8
0,2

$=$

y_1
0,3 ∧ 0,8
0,7 ∧ 0,8
1 ∧ 0,2

$=$

y_1
0,3
0,7
0,2

.

Carrying out the operation max on the elements of the column obtained, we have

$$0{,}3 \vee 0{,}7 \vee 0{,}2 = 0{,}7 . \tag{15.22}$$

Thus,

(15.23) $$\mu_{\underset{\sim}{B}}(y_1) = 0,7 \ .$$

Doing the same between the elements of (15.20) and the other columns of (15.19), we have

(15.24) $$\mu_{\underset{\sim}{B}}(y_2) = 0,3 \quad , \quad \mu_{\underset{\sim}{B}}(y_3) = 0,7 \quad , \quad \mu_{\underset{\sim}{B}}(y_4) = 0,4 \quad , \quad \mu_{\underset{\sim}{B}}(y_5) = 1 \ .$$

And finally

(15.25) $$\underset{\sim}{B} = \{(y_1 \mid 0,7) \ , \ (y_2 \mid 0,3) \ , \ (y_3 \mid 0,7) \ , (y_4 \mid 0,4) \ , (y_5 \mid 1)\} \ ,$$

or what is the same

(15.26) $\underset{\sim}{B}$ =

y_1	y_2	y_3	y_4	y_5
0,7	0,3	0,7	0,4	1

.

Example 2. Evidently, formula (15.15) or (15.2) may also be applied to the case where the subsets are ordinary and the relation $\mathcal{R}$ is boolean (that is, formal). In this case, the formulas become

(15.27) $$\mu_B(y) = \dot{\sum_x} \mu_A(x) \cdot \mu_{\mathcal{R}}(x\,,\,y) \ ,$$

where $\dot{\sum_x}$ is the boolean sum.

Let

(15.28) $$E_1 = \{x_1\,,x_2\,,x_3\} \ ,$$

(15.29) $$A = \{(x_1 \mid 1) \ , \ (x_2 \mid 0) \ , \ (x_3 \mid 1)\} \ ,$$

(15.30) $$E_2 = \{y_1\,,y_2\,,y_3\,,y_4\,,y_5\} \ ,$$

(15.31)

$\mathcal{R}$	y_1	y_2	y_3	y_4	y_5
x_1	1	0	0	1	0
x_2	0	1	1	1	0
x_3	1	0	0	0	1

Then, in carrying out the boolean operations indicated by (15.27) for

(15.32) A =

x_1	x_2	x_3
1	0	1

,

with (15.31) one finds

(15.33)

	y_1	y_2	y_3	y_4	y_5
B =	1	0	0	1	1

Example 3. We now see a case where the reference set is continuous.

(15.34) Let $$\mathbf{E}_1 = \mathbf{R}^+ ,$$

(15.35) $$\underset{\sim}{\mathbf{A}} = \{ x \mid \mu_{\underset{\sim}{\mathbf{A}}}(x) = e^{-k_1 x} \} , \; k_1 \in \mathbf{R}^+,$$

(15.36) $$\underset{\sim}{\mathcal{R}} = \{ (x , y) \mid \mu_{\underset{\sim}{\mathcal{R}}}(x , y) = e^{-k_2 |x-y|} \} , \; k_2 \in \mathbf{R}^+ ,$$

with $k_2 > k_1$.

We now determine the minimum with respect to x among the $\mu_{\underset{\sim}{\mathbf{A}}}(x)$ (Figure 15.1a) and $\mu_{\underset{\sim}{\mathcal{R}}}(x, y)$ (Figure 15.1b). The two curves intersect at two points:

(15.37)

$0 \leqslant x \leqslant y \quad e^{-k_1 x} = e^{-k_2 (y - x)}$

gives $$x = \frac{k_2}{k_2 + k_1} y ,$$

(15.38) $y \leqslant x \quad e^{-k_1 x} = e^{-k_2 (x-y)}$

gives $$x = \frac{k_2}{k_2 - k_1} y .$$

In Figure 15.1c we have traced with a solid curve

(15.39)
$$\mu(x , y) = \mu_{\underset{\sim}{\mathbf{A}}}(x) \wedge \mu_{\underset{\sim}{\mathcal{R}}}(x , y) ,$$

whose maximum holds for

(15.40) $$x = \frac{k_2}{k_2 + k_1} y ;$$

thus

(15.41) $$\mu_{\underset{\sim}{\mathbf{B}}}(y) = e^{-k_1 \left(\frac{k_2}{k_2 + k_1} \right) y}$$
$$= e^{-\frac{k_1 k_2}{k_2 + k_1} y} .$$

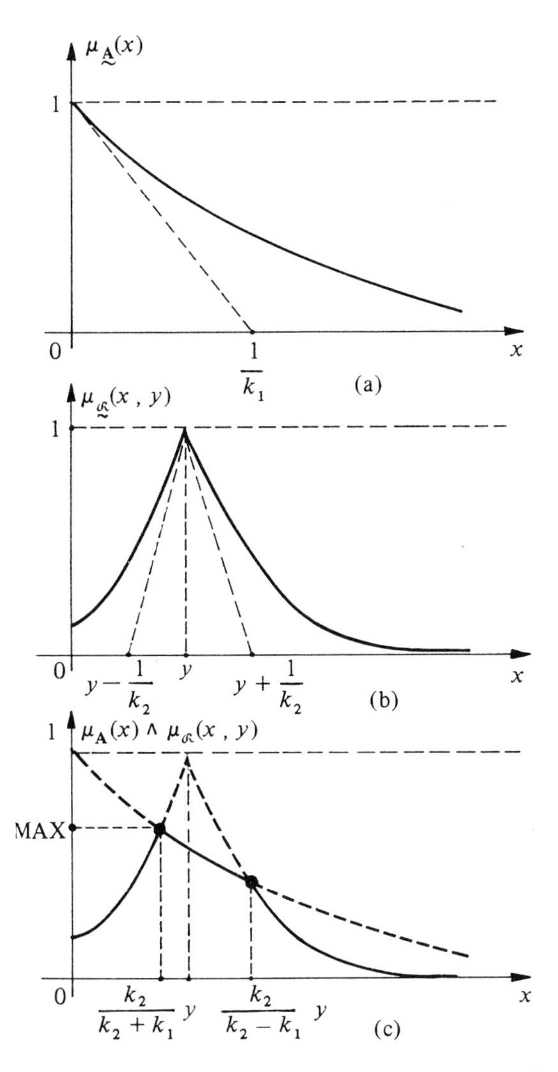

Fig. 15.1

General remark. One may evidently ask the following important question: Knowing that $\underset{\sim}{\mathbf{Y}} = \underset{\sim}{\mathbf{B}}$ through $\underset{\sim}{\mathcal{R}}$ if $\underset{\sim}{\mathbf{X}} = \underset{\sim}{\mathbf{A}}$, may one conclude $\underset{\sim}{\mathbf{X}} = \underset{\sim}{\mathbf{A}}$ through $\underset{\sim}{\mathcal{R}}'$ knowing that $\underset{\sim}{\mathbf{Y}} = \underset{\sim}{\mathbf{B}}$, where $\underset{\sim}{\mathcal{R}}'$ is the reciprocal fuzzy relation?† Except in particular cases, this reciprocal property does not hold. The relation $\underset{\sim}{\mathcal{R}}'$ is not an inverse relation $\underset{\sim}{\mathcal{R}}$.

On this subject, we consider an example, that given by (15.16)–(15.20) and (15.27):

(15.42)

	x_1	x_2	x_3
$\underset{\sim}{\mathbf{A}}$ =	0,3	0,7	1

and

(15.43)

$\underset{\sim}{\mathcal{R}}$ =	y_1	y_2	y_3	y_4	y_5
x_1	0,8	1	0	0,3	0,7
x_2	0,8	0,3	0,8	0,4	0,7
x_3	0,2	0,3	0	0,2	1

,

giving

(15.44)

	y_1	y_2	y_3	y_4	y_5
$\underset{\sim}{\mathbf{B}}$ =	0,7	0,3	0,7	0,4	1

Whereas $\underset{\sim}{\mathbf{B}}$ and

(15.45)

$\underset{\sim}{\mathcal{R}}'$ =	x_1	x_2	x_3
y_1	0,8	0,8	0,2
y_2	1	0,3	0,3
y_3	0	0,8	0
y_4	0,3	0,4	0,2
y_5	0,7	0,7	1

would give‡

(15.46)

	x_1	x_2	x_3
$\underset{\sim}{\mathbf{A}}'$ =	0,7	0,7	1

.

†By reciprocal relation one means here the relation obtained by interchanging the rows and columns of the table constituting the relation.

‡Thus, it is the same as in matrix calculations (linear vector space) where $[M]\{x\} = \{y\}$ and $[M]'\ \{y\}\ \{x\}$. If $[M]$ is square and nonsingular, then $[M]$ has an inverse $[M]^{-1}$ such that $[M] \cdot [M]^{-1} = [1]$ and one has

$$[M] \cdot \{x\} = \{y\} \quad \text{and} \quad \{x\} = [M]^{-1} \cdot \{y\}$$

where here $\{x\}$ and $\{y\}$ are column matrices.

Fuzzy subsets successively conditioning one another. If $\underset{\sim}{\mathbf{A}}_1$ induces $\underset{\sim}{\mathbf{A}}_2$ through $\underset{\sim}{\mathcal{R}}_1$, $\underset{\sim}{\mathbf{A}}_2$ induces $\underset{\sim}{\mathbf{A}}_3$ through $\underset{\sim}{\mathcal{R}}_2, \ldots,$ and $\underset{\sim}{\mathbf{A}}_{n-1}$ induces $\underset{\sim}{\mathbf{A}}_n$ through $\underset{\sim}{\mathcal{R}}_{n-1}$, then $\underset{\sim}{\mathbf{A}}_1$ induces $\underset{\sim}{\mathbf{A}}_n$ through $\underset{\sim}{\mathcal{R}}_{n-1} \circ \underset{\sim}{\mathcal{R}}_{n-2} \circ \ldots \circ \underset{\sim}{\mathcal{R}}_1$.

Example.

(15.47) $$\mathbf{E}_1 = \{x_1, x_2\},$$

(15.48) $\underset{\sim}{\mathbf{A}} =$

x_1	x_2
0,8	0,3

,

(15.49) $$\mathbf{E}_2 = \{y_1, y_2, y_3\},$$

(15.50) $\underset{\sim}{\mathcal{R}}_1 =$

	y_1	y_2	y_3
x_1	0,3	1	0
x_2	0,8	0	0,7

,

(15.51) $$\mathbf{E}_3 = \{z_1, z_2, z_3\},$$

(15.52) $\underset{\sim}{\mathcal{R}}_2 =$

	z_1	z_2	z_3
y_1	0,7	0,4	1
y_2	0,2	0	0,8
y_3	0	0,3	0,9

.

(15.53)

x_1	x_2
0,8	0,3

$\circ$ $\underset{\sim}{\mathcal{R}}_1$

y_1	y_2	y_3
0,3	1	0
0,8	0	0,7

$\circ$ $\underset{\sim}{\mathcal{R}}_2$

z_1	z_2	z_3
0,7	0,4	1
0,2	0	0,8
0	0,3	0,9

$=$

y_1	y_2	y_3
0,3	0,8	0,3

$\circ$ $\underset{\sim}{\mathcal{R}}_2$

z_1	z_2	z_3
0,7	0,4	1
0,2	0	0,8
0	0,3	0,9

$=$

z_1	z_2	z_3
0,3	0,3	0,8

$\subset \mathbf{E}_3$.

Nearest ordinary subsets conditioning one another. It is easy to show [it suffices to refer to (13.38)] that

(15.54)
$$\underset{\sim}{\mathbf{A}} \underset{\underset{\sim}{\mathcal{R}}}{\leadsto} \underset{\sim}{\mathbf{B}} \quad \Rightarrow \quad \underset{\approx}{\mathbf{A}} \underset{\underset{\approx}{\mathcal{R}}}{\leadsto} \underset{\approx}{\mathbf{B}} .$$

Example.

(15.55)
$$\begin{array}{ccc} x_1 & x_2 & x_3 \\ \hline 0,3 & 0,7 & 1 \end{array} \circ \begin{array}{c|ccccc} \underset{\sim}{\mathcal{R}} & y_1 & y_2 & y_3 & y_4 & y_5 \\ \hline x_1 & 0,8 & 1 & 0 & 0,3 & 0,7 \\ x_2 & 0,8 & 0,3 & 0,8 & 0,4 & 0,7 \\ x_3 & 0,2 & 0,3 & 0 & 0,2 & 1 \end{array} = \begin{array}{ccccc} y_1 & y_2 & y_3 & y_4 & y_5 \\ \hline 0,7 & 0,3 & 0,7 & 0,4 & 1 \end{array} ,$$

(15.56)
$$\begin{array}{ccc} x_1 & x_2 & x_3 \\ \hline 0 & 1 & 1 \end{array} \circ \begin{array}{c|ccccc} \underset{\approx}{\mathcal{R}} & y_1 & y_2 & y_3 & y_4 & y_5 \\ \hline x_1 & 1 & 1 & 0 & 0 & 1 \\ x_2 & 1 & 0 & 1 & 0 & 1 \\ x_3 & 0 & 0 & 0 & 0 & 1 \end{array} = \begin{array}{ccccc} y_1 & y_2 & y_3 & y_4 & y_5 \\ \hline 1 & 0 & 1 & 0 & 1 \end{array} .$$

This property remains valid whatever the nature of the reference sets $\mathbf{E}_1$ and $\mathbf{E}_2$, where $x_i \in \mathbf{E}_1$ and $y_j \in \mathbf{E}_2$, whether or not $\mathbf{E}_1$ and $\mathbf{E}_2$ are finite.

16. PROPERTIES OF FUZZY BINARY RELATIONS

We shall consider the case where

(16.1)
(16.2)
$$\mathbf{E}_1 = \mathbf{E}_2 = \mathbf{E} \quad \text{and} \quad \mathbf{M} = [0\,,\,1]\,,$$

and occupy ourselves with some properties of fuzzy binary relations in $\mathbf{E} \times \mathbf{E}$.

Example 1. Let

(16.3) $\mathbf{E} = \{A\,,B\,,C\,,D\,,E\}\,,$

(16.4) $\mathbf{M} = [0\,,\,1]\;;$

the table or matrix in Figure 16.1 represents a fuzzy relation in $\mathbf{E} \times \mathbf{E}$.

$\underset{\sim}{\mathcal{R}}$	A	B	C	D	E
A	0,1	0	0	1	0,8
B	0,8	0,3	0	0,7	1
C	0,8	0,3	0,2	0	0,9
D	0,6	0	1	0,5	0
E	0,2	0,5	1	0,6	0,4

Fig. 16.1

Example 2. Let **R** be the set of real numbers, and let $x \in \mathbf{R}, y \in \mathbf{R}$, then

$$|y| \gg |x| , \tag{16.5}$$

is a fuzzy binary relation $\underset{\sim}{\mathcal{R}}$ in $\mathbf{R} \times \mathbf{R}$ provided one is given $\mu_{\underset{\sim}{\mathcal{R}}}(x, y)$ defining (16.5) for all (x, y).

We proceed to examine the principal properties of fuzzy relations. In order to represent the membership function defining the fuzzy relation, we shall use indifferently the notation $\mu_{\underset{\sim}{\mathcal{R}}}(x, y)$ or $\mu_{\underset{\sim}{G}}(x, y)$ since a fuzzy relation is a fuzzy graph.

Symmetry. A symmetric fuzzy binary relation is defined by

$$\forall (x, y) \in E \times E : \quad (\mu_{\underset{\sim}{\mathcal{R}}}(x, y) = \mu) \Rightarrow (\mu_{\underset{\sim}{\mathcal{R}}}(y, x) = \mu) . \tag{16.6}$$

Example. See Figure 16.2.

Another example. Let **R** be the set of real numbers, and let $x \in \mathbf{R}$, $y \in \mathbf{R}$; then the relation

y is near x

intuitively is a fuzzy symmetric relation in $\mathbf{R} \times \mathbf{R}$.

$\underset{\sim}{\mathcal{R}}$	A	B	C	D	E
A	0	0,1	0	0,1	0,9
B	0,1	1	0,2	0,3	0,4
C	0	0,2	0,8	0,8	1
D	0,1	0,3	0,8	0,7	1
E	0,9	0,4	1	1	0

Fig. 16.2

Reflexivity. This property is defined by

$$\forall (x, x) \in E \times E \quad : \quad \mu_{\underset{\sim}{\mathcal{R}}}(x, x) = 1 . \tag{16.7}$$

Example. See Figure 16.3.

Another example. y is near x, in the example given for symmetry, is reflexive.

$\underset{\sim}{\mathcal{R}}$	A	B	C	D
A	1	0	0,2	0,3
B	0	1	0,1	1
C	0,2	0,7	1	0,4
D	0	1	0,4	1

Fig. 16.3

Transitivity. Let $x, y, z \in E$; then

$$\forall (x, y), (y, z), (x, z) \in E \times E :$$

$$\mu_{\underset{\sim}{\mathcal{R}}}(x, z) \geqslant \underset{y}{\mathrm{MAX}} [\mathrm{MIN}(\mu_{\underset{\sim}{\mathcal{R}}}(x, y), \mu_{\underset{\sim}{\mathcal{R}}}(y, z))] . \tag{16.8}$$

This relation defines the property of transitivity of a fuzzy relation. Such a relation may also be written using the notation

$$(16.9) \qquad \mu_{\underset{\sim}{\mathcal{R}}}(x, z) \geqslant \bigvee_{y} [\mu_{\underset{\sim}{\mathcal{R}}}(x, y) \wedge \mu_{\underset{\sim}{\mathcal{R}}}(y, z)]$$

where, we recall, $\vee$ is a symbol signifying "maximum of" and $\wedge$ is a symbol signifying "minimum of."

Before giving some examples, it is interesting to verify that definition (16.9) in fact generalizes the following in the case of the notion of transitivity in formal relations. For such relations, one knows that transitivity is defined by the property

$\forall (x, y), (y, z), (x, z) \in \mathbf{E} \times \mathbf{E}$:

$$(16.10) \qquad ((x, y) \in \mathbf{G} \text{ and } (y, z) \in \mathbf{G}) \Rightarrow (x, z) \in \mathbf{G} .$$

This relation expresses the fact that, if there exists at least one y such that $(x, y) \in \mathbf{G}$, and $(y, z) \in \mathbf{G}$, that is, $\mu(x, y) = 1$ and $\mu(x, z) = 1$, then $\mu(x, z) = 1$ and $(x, z) \in \mathbf{G}$.

The operation $\wedge$ (min) corresponds well to "and" in propositional logic and the operation $\bigvee_y$ (max with respect to all y's) corresponds well to the result that may be obtained through implication $\Rightarrow$.

We shall see several examples of how to apply formula (16.8) [that is, what is the same thing, (16.9)].

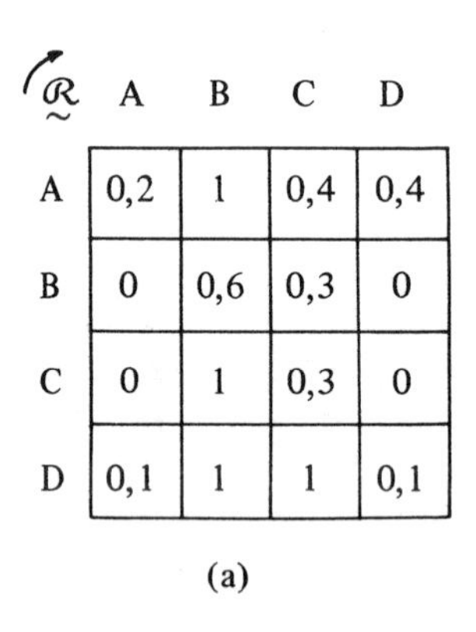

$\underset{\sim}{\mathcal{R}}$	A	B	C	D
A	0,2	1	0,4	0,4
B	0	0,6	0,3	0
C	0	1	0,3	0
D	0,1	1	1	0,1

(a)

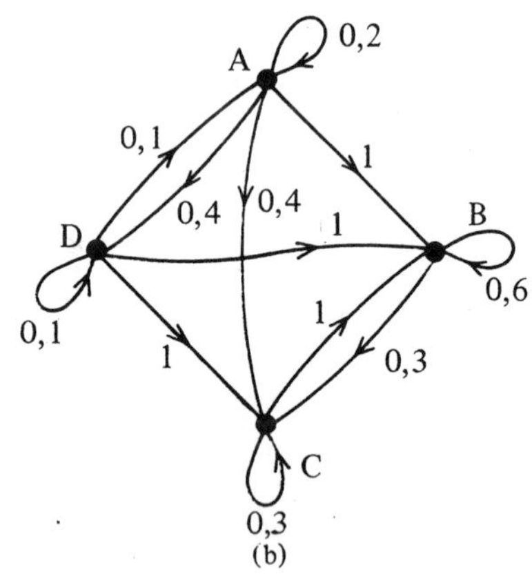

(b)

(arcs of zero membership have not been traced)

Fig. 16.4

Example 1. Figure (16.4). This relation is transitive. As an exercise we carry out the complete verification. There are 16 × 4 operations to perform:†

†For a finite set E whose cardinality is n, there would be n^2 times n operations to perform, thus n^3, in order to verify transitivity if no rule allows reasoning on the membership function.

Arc (A , A).

$\mu(A, A) \wedge \mu(A, A) = 0{,}2 \wedge 0{,}2 = 0{,}2$,
$\mu(A, B) \wedge \mu(B, A) = 1 \wedge 0 = 0$,
$\mu(A, C) \wedge \mu(C, A) = 0{,}4 \wedge 0 = 0$,
$\mu(A, D) \wedge \mu(D, A) = 0{,}4 \wedge 0{,}1 = 0{,}1$,
$\mathrm{MAX}[0{,}2 , 0 , 0 , 0{,}1] = 0{,}2$,
$\mu(A, A) = 0{,}2 \geqslant 0{,}2$.

Arc (A , B).

$\mu(A, A) \wedge \mu(A, B) = 0{,}2 \wedge 1 = 0{,}2$,
$\mu(A, B) \wedge \mu(B, B) = 1 \wedge 0{,}6 = 0{,}6$,
$\mu(A, C) \wedge \mu(C, B) = 0{,}4 \wedge 1 = 0{,}4$,
$\mu(A, D) \wedge \mu(D, B) = 0{,}4 \wedge 1 = 0{,}4$,
$\mathrm{MAX}[0{,}2 , 0{,}6 , 0{,}4 , 0{,}4] = 0{,}6$,
$\mu(A, B) = 1 \geqslant 0{,}6$.

Arc (A , C).

$\mu(A, A) \wedge \mu(A, C) = 0{,}2 \wedge 0{,}4 = 0{,}2$,
$\mu(A, B) \wedge \mu(B, C) = 1 \wedge 0{,}3 = 0{,}3$,
$\mu(A, C) \wedge \mu(C, C) = 0{,}4 \wedge 0{,}3 = 0{,}3$,
$\mu(A, D) \wedge \mu(D, C) = 0{,}4 \wedge 1 = 0{,}4$,
$\mathrm{MAX}[0{,}2 , 0{,}3 , 0{,}3 , 0{,}4] = 0{,}4$,
$\mu(A, C) = 0{,}4 \geqslant 0{,}4$.

Arc (A , D).

$\mu(A, A) \wedge \mu(A, D) = 0{,}2 \wedge 0{,}4 = 0{,}2$,
$\mu(A, B) \wedge \mu(B, D) = 1 \wedge 0 = 0$,
$\mu(A, C) \wedge \mu(C, D) = 0{,}4 \wedge 0 = 0$,
$\mu(A, D) \wedge \mu(D, D) = 0{,}4 \wedge 0{,}1 = 0{,}1$,
$\mathrm{MAX}[0{,}2 , 0 , 0 , 0{,}1] = 0{,}2$,
$\mu(A, D) = 0{,}4 \geqslant 0{,}2$.

Arc (B , A).

$\mu(B, A) \wedge \mu(A, A) = 0 \wedge 0{,}2 = 0$,
$\mu(B, B) \wedge \mu(B, A) = 0{,}6 \wedge 0 = 0$,
$\mu(B, C) \wedge \mu(C, A) = 0{,}3 \wedge 0 = 0$,
$\mu(B, D) \wedge \mu(D, A) = 0 \wedge 0{,}1 = 0$,
$\mathrm{MAX}[0 , 0 , 0 , 0] = 0$,
$\mu(B, A) = 0 \geqslant 0$.

Arc (B , B).

$\mu(B, A) \wedge \mu(A, B) = 0 \wedge 1 = 0$,
$\mu(B, B) \wedge \mu(B, B) = 0{,}6 \wedge 0{,}6 = 0{,}6$,
$\mu(B, C) \wedge \mu(C, B) = 0{,}3 \wedge 1 = 0{,}3$,
$\mu(B, D) \wedge \mu(D, B) = 0 \wedge 1 = 0$,
$\mathrm{MAX}[0 , 0{,}6 , 0{,}3 , 0] = 0{,}6$,
$\mu(B, B) = 0{,}6 \geqslant 0{,}6$.

Arc (B , C)

$$\begin{aligned}
&\mu(B , A) \wedge \mu(A , C) = 0 \wedge 0{,}4 = 0 , \\
&\mu(B , B) \wedge \mu(B , C) = 0{,}6 \wedge 0{,}3 = 0{,}3 , \\
&\mu(B , C) \wedge \mu(C , C) = 0{,}3 \wedge 0{,}3 = 0{,}3 , \\
&\mu(B , D) \wedge \mu(D , C) = 0 \wedge 1 = 0 , \\
&\mathrm{MAX}[0 , 0{,}3 , 0{,}3 , 0] = 0{,}3 , \\
&\mu(B , C) = 0{,}3 \geqslant 0{,}3 .
\end{aligned}$$

Arc (B , D).

$$\begin{aligned}
&\mu(B , A) \wedge \mu(A , D) = 0 \wedge 0{,}4 = 0 , \\
&\mu(B , B) \wedge \mu(B , D) = 0{,}6 \wedge 0 = 0 , \\
&\mu(B , C) \wedge \mu(C , D) = 0{,}3 \wedge 0 = 0 , \\
&\mu(B , D) \wedge \mu(D , D) = 0 \wedge 0{,}1 = 0 , \\
&\mathrm{MAX}[0 , 0 , 0 , 0] = 0 , \\
&\mu(B , D) = 0 \geqslant 0 .
\end{aligned}$$

Arc (C , A).

$$\begin{aligned}
&\mu(C , A) \wedge \mu(A , A) = 0 \wedge 0{,}2 = 0 , \\
&\mu(C , B) \wedge \mu(B , A) = 1 \wedge 0 = 0 , \\
&\mu(C , C) \wedge \mu(C , A) = 0{,}3 \wedge 0 = 0 , \\
&\mu(C , D) \wedge \mu(D , A) = 0 \wedge 0{,}1 = 0 , \\
&\mathrm{MAX}[0 , 0 , 0 , 0] = 0 , \\
&\mu(C , A) = 0 \geqslant 0 .
\end{aligned} \tag{16.11}$$

Arc (C , B).

$$\begin{aligned}
&\mu(C , A) \wedge \mu(A , B) = 0 \wedge 1 = 0 , \\
&\mu(C , B) \wedge \mu(B , B) = 1 \wedge 0{,}6 = 0{,}6 , \\
&\mu(C , C) \wedge \mu(C , B) = 0{,}3 \wedge 1 = 0{,}3 , \\
&\mu(C , D) \wedge \mu(D , B) = 0 \wedge 1 = 0 , \\
&\mathrm{MAX}[0 , 0{,}6 , 0{,}3 , 0] = 0{,}6 , \\
&\mu(C , B) = 1 \geqslant 0{,}6 .
\end{aligned}$$

Arc (C , C).

$$\begin{aligned}
&\mu(C , A) \wedge \mu(A , C) = 0 \wedge 0{,}4 = 0 , \\
&\mu(C , B) \wedge \mu(B , C) = 1 \wedge 0{,}3 = 0{,}3 , \\
&\mu(C , C) \wedge \mu(C , C) = 0{,}3 \wedge 0{,}3 = 0{,}3 , \\
&\mu(C , D) \wedge \mu(D , C) = 0 \wedge 1 = 0 , \\
&\mathrm{MAX}[0 , 0{,}3 , 0{,}3 , 0] = 0{,}3 , \\
&\mu(C , C) = 0{,}3 \geqslant 0{,}3 .
\end{aligned}$$

Arc (C , D).

$$\begin{aligned}
&\mu(C , A) \wedge \mu(A , D) = 0 \wedge 0{,}4 = 0 , \\
&\mu(C , B) \wedge \mu(B , D) = 1 \wedge 0 = 0 , \\
&\mu(C , C) \wedge \mu(C , D) = 0{,}3 \wedge 0 = 0 , \\
&\mu(C , D) \wedge \mu(D , D) = 0 \wedge 0{,}1 = 0 , \\
&\mathrm{MAX}[0 , 0 , 0 , 0] = 0 , \\
&\mu(C , D) = 0 \geqslant 0 .
\end{aligned}$$

Arc (D , A).

$\mu(D, A) \wedge \mu(A, A) = 0{,}1 \wedge 0{,}2 = 0{,}1$,
$\mu(D, B) \wedge \mu(B, A) = 1 \wedge 0 = 0$,
$\mu(D, C) \wedge \mu(C, A) = 1 \wedge 0 = 0$,
$\mu(D, D) \wedge \mu(D, A) = 0{,}1 \wedge 0{,}1 = 0{,}1$,
$\mathrm{MAX}[0{,}1 , 0 , 0 , 0{,}1] = 0{,}1$,
$\mu(D, A) = 0{,}1 \geqslant 0{,}1$.

Arc (D , B).

$\mu(D, A) \wedge \mu(A, B) = 0{,}1 \wedge 1 = 0{,}1$,
$\mu(D, B) \wedge \mu(B, B) = 1 \wedge 0{,}6 = 0{,}6$,
$\mu(D, C) \wedge \mu(C, B) = 1 \wedge 1 = 1$,
$\mu(D, D) \wedge \mu(D, B) = 0{,}1 \wedge 1 = 0{,}1$,
$\mathrm{MAX}[0{,}1 , 0{,}6 , 1 , 0{,}1] = 1$,
$\mu(D, B) = 1 \geqslant 1$.

Arc (D , C).

$\mu(D, A) \wedge \mu(A, C) = 0{,}1 \wedge 0{,}4 = 0{,}1$,
$\mu(D, B) \wedge \mu(B, C) = 1 \wedge 0{,}3 = 0{,}3$,
$\mu(D, C) \wedge \mu(C, C) = 1 \wedge 0{,}3 = 0{,}3$,
$\mu(D, D) \wedge \mu(D, C) = 0{,}1 \wedge 1 = 0{,}1$,
$\mathrm{MAX}[0{,}1 0{,}3 , 0{,}3 , 0{,}1] = 0{,}3$,
$\mu(D, C) = 1 \geqslant 0{,}3$.

Arc (D , D).

$\mu(D, A) \wedge \mu(A, D) = 0{,}1 \wedge 0{,}4 = 0{,}1$,
$\mu(D, B) \wedge \mu(B, D) = 1 \wedge 0 = 0$,
$\mu(D, C) \wedge \mu(C, D) = 1 \wedge 0 = 0$,
$\mu(D, D) \wedge \mu(D, D) = 0{,}1 \wedge 0{,}1 = 0{,}1$,
$\mathrm{MAX}[0{,}1 , 0 , 0 , 0{,}1] = 0{,}1$,
$\mu(D, D) = 0{,}1 \geqslant 0.1$.

Example 2. The following are transitive fuzzy relations:

Y is much larger than **X**
A is clearer than **B**
X is a distant relative of **Y**

on the contrary, the relation

X resembles **Y**

is not transitive. One may have that **X** resembles **Y** and that **Y** resembles **Z**, without necessarily having that **X** resembles **Z**; however, all depends on the nature of the function $\mu_{\underset{\sim}{\mathcal{R}}}(x, y)$, which evaluates the resemblance. This will lead us later to define with more precision what is intended in the present theory by "resemblance."

Example 3. Consider the relation $x \underset{\sim}{\mathcal{R}} y$, where $x, y \in \mathbf{N}$, with

(16.12) $$\mu_{\underset{\sim}{\mathcal{R}}}(x\,,\,y) = e^{-k(x-y)^2},$$

with $k > 1$ sufficiently large so that this membership function expresses the relation "x and y are very near to one another." We shall show that the fuzzy relation defined by (16.12) is not transitive.

We write the matrix representing relation (16.12), obtaining Figure 16.5.

$\underset{\sim}{\mathcal{R}}$	0	1	2	3	4	5	6	7	...
0	1	e^{-k}	e^{-4k}	e^{-9k}	e^{-16k}	e^{-25k}	e^{-36k}	e^{-49k}	...
1	e^{-k}	1	e^{-k}	e^{-4k}	e^{-9k}	e^{-16k}	e^{-25k}	e^{-36k}	...
2	e^{-4k}	e^{-k}	1	e^{-k}	e^{-4k}	e^{-9k}	e^{-16k}	e^{-25k}	...
3	e^{-9k}	e^{-4k}	e^{-k}	1	e^{-k}	e^{-4k}	e^{-9k}	e^{-16k}	...
4	e^{-16k}	e^{-9k}	e^{-4k}	e^{-k}	1	e^{-k}	e^{-4k}	e^{-9k}	...
5	e^{-25k}	e^{-16k}	e^{-9k}	e^{-4k}	e^{-k}	1	e^{-k}	e^{-4k}	...
6	e^{-36k}	e^{-25k}	e^{-16k}	e^{-9k}	e^{-4k}	e^{-k}	1	e^{-k}	...
7	e^{-49k}	e^{-36k}	e^{-25k}	e^{-16k}	e^{-9k}	e^{-4k}	e^{-k}	1	...
⋮	⋮	⋮	⋮	⋮	⋮	⋮	⋮		⋱

Fig. 16.5

In Figure 16.6 we have carried out the calculations giving the right-hand member of the transitivity relation (16.9). One may establish that (16.9) is not verified for all pairs; thus the relation $\underset{\sim}{\mathcal{R}}$ defined by (16.12) is not transitive.

In Section 29, we shall return in detail to the case where **E** is not finite. Transitivity in such cases deserves particular attention.

Now we consider an example where **E** is denumerable, but this time the relation is transitive.

$\underset{\sim}{\mathcal{R}} \circ \underset{\sim}{\mathcal{R}}$	0	1	2	3	4	5	6	7	
0	1	e^{-k}	e^{-k}	e^{-4k}	e^{-4k}	e^{-9k}	e^{-9k}	e^{-16k}	...
1	e^{-k}	1	e^{-k}	e^{-k}	e^{-4k}	e^{-4k}	e^{-9k}	e^{-9k}	...
2	e^{-k}	e^{-k}	1	e^{-k}	e^{-k}	e^{-4k}	e^{-4k}	e^{-9k}	...
3	e^{-4k}	e^{-k}	e^{-k}	1	e^{-k}	e^{-k}	e^{-4k}	e^{-4k}	...
4	e^{-4k}	e^{-4k}	e^{-k}	e^{-k}	1	e^{-k}	e^{-k}	e^{-4k}	...
5	e^{-9k}	e^{-4k}	e^{-4k}	e^{-k}	e^{-k}	1	e^{-k}	e^{-k}	...
6	e^{-9k}	e^{-9k}	e^{-4k}	e^{-4k}	e^{-k}	e^{-k}	1	e^{-k}	...
7	e^{-16k}	e^{-9k}	e^{-9k}	e^{-4k}	e^{-4k}	e^{-k}	e^{-k}	1	...
⋮	⋮	⋮	⋮	⋮	⋮	⋮	⋮	⋮	⋱

Fig. 16.6

Example 4. Consider the relation $x \underset{\sim}{\mathcal{R}} y$, where $x, y \in \mathbf{M}$, with

$$\text{(16.13)} \qquad \begin{aligned} \mu_{\underset{\sim}{\mathcal{R}}}(x, y) &= 0 \qquad y < x\,, \\ &= e^{-x} \qquad y \geqslant x\,. \end{aligned}$$

The matrix of this relation is presented in Figure 16.7. Figure 16.8 shows the results of calculating the right-hand member of (16.9). By comparing the two figures one may verify that (16.9) is satisfied for all pairs. This relation is transitive.

One may also verify that this remains true if $x, y \in \mathbf{R}^+$. This relation may represent "x is small and independent of y." Sometimes this is a trivial relation.

$\underset{\sim}{\mathcal{R}}$ $x \backslash y$	0	1	2	3	4	5	6	7	...
0	1	1	1	1	1	1	1	1	...
1	0	e^{-1}	e^{-1}	e^{-1}	e^{-1}	e^{-1}	e^{-1}	e^{-1}	...
2	0	0	e^{-2}	e^{-2}	e^{-2}	e^{-2}	e^{-2}	e^{-2}	...
3	0	0	0	e^{-3}	e^{-3}	e^{-3}	e^{-3}	e^{-3}	...
4	0	0	0	0	e^{-4}	e^{-4}	e^{-4}	e^{-4}	...
5	0	0	0	0	0	e^{-5}	e^{-5}	e^{-5}	...
6	0	0	0	0	0	0	e^{-6}	e^{-6}	...
7	0	0	0	0	0	0	0	e^{-7}	...
⋮	⋮	⋮	⋮	⋮	⋮	⋮	⋮	⋮	⋱

Fig. 16.7

$\underset{\sim}{\mathcal{R}} \circ \underset{\sim}{\mathcal{R}}$ $x \backslash y$	0	1	2	3	4	5	6	7	...
0	1	1	1	1	1	1	1	1	...
1	0	e^{-1}	e^{-1}	e^{-1}	e^{-1}	e^{-1}	e^{-1}	e^{-1}	...
2	0	0	e^{-2}	e^{-2}	e^{-2}	e^{-2}	e^{-2}	e^{-2}	...
3	0	0	0	e^{-3}	e^{-3}	e^{-3}	e^{-3}	e^{-3}	...
4	0	0	0	0	e^{-4}	e^{-4}	e^{-4}	e^{-4}	...
5	0	0	0	0	0	e^{-5}	e^{-5}	e^{-5}	...
6	0	0	0	0	0	0	e^{-6}	e^{-6}	...
7	0	0	0	0	0	0	0	e^{-7}	...
⋮	⋮	⋮	⋮	⋮	⋮	⋮	⋮	⋮	⋱

Fig. 16.8

Remark on finite relations. The operation defined by (16.8) or (16.9) is carried out "rows by columns" as one ordinarily does in matrix calculations. Mnemonically one may say "roco" (rows–columns). Figure 16.9 indicates how to proceed to obtain

$$\vee\, [(x_{i1} \wedge x_{1j}),\, (x_{i2} \wedge x_{2j}),\, (x_{i3} \wedge x_{3j}),\, \ldots,\, (x_{i,n-1} \wedge x_{n-1\,j}),\, (x_{in} \wedge x_{nj})]. \tag{16.14}$$

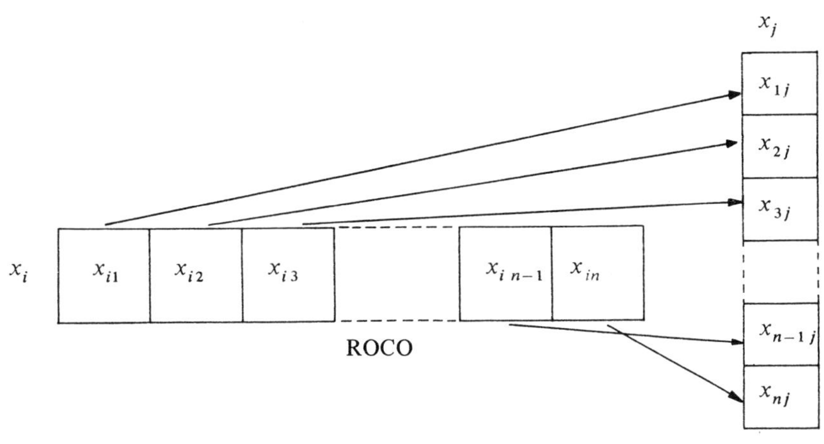

Fig. 16.9

The composition of fuzzy binary relations is a kind of matrix calculation, and at the same time one of the theory of graphs† of a different type than is classical in these calculations and in this theory. The composition of binary relations is, moreover, a particular case of the very general theory of monoids.

17. TRANSITIVE CLOSURE OF A FUZZY BINARY RELATION

Let $\underset{\sim}{\mathcal{R}}$ be a fuzzy relation in $\mathbf{E} \times \mathbf{E}$; then define

$$\underset{\sim}{\mathcal{R}}^2 = \underset{\sim}{\mathcal{R}} \circ \underset{\sim}{\mathcal{R}}\,, \tag{17.1}$$

by

$$\mu_{\underset{\sim}{\mathcal{R}}}(x, z) = \underset{y}{\mathrm{MAX}}\, [\mathrm{MIN}\, (\mu_{\underset{\sim}{\mathcal{R}}}(x, y), \mu_{\underset{\sim}{\mathcal{R}}}(y, z)),] \tag{17.2}$$

where $x, y, z \in \mathbf{E}$; The expression (17.2) may be rewritten

$$\mu_{\underset{\sim}{\mathcal{R}}^2}(x, z) = \underset{y}{\vee}\, [\mu_{\underset{\sim}{\mathcal{R}}}(x, y) \wedge \mu_{\underset{\sim}{\mathcal{R}}}(y, z)]\,. \tag{17.3}$$

Property (16.8) or (16.9) defining transitivity may also be presented in the following fashion:

$$\underset{\sim}{\mathcal{R}} \circ \underset{\sim}{\mathcal{R}} \subset \underset{\sim}{\mathcal{R}}\,, \tag{17.4}$$

†See for example, references [1K, 2K, 3K, 5K].

Suppose

(17.5) $$\underset{\sim}{\mathcal{R}}^2 \subset \underset{\sim}{\mathcal{R}} ,$$

and also that

(17.6) $$\underset{\sim}{\mathcal{R}}^{k+1} \subset \underset{\sim}{\mathcal{R}}^k , \quad k = 1, 2, 3, \ldots .$$

Then also, evidently,

(17.7) $$\underset{\sim}{\mathcal{R}}^{k} \subset \underset{\sim}{\mathcal{R}} \quad k = 1, 2, 3, \ldots .$$

We shall call the *transitive closure* of a fuzzy binary relation the relation

(17.8) $$\hat{\underset{\sim}{\mathcal{R}}} = \underset{\sim}{\mathcal{R}} \cup \underset{\sim}{\mathcal{R}}^2 \cup \underset{\sim}{\mathcal{R}}^3 \cup \ldots .$$

Theorem I. The transitive closure of any fuzzy binary relation is a transitive binary relation.

Proof. According to (17.8), we may write

(17.9) $$\hat{\underset{\sim}{\mathcal{R}}}^2 = \hat{\underset{\sim}{\mathcal{R}}} \circ \hat{\underset{\sim}{\mathcal{R}}} = \underset{\sim}{\mathcal{R}}^2 \cup \underset{\sim}{\mathcal{R}}^3 \cup \underset{\sim}{\mathcal{R}}^4 \cup \ldots .$$

Then, comparing (17.8) and (17.9), we may write

(17.10) $$\hat{\underset{\sim}{\mathcal{R}}}^2 \subset \hat{\underset{\sim}{\mathcal{R}}} ;$$

which proves that $\underset{\sim}{\mathcal{R}}$ is transitive.

To summarize, we have the following properties:

(17.11) $$(\underset{\sim}{\mathcal{R}} \supset \underset{\sim}{\mathcal{R}}^2) \Leftrightarrow (\underset{\sim}{\mathcal{R}} = \hat{\underset{\sim}{\mathcal{R}}}) \Leftrightarrow (\underset{\sim}{\mathcal{R}} \text{ is transitive}),$$

(17.12) $$(\underset{\sim}{\mathcal{R}} = \underset{\sim}{\mathcal{R}}^2) \Rightarrow (\underset{\sim}{\mathcal{R}} = \hat{\underset{\sim}{\mathcal{R}}}) \Leftrightarrow (\underset{\sim}{\mathcal{R}} \text{ is transitive}).$$

Remark. Theorem I gives the means for constructing a transitive relation from any relation. Such a synthesis may have interest.

Theorem II. Let $\underset{\sim}{\mathcal{R}}$ be any fuzzy binary relation. If, for some k, one has

(17.13) $$\underset{\sim}{\mathcal{R}}^{k+1} = \underset{\sim}{\mathcal{R}}^k ,$$

then

(17.14) $$\hat{\underset{\sim}{\mathcal{R}}} = \underset{\sim}{\mathcal{R}} \cup \underset{\sim}{\mathcal{R}}^2 \cup \ldots \cup \underset{\sim}{\mathcal{R}}^k .$$

We shall note that the reverse is not true.

Proof. The proof is almost trivial. One has

(17.15) $$\begin{aligned} \hat{\underset{\sim}{\mathcal{R}}} &= \underset{\sim}{\mathcal{R}} \cup \underset{\sim}{\mathcal{R}}^2 \cup \ldots \cup \underset{\sim}{\mathcal{R}}^k \cup \underset{\sim}{\mathcal{R}}^{k+1} \cup \underset{\sim}{\mathcal{R}}^{k+2} \cup \ldots \\ &= \underset{\sim}{\mathcal{R}} \cup \underset{\sim}{\mathcal{R}}^2 \cup \ldots \cup \underset{\sim}{\mathcal{R}}^k \cup \underset{\sim}{\mathcal{R}}^k \cup \underset{\sim}{\mathcal{R}}^k \cup \ldots \\ &= \underset{\sim}{\mathcal{R}} \cup \underset{\sim}{\mathcal{R}}^2 \cup \ldots \cup \underset{\sim}{\mathcal{R}}^k . \end{aligned}$$

We shall prove later that, if $\underset{\sim}{\mathcal{R}} \subset \mathbf{E} \times \mathbf{E}$, where $\mathbf{E}$ is finite and card(E) = n, then

$$\hat{\underset{\sim}{\mathcal{R}}} = \underset{\sim}{\mathcal{R}} \cup \underset{\sim}{\mathcal{R}}^2 \cup \ldots \cup \underset{\sim}{\mathcal{R}}^n . \tag{17.16}$$

and there exists a k defined by (17.14) such that $k \leqslant n$;

We consider several examples.

Example 1. Consider the relation $\underset{\sim}{\mathcal{R}}$ given in Figure 17.1a. One may calculate $\underset{\sim}{\mathcal{R}}^2$ (Figure 17.1b), then $\underset{\sim}{\mathcal{R}}^3$(Figure 17.1c). We see that $\underset{\sim}{\mathcal{R}}^3 = \underset{\sim}{\mathcal{R}}^2$; one may then stop there, and $\hat{\underset{\sim}{\mathcal{R}}}$ is given in Figure 17.1d.

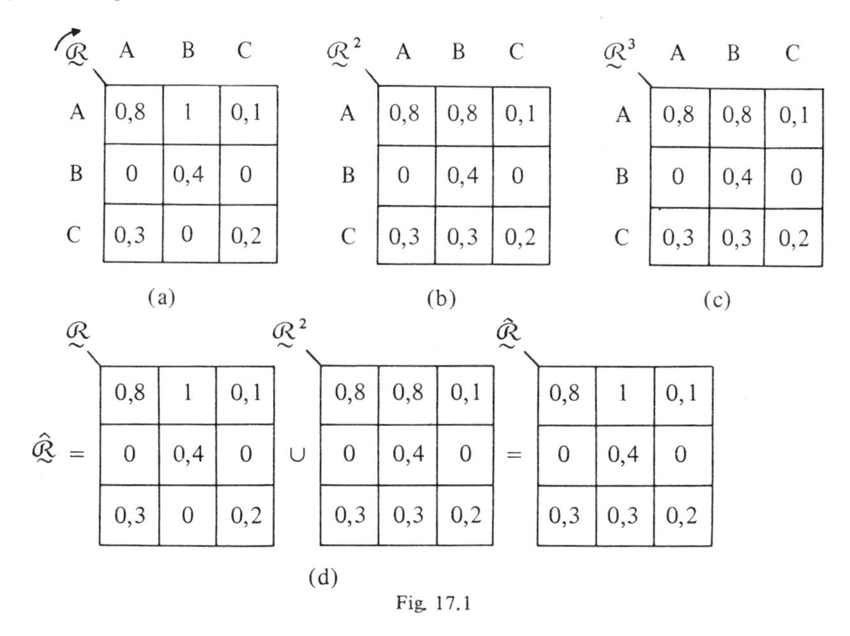

(a) $\underset{\sim}{\mathcal{R}}$

	A	B	C
A	0,8	1	0,1
B	0	0,4	0
C	0,3	0	0,2

(b) $\underset{\sim}{\mathcal{R}}^2$

	A	B	C
A	0,8	0,8	0,1
B	0	0,4	0
C	0,3	0,3	0,2

(c) $\underset{\sim}{\mathcal{R}}^3$

	A	B	C
A	0,8	0,8	0,1
B	0	0,4	0
C	0,3	0,3	0,2

(d) $\hat{\underset{\sim}{\mathcal{R}}} = \underset{\sim}{\mathcal{R}} \cup \underset{\sim}{\mathcal{R}}^2 =$

0,8	1	0,1
0	0,4	0
0,3	0,3	0,2

Fig. 17.1

In Figure 17.2 we have verified that

$$\hat{\underset{\sim}{\mathcal{R}}}^2 \subset \hat{\underset{\sim}{\mathcal{R}}} . \tag{17.17}$$

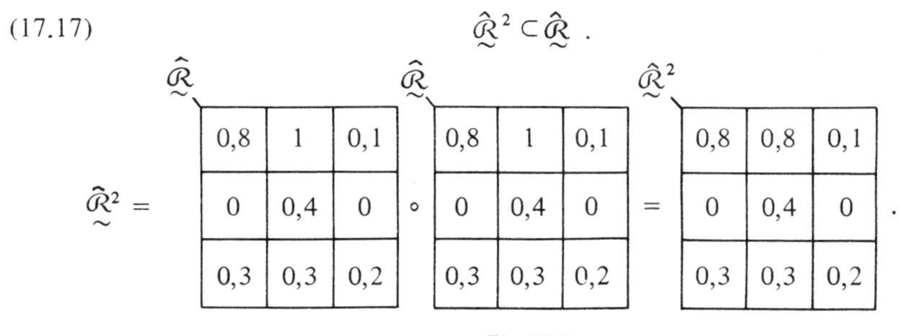

$\hat{\underset{\sim}{\mathcal{R}}}^2 = \hat{\underset{\sim}{\mathcal{R}}} \circ \hat{\underset{\sim}{\mathcal{R}}}$, where $\hat{\underset{\sim}{\mathcal{R}}}$:

0,8	1	0,1
0	0,4	0
0,3	0,3	0,2

$\hat{\underset{\sim}{\mathcal{R}}}^2$:

0,8	0,8	0,1
0	0,4	0
0,3	0,3	0,2

Fig. 17.2

Example 2. In the example presented in Figure 17.3, we have a relation $\underset{\sim}{\mathcal{R}}$ that is transitive. By carrying out the calculations in the same order as the above, one sees that

$$\hat{\underset{\sim}{\mathcal{R}}} = \underset{\sim}{\mathcal{R}} . \tag{17.18}$$

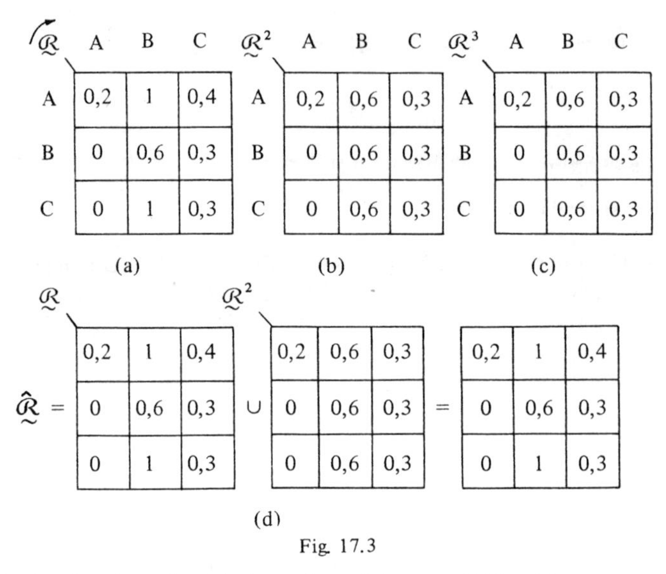

(a)

$\underset{\sim}{\mathcal{R}}$	A	B	C
A	0,2	1	0,4
B	0	0,6	0,3
C	0	1	0,3

(b)

$\underset{\sim}{\mathcal{R}}^2$	A	B	C
A	0,2	0,6	0,3
B	0	0,6	0,3
C	0	0,6	0,3

(c)

$\underset{\sim}{\mathcal{R}}^3$	A	B	C
A	0,2	0,6	0,3
B	0	0,6	0,3
C	0	0,6	0,3

(d)

$\hat{\underset{\sim}{\mathcal{R}}}$ = $\underset{\sim}{\mathcal{R}}$ ∪ $\underset{\sim}{\mathcal{R}}^2$ =

0,2	1	0,4
0	0,6	0,3
0	1	0,3

∪

0,2	0,6	0,3
0	0,6	0,3
0	0,6	0,3

=

0,2	1	0,4
0	0,6	0,3
0	1	0,3

Fig. 17.3

Example 3. Consider the relation $x\underset{\sim}{\mathcal{R}}y$, where $x, y \in \mathbf{N}$, with

(17.19) $$\mu_{\underset{\sim}{\mathcal{R}}}(x, y) = e^{-kxy},$$

with $k > 1$ sufficiently large so that this membership function expresses the relation "x and y are both rather small nonnegative integers."† For a matrix representation of this relation, one has

(17.20) $\underset{\sim}{\mathcal{R}}$ =

	0	1	2	3	4	5	...
0	1	1	1	1	1	1	...
1	1	e^{-k}	e^{-2k}	e^{-3k}	e^{-4k}	e^{-5k}	...
2	1	e^{-2k}	e^{-4k}	e^{-6k}	e^{-8k}	e^{-10k}	...
3	1	e^{-3k}	e^{-6k}	e^{-9k}	e^{-12k}	e^{-15k}	...
4	1	e^{-4k}	e^{-8k}	e^{-12k}	e^{-16k}	e^{-20k}	...
5	1	e^{-5k}	e^{-10k}	e^{-15k}	e^{-20k}	e^{-25k}	...
⋮	⋮	⋮	⋮	⋮	⋮	⋮	⋱

†One may say for this that among the two elements of the ordered pair (x, y) there is at least one that is rather small.

Calculation of $\underset{\sim}{\mathcal{R}}^2$ gives

(17.21)

$\underset{\sim}{\mathcal{R}}^2 =$

	0	1	2	3	4	5	
0	1	1	1	1	1	1	...
1	1	1	1	1	1	1	...
2	1	1	1	1	1	1	...
3	1	1	1	1	1	1	...
4	1	1	1	1	1	1	...
5	1	1	1	1	1	1	...
⋮	⋮	⋮	⋮	⋮	⋮	⋮	⋱

Thus, since $\underset{\sim}{\mathcal{R}}^2 \supset \underset{\sim}{\mathcal{R}}$ instead of $\underset{\sim}{\mathcal{R}}^2 \subset \underset{\sim}{\mathcal{R}}$, this fuzzy relation is not transitive.

A similar and easy proof would show that it is the same if $x, y \in \mathbf{R}^+$ instead of $\mathbf{N}$.

We shall return in Section 29, as promised in Section 16, to the case where $\mathbf{E}$ is not finite.

Example 4. We return to the case of a relation $\underset{\sim}{\mathcal{R}} \subset \mathbf{E} \times \mathbf{E}$, where $\mathbf{E}$ is finite. This is done in order to make it clear that one does not always have the favorable case (17.13). But we shall go on to show also from this example a very interesting phenomenon.

In Figure 17.4 we have given a relation $\underset{\sim}{\mathcal{R}}$ and successively calculated $\underset{\sim}{\mathcal{R}}^2, \underset{\sim}{\mathcal{R}}^3, \ldots$. One notices that this does not converge; there does not exist a fixed k after which $\underset{\sim}{\mathcal{R}}^{k+1} = \underset{\sim}{\mathcal{R}}^k$.

Fortunately, thanks to (17.16) we know that we may stop at $k = 3$. And then one obtains $\hat{\underset{\sim}{\mathcal{R}}}$ easily.

But, if the reader considers attentively all the relations obtained, he sees that for $k > 3$, we have

$$(17.22) \qquad \underset{\sim}{\mathcal{R}}^4 = \underset{\sim}{\mathcal{R}}^6 = \cdots = \underset{\sim}{\mathcal{R}}^{2\nu} = \underset{\sim}{\mathcal{R}}^{2\nu+2} = \cdots = \underset{\sim}{\mathcal{R}}_p$$

$$(17.23) \qquad \underset{\sim}{\mathcal{R}}^5 = \underset{\sim}{\mathcal{R}}^7 = \cdots = \underset{\sim}{\mathcal{R}}^{2\nu+1} = \underset{\sim}{\mathcal{R}}^{2\nu+3} = \cdots = \underset{\sim}{\mathcal{R}}_i$$

Thus there appears a cyclic phenomenon that would be interesting to study. We lack room here to study "cyclic fuzzy relations," which we leave with these remarks; but we commend these to the reader who perhaps finds himself interested.

$\underset{\sim}{\mathcal{R}}$	A	B	C
A	0,3	0,6	0,2
B	1	0,2	0,7
C	0	0,8	0,1

$\underset{\sim}{\mathcal{R}}^2$	A	B	C
A	0,6	0,3	0,6
B	0,3	0,7	0,2
C	0,8	0,2	0,7

$\underset{\sim}{\mathcal{R}}^3$	A	B	C
A	0,3	0,6	0,3
B	0,7	0,3	0,7
C	0,3	0,7	0,2

$\underset{\sim}{\mathcal{R}}^4$	A	B	C
A	0,6	0,3	0,6
B	0,3	0,7	0,3
C	0,7	0,3	0,7

$\underset{\sim}{\mathcal{R}}^5$	A	B	C
A	0,3	0,6	0,3
B	0,7	0,3	0,7
C	0,3	0,7	0,3

$\underset{\sim}{\mathcal{R}}^6$	A	B	C
A	0,6	0,3	0,6
B	0,3	0,7	0,3
C	0,7	0,3	0,7

$\underset{\sim}{\mathcal{R}}^7$	A	B	C
A	0,3	0,6	0,3
B	0,7	0,3	0,7
C	0,3	0,7	0,3

$\underset{\sim}{\mathcal{R}}^8$	A	B	C
A	0,6	0,3	0,6
B	0,3	0,7	0,3
C	0,7	0,3	0,7

$\underset{\sim}{\mathcal{R}}^9$	A	B	C
A	0,3	0,6	0,3
B	0,7	0,3	0,7
C	0,3	0,7	0,3

$\underset{\sim}{\mathcal{R}}^{10}$	A	B	C
A	0,6	0,3	0,6
B	0,3	0,7	0,3
C	0,7	0,3	0,7

. . . .

$\hat{\underset{\sim}{\mathcal{R}}}$	A	B	C
A	0,6	0,6	0,6
B	1	0,7	0,7
C	0,8	0,8	0,7

Fig. 17.4

Remark. One may ask the following interesting question: Does the composition of two transitive relations $\underset{\sim}{\mathcal{R}}_1$ and $\underset{\sim}{\mathcal{R}}_2$ always give a relation $\underset{\sim}{\mathcal{R}}_1 \circ \underset{\sim}{\mathcal{R}}_2$ and/or $\underset{\sim}{\mathcal{R}}_2 \circ \underset{\sim}{\mathcal{R}}_1$ that is transitive? This is, unfortunately, not the case, as the following counterexample shows:

Example. Let $\underset{\sim}{\mathcal{R}}_1$ be as given in (17.24); by checking the property $\underset{\sim}{\mathcal{R}}_1^2 \subset \underset{\sim}{\mathcal{R}}_1$, one may verify that this relation is indeed transitive:

(17.24)

$\underset{\sim}{\mathcal{R}}_1$	A	B	C	D	E
A	0,5	0,9	0	0	0,5
B	0	0,7	0	0	0
C	0	1	0,1	0,1	0
D	0	1	0,4	1	0
E	0,7	0,9	0	0	0,5

$\circ$

$\underset{\sim}{\mathcal{R}}_1$	A	B	C	D	E
A	0,5	0,9	0	0	0,5
B	0	0,7	0	0	0
C	0	1	0,1	0,1	0
D	0	1	0,4	1	0
E	0,7	0,9	0	0	0,5

$=$

$\underset{\sim}{\mathcal{R}}_1^2$	A	B	C	D	E
A	0,5	0,7	0	0	0,5
B	0	0,7	0	0	0
C	0	0,7	0,1	0,1	0
D	0	1	0,4	1	0
E	0,5	0,7	0	0	0,5

Let $\underset{\sim}{\mathcal{R}}_2$ be as in (17.25); by checking the property $\underset{\sim}{\mathcal{R}}_2^2 \subset \underset{\sim}{\mathcal{R}}_2$, one verifies that this relation also is transitive:

(17.25)

$\underset{\sim}{\mathcal{R}}_2$	A	B	C	D	E
A	0,7	0	0	0	0
B	0,8	1	0,6	0,6	1
C	0	0	0,5	0,5	0
D	0	0	0,2	0,4	0
E	0,8	1	0,6	0,6	1

$\circ$

$\underset{\sim}{\mathcal{R}}_2$	A	B	C	D	E
A	0,7	0	0	0	0
B	0,8	1	0,6	0,6	1
C	0	0	0,5	0,5	0
D	0	0	0,2	0,4	0
E	0,8	1	0,6	0,6	1

$=$

$\underset{\sim}{\mathcal{R}}_2^2$	A	B	C	D	E
A	0,7	0	0	0	0
B	0,8	1	0,6	0,6	1
C	0	0	0,5	0,5	0
D	0	0	0,2	0,4	0
E	0,8	1	0,6	0,6	1

We now calculate $\underset{\sim}{\mathcal{R}}_2 \circ \underset{\sim}{\mathcal{R}}_1$:

(17.26)

$\underset{\sim}{\mathcal{R}}_1$	A	B	C	D	E
A	0,5	0,9	0	0	0,5
B	0	0,7	0	0	0
C	0	1	0,1	0,1	0
D	0	1	0,4	1	0
E	0,7	0,9	0	0	0,5

$\circ$

$\underset{\sim}{\mathcal{R}}_2$	A	B	C	D	E
A	0,7	0	0	0	0
B	0,8	1	0,6	0,6	1
C	0	0	0,5	0,5	0
D	0	0	0,2	0,4	0
E	0,8	1	0,6	0,6	1

$=$

$\underset{\sim}{\mathcal{R}}_2 \circ \underset{\sim}{\mathcal{R}}_1$	A	B	C	D	E
A	0,8	0,9	0,6	0,6	0,9
B	0,7	0,7	0,6	0,6	0,7
C	0,8	1	0,6	0,6	1
D	0,8	1	0,6	0,6	1
E	0,8	0,9	0,6	0,6	0,9

and $(\underset{\sim}{\mathcal{R}}_2 \circ \underset{\sim}{\mathcal{R}}_1)^2$:

(17.27)

$\underset{\sim}{\mathcal{R}}_2 \circ \underset{\sim}{\mathcal{R}}_1$	A	B	C	D	E
A	0,8	0,9	0,6	0,6	0,9
B	0,7	0,7	0,6	0,6	0,7
C	0,8	1	0,6	0,6	1
D	0,8	1	0,6	0,6	1
E	0,8	0,9	0,6	0,6	0,9

$\circ$

$\underset{\sim}{\mathcal{R}}_2 \circ \underset{\sim}{\mathcal{R}}_1$	A	B	C	D	E
A	0,8	0,9	0,6	0,6	0,9
B	0,7	0,7	0,6	0,6	0,7
C	0,8	1	0,6	0,6	1
D	0,8	1	0,6	0,6	1
E	0,8	0,9	0,6	0,6	0,9

$=$

$(\underset{\sim}{\mathcal{R}}_2 \circ \underset{\sim}{\mathcal{R}}_1)^2$	A	B	C	D	E
A	0,8	0,9	0,6	0,6	0,9
B	0,7	0,7	0,6	0,6	0,7
C	0,8	0,9	0,6	0,6	0,9
D	0,8	0,9	0,6	0,6	0,9
E	0,8	0,9	0,6	0,6	0,9

The relation $(\underset{\sim}{\mathcal{R}}_2 \circ \underset{\sim}{\mathcal{R}}_1)^2 \subset \underset{\sim}{\mathcal{R}}_2 \circ \underset{\sim}{\mathcal{R}}_1$ is certainly verified.

We now calculate $\underset{\sim}{\mathcal{R}}_1 \circ \underset{\sim}{\mathcal{R}}_2$:

(17.28)

$\underset{\sim}{\mathcal{R}}_2$	A	B	C	D	E
A	0,7	0	0	0	0
B	0,8	1	0,6	0,6	1
C	0	0	0,5	0,5	0
D	0	0	0,2	0,4	0
E	0,8	1	0,6	0,6	1

$\circ$

$\underset{\sim}{\mathcal{R}}_1$	A	B	C	D	E
A	0,5	0,9	0	0	0,5
B	0	0,7	0	0	0
C	0	1	0,1	0,1	0
D	0	1	0,4	1	0
E	0,7	0,9	0	0	0,5

$=$

$\underset{\sim}{\mathcal{R}}_1 \circ \underset{\sim}{\mathcal{R}}_2$	A	B	C	D	E
A	0,5	0,7	0	0	0,5
B	0,7	0,9	0,4	0,6	0,5
C	0	0,5	0,4	0,5	0
D	0	0,4	0,4	0,4	0
E	0,7	0,9	0,4	0,6	0,5

and $(\mathcal{R}_1 \circ \mathcal{R}_2)^2$:

(17.29)

$\underset{\sim}{\mathcal{R}}_1 \circ \underset{\sim}{\mathcal{R}}_2$	A	B	C	D	E
A	0,5	0,7	0	0	0,5
B	0,7	0,9	0,4	0,6	0,5
C	0	0,5	0,4	0,5	0
D	0	0,4	0,4	0,4	0
E	0,7	0,9	0,4	0,6	0,5

$\circ$

$\underset{\sim}{\mathcal{R}}_1 \circ \underset{\sim}{\mathcal{R}}_2$	A	B	C	D	E
A	0,5	0,7	0	0	0,5
B	0,7	0,9	0,4	0,6	0,5
C	0	0,5	0,4	0,5	0
D	0	0,4	0,4	0,4	0
E	0,7	0,9	0,4	0,6	0,5

$=$

$(\underset{\sim}{\mathcal{R}}_1 \circ \underset{\sim}{\mathcal{R}}_2)^2$	A	B	C	D	E
A	0,7	0,7	0,4	0,6	0,5
B	0,7	0,9	0,4	0,6	0,5
C	0,5	0,5	0,4	0,5	0,5
D	0,4	0,4	0,4	0,4	0,4
E	0,7	0,9	0,4	0,6	0,5

One sees that we do not have $(\underset{\sim}{\mathcal{R}}_1 \circ \underset{\sim}{\mathcal{R}}_2)^2 \subset \underset{\sim}{\mathcal{R}}_1 \circ \underset{\sim}{\mathcal{R}}_2$; and it follows that $\underset{\sim}{\mathcal{R}}_1 \circ \underset{\sim}{\mathcal{R}}_2$ is not transitive.

Thus, the composition of two transitive relations will not always give a transitive relation.

18. A PATH IN A FINITE FUZZY GRAPH

We shall consider in the finite graph $\mathbf{G} \subset \mathbf{E} \times \mathbf{E}$ an ordered r-tuple with or without repetition†,

$$C = (x_{i_1}, x_{i_2}, \ldots, x_{i_r}) , \tag{18.1}$$

where the $x_{i_k} \in \mathbf{E}$, $k = 1, 2, \ldots, r$, and with the condition

$$\forall (x_{i_k}, x_{i_{k+1}}) : \mu_{\underset{\sim}{\mathcal{R}}}(x_{i_k}, x_{i_{k+1}}) > 0 , \quad k = 1, 2, \ldots, r-1 . \tag{18.2}$$

Such an ordered r-tuple will be called a *path from* x_{i_1} to x_{i_r} in the graph $\mathbf{G}$ (one also says in the relation $\underset{\sim}{\mathcal{R}}$).

With each path $(x_{i_1}, x_{i_2}, \ldots, x_{i_r})$ we shall associate a value defined by

$$l(x_{i_1}, x_{i_2}, \ldots, x_{i_r}) = \mu_{\underset{\sim}{\mathcal{R}}}(x_{i_1}, x_{i_2}) \wedge \mu_{\underset{\sim}{\mathcal{R}}}(x_{i_2}, x_{i_3}) \wedge \ldots \wedge \mu\ (x_{i_{r-1}}, x_{i_r}) . \tag{18.3}$$

We now consider all possible paths existing between x_i and x_j, two arbitrary elements of $\mathbf{E}$; let $\mathbf{C}(x_i, x_j)$ be the ordinary set of all such paths:

$$\mathbf{C}(x_i, x_j) = \{\ c(x_i, x_j)\ c(x_i, x_j) = (x_{i_1} = x_i, x_{i_2}, .x_{i_{r-1}}, x_{i_r} = x_j)\ \}.$$

We shall define the strongest path $C^*(x_i, x_j)$ from x_i to x_j by

$$l^*(x_i, x_j) = \bigvee_{C(x_i, x_j)} l(x_{i_1} = x_i, x_{i_2}, \ldots, x_{i_{r-1}}, x_{i_r} = x_j) . \tag{18.4}$$

Also, the number of elements less one constituting a path will be called the length of the path.

Before giving several examples, we consider various theorems.

Theorem I. Let $\underset{\sim}{\mathcal{R}} \subset \mathbf{E} \times \mathbf{E}$; then one has

$$\forall (x, y) \in \mathbf{E} \times \mathbf{E} : \quad \mu_{\underset{\sim}{\mathcal{R}}^k}(x, y) = l_k^*(x, y) , \tag{18.5}$$

where $l_k^*(x, y)$ is the strongest path existing from x to y of length k.

Proof. The result is immediate; it suffices to consider (18.4) and (18.3) on the one hand, and from there the composition $\underset{\sim}{\mathcal{R}} \circ \underset{\sim}{\mathcal{R}} \circ \cdots \circ \underset{\sim}{\mathcal{R}}$. It is in fact the same max–min operation presented in two different fashions.

†In other words, r may be less than, equal to, or greater than n = card E, depending on the case.

Theorem II. Let $\underset{\sim}{\mathcal{R}} \subset \mathbf{E} \times \mathbf{E}$ and $\hat{\underset{\sim}{\mathcal{R}}}$ be the transitive closure of $\underset{\sim}{\mathcal{R}}$; then one has

$$\forall (x, y) \in \mathbf{E} \times \mathbf{E} : \mu_{\hat{\underset{\sim}{\mathcal{R}}}}(x, y) = l^*(x, y) . \tag{18.6}$$

Proof. It suffices to review the definitions of $\hat{\underset{\sim}{\mathcal{R}}}$ and of $l^*(x, y)$.

Theorem III. Let n = card $\mathbf{E}$; if k is the length of a path from x_i to x_j with $k > n$ = card $\mathbf{E}$, then all the elements of the chain are not unique; there is at least one "circuit" (closed path) in the path. If one removes this (or these) circuit(s), the resulting path has a length less than or equal to n; one may also state

$$l_k^*(x, y) = l_{j \leqslant n}^*(x, y), \tag{18.7}$$

where $l_{i \leqslant n}^*(x, y)$ is the value of the strongest path of length less than or equal to n from x to y.

Proof. After removing the circuits there remains a chain that has at most length n; relation (18.7) is then verified.

Theorem IV.† If $\underset{\sim}{\mathcal{R}} \subset \mathbf{E} \times \mathbf{E}$ and n = card $\mathbf{E}$, then

$$\hat{\underset{\sim}{\mathcal{R}}} = \underset{\sim}{\mathcal{R}} \cup \underset{\sim}{\mathcal{R}}^2 \cup \ldots \cup \underset{\sim}{\mathcal{R}}^n . \tag{18.8}$$

Proof. This follows immediately from Theorem II [see (18.6)].

Example. We consider the relation $\underset{\sim}{\mathcal{R}}$ represented in Figure 18.1. The results presented in Figure 18.2 will be used in our explanations. Let (B, C, A, D) be a path. We calculate its value:

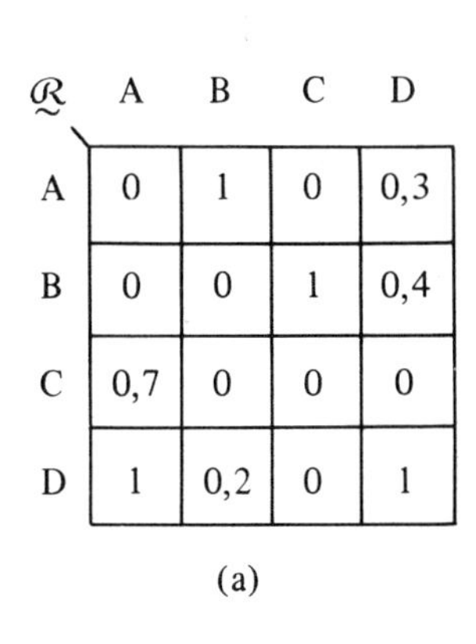

$\underset{\sim}{\mathcal{R}}$	A	B	C	D
A	0	1	0	0,3
B	0	0	1	0,4
C	0,7	0	0	0
D	1	0,2	0	1

(a)

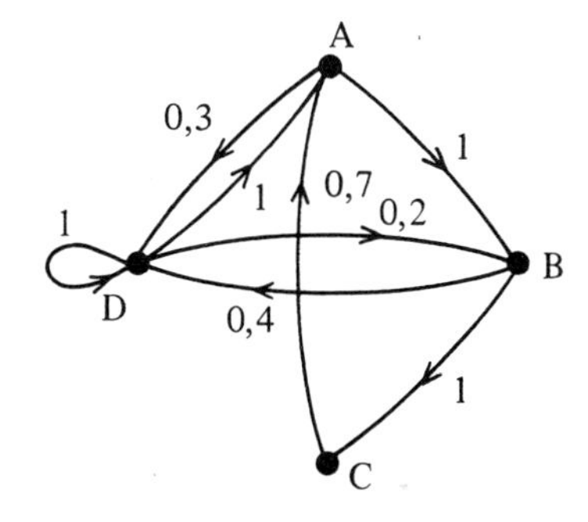

(the ordered pairs (x, y) such that $\mu_{\underset{\sim}{\mathcal{R}}}(x, y) = 0$ have not been represented)

(b)

Fig. 18.1

†Recall that in the contents of Section 18 all reference sets were finite.

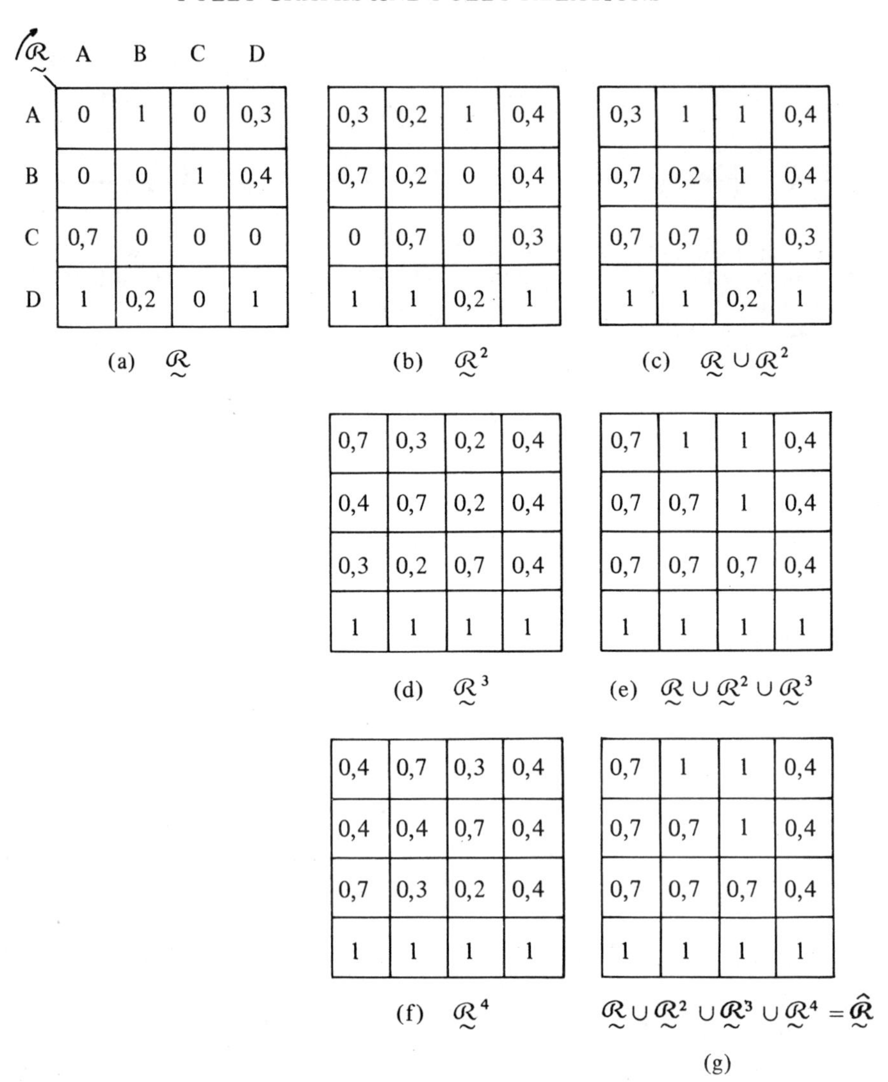

$\underset{\sim}{\mathcal{R}}$	A	B	C	D
A	0	1	0	0,3
B	0	0	1	0,4
C	0,7	0	0	0
D	1	0,2	0	1

(a) $\underset{\sim}{\mathcal{R}}$

0,3	0,2	1	0,4
0,7	0,2	0	0,4
0	0,7	0	0,3
1	1	0,2	1

(b) $\underset{\sim}{\mathcal{R}}^2$

0,3	1	1	0,4
0,7	0,2	1	0,4
0,7	0,7	0	0,3
1	1	0,2	1

(c) $\underset{\sim}{\mathcal{R}} \cup \underset{\sim}{\mathcal{R}}^2$

0,7	0,3	0,2	0,4
0,4	0,7	0,2	0,4
0,3	0,2	0,7	0,4
1	1	1	1

(d) $\underset{\sim}{\mathcal{R}}^3$

0,7	1	1	0,4
0,7	0,7	1	0,4
0,7	0,7	0,7	0,4
1	1	1	1

(e) $\underset{\sim}{\mathcal{R}} \cup \underset{\sim}{\mathcal{R}}^2 \cup \underset{\sim}{\mathcal{R}}^3$

0,4	0,7	0,3	0,4
0,4	0,4	0,7	0,4
0,7	0,3	0,2	0,4
1	1	1	1

(f) $\underset{\sim}{\mathcal{R}}^4$

0,7	1	1	0,4
0,7	0,7	1	0,4
0,7	0,7	0,7	0,4
1	1	1	1

$\underset{\sim}{\mathcal{R}} \cup \underset{\sim}{\mathcal{R}}^2 \cup \underset{\sim}{\mathcal{R}}^3 \cup \underset{\sim}{\mathcal{R}}^4 = \hat{\underset{\sim}{\mathcal{R}}}$

(g)

Fig. 18.2

$$(18.9) \qquad l(\mathrm{B}, \mathrm{C}, \mathrm{A}, \mathrm{D}) = \mu_{\underset{\sim}{\mathcal{R}}}(\mathrm{B}, \mathrm{C}) \wedge \mu_{\underset{\sim}{\mathcal{R}}}(\mathrm{C}, \mathrm{A}) \wedge \mu_{\underset{\sim}{\mathcal{R}}}(\mathrm{A}, \mathrm{D})$$
$$= 1 \wedge 0{,}7 \wedge 0{,}3 = 0{,}3 \; .$$

Now we examine all paths‡ from B to D whose length is less than or equal to 3; these are only the paths (B, D), (B, D, D), (B, D, D, D), for which we have

$$(18.10) \quad l(\mathrm{B}, \mathrm{D}) = \mu_{\underset{\sim}{\mathcal{R}}}(\mathrm{B}, \mathrm{D}) = 0{,}4 \, , \; l(\mathrm{B}, \mathrm{D}, \mathrm{D}) = \mu_{\underset{\sim}{\mathcal{R}}}(\mathrm{B}, \mathrm{D}) \wedge \mu_{\underset{\sim}{\mathcal{R}}}(\mathrm{D}, \mathrm{D}) = 0{,}4 \wedge 1 = 0{,}4$$

†One knows how to carry out such an enumeration in an automatic fashion, without omission and without repetition. See, for example, the references [1K, 2K].

$$l(\mathrm{B}, \mathrm{D}, \mathrm{D}, \mathrm{D}) = \mu_{\underset{\sim}{\mathcal{R}}}(\mathrm{B}, \mathrm{D}) \wedge \mu_{\underset{\sim}{\mathcal{R}}}(\mathrm{D}, \mathrm{D}) \wedge \mu_{\underset{\sim}{\mathcal{R}}}(\mathrm{D}, \mathrm{D}) = 0{,}4 \wedge 1 \wedge 1 = 0{,}4.$$

One then has

$$(18.11) \quad l^*(\mathrm{B}, \mathrm{D}) = l(\mathrm{B}, \mathrm{C}, \mathrm{A}, \mathrm{D}) \vee l(\mathrm{B}, \mathrm{D}) \vee l(\mathrm{B}, \mathrm{D}, \mathrm{D}) \vee l(\mathrm{B}, \mathrm{D}, \mathrm{D}, \mathrm{D})$$
$$= 0{,}3 \vee 0{,}4 \vee 0{,}4 \vee 0{,}4 = 0{,}4.$$

If we locate $\hat{\underset{\sim}{\mathcal{R}}}$ in Figure 18.2g, we find

$$(18.12) \quad \mu_{\hat{\underset{\sim}{\mathcal{R}}}}(\mathrm{B}, \mathrm{D}) = 0{,}4 . \quad \text{(Theorem II - 18.6)}$$

On the other hand, there are two paths of length 3 between B and D; these are (B, C, A, D) and (B, D, D, D). One then has

$$(18.13) \quad l_3^*(\mathrm{B}, \mathrm{D}) = l(\mathrm{B}, \mathrm{C}, \mathrm{A}, \mathrm{D}) \vee l(\mathrm{B}, \mathrm{D}, \mathrm{D}, \mathrm{D})$$
$$= 0{,}3 \vee 0{,}4 = 0{,}4.$$

One verifies with†

$$(18.14) \quad \mu_{\underset{\sim}{\mathcal{R}}^3}(\mathrm{B}, \mathrm{D}) = 0{,}4 \quad \text{(Theorem I– 18.5)} .$$

Now consider the path (C, A, B, D, A, D). This path possesses a circuit (D, A, D); eliminating this we see

$$(18.15) \quad l_5^*(\mathrm{C}, \mathrm{D}) = l^*_{j \leqslant 4}(\mathrm{C}, \mathrm{D})$$
$$= l_1^*(\mathrm{C}, \mathrm{D}) \vee l_2^*(\mathrm{C}, \mathrm{D}) \vee l_3^*(\mathrm{C}, \mathrm{D}) \vee l_4^*(\mathrm{C}, \mathrm{D})$$
$$= \mu_{\underset{\sim}{\mathcal{R}}}(\mathrm{C}, \mathrm{D}) \vee \mu_{\underset{\sim}{\mathcal{R}}^2}(\mathrm{C}, \mathrm{D}) \vee \mu_{\underset{\sim}{\mathcal{R}}^3}(\mathrm{C}, \mathrm{D}) \vee \mu_{\underset{\sim}{\mathcal{R}}^4}(\mathrm{C}, \mathrm{D})$$
$$= 0 \vee 0{,}3 \vee 0{,}4 \vee 0{,}4 = 0{,}4 .$$

One may have expected to find 0,3; but the strongest path of length 5 between C and D is not (C, A, B, D, A, D) but (C, A, B, D, D, D); these two, moreover, reduce to (C, A, B, D) when the circuits are eliminated. All this may be seen clearly in Figure 18.1b.

Notion of a path defined with respect to other operators. Max–star transitivity. The value defined with the aid of expression (18.3) may be extended, in its definition, to operators other than $\wedge$, under the restriction that those considered have the properties of associativity and monotonicity. If $*$ is such an operator, one then sees

$$(18.16) \quad l(x_{i_1}, x_{i_2}, \ldots, x_{i_r}) = \mu_{\underset{\sim}{\mathcal{R}}}(x_{i_1}, x_{i_2}) * \mu_{\underset{\sim}{\mathcal{R}}}(x_{i_2}, x_{i_3}) * \cdots * \mu_{\underset{\sim}{\mathcal{R}}}(x_{i_{r-1}}, x_{i_r}) .$$

In particular, if $*$ is the product operator, denoted $\cdot$ and defined by (12.35), one sees

$$(18.17) \quad l(x_{i_1}, x_{i_2}, \ldots, x_{i_r}) = \mu_{\underset{\sim}{\mathcal{R}}}(x_{i_1}, x_{i_2}) \cdot \mu_{\underset{\sim}{\mathcal{R}}}(x_{i_2}, x_{i_3}) \cdot \ldots \cdot \mu_{\underset{\sim}{\mathcal{R}}}(x_{i_{r-1}}, x_{i_r}) .$$

Due to the property

$$(18.18) \quad a \cdot b \leqslant a \wedge b \quad \text{if} \quad a, b \in [0, 1] ,$$

†One finds here that $l_3{}^*(\mathrm{B}, \mathrm{D}) = l^*(\mathrm{B}, \mathrm{D})$; but this is fortuitous—the relations $\underset{\sim}{\mathcal{R}}^3$ and $\hat{\underset{\sim}{\mathcal{R}}}$ being different (see Figures 18.2d and 18.2g).

it is easy to verify that the transitivity of the operator $\wedge$ entails the transitivity of the operator $\cdot$. Thus

$$\underset{\sim}{\mathcal{R}} \circ \underset{\sim}{\mathcal{R}} \subset \underset{\sim}{\mathcal{R}} \;\Rightarrow\; \underset{\sim}{\mathcal{R}} \cdot \underset{\sim}{\mathcal{R}} \subset \underset{\sim}{\mathcal{R}} \,. \tag{18.19}$$

The reverse implication is not true by all evidence.

When one has shown max–min transitivity [according to (16.9)] of a relation, it is unnecessary to check max–product transitivity; it is implied by the first.

It is interesting, for some applications, to have operators other than $\wedge$ at one's disposal, with a view to defining transitivities and paths for some particular uses.

19. RELATION OF FUZZY PREORDER

A binary fuzzy relation that is

(1) transitive [(16.9)]
(2) reflexive [(16.7)]

is a relation of fuzzy preorder.

First we consider an important theorem.

Theorem I. If $\underset{\sim}{\mathcal{R}}$ is transitive and reflexive (that is, is a preorder), then

$$\underset{\sim}{\mathcal{R}}^k = \underset{\sim}{\mathcal{R}} \quad , \quad k = 1, 2, 3, \ldots . \tag{19.1}$$

Proof. It suffices to review the definition of transitivity [(16.9) and (17.5)] and to show

$$\underset{\sim}{\mathcal{R}}^2 = \underset{\sim}{\mathcal{R}} , \tag{19.2}$$

if one asserts that

$$\forall x : \quad \mu_{\underset{\sim}{\mathcal{R}}}(x, x) = 1 . \tag{19.3}$$

Since

$$\underset{\sim}{\mathcal{R}}^2 = \underset{\sim}{\mathcal{R}} \circ \underset{\sim}{\mathcal{R}} ,$$

one has, according to (13.2)

$$\mu_{\underset{\sim}{\mathcal{R}}^2}(x, z) = \bigvee_y \,[\mu_{\underset{\sim}{\mathcal{R}}}(x, y) \wedge \mu_{\underset{\sim}{\mathcal{R}}}(y, z)] . \tag{19.4}$$

The right-hand member of (19.4) contains two equal terms

$$\mu_{\underset{\sim}{\mathcal{R}}}(x, x) \wedge \mu_{\underset{\sim}{\mathcal{R}}}(x, z) = \mu_{\underset{\sim}{\mathcal{R}}}(x, z) \wedge \mu_{\underset{\sim}{\mathcal{R}}}(z, z) = \mu_{\underset{\sim}{\mathcal{R}}}(x, z) \tag{19.5}$$

because

$$\mu_{\underset{\sim}{\mathcal{R}}}(x, x) - \mu_{\underset{\sim}{\mathcal{R}}}(z, z) = 1 \qquad \text{reflexivity} . \tag{19.6}$$

Recall that $\underset{\sim}{\mathcal{R}}$ is transitive (16.9), that is,

$$\mu_{\underset{\sim}{\mathcal{R}}}(x, z) \geqslant \bigvee_y \,[\mu_{\underset{\sim}{\mathcal{R}}}(x, y) \wedge \mu_{\underset{\sim}{\mathcal{R}}}(y, z)] ; \tag{19.7}$$

it then results that $\mu_{\underset{\sim}{\mathcal{R}}}(x, y)$ is greater than or equal to the terms $\mu_{\underset{\sim}{\mathcal{R}}}(x, y) \wedge \mu_{\underset{\sim}{\mathcal{R}}}(y, z)$; this is then the value of the member on the right of (19.4), and one indeed has

$$\underset{\sim}{\mathcal{R}}^2 = \underset{\sim}{\mathcal{R}} . \tag{19.8}$$

Q.E.D.

Theorem II. If $\underset{\sim}{\mathcal{R}}$ is a preorder, then

$$\underset{\sim}{\mathcal{R}} = \underset{\sim}{\mathcal{R}}^2 = \cdots = \underset{\sim}{\mathcal{R}}^k = \cdots = \hat{\underset{\sim}{\mathcal{R}}} . \tag{19.9}$$

Proof. This is a corollary to Theorem I. It suffices to consider (17.8) and (19.8) together.

Example 1. Figure 19.1 represents a preorder on

$$\mathbf{E} = \{A , B , C , D , E\} . \tag{19.10}$$

One may verify transitivity with the aid of the relation

$$\underset{\sim}{\mathcal{R}}^2 \subset \underset{\sim}{\mathcal{R}} . \tag{19.11}$$

Reflexivity is directly apparent from the presence of the ones on the principal diagonal.

Finally, one may verify that one indeed has

$$\underset{\sim}{\mathcal{R}}^2 = \underset{\sim}{\mathcal{R}} . \tag{19.12}$$

$\underset{\sim}{\mathcal{R}}$	A	B	C	D	E
A	1	0,7	0,8	0,5	0,5
B	0	1	0,3	0	0,2
C	0	0,7	1	0	0,2
D	0,6	1	0,9	1	0,6
E	0	0	0	0	1

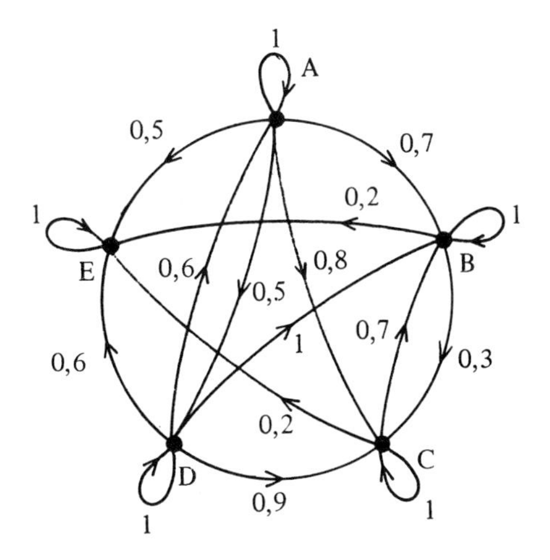

Fig. 19.1

Example 2. Consider a graph $\mathbf{G} \subset \mathbf{E} \times \mathbf{E}$, where $\mathbf{E}$ is finite, and suppose that $\mathbf{G}$ is reflexive. Then the binary fuzzy relation "there exists a path from x to y in $\mathbf{G}$" (in the sense of the word *path* given in Section 18) is a preorder.

Example 3. The fuzzy binary relation $x \underset{\sim}{\mathcal{R}} y$ where $x, y \in \mathbf{N}$ with

$$\mu_{\underset{\sim}{\mathcal{R}}}(x, y) = e^{-k(x-y)^2}, \quad \text{with } k > 1, \tag{19.13}$$

is not a preorder because it is not transitive [see (16.12)].

Example 4. (Figure 19.2).

$$0 \leqslant a_1 \leqslant a_2 \leqslant \cdots \cdots \leqslant a_k \leqslant \cdots \leqslant 1 \tag{19.14}$$

this relation on a denumberably infinite set **E** is a preorder.

$\underset{\sim}{\mathcal{R}}$	x_1	x_2	x_3	x_4	x_5	x_6	...
x_1	1	a_1	a_1	a_1	a_1	a_1	...
x_2	0	1	a_2	a_2	a_2	a_2	...
x_3	0	a_1	1	a_3	a_3	a_3	...
x_4	0	a_1	a_2	1	a_4	a_4	...
x_5	0	a_1	a_2	a_3	1	a_5	...
x_6	0	a_1	a_2	a_3	a_4	1	...
	⋮	⋮	⋮	⋮	⋮	⋮	⋱

Fig. 19.2

Fuzzy semipreorder. A transitive fuzzy relation that is not reflexive is called a *semipreorder*, or what is the same thing, a *nonreflexive fuzzy preorder.*

Example 1. The relation represented in Figure 19.3 is transitive but not reflexive; it is a semipreorder.

$\underset{\sim}{\mathcal{R}}$	A	B	C
A	0,2	1	0,4
B	0	0,6	0,3
C	0	1	0,3

Fig. 19.3

Example 2. The relation presented in Figure 16.7 is a semipreorder.

Antireflexive fuzzy preorder. A particular case of a fuzzy semipreorder is that where

$$\forall x \in \mathbf{E} : \mu_{\underset{\sim}{\mathcal{R}}}(x, x) = 0 \tag{19.15}$$

One says then that the fuzzy preorder is antireflexive.

Thus, the preorder relation presented in Figure 19.4 is antireflexive.

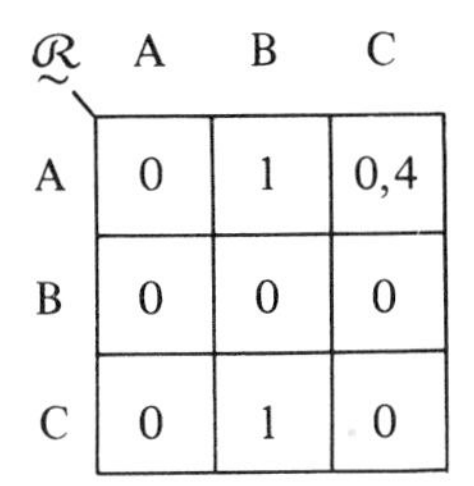

$\underset{\sim}{\mathcal{R}}$	A	B	C
A	0	1	0,4
B	0	0	0
C	0	1	0

Fig. 19.4

20. RELATION OF SIMILITUDE

A fuzzy binary relation that is

(1) transitive (16.9)
(2) reflexive (16.7)
(3) symmetric (16.6)

is called a *relation of similitude* or a *fuzzy equivalence relation*. It is evidently a preorder.

First, we give several examples.

Example 1. An example is presented in Figure 20.1. One may verify reflexivity and symmetry directly. In order to verify transitivity, it suffices to calculate $\underset{\sim}{\mathcal{R}}^2$. One must then have, according to (19.9),

$$\underset{\sim}{\mathcal{R}}^2 = \underset{\sim}{\mathcal{R}} \tag{20.1}$$

$\underset{\sim}{\mathcal{R}}$	A	B	C	D	E
A	1	0,8	0,7	1	0,9
B	0,8	1	0,7	0,8	0,8
C	0,7	0,7	1	0,7	0,7
D	1	0,8	0,7	1	0,9
E	0,9	0,8	0,7	0,9	1

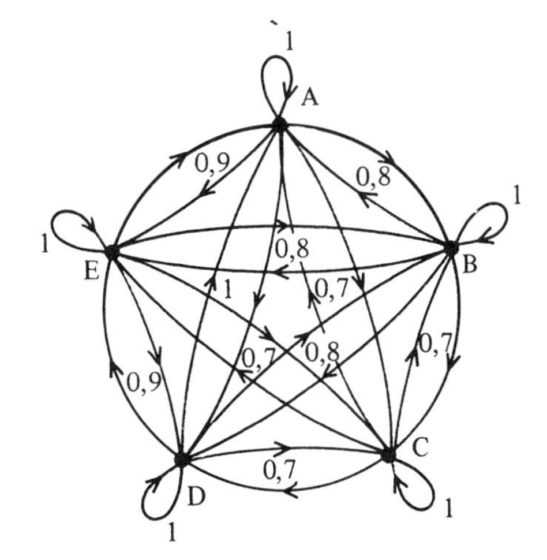

Fig. 20.1

Example 2. (Figure 20.2). If one takes $0 \leqslant a \leqslant 1$, then one has a similitude relation.

$\underset{\sim}{\mathcal{R}}$	A	B	C	D	E
A	1	a	a	a	a
B	a	1	a	a	a
C	a	a	1	a	a
D	a	a	a	1	a
E	a	a	a	a	1

Fig. 20.2

Example 3. (Figure 20.3). If we suppose

(20.2) $0 \leqslant a_1 \leqslant a_2 \leqslant \cdots \cdots \leqslant a_k \leqslant \cdots \leqslant 1$;

this is a similitude relation in an infinite set $\mathbf{E}$.

$\underset{\sim}{\mathcal{R}}$	x_1	x_2	x_3	x_4	x_5	x_6	x_7	...
x_1	1	a_1	a_1	a_1	a_1	a_1	a_1	...
x_2	a_1	1	a_2	a_2	a_2	a_2	a_2	...
x_3	a_1	a_2	1	a_3	a_3	a_3	a_3	...
x_4	a_1	a_2	a_3	1	a_4	a_4	a_4	...
x_5	a_1	a_2	a_3	a_4	1	a_5	a_5	...
x_6	a_1	a_2	a_3	a_4	a_5	1	a_6	...
x_7	a_1	a_2	a_3	a_4	a_5	a_6	1	...
	⋮	⋮	⋮	⋮	⋮	⋮	⋮	⋱

Fig. 20.3

Example 4. The fuzzy relation $x \underset{\sim}{\mathcal{R}} y$, where $x, y \in \mathbf{R}^+$ with

$$\begin{aligned} \mu_{\underset{\sim}{\mathcal{R}}}(x, y) &= e^{-k(y+1)} \quad && y < x \,, \; k > 1 \,, \\ &= 1 && y = x \\ &= e^{-k(x+1)} && y > x \,, \; k > 1 \,, \end{aligned} \tag{20.3}$$

is a similitude relation, as the reader is asked to verify in the Exercises (see also Section 29 a little later).

Theorem I. Let $\underset{\sim}{\mathcal{R}} \subset \mathbf{E} \times \mathbf{E}$ be a similitude relation. Let x, y, z be three elements of $\mathbf{E}$. Put

$$\begin{aligned} c &= \mu_{\underset{\sim}{\mathcal{R}}}(x, z) = \mu_{\underset{\sim}{\mathcal{R}}}(z, x) \,, \\ a &= \mu_{\underset{\sim}{\mathcal{R}}}(x, y) = \mu_{\underset{\sim}{\mathcal{R}}}(y, x) \,, \\ b &= \mu_{\underset{\sim}{\mathcal{R}}}(y, z) = \mu_{\underset{\sim}{\mathcal{R}}}(z, y) \,. \end{aligned} \tag{20.4}$$

Then

$$c \geqslant a = b \quad \text{or} \quad a \geqslant b = c \quad \text{or} \quad b \geqslant c = a \,. \tag{20.5}$$

In other words, of these three quantities a, b, and c at least two are equal and the third is greater than the other two.

Proof. One already has by hypothesis

$$c \geqslant a \wedge b \,, \tag{20.6}$$

$$b \geqslant c \wedge a \,, \tag{20.7}$$

$$a \geqslant b \wedge c \,. \tag{20.8}$$

We suppose that we have

$$c \geqslant b > a \,, \tag{20.9}$$

then (20.6) and (20.7) are verified, but (20.8) is not; and if one takes $b = a$, the three relations are verified.

Suppose that we have

$$c \geqslant a > b \,, \tag{20.10}$$

then (20.6) and (20.8) are verified, but (20.8) is not; and if one takes $a = b$, the three relations are verified.

One then may not have (20.9) nor (20.10), but on the contrary

$$c \geqslant a = b \quad \text{holds.} \tag{20.11}$$

One could show in the same manner that one may not have $a \geqslant b > c$ or $a \geqslant c > b$.

But

$$a \geqslant b = c \quad \text{holds.} \tag{20.12}$$

One could show again in the same manner that one may not have $b \geqslant c > a$ or $b \geqslant a > c$, but

$$b \geqslant a = c \quad \text{holds.} \tag{20.13}$$

Thus it is necessary that one always have at least two of the values equal.

The inequalities (20.6)–(20.8) then give:

If $a = b$:

$$c \geqslant a \wedge b \,, \quad b = c \wedge a \,, \quad a = b \wedge c \,. \tag{20.14}$$

If $b = c$:

$$c = a \wedge b \,, \quad b = c \wedge a \,, \quad a \geqslant b \wedge c \,. \tag{20.15}$$

If $c = a$:

$$c = a \wedge b \,, \quad b \geqslant c \wedge a \,, \quad a = b \wedge c \,. \tag{20.16}$$

21. SUBRELATION OF SIMILITUDE IN A FUZZY PREORDER

Let $\underset{\sim}{\mathcal{R}} \subset \mathbf{E} \times \mathbf{E}$ be a relation of fuzzy preorder. If there exists an ordinary subset $\mathbf{E}_1 \subset \mathbf{E}$ such that $\forall x, y \in \mathbf{E}_1 : \mu_{\underset{\sim}{\mathcal{R}}}(x, y) = \mu_{\underset{\sim}{\mathcal{R}}}(y, x)$, the elements of $\mathbf{E}_1$ form among themselves a similitude relation that we shall call a *similitude subrelation* in the preorder $\underset{\sim}{\mathcal{R}}$.

We shall say that a similitude subrelation is maximal if there is no other similitude relation of the same nature in the relation being considered.

Suppose now that a preorder relation is such that each of the elements of the reference set involved belongs to a maximal similitude subrelation and does not belong to any other. This may be rephrased: all the maximal subrelations are disjoint. In this case we call the subsets for which one has such disjoint maximal similitude subrelations *similitude classes* of the preorder.

Thus, not all fuzzy preorders are decomposable into similitude classes. We shall consider several examples.

Example 1. The relation shown in Figure 21.1 is certainly a preorder [one may verify this with reference to (19.2)]. But this preorder is not a symmetric relation. However, note that the relation $\underset{\sim}{\mathcal{R}}$ may be decomposed into three subrelations: $\underset{\sim}{\mathcal{R}}_1$ relative to {A, B, C, E, F}, $\underset{\sim}{\mathcal{R}}_2$ relative to {D}, and $\underset{\sim}{\mathcal{R}}_3$ relative to {G}. The ordinary subsets $\mathbf{K}_1$ = {A, B, C, E, F}, $\mathbf{K}_2$ = {D}, $\mathbf{K}_3$ = {G} are clearly maximal for the property of similitude [this is not the case, for example, for {B, C, F} or {A, C, E}]. We shall say that the relation $\underset{\sim}{\mathcal{R}}$ of fuzzy preorder is decomposable into maximal disjoint similitude subrelations relative to $\mathbf{K}_1$, $\mathbf{K}_2$, and $\mathbf{K}_3$, forming the similitude classes existing in the preordered set.

$\underset{\sim}{\mathcal{R}}$	A	B	C	E	F	D	G
A	1	0,2	0,2	0,2	0,2	0,3	0,4
B	0,2	1	0,5	0,2	0,2	0,3	0,5
C	0,2	0,5	1	0,2	0,2	0,3	0,5
E	0,2	0,2	0,2	1	0,8	0,3	0,5
F	0,2	0,2	0,2	0,8	1	0,3	0,5
D	0,2	0,2	0,2	0,2	0,2	1	0,4
G	0,2	0,2	0,2	0,2	0,2	0,2	1

Fig. 21.1

If we now consider the strongest paths existing between these classes [see the definition, (18.4)], these classes then form among themselves (Figure 21.2) a transitive

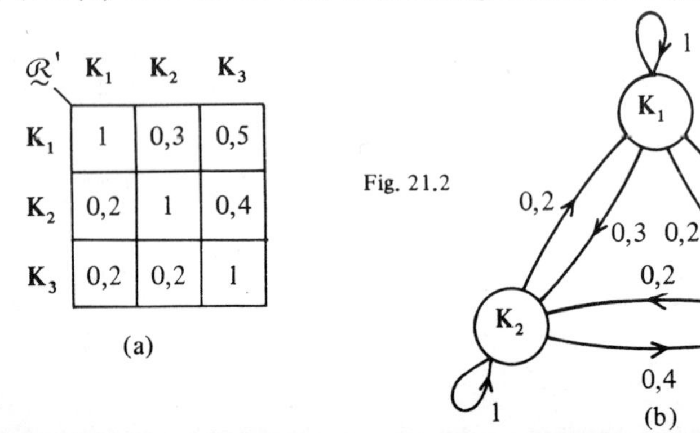

$\underset{\sim}{\mathcal{R}}'$	$\mathbf{K}_1$	$\mathbf{K}_2$	$\mathbf{K}_3$
$\mathbf{K}_1$	1	0,3	0,5
$\mathbf{K}_2$	0,2	1	0,4
$\mathbf{K}_3$	0,2	0,2	1

(a)

(b)

Fig. 21.2

nonsymmetric fuzzy relation; we shall see in Section 23 that this is a relation of fuzzy order.

Example 2. Figure 21.3a represents a fuzzy preorder relation. We may find three similitude subrelations $\underset{\sim}{\mathcal{R}}_1, \underset{\sim}{\mathcal{R}}_2$, and $\underset{\sim}{\mathcal{R}}_3$ (Figure 21.3b); but if these are maximal, they are not disjoint and thus do not constitute similitude classes.

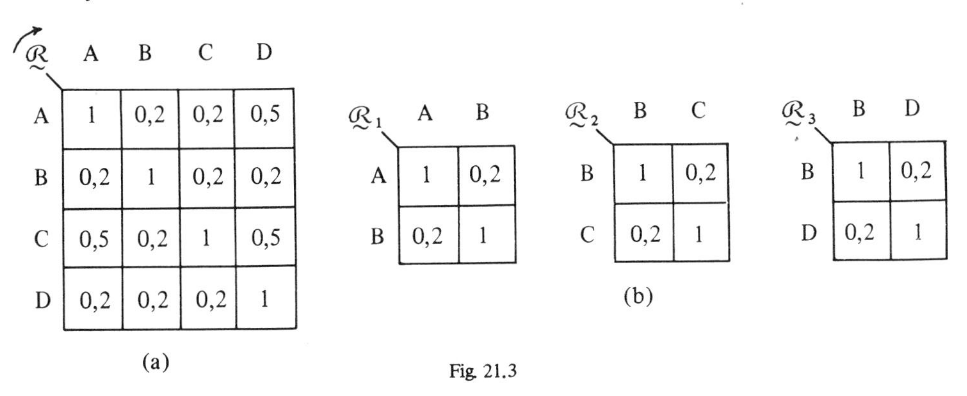

$\underset{\sim}{\mathcal{R}}$	A	B	C	D
A	1	0,2	0,2	0,5
B	0,2	1	0,2	0,2
C	0,5	0,2	1	0,5
D	0,2	0,2	0,2	1

(a)

$\underset{\sim}{\mathcal{R}}_1$	A	B
A	1	0,2
B	0,2	1

$\underset{\sim}{\mathcal{R}}_2$	B	C
B	1	0,2
C	0,2	1

$\underset{\sim}{\mathcal{R}}_3$	B	D
B	1	0,2
D	0,2	1

(b)

Fig. 21.3

Reducible fuzzy preorder. A fuzzy preorder decomposable into similitude classes will be called a *reducible fuzzy preorder.* Thus, the fuzzy preorder in Figure 21.1 is reducible, but that of Figure 21.3a is not.

The examples given above have involved finite sets **E**; but decomposition into similitude classes, such as that which has been explained, remains valid if **E** is infinite, denumerably or not. The classes may then be finite or not, and their number finite or infinite. But, of course, representations using matrices or Berge graphs may be made only in cases where **E** is denumerable.

The search for maximal similitude subrelations of a preorder (E finite). In certain simple cases, by examining the pairs of elements for which one has symmetry, one obtains immediately the maximal similitude subrelations, which may or may not be disjoint. But it is convenient to have at one's disposal a general procedure. We give in Appendix B, page 387, some appropriate algorithms.

22. ANTISYMMETRY

A fuzzy binary relation is *antisymmetric* if

$\forall (x, y) \in \mathbf{E} \times \mathbf{E}$ with $x \neq y$:

$$(\mu_{\underset{\sim}{\mathcal{R}}}(x, y) \neq \mu_{\underset{\sim}{\mathcal{R}}}(y, x)) \quad \text{or} \quad (\mu_{\underset{\sim}{\mathcal{R}}}(x, y) = \mu_{\underset{\sim}{\mathcal{R}}}(y, x) = 0) \ . \tag{22.1}$$

Examples. Figures 22.1–22.3 give some examples of antisymmetric fuzzy binary relations. Thus (Figure 22.1),

$$\begin{aligned} &\mu_{\underset{\sim}{\mathcal{R}}}(\mathrm{A}, \mathrm{B}) < \mu_{\underset{\sim}{\mathcal{R}}}(\mathrm{B}, \mathrm{A}) \ , \\ &\mu_{\underset{\sim}{\mathcal{R}}}(\mathrm{A}, \mathrm{C}) = \mu_{\underset{\sim}{\mathcal{R}}}(\mathrm{C}, \mathrm{A}) = 0 \ , \end{aligned} \tag{22.2}$$

$\underset{\sim}{\mathcal{R}}$	A	B	C	D	E
A	0	0,3	0	0,9	1
B	0,5	0,8	0,6	0,8	0
C	0	0,5	1	0	1
D	0,5	1	0,2	1	0,3
E	0	0	0	0,2	0

Fig. 22.1

$\underset{\sim}{\mathcal{R}}$	A	B	C	D
A	0	0	0	0,8
B	0	0	0,6	0
C	1	0,2	0,3	1
D	1	0	0	1

Fig. 22.2

$\underset{\sim}{\mathcal{R}}$	A	B	C	D	E	F
A	0	0,3	0	0,2	0	0,8
B	0	1	1	0,8	0	0,5
C	0	0	0,3	0	0	0,6
D	0	0	0	0	0	0
E	0	0,3	0	0,2	0,5	0,5
F	0,7	0	0	0	0	0,1

Fig. 22.3

(22.2)
$$\mu_{\underset{\sim}{\mathcal{R}}}(\mathrm{A}, \mathrm{D}) > \mu_{\underset{\sim}{\mathcal{R}}}(\mathrm{D}, \mathrm{A}),$$
$$\mu_{\underset{\sim}{\mathcal{R}}}(\mathrm{A}, \mathrm{E}) > \mu_{\underset{\sim}{\mathcal{R}}}(\mathrm{E}, \mathrm{A}),$$

and so on.

Another example. Let $x \underset{\sim}{\mathcal{R}} y$ where $x, y \in \mathbf{R}^+$; the relation $\underset{\sim}{\mathcal{R}}$ such that

(22.3)
$$\mu_{\underset{\sim}{\mathcal{R}}}(x, y) = e^{-(ax+by)} \qquad a > b > 1,$$

is antisymmetric.

Remark. One should not confuse a nonsymmetric graph with an antisymmetric graph. For the first, one may write

$\exists (x, y) \in \mathbf{E} \times \mathbf{E}$ with $x \neq y$:

(22.4)
$$\mu_{\underset{\sim}{\mathcal{R}}}(x, y) \neq \mu_{\underset{\sim}{\mathcal{R}}}(y, x).$$

Thus, the graph in Figure 22.4 is nonsymmetric [there exists at least one ordered pair (x, y) for which (22.4) is satisfied]. But this graph is not antisymmetric [there is at least one ordered pair (x, y) for which $\mu_{\underset{\sim}{\mathcal{R}}}(x, y) = \mu_{\underset{\sim}{\mathcal{R}}}(y, x) \neq 0$, for example the ordered pair (C, D)].

$\underset{\sim}{\mathcal{R}}$	A	B	C	D	E	F
A	0	0,5	0,3	1	0	0
B	0,3	0	0,4	1	0,2	0
C	0	0,4	0,6	1	0,3	0,4
D	0	0	0	1	0,2	0,6
E	0	0,2	0,3	0,3	0,7	1
F	0	0,6	0,4	0,6	1	0

Fig. 22.4

Ordinary antisymmetric graph associated with an antisymmetric fuzzy relation. To any antisymmetric fuzzy relation $\underset{\sim}{\mathcal{R}}$ one will associate one (and only one) ordinary antisymmetric graph **G** such that

$\forall (x, y) \in \mathbf{E} \times \mathbf{E}$:

(22.5)

1) $x \neq y$ and $\mu_{\underset{\sim}{\mathcal{R}}}(x, y) > \mu_{\underset{\sim}{\mathcal{R}}}(y, x) \Rightarrow (x, y) \in \mathbf{G}$ and $(y, x) \notin \mathbf{G}$,

2) $x \neq y$ and $\mu_{\underset{\sim}{\mathcal{R}}}(x, y) = \mu_{\underset{\sim}{\mathcal{R}}}(x, y) = 0 \Rightarrow (x, y) \notin \mathbf{G}$ and $(y, x) \notin \mathbf{G}$,

We shall take (arbitrarily) for **G**

$$\forall (x, x) \in \mathbf{E} \times \mathbf{E} : \quad (x, x) \in \mathbf{G} . \tag{22.6}$$

This will prove convenient later for the study of nonstrict relations of order.

Example 1. Figure 22.5 and 22.6 represent ordinary antisymmetric graphs associated with the relations in Figures 22.1 and 22.2.

G	A	B	C	D	E
A	1	0	0	1	1
B	1	1	1	0	0
C	0	0	1	0	1
D	0	1	1	1	1
E	0	0	0	0	1

(a)

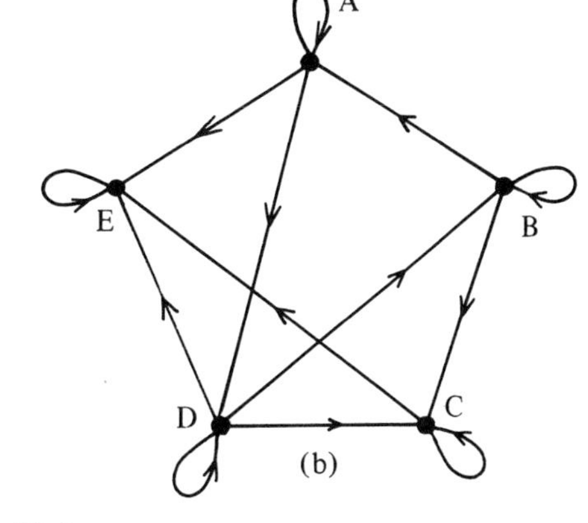

Fig. 22.5

G	A	B	C	D
A	1	0	0	0
B	0	1	1	0
C	1	0	1	1
D	1	0	0	1

(a)

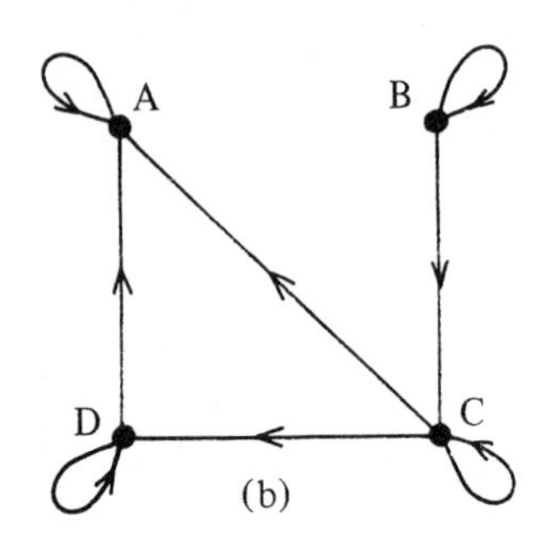

Fig. 22.6

Example 2. We recall that the notion of an ordinary graph encompasses all ordinary sets, finite or not. Thus, to any antisymmetric fuzzy relation defined on a finite or an infinite set, one may associate an ordinary antisymmetric graph. Thus, to the fuzzy antisymmetric relation defined by (22.3), we shall associate the ordinary graph

$$\text{(22.7)} \qquad G = \{(x, y) \mid y \geqslant x\}.$$

this graph is represented in Figure 22.7.

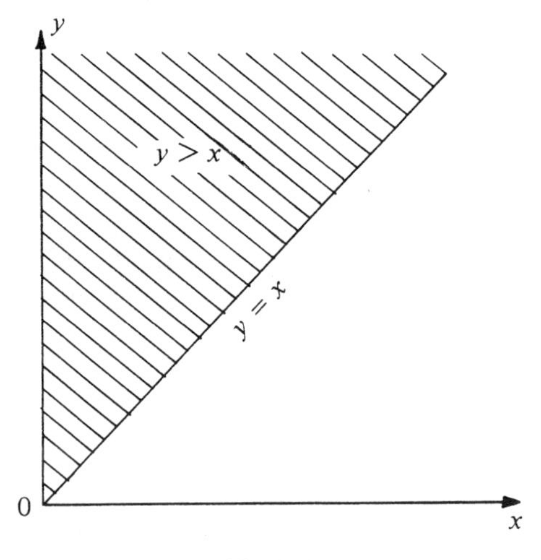

Fig. 22.7

Remark. One ought not to confuse the concept of the ordinary antisymmetric graph associated with an antisymmetric fuzzy relation with that of the ordinary graph nearest to this fuzzy relation; these two graphs have no direct relationship.

Perfect antisymmetry. L. A. Zadeh defines antisymmetry more restrictively, but in a way having some further interesting properties; we shall call this *perfect antisymmetry*. A perfect antisymmetric relation is one such that†

$\forall (x, y) \in E \times E$ with $x \neq y$:

$$\text{(22.8)} \qquad \mu_{\underset{\sim}{\mathcal{R}}}(x, y) > 0 \quad \Rightarrow \quad \mu_{\underset{\sim}{\mathcal{R}}}(y, x) = 0 .$$

†L. A. Zadeh gives another definition:

$$(\mu_{\underset{\sim}{\mathcal{R}}}(x, y) > 0 \quad \text{and} \quad \mu_{\underset{\sim}{\mathcal{R}}}(y, x) > 0) \quad \Rightarrow \quad (x = y) .$$

We shall return later to several interesting properties of perfect antisymmetry in a discussion of the idea of perfect order.

Remark. Any perfect antisymmetric relation is evidently antisymmetric.

Example 1. Figure 22.8 represents a perfect antisymmetric relation. Figure 22.9 shows the ordinary antisymmetric graph associated with this relation.

$\underset{\sim}{\mathcal{R}}$	A	B	C	D	E	F
A	0	0,8	0,4	0,6	0	0
B	0	0,3	0	0,6	0	0,7
C	0	0,3	1	0,2	1	0,6
D	0	0	0	0,8	0	0,3
E	0	0,5	0	0	0	1
F	0,7	0	0	0	0	1

Fig. 22.8

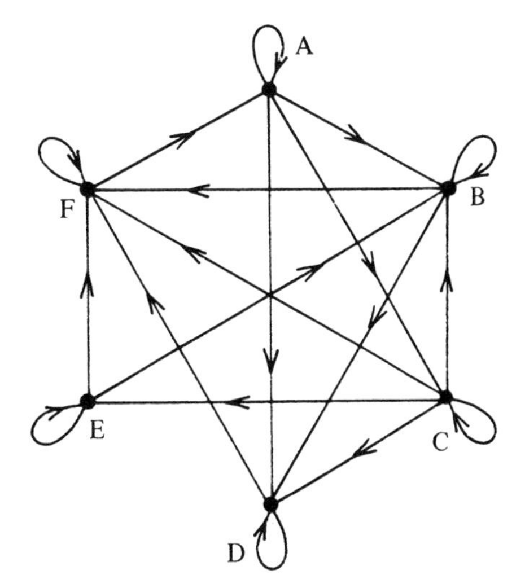

Fig. 22.9

Example 2. Consider the two domains $\mathbf{D}_1$ and $\mathbf{D}_2$ of $\mathbf{R}^+ + \mathbf{R}^+$ indicated in Figure 22.10. The relation $x \underset{\sim}{\mathcal{R}} y$ defined on $\mathbf{R}^+$.

$$(22.9)\quad \mu_{\underset{\sim}{\mathcal{R}}}(x, y) = \mu_1(x, y) \quad \text{if} \quad (x, y) \in \mathbf{D}_1,$$
$$= \mu_2(x, y) \quad \text{if} \quad (x, y) \in \mathbf{D}_2,$$
$$= 0 \quad \text{if} \quad (x, y) \notin \mathbf{D}_1 \cup \mathbf{D}_2,$$

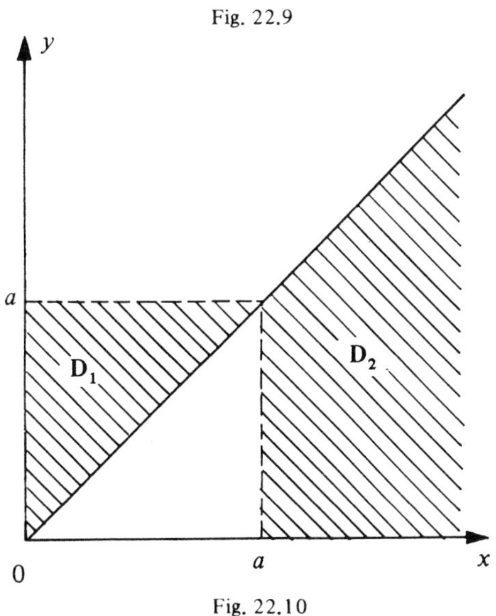

Fig. 22.10

is a perfect antisymmetric relation. Moreover, Figure 22.10 represents the ordinary antisymmetric graph associated with the expression (22.9).

23. FUZZY ORDER RELATIONS

A binary relation that is

(1) reflexive [according to (16.7)]
(2) transitive [according to (16.8) or (16.9)]
(3) antisymmetric [according to (22.1)]

is a *fuzzy order relation* (we shall also say simply *order relation* if no confusion is possible).

One may also define this property in the following fashion: A fuzzy preorder relation that is antisymmetric† is a fuzzy order relation.

Example 1. Figures 23.1 and 23.2 represent fuzzy order relations. One may verify that these are indeed reflexive, transitive, and antisymmetric.

$\underset{\sim}{\mathcal{R}}$	A	B	C	D
A	1	0,8	0	0
B	0,2	1	0	0
C	0,3	0,4	1	0,1
D	0	0	0	1

Fig. 23.1

$\underset{\sim}{\mathcal{R}}$	A	B	C	D
A	1	0,8	0,8	0,8
B	0,5	1	0,6	1
C	0,5	1	1	1
D	0,5	0,6	0,6	1

Fig. 23.2

Example 2. The relation defined by (19.14) and Figure 19.2 is a fuzzy order relation.

Example 3. The relation $x \underset{\sim}{\mathcal{R}} y$ where $x, y \in \mathbf{N}$ (Figure 23.3) is a fuzzy order relation.

†This is then reducible and each similitude class contains only one element.

$\underset{\sim}{\mathcal{R}}$	0	1	2	3	4	5	
0	1	e^{-1}	e^{-2}	e^{-3}	e^{-4}	e^{-5}	...
1	0	1	e^{-3}	e^{-4}	e^{-5}	e^{-6}	...
2	0	0	1	e^{-5}	e^{-6}	e^{-7}	...
3	0	0	0	1	e^{-7}	e^{-8}	...
4	0	0	0	0	1	e^{-9}	...
5	0	0	0	0	0	1	...
	⋮	⋮	⋮	⋮	⋮	⋮	⋱

Fig. 23.3

Theorem I. Every fuzzy order relation induces an order (in the sense of the theory of sets) on its reference set through the relation

$$\mu_{\underset{\sim}{\mathcal{R}}}(x, y) \geqslant \mu_{\underset{\sim}{\mathcal{R}}}(y, x) . \tag{23.1}$$

This order will be denoted $y \succcurlyeq x$.

Proof. It suffices to consider the ordinary antisymmetric graph associated with the fuzzy order relation.

Examples. Figures 23.4 and 23.5 represent, respectively, the ordinary antisymmetric graphs associated with the fuzzy order relations given in Figures 23.1 and 23.2.

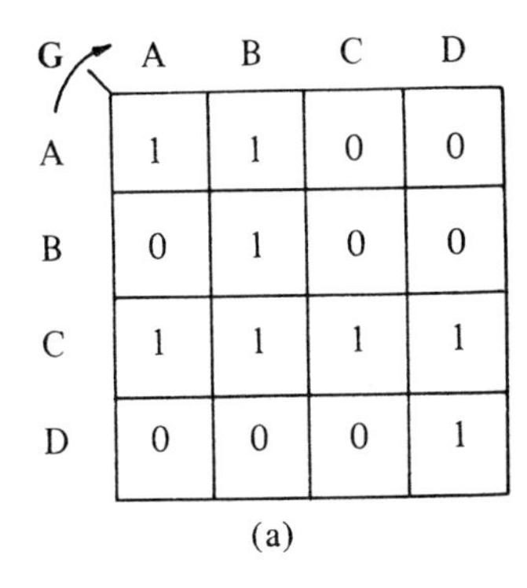

G	A	B	C	D
A	1	1	0	0
B	0	1	0	0
C	1	1	1	1
D	0	0	0	1

(a)

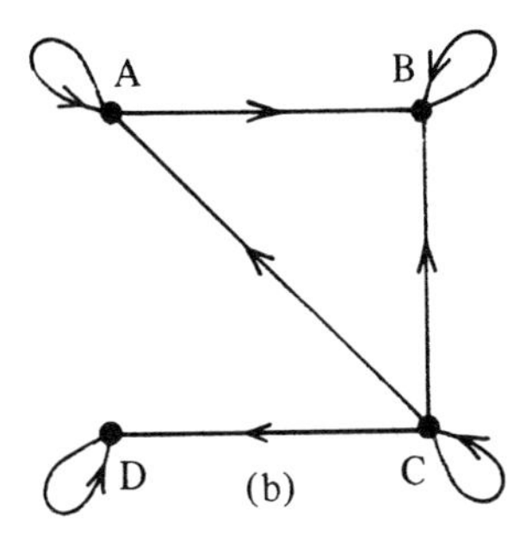

(b)

Fig. 23.4

G	A	B	C	D
A	1	1	1	1
B	0	1	0	1
C	0	1	1	1
D	0	0	0	1

(a)

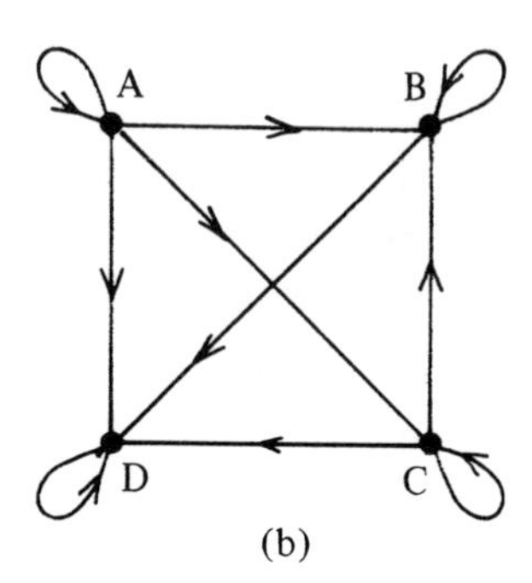

(b)

Fig. 23.5

Figure 23.6 represents the denumerably infinite ordinary graph associated with the relation presented in Figure 23.3.

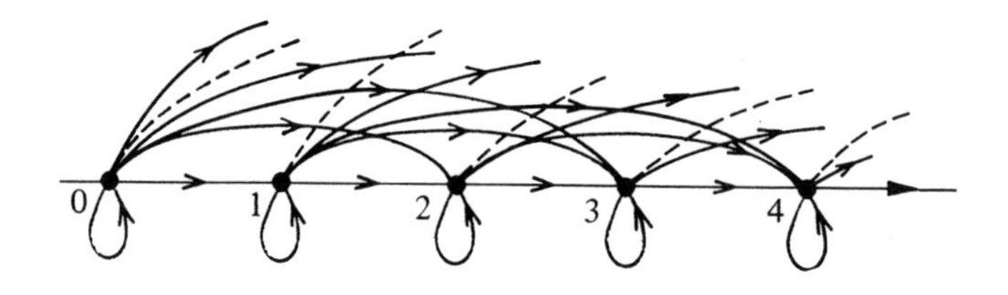

Fig. 23.6

Fuzzy relation of total order.† A fuzzy relation is of total order (a totally ordered fuzzy relation) if its associated ordinary graph represents a total order.

An example is given in Figure 23.5. Using the notation

(23.2) $$y \succcurlyeq x \quad \text{if} \quad \mu_{\underset{\sim}{\mathcal{R}}}(x, y) > \mu_{\underset{\sim}{\mathcal{R}}}(y, x)$$

[that is, if $(x, y) \in \mathbf{G}$ and $(y, x) \in \mathbf{G}$], one then has

(23.3) $$\mathrm{D} \geqslant \mathrm{B} \geqslant \mathrm{C} \geqslant \mathrm{A} .$$

Partially ordered fuzzy relations. A fuzzy relation is of partial order (a partially ordered fuzzy relation) if its associated ordinary graph is partially ordered, that is, is ordered but not totally ordered.

This is the case in the example of Figure 23.4. One has,

(23.4) $$\begin{gathered} \mathrm{B} \geqslant \mathrm{A} \geqslant \mathrm{C} , \\ \mathrm{D} \geqslant \mathrm{C} . \end{gathered}$$

†This is called a linear order relation by L. A. Zadeh when this order is perfect. One may define a linear order with the more restrictive condition of antisymmetry:

$$x \neq y, \quad \mu_{\underset{\sim}{\mathcal{R}}}(x, y) > 0 \qquad \underset{\text{(exclusive)}}{\text{or}} \qquad \mu_{\underset{\sim}{\mathcal{R}}}(y, x) > 0 .$$

Perfect order relations. If one takes the notion of perfect antisymmetry [according to (22.8)] in place of the notion defined by (22.1), one will then have a perfect order relation.

All of these order relations have particularly interesting properties, which we shall examine later.

Nonstrict and strict order relations. As in the theory of ordinary sets, one may distinguish between nonstrict (transitive, reflexive, antisymmetric) order relations and strict (transitive, reflexive, antisymmetric) order relations. A nonstrict order relation will generally be called an *order relation*, and a strict order relation will have to be made precise by its adjective. Such a relation may also be called a *nonreflexive order relation.*

A nonstrict order being denoted, as we have indicated,

(23.5) $$y \succcurlyeq x\,,$$

then a strict order will be denoted

(23.6) $$y \succ x\,.$$

We shall give several examples of fuzzy order relations that are strict.

Example 1. Figure 23.7 gives an example of a strict order relation; it is also a perfect order relation. Further, the order is total. One may verify that one has

(23.7) $$\mathrm{A} < \mathrm{B} < \mathrm{C} < \mathrm{D}\,.$$

$\underset{\sim}{\mathcal{R}}$	A	B	C	D
A	0	0,8	0,7	0,7
B	0	0	0,6	0,4
C	0	0	0	0,5
D	0	0	0	0

Fig. 23.7

Example 2. Consider $x \underset{\sim}{\mathcal{R}} y$, where $x, y \in \mathbf{R}$ and

$$\mu_{\underset{\sim}{\mathcal{R}}}(x\,,y) = 0 \;\underset{\sim}{\mathcal{R}}\; y < x\,,$$

(23.8) $$= \frac{1}{1 + \dfrac{1}{(y-x)^2}}\,,\; y \geqslant x\,.$$

This is a relation of strict and perfect order [one may check that for $y = x$, $\mu_{\underset{\sim}{\mathcal{R}}}(x, y) = 0$]. Further, the order is total. This fuzzy relation may be taken to represent (rather poorly) the proposition $y \gg x$.

Important general remark. All definitions associated with order relations in ordinary sets† are directly transposable to fuzzy order relations; it is sufficient to pass through the notion of the associated ordinary graph. It is thus that one may study for fuzzy order relations the classical concepts:

greatest and least element;
majorant and minorant;
limit superior and limit inferior;
maximal chain;
filtering set;
Hasse diagram;
semilattice and lattice.

We shall take up again, when necessary, certain of these concepts for particular uses.

We return now to the concept of a reducible fuzzy preorder for an important theroem.

Theorem II. In a reducible fuzzy preorder relation, there exists at least one similitude class, and the similitude classes form among themselves a fuzzy order relation if one considers the concept of the strongest path from one class to another.

Proof. The relation formed by the similitude classes is necessarily antisymmetric, otherwise certain classes would not be disjoint.

Example 1. Return now to the example of Figure 21.2. For the order relation between these classes, one has the graph in Figure 23.8 for the associated ordinary graph.

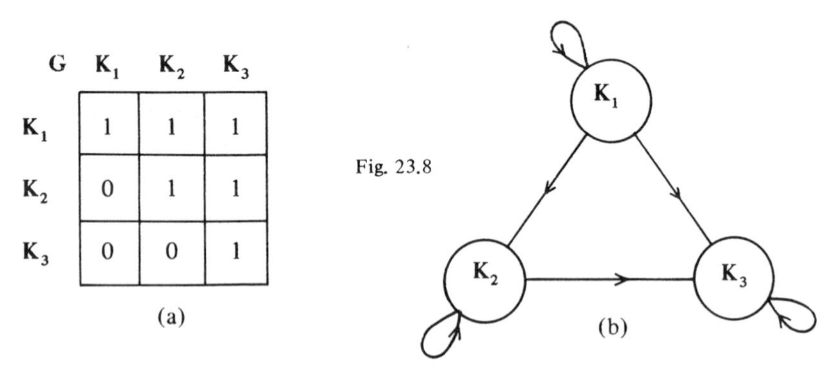

G	K_1	K_2	K_3
K_1	1	1	1
K_2	0	1	1
K_3	0	0	1

Fig. 23.8

Thus, for these classes, there exists a total order

$$K_3 \succ K_2 \succ K_1 \, . \tag{23.9}$$

Having thus presented the ordinary graph constituting the ordinary order relation between the similitude classes, one may then give the fuzzy order relation existing between

†For all that concerns definitions with respect to the theory of (ordinary) sets, we refer to the work of Kaufmann and Precigout [3K].

the classes, obtaining the relation determined by the strongest path existing between each class. For the example of Figures 21.1, 21.2, and 23.8, these results have been given in Figure 21.2; similarly, we reproduce here (Figure 23.9) those holding for Figure 23.8.

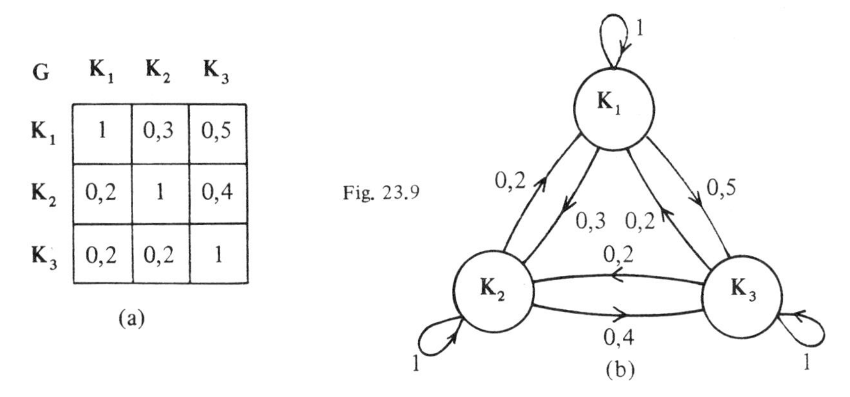

G	K_1	K_2	K_3
K_1	1	0,3	0,5
K_2	0,2	1	0,4
K_3	0,2	0,2	1

(a)

Fig. 23.9

In the case of the example presented in Figure 21.1 determination of the strongest paths existing between $\mathbf{K}_1$ and $\mathbf{K}_2$, $\mathbf{K}_1$ and $\mathbf{K}_3$, and finally between $\mathbf{K}_2$ and $\mathbf{K}_3$ is very easy. The class $\mathbf{K}_1 = \{A, B, C, E, F\}$, and the class $\mathbf{K}_2 = \{D\}$, are joined by paths whose values are given by

A	0,3
B	0,3
C	0,3
E	0,3
F	0,3

The strongest path (not unique) has value 0,3. For $\mathbf{K}_2$ toward $\mathbf{K}_1$, one sees

	A	B	C	E	F
D	0,2	0,2	0,2	0,2	0,2

and the nonunique strongest path has value 0,2.

For $\mathbf{K}_1$ toward $\mathbf{K}_3 = \{G\}$, one may see

A	0,4
B	0,5
C	0,5
E	0,5
F	0,5

and the nonunique strongest path has value 0,5. In the same manner one finds that the nonunique strongest path for $\mathbf{K}_3$ toward $\mathbf{K}_1$ has value 0,2.

One also obtains 0,4 for $\mathbf{K}_2 \to \mathbf{K}_3$ and 0,2 for $\mathbf{K}_3 \to \mathbf{K}_2$, where the determination is trivial since these classes each have only one unique element. It is thus that Figure 23.9 has been obtained.

More generally, to construct the fuzzy order relation existing between the classes, one proceeds in the following fashion:

(1) Find the similitude classes $\mathbf{K}_i$ in the reducible fuzzy preorder. For these, consider the ordered pairs (x, y) for which one has

$$\mu_{\underset{\sim}{\mathcal{R}}}(x\,,\,y) = \mu_{\underset{\sim}{\mathcal{R}}}(y\,,\,x)\ .$$

With respect to these ordered pairs construct the maximal similitude subrelations.† If these are all disjoint, one has obtained the similitude classes. If there exist at least two that are not disjoint, we do not have a reducible fuzzy preorder.

(2) For each ordered pair $(\mathbf{K}_i, \mathbf{K}_j)$, $i \neq j$, examine the fuzzy subrelation $\underset{\sim}{\mathcal{R}}_{ij}$ existing between $\mathbf{K}_i$ and $\mathbf{K}_j$ (rows of $\mathbf{K}_i$ and columns of $\mathbf{K}_j$). Determine the global projection of $\underset{\sim}{\mathcal{R}}_{ij}$ [see (12.13)]; thus,

$$h(\underset{\sim}{\mathcal{R}}_{ij}) = \bigvee_x \bigvee_y \mu_{\underset{\sim}{\mathcal{R}}_{ij}}(x\,,\,y)\ ,\ x \in \mathbf{K}_1\ ,\ y \in \mathbf{K}_2\ .$$

(3) Assign the value $h(\underset{\sim}{\mathcal{R}}_{ij})$ to the membership function of the pair $(\mathbf{K}_i, \mathbf{K}_j)$.

		$\mathbf{K}_1$		$\mathbf{K}_2$			$\mathbf{K}_3$	$\mathbf{K}_4$		$\mathbf{K}_5$
	$\underset{\sim}{\mathcal{R}}$	B	D	H	A	C	E	F	G	I
$\mathbf{K}_1$	B	1	0,7	0,7	0,7	0,7	0,4	0,7	0,7	0,1
	D	0,7	1	0,8	0,8	0,9	0,4	0,7	0,8	0,1
$\mathbf{K}_2$	H	0	0	1	0,8	0,6	0,3	0	0	0
	A	0	0	0,8	1	0,6	0,3	0	0	0
	C	0	0	0,6	0,6	1	0,3	0	0	0
$\mathbf{K}_3$	E	0	0	0,2	0,2	0,2	1	0	0	0
$\mathbf{K}_4$	F	0	0	0	0	0	0	1	0,3	0
	G	0	0	0	0	0	0	0,3	1	0
$\mathbf{K}_5$	I	0	0	0	0	0	0	0	0	1

Fig. 23.10

†Use if necessary one of the algorithms given in Appendix B, p. 387.

Example 2. The example given in Figure 23.10 is a little more complicated. One may notice in this reducible fuzzy preorder several particularities that have not appeared in the preceding examples. The existence of a partial order between the classes is evident in Figure 23.11. In the preceding example, we had a total order (see Figure 23.8).

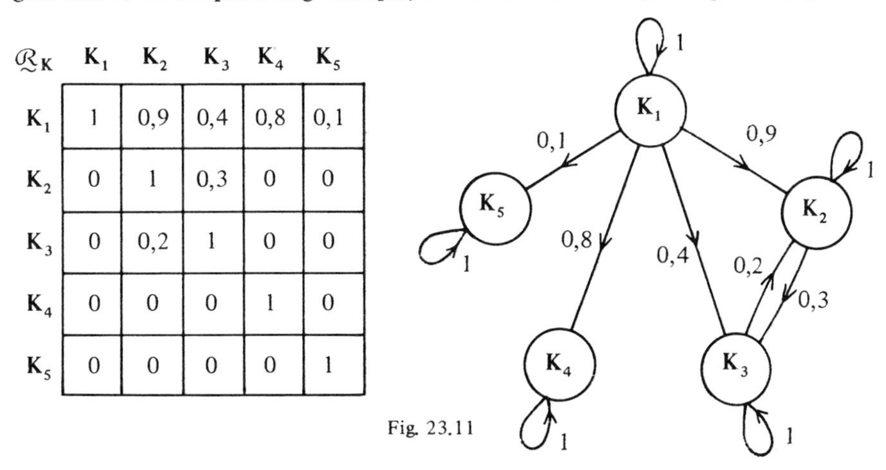

$\underset{\sim}{\mathcal{R}}_K$	K_1	K_2	K_3	K_4	K_5
K_1	1	0,9	0,4	0,8	0,1
K_2	0	1	0,3	0	0
K_3	0	0,2	1	0	0
K_4	0	0	0	1	0
K_5	0	0	0	0	1

Fig. 23.11

Example 3. The fuzzy relation presented in Figure 23.12 is a reducible fuzzy preorder relation if one imposes

$$0 \leqslant a_1 \leqslant a_2 \leqslant \cdots \leqslant a_k \leqslant \cdots \leqslant 1 \ .$$

We may see that this decomposes into an infinity of similitude classes forming among themselves a total order

$$C_1 \prec C_2 \prec C_3 \prec \ldots .$$

	C_1		C_2		C_2		C_3		
	x_1	x_2	x_3	x_4	x_5	x_6	x_7	x_8	...
x_1	1	a_1	a_1	a_1	a_1	a_1	a_1	a_1	...
x_2	a_1	1	a_2	a_2	a_2	a_2	a_2	a_2	...
x_3	0	0	1	a_3	a_3	a_3	a_3	a_3	...
x_4	0	0	a_3	1	a_4	a_4	a_4	a_4	...
x_5	0	0	0	0	1	a_5	a_5	a_5	...
x_6	0	0	0	0	a_5	1	a_6	a_6	...
x_7	0	0	0	0	0	0	1	a_7	...
x_8	0	0	0	0	0	0	a_7	1	...
⋮	⋮	⋮	⋮	⋮	⋮	⋮	⋮	⋮	⋱

Fig. 23.12

24. ANTISYMMETRIC RELATIONS WITHOUT CIRCUITS, ORDINAL RELATIONS, ORDINAL FUNCTION IN A FUZZY ORDER RELATION

We shall consider a fuzzy relation (**E** finite) that possesses the three properties:

(1) reflexivity [according to (16.7)]

(2) antisymmetry [according to (22.1)]

(3) does not have an ordinary circuit in its associated ordinary graph, other than loops, that is, other than circuits of length1, such as (x, x).

Such a relation will be called a *fuzzy ordinal relation.*

Example 1. The fuzzy relation in Figure 24.1 is an ordinal relation. By constructing the associated ordinary graph (Figure 24.2), one may verify that this relation is indeed reflexive, antisymmetric, and without circuits other than loops.

$\underset{\sim}{\mathcal{R}}$	A	B	C	D	E
A	1	0,8	0	0	0,7
B	0,4	1	0,6	1	0,3
C	0,9	0,8	1	0	0
D	0	0	0	1	0
E	0,2	0	0	0	1

Fig. 24.1

G	A	B	C	D	E
A	1	1	0	0	1
B	0	1	0	1	1
C	1	1	1	0	0
D	0	0	0	1	0
E	0	0	0	0	1

(a)

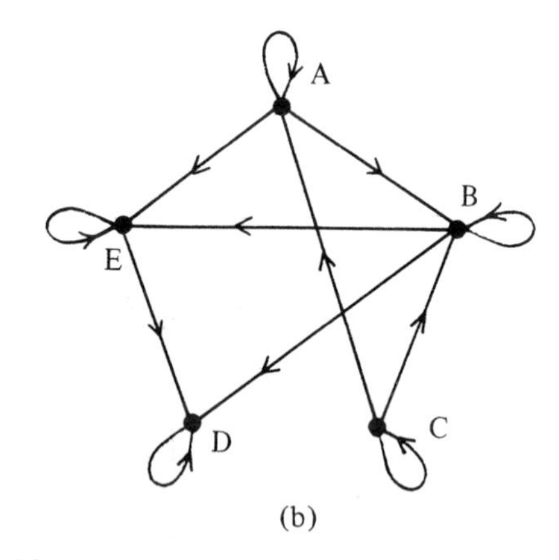

(b)

Fig. 24.2

Review of the notion of the ordinal function of an ordinary antisymmetric finite graph without circuits. We shall consider an ordinary graph without circuits $\mathbf{G} \subset \mathbf{E} \times \mathbf{E}$, **E** finite. We shall again designate **G** by the ordered pair $(\mathbf{E}, \Gamma)$, where $\mathbf{E} \underset{\Gamma}{\rightsquigarrow} \mathbf{E}$, Γ representing the mapping of **E** into **E**, which in general is multivalued.

We define the ordinary subsets $\mathbf{N}_0, \mathbf{N}_1, \ldots, \mathbf{N}_r$ such that†

†Some authors prefer to define the ordinal function of an ordinary graph by replacing the inverse mapping Γ^{-1} with the direct mapping Γ in formulas (24.1). This has the effect of eventually giving another order of levels.

$$
\begin{aligned}
N_0 &= \{X_i \mid \Gamma^{-1}\{X_i\} = \phi\}, \\
N_1 &= \{X_i \mid \Gamma^{-1}\{X_i\} \subset N_0\}, \\
N_2 &= \{X_i \mid \Gamma^{-1}\{X_i\} \subset N_0 \cup N_1, \\
&\vdots \\
N_r &= \{X_i \mid \Gamma^{-1}\{X_i\} \subset \bigcup_{k=0}^{r-1} N_k,
\end{aligned}
\tag{24.1}
$$

where r is the least integer such that

$$\Gamma N_r = \phi \ . \tag{24.2}$$

One may easily show that the ordinary subsets $\mathbf{N}_k$, $k = 0, 1, 2, \ldots, r$, form a partition of **E** and are totally and strictly ordered by the relation

$$N_k < N'_k \quad \Leftrightarrow \quad k < k' . \tag{24.3}$$

The function $O(X_i)$ defined by

$$X_i \in N_k \quad \Rightarrow \quad O(X_i) = k , \tag{24.4}$$

is called the *ordinal function of an ordinary graph without circuits.*

In other words, less precise but more concise: One has in mind the decomposition of the set of vertices of the ordinary graph **G** without circuits into ordinary subsets, disjoint and ordered so that if one of these vertices belongs to one of the subsets carrying the number k, all vertices following the vertex being considered must be placed in a subset carrying a number larger than k.

The ordinary subsets of the partiion are called *levels.*†

Example. The ordinary graph without circuits in Figure 24.3 has been decomposed into levels in Figure 24.4. If X_i is a vertex of the graph, to each X_i there corresponds an N_k, or more simply a $k \in \{0, 1, \ldots, 5\}$. The function $X_i \rightsquigarrow k$ represented in Figure 24.5 is the ordinal function of the graph. An enumeration of the vertices is presented in Figure 24.6.

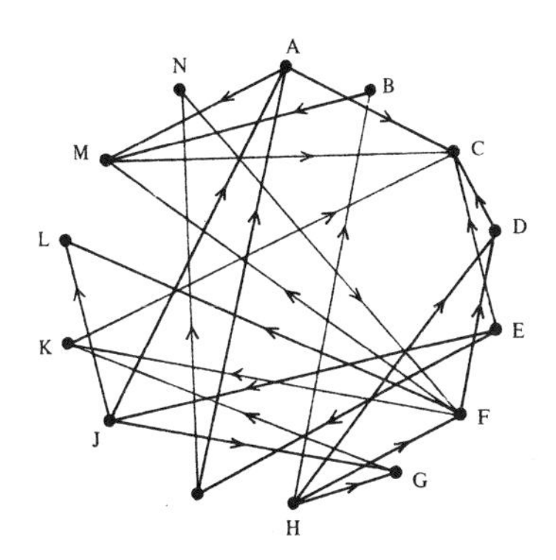

Fig. 24.3

†Some authors call these ordinary subsets *ranks.*

The ordinal function of a graph is not in general unique; it may be defined with respect to the largest elements† of the ordered set instead of with respect to the smallest, that is to say, ordered from right to left instead of toward the right as we have done in the examples of Figure 24.3–24.6.

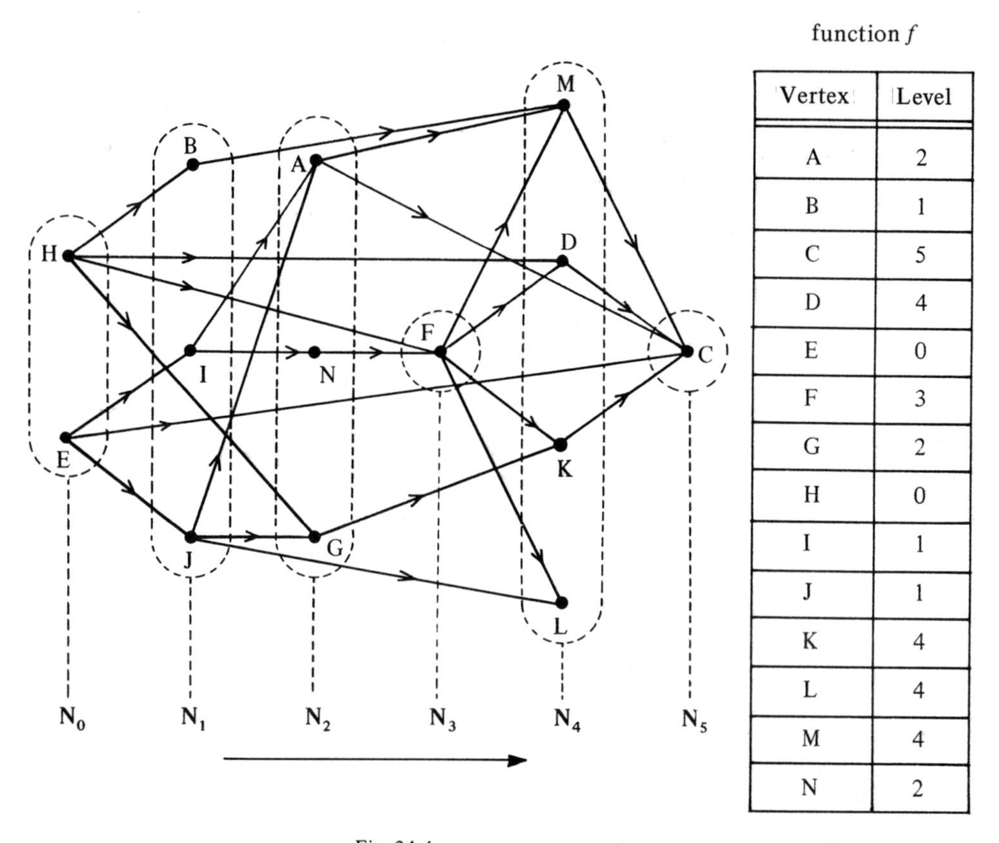

Vertex	Level
A	2
B	1
C	5
D	4
E	0
F	3
G	2
H	0
I	1
J	1
K	4
L	4
M	4
N	2

Fig. 24.4

Fig. 24.5

The notion of ordinal function plays an important role in a large number of theoretical combinatorial problems and practical applications.

Extension of the notion of ordinal function to an ordinary graph having circuits. For this, it suffices to consider equivalence classes (with respect to the relation, "there exists a path from X_i to X_j, and vice versa") of the ordinary graph.

These classes are the maximal ordinary subsets for the equivalence relation. These classes then form an order (total or partial, depending on the case). If the order is total, one has the ordinal function; if it is partial, one will seek the ordinal function of the ordinary graph without circuits formed by these classes.

†In the sense given to the words greatest element and least element in the theory of ordinary ordered sets.

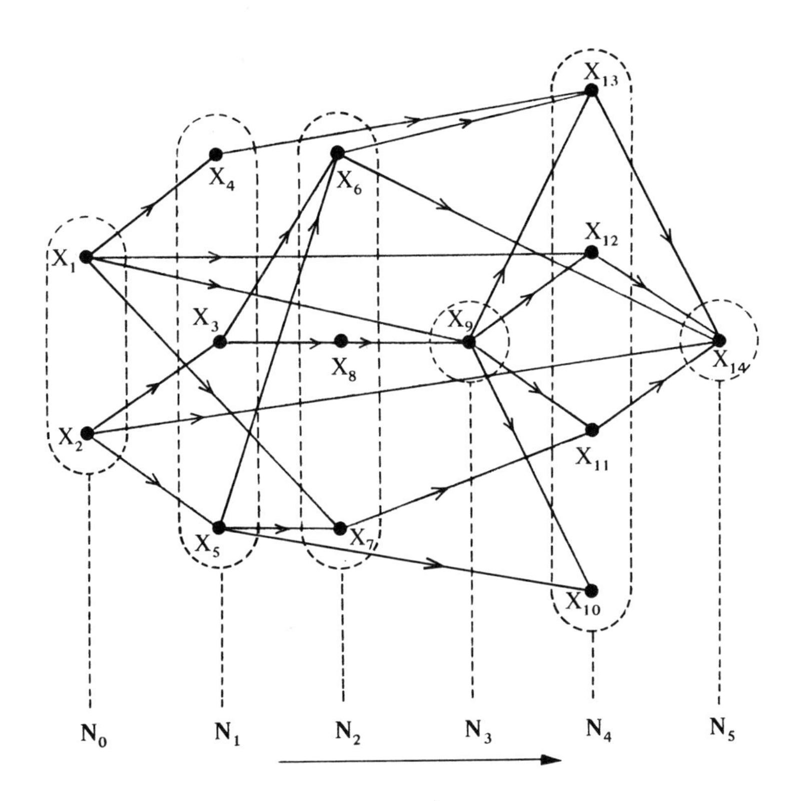

Fig. 24.6

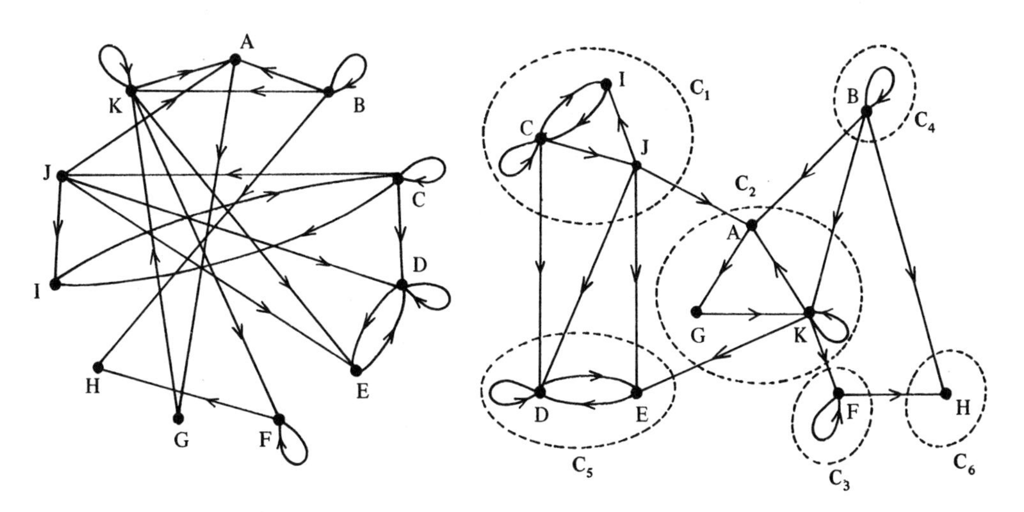

Fig. 24.7

Fig. 24.8

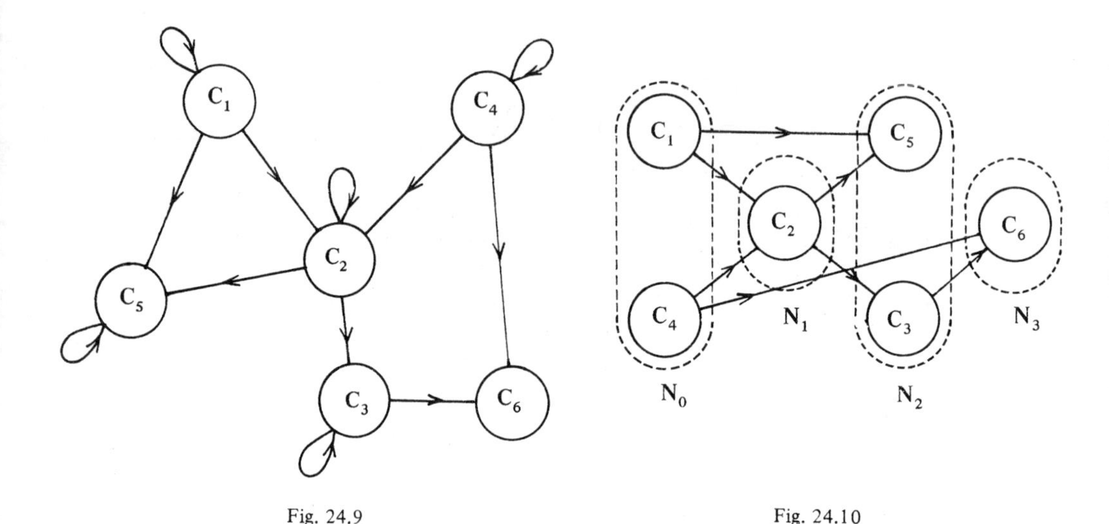

Fig. 24.9

Fig. 24.10

An example is given in Figures 24.7–24.10.

Method*† *for determining the level of a graph without circuits. We shall consider the boolean matrix of the ordinary graph in Figure 24.3; this matrix is presented in Figure 24.11. We form a row Λ_0 in which appears the sum of the rows of the matrix.‡ The zeros of Λ_0 give the vertices that are not precedents of one another; thus E and H form level $\mathbf{N}_0$. Eliminating the sum of the E and H rows from the Λ_0 row, one obtains the row Λ_1, where the zeros of the row Λ_0 have been replaced by a $\times$ (cross). The zeros that appear in the row Λ_1 give the vertices that are not precedents of one another whenever E and H have been eliminated; these are B, I, and J, which form $\mathbf{N}_1$. We eliminate from row Λ_1 the sum of rows B, I, and J after having replaced all the zeros previously appearing with an $\times$; the new zeros that appear in Λ_2 give the vertices that are not precedents of one another whenever E, H, B, I, and J have been eliminated; these are A, G, and N, which form $\mathbf{N}_2$. And we continue thus until exhaustion. Afterward, it remains only to construct the ordinary graph (Figure 24.4) where the vertices appear with their respective levels. An arbitrary enumeration of the vertices is represented in Figure 24.6; it respects the ordinal function.

When the graph contains at least one circuit, there exists a row Λ_i in which it is impossible to make a new zero appear. This also therefore gives us an automatic means of checking whether a graph is without circuits.

If one has seen how to obtain the ordinal function in the inverse sense, by taking the greatest elements of the order (that is, from the right to the left in our representation), one may utilize exactly the same procedure by taking the transpose of the boolean matrix

†A method due to M. Demoucron of Honeywell-Bull Cie., Paris.

‡That is, the sum calculated in each column.

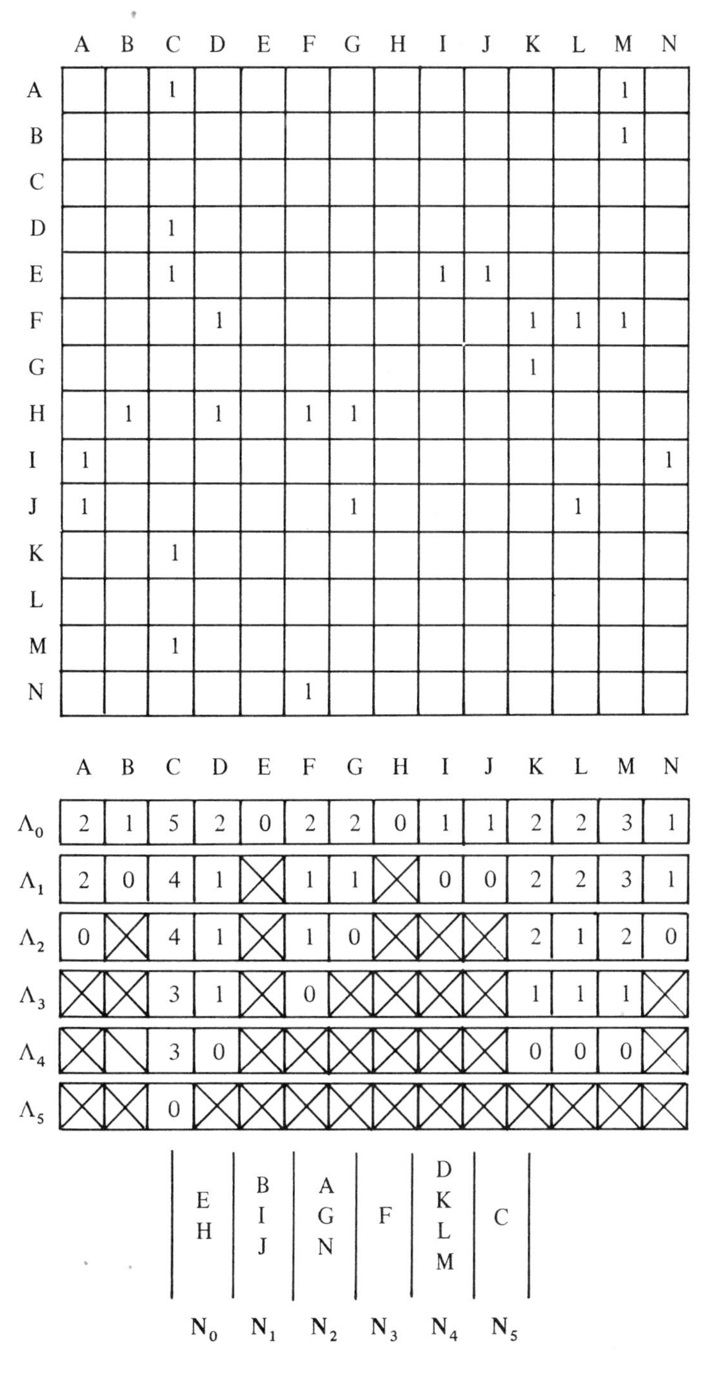

	A	B	C	D	E	F	G	H	I	J	K	L	M	N
A			1										1	
B													1	
C														
D			1											
E			1						1	1				
F				1							1	1	1	
G											1			
H		1		1		1	1							
I	1													1
J	1						1					1		
K			1											
L														
M			1											
N						1								

	A	B	C	D	E	F	G	H	I	J	K	L	M	N
Λ_0	2	1	5	2	0	2	2	0	1	1	2	2	3	1
Λ_1	2	0	4	1	X	1	1	X	0	0	2	2	3	1
Λ_2	0	X	4	1	X	1	0	X	X	X	2	1	2	0
Λ_3	X	X	3	1	X	0	X	X	X	X	1	1	1	X
Λ_4	X	X	3	0	X	X	X	X	X	X	0	0	0	X
Λ_5	X	X	0	X	X	X	X	X	X	X	X	X	X	X

N_0	N_1	N_2	N_3	N_4	N_5
E H	B I J	A G N	F	D K L M	C

Fig. 24.11

(the rows become the columns and vice versa). Thus, reconsidering the example in Figures 24.3–24.6, we seek this time an ordinal function from right to left. The result is presented in Figure 24.12.

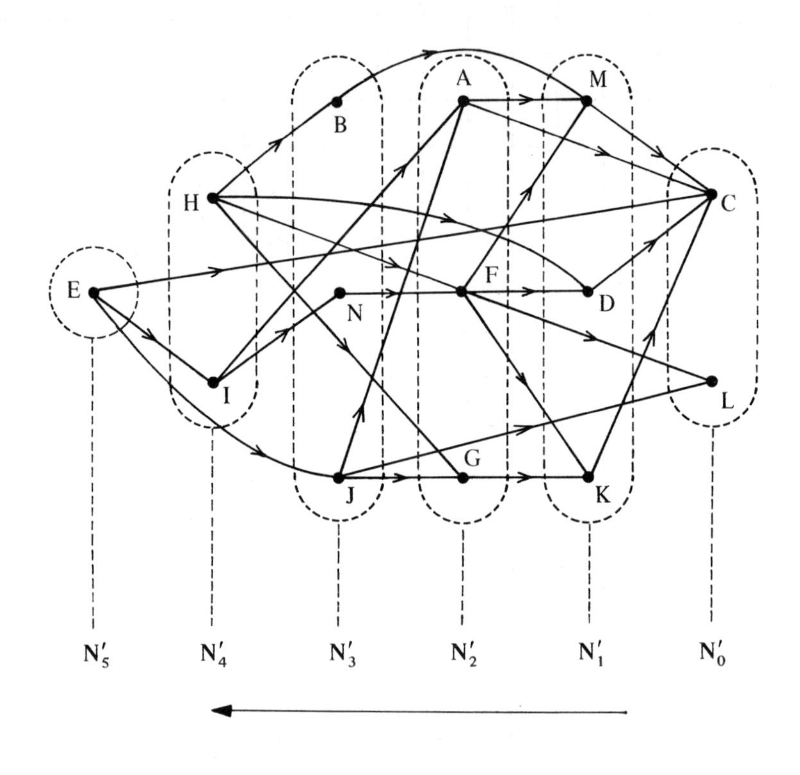

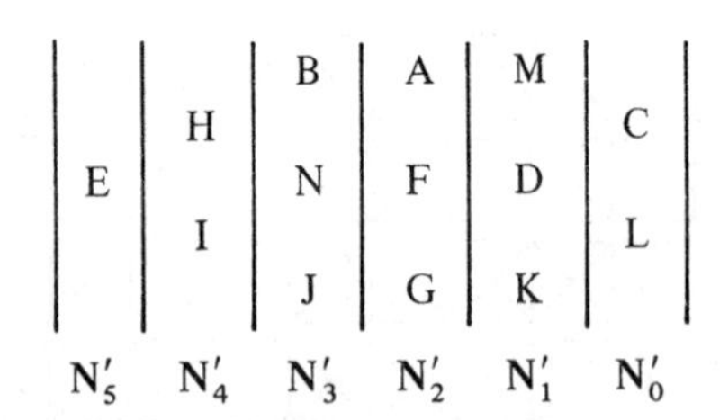

Fig. 24.12

Ordinal function of a fuzzy order relation. An order relation is an ordinal relation; it is reflexive, antisymmetric, and without circuits; it is moreover transitive. One may then define an ordinal function for it.

An example will serve to illustrate.

In Figure 24.13 we have presented a fuzzy order relation that constitutes a partial order. In Figure 24.14 we have presented the ordinal function of the associated ordinary graph with respect to the smallest elements. In this graph we have intentionally omitted the loops.

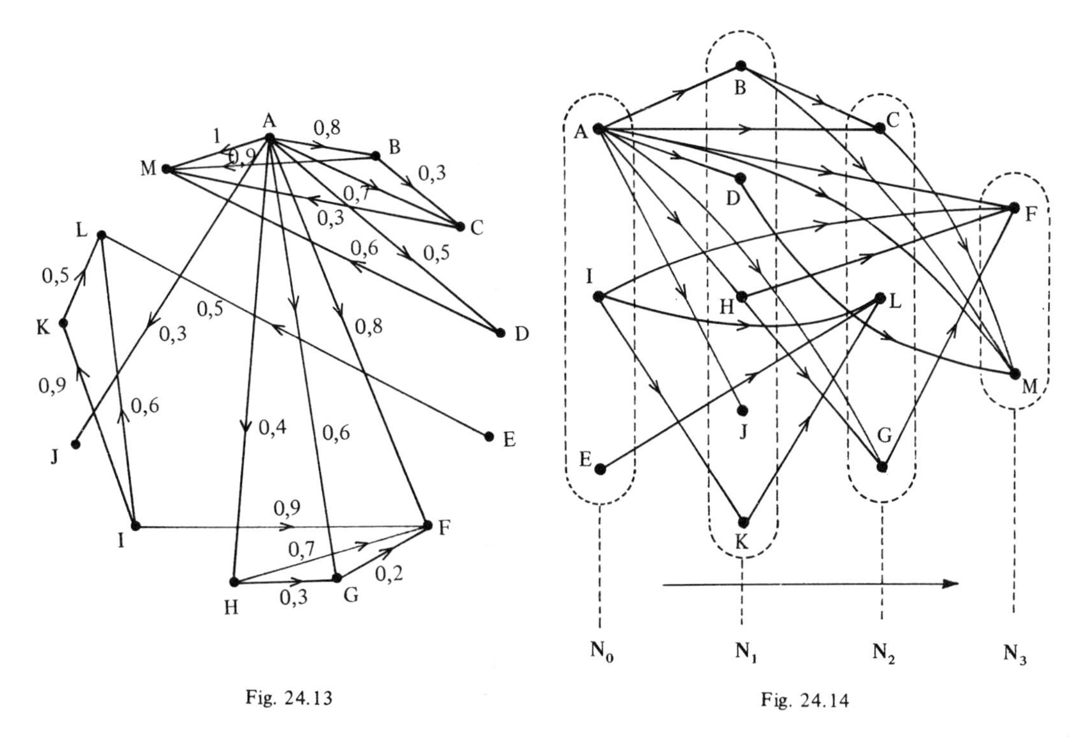

Fig. 24.13

Fig. 24.14

We now consider the fuzzy order relation† presented in Figure 24.15a. Its associated ordinary graph has been given in Figure 24.15b. By permuting the elements in a manner to satisfy the ordinal function given in Figure 24.14 (Figures 24.14 and 24.15b represent the same ordinary graph), one sees a triangular matrix appear. By reconsidering the fuzzy order relation in the total order of its elements chosen to conform to the ordinal function, one obtains a fuzzy order relation that will be said to be triangular (Figure 24.15d). One knows that it is important, for whatever calculations, to know how to reduce a matrix to triangular form.

†We have taken as an example a perfect order relation with the desire of presenting a simple example; but the considerations that follow would remain valid for a fuzzy order relation that is not perfect, and the property that gives a triangular matrix is verified only for ordered pairs (x, y) such that $\mu_{\underset{\sim}{\mathcal{R}}}(x, y) > \mu_{\underset{\sim}{\mathcal{R}}}(y, x)$.

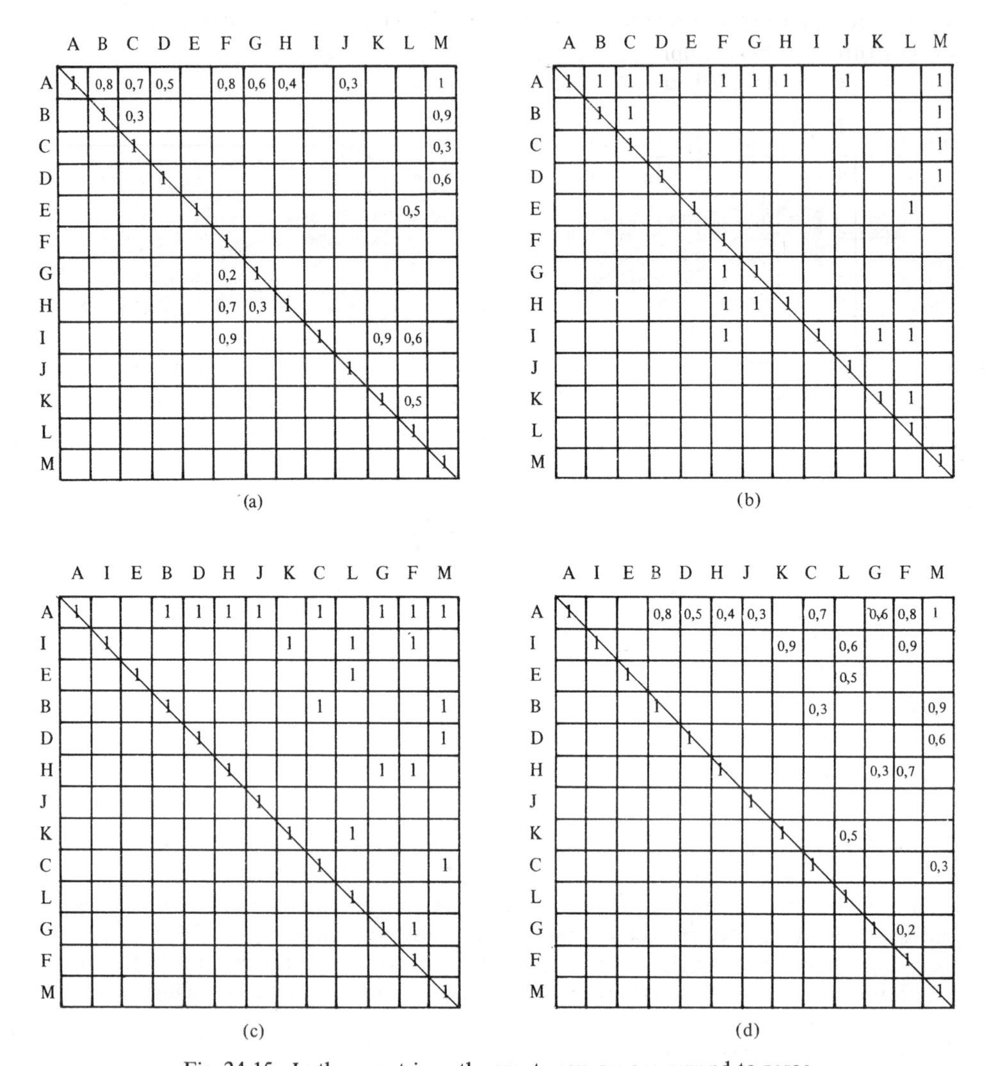

(a)

	A	B	C	D	E	F	G	H	I	J	K	L	M
A	1	0,8	0,7	0,5		0,8	0,6	0,4		0,3			1
B		1	0,3										0,9
C			1										0,3
D				1									0,6
E					1							0,5	
F						1							
G						0,2	1						
H						0,7	0,3	1					
I						0,9			1		0,9	0,6	
J										1			
K											1	0,5	
L												1	
M													1

(b)

	A	B	C	D	E	F	G	H	I	J	K	L	M
A	1	1	1	1		1	1	1		1			1
B		1	1										1
C			1										1
D				1									1
E					1							1	
F						1							
G						1	1						
H						1	1	1					
I						1			1		1	1	
J										1			
K											1	1	
L												1	
M													1

(c)

	A	I	E	B	D	H	J	K	C	L	G	F	M
A	1			1	1	1	1		1		1	1	1
I		1						1		1		1	
E			1							1			
B				1					1				1
D					1								1
H						1					1	1	
J							1						
K								1		1			
C									1				1
L										1			
G											1	1	
F												1	
M													1

(d)

	A	I	E	B	D	H	J	K	C	L	G	F	M
A	1			0,8	0,5	0,4	0,3		0,7		0,6	0,8	1
I		1						0,9		0,6		0,9	
E			1							0,5			
B				1					0,3				0,9
D					1								0,6
H						1					0,3	0,7	
J							1						
K								1		0,5			
C									1				0,3
L										1			
G											1	0,2	
F												1	
M													1

Fig. 24.15. In these matrices, the empty squares correspond to zeros.

Utility of the notion of ordinal function in fuzzy preorder relations. We have seen in Section 23 that the notion of similitude class induces in a fuzzy preorder relation an order (total or partial) of the similitude classes (in the case of a reducible preorder).

The associated ordinary graph of this order is evidently reflexive and antisymmetric; it is also transitive. If the preorder is an order, it may be reduced, as we have just seen, to a triangular form in its matrix representation. If the preorder is not an order, it may then always be reduced to a block-triangular form. Such a block-triangular form has already been presented in the example given in Figure 23.10, which we reproduce here associated with its boolean matrix (Figure 24.16) in order to show that it is a block-triangular form.

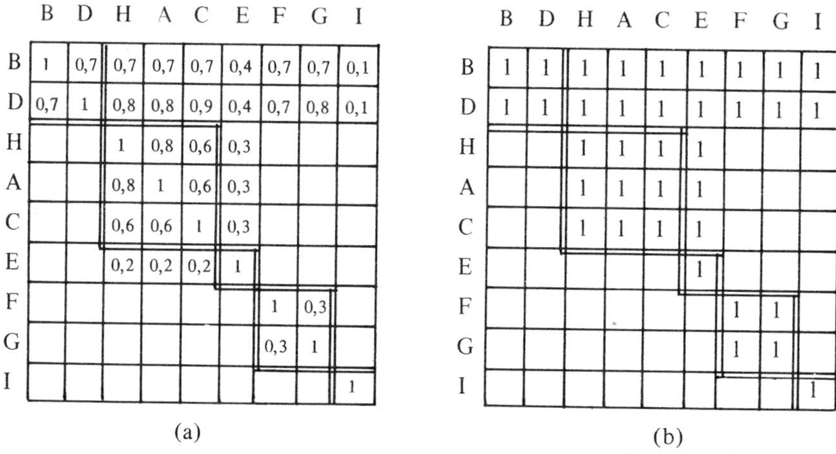

(a)

	B	D	H	A	C	E	F	G	I
B	1	0,7	0,7	0,7	0,7	0,4	0,7	0,7	0,1
D	0,7	1	0,8	0,8	0,9	0,4	0,7	0,8	0,1
H			1	0,8	0,6	0,3			
A			0,8	1	0,6	0,3			
C			0,6	0,6	1	0,3			
E			0,2	0,2	0,2	1			
F							1	0,3	
G							0,3	1	
I									1

(b)

	B	D	H	A	C	E	F	G	I
B	1	1	1	1	1	1	1	1	1
D	1	1	1	1	1	1	1	1	1
H			1	1	1	1			
A			1	1	1	1			
C			1	1	1	1			
E						1			
F							1	1	
G							1	1	
I									1

Fig. 24.16. The empty squares represent zeros.

Further, the construction of the ordinal function permits the automatic realization of the search for the Hasse diagram† corresponding to the order relation and to the determination of the levels of this diagram.

25. DISSIMILITUDE RELATIONS

We shall consider a similitude relation $\underset{\sim}{\mathcal{R}}$ such as that defined in Section 20. For convenience, we recall here the three properties of similitude:

(25.1) 1) $\forall (x, y), (y, z), (z, x) \in \mathbf{E} \times \mathbf{E}$:

$$\mu_{\underset{\sim}{\mathcal{R}}}(x, z) \geqslant \bigvee_{y} [\mu_{\underset{\sim}{\mathcal{R}}}(x, y) \wedge \mu_{\underset{\sim}{\mathcal{R}}}(y, z)] , \qquad \text{transitivity} .$$

(25.2) 2) $\forall (x, x) \in \mathbf{E} \times \mathbf{E}$: $\quad \mu_{\underset{\sim}{\mathcal{R}}}(x, x) = 1$, $\qquad$ reflexivity .

(25.3) 3) $\forall (x, y) \in \mathbf{E} \times \mathbf{E}$: $\quad \mu_{\underset{\sim}{\mathcal{R}}}(x, y) = \mu_{\underset{\sim}{\mathcal{R}}}(y, x)$, $\qquad$ symmetry .

Now we associate with $\underset{\sim}{\mathcal{R}}$ a relation $\overline{\underset{\sim}{\mathcal{R}}}$ such that

(25.4) $\quad \forall (x, y) \in \mathbf{E} \times \mathbf{E}$: $\quad \mu_{\overline{\underset{\sim}{\mathcal{R}}}}(x, y) = 1 - \mu_{\underset{\sim}{\mathcal{R}}}(x, y)$.

†Those who are not familiar with what is called a Hasse diagram in the ordinary theory of sets may consult, for example, references [K_1 , K_2].

Knowing that $\underset{\sim}{\mathcal{R}}$ has properties (25.1)–(25.3), these are then properties of $\overline{\underset{\sim}{\mathcal{R}}}$. Beginning with (25.1), one has

(25.5) $$1 - \mu_{\overline{\underset{\sim}{\mathcal{R}}}}(x, z) \geqslant \bigvee_y \Big[[1 - \mu_{\overline{\underset{\sim}{\mathcal{R}}}}(x, y)] \wedge [1 - \mu_{\overline{\underset{\sim}{\mathcal{R}}}}(y, z)]\Big].$$

But, according to (7.32),

(25.6) $$[1 - \mu_{\overline{\underset{\sim}{\mathcal{R}}}}(x, y)] \wedge [1 - \mu_{\overline{\underset{\sim}{\mathcal{R}}}}(y, z)] = 1 - \mu_{\overline{\underset{\sim}{\mathcal{R}}}}(x, y) \vee \mu_{\overline{\underset{\sim}{\mathcal{R}}}}(y, z).$$

Thus, (25.5) may be written

(25.7) $$1 - \mu_{\overline{\underset{\sim}{\mathcal{R}}}}(x, z) \geqslant \bigvee_y [1 - (\mu_{\overline{\underset{\sim}{\mathcal{R}}}}(x, y) \vee \mu_{\overline{\underset{\sim}{\mathcal{R}}}}(y, z))],$$

or

(25.8) $$\mu_{\overline{\underset{\sim}{\mathcal{R}}}}(x, z) \leqslant \bigwedge_y [\mu_{\overline{\underset{\sim}{\mathcal{R}}}}(x, y) \vee \mu_{\overline{\underset{\sim}{\mathcal{R}}}}(y, z)].$$

This property will be called *min–max transitivity*.†

Concerning (25.2), one may see

(25.9) $$\mu_{\overline{\underset{\sim}{\mathcal{R}}}}(x, x) = 1 - \mu_{\underset{\sim}{\mathcal{R}}}(x, x) = 1 - 1 = 0.$$

And finally, symmetry is preserved. Thus we have

(25.10) 1) $\forall (x, y), (y, z), (z, x) \in \mathbf{E} \times \mathbf{E}$:

$$\mu_{\overline{\underset{\sim}{\mathcal{R}}}}(x, z) \leqslant \bigwedge_y [\mu_{\overline{\underset{\sim}{\mathcal{R}}}}(x, y) \vee \mu_{\overline{\underset{\sim}{\mathcal{R}}}}(y, z)],$$ min–max transitivity,

(25.11) 2) $\forall (x, x) \in \mathbf{E} \times \mathbf{E}$: $\mu_{\overline{\underset{\sim}{\mathcal{R}}}}(x, x) = 0$, antireflexivity,

(25.12) 3) $\forall (x, y) \in \mathbf{E} \times \mathbf{E}$: $\mu_{\overline{\underset{\sim}{\mathcal{R}}}}(x, y) = \mu_{\overline{\underset{\sim}{\mathcal{R}}}}(y, x)$, symmetry.

A fuzzy binary relation that possesses properties (25.10)–(25.12) is called a *dissimilitude relation.*

Example 1. Figure 25.1 represents a dissimilitude relation (it is, moreover, the relation $\overline{\underset{\sim}{\mathcal{R}}}$ corresponding to the similitude relation $\underset{\sim}{\mathcal{R}}$ presented in Figure 20.1). As an exercise, we verify (25.10) for several pairs of elements.

$\underset{\sim}{\mathcal{R}}$	A	B	C	D	E
A	0	0,2	0,3	0	0,1
B	0,2	0	0,3	0,2	0,2
C	0,3	0,3	0	0,3	0,3
D	0	0,2	0,3	0	0,1
E	0,1	0,2	0,3	0,1	0

Fig. 25.1

†One may also call this min–max cotransitivity.

Arc (A, B)

$$\mu(A, A) \vee \mu(A, B) = 0 \vee 0{,}2 = 0{,}2 ,$$
$$\mu(A, B) \vee \mu(B, B) = 0{,}2 \vee 0 = 0{,}2 ,$$
$$\mu(A, C) \vee \mu(C, B) = 0{,}3 \vee 0{,}3 = 0{,}3 ,$$
$$\mu(A, D) \vee \mu(D, B) = 0 \vee 0{,}2 = 0{,}2 ,$$
$$\mu(A, E) \vee \mu(E, B) = 0{,}1 \vee 0{,}2 = 0{,}2 ,$$
$$\text{MIN}[0{,}2 , 0{,}2 , 0{,}3 , 0{,}2 , 0{,}2] = 0{,}2 , \quad (25.13)$$
$$\mu(A, B) = 0{,}2 \leqslant 0{,}2 .$$

Arc (A, C)

$$\mu(A, A) \vee \mu(A, C) = 0 \vee 0{,}3 = 0{,}3 ,$$
$$\mu(A, B) \vee \mu(B, C) = 0{,}2 \vee 0{,}3 = 0{,}3 ,$$
$$\mu(A, C) \vee \mu(C, C) = 0{,}3 \vee 0 = 0{,}3 ,$$
$$\mu(A, D) \vee \mu(D, C) = 0 \vee 0{,}3 = 0{,}3 ,$$
$$\mu(A, E) \vee \mu(E, C) = 0{,}1 \vee 0{,}3 = 0{,}3 ,$$
$$\text{MIN}[0{,}3 , \ldots] = 0{,}3 ,$$
$$\mu(A, C) = 0{,}3 \leqslant 0{,}3 .$$

And so on.

Example 2. The relation presented in Figure 25.2 is a dissimilitude relation if

$$1 \geqslant b_1 \geqslant b_2 \geqslant \cdots \geqslant b_l \geqslant \cdots \geqslant 0 . \quad (25.14)$$

$\underset{\sim}{\mathcal{R}}$	x_1	x_2	x_3	x_4	x_5	x_6	x_7	...
x_1	0	b_1	b_1	b_1	b_1	b_1	b_1	...
x_2	b_1	0	b_2	b_2	b_2	b_2	b_2	...
x_3	b_1	b_2	0	b_3	b_3	b_3	b_3	...
x_4	b_1	b_2	b_3	0	b_4	b_4	b_4	...
x_5	b_1	b_2	b_3	b_4	0	b_5	b_5	...
x_6	b_1	b_2	b_3	b_4	b_5	0	b_6	...
x_7	b_1	b_2	b_3	b_4	b_5	b_6	0	...
⋮	⋮	⋮	⋮	⋮	⋮	⋮	⋮	⋱

Fig. 25.2

This relation has been obtained from that presented in Figure 20.3 by setting

(25.15) $\mu_{\bar{\underset{\sim}{\mathcal{R}}}}(x, y) = 1 - \mu_{\underset{\sim}{\mathcal{R}}}(x, y)$ letting $b_i = 1 - a_i \quad i = 1, 2, 3, \ldots$.

Example 3. The fuzzy relation

$$\begin{aligned} \mu_{\bar{\underset{\sim}{\mathcal{R}}}}(x, y) &= 1 - e^{-k(y+1)}, \quad y < x, \quad k > 1, \\ &= 0, \quad y = x, \\ &= 1 - e^{-k(x+1)}, \quad y > x, \quad k > 1, \end{aligned} \tag{25.16}$$

is a dissimilitude relation. It has been obtained from (20.3) by setting

$$\mu_{\bar{\underset{\sim}{\mathcal{R}}}}(x, y) = 1 - \mu_{\underset{\sim}{\mathcal{R}}}(x, y) .$$

We shall see several examples; but first we recall here, in order to have them nearby, axioms (5.49)–(5.52) concerning the notion of distance between two elements of a set.

If $d(\mathrm{X}, \mathrm{Y})$ is this distance between X and Y:

$\forall \mathrm{X}, \mathrm{Y}, \mathrm{Z} \in \mathrm{E}$, one must have

(25.17) 1) $$d(\mathrm{X}, \mathrm{Y}) \geqslant 0 ,$$

(25.18) 2) $$d(\mathrm{X}, \mathrm{Y}) = d(\mathrm{Y}, \mathrm{X}) ,$$

(25.19) 3) $$d(\mathrm{X}, \mathrm{Y}) * d(\mathrm{Y}, \mathrm{Z}) \geqslant d(\mathrm{X}, \mathrm{Z}) ,$$

where $*$ is the operation considered among the distances $d(\mathrm{X}, \mathrm{Y})$.

To these three conditions, one may logically introduce a fourth

$$d(\mathrm{X}, \mathrm{X}) = 0 . \tag{25.20}$$

Then considering $\mu_{\bar{\underset{\sim}{\mathcal{R}}}}(x, y)$, one has indeed, by definition, that (25.17) is satisfied since $0 \leqslant \mu_{\bar{\underset{\sim}{\mathcal{R}}}}(x, y) \leqslant 1$. Relation (25.18) is satisfied [see (25.12)]. Relation (25.19) is satisfied [see (25.10), where the operation $*$ is the min–max operation]. Finally, (25.20) is verified [see (25.11)]. Thus, one may put

$$d(x, y) = \mu_{\bar{\underset{\sim}{\mathcal{R}}}}(x, y) \tag{25.21}$$

and consider $\mu_{\bar{\underset{\sim}{\mathcal{R}}}}(x, y)$ as a distance† existing between x and y.

Min–max distance between two elements in a similitude relation. Let $\underset{\sim}{\mathcal{R}}$ be a similitude relation. We shall call the *min–max distance* between x and y, $y \subset \mathrm{E}$ $\underset{\sim}{\mathcal{R}} \subset \mathrm{E} \times \mathrm{E}$.

$$d_{\underset{\sim}{\mathcal{R}}}(x, y) = 1 - \mu_{\underset{\sim}{\mathcal{R}}}(x, y) . \tag{25.22}$$

Example 1. We take up again the example of Figure 20.1 (seen again in Figure 25.3). This is a similitude relation $\underset{\sim}{\mathcal{R}}$. Figure 25.4 represents the dissimilitude relation associated with that of Figure 25.3. One thus has

†In this case, one may also call $\mu_{\underset{\sim}{\mathcal{R}}}(x, y)$ the codistance between x and y.

$\underset{\sim}{\mathcal{R}}$	A	B	C	D	E
A	1	0,8	0,7	1	0,9
B	0,8	1	0,7	0,8	0,8
C	0,7	0,7	1	0,7	0,7
D	1	0,8	0,7	1	0,9
E	0,9	0,8	0,7	0,9	1

Fig. 25.3

$\overline{\underset{\sim}{\mathcal{R}}}$	A	B	C	D	E
A	0	0,2	0,3	0	0,1
B	0,2	0	0,3	0,2	0,2
C	0,3	0,3	0	0,3	0,3
D	0	0,2	0,3	0	0,1
E	0,1	0,2	0,3	0,1	0

Fig. 25.4

(25.23)
$$d_{\underset{\sim}{\mathcal{R}}}(\text{A}, \text{B}) = 0{,}2 ,$$
$$d_{\underset{\sim}{\mathcal{R}}}(\text{A}, \text{C}) = 0{,}3 ,$$
$$d_{\underset{\sim}{\mathcal{R}}}(\text{A}, \text{D}) = 0 ,$$
$$\ldots \text{ etc.}$$

Example 2. Consider again example (20.3); one then has

(25.24)
$$\begin{aligned} d(x, y) &= 1 - e^{-k(y+1)} , \quad y < x , \quad k > 1 , \\ &= 0 \qquad\qquad\quad y = x , \\ &= 1 - e^{-k(x+1)} , \quad y > x , \quad k > 1 . \end{aligned}$$

26. RESEMBLANCE RELATIONS†

A relation $\underset{\sim}{\mathcal{R}}$ such that

(26.1) 1) $\forall (x, x) \in \text{E} \times \text{E} : \quad \mu_{\underset{\sim}{\mathcal{R}}}(x, x) = 1$ (reflexivity),

†In the theory of ordinary sets, the fact that this binary relation does not inherit the property of transitivity has provoked an almost total disinterest on the part of mathematicians in this property (an exception being C. Flament, Analyse des structures preferentielles intrasitives, *Proc. Sec. Intern. Conf. O.R.*, p. 150, 1960). Just as humerous cartoonists of all times, they have engaged in a very common error, that of believing that resemblance is transitive. Recall those caricatures that one has seen in which modified images appear one after the other, as King Louis-Philippe has been transformed in countenance or the emperor Napoleon III transformed into a mackerel.The talent of these humorists must not obscure their logical error. Writing, in the sense of the theory of ordinary sets, A resembles B, B resembles C, C resembles D, . . ., K resembles L, therefore A resembles L, indeed A = L, constitutes a sequence of deductions without validity. Often enough, moreover, false deductions of this nature are used by men in the spirit of making a joke or by political men to make the best of the stupidity of certain voters. The sophists have a particular habit of making us believe in the existence of transitivity where its existence may well be doubted.

But, with the theory of fuzzy subsets, one may measure several sorts of resemblance with the aid of the notion of distance in the transitive closure. The notion of similitude then constitutes the bridge existing between equivalence and resemblance.

(26.2) 2) $\forall (x, y) \in \mathbf{E} \times \mathbf{E}: \quad \mu_{\underset{\sim}{\mathcal{R}}}(x, y) = \mu_{\underset{\sim}{\mathcal{R}}}(y, x)$ symmetry ,

is called a *resemblance relation*.†

Example 1. Figure 26.1 gives an example of a resemblance relation.

Example 2. The relation (16.12),

(26.3) $\mu_{\underset{\sim}{\mathcal{R}}}(x, y) = e^{-k(x-y)^2}, \quad x, y \in \mathbf{N},$

is not, as we have seen, transitive; but it is reflexive and symmetric; it is a fuzzy relation of resemblance.

$\underset{\sim}{\mathcal{R}}$	A	B	C	D	E
A	1	0,1	0,8	0,2	0,3
B	0,1	1	0	0,3	1
C	0,8	0	1	0,7	0
D	0,2	0,3	0,7	1	0,6
E	0,3	1	0	0,6	1

Fig. 26.1

Min–max distance in a resemblance relation. If $\underset{\sim}{\mathcal{R}}$ is a resemblance relation,‡ then $\hat{\underset{\sim}{\mathcal{R}}}$, its transitive closure, is a similitude relation. One may then define the notion of min–max distance in $\underset{\sim}{\mathcal{R}}$ by that in $\hat{\underset{\sim}{\mathcal{R}}}$. Thus,

(26.4) $$d_{\underset{\sim}{\mathcal{R}}}(x, y) = 1 - \mu_{\hat{\underset{\sim}{\mathcal{R}}}}(x, y) .$$

Example 1. We reconsider the example of Figure 26.1. With the aid of the composition formula (17.3) we have calculated $\underset{\sim}{\mathcal{R}}$, the transitive closure of $\hat{\underset{\sim}{\mathcal{R}}}$.

$\hat{\underset{\sim}{\mathcal{R}}}$	A	B	C	D	E
A	1	0,6	0,8	0,7	0,6
B	0,6	1	0,6	0,6	1
C	0,8	0,6	1	0,7	0,6
D	0,7	0,6	0,7	1	0,6
E	0,6	1	0,6	0,6	1

Fig. 26.2

$\overline{\hat{\underset{\sim}{\mathcal{R}}}}$	A	B	C	D	E
A	0	0,4	0,2	0,3	0,4
B	0,4	0	0,4	0,4	0
C	0,2	0,4	0	0,3	0,4
D	0,3	0,4	0,3	0	0,4
E	0,4	0	0,4	0,4	0

Fig. 26.3

This result is presented in Figure 26.2. Next we have calculated $\overline{\hat{\underset{\sim}{\mathcal{R}}}}$ such that

†See the previous footnote.

‡The composition of $\underset{\sim}{\mathcal{R}}$ with $\underset{\sim}{\mathcal{R}}$ conserves reflexivity and symmetry.

$$\mu_{\bar{\hat{\underset{\sim}{\mathcal{R}}}}}(x\,,\,y) = 1 - \mu_{\hat{\underset{\sim}{\mathcal{R}}}}(x\,,\,y)\;; \tag{26.5}$$

the result is presented in Figure 26.3.

Finally one has

$$
\begin{aligned}
&d_{\bar{\hat{\underset{\sim}{\mathcal{R}}}}}(\mathrm{A}\,,\,\mathrm{B}) = 0{,}4\;,\\
&d_{\bar{\hat{\underset{\sim}{\mathcal{R}}}}}(\mathrm{A}\,,\,\mathrm{C}) = 0{,}2\;,\\
&\quad\vdots\\
&d_{\bar{\hat{\underset{\sim}{\mathcal{R}}}}}(\mathrm{B}\,,\,\mathrm{D}) = 0{,}4\;,\\
&\ldots\ \text{etc.}
\end{aligned} \tag{26.6}
$$

Example 2. Consider the resemblance relation $\underset{\sim}{\mathcal{R}}$ defined by

$$\mu_{\underset{\sim}{\mathcal{R}}}(x\,,\,y) = \frac{1}{1 + |x - y|}\,, \quad x \in \mathbf{N}\;,\; y \in \mathbf{N}\;. \tag{26.7}$$

This relation is represented in Figure 26.4.

$\underset{\sim}{\mathcal{R}}$	0	1	2	3	4	5	6	7	8	...
0	1	$\frac{1}{2}$	$\frac{1}{3}$	$\frac{1}{4}$	$\frac{1}{5}$	$\frac{1}{6}$	$\frac{1}{7}$	$\frac{1}{8}$	$\frac{1}{9}$	...
1	$\frac{1}{2}$	1	$\frac{1}{2}$	$\frac{1}{3}$	$\frac{1}{4}$	$\frac{1}{5}$	$\frac{1}{6}$	$\frac{1}{7}$	$\frac{1}{8}$	...
2	$\frac{1}{3}$	$\frac{1}{2}$	1	$\frac{1}{2}$	$\frac{1}{3}$	$\frac{1}{4}$	$\frac{1}{5}$	$\frac{1}{6}$	$\frac{1}{7}$	...
3	$\frac{1}{4}$	$\frac{1}{3}$	$\frac{1}{2}$	1	$\frac{1}{2}$	$\frac{1}{3}$	$\frac{1}{4}$	$\frac{1}{5}$	$\frac{1}{6}$	...
4	$\frac{1}{5}$	$\frac{1}{4}$	$\frac{1}{3}$	$\frac{1}{2}$	1	$\frac{1}{2}$	$\frac{1}{3}$	$\frac{1}{4}$	$\frac{1}{5}$	...
5	$\frac{1}{6}$	$\frac{1}{5}$	$\frac{1}{4}$	$\frac{1}{3}$	$\frac{1}{2}$	1	$\frac{1}{2}$	$\frac{1}{3}$	$\frac{1}{4}$	...
6	$\frac{1}{7}$	$\frac{1}{6}$	$\frac{1}{5}$	$\frac{1}{4}$	$\frac{1}{3}$	$\frac{1}{2}$	1	$\frac{1}{2}$	$\frac{1}{3}$	...
7	$\frac{1}{8}$	$\frac{1}{7}$	$\frac{1}{6}$	$\frac{1}{5}$	$\frac{1}{4}$	$\frac{1}{3}$	$\frac{1}{2}$	1	$\frac{1}{2}$	...
8	$\frac{1}{9}$	$\frac{1}{8}$	$\frac{1}{7}$	$\frac{1}{6}$	$\frac{1}{5}$	$\frac{1}{4}$	$\frac{1}{3}$	$\frac{1}{2}$	1	...
⋮	⋮	⋮	⋮	⋮	⋮	⋮	⋮	⋮	⋮	⋱

Fig. 26.4

$\hat{\underset{\sim}{\mathcal{R}}}$	0	1	2	3	4	5	6	7	8	...
0	1	$\frac{1}{2}$	$\frac{1}{2}$	$\frac{1}{2}$	$\frac{1}{2}$	$\frac{1}{2}$	$\frac{1}{2}$	$\frac{1}{2}$	$\frac{1}{2}$	...
1	$\frac{1}{2}$	1	$\frac{1}{2}$	$\frac{1}{2}$	$\frac{1}{2}$	$\frac{1}{2}$	$\frac{1}{2}$	$\frac{1}{2}$	$\frac{1}{2}$	...
2	$\frac{1}{2}$	$\frac{1}{2}$	1	$\frac{1}{2}$	$\frac{1}{2}$	$\frac{1}{2}$	$\frac{1}{2}$	$\frac{1}{2}$	$\frac{1}{2}$	...
3	$\frac{1}{2}$	$\frac{1}{2}$	$\frac{1}{2}$	1	$\frac{1}{2}$	$\frac{1}{2}$	$\frac{1}{2}$	$\frac{1}{2}$	$\frac{1}{2}$	...
4	$\frac{1}{2}$	$\frac{1}{2}$	$\frac{1}{2}$	$\frac{1}{2}$	1	$\frac{1}{2}$	$\frac{1}{2}$	$\frac{1}{2}$	$\frac{1}{2}$	...
5	$\frac{1}{2}$	$\frac{1}{2}$	$\frac{1}{2}$	$\frac{1}{2}$	$\frac{1}{2}$	1	$\frac{1}{2}$	$\frac{1}{2}$	$\frac{1}{2}$	...
6	$\frac{1}{2}$	$\frac{1}{2}$	$\frac{1}{2}$	$\frac{1}{2}$	$\frac{1}{2}$	$\frac{1}{2}$	1	$\frac{1}{2}$	$\frac{1}{2}$	...
7	$\frac{1}{2}$	$\frac{1}{2}$	$\frac{1}{2}$	$\frac{1}{2}$	$\frac{1}{2}$	$\frac{1}{2}$	$\frac{1}{2}$	1	$\frac{1}{2}$	...
8	$\frac{1}{2}$	$\frac{1}{2}$	$\frac{1}{2}$	$\frac{1}{2}$	$\frac{1}{2}$	$\frac{1}{2}$	$\frac{1}{2}$	$\frac{1}{2}$	1	...
⋮	⋮	⋮	⋮	⋮	⋮	⋮	⋮	⋮	⋮	⋱

Fig. 26.5

Calculating†

(26.8) $$\hat{\underset{\sim}{\mathcal{R}}} = \underset{\sim}{\mathcal{R}} \cup \underset{\sim}{\mathcal{R}}^2 \cup \underset{\sim}{\mathcal{R}}^3 \cup \ldots ,$$

one obtains the relation given in Figure 26.5. One then has

(26.9) $$\begin{aligned} \mu_{\hat{\underset{\sim}{\mathcal{R}}}}(x, y) &= \tfrac{1}{2} \quad x \neq y \\ &= 1 \quad x = y . \end{aligned}$$

Hence, in conclusion

(26.10) $$\begin{aligned} d_{\underset{\sim}{\mathcal{R}}}(x, y) &= \tfrac{1}{2} \quad x \neq y \\ &= 0 \quad x = y . \end{aligned}$$

We note that if one reconsiders (26.7) but this time with

(26.11) $$x \in \mathbf{R}^+ \quad \text{and} \quad y \in \mathbf{R}^+ ,$$

one would find

(26.12) $$d_{\underset{\sim}{\mathcal{R}}}(x, y) = 0 ,$$

for all x and all y. This is not paradoxical since the distance between x and $y = x + dx$ is infinitely small and of the same order as dx. Of course, if one would give the distance some other significance than the min–max distance considered here, it would be proper to review this conclusion.

Max–product transitive closure for a resemblance relation. Let $\underset{\sim}{\mathcal{R}}$ be a resemblance relation. In certain cases it is preferable to measure the distance existing between elements with the aid of the max–product operation instead of the max–min operation, that is, to use (13.19) instead of (13.2); thus

(26.13) $$\mu_{\underset{\sim}{\mathcal{R}}^2}(x, z) = \bigvee_y [\mu_{\underset{\sim}{\mathcal{R}}}(x, y) \cdot \mu_{\underset{\sim}{\mathcal{R}}}(y, z)] .$$

The max–product transitive closure of a relation is

(26.14) $$\dot{\hat{\underset{\sim}{\mathcal{R}}}} = \underset{\sim}{\mathcal{R}} \cup \underset{\sim}{\mathcal{R}}^{\dot{2}} \cup \underset{\sim}{\mathcal{R}}^{\dot{3}} \cup \ldots$$

where

(26.15) $$\underset{\sim}{\mathcal{R}}^{\dot{k}} = \underbrace{\underset{\sim}{\mathcal{R}} \cdot \underset{\sim}{\mathcal{R}} \cdot \ldots \cdot \underset{\sim}{\mathcal{R}}}_{k \text{ times}} , \qquad k = 1, 2, 3, \ldots .$$

The points on $\dot{\hat{}}$ and $\dot{k}$ remind us that we have a max–product composition.

†In order to obtain $\hat{\underset{\sim}{\mathcal{R}}}$ it is necessary to take $\underset{\sim}{\mathcal{R}} \cup \underset{\sim}{\mathcal{R}}^2 \cup \underset{\sim}{\mathcal{R}}^3 \cup \cdots$; it is clear that all the elements of $\hat{\underset{\sim}{\mathcal{R}}}$ tend toward 1/2, except those on the principal diagonal which remain equal to 1.

We see an example. Recall that for Figure 26.1, we have calculated $\hat{\underset{\sim}{\mathcal{R}}}$ and $\bar{\hat{\underset{\sim}{\mathcal{R}}}}$ in Figures 26.2 and 26.3. In Figure 26.5 one may observe how we have calculated $\underset{\sim}{\mathcal{R}}^{\dot{2}}$, $\underset{\sim}{\mathcal{R}}^{\dot{3}}$, $\underset{\sim}{\mathcal{R}}^{\dot{4}}$, $\underset{\sim}{\mathcal{R}}^{\dot{5}}$, $\dot{\hat{\underset{\sim}{\mathcal{R}}}}$.

$\underset{\sim}{\mathcal{R}}$	A	B	C	D	E
A	1	0,1	0,8	0,2	0,3
B	0,1	1	0	0,3	1
C	0,8	0	1	0,7	0
D	0,2	0,3	0,7	1	0,6
E	0,3	1	0	0,6	1

,

$\underset{\sim}{\mathcal{R}}^{\dot{2}}$	A	B	C	D	E
A	1	0,3	0,8	0,56	0,3
B	0,3	1	0,21	0,6	1
C	0,8	0,21	1	0,7	0,42
D	0,56	0,6	0,7	1	0,6
E	0,3	1	0,42	0,6	1

,

$\underset{\sim}{\mathcal{R}}^{\dot{3}}$	A	B	C	D	E
A	1	0,3	0,8	0,56	0,336
B	0,3	1	0,42	0,6	1
C	0,8	0,42	1	0,7	0,42
D	0,56	0,6	0,7	1	0,6
E	0,336	1	0,42	0,6	1

,

$\underset{\sim}{\mathcal{R}}^{\dot{4}}$	A	B	C	D	E
A	1	0,336	0,8	0,56	0,336
B	0,336	1	0,42	0,6	1
C	0,8	0,42	1	0,7	0,42
D	0,56	0,6	0,7	1	0,6
E	0,336	1	0,42	0,6	1

,

$\underset{\sim}{\mathcal{R}}^{\dot{5}}$	A	B	C	D	E
A	1	0,336	0,8	0,56	0,336
B	0,336	1	0,42	0,6	1
C	0,8	0,42	1	0,7	0,42
D	0,56	0,6	0,7	1	0,6
E	0,336	1	0,42	0,6	1

,

$\dot{\hat{\underset{\sim}{\mathcal{R}}}}$	A	B	C	D	E
A	1	0,336	0,8	0,56	0,336
B	0,336	1	0,42	0,6	1
C	0,8	0,42	1	0,7	0,42
D	0,56	0,6	0,7	1	0,6
E	0,336	1	0,42	0,6	1

,

Fig. 26.6

Remarks on the calculation of $\dot{\hat{\underset{\sim}{\mathcal{R}}}}$. We have seen in (18.19) that

(26.16) $$\underset{\sim}{\mathcal{R}} \circ \underset{\sim}{\mathcal{R}} \subset \underset{\sim}{\mathcal{R}} \quad \Rightarrow \quad \underset{\sim}{\mathcal{R}} \cdot \underset{\sim}{\mathcal{R}} \subset \underset{\sim}{\mathcal{R}}$$

without having the reverse be true.

Theorem II of Section 17, that is, (17.13), is also verified for the max–product. With respect to a particular k

(26.17) $$\underset{\sim}{\mathcal{R}}^{\widetilde{\dot{k+1}}} = \underset{\sim}{\mathcal{R}}^{\dot{k}} \quad \Rightarrow \quad \dot{\hat{\underset{\sim}{\mathcal{R}}}} = \underset{\sim}{\mathcal{R}} \cup \underset{\sim}{\mathcal{R}}^{\dot{2}} \cup \ldots \cup \underset{\sim}{\mathcal{R}}^{\dot{k}} .$$

And in the case where $\underset{\sim}{\mathcal{R}}$ is a resemblance relation, one has likewise

(26.18) $$\underset{\sim}{\mathcal{R}}^{\widetilde{\dot{k+1}}} = \underset{\sim}{\mathcal{R}}^{\dot{k}} \quad \Rightarrow \quad \dot{\hat{\underset{\sim}{\mathcal{R}}}} = \underset{\sim}{\mathcal{R}}^{\dot{k}} .$$

Min–sum distance in a resemblance relation. We shall call

$$(26.19) \qquad \gamma_{\underset{\sim}{\mathcal{R}}}(x\,,\,y) = \mu_{\overline{\hat{\underset{\sim}{\mathcal{R}}}}}(x\,,\,y)\,,$$

the *min–sum distance*; but first we must determine whether the distance axioms (25.17)–(25.20) are satisfied.

(25.17) is verified a priori since $\qquad \mu_{\hat{\underset{\sim}{\mathcal{R}}}}(x\,,\,y) \in [0\,,\,1]\,,$

(25.18) is verified a priori since the relation $\dot{\hat{\underset{\sim}{\mathcal{R}}}}$ is symmetric.

(25.20) is verified a priori since the relation $\dot{\hat{\underset{\sim}{\mathcal{R}}}}$ is reflexive, which entails $\mu_{\hat{\underset{\sim}{\mathcal{R}}}}(x, x) = 0$.

It remains to show that one indeed has property (25.19). We shall operate as for (25.5)–(25.9).

One then has

$$(26.20) \qquad \mu_{\dot{\hat{\underset{\sim}{\mathcal{R}}}}}(x\,,\,z) \geqslant \bigvee_{y} \left[\mu_{\dot{\hat{\underset{\sim}{\mathcal{R}}}}}(x\,,\,y) \cdot \mu_{\dot{\hat{\underset{\sim}{\mathcal{R}}}}}(y\,,\,z)\right]\,,$$

and from there, following (8.23),

$$(26.21) \qquad 1 - \mu_{\overline{\hat{\underset{\sim}{\mathcal{R}}}}}(x\,,\,z) \geqslant \bigvee_{y} \left[[1 - \mu_{\overline{\hat{\underset{\sim}{\mathcal{R}}}}}(x\,,\,y)] \cdot [1 - \mu_{\overline{\hat{\underset{\sim}{\mathcal{R}}}}}(y\,,\,z)]\right]$$

$$\geqslant \bigvee_{y} \left[1 - \mu_{\overline{\hat{\underset{\sim}{\mathcal{R}}}}}(x\,,\,y) - \mu_{\overline{\hat{\underset{\sim}{\mathcal{R}}}}}(y\,,\,z) + \mu_{\overline{\hat{\underset{\sim}{\mathcal{R}}}}}(x\,,\,y) \cdot \mu_{\overline{\hat{\underset{\sim}{\mathcal{R}}}}}(y\,,\,z)\right]\,.$$

This gives

$$(26.22) \qquad \mu_{\overline{\hat{\underset{\sim}{\mathcal{R}}}}}(x\,,\,z) \leqslant \bigwedge_{y} \left[\mu_{\overline{\hat{\underset{\sim}{\mathcal{R}}}}}(x\,,\,y) + \mu_{\overline{\hat{\underset{\sim}{\mathcal{R}}}}}(y\,,\,z) - \mu_{\overline{\hat{\underset{\sim}{\mathcal{R}}}}}(x\,,\,y) \cdot \mu_{\overline{\hat{\underset{\sim}{\mathcal{R}}}}}(y\,,\,z)\right]\,.$$

That is

$$(26.23) \qquad \mu_{\overline{\hat{\underset{\sim}{\mathcal{R}}}}}(x\,,\,z) \leqslant \bigwedge_{y} \left[\mu_{\overline{\hat{\underset{\sim}{\mathcal{R}}}}}(x\,,\,y) \,\hat{+}\, \mu_{\overline{\hat{\underset{\sim}{\mathcal{R}}}}}(y\,,\,z)\right]\,,$$

where $\hat{+}$ is the algebraic sum defined by (12.42). Then we certainly have property (25.19) for the min–sum operation.

Example 1. Consider again the example of Figure 26.1. In Figure 26.6 we have calculated the max–product transitive closure, that is, $\dot{\hat{\underset{\sim}{\mathcal{R}}}}$ The min–sum distances will then be given by the relation $\overline{\dot{\hat{\underset{\sim}{\mathcal{R}}}}}$ for which one has

$$(26.24) \qquad \gamma(x\,,\,y) = \mu_{\overline{\hat{\underset{\sim}{\mathcal{R}}}}}(x\,,\,y) = 1 - \mu_{\dot{\hat{\underset{\sim}{\mathcal{R}}}}}(x\,,\,y)\,.$$

Figure 26.7 gives the min–sum distances between the various elements. Thus,

$$\gamma(\mathrm{C}\,,\,\mathrm{E}) = 0{,}58\,.$$

$$\gamma(\mathrm{D}\,,\,\mathrm{B}) = 0{,}4\,.$$

$\overline{\hat{\underset{\sim}{\mathcal{R}}}}$	A	B	C	D	E
A	0	0,664	0,2	0,44	0,664
B	0,664	0	0,58	0,4	0
C	0,2	0,58	0	0,3	0,58
D	0,44	0,4	0,3	0	0,4
E	0,664	0	0,58	0,4	0

Fig. 26.7

$\overline{\hat{\underset{\sim}{\mathcal{R}}}}$	0	1	2	3	4	5	6	7	8	...
0	0	$\frac{1}{2}$	$\frac{2}{3}$	$\frac{3}{4}$	$\frac{4}{5}$	$\frac{5}{6}$	$\frac{6}{7}$	$\frac{7}{8}$	$\frac{8}{9}$	...
1	$\frac{1}{2}$	0	$\frac{1}{2}$	$\frac{2}{3}$	$\frac{3}{4}$	$\frac{4}{5}$	$\frac{5}{6}$	$\frac{6}{7}$	$\frac{7}{8}$	...
2	$\frac{2}{3}$	$\frac{1}{2}$	0	$\frac{1}{2}$	$\frac{2}{3}$	$\frac{3}{4}$	$\frac{4}{5}$	$\frac{5}{6}$	$\frac{6}{7}$	...
3	$\frac{3}{4}$	$\frac{2}{3}$	$\frac{1}{2}$	0	$\frac{1}{2}$	$\frac{2}{3}$	$\frac{3}{4}$	$\frac{4}{5}$	$\frac{5}{6}$	...
4	$\frac{4}{5}$	$\frac{3}{4}$	$\frac{2}{3}$	$\frac{1}{2}$	0	$\frac{1}{2}$	$\frac{2}{3}$	$\frac{3}{4}$	$\frac{4}{5}$	...
5	$\frac{5}{6}$	$\frac{4}{5}$	$\frac{3}{4}$	$\frac{2}{3}$	$\frac{1}{2}$	0	$\frac{1}{2}$	$\frac{2}{3}$	$\frac{3}{4}$	...
6	$\frac{6}{7}$	$\frac{5}{6}$	$\frac{4}{5}$	$\frac{3}{4}$	$\frac{2}{3}$	$\frac{1}{2}$	0	$\frac{1}{2}$	$\frac{2}{3}$	...
7	$\frac{7}{8}$	$\frac{6}{7}$	$\frac{5}{6}$	$\frac{4}{5}$	$\frac{3}{4}$	$\frac{2}{3}$	$\frac{1}{2}$	0	$\frac{1}{2}$	...
8	$\frac{8}{9}$	$\frac{7}{8}$	$\frac{6}{7}$	$\frac{5}{6}$	$\frac{4}{5}$	$\frac{3}{4}$	$\frac{2}{3}$	$\frac{1}{2}$	0	...
⋮	⋮	⋮	⋮	⋮	⋮	⋮	⋮	⋮	⋮	⋱

Fig. 26.8

Example 2. We take up again the example of Figure 26.5. A max–product composition shows immediately that

(26.25) $$\hat{\underset{\sim}{\mathcal{R}}}^{\cdot} = \underset{\sim}{\mathcal{R}} \, .$$

The relation $\overline{\hat{\underset{\sim}{\mathcal{R}}}^{\cdot}}$ is given in Figure 26.8.

One sees that

(26.26) $$\gamma(n_1\,, n_2) = \frac{|n_2 - n_1|}{|n_2 - n_1| + 1} \, ,$$

and that, as a consequence,

(26.27) $$\lim_{|n_2 - n_1| \to \infty} \gamma(n_1\,, n_2) = 1 \, .$$

Remark. It appears that $\gamma(x, y)$ gives a better practical idea of distance than $d(x, y)$; this may be very important for all concerned with problems of resemblance, hence the interest that we have shown in the min–sum distance. But, as we shall go on to see in Figure 27.10 of the next section, decomposition into ordinary partial graphs is no longer possible.

Theorem I. Let $\underset{\sim}{\mathcal{R}}$ be a resemblance relation. Then one always has

(26.28) $$\overline{\hat{\underset{\sim}{\mathcal{R}}}} \subset \overline{\hat{\underset{\sim}{\mathcal{R}}}^{\cdot}} \, .$$

that is,

(26.29) $$\forall (x\,, y) : \quad d(x\,, y) \leqslant \gamma(x\,, y) \, .$$

Proof. On account of max–min transitivity one has

(26.30) $$\mu_{\underset{\sim}{\mathcal{R}}}(x\,, z) \geqslant \bigvee_y \, [\mu_{\underset{\sim}{\mathcal{R}}}(x\,, y) \wedge \mu_{\underset{\sim}{\mathcal{R}}}(y\,, z)] \, .$$

From max–product transitivity one has

(26.31) $$\mu_{\underset{\sim}{\mathcal{R}}}(x\,, z) \geqslant \bigvee_y \, [\mu_{\underset{\sim}{\mathcal{R}}}(x\,, y) \cdot \mu_{\underset{\sim}{\mathcal{R}}}(y, z)] .$$

But, according to (18.18),

(26.32) $$\mu_{\underset{\sim}{\mathcal{R}}}(x\,, y) \wedge \mu_{\underset{\sim}{\mathcal{R}}}(y\,, z) \geqslant \mu_{\underset{\sim}{\mathcal{R}}}(x\,, y) \cdot \mu_{\underset{\sim}{\mathcal{R}}}\,(y, z) \, ,$$

which implies

(26.33) $$\underbrace{\bigvee_y \, [\mu_{\underset{\sim}{\mathcal{R}}}(x\,, y) \wedge \mu_{\underset{\sim}{\mathcal{R}}}(y\,, z)]}_{\text{max–min}} \geqslant \underbrace{\bigvee_y \, [\mu_{\underset{\sim}{\mathcal{R}}}(x\,, y) \cdot \mu_{\underset{\sim}{\mathcal{R}}}(y\,, z)]}_{\text{max–product}} \, ,$$

that is,

(26.34) $$\underset{\sim}{\mathcal{R}} \cdot \underset{\sim}{\mathcal{R}} \subset \underset{\sim}{\mathcal{R}} \circ \underset{\sim}{\mathcal{R}}$$

where, we recall, $\cdot$ indicates max–product composition and $\circ$ max–min composition.

From here

$$\dot{\hat{\underset{\sim}{\mathcal{R}}}} \subset \hat{\underset{\sim}{\mathcal{R}}}\,, \tag{26.35}$$

and thus

$$\bar{\hat{\underset{\sim}{\mathcal{R}}}} \subset \bar{\dot{\hat{\underset{\sim}{\mathcal{R}}}}}\,. \tag{26.36}$$

Dissemblance relation. A relation $\underset{\sim}{\mathcal{R}}$ such that

(26.37) 1) $\forall (x, x) \in E \times E: \quad \mu_{\underset{\sim}{\mathcal{R}}}(x, x) = 0$ antireflexivity.

(26.38) 2) $\forall (x, y) \in E \times E: \quad \mu_{\underset{\sim}{\mathcal{R}}}(x, y) = \mu_{\underset{\sim}{\mathcal{R}}}(y, x)$ symmetry.

is called a *dissemblance relation*. Figure 26.9 gives an example.

We consider some evident properties. If $\underset{\sim}{\mathcal{R}}$ is a resemblance relation, $\bar{\underset{\sim}{\mathcal{R}}}$ is a dissemblance relation, and vice versa.

Theorem II. If $\hat{\underset{\sim}{\mathcal{R}}}$ is the max–min transitive closure of the resemblance relation $\underset{\sim}{\mathcal{R}}$, then $\bar{\hat{\underset{\sim}{\mathcal{R}}}}$ is the min–max transitive closure of the corresponding dissemblance relation.

$\underset{\sim}{\mathcal{R}}$	A	B	C	D	E
A	0	0,3	0,9	1	0,2
B	0,3	0	0,4	0,1	0
C	0,9	0,4	0	0,8	0,1
D	1	0,1	0,8	0	1
E	0,2	0	0,1	1	0

Fig. 26.9

Proof. The max–min transitive closure is expressed by (17.8) and (17.3); thus,

$$\hat{\underset{\sim}{\mathcal{R}}} = \underset{\sim}{\mathcal{R}} \cup \underset{\sim}{\mathcal{R}}^2 \cup \underset{\sim}{\mathcal{R}}^3 \cup \ldots\,. \tag{26.39}$$

and

$$\mu_{\underset{\sim}{\mathcal{R}} \circ \underset{\sim}{\mathcal{R}}}(x, z) = \bigvee_y \left[\mu_{\underset{\sim}{\mathcal{R}}}(x, y) \wedge \mu_{\underset{\sim}{\mathcal{R}}}(y, z)\right] \tag{26.40}$$

The min–max transitive closure will then be expressed by†

$$\check{\underset{\sim}{\mathcal{R}}} = \underset{\sim}{\mathcal{R}} \cap (\underset{\sim}{\mathcal{R}} * \underset{\sim}{\mathcal{R}}) \cap (\underset{\sim}{\mathcal{R}} * \underset{\sim}{\mathcal{R}} * \underset{\sim}{\mathcal{R}}) \cap \ldots\,.(^1) \tag{26.41}$$

and

$$\mu_{\underset{\sim}{\mathcal{R}} * \underset{\sim}{\mathcal{R}}}(x, z) = \bigwedge_y \left[\mu_{\underset{\sim}{\mathcal{R}}}(x, y) \vee \mu_{\underset{\sim}{\mathcal{R}}}(y, z)\right]\,. \tag{26.42}$$

Let $\underset{\sim}{\mathcal{R}}$ be a resemblance relation; $\hat{\underset{\sim}{\mathcal{R}}}$ is a similitude relation; $\bar{\underset{\sim}{\mathcal{R}}}$ is a dissemblance relation; and $\check{\underset{\sim}{\mathcal{R}}}$ is a dissimilitude relation. We show that

$$\bar{\hat{\underset{\sim}{\mathcal{R}}}} = \check{\bar{\underset{\sim}{\mathcal{R}}}}\,. \tag{26.43}$$

We have already shown in (25.4)–(25.8) that if $\underset{\sim}{\mathcal{R}}$ is max–min transitive, then $\bar{\underset{\sim}{\mathcal{R}}}$ is min–max transitive.

†One may denote $\underset{\sim}{\mathcal{R}} * \underset{\sim}{\mathcal{R}} = \underset{\sim}{\mathcal{R}}^2$ if there is no danger of confusion with the max–min operation and $\underset{\sim}{\mathcal{R}} * \underset{\sim}{\mathcal{R}} * \ldots * \underset{\sim}{\mathcal{R}} = \underset{\sim}{\mathcal{R}}^n$.

We show now that

$$\overline{\underset{\sim}{\mathcal{R}} \circ \underset{\sim}{\mathcal{R}}} = \overline{\underset{\sim}{\mathcal{R}}} * \overline{\underset{\sim}{\mathcal{R}}} \,. \tag{26.44}$$

max-min min-max

In order to verify this, one proceeds as we did in (25.4)–(25.8):

$$\mu_{\underset{\sim}{\mathcal{R}} \circ \underset{\sim}{\mathcal{R}}}(x, z) = \bigvee_y [\mu_{\underset{\sim}{\mathcal{R}}}(x, y) \wedge \mu_{\underset{\sim}{\mathcal{R}}}(y, z)] \tag{26.45}$$

$$\begin{aligned} \mu_{\overline{\underset{\sim}{\mathcal{R}} \circ \underset{\sim}{\mathcal{R}}}}(x, z) &= 1 - \mu_{\underset{\sim}{\mathcal{R}} \circ \underset{\sim}{\mathcal{R}}}(x, z) \\ &= 1 - \bigvee_y [\mu_{\underset{\sim}{\mathcal{R}}}(x, y) \wedge \mu_{\underset{\sim}{\mathcal{R}}}(y, z)] \\ &= \bigwedge_y [\mu_{\overline{\underset{\sim}{\mathcal{R}}}}(x, y) \vee \mu_{\overline{\underset{\sim}{\mathcal{R}}}}(y, z) \\ &= \mu_{\overline{\underset{\sim}{\mathcal{R}}} * \overline{\underset{\sim}{\mathcal{R}}}}(x, z) \,. \end{aligned} \tag{26.46}$$

This proves (26.44).

Now we write

$$\begin{aligned} \bar{\hat{\underset{\sim}{\mathcal{R}}}} &= \overline{\underset{\sim}{\mathcal{R}} \cup \underset{\sim}{\mathcal{R}}^2 \cup \underset{\sim}{\mathcal{R}}^3 \cup \ldots} \\ &= \overline{\underset{\sim}{\mathcal{R}} \cup (\underset{\sim}{\mathcal{R}} \circ \underset{\sim}{\mathcal{R}}) \cup (\underset{\sim}{\mathcal{R}} \circ \underset{\sim}{\mathcal{R}} \circ \underset{\sim}{\mathcal{R}}) \cup \ldots} \\ &= \overline{\underset{\sim}{\mathcal{R}}} \cap \overline{\underset{\sim}{\mathcal{R}} \circ \underset{\sim}{\mathcal{R}}} \cap \overline{\underset{\sim}{\mathcal{R}} \circ \underset{\sim}{\mathcal{R}} \circ \underset{\sim}{\mathcal{R}}} \cap \ldots \end{aligned} \tag{26.47}$$

(applying De Morgan's theorem)

$$= \overline{\underset{\sim}{\mathcal{R}}} \cap \overline{\underset{\sim}{\mathcal{R}}} * \overline{\underset{\sim}{\mathcal{R}}} \cap \overline{\underset{\sim}{\mathcal{R}}} * \overline{\underset{\sim}{\mathcal{R}}} * \overline{\underset{\sim}{\mathcal{R}}} \cap \ldots$$

[according to (26.44)]

$$= \bar{\check{\underset{\sim}{\mathcal{R}}}} \,.$$

We shall see an example. We take again the resemblance relation given by Figure 26.1, whose corresponding similitude relation has been given in Figure 26.2 and whose matrix of distances in Figure 26.3. We meet these relations again in the calculations that end in $\bar{\check{\underset{\sim}{\mathcal{R}}}}$ in Figures 26.10d–h.

$\underset{\sim}{\mathcal{R}}$	A	B	C	D	E
A	1	0,1	0,8	0,2	0,3
B	0,1	1	0	0,3	1
C	0,8	0	1	0,7	0
D	0,2	0,3	0,7	1	0,6
E	0,3	1	0	0,6	1

(a)

$\hat{\underset{\sim}{\mathcal{R}}}$	A	B	C	D	E
A	1	0,6	0,8	0,7	0,6
B	0,6	1	0,6	0,6	1
C	0,8	0,6	1	0,7	0,6
D	0,7	0,6	0,7	1	0,6
E	0,6	1	0,6	0,6	1

(b)

$\bar{\hat{\underset{\sim}{\mathcal{R}}}}$	A	B	C	D	E
A	0	0,4	0,2	0,3	0,4
B	0,4	0	0,4	0,4	0
C	0,2	0,4	0	0,3	0,4
D	0,3	0,4	0,3	0	0,4
E	0,4	0	0,4	0,4	0

(c)

$\underset{\sim}{\bar{\mathcal{R}}}$	A	B	C	D	E
A	0	0,9	0,2	0,8	0,7
B	0,9	0	1	0,7	0
C	0,2	1	0	0,3	1
D	0,8	0,7	0,3	0	0,4
E	0,7	0	1	0,4	0

(d)

$\underset{\sim}{\bar{\mathcal{R}}} * \underset{\sim}{\bar{\mathcal{R}}}$	A	B	C	D	E
A	0	0,7	0,2	0,3	0,7
B	0,7	0	0,7	0,4	0
C	0,2	0,7	0	0,3	0,4
D	0,3	0,4	0,3	0	0,4
E	0,7	0	0,4	0,4	0

$\underset{\sim}{\bar{\mathcal{R}}} * \underset{\sim}{\bar{\mathcal{R}}}$ through min–max composition

(e)

$\underset{\sim}{\bar{\mathcal{R}}}^3_*$	A	B	C	D	E
A	0	0,7	0,2	0,3	0,4
B	0,7	0	0,4	0,4	0
C	0,2	0,4	0	0,3	0,4
D	0,3	0,4	0,3	0	0,4
E	0,4	0	0,4	0,4	0

min–max composition

(f)

$\underset{\sim}{\bar{\mathcal{R}}}^4_*$	A	B	C	D	E
A	0	0,4	0,2	0,3	0,4
B	0,4	0	0,4	0,4	0
C	0,2	0,4	0	0,3	0,4
D	0,3	0,4	0,3	0	0,4
E	0,4	0	0,4	0,4	0

min–max composition

(g)

$\underset{\sim}{\check{\bar{\mathcal{R}}}}$	A	B	C	D	E
A	0	0,4	0,2	0,3	0,4
B	0,4	0	0,4	0,4	0
C	0,2	0,4	0	0,3	0,4
D	0,3	0,4	0,3	0	0,4
E	0,4	0	0,4	0,4	0

intersection of min–max compositions

(h)

Fig. 26.10

Theorem II may be extended to the case of any relation, without imposing that it be a resemblance relation; the proof remains valid. Thus we may announce a more general theorem.

Theorem III. ‡ Let $\underset{\sim}{\hat{\mathcal{R}}}$ be the max–min transitive closure of any fuzzy relation $\underset{\sim}{\mathcal{R}} \subset \mathbf{E} \times \mathbf{E}$ whatever, and let $\underset{\sim}{\check{\bar{\mathcal{R}}}}$ be the min–max transitive closure of $\underset{\sim}{\bar{\mathcal{R}}}$. Then

$$\bar{\hat{\underset{\sim}{\mathcal{R}}}} = \check{\bar{\underset{\sim}{\mathcal{R}}}} . \tag{26.48}$$

This may also be expressed by writing: One may permute the order of the operations ˆ and ¯, but ˆ becomes ˇ (or vice versa) in the permutation.

†We might have introduced this theorem earlier, in Section 17, but with a didactic aim (not to overload any section, operating progressively), we report this useful and important theorem in Section 26, where we have a true need for the notion of distance.

With respect to these, the reader may seek other interesting properties concerning the max–min and min–max transitive closures, which one may characterize as duals without being the object of reproach for using that word.

27. VARIOUS PROPERTIES CONCERNING SIMILITUDE AND RESEMBLANCE

Theorem of decomposition for a similitude relation. Let $\underset{\sim}{\mathcal{R}}$ be a similitude relation in $\mathbf{E} \times \mathbf{E}$. Then $\underset{\sim}{\mathcal{R}}$ may be decomposed in the form

(27.1) $$\underset{\sim}{\mathcal{R}} = \bigvee_{\alpha} \alpha \cdot \mathcal{R}_\alpha \qquad 0 < \alpha \leqslant 1$$

$$\text{with } \alpha_1 > \alpha_2 \Rightarrow \mathcal{R}_2 \supset \mathcal{R}_1,$$

where the $\mathcal{R}_\alpha$ are equivalence relations in the sense of ordinary set theory and $\alpha . \mathcal{R}_\alpha$ indicates that all the elements of the ordinary relation $\mathcal{R}_\alpha$ are multiplied by α.

Proof. First, $\mu_{\underset{\sim}{\mathcal{R}}}(x, x) = 1$; it follows that $(x, x) \in \mathcal{R}_\alpha$ for $\alpha \in [0, 1]$; and thus $\mathcal{R}_\alpha$ has the property of reflexivity.

Then, letting $(x, y) \in \mathcal{R}_\alpha$ $\alpha \in [0, 1]$, this implies that $\mu_{\underset{\sim}{\mathcal{R}}}(x, y) \geqslant \alpha$ and, by the symmetry of $\underset{\sim}{\mathcal{R}}, \mu_{\underset{\sim}{\mathcal{R}}}(y, x) \geqslant \alpha$. Then $\mathcal{R}_\alpha$ has the property of symmetry.

Finally, for all $\alpha \in [0, 1]$, suppose that $(x, y) \in \mathcal{R}_\alpha$ and $(y, z) \in \mathcal{R}_\alpha$; then $\mu_{\underset{\sim}{\mathcal{R}}}(x, y) \geqslant \alpha$ and $\mu_{\underset{\sim}{\mathcal{R}}}(y, z) \geqslant \alpha$; then by transitivity, $\mu_{\underset{\sim}{\mathcal{R}}}(x, z) \geqslant \alpha$ and also $\mathcal{R}_\alpha$ is transitive.

Then $\mathcal{R}_\alpha$, being reflexive, symmetric, and transitive, is an equivalence relation.

The converse theorem is equally true.

Converse. $\mathcal{R}_1$ is nonempty, $(x, x) \in \mathcal{R}_1$, and also

(27.2) $$\mu_{\underset{\sim}{\mathcal{R}}}(x, x) = 1 \quad , \quad \forall x \in \mathbf{E},$$

then $\underset{\sim}{\mathcal{R}}$ is a reflexive fuzzy relation.

On the other hand, referring to (13.31), one may write

(27.3) $$\forall (x, y) \in \mathbf{E} \times \mathbf{E} : \quad \mu_{\underset{\sim}{\mathcal{R}}}(x, y) = \bigvee_{\alpha} \alpha \cdot \mu_{\mathcal{R}_\alpha}(x, y).$$

It is evident that the symmetry of each $\mathcal{R}_\alpha$ implies the symmetry of $\underset{\sim}{\mathcal{R}}$.

Finally, let

(27.4) $$\mu_{\underset{\sim}{\mathcal{R}}}(x, y) = \alpha \quad \text{and} \quad \mu_{\underset{\sim}{\mathcal{R}}}(y, z) = \beta ;$$

then

(27.5) $$(x, y) \in \mathcal{R}_{\alpha \wedge \beta} \quad \text{and} \quad (y, z) \in \mathcal{R}_{\alpha \wedge \beta} .$$

As a consequence

(27.6) $$(x, z) \in \mathcal{R}_{\alpha \wedge \beta} ;$$

because $\mathcal{R}_{\alpha > \beta}$ is transitive.

It follows that

$$\forall x, y, z \in \mathbf{E} : \quad \mu_{\underset{\sim}{\mathcal{R}}}(x, z) \geqslant \alpha \wedge \beta ; \tag{27.7}$$

and also

$$\mu_{\underset{\sim}{\mathcal{R}}}(x, z) \geqslant \bigvee_{y} (\mu_{\underset{\sim}{\mathcal{R}}}(x, y) \wedge \mu_{\underset{\sim}{\mathcal{R}}}(y, z)) . \tag{27.8}$$

This with (27.2) and (27.3) proves the transitivity of $\underset{\sim}{\mathcal{R}}$.

This converse allows the synthesis of similitude relations, as the direct theorem permits analysis.

Interesting remark. It follows from this theorem that the ordinary relation closest to a similitude relation is an equivalence relation. This one may see immediately by considering what represents $\mathcal{R}_\alpha$ when $\alpha > 0{,}5$.

Examples. We now see the analysis of the relation given in Figure 20.1. The decomposition has been presented in Figure 27.1.

$\underset{\sim}{\mathcal{R}}$	A	B	C	D	E
A	1	0,8	0,7	1	0,9
B	0,8	1	0,7	0,8	0,8
C	0,7	0,7	1	0,7	0,7
D	1	0,8	0,7	1	0,9
E	0,9	0,8	0,7	0,9	1

= V ((0,7) · $\mathcal{R}_{0,7}$, (0,8) · $\mathcal{R}_{0,8}$, (0,9) · $\mathcal{R}_{0,9}$, (1) · $\mathcal{R}_1$)

$\mathcal{R}_{0,7}$

1	1	1	1	1
1	1	1	1	1
1	1	1	1	1
1	1	1	1	1
1	1	1	1	1

$\mathcal{R}_{0,8}$

1	1	0	1	1
1	1	0	1	1
0	0	1	0	0
1	1	0	1	1
1	1	0	1	1

$\mathcal{R}_{0,9}$

1	0	0	1	1
0	1	0	0	0
0	0	1	0	0
1	0	0	1	1
1	0	0	1	1

$\mathcal{R}_1$

1	0	0	1	0
0	1	0	0	0
0	0	1	0	0
1	0	0	1	0
0	0	0	0	1

Fig. 27.1

Next we consider an example of synthesis. Let the four equivalence relations be successively included in one another (Figure 27.2):

$\mathcal{R}_{0,2}$	A	B	C	D
A	1	1	1	1
B	1	1	1	1
C	1	1	1	1
D	1	1	1	1

(a)

$\mathcal{R}_{0,6}$	A	B	C	D
A	1	1	0	0
B	1	1	0	0
C	0	0	1	1
D	0	0	1	1

(b)

$\mathcal{R}_{0,8}$	A	B	C	D
A	1	1	0	0
B	1	1	0	0
C	0	0	1	0
D	0	0	0	1

(c)

$\mathcal{R}_{1}$	A	B	C	D
A	1	0	0	0
B	0	1	0	0
C	0	0	1	0
D	0	0	0	1

(d)

Fig. 27.2

One then has

(27.9) $$\underset{\sim}{\mathcal{R}} = \vee\,(0{,}2 \cdot \mathcal{R}_{0,2}\ ,\ 0{,}6 \cdot \mathcal{R}_{0,6}\ ,\ 0{,}8 \cdot \mathcal{R}_{0,8}\ ,\ 1 \cdot \mathcal{R}_{1})\ .$$

The result is shown in Figure 27.3.

Another example is shown in Figure 27.4, where we have supposed that a and $b \in [0, 1]$ with $a < b$.

$\underset{\sim}{\mathcal{R}}$	A	B	C	D
A	1	0,8	0,2	0,2
B	0,8	1	0,2	0,2
C	0,2	0,2	1	0,6
D	0,2	0,2	0,6	1

Fig. 27.3

$\vee\Big($ $0 \cdot$

1	1	1	1	1
1	1	1	1	1
1	1	1	1	1
1	1	1	1	1
1	1	1	1	1

$,\ a \cdot$

1	0	0	1	1
0	1	1	0	0
0	1	1	0	0
1	0	0	1	1
1	0	0	1	1

$,\ b \cdot$

1	0	0	0	1
0	1	1	0	0
0	1	1	0	0
0	0	0	1	0
1	0	0	0	1

$,\ 1 \cdot$

1	0	0	0	0
0	1	0	0	0
0	0	1	0	0
0	0	0	1	0
0	0	0	0	1

$\Big)$

=

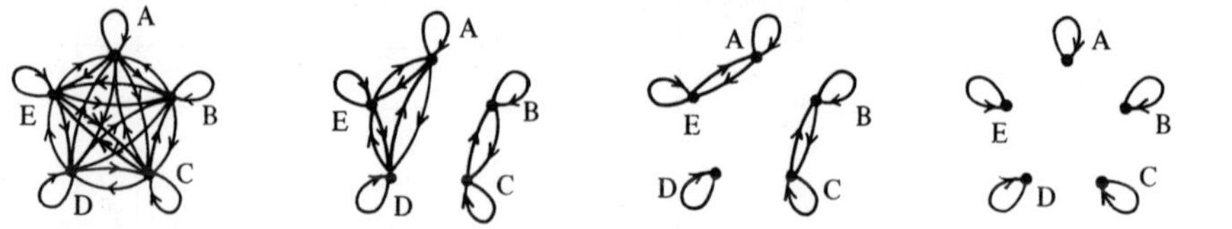

	A	B	C	D	E
A	1	0	0	a	b
B	0	1	b	0	0
C	0	b	1	0	0
D	a	0	0	1	a
E	b	0	0	a	1

Fig. 27.4

Transitive graphs of distances. It is interesting to present for each similitude relation the transitive graphs corresponding to the min–max distances. Some examples serve to demonstrate the interest.

Example 1. Figure 27.5 gives an example of a dissimilitude relation. In Figure 27.6 we have represented the transitive graphs corresponding to various distances.

$\underset{\sim}{\mathcal{R}}$	A	B	C	D	E
A	0	0,2	0,8	0	0,1
B	0,2	0	0,8	0,2	0,2
C	0,8	0,8	0	0,8	0,8
D	0	0,2	0,8	0	0,1
E	0,1	0,2	0,8	0,1	0

Fig. 27.5

distances equal to 0:

	A	B	C	D	E
A	1	0	0	1	0
B	0	1	0	0	0
C	0	0	1	0	0
D	1	0	0	1	0
E	0	0	0	0	1

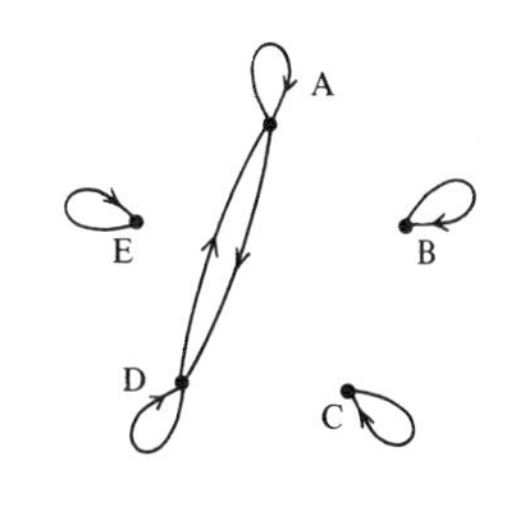

distances less than or equal to 0,1:

	A	B	C	D	E
A	1	0	0	1	1
B	0	1	0	0	0
C	0	0	1	0	0
D	1	0	0	1	1
E	1	0	0	1	1

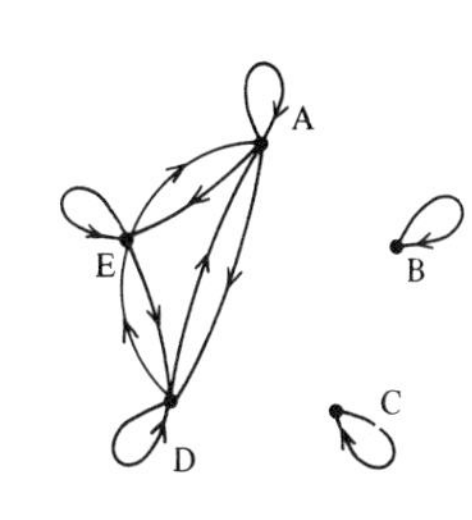

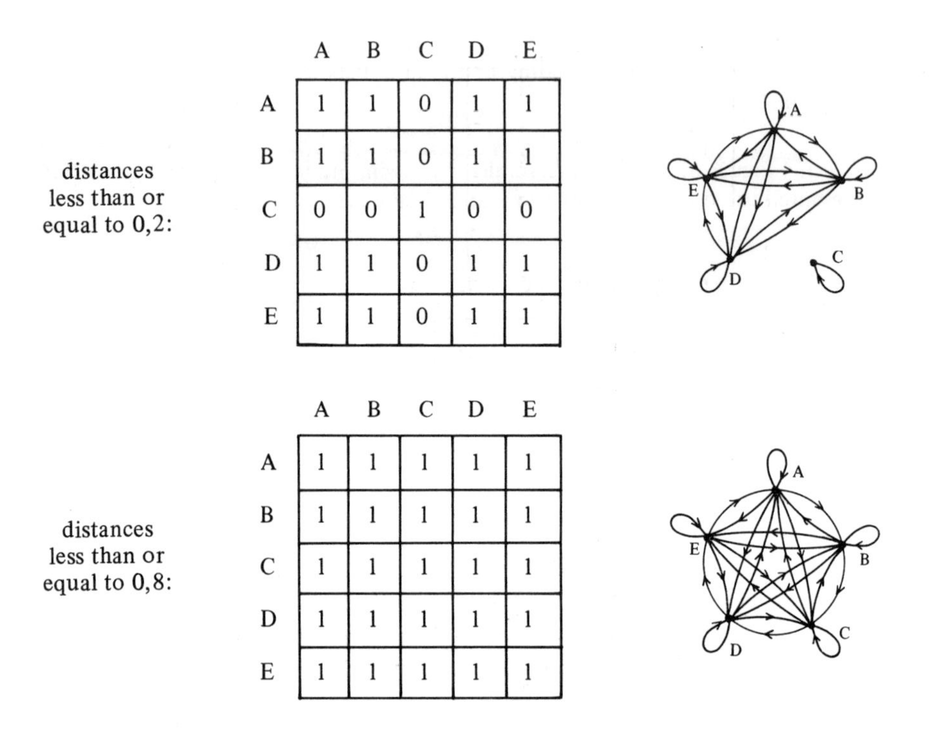

distances less than or equal to 0,2:

	A	B	C	D	E
A	1	1	0	1	1
B	1	1	0	1	1
C	0	0	1	0	0
D	1	1	0	1	1
E	1	1	0	1	1

distances less than or equal to 0,8:

	A	B	C	D	E
A	1	1	1	1	1
B	1	1	1	1	1
C	1	1	1	1	1
D	1	1	1	1	1
E	1	1	1	1	1

Fig. 27.6 Transitive graphs of distance

Example 2. (Figures 27.7 and 27.8). This example is relative to the transitive closure (Figure 26.2) of the resemblance relation (Figure 26.1). The decomposition obtained will be compared to that of the following example (Figures 27.9 and 27.10).

$\bar{\hat{\underset{\sim}{\mathcal{R}}}}$	A	B	C	D	E
A	0	0,4	0,2	0,3	0,4
B	0,4	0	0,4	0,4	0
C	0,2	0,4	0	0,3	0,4
D	0,3	0,4	0,3	0	0,4
E	0,4	0	0,4	0,4	0

Fig. 27.7

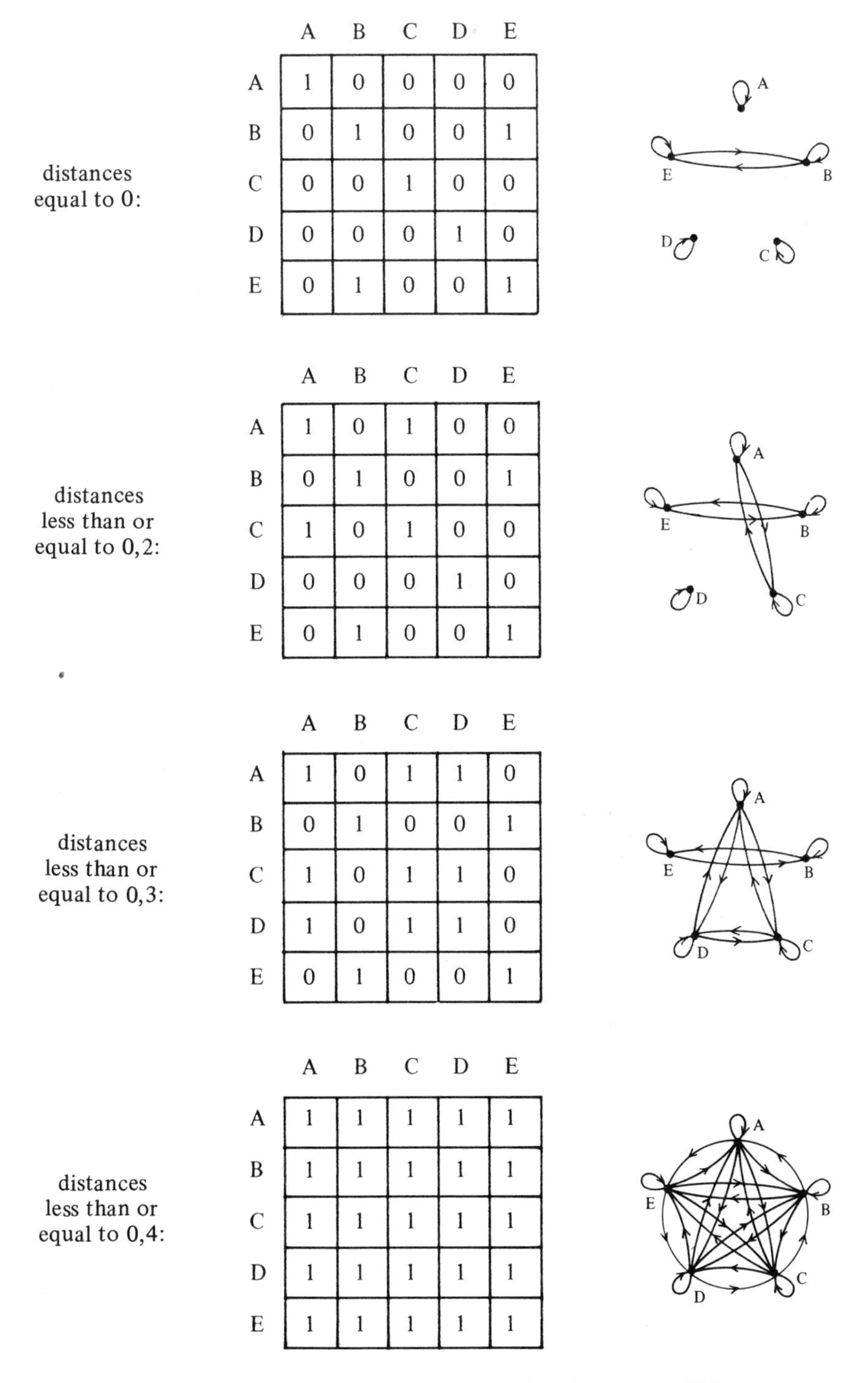

distances equal to 0:

	A	B	C	D	E
A	1	0	0	0	0
B	0	1	0	0	1
C	0	0	1	0	0
D	0	0	0	1	0
E	0	1	0	0	1

distances less than or equal to 0,2:

	A	B	C	D	E
A	1	0	1	0	0
B	0	1	0	0	1
C	1	0	1	0	0
D	0	0	0	1	0
E	0	1	0	0	1

distances less than or equal to 0,3:

	A	B	C	D	E
A	1	0	1	1	0
B	0	1	0	0	1
C	1	0	1	1	0
D	1	0	1	1	0
E	0	1	0	0	1

distances less than or equal to 0,4:

	A	B	C	D	E
A	1	1	1	1	1
B	1	1	1	1	1
C	1	1	1	1	1
D	1	1	1	1	1
E	1	1	1	1	1

Fig 27.8. Transitivity graphs of min–max distances

Example 3. (Figures 27.9 and 27.10). The max–product transitive closure of the resemblance relation of Figure 26.1 has been obtained in Figure 26.6. For this, we have drawn in Figure 26.7 the matrix of min–sum distances. The decomposition into ordinary graphs of distances that will not all be transitive appears in this example. It is an inconvenience to use the max–product transitive closure in a resemblance relation in comparison to the use of the max–min transitive closure.

$\hat{\bar{\underset{\sim}{R}}}$	A	B	C	D	E
A	0	0,664	0,2	0,44	0,664
B	0,664	0	0,58	0,4	0
C	0,2	0,58	0	0,3	0,58
D	0,44	0,4	0,3	0	0,4
E	0,664	0	0,58	0,4	0

Fig. 27.9

distances equal to 0:

1	0	0	0	0
0	1	0	0	1
0	0	1	0	0
0	0	0	1	0
0	1	0	0	1

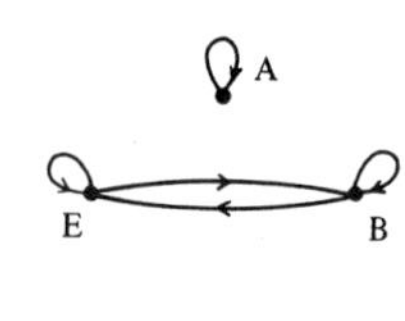

distances less than or equal to 0,2:

1	0	1	0	0
0	1	0	0	1
1	0	1	0	0
0	0	0	1	0
0	1	0	0	1

distances less than or equal to 0,3:

1	0	1	0	0
0	1	0	0	1
1	0	1	1	0
0	0	1	1	0
0	1	0	0	1

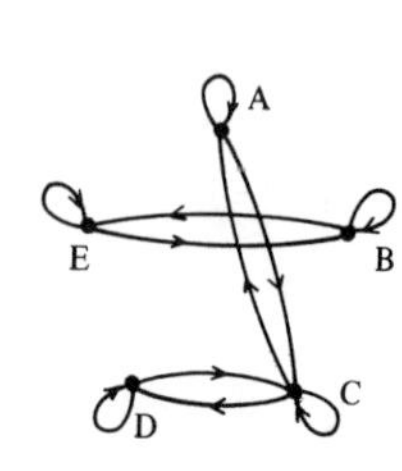

distances less than or equal to 0,4:

1	0	1	0	0
0	1	0	1	1
1	0	1	1	0
0	1	1	1	1
0	1	0	1	1

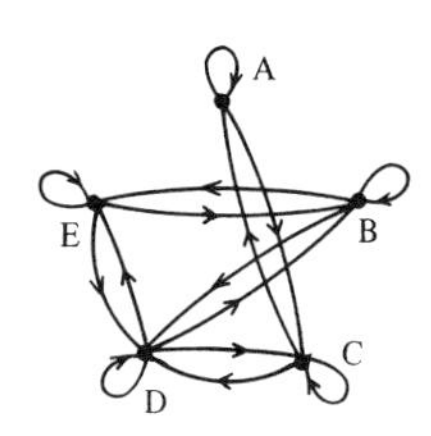

distances less than or equal to 0,44:

1	0	1	1	0
0	1	0	1	1
1	0	1	1	0
1	1	1	1	1
0	1	0	1	1

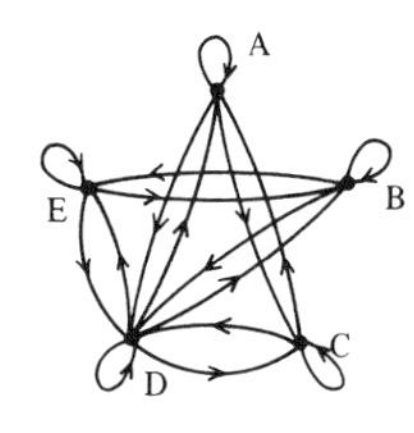

distances less than or equal to 0,58:

1	0	1	1	0
0	1	1	1	1
1	1	1	1	1
1	1	1	1	1
0	1	1	1	1

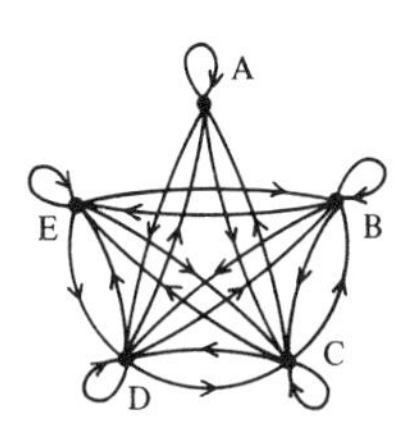

distances less than or equal to 0,664:

1	1	1	1	1
1	1	1	1	1
1	1	1	1	1
1	1	1	1	1
1	1	1	1	1

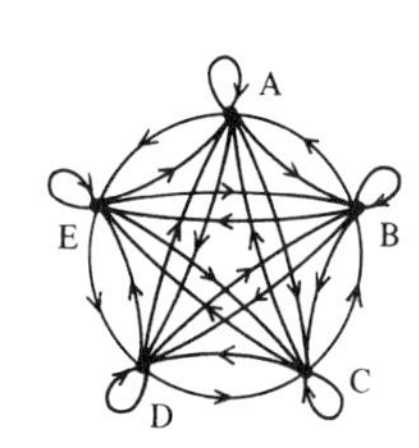

Fig. 27.10

Tree decomposition. A reader who examines Figure 27.1 is led to note that gradually as α takes the values 0,7, 0,8, 0,9, and 1, the partition of **E** into equivalence classes includes more and more parts. This decomposition has been carried out according to a tree scheme, which has been represented in Figure 27.11. An ordered scheme such as this is called *tree decomposition.*

Another example relative to Figure 27.4 is given in Figure 27.12.

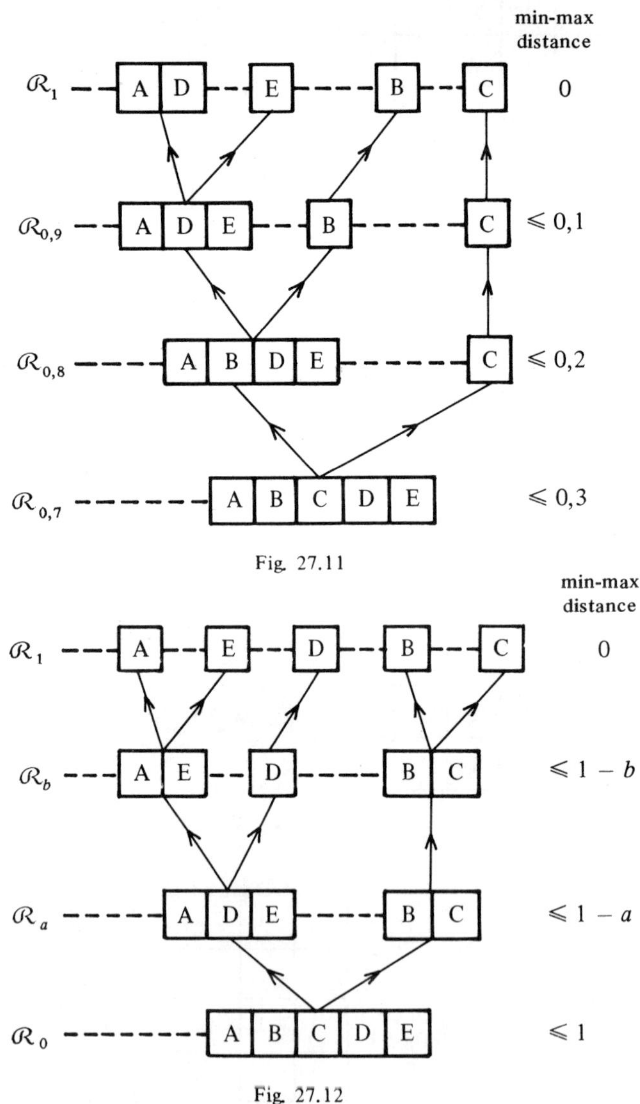

Fig. 27.11

Fig. 27.12

One may verify that two elements x and y belonging to **E** would belong to the same class of level α if and only if

$$\mu_{\mathcal{R}}(x, y) \geqslant \alpha . \tag{27.10}$$

This decomposition tree reflects the structure of the similitude relation well, or if one prefers, the groupings of elements by their transitive distances from one another.

One may represent a tree in various manners. Using the notation of linguistics, one may write sequentially following the tree of Figure 27.11

(27.11) 0,7(0,8(0,9(1{A ,D}, 1{E}), 0,9(1{B})) , 0,8(0,9(1{C}))).

Such a use of parentheses is not convenient.

One may also use the notion of a pile and represent the tree (27.11) with the sequence: 0,7 (ABCDE) 0,8 (ABDE) 0,9 (ADE) 1 (AE) 0,9 (ADE) 1 (D) 0,9 (ADE) 0,8 (ABDE) 0,9 (B) 1 (B) 0,9 (B) 0,8 (ABDE) 0,7 (ABCDE) 0,8 (C) 0,9 (C) 1 (C) 0,9 (C) 0,8 (C) 0,7 (ABCDE). This notation is associated with the scheme known as Polish notation.

It is easy to follow the sequence in Figure 27.11.

Selection of the transitively nearest messages. One may consider a fuzzy subset as a message that is fuzzy instead of being binary.

Consider an ordinary set **F** of fuzzy subsets $\underset{\sim}{A}_i$ belonging to the same reference set **E**:

(27.12) $$F = \{\underset{\sim}{A}_1 , \underset{\sim}{A}_2 , \ldots , \underset{\sim}{A}_n\} .$$

We have in mind the determination of which fuzzy subsets or fuzzy messages are transitively nearest. We shall make precise a little later the inconveniences of the notion of transitivity that will be considered; the advantages are at once apparent.

We shall proceed as follows (and shall explain at the same time what is meant by transitively nearest):

(1) For each pair $(\underset{\sim}{A}_i, \underset{\sim}{A}_j)$, $i, j = 1, 2, \ldots, n$, evaluate the relative generalized Hamming distance† $\delta(\underset{\sim}{A}_i, \underset{\sim}{A}_j)$; this gives a dissemblance relation $\underset{\sim}{\mathfrak{D}}$.

(2) Take the min–max transitive closure [that defined by (26.41)]. The relation $\underset{\sim}{\check{\mathfrak{D}}}$ obtained gives the min–max transitive distance:

(27.13) $$\check{\delta}(\underset{\sim}{A}_i , \underset{\sim}{A}_j) .$$

(3) Then decompose $\underset{\sim}{\check{\mathfrak{D}}}$ according to (27.1) and obtain the following ordinary subsets of **F**:

transitively nearest messages for which one has

(27.14) $$\check{\delta}(\underset{\sim}{A}_i , \underset{\sim}{A}_j) = 0 .$$

transitively nearest messages for which one has

(27.15) $$0 < \check{\delta}(\underset{\sim}{A}_i , \underset{\sim}{A}_j) = \alpha_1 < \alpha_2 < \cdots$$

transitively nearest messages for which one has

(27.16) $$0 < \alpha_1 < \check{\delta}(\underset{\sim}{A}_i , \underset{\sim}{A}_j) = \alpha_2 < \alpha_3 < \cdots$$

†Or relative euclidean distance $\epsilon(A_i, A_j)$, this depending on the nature of the problem, or even some other notion of distance.

and so on.

(4) Construct the corresponding composition tree.

Example. Let **E** be a finite reference set with card(**E**) = 7 and consider six subsets or messages $\underset{\sim}{\mathbf{A}}_i$, $i = 1, 2, \ldots, 6$,

(27.17)

	x_1	x_2	x_3	x_4	x_5	x_6	x_7	
$\underset{\sim}{\mathbf{A}}_1 =$	0,1	0,8	0,3	1	0,1	0	1	,
$\underset{\sim}{\mathbf{A}}_2 =$	0,3	0,8	0,1	1	0	1	0,7	,
$\underset{\sim}{\mathbf{A}}_3 =$	0,7	1	0	1	0,8	0,3	0,7	,
$\underset{\sim}{\mathbf{A}}_4 =$	0,1	0,8	0,7	0	0,1	1	0,3	,
$\underset{\sim}{\mathbf{A}}_5 =$	0,6	1	0	0,7	0,8	0	1	,
$\underset{\sim}{\mathbf{A}}_6 =$	0	0,3	0,5	0,1	0,1	0,5	0,8	.

Then calculate the relative generalized Hamming distance:

(27.18)
$$\delta(\underset{\sim}{\mathbf{A}}_i, \underset{\sim}{\mathbf{A}}_j) = \frac{d(\underset{\sim}{\mathbf{A}}_i, \underset{\sim}{\mathbf{A}}_j)}{7}.$$

This gives the dissemblance relation $\underset{\sim}{\mathcal{D}}$ (Figure 27.13a). One then calculates with the aid of (26.41) the min–max transitive closure $\check{\underset{\sim}{\mathcal{D}}}$, which gives the transitive distances δ. (See Figures 27.14 and 27.15.)

$\underset{\sim}{\mathcal{D}}$	$\underset{\sim}{\mathbf{A}}_1$	$\underset{\sim}{\mathbf{A}}_2$	$\underset{\sim}{\mathbf{A}}_3$	$\underset{\sim}{\mathbf{A}}_4$	$\underset{\sim}{\mathbf{A}}_5$	$\underset{\sim}{\mathbf{A}}_6$
$\underset{\sim}{\mathbf{A}}_1$	0	0,25	0,34	0,44	0,28	0,34
$\underset{\sim}{\mathbf{A}}_2$	0,25	0	0,31	0,32	0,42	0,40
$\underset{\sim}{\mathbf{A}}_3$	0,34	0,31	0	0,61	0,14	0,54
$\underset{\sim}{\mathbf{A}}_4$	0,44	0,32	0,61	0	0,64	0,27
$\underset{\sim}{\mathbf{A}}_5$	0,28	0,42	0,14	0,64	0	0,54
$\underset{\sim}{\mathbf{A}}_6$	0,34	0,40	0,54	0,27	0,54	0

(a)

$\check{\underset{\sim}{\mathcal{D}}}$	$\underset{\sim}{\mathbf{A}}_1$	$\underset{\sim}{\mathbf{A}}_2$	$\underset{\sim}{\mathbf{A}}_3$	$\underset{\sim}{\mathbf{A}}_4$	$\underset{\sim}{\mathbf{A}}_5$	$\underset{\sim}{\mathbf{A}}_6$
$\underset{\sim}{\mathbf{A}}_1$	0	0,25	0,28	0,32	0,28	0,32
$\underset{\sim}{\mathbf{A}}_2$	0,25	0	0,28	0,32	0,28	0,32
$\underset{\sim}{\mathbf{A}}_3$	0,28	0,28	0	0,32	0,14	0,32
$\underset{\sim}{\mathbf{A}}_4$	0,32	0,32	0,32	0	0,32	0,27
$\underset{\sim}{\mathbf{A}}_5$	0,28	0,28	0,14	0,32	0	0,32
$\underset{\sim}{\mathbf{A}}_6$	0,32	0,32	0,32	0,27	0,32	0

(b)

Fig. 27.13

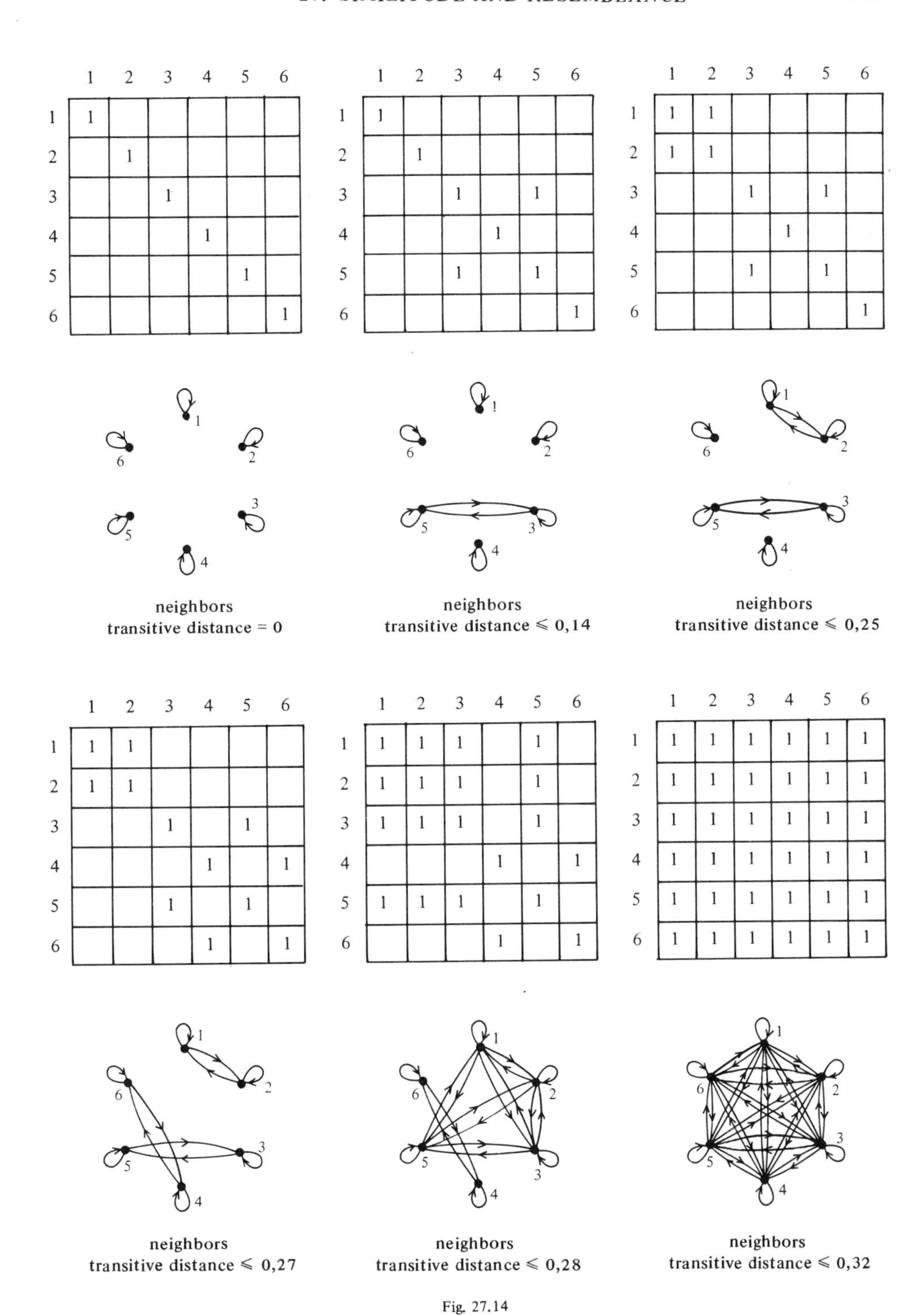

	1	2	3	4	5	6
1	1					
2		1				
3			1			
4				1		
5					1	
6						1

neighbors
transitive distance = 0

	1	2	3	4	5	6
1	1					
2		1				
3			1		1	
4				1		
5			1		1	
6						1

neighbors
transitive distance ⩽ 0,14

	1	2	3	4	5	6
1	1	1				
2	1	1				
3			1		1	
4				1		
5			1		1	
6						1

neighbors
transitive distance ⩽ 0,25

	1	2	3	4	5	6
1	1	1				
2	1	1				
3			1		1	
4				1		1
5			1		1	
6				1		1

neighbors
transitive distance ⩽ 0,27

	1	2	3	4	5	6
1	1	1	1		1	
2	1	1	1		1	
3	1	1	1		1	
4				1		1
5	1	1	1		1	
6				1		1

neighbors
transitive distance ⩽ 0,28

	1	2	3	4	5	6
1	1	1	1	1	1	1
2	1	1	1	1	1	1
3	1	1	1	1	1	1
4	1	1	1	1	1	1
5	1	1	1	1	1	1
6	1	1	1	1	1	1

neighbors
transitive distance ⩽ 0,32

Fig. 27.14

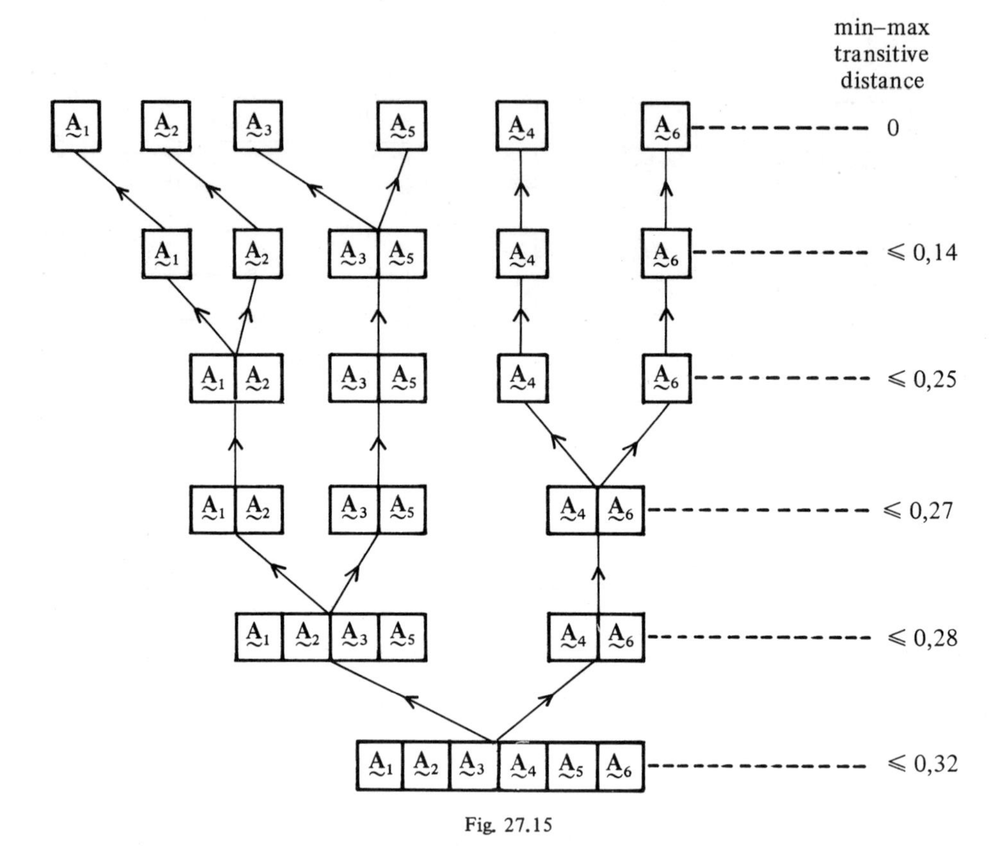

Fig. 27.15

Important remark on the subject of transitive distance. Depending on the nature of the problem being treated, the min–max transitive closure of a distance matrix may not be significant in its practical employment. We consider an example. Consider the following four messages:

$$\underset{\sim}{A}_1 = \begin{array}{c|c|c|} & x_1 & x_2 \\ \hline & 0 & 0 \\ \hline \end{array}$$

$$\underset{\sim}{A}_2 = \begin{array}{c|c|c|} & x_1 & x_2 \\ \hline & 0 & \frac{1}{2} \\ \hline \end{array}$$

$$\underset{\sim}{A}_3 = \begin{array}{c|c|c|} & x_1 & x_2 \\ \hline & \frac{1}{2} & 0 \\ \hline \end{array}$$

$$\underset{\sim}{A}_4 = \begin{array}{c|c|c|} & x_1 & x_2 \\ \hline & 0 & 1 \\ \hline \end{array}$$

$\bar{\mathcal{R}}$	(0, 0)	(0, ½)	(½, 0)	(0, 1)
(0, 0)	0	$\frac{1}{4}$	$\frac{1}{4}$	$\frac{1}{2}$
(0, ½)	$\frac{1}{4}$	0	$\frac{1}{2}$	$\frac{1}{4}$
(½, 0)	$\frac{1}{4}$	$\frac{1}{2}$	0	$\frac{3}{4}$
(0, 1)	$\frac{1}{2}$	$\frac{1}{4}$	$\frac{3}{4}$	0

Fig. 27.16

$\check{\bar{\mathcal{R}}}$	(0, 0)	(0, ½)	(½, 0)	(0, 1)
(0, 0)	0	$\frac{1}{4}$	$\frac{1}{4}$	$\frac{1}{4}$
(0, ½)	$\frac{1}{4}$	0	$\frac{1}{4}$	$\frac{1}{4}$
(½, 0)	$\frac{1}{4}$	$\frac{1}{4}$	0	$\frac{1}{4}$
(0, 1)	$\frac{1}{4}$	$\frac{1}{4}$	$\frac{1}{4}$	0

Fig. 27.17

The relative generalized Hamming distances for these messages are given in Figure 27.16, which then constitutes a dissemblance matrix $\bar{\mathcal{R}}$. In Figure 27.17 we have calculated the min–max closure of $\bar{\mathcal{R}}$, that is, $\check{\bar{\mathcal{R}}}$. One sees then that all these messages are transitively equidistant.

This conception of min–max transitive distance may seem to be unacceptable in numerous applications. But the relative generalized Hamming distance is transitive for the ordinary min–addition operation, that is,

$$\delta(x, z) \leqslant \underset{y}{\text{MIN}} [\delta(x, y) + \delta(y, z)] , \tag{27.19}$$

since this is a distance, that is,

$$\forall y : \qquad \delta(x, z) \leqslant \delta(x, y) + \delta(y, z) . \tag{27.20}$$

One comes to the same conclusions for relative euclidean distance.

Thus, any relation $\mathcal{Q}$ giving the relative generalized Hamming distance (or relative euclidean distance) is a relation that is its own ordinary min–addition transitive closure. Note that the member on the right-hand side of (27.19) may give a sum greater than 1, since it is an ordinary addition; but this constrains nothing since the member on the left, by construction, always belongs to [0, 1].

The decomposition by levels relative to values contained in the dissemblance relation will no longer give equivalence classes, but maximal subrelations, as we explain hereafter.

Ordinary min–addition dissimilitude. Decomposition into maximal subrelations. The relation (27.19) may be considered as a dissimilitude relation, which we may call ordinary min–addition dissimilitude. As may be seen in the example given in Figure 27.19,

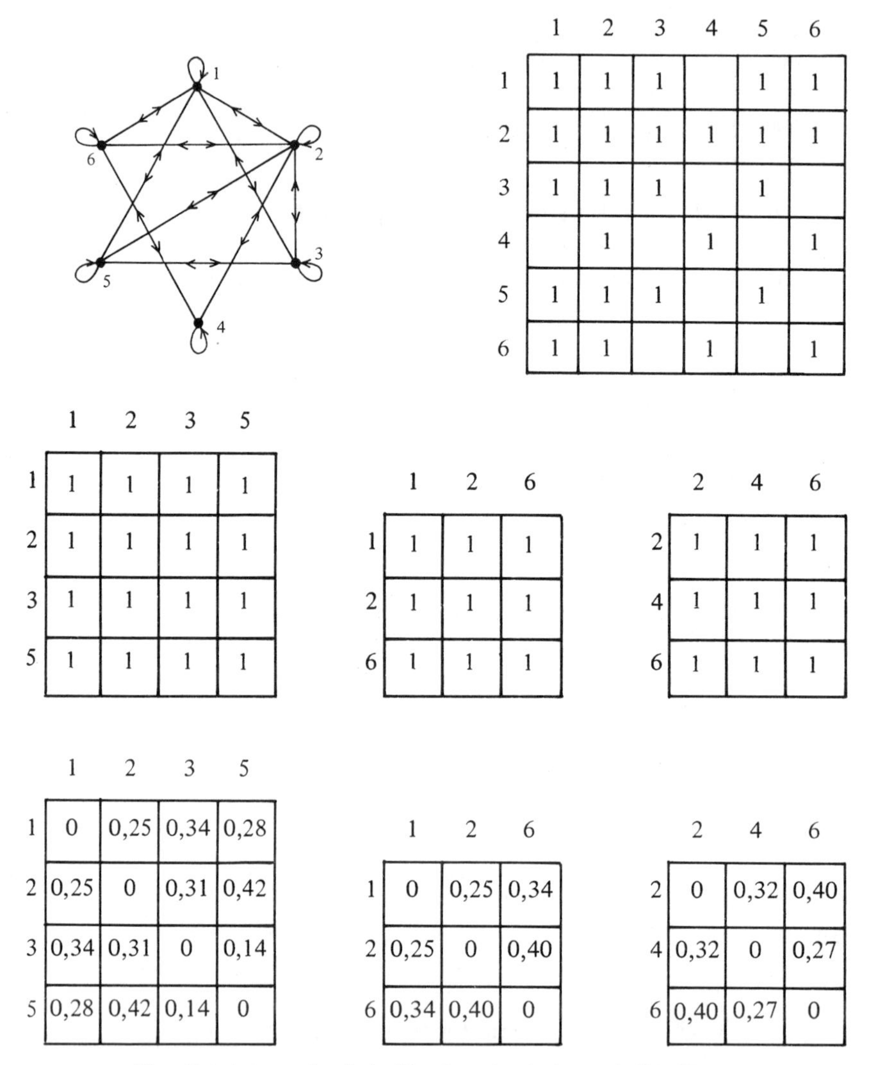

	1	2	3	4	5	6
1	1	1	1		1	1
2	1	1	1	1	1	1
3	1	1	1		1	
4		1		1		1
5	1	1	1		1	
6	1	1		1		1

	1	2	3	5
1	1	1	1	1
2	1	1	1	1
3	1	1	1	1
5	1	1	1	1

	1	2	6
1	1	1	1
2	1	1	1
6	1	1	1

	2	4	6
2	1	1	1
4	1	1	1
6	1	1	1

	1	2	3	5
1	0	0,25	0,34	0,28
2	0,25	0	0,31	0,42
3	0,34	0,31	0	0,14
5	0,28	0,42	0,14	0

	1	2	6
1	0	0,25	0,34
2	0,25	0	0,40
6	0,34	0,40	0

	2	4	6
2	0	0,32	0,40
4	0,32	0	0,27
6	0,40	0,27	0

Nondisjoint maximal similitude subrelations, $d \leqslant 0{,}42$

Fig. 27.18

one does not obtain for a distance $d \leqslant k$ (k arbitrary) ordinary graphs whose subgraphs constitute equivalence classes. Sometimes one may use a less strong concept, which is rather interesting for various operations, that of maximal subrelations–which may be or may not be disjoint.

Take the case of Figure 27.19 and more particularly that of the ordinary symmetric graph corresponding to $d \leqslant 0{,}42$. In Figure 27.18 we have reproduced this ordinary graph and made evident three maximal subrelations or complete ordinary graphs, each constituting an equivalence relation. For each of these subrelations, the distance of each element to another is less than or equal to 0,42 and property (27.19) is verified. In general, such a decomposition may not be made without an appropriate algorithm; we give two of these in Appendix B, page 387.

Remark. Ordinary min–addition dissimilitude is not dual to that of max–product similitude; it is algebraic min–sum dissimilitude that corresponds in this duality [see (26.33)].

We shall see a completely developed example where there appear maximal subrelations.

Example. We decompose the dissemblance relation (27.13a) (Figure 27.19).

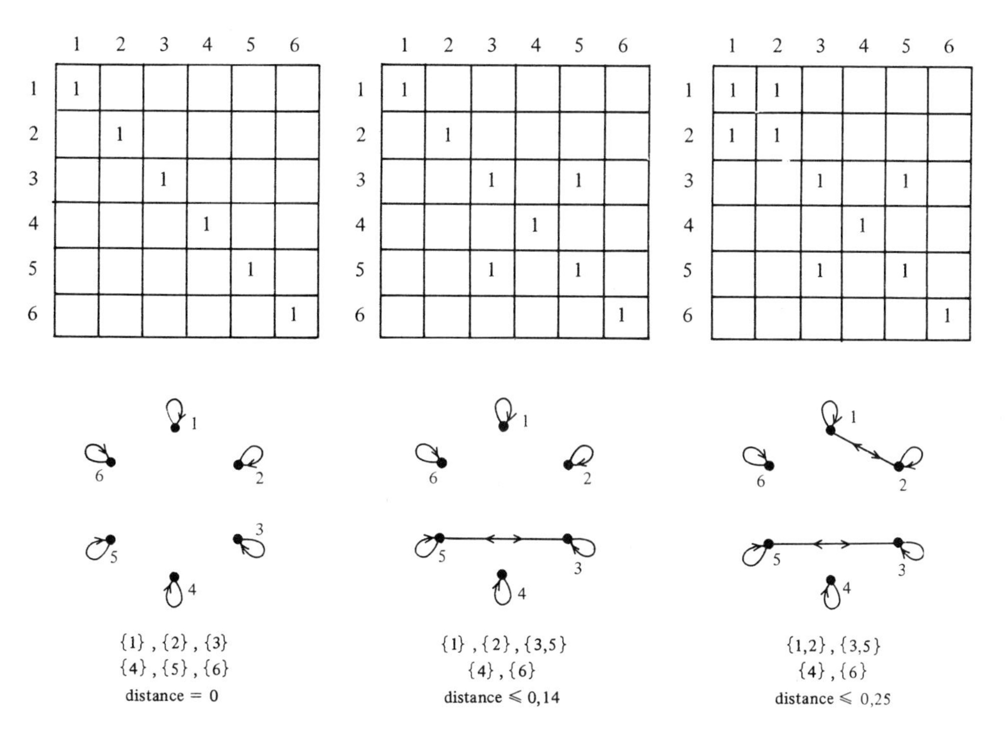

Fig. 27.19

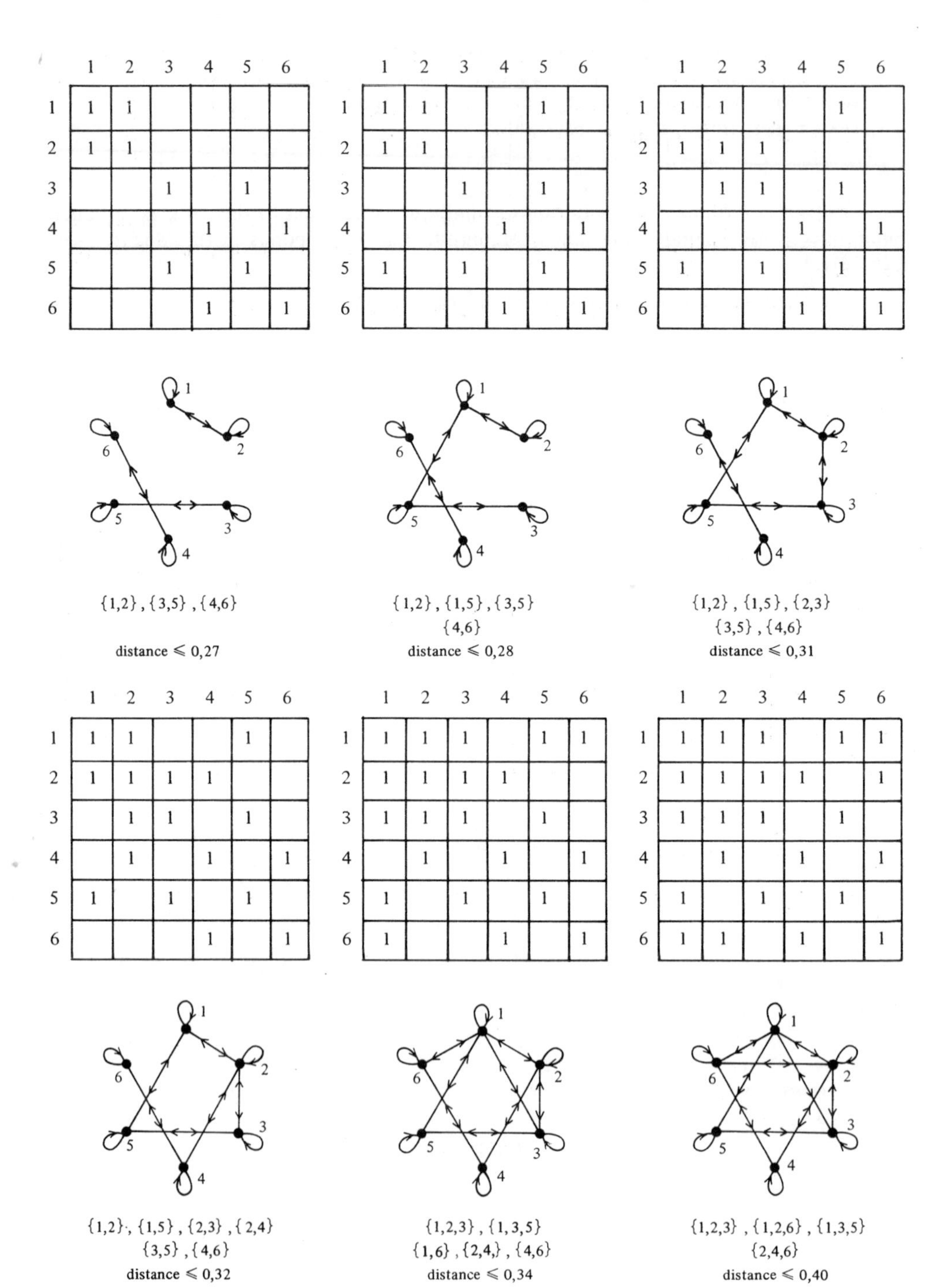

	1	2	3	4	5	6
1	1	1				
2	1	1				
3			1		1	
4				1		1
5			1		1	
6				1		1

{1,2}, {3,5}, {4,6}

distance ⩽ 0,27

	1	2	3	4	5	6
1	1	1			1	
2	1	1				
3			1		1	
4				1		1
5	1		1		1	
6				1		1

{1,2}, {1,5}, {3,5}
{4,6}

distance ⩽ 0,28

	1	2	3	4	5	6
1	1	1			1	
2	1	1	1			
3		1	1		1	
4				1		1
5	1		1		1	
6				1		1

{1,2}, {1,5}, {2,3}
{3,5}, {4,6}

distance ⩽ 0,31

	1	2	3	4	5	6
1	1	1			1	
2	1	1	1	1		
3		1	1		1	
4		1		1		1
5	1		1		1	
6				1		1

{1,2}, {1,5}, {2,3}, {2,4}
{3,5}, {4,6}

distance ⩽ 0,32

	1	2	3	4	5	6
1	1	1	1		1	1
2	1	1	1	1		
3	1	1	1		1	
4		1		1		1
5	1		1		1	
6	1			1		1

{1,2,3}, {1,3,5}
{1,6}, {2,4,}, {4,6}

distance ⩽ 0,34

	1	2	3	4	5	6
1	1	1	1		1	1
2	1	1	1	1		1
3	1	1	1		1	
4		1		1		1
5	1		1		1	
6	1	1		1		1

{1,2,3}, {1,2,6}, {1,3,5}
{2,4,6}

distance ⩽ 0,40

Fig. 27-19 (suite)

	1	2	3	4	5	6
1	1	1	1		1	1
2	1	1	1	1	1	1
3	1	1	1		1	
4		1		1		1
5	1	1	1		1	
6	1	1		1		1

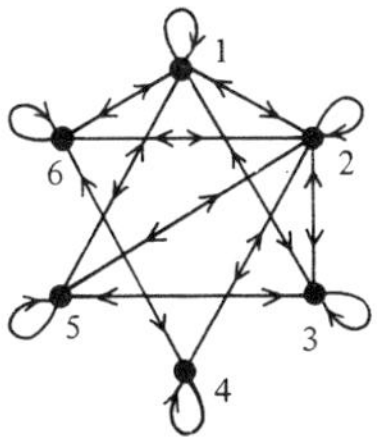

{1,2,3,5}, {1,2,6}
{2,4,6}
distance ⩽ 0,42

	1	2	3	4	5	6
1	1	1	1	1	1	1
2	1	1	1	1	1	1
3	1	1	1		1	
4	1	1		1		1
5	1	1	1		1	
6	1	1		1		1

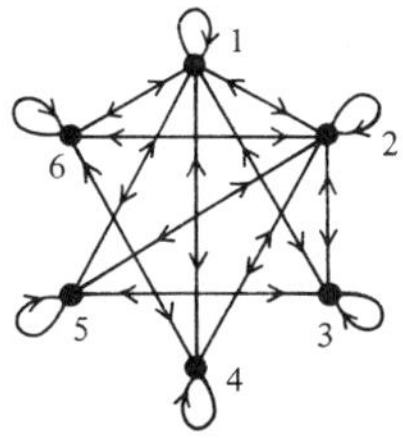

{1,2,4,6}, {1,2,3,5}
distance ⩽ 0,44

	1	2	3	4	5	6
1	1	1	1	1	1	1
2	1	1	1	1	1	1
3	1	1	1		1	1
4	1	1		1		1
5	1	1	1		1	1
6	1	1	1	1	1	1

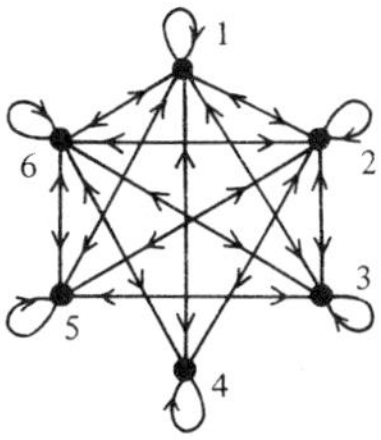

{1,2,3,5,6}, {1,2,4,6}
distance ⩽ 0,54

	1	2	3	4	5	6
1	1	1	1	1	1	1
2	1	1	1	1	1	1
3	1	1	1	1	1	1
4	1	1	1	1		1
5	1	1	1		1	1
6	1	1	1	1	1	1

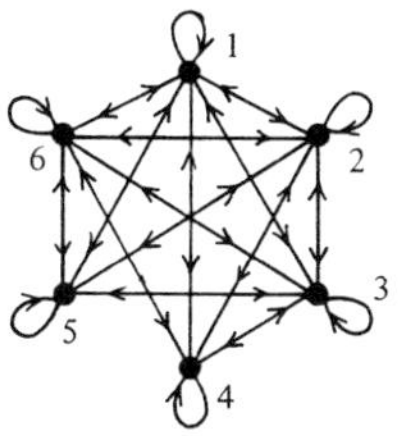

{1,2,3,5,6}, {1,2,3,4,6}
distance ⩽ 0,61

	1	2	3	4	5	6
1	1	1	1	1	1	1
2	1	1	1	1	1	1
3	1	1	1	1	1	1
4	1	1	1	1	1	1
5	1	1	1	1	1	1
6	1	1	1	1	1	1

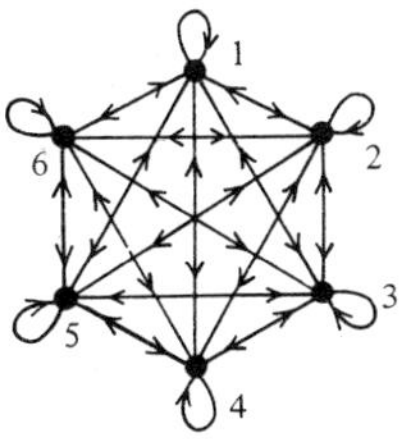

{1,2,3,4,5,6}
distance ⩽ 0,64

Fig. 27.19

Finally, one may also use the algebraic min–sum ($a \hat{+} b = a + b - ab$) transitivity to obtain the decomposition into maximal subrelations.

By comparing Figures 27.14 and 27.19, one may see the advantages and inconveniences of using min–max transitivity on the one hand and min–addition transitivity on the other. The first gives equivalence classes that are formed gradually depending on α; in contrast, interpretation is very debatable. The other gives only maximal subrelations that are not generally disjoint; but the interpretation is incontestable, particularly as concerns applications in the domain of classification of structures.

28. VARIOUS PROPERTIES OF FUZZY PERFECT ORDER RELATIONS

Decomposition theorem for a fuzzy perfect order relation. Let $\underset{\sim}{\mathcal{R}}$ be a fuzzy perfect order relation in $\mathbf{E} \times \mathbf{E}$. The $\underset{\sim}{\mathcal{R}}$ may be decomposed in the form

$$\underset{\sim}{\mathcal{R}} = \bigvee_{\alpha} \alpha \cdot \mathcal{R}_\alpha , \qquad 0 < \alpha \leqslant 1 , \tag{28.1}$$

with

$$\alpha_1 \geqslant \alpha_2 \quad \Rightarrow \quad \mathcal{R}_{\alpha_1} \subset \mathcal{R}_{\alpha_2} ,$$

where the $\mathcal{R}_\alpha$ are order relations in the sense of the theory of ordinary sets, and $\alpha \cdot \mathcal{R}_\alpha$ expresses the product of all elements of $\mathcal{R}_\alpha$ by the quantity α.

Proof. Reflexivity and transitivity of $\mathcal{R}_\alpha$ are proved as was (27.1) in Section 27. We shall see that this happens also for perfect antisymmetry according to (22.8).

In order to show the antisymmetry of $\mathcal{R}_\alpha$ we remark first that, since $\underset{\sim}{\mathcal{R}}$ is reflexive, one may replace the definition

$$\mu_{\mathcal{R}}(x, y) > 0 \quad \Rightarrow \quad \mu_{\mathcal{R}}(y, x) = 0 \tag{28.2}$$

by

$$(\mu_{\mathcal{R}}(x, y) > 0 \ \text{ and } \ \mu_{\mathcal{R}}(y, x) > 0) \quad \Rightarrow \quad (x = y) . \tag{28.3}$$

We shall reason by contradiction.

Suppose that $(x, y) \in \mathcal{R}_\alpha$ and $(y, x) \in \mathcal{R}_\alpha$. Then $\mu_{\underset{\sim}{\mathcal{R}}}(x, y) \geqslant \alpha$ and $\mu_{\underset{\sim}{\mathcal{R}}}(y, x) \geqslant \alpha$; thus by the antisymmetry of $\underset{\sim}{\mathcal{R}}$, $x = y$. Conversely, suppose $\mu_{\underset{\sim}{\mathcal{R}}}(x, y) = \alpha > 0$ and $\mu_{\underset{\sim}{\mathcal{R}}}(y, x) = \beta \geqslant 0$. Put $\gamma = \alpha \geqslant \beta$. Then $(x, y) \in \mathcal{R}_\gamma$ and $(y, x) \in \mathcal{R}_\gamma$, and from the antisymmetry of $\mathcal{R}_\gamma$ it follows that $x = y$. But one may not have $x \neq y$ with these hypotheses.

Example 1. Figure 28.1 represents a decomposition of a fuzzy perfect order relation. To simplify reading of the results, we have omitted the zeros. Beneath each $\mathcal{R}_\alpha$ we have placed a sketch representing the ordinary antisymmetric graph.

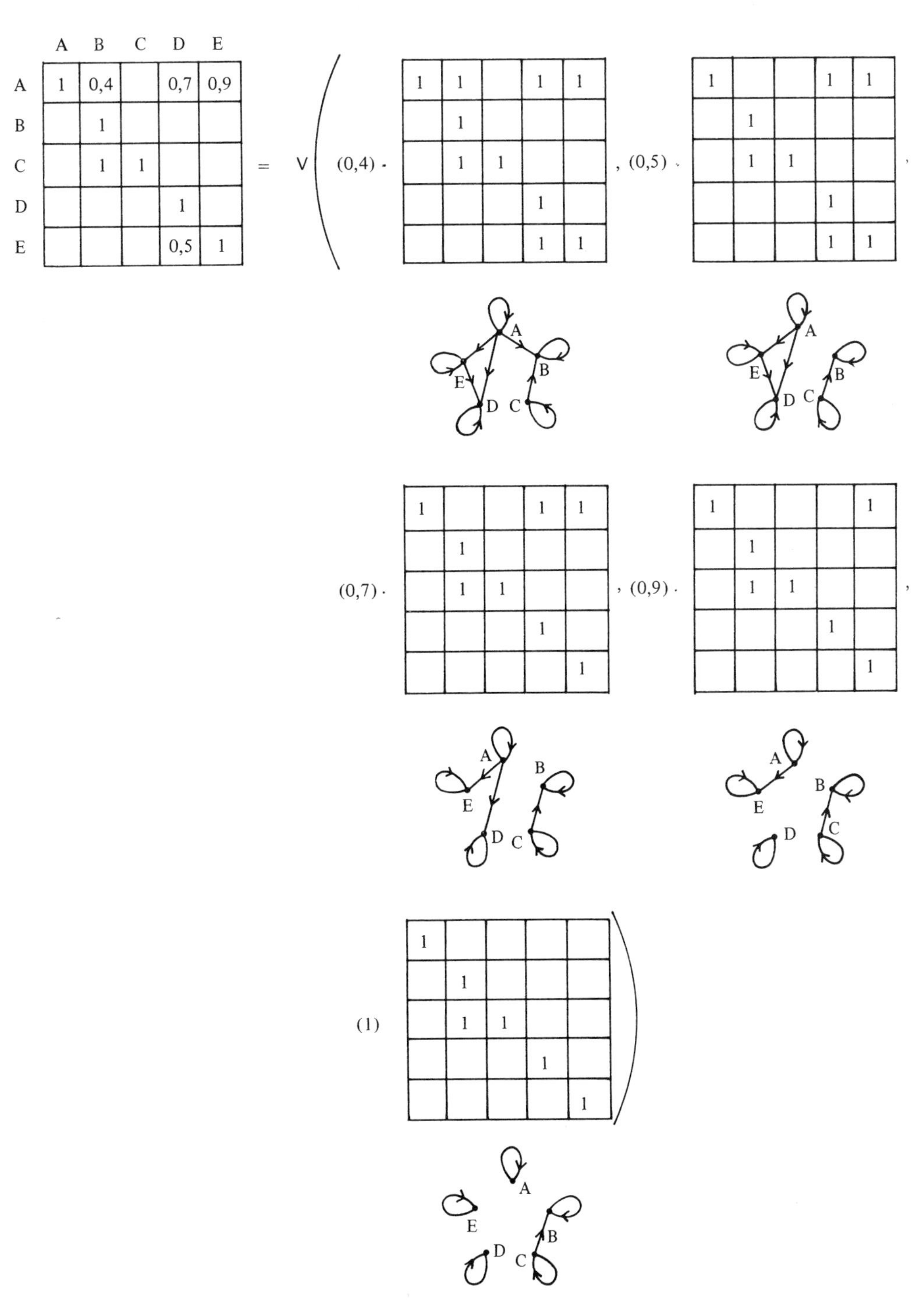

Fig. 28.1

Example 2. We see how to realize a synthesis of a perfect order relation (Figure 28.2).

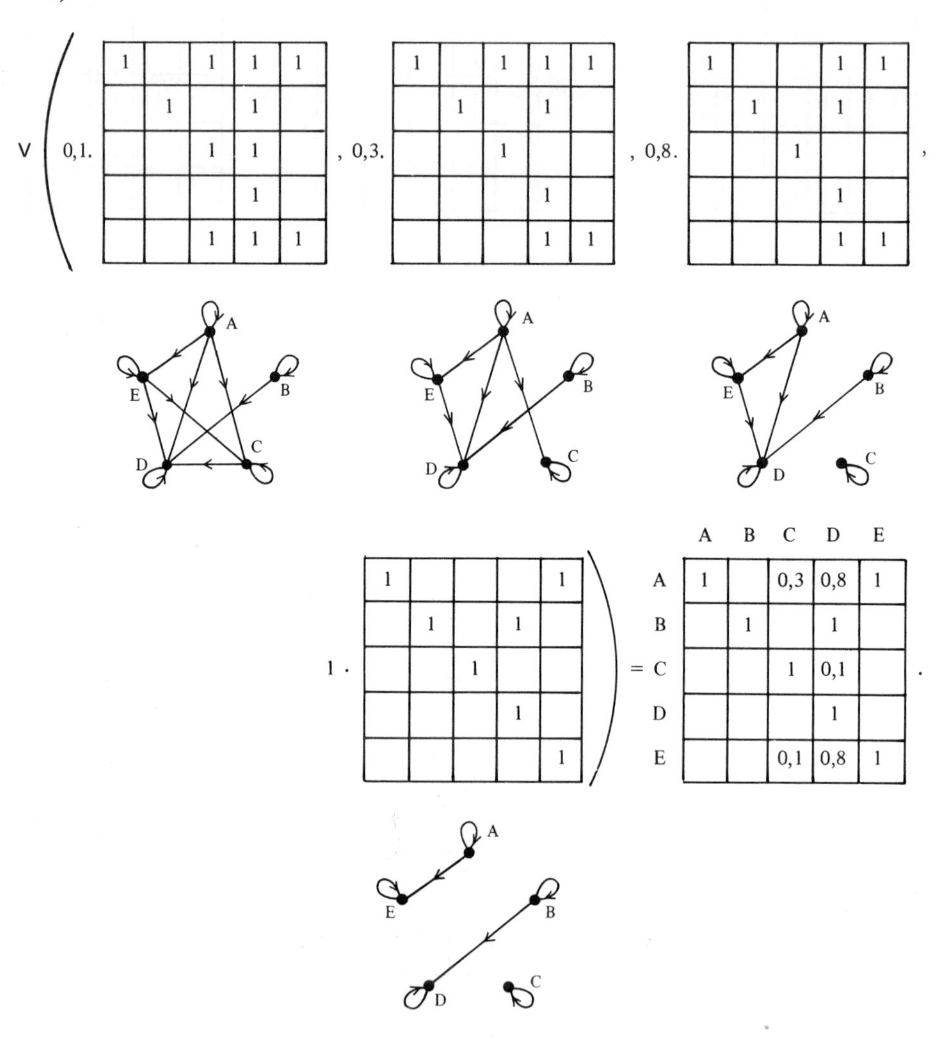

Fig. 28.2

Extension of the decomposition property to the case of a reducible preorder whose similitude classes are perfectly ordered. Properties (27.1) and (28.1) are combined whenever one considers a reducible preorder whose similitude classes constitute a perfect order.

Example. Figure 28.3 (pp. 163–164) gives an example of such a decomposition. In this figure the zeros are omitted to allow rapid examination. On the other hand, there are numerous elements and similitude classes for which the properties are readily apparent.

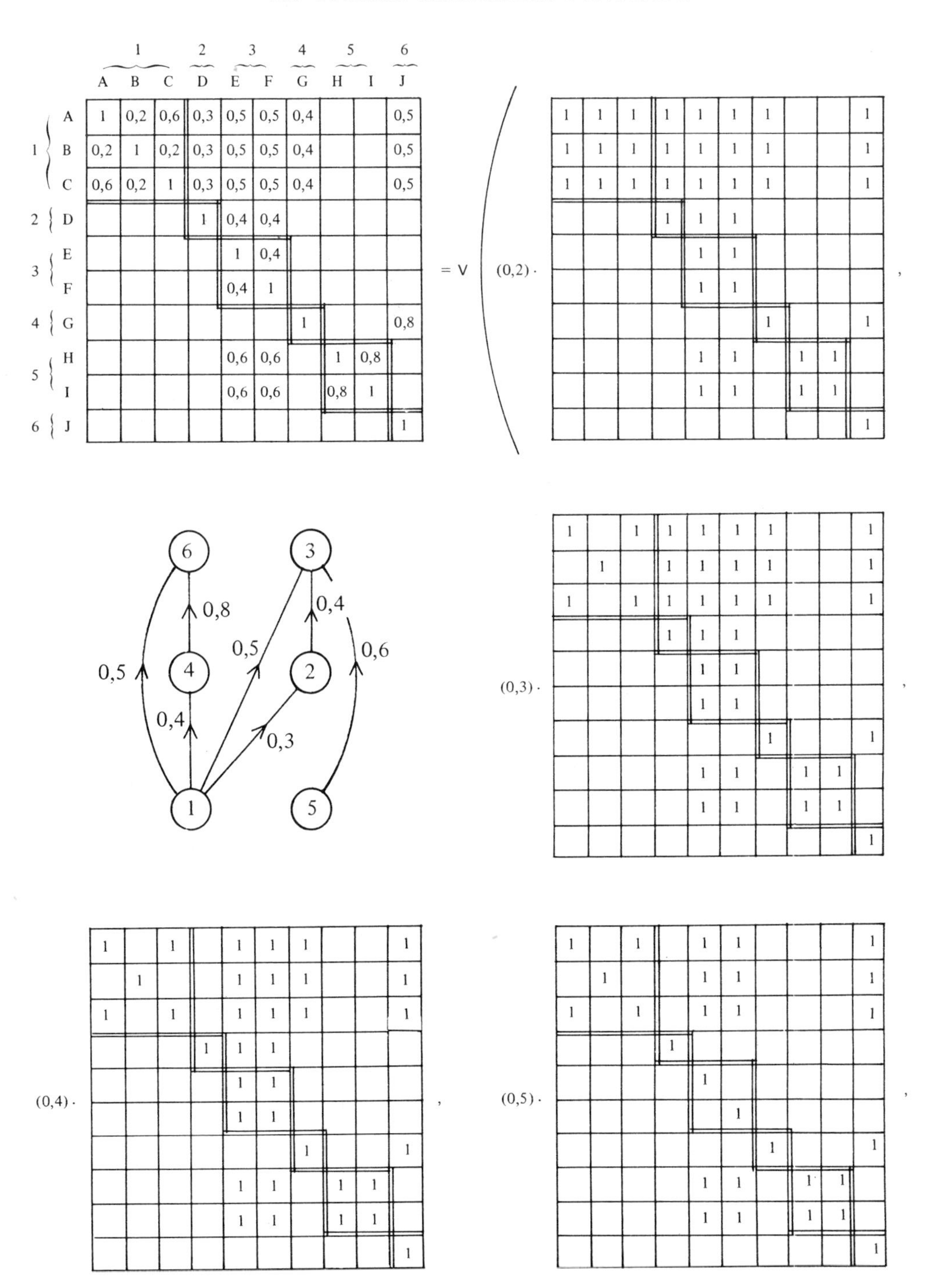

Fig. 28.3

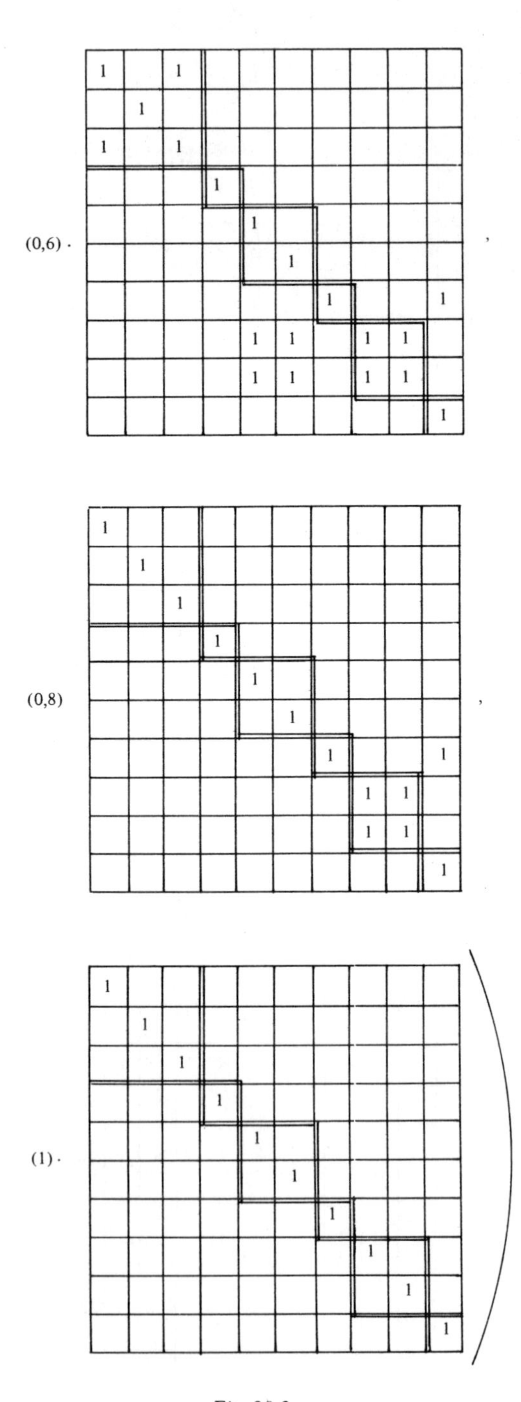

Fig. 28.3

Another example of synthesis. See Figure 28.4. Figure 28.5 illustrates the block-triangular form of the preorder.

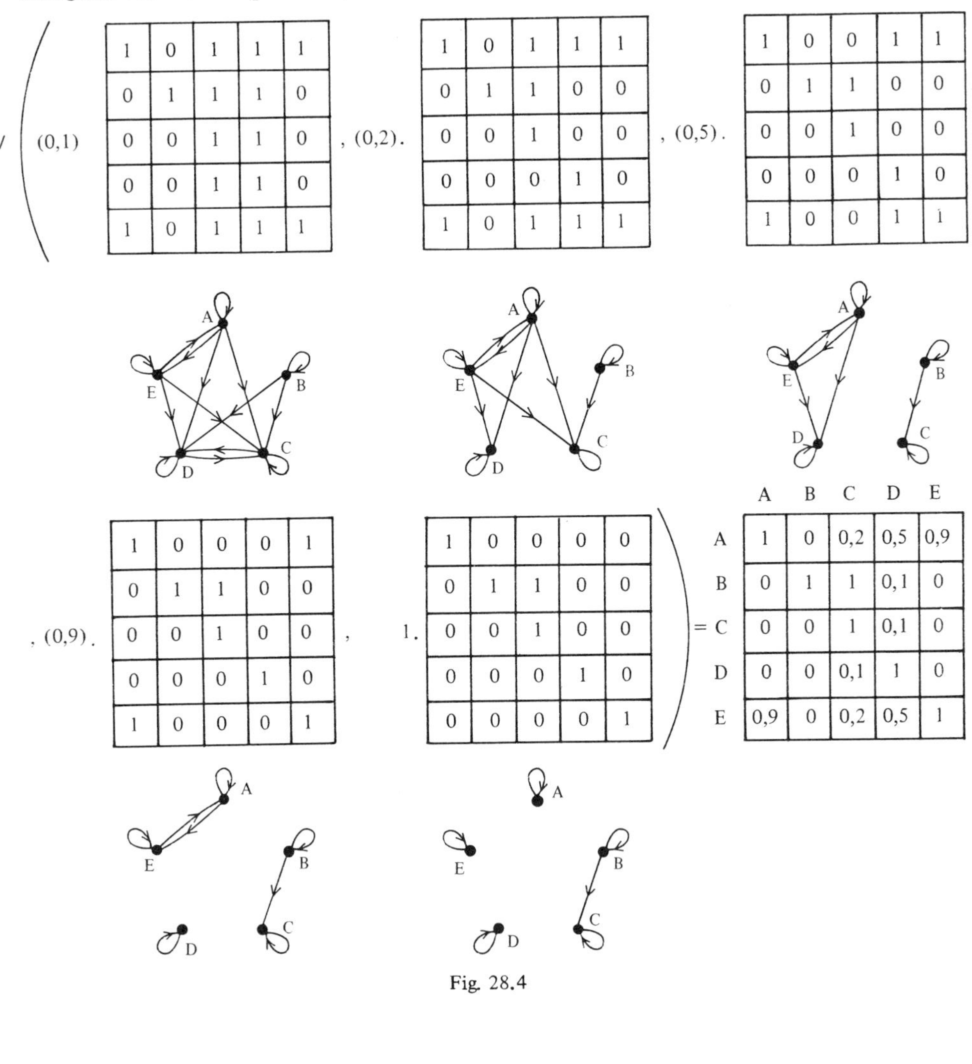

Fig. 28.4

$\underset{\sim}{\mathcal{R}}$	A	E	B	C	D
A	1	0,9	0	0,2	0,5
E	0,9	1	0	0,2	0,5
B	0	0	1	1	0,1
C	0	0	0	1	0,1
D	0	0	0	0,1	1

Fig. 28.5

Perfect total order induced in a perfect partial order by the ordinal function (case where E is finite). † Recalling what we have seen in Section 24, we take up again the example of Figure 28.3 and seek an ordinal function for the ordinary graph representing the order of the classes; consider the graph of Figure 28.6 in which appear three levels $\mathbf{N}_0, \mathbf{N}_1, \mathbf{N}_2$.

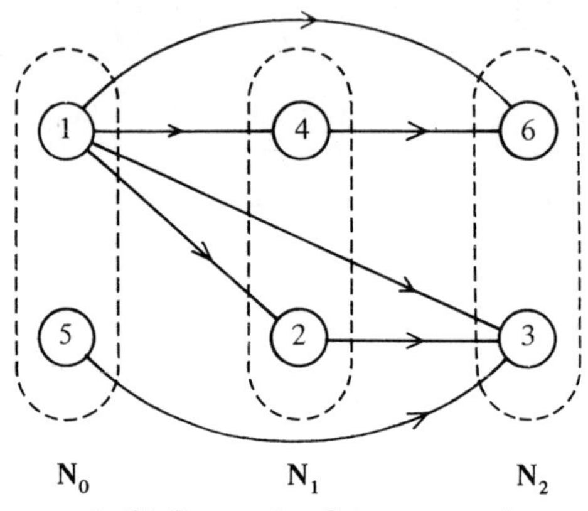

In this figure a class $\mathbf{C}_i$ is represented by its index $i = 1, 2, \ldots, 6$.

Fig. 28.6

These levels induce in the set of classes $\{\mathbf{C}_1, \mathbf{C}_2, \ldots, \mathbf{C}_6\}$ a (nonunique) total order such that, with respect to this order, the fuzzy relation takes a block-triangular form.

		1			5		4	2	6	3	
		A	B	C	H	I	G	D	J	E	F
1	A	1	0,2	0,6	0	0	0,4	0,3	0,5	0,5	0,5
	B	0,2	1	0,2	0	0	0,4	0,3	0,5	0,5	0,5
	C	0,6	0,2	1	0	0	0,4	0,3	0,5	0,5	0,5
5	H	0	0	0	1	0,8	0	0	0	0,6	0,6
	I	0	0	0	0,8	1	0	0	0	0,6	0,6
4	G	0	0	0	0	0	1	0	0,8	0	0
2	D	0	0	0	0	0	0	1	0	0,4	0,4
6	J	0	0	0	0	0	0	0	1	0	0
3	E	0	0	0	0	0	0	0	0	1	0,4
	F	0	0	0	0	0	0	0	0	0,4	1

In this figure a class $\mathbf{C}_i$ is represented by its index $i = 1, 2, \ldots, 6$.

Fig. 28.7

†This property is presented by certain authors under the name of the theorem of Szpilrajn. The introduction of the notion of the ordinal function of a graph avoids the rather delicate proof of this theorem. This is one of the advantages among many others of this important notion of ordinal function.

Figure 28.7 represents the results obtained in taking the total order

$$C_1 \succ C_5 \succ C_4 \succ C_2 \succ C_6 \succ C_3 \;,$$

with which one obtains a half-matrix of zeros below the diagonal of these blocks.

By choosing a total order in an ordinal function numbered from right to left, one would obtain a half-matrix of zeros above the diagonal of the blocks.

This we summarize in an example generalizable to all cases conforming to the title of this subsection.

PROPERTIES OF THE PRINCIPAL FUZZY RELATIONS

	Reflexivity	*Anti reflexivity*	*Max–min transitivity*	*Min–max transitivity*	*Sym-metry*	*Anti sym-metry*	*Does not possess cir-cuits other than loops*
Preorder	yes		yes				
Similitude	yes		yes		yes		
Dissimilitude		yes		yes	yes		
Resemblance	yes				yes		
Dissemblance		yes			yes		
Ordinate	yes					yes	yes
Nonstrict order	yes		yes			yes	yes
Strict order		yes	yes			yes	yes

29. COMMON MEMBERSHIP FUNCTIONS

In the tables that follow we have presented various continuous membership functions that are useful for representing numerical fuzzy subsets corresponding to the following fuzzy propositions:

x is small (29.1)–(29.7)

x is large (29.8)–(29.14)

$|x|$ is small (29.15)–(29.21)

$|x|$ is large (29.22)–(29.28)

With respect to these one may construct numerical fuzzy subsets relative to two variables. We shall show how to proceed. Also in the same section we shall show how to analyze or synthesize transitive fuzzy relations.

REFERENCE SETS: $\mathbf{R}^+$, $\mathbf{N}$

MEMBERSHIP FUNCTION CORRESPONDING TO "x IS SMALL"

	Domain	Curve	Function
(29.1)	$\mathbf{R}^+$ $\mathbf{N}$	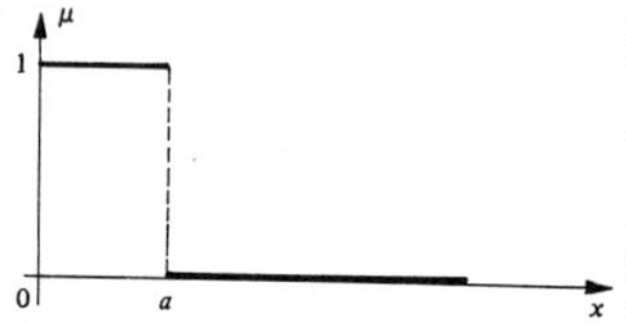	$\mu(x) = 1 \;,\; 0 \leqslant x \leqslant a \;,$ $= 0 \;,\; x > a \;.$
(29.2)	$\mathbf{R}^+$ $\mathbf{N}$	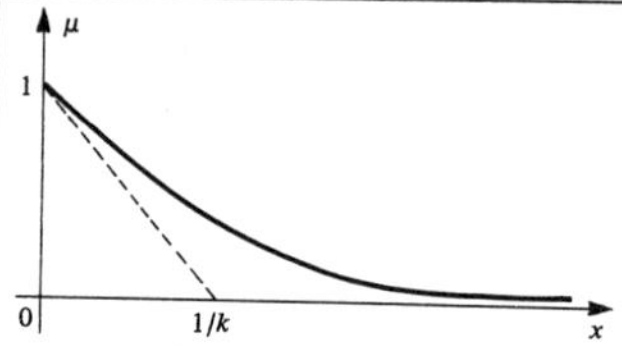	$\mu(x) = e^{-kx} \;,\; k > 0 \;.$
(29.3)	$\mathbf{R}^+$ $\mathbf{N}$	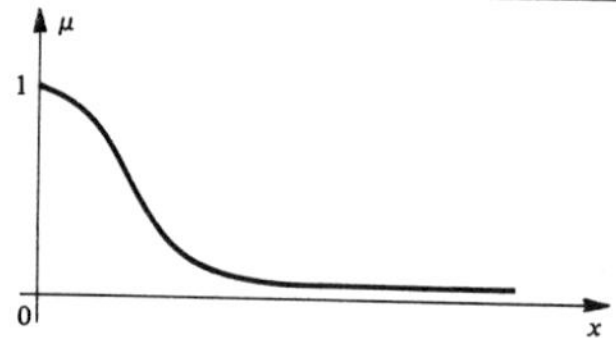	$\mu(x) = e^{-kx^2} \;,\; k > 0 \;.$
(29.4)	$\mathbf{R}^+$ $\mathbf{N}$	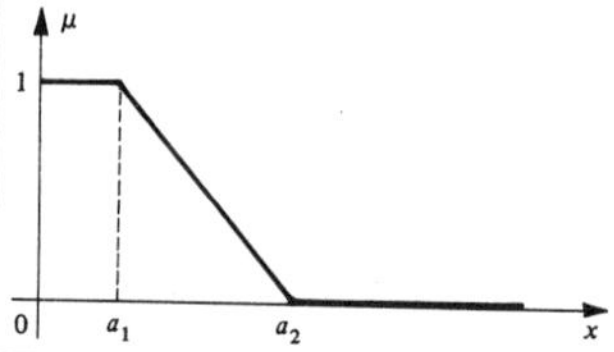	$\mu(x) = 1 \;,\; 0 \leqslant x \leqslant a_1 \;,$ $= \dfrac{a_2 - x}{a_2 - a_1} \;,\; a_1 \leqslant x \leqslant a_2 \;,$ $= 0 \;,\; a_2 \leqslant x \;.$
(29.5)	$\mathbf{R}^+$ $\mathbf{N}$	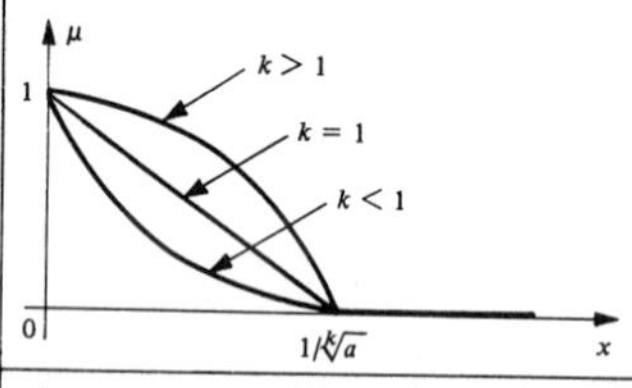	$\mu(x) = 1 - ax^k \;,\; 0 \leqslant x \leqslant \dfrac{1}{\sqrt[k]{a}} \;,$ $= 0 \;,\; \dfrac{1}{\sqrt[k]{a}} \leqslant x \;.$
(29.6)	$\mathbf{R}^+$ $\mathbf{N}$	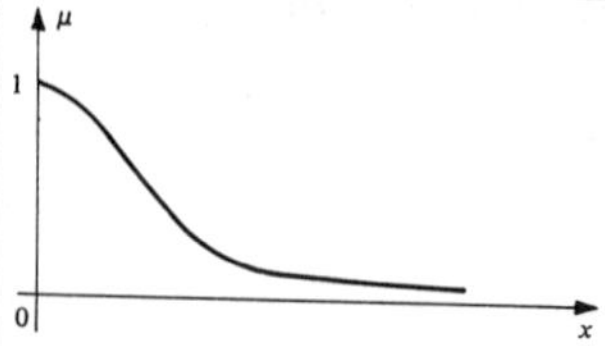	$\mu(x) = \dfrac{1}{1 + kx^2} \;,$ $k > 1 \;.$
(29.7)	$\mathbf{R}^+$ $\mathbf{N}$	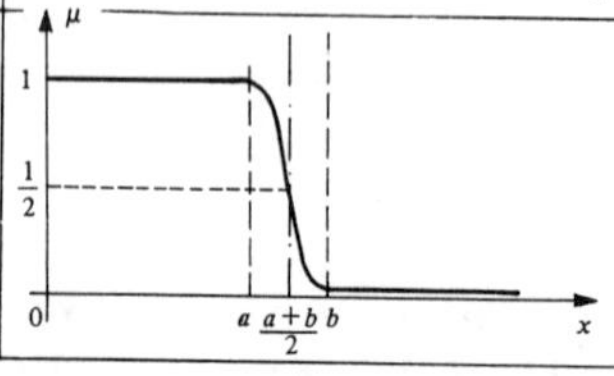	$\mu(x) = 1 \;,\; 0 \leqslant x \leqslant a,$ $= \dfrac{1}{2} - \dfrac{1}{2} \sin \dfrac{\pi}{b - a}\left(x - \dfrac{a + b}{2}\right),$ $a \leqslant x \leqslant b,$ $= 0 \;,\; b \leqslant x \;.$

REFERENCE SETS: $\mathbf{R}^+$, $\mathbf{N}$

MEMBERSHIP FUNCTION CORRESPONDING TO "a IS LARGE"

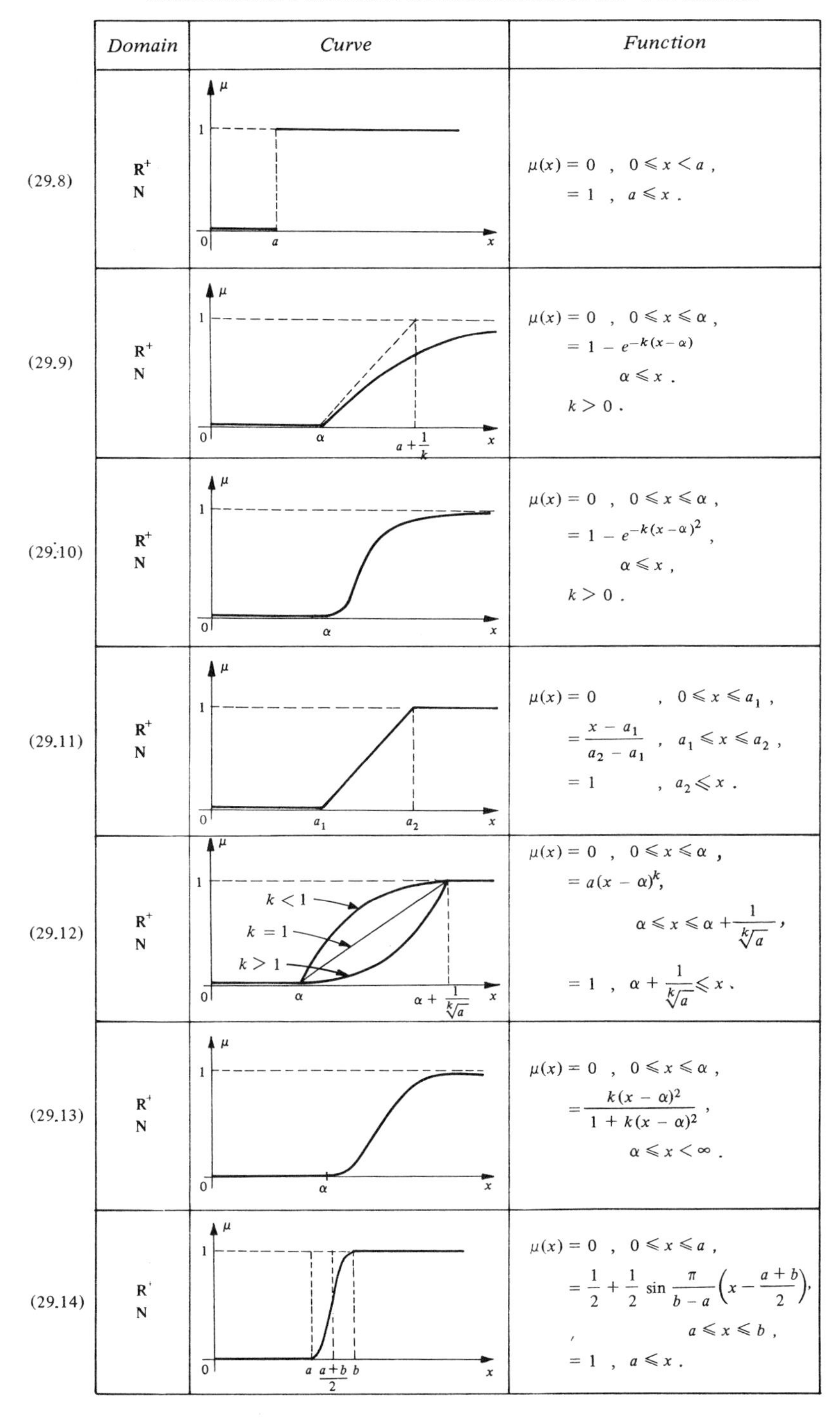

	Domain	*Curve*	*Function*
(29.8)	$\mathbf{R}^+$ $\mathbf{N}$		$\mu(x) = 0$, $0 \leqslant x < a$, $= 1$, $a \leqslant x$.
(29.9)	$\mathbf{R}^+$ $\mathbf{N}$		$\mu(x) = 0$, $0 \leqslant x \leqslant \alpha$, $= 1 - e^{-k(x-\alpha)}$ $\alpha \leqslant x$. $k > 0$.
(29.10)	$\mathbf{R}^+$ $\mathbf{N}$		$\mu(x) = 0$, $0 \leqslant x \leqslant \alpha$, $= 1 - e^{-k(x-\alpha)^2}$, $\alpha \leqslant x$, $k > 0$.
(29.11)	$\mathbf{R}^+$ $\mathbf{N}$		$\mu(x) = 0$, $0 \leqslant x \leqslant a_1$, $= \dfrac{x - a_1}{a_2 - a_1}$, $a_1 \leqslant x \leqslant a_2$, $= 1$, $a_2 \leqslant x$.
(29.12)	$\mathbf{R}^+$ $\mathbf{N}$		$\mu(x) = 0$, $0 \leqslant x \leqslant \alpha$, $= a(x - \alpha)^k$, $\alpha \leqslant x \leqslant \alpha + \dfrac{1}{\sqrt[k]{a}}$, $= 1$, $\alpha + \dfrac{1}{\sqrt[k]{a}} \leqslant x$.
(29.13)	$\mathbf{R}^+$ $\mathbf{N}$		$\mu(x) = 0$, $0 \leqslant x \leqslant \alpha$, $= \dfrac{k(x-\alpha)^2}{1 + k(x-\alpha)^2}$, $\alpha \leqslant x < \infty$.
(29.14)	$\mathbf{R}^+$ $\mathbf{N}$		$\mu(x) = 0$, $0 \leqslant x \leqslant a$, $= \dfrac{1}{2} + \dfrac{1}{2} \sin \dfrac{\pi}{b-a}\left(x - \dfrac{a+b}{2}\right)$, $a \leqslant x \leqslant b$, $= 1$, $a \leqslant x$.

REFERENCE SETS: **R, Z**
MEMBERSHIP FUNCTION CORRESPONDING TO "$|x|$ IS SMALL"

	Domain	*Curve*	*Function*
(29.15)	**R** **Z**	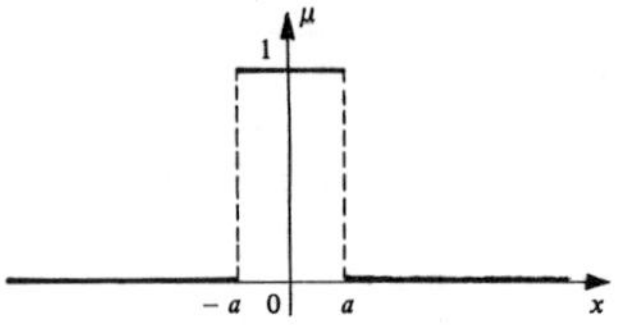	$\mu(x) = 0 \quad , \quad -\infty < x < a \,,$ $= 1 \quad , \quad -a \leqslant x \leqslant a \,,$ $= 0 \quad , \quad a \leqslant x \,.$
(29.16)	**R** **Z**	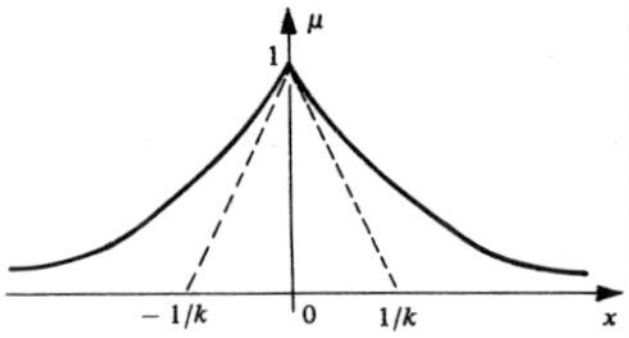	$\mu(x) = e^{kx} \quad , \quad -\infty < x \leqslant 0 \,,$ $= e^{-kx} \,, \quad 0 \leqslant x < \infty ,$ $k > 1 \,.$
(29.17)	**R** **Z**	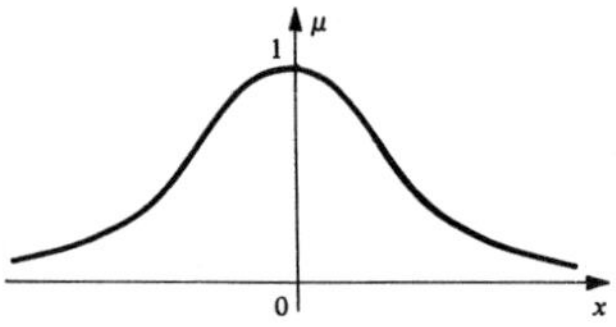	$\mu(x) = e^{-kx^2} \,.$
(29.18)	**R** **Z**	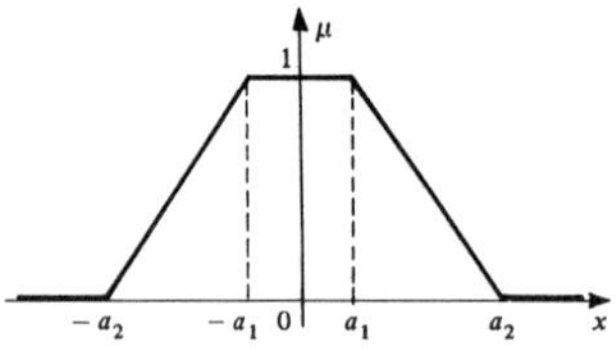	$\mu(x) = 0 \quad , \quad -\infty < x \leqslant -a_2 \,,$ $= \dfrac{a_2 + x}{a_2 - a_1} \,, \quad -a_2 \leqslant x \leqslant -a_1 \,,$ $= 1 \quad , \quad -a_1 \leqslant x \leqslant a_1 \,,$ $= \dfrac{a_2 - x}{a_2 - a_1} \,, \quad a_1 \leqslant x \leqslant a_2 \,,$ $= 0 \quad , \quad a_2 \leqslant x < \infty \,.$
(29.19)	**R** **Z**	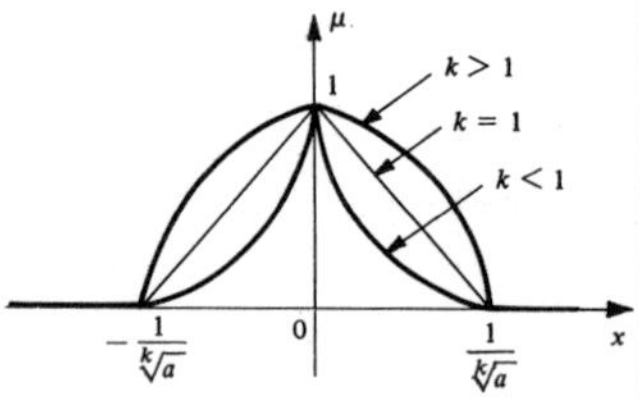	$\mu(x) = 0 \qquad -\infty < x \leqslant -\dfrac{1}{\sqrt[k]{a}}$ $= 1 - a(-x)^k \,, \quad -\dfrac{1}{\sqrt[k]{a}} \leqslant x \leqslant 0 \,,$ $= 1 - a(x)^k \quad , \quad 0 \leqslant x \leqslant \dfrac{1}{\sqrt[k]{a}} \,,$ $= 0 \quad , \quad \dfrac{1}{\sqrt[k]{a}} \leqslant x < \infty \,.$
(29.20)	**R** **Z**	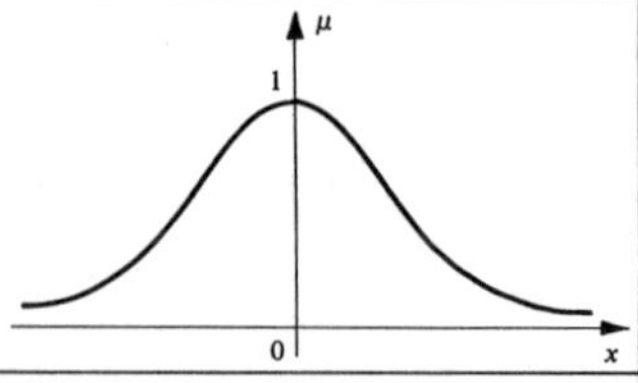	$\mu(x) = \dfrac{1}{1 + kx^2} \,,$ $k > 1 \,.$
(29.21)	**R** **Z**	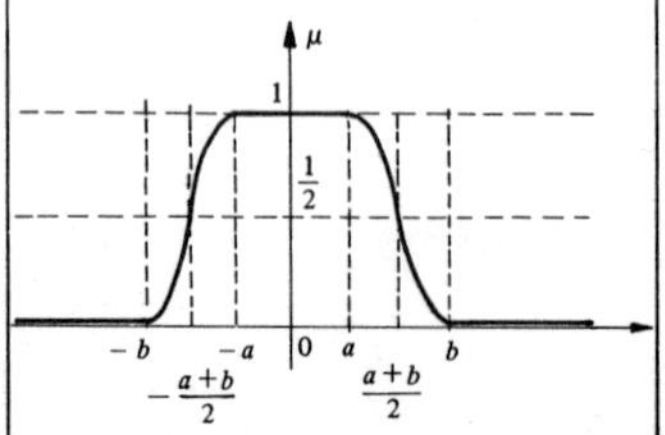	$\mu(x) = 0 \quad , \quad -\infty < x \leqslant -b \,,$ $= \dfrac{1}{2} + \dfrac{1}{2} \sin \dfrac{\pi}{b - a} \left(x + \dfrac{a + b}{2} \right),$ $-b \leqslant x \leqslant -a \,,$ $= 1 \quad , \quad -a \leqslant x \leqslant a \,,$ $= \dfrac{1}{2} - \dfrac{1}{2} \sin \dfrac{\pi}{b - a} \left(x - \dfrac{a + b}{2} \right),$ $a \leqslant x \leqslant b \,,$ $= 0 \quad , \quad b \leqslant x < \infty \,.$

REFERENCE SETS: **R, Z**
MEMBERSHIP FUNCTION CORRESPONDING TO "$|x|$ IS LARGE"

	Domain	*Curve*	*Function*
(29.22)	**R** **Z**	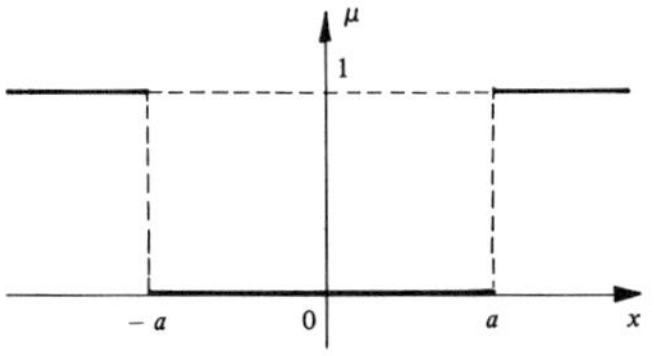	$\mu(x) = 1 \;,\; -\infty < x < -a \;,$ $= 0 \;,\; -a < x < a \;,$ $= 1 \;,\; a < x < \infty \;.$
(29.23)	**R** **Z**	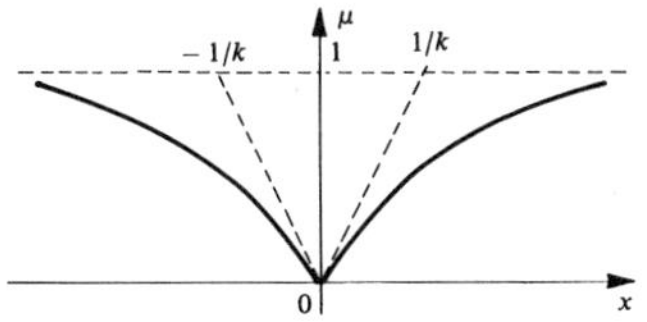	$\mu(x) = 1 - e^{kx} \;,\; -\infty < x \leqslant 0 \;,$ $= 1 - e^{-kx} \;,\; 0 \leqslant x < \infty \;,$ $k > 1 \;,$
(29.24)	**R** **Z**	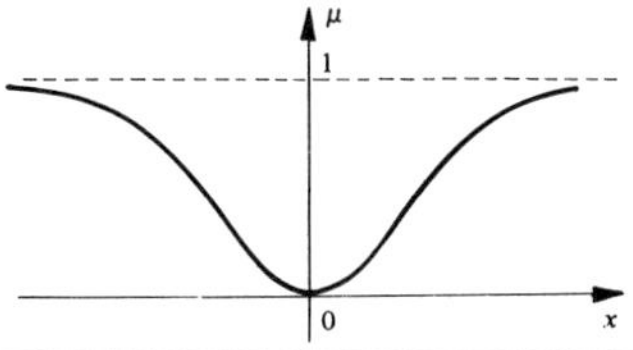	$\mu(x) = 1 - e^{-kx^2} \;,$ $k > 1 \;.$
(29.25)	**R** **Z**	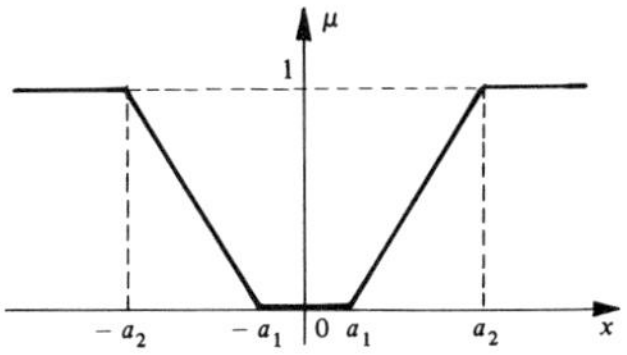	$\mu(x) = 1 \;,\; -\infty < x < -a_2 \;,$ $= -\dfrac{x + a_1}{a_2 - a_1} \;,\; -a_2 \leqslant x \leqslant -a_1 \;,$ $= 0 \;,\; -a_1 \leqslant x \leqslant a_1 \;,$ $= \dfrac{x - a_1}{a_2 - a_1} \;,\; a_1 \leqslant x \leqslant a_2 \;,$ $= 1 \;,\; a_2 \leqslant x < \infty \;.$
(29.26)	**R** **Z**	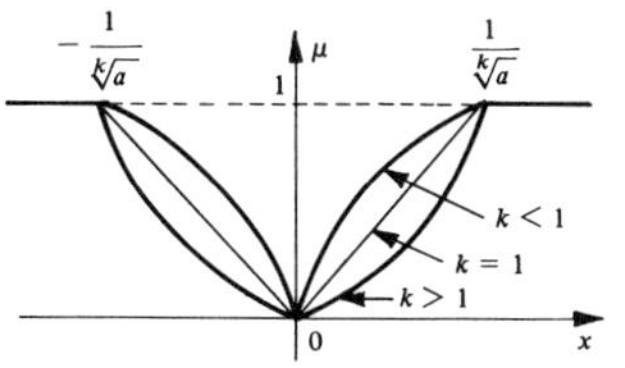	$\mu(x) = 1 \;,\; -\infty < x < -\dfrac{1}{\sqrt[k]{a}} \;,$ $= a(-x)^k \;,\; -\dfrac{1}{\sqrt[k]{a}} \leqslant x \leqslant 0 \;,$ $= ax^k \;,\; 0 \leqslant x \leqslant \dfrac{1}{\sqrt[k]{a}} \;,$ $= 1 \;,\; \dfrac{1}{\sqrt[k]{a}} \leqslant x < \infty \;.$
(29.27)	**R** **Z**	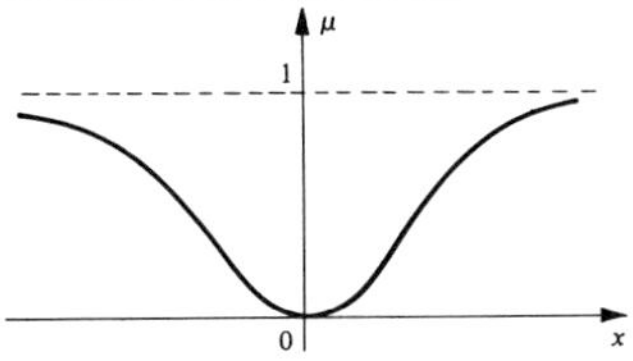	$\mu(x) = \dfrac{kx^2}{1 + kx^2} = \dfrac{1}{1 + \dfrac{1}{kx^2}} \;,$ $k > 1 \;.$
(29.28)	**R** **Z**	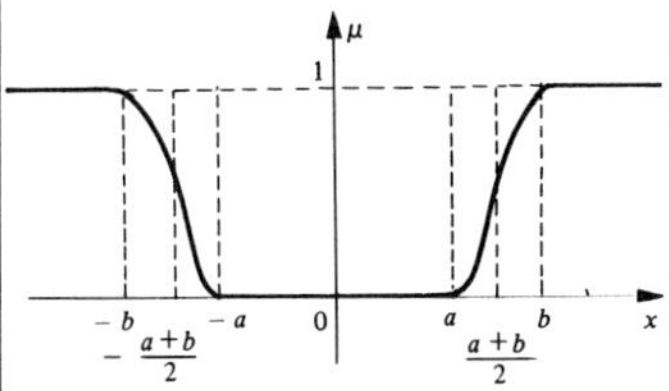	$\mu(x) = 1 \;,\; -\infty < x \leqslant -b \;,$ $= \dfrac{1}{2} - \dfrac{1}{2} \sin \dfrac{\pi}{b - a}\left(x + \dfrac{a + b}{2}\right),$ $-b \leqslant x \leqslant -a \;,$ $= 0 \;,\; -a \leqslant x \leqslant a \;,$ $= \dfrac{1}{2} + \dfrac{1}{2} \sin \dfrac{\pi}{b - a}\left(x - \dfrac{a + b}{2}\right),$ $a \leqslant x \leqslant b \;,$ $= 1 \;,\; b \leqslant x < \infty \;.$

REFERENCE SETS: $\mathbf{R}^+ \times \mathbf{R}^+$, $\mathbf{R} \times \mathbf{R}$, $\mathbf{N} \times \mathbf{N}$, $\mathbf{Z} \times \mathbf{Z}$

A. Cylindrical membership functions,† of the type

(29.29) $$\mu(x, y) = f(x^2 + y^2)$$

corresponding to "$x^2 + y^2$ has property $\mathscr{P}$".

Take the curves and functions (29.1)–(29.14) and replace

$$x \quad \text{by} \quad \rho = \sqrt{x^2 + y^2}$$

For (29.1)–(29.14), property $\mathscr{P}$ will be

$x^2 + y^2$ is small

or x and y are small

For (28.8)–(29.14), property $\mathscr{P}$ will be

$x^2 + y^2$ is large

or x and y are large

Example. – With reference to (29.6), one may see

(29.30) $$\mu(x, y) = \frac{1}{1 + k(x^2 + y^2)}$$

B. Hyperbolic membership functions, of the type

(29.31) $$\mu(x, y) = f(|y - x|)$$

or

(29.32) $$\mu(x, y) = f(|y^2 - x^2|)$$

corresponding to $|y - x|$ or $|y^2 - x^2|$ have property $\mathscr{P}$.

Take the curves and functions of (29.1)–(29.14) and replace

(29.33) $$x \quad \text{by} \quad \rho = |y - x|$$

or

$$\rho = \sqrt{|y^2 - x^2|}\,.$$

For (29.1)–(29.7), property $\mathscr{P}$ will be

y is very near x

For (29.8)–(29.14), it will be

y is very different than x

One may also use $\rho = |y - kx|$ with k sufficiently large and reversing the opposite properties above.

Remark. – One knows that

(29.34) $$e^{-u} = \frac{1}{1 + u + \dfrac{u^2}{2!} + \dfrac{u^3}{3!} + \cdots}\,.$$

Thus, the functions

(29.35) $$e^{-u} \quad \text{et} \quad \frac{1}{1 + u}$$

†We shall find it convenient to describe it so.

will give similar results, taking these as membership functions when $u = \phi(x)$, $u = (\sqrt{x^2 + y^2})$, $u = \phi(\sqrt{|y^2 - x^2|})$ and similarly for the other variables.

Determination of the property of max–min transitivity in the case of continuous membership function of a relation. It is generally very easy to evaluate a membership function $\mu_{\underset{\sim}{\mathcal{R}}}(x, y)$ if it presents one of the following properties:

reflexivity,
symmetry,
antisymmetry.

But it is generally much less easy to be concerned with transitivity. We shall first consider max–min transitivity, then max–product transitivity.†

Recall that max–min transitivity is expressed by the property

$$\mu_{\underset{\sim}{\mathcal{R}}}(x, z) \geqslant \bigvee_{y} [\mu_{\underset{\sim}{\mathcal{R}}}(x, y) \wedge \mu_{\underset{\sim}{\mathcal{R}}}(y, z)] . \tag{29.36}$$

In Figure 29.1 we have shown how to obtain the member on the right of (29.36). In this example there is a single intersection point between $\mu_{\underset{\sim}{\mathcal{R}}}(x, y)$ and $\mu_{\underset{\sim}{\mathcal{R}}}(y, z)$ when x and z are taken as parameters; there may exist several in other cases, but each time one determines the maximum y_{M}. In the sequel it is convenient to proceed in the following fashion:

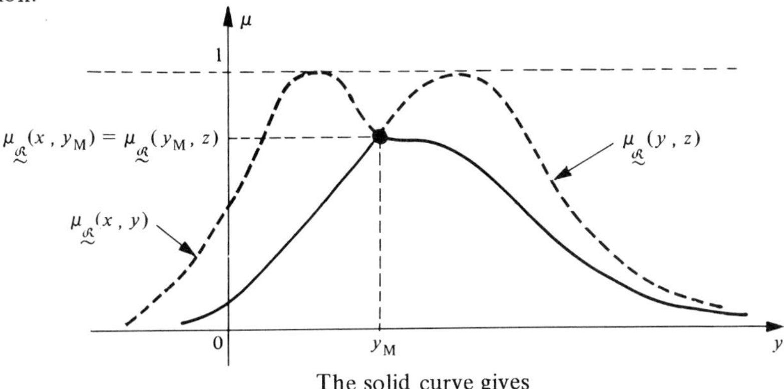

The solid curve gives
$\mu_{\underset{\sim}{\mathcal{R}}}(x, y) \wedge \mu_{\underset{\sim}{\mathcal{R}}}(y, z)$

Fig. 29.1

(1) Determine the point y_{M} as a function of x and z such that

$$\mu_{\underset{\sim}{\mathcal{R}}}(x, y_{\mathrm{M}}) = \mu_{\underset{\sim}{\mathcal{R}}}(y_{\mathrm{M}}, z) . \tag{29.37}$$

(2) Substitute the value of y_{M} as a function of x and z into $\mu_{\underset{\sim}{\mathcal{R}}}(x, y_{\mathrm{M}})$ or in $\mu_{\underset{\sim}{\mathcal{R}}}(y_{\mathrm{M}}, z)$; this gives a function $\lambda(x, z)$.

(3) Compare $\lambda(x, z)$ to $\mu_{\underset{\sim}{\mathcal{R}}}(x, z)$. If

$$\forall (x, z) \quad : \quad \mu_{\underset{\sim}{\mathcal{R}}}(x, z) \geqslant \lambda(x, z) , \tag{29.38}$$

then the relation $\underset{\sim}{\mathcal{R}}$ is transitive. If

†One may easily pass to proofs involving min–max transitivity and min–sum transitivity, or even ordinary min–addition.

(29.39)
$$\exists (x, z) \quad : \quad \mu_{\underset{\sim}{\mathcal{R}}}(x, z) < \lambda(x, z) ,$$

then the relation $\underset{\sim}{\mathcal{R}}$ is not transitive.

We consider several examples.

Example 1. Consider the fuzzy relation $\underset{\sim}{\mathcal{R}}$ defined for $x \in \mathbf{R}^+$ and $y \in \mathbf{R}^+$:

(29.40)
$$\begin{aligned} \mu_{\underset{\sim}{\mathcal{R}}}(x, y) &= e^{-x} & y < x \\ &= 1 & y = x \\ &= e^{-y} & y > x . \end{aligned}$$

In Figure 29.2 we have represented, as a function of y, the function $\mu_{\underset{\sim}{\mathcal{R}}}(x, y)$ (x taken as a parameter) and $\mu_{\underset{\sim}{\mathcal{R}}}(y, z)$ (z taken as a parameter) in the cases where $x < z$ (Figure 29.2a) and $x > z$ (Figure 29.2b). In these figures ABCD represents $\mu_{\underset{\sim}{\mathcal{R}}}(x, y)$ (with x as parameter) and A'B'CD represents $\mu_{\underset{\sim}{\mathcal{R}}}(y, z)$ (with z as parameter).

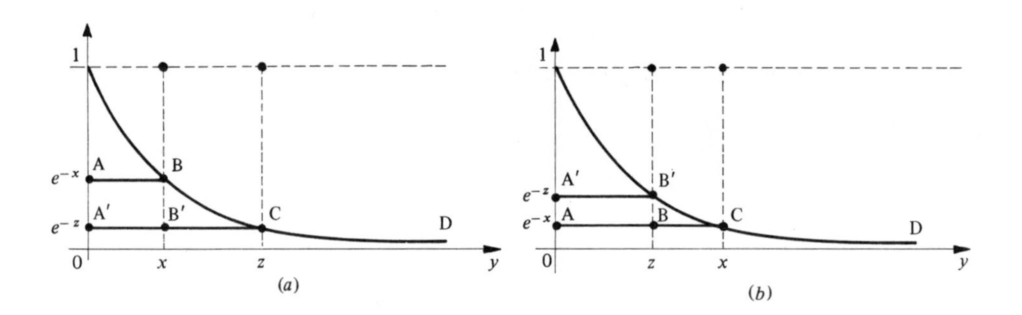

Fig. 29.2

In the $x < z$ case the max–min is equal to e^{-z}, and in the $x > z$ case the max–min is equal to e^{-x}. Thus, one may write

(29.41)
$$\begin{aligned} \lambda(x, z) &= e^{-z} \quad , \quad x \leqslant z , \\ &= e^{-x} \quad , \quad x \geqslant z . \end{aligned}$$

Compare $\lambda(x, z)$ with $\mu_{\underset{\sim}{\mathcal{R}}}(x, z)$ given by (29.40) where z replaces y:

(29.42)
$$\begin{aligned} \mu_{\underset{\sim}{\mathcal{R}}}(x ; z) &= e^{-z} , \quad x < z , \\ &= 1 \quad , \quad x = z , \\ &= e^{-x} , \quad x > z . \end{aligned}$$

By comparing one sees that

(29.43)
$$\mu_{\underset{\sim}{\mathcal{R}}}(x, z) = \lambda(x, z) , \quad x \neq z ,$$

(29.44)
$$\mu_{\underset{\sim}{\mathcal{R}}}(x, z) > \lambda(x, z) , \quad x = z ,$$

Then $\underset{\sim}{\mathcal{R}}$ is a transitive relation. We note that this relation is a similitude relation.

In Figure 29.3 we have represented with the aid of a matrix the fuzzy relation corresponding to (29.40) but using **N** instead of $\mathbf{R}^+$. This shows the particular arrangement of the values of the membership function. The reader should involve himself in applying and verifying transitivity in this case using (29.36). The row-by-column max–min operation will permit one to check (29.43) and (29.44).

$x \backslash y$	0	1	2	3	4	5	6	7	...
0	1	e^{-1}	e^{-2}	e^{-3}	e^{-4}	e^{-5}	e^{-6}	e^{-7}	...
1	e^{-1}	1	e^{-2}	e^{-3}	e^{-4}	e^{-5}	e^{-6}	e^{-7}	...
2	e^{-2}	e^{-2}	1	e^{-3}	e^{-4}	e^{-5}	e^{-6}	e^{-7}	...
3	e^{-3}	e^{-3}	e^{-3}	1	e^{-4}	e^{-5}	e^{-6}	e^{-7}	...
4	e^{-4}	e^{-4}	e^{-4}	e^{-4}	1	e^{-5}	e^{-6}	e^{-7}	...
5	e^{-5}	e^{-5}	e^{-5}	e^{-5}	e^{-5}	1	e^{-6}	e^{-7}	...
6	e^{-6}	e^{-6}	e^{-6}	e^{-6}	e^{-6}	e^{-6}	1	e^{-7}	...
7	e^{-7}	e^{-7}	e^{-7}	e^{-7}	e^{-7}	e^{-7}	e^{-7}	1	...
⋮	⋮	⋮	⋮	⋮	⋮	⋮	⋮	⋮	⋱

Fig. 29.3

Example 2. Consider the fuzzy relation $\underset{\sim}{\mathcal{R}}$ defined for $x \in \mathbf{R}$ and $y \in \mathbf{R}$:

$$\mu_{\underset{\sim}{\mathcal{R}}}(x, y) = e^{-(x-y)^2} : \tag{29.45}$$

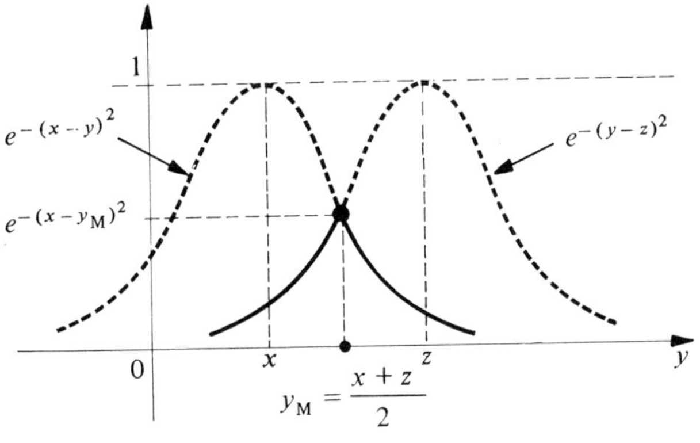

Fig. 29.4

One finds easily

$$e^{-(x-y_M)^2} = e^{-(y_M-z)^2}, \tag{29.46}$$

Thus

$$x - y_M = y_M - z, \tag{29,47}$$

or

$$y_M = \frac{x+z}{2}. \tag{29.48}$$

See Figure 29.4. Putting this value in the member on the right of (29.45), we have

$$\begin{aligned} \lambda(x, z) &= e^{-(x-y_M)^2} \\ &= e^{\frac{-(x-z)^2}{4}}. \end{aligned} \tag{29.49}$$

And we see that

$$\forall (x, z) \quad : \quad e^{-(x-z)^2} \leqslant e^{\frac{-(x-z)^2}{4}}, \tag{29.50}$$

that is

$$\forall (x, z) \quad : \quad \mu_{\underset{\sim}{\mathcal{R}}}(x, z) \leqslant \lambda(x, z). \tag{29.51}$$

Thus, this relation $\underset{\sim}{\mathcal{R}}$ is not transitive. We note that sometimes this is a resemblance relation.

If Figure 29.5 we have represented the corresponding relation $\underset{\sim}{\mathcal{R}}$ but using **N** instead of $\mathbf{R}^+$.

$x \backslash y$	0	1	2	3	4	5	6	7	...
0	1	e^{-1}	e^{-4}	e^{-9}	e^{-16}	e^{-25}	e^{-36}	e^{-49}	...
1	e^{-1}	1	e^{-1}	e^{-4}	e^{-9}	e^{-16}	e^{-25}	e^{-36}	...
2	e^{-4}	e^{-1}	1	e^{-1}	e^{-4}	e^{-9}	e^{-16}	e^{-25}	...
3	e^{-9}	e^{-4}	e^{-1}	1	e^{-1}	e^{-4}	e^{-9}	e^{-16}	...
4	e^{-16}	e^{-9}	e^{-4}	e^{-1}	1	e^{-1}	e^{-4}	e^{-9}	...
5	e^{-25}	e^{-16}	e^{-9}	e^{-4}	e^{-1}	1	e^{-1}	e^{-4}	...
6	e^{-36}	e^{-25}	e^{-16}	e^{-9}	e^{-4}	e^{-1}	1	e^{-1}	...
7	e^{-49}	e^{-36}	e^{-25}	e^{-16}	e^{-9}	e^{-4}	e^{-1}	1	...
⋮	⋮	⋮	⋮	⋮	⋮	⋮	⋮	⋮	⋱

Fig. 29.5

Example 3. Consider the fuzzy relation $\underset{\sim}{\mathcal{R}}$ defined for $x \in \mathbf{R}^+$ and $y \in \mathbf{R}^+$:

$$\mu_{\underset{\sim}{\mathcal{R}}}(x, y) = \frac{xy}{1 + xy} \qquad y > x , \tag{29.52}$$

$$= 0 \qquad y \leqslant x .$$

Figure 29.6 shows that the min–max corresponds to $y_{\mathbf{M}} = z$, whence

$$\lambda(x, y_{\mathrm{M}}) = \lambda(x, z), \tag{29.53}$$

$$= \frac{xz}{1 + xz}, \; z > x ,$$

$$= 0 , \qquad z < x .$$

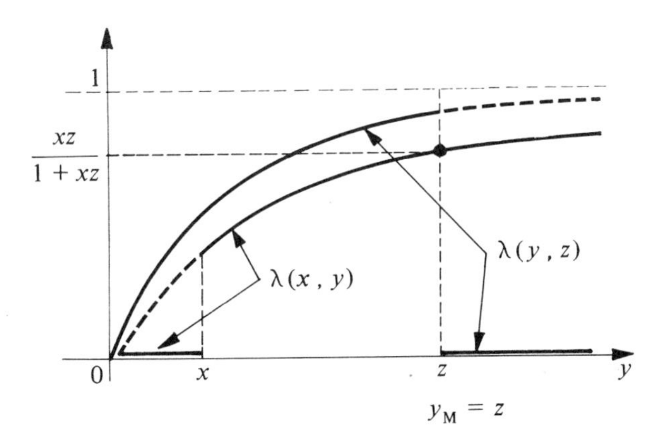

Fig. 29.6

We see that

$$\mu_{\underset{\sim}{\mathcal{R}}}(x, z) = \lambda(x, z) . \tag{29.54}$$

Thus the relation $\underset{\sim}{\mathcal{R}}$ is indeed transitive. One may also verify that this relation is a total fuzzy order.

Figure 29.7 represents the corresponding relation with $\mathbf{R}^+$ replaced by $\mathbf{N}$.

Remark on the synthesis of a transitive fuzzy relation. If the analysis of a fuzzy relation in $\mathbf{R}$ or $\mathbf{R}^+$ is not very easy, as we have seen, its synthesis is even more difficult except in certain very simple particular cases. Thus, a good method consists in carrying out the synthesis of the relation in $\mathbf{N}$, and then passing from there to $\mathbf{R}^+$ or $\mathbf{R}$.

The decomposition theorem for a similitude relation (Section 27) and that for a perfect order relation (Section 28) allow the easy synthesis of the corresponding relations. One may provide an algorithm.

Algorithm for construction of a fuzzy transitive relation in a denumerable set.

(1) We are given a sequence, finite or not, of numbers $x_i \in [0, 1]$, strictly ordered on i:

$$1 > a_1 > a_2 > \cdots > a_r > \cdots > 0 . \tag{29.55}$$

x \ y	0	1	2	3	4	5	6	7	...
0	0	0	0	0	0	0	0	0	...
1	0	0	$\frac{2}{3}$	$\frac{3}{4}$	$\frac{4}{5}$	$\frac{5}{6}$	$\frac{6}{7}$	$\frac{7}{8}$	...
2	0	0	0	$\frac{6}{7}$	$\frac{8}{9}$	$\frac{10}{11}$	$\frac{12}{13}$	$\frac{14}{15}$	...
3	0	0	0	0	$\frac{12}{13}$	$\frac{15}{16}$	$\frac{18}{19}$	$\frac{21}{22}$	...
4	0	0	0	0	0	$\frac{20}{21}$	$\frac{24}{25}$	$\frac{28}{29}$	...
5	0	0	0	0	0	0	$\frac{30}{31}$	$\frac{35}{36}$	...
6	0	0	0	0	0	0	0	$\frac{42}{43}$	...
7	0	0	0	0	0	0	0	0	...
⋮	⋮	⋮	⋮	⋮	⋮	⋮	⋮	⋮	⋱

Fig. 29.7

Step-by-step construct a transitive ordinary graph by enriching the arcs, always maintaining transitivity. To each passage from a transitive graph to a transitive graph richer in arcs, this constituting a step, associate with the corresponding arcs of the fuzzy graph the value a_{i+1} that follows the preceding a_i value. The finite fuzzy graph obtained by stopping at step i is transitive; if the procedure is not stopped at a finite number, one obtains an infinite graph that is transitive.

Example. Consider the infinite sequence:

$$1 > \frac{1}{2} > \frac{1}{3} > \frac{1}{4} > \cdots > \frac{1}{r} > \cdots > 0 . \tag{29.56}$$

We propose to construct a transitive and antireflexive fuzzy graph having perfect antisymmetry. The construction will be carried out according to the order indicated in Figure 29.8, where the addition of arcs is arbitrary except that, at each step, transitivity must be maintained. In Figure 29.9 one may see how to construct the fuzzy graph.

The same procedure may be used for preorders, similitudes, orders, etc.

From the matrix obtained one may see the corresponding relation $\underset{\sim}{\mathcal{R}}$ to be obtained for $\mathbf{R}^+$ and eventually for $\mathbf{R}$. This is not, evidently, always easy.

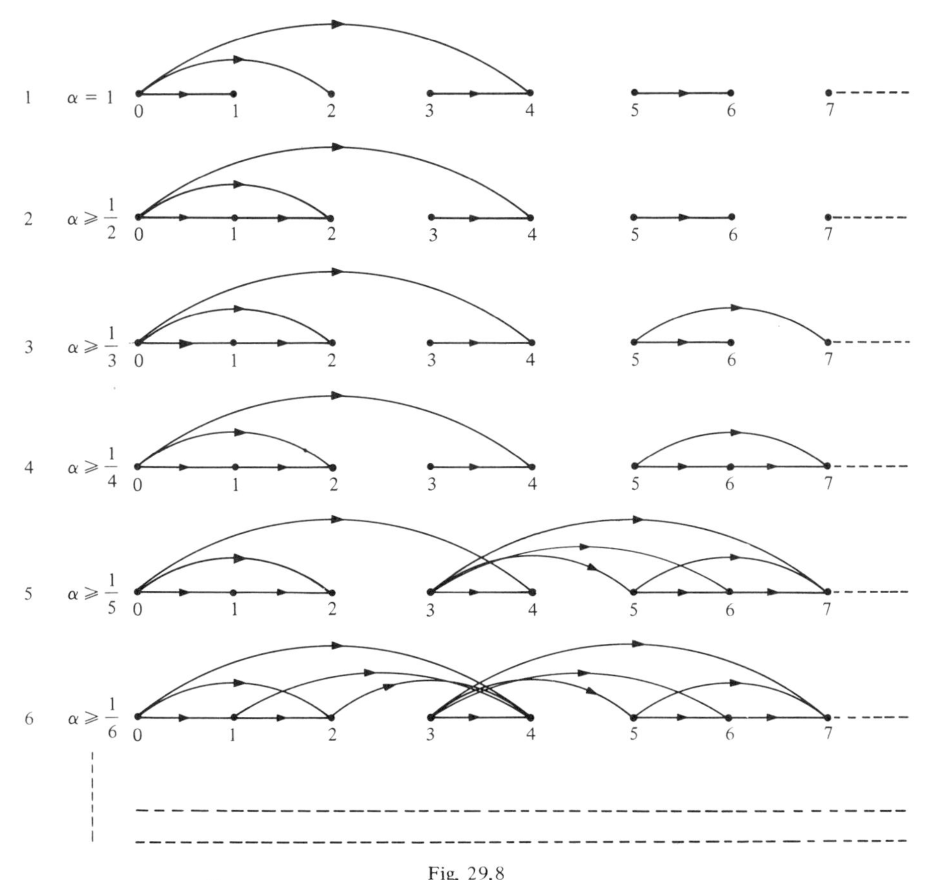

Fig. 29.8

$x \backslash y$	0	1	2	3	4	5	6	7	...
0	0	1 ①	1 ①		1 ①				...
1	0	0	$\frac{1}{2}$ ②		$\frac{1}{6}$ ⑥				...
2	0	0	0		$\frac{1}{6}$ ⑥				...
3	0	0	0	0	1 ①	$\frac{1}{5}$ ⑤	$\frac{1}{5}$ ⑤	$\frac{1}{5}$ ⑤	...
4	0	0	0	0	0				...
5	0	0	0	0	0	0	1 ①	$\frac{1}{3}$ ③	...
6	0	0	0	0	0	0	0	$\frac{1}{4}$ ④	...
7	0	0	0	0	0	0	0	0	...
⋮	⋮	⋮	⋮	⋮	⋮	⋮	⋮	⋮	⋱

Fig. 29.9

30. EXERCISES

II1. Consider the following fuzzy graphs; for each of them give the first projection, the second projection, and the global projection.

a)

$\underset{\sim}{\mathcal{R}}$	a	b	c	d
A	0,3	1	0,2	0,1
B	0,9	0,2	0	0,5
C	0,8	0,1	0,8	0,9
D	0,9	0,5	1	0,9
E	0,5	0	0,7	0,7

b)

$\underset{\sim}{\mathcal{R}}$	a	b	c	d
α	0	$\frac{1}{2}$	0	$\frac{1}{2}$
β	1	0	$\frac{1}{2}$	$\frac{1}{2}$
γ	1	0	0	1
δ	0	0	0	0

c)

$\underset{\sim}{\mathcal{R}}$	A	B	C	D	E	F
M	0	0,3	0,2	0,4	0,5	0,9
N	0,8	0	0	0,2	0,3	0,8

II2. Carry out the instruction for II1 for the following graphs:

a) $x \underset{\sim}{\mathcal{R}} y\,;\, x,\, y \in \mathbf{R}$ with :

$$\mu_{\underset{\sim}{\mathcal{R}}}(x\,,\,y) = \frac{1}{1 + k(x - y)^2}\ , \qquad k > 1\ .$$

b) $x \underset{\sim}{\mathcal{R}} y\,;\, x,\, y \in \mathbf{R}$ with :

$$\mu_{\underset{\sim}{\mathcal{R}}}(x\,,\,y) = \frac{1}{1 + k\,|x - y|}\ , \qquad k > 1\ .$$

II3. For Exercises II1a, b, c and II2a, b, calculate the support of each graph.

II4. Given the following relations

$\underset{\sim}{\mathcal{R}}_1$	y_1	y_2	y_3	y_4
x_1	0	0,1	0	0,4
x_2	0,5	1	0	0,7
x_3	0,8	0,9	0,9	1

$\underset{\sim}{\mathcal{R}}_2$	y_1	y_2	y_3	y_4
x_1	0,1	0	0,2	0,5
x_2	0	1	0,1	1
x_3	0,9	0,4	0,7	0

$\underset{\sim}{\mathcal{R}}_3$	y_1	y_2	y_3	y_4
x_1	0,5	0	0,2	0
x_2	0	1	0,1	0,2
x_3	0,9	0,4	0	1

Calculate:

(a) $\underset{\sim}{\mathcal{R}}_1 \cap \underset{\sim}{\mathcal{R}}_2$, b) $\underset{\sim}{\mathcal{R}}_1 \cup \underset{\sim}{\mathcal{R}}_3$, c) $\underset{\sim}{\mathcal{R}}_1 \cap \underset{\sim}{\mathcal{R}}_2 \cap \underset{\sim}{\mathcal{R}}_3$,

(d) $\underset{\sim}{\mathcal{R}}_1 \cap (\underset{\sim}{\mathcal{R}}_2 \cup \underset{\sim}{\mathcal{R}}_3)$, e) $\underset{\sim}{\mathcal{R}}_1 \cdot \underset{\sim}{\mathcal{R}}_2$, f) $\underset{\sim}{\mathcal{R}}_1 \hat{+} \underset{\sim}{\mathcal{R}}_2$

(g) $\overline{\underset{\sim}{\mathcal{R}}}_1 \cap \overline{\underset{\sim}{\mathcal{R}}}_2$, h) $\underset{\sim}{\mathcal{R}}_1 \oplus \underset{\sim}{\mathcal{R}}_3$, i) $\underset{\sim}{\mathcal{R}}_1 \hat{+} (\underset{\sim}{\mathcal{R}}_2 \oplus \underset{\sim}{\mathcal{R}}_3)$.

II5. Considering the example proposed in (23.35) in the text, give:

a) $\underset{\sim}{\mathcal{R}}_1 \cup \underset{\sim}{\mathcal{R}}_2$, b) $\overline{\underset{\sim}{\mathcal{R}}}_1$, c) $\underset{\sim}{\mathcal{R}}_1 \oplus \underset{\sim}{\mathcal{R}}_2$.

The results should be given infigures like thosewe did in Figure 12.11.

II6. For each of the fuzzy relations given in Exercise II4, give the nearest ordinary relation.

II7. For the following relations

$\underset{\sim}{\mathcal{R}}_1$	y_1	y_2	y_3	y_4	y_5
x_1	1	0,2	0	0,2	1
x_2	1	0,5	0,4	1	0,4
x_3	0,7	0	0,5	0	0,9

$\underset{\sim}{\mathcal{R}}_2$	z_1	z_2	z_3	z_4
y_1	0,5	0,8	0	0,7
y_2	0,7	0	0,5	0,8
y_3	1	1	1	0
y_4	0,5	0,2	0	0,4
y_5	0,9	0,7	0,8	0,7

$\underset{\sim}{\mathcal{R}}_3$	t_1	t_2	t_3	t_4	t_5
z_1	0,8	0	0,2	0,8	1
z_2	0,9	0,2	0	1	0
z_3	1	0,5	0,7	0	0,4
z_4	0,4	1	0	0,4	0,9

Carry out the composition (max–min composition):

a) $\underset{\sim}{\mathcal{R}}_2 \circ \underset{\sim}{\mathcal{R}}_1$, b) $\underset{\sim}{\mathcal{R}}_3 \circ \underset{\sim}{\mathcal{R}}_2 \circ \underset{\sim}{\mathcal{R}}_1$, c) $\overline{\underset{\sim}{\mathcal{R}}}_2 \circ \overline{\underset{\sim}{\mathcal{R}}}_1$

d) $(\underset{\sim}{\mathcal{R}}_3 \circ \underset{\sim}{\mathcal{R}}_2) \cap (\overline{\underset{\sim}{\mathcal{R}}}_3 \circ \overline{\underset{\sim}{\mathcal{R}}}_2)$, e) $\underset{\sim}{\mathcal{R}}_2 \cdot \underset{\sim}{\mathcal{R}}_1$ max–product composition.

II8. Consider the fuzzy relations (13.3) and (13.4). Give

a) $\underset{\sim}{\mathcal{R}}_2 \circ \overline{\underset{\sim}{\mathcal{R}}}_1$, b) $\overline{\underset{\sim}{\mathcal{R}}}_2 \circ \overline{\underset{\sim}{\mathcal{R}}}_1$.

II9. As we have done for example (13.34), decompose each of the relations $\underset{\sim}{\mathcal{R}}_1, \underset{\sim}{\mathcal{R}}_2$ and $\underset{\sim}{\mathcal{R}}_3$ of Exercise II7.

II10. Let

$$E_1 = \{x_1, x_2, x_3, x_4\},$$

$$E_2 = \{y_1, y_2, y_3, y_4, y_5\},$$

$$\underset{\sim}{A} = \{(x_1 \mid 0{,}3), (x_2 \mid 0{,}7), (x_3 \mid 0), (x_4 \mid 0{,}9)\}.$$

(1) Give the fuzzy subsets $\underset{\sim}{B}_1 \subset E_2$ and $\underset{\sim}{B}_2 \subset E_2$ induced by the following mappings Γ_1 and Γ_2:

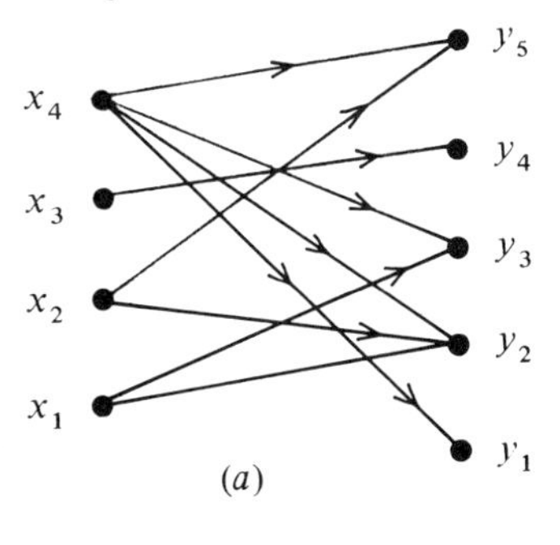

(a)

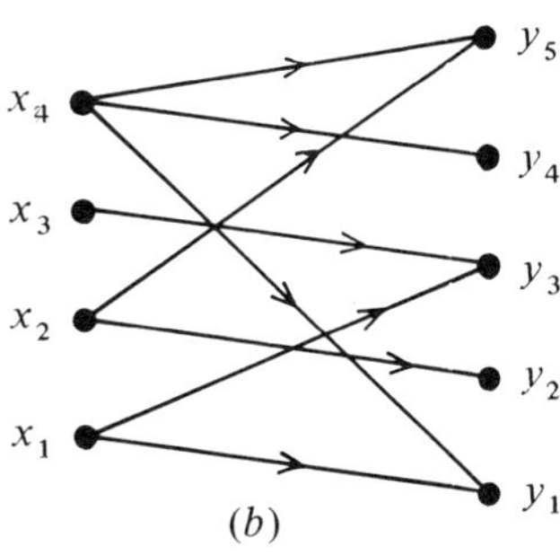

(b)

(2) Let

$$\underset{\sim}{\mathbf{B}} = \{(y_1 \mid 0{,}3)\ ,\ (y_2 \mid 0)\ ,\ (y_3 \mid 0{,}1)\ ,\ (y_4 \mid 1)\ ,\ (y_5 \mid 0{,}8)\}\ ,$$

Then give the fuzzy subsets $\underset{\sim}{\mathbf{A}} \subset \mathbf{E}_1$ and $\underset{\sim}{\mathbf{A}} \subset \mathbf{E}_2$ induced by Γ_1^{-1} and Γ_2^{-1}.

II11. Let $\mathbf{E}_1$, $\mathbf{E}_2$, and $\underset{\sim}{\mathbf{A}}$ be as given in Exercise II10. Let $\underset{\sim}{\mathcal{R}}$ be the following fuzzy relation:

$\underset{\sim}{\mathcal{R}}$	y_1	y_2	y_3	y_4	y_5
x_1	0,2	1	0	0,7	0,3
x_2	0	0	0,2	1	0,7
x_3	1	0,4	1	0,5	0,2
x_4	0	0,2	0,4	0,8	0

What is the fuzzy subset $\underset{\sim}{\mathbf{B}} \subset \mathbf{E}_2$ conditioned by $\underset{\sim}{\mathcal{R}}$ with respect to $\underset{\sim}{\mathbf{A}}$?

II12. Consider the three fuzzy relations $\underset{\sim}{\mathcal{R}}_1$, $\underset{\sim}{\mathcal{R}}_2$ and $\underset{\sim}{\mathcal{R}}_3$ below:

$\underset{\sim}{\mathcal{R}}_1$	y_1	y_2	y_3
x_1	0,3	1	0
x_2	0,4	0,2	0,8

$\underset{\sim}{\mathcal{R}}_2$	z_1	z_2	z_3	z_4
y_1	0,3	0,2	1	0
y_2	0,8	0	0,2	0
y_3	0,9	1	0,9	0,4

$\underset{\sim}{\mathcal{R}}_3$	t_1	t_2	t_3
z_1	0,4	0	1
z_2	1	0,3	0,2
z_3	1	1	0,3
z_4	0,1	0,3	0,2

What is the fuzzy subset $\underset{\sim}{\mathbf{D}} \subset \mathbf{T} = \{t_1, t_2, t_3\}$ conditioned by $\underset{\sim}{\mathcal{R}}_3 \circ \underset{\sim}{\mathcal{R}}_2 \circ \underset{\sim}{\mathcal{R}}_1$ with respect to

$$\underset{\sim}{\mathbf{A}} = \{(x_1 \mid 0{,}3)\ ,\ (x_2 \mid 0{,}7)\}\ ?$$

II13. Among the fuzzy binary relations below, which are

(a) symmetric, (b) reflexive, (c) transitive?

$\underset{\sim}{\mathcal{R}}_1$	A	B	C	D	E
A	0	1	1	1	1
B	0	0	0,9	0,7	0,3
C	0	0	0	0,7	0,3
D	0	0	0	0	0,3
E	0	0	0	0	0

$\underset{\sim}{\mathcal{R}}_2$	A	B	C	D	E
A	0	0,3	1	0	0,5
B	0,3	0,2	0	0,8	0,1
C	1	0	0	0,2	1
D	0	0,8	0,2	1	0,4
E	0,5	0,1	1	0,4	0,4

$\underset{\sim}{\mathcal{R}}_3$	A	B	C	D	E
A	1	0,5	0,5	0	0,7
B	0	1	0,7	0	0
C	0	1	1	0	0
D	0	0,3	0,3	1	0
E	1	0,5	0,5	0	1

$\underset{\sim}{\mathcal{R}}_4$	A	B	C	D	E
A	0	0	0,3	0,2	0
B	0,6	1	0,8	1	0,2
C	0,2	0	1	0,8	0,3
D	0	0	0	1	0
E	1	0	0,2	0,6	0

$\underset{\sim}{\mathcal{R}}_5$	A	B	C	D	E
A	1	0	0	0	0,4
B	0	1	0,9	1	0
C	0	0,9	1	0,9	0
D	0	1	0,9	1	0
E	0,4	0	0	0	1

$\underset{\sim}{\mathcal{R}}_6$	A	B	C	D	E
A	0	0,1	0	1	0,8
B	0,1	0	0	0	0,6
C	0	0	0	0,4	1
D	1	0	0,4	0	1
E	0,8	0,6	1	1	0

II14. Give the transitive closure of each of the relations $\underset{\sim}{\mathcal{R}}_i, i = 1, 2, 3, 4, 5, 6$, of Exercise II13.

II15. Carry out Exercise II14 but using max–product transitivity in place of max–min transitivity.

II16. Calculate the max–min transitive closure of each of the relation $\underset{\sim}{\mathcal{R}}$ below:

$\underset{\sim}{\mathcal{R}}_1$	x_1	x_2	x_3
x_1	0,3	0,2	0,8
x_2	1	0,7	0
x_3	0	1	0,8

$\underset{\sim}{\mathcal{R}}_2$	x_1	x_2	x_3	x_4	x_5
x_1	0	b	a	1	c
x_2	1	a	b	b	1
x_3	c	c	0	a	0
x_4	1	a	1	c	b
x_5	0	1	b	a	c

$0 < a < b < c < 1$.

$\underset{\sim}{\mathcal{R}}_3$	x_1	x_2	x_3	x_4	x_5
x_1	1	a	c	0	c
x_2	0	1	0	b	b
x_3	0	b	1	a	0
x_4	c	0	0	1	c
x_5	0	0	c	0	1

$0 < a < b < c < 1$.

$\underset{\sim}{\mathcal{R}}_4$	x_1	x_2	x_3	x_4	x_5	x_6	x_7	x_8	...
x_1	1	$\frac{1}{2}$	$\frac{1}{3}$	$\frac{1}{4}$	$\frac{1}{5}$	$\frac{1}{6}$	$\frac{1}{7}$	$\frac{1}{8}$	...
x_2	$\frac{1}{2}$	$\frac{1}{3}$	$\frac{1}{4}$	$\frac{1}{5}$	$\frac{1}{6}$	$\frac{1}{7}$	$\frac{1}{8}$	$\frac{1}{9}$	...
x_3	$\frac{1}{3}$	$\frac{1}{4}$	$\frac{1}{5}$	$\frac{1}{6}$	$\frac{1}{7}$	$\frac{1}{8}$	$\frac{1}{9}$	$\frac{1}{10}$	...
x_4	$\frac{1}{4}$	$\frac{1}{5}$	$\frac{1}{6}$	$\frac{1}{7}$	$\frac{1}{8}$	$\frac{1}{9}$	$\frac{1}{10}$	$\frac{1}{11}$	...
x_5	$\frac{1}{5}$	$\frac{1}{6}$	$\frac{1}{7}$	$\frac{1}{8}$	$\frac{1}{9}$	$\frac{1}{10}$	$\frac{1}{11}$	$\frac{1}{12}$	...
x_6	$\frac{1}{6}$	$\frac{1}{7}$	$\frac{1}{8}$	$\frac{1}{9}$	$\frac{1}{10}$	$\frac{1}{11}$	$\frac{1}{12}$	$\frac{1}{13}$	...
x_7	$\frac{1}{7}$	$\frac{1}{8}$	$\frac{1}{9}$	$\frac{1}{10}$	$\frac{1}{11}$	$\frac{1}{12}$	$\frac{1}{13}$	$\frac{1}{14}$	...
x_8	$\frac{1}{8}$	$\frac{1}{9}$	$\frac{1}{10}$	$\frac{1}{11}$	$\frac{1}{12}$	$\frac{1}{13}$	$\frac{1}{14}$	$\frac{1}{15}$	...
⋮	⋮	⋮	⋮	⋮	⋮	⋮	⋮	⋮	⋱

$\underset{\sim}{\mathcal{R}}_5$	x_1	x_2	x_3	x_4	x_5	x_6	x_7	x_8	...
x_1	1	$\frac{1}{2}$	$\frac{1}{3}$	$\frac{1}{4}$	$\frac{1}{5}$	$\frac{1}{6}$	$\frac{1}{7}$	$\frac{1}{8}$	...
x_2	$\frac{1}{2}$	$\frac{1}{2}$	$\frac{1}{3}$	$\frac{1}{4}$	$\frac{1}{5}$	$\frac{1}{6}$	$\frac{1}{7}$	$\frac{1}{8}$	...
x_3	$\frac{1}{3}$	$\frac{1}{3}$	$\frac{1}{3}$	$\frac{1}{4}$	$\frac{1}{5}$	$\frac{1}{6}$	$\frac{1}{7}$	$\frac{1}{8}$	...
x_4	$\frac{1}{4}$	$\frac{1}{4}$	$\frac{1}{4}$	$\frac{1}{4}$	$\frac{1}{5}$	$\frac{1}{6}$	$\frac{1}{7}$	$\frac{1}{8}$	...
x_5	$\frac{1}{5}$	$\frac{1}{5}$	$\frac{1}{5}$	$\frac{1}{5}$	$\frac{1}{5}$	$\frac{1}{6}$	$\frac{1}{7}$	$\frac{1}{8}$	...
x_6	$\frac{1}{6}$	$\frac{1}{6}$	$\frac{1}{6}$	$\frac{1}{6}$	$\frac{1}{6}$	$\frac{1}{6}$	$\frac{1}{7}$	$\frac{1}{8}$	...
x_7	$\frac{1}{7}$	$\frac{1}{7}$	$\frac{1}{7}$	$\frac{1}{7}$	$\frac{1}{7}$	$\frac{1}{7}$	$\frac{1}{7}$	$\frac{1}{8}$	...
x_8	$\frac{1}{8}$	$\frac{1}{8}$	$\frac{1}{8}$	$\frac{1}{8}$	$\frac{1}{8}$	$\frac{1}{8}$	$\frac{1}{8}$	$\frac{1}{8}$	...
⋮	⋮	⋮	⋮	⋮	⋮	⋮	⋮	⋮	⋱

II17. Consider the three relations $\underset{\sim}{\mathcal{R}}_1$, $\underset{\sim}{\mathcal{R}}_2$, $\underset{\sim}{\mathcal{R}}_3 \subset \mathbf{E} \times \mathbf{E}$:

$\underset{\sim}{\mathcal{R}}_1$	A	B	C	D
A	0,3	1	0	0
B	0,2	0	0,4	0
C	0,3	0	1	0
D	0,2	1	0,8	0

$\underset{\sim}{\mathcal{R}}_2$	A	B	C	D
A	0	0	0,5	0
B	0	1	0,4	1
C	0,6	0,3	0,9	0,2
D	0,5	0	1	0,2

$\underset{\sim}{\mathcal{R}}_3$	A	B	C	D
A	0,6	0,3	0	0,2
B	0,7	1	0	0
C	0	0,8	0	0
D	0,6	0,2	0,5	0

calculate:

a) $\widehat{\underset{\sim}{\mathcal{R}}_1 \circ \underset{\sim}{\mathcal{R}}_2}$, b) $\widehat{\underset{\sim}{\mathcal{R}}_2 \circ \underset{\sim}{\mathcal{R}}_3}$, c) $\widehat{\underset{\sim}{\mathcal{R}}_1 \circ \underset{\sim}{\mathcal{R}}_2 \circ \underset{\sim}{\mathcal{R}}_3}$

d) $\widehat{\underset{\sim}{\mathcal{R}}_1 \circ \underset{\sim}{\mathcal{R}}_2 \circ \underset{\sim}{\mathcal{R}}_1 \circ \underset{\sim}{\mathcal{R}}_2}$, e) $\widehat{\underset{\sim}{\mathcal{R}}_1^2 \circ \underset{\sim}{\mathcal{R}}_2^2}$.

f) $\check{\underset{\sim}{\mathcal{R}}}_1$, g) $\hat{\underset{\sim}{\mathcal{R}}}_1 \cap \check{\underset{\sim}{\mathcal{R}}}_2$, h) $\hat{\underset{\sim}{\mathcal{R}}}_1 \cup \check{\underset{\sim}{\mathcal{R}}}_2$.

II18. Prove that the fuzzy relation $\underset{\sim}{\mathcal{R}}_1$ below is a fuzzy preorder.

$\underset{\sim}{\mathcal{R}}_1$	A	B	C	D	E	F
A	1	0,7	0,2	0	0,8	1
B	0	1	0,2	0	0	0
C	0	0,5	1	0	0	0
D	0	0,1	0,1	1	0,1	0,1
E	0	0	0	0	1	0,8
F	0	0	0	0	0,8	1

II19. Verify that the relations $\underset{\sim}{\mathcal{R}}_2$, $\underset{\sim}{\mathcal{R}}_3$, $\underset{\sim}{\mathcal{R}}_4$ below are similitude relations.

$\underset{\sim}{\mathcal{R}}_2$	A	B	C	D	E	F
A	1	1	0,8	0,3	0,7	1
B	1	1	0,8	0,3	0,7	1
C	0,8	0,8	1	0,3	0,7	0,8
D	0,3	0,3	0,3	1	0,3	0,3
E	0,7	0,7	0,7	0,3	1	0,7
F	1	1	0,8	0,3	0,7	1

$\underset{\sim}{\mathcal{R}}_3$	A	B	C	D	E	F
A	1	1	0	0	c	0
B	1	1	0	0	c	0
C	0	0	1	b	0	a
D	0	0	b	1	0	a
E	c	c	0	0	1	0
F	0	0	a	a	0	1

$0 \leqslant a \leqslant b \leqslant c \leqslant 1$

$\underset{\sim}{\mathcal{R}}_4$	A	B	C	D	E	F
A	1	0,1	0,1	0	0	0,5
B	0,1	1	0,6	0	0	0,1
C	0,1	0,6	1	0	0	0,1
D	0	0	0	1	0,3	0
E	0	0	0	0,3	1	0
F	0,5	0,1	0,1	0	0	1

II20. If $\underset{\sim}{\mathcal{R}}_1$, and $\underset{\sim}{\mathcal{R}}_2$ are preorder relations in the same set **E**, may one say that $\underset{\sim}{\mathcal{R}}_2 \circ \underset{\sim}{\mathcal{R}}_1$ is also a preorder relation?

Answer the same question considering $\underset{\sim}{\mathcal{R}}_1 \cap \underset{\sim}{\mathcal{R}}_2$, $\underset{\sim}{\mathcal{R}}_1 \cup \underset{\sim}{\mathcal{R}}_2$, and $\overline{\underset{\sim}{\mathcal{R}}}_1$.

II21. Give the maximal similitude subrelations for the relation $\underset{\sim}{\mathcal{R}}_1$ (II18), $\underset{\sim}{\mathcal{R}}_3$ $\underset{\sim}{\mathcal{R}}_4$ (II19) (possibly using one of the algorithms given in Appendix B, p. 387.

II22. Among the six relations given in Exercise II13, which are antisymmetric? Which are perfectly antisymmetric?

II23. Verify that the relations $\underset{\sim}{\mathcal{R}}_1$, $\underset{\sim}{\mathcal{R}}_2$, $\underset{\sim}{\mathcal{R}}_3$ given below are indeed fuzzy order relations. Which of these are perfect fuzzy order relations? Which of these constitute a total order? Which do not constitute a total order?

$\underset{\sim}{\mathcal{R}}_1$	A	B	C	D	E
A	1	0	0	0	0,6
B	0,5	1	0,9	0,5	0,8
C	0	0	1	0,2	0
D	0	0	0,1	1	0
E	0,3	0	0	0	1

,

$\underset{\sim}{\mathcal{R}}_2$	A	B	C	D	E
A	1	1	0,9	0,7	0,8
B	0	1	0,9	0	0
C	0	0	1	0	0
D	0	0	0	1	0
E	0	0	0	0,6	1

,

$\underset{\sim}{\mathcal{R}}_3$	A	B	C	D	E
A	1	0,5	0,5	0,5	0,5
B	0	1	1	1	1
C	0	0	1	0,9	0
D	0	0	0	1	0
E	0	0	0,8	0,8	1

.

II24. Using the notion of the ordinal function of the associated ordinary graph, put each of the fuzzy order relations below into triangular form.

$\underset{\sim}{\mathcal{R}}_1$	A	B	C	D	E
A	1	0,7	0,8	0,5	0,5
B	0	1	0,3	0	0,2
C	0	0,7	1	0	0,2
D	0,6	1	0,9	1	0,6
E	0	0	0	0	1

,

$\underset{\sim}{\mathcal{R}}_2$	A	B	C	D	E	F
A	1	0,7	0,2	0	0,8	1
B	0	1	0,2	0	0	0
C	0	0,5	1	0	0	0
D	0	0,1	0,1	1	0,1	0,1
E	0	0	0	0	1	0,8
F	0	0	0	0	0	1

.

II25. For each of the reflexive relations below, calculate the max–min transitive closure. One thus obtains preorder relations.

(a) Determine the set of maximal similitude subrelations.

(b) Are these subrelations disjoint?

(c) May one present $\underset{\sim}{\mathcal{R}}_1$ and/or $\underset{\sim}{\mathcal{R}}_2$ in block-triangular form?

$\underset{\sim}{\mathcal{R}}_1$	A	B	C	D	E	F
A	1	0,5	0,4	0,6	0	0
B	0,5	1	0	0,6	0,2	0,7
C	0,5	0,3	1	0,1	0,5	0
D	0,2	0,6	0	1	0	0,8
E	0	0,2	0	0	1	0,1
F	0	0,7	0,2	0,8	0	1

,

$\underset{\sim}{\mathcal{R}}_2$	A	B	C	D	E	F	G	H
A	1	0,2	0,3	0,8	0,6	0,7	0	0
B	0,1	1	0	0,4	0,5	0,7	0	0
C	0,4	0,5	1	0,3	0	1	0,9	1
D	0,3	0,4	0,2	1	0,5	0	0	0
E	0,6	0,5	0	0,5	1	0,8	0	0
F	0,7	0,7	1	1	0	1	0,8	0
G	0,3	0	0,8	0	0,9	0	1	0,8
H	0	0	0	0,9	0	0	0,8	1

II26. Consider the resemblance relations below.

(1) Find the corresponding similitude relations by taking their transitive closure.

(2) Give the corresponding dissimilitude relations.

(3) Give the classes of pairs {X, Y}whose distances $d(X, Y)$ are equal to: 0; 0,1; 0,2; . . . ; 0,9; 1.

$\underset{\sim}{\mathcal{R}}_1$	A	B	C	D	E	F	G
A	1	0	0,1	0	0,8	1	0,6
B	0	1	0	1	0	0,8	1
C	0,1	0	1	0,7	0,6	0	0,1
D	0	1	0,7	1	0	0,9	0
E	0,8	0	0,6	0	1	0,7	0,5
F	1	0,8	0	0,9	0,7	1	0,4
G	0,6	1	0,1	0	0,5	0,4	1

$\underset{\sim}{\mathcal{R}}_2$	A	B	C	D	E	F	G
A	1	0,9	0,8	0,7	0,6	0,5	0,4
B	0,9	1	0,7	0,6	0,5	0,4	0,3
C	0,8	0,7	1	0,5	0,4	0,3	0,2
D	0,7	0,6	0,5	1	0,3	0,2	0,1
E	0,6	0,5	0,4	0,3	1	0,1	0
F	0,5	0,4	0,3	0,2	0,1	1	0
G	0,4	0,3	0,2	0,1	0	0	1

II27. In Exercise II26 we have obtained two similitude relations and from there two dissimilitude relations.

(1) For each similitude relations, give a decomposition according to formula (27.1 The result should be presented in a form like that given in Figure 27.1.

(2) For each of the corresponding dissimilitude relations, give the graphs of min–max distances in the fashion that has been realized in Figure 27.9.

II28. Given the following seven fuzzy messages

	x_1	x_2	x_3	x_4	x_5	x_6	x_7	x_8	
$\underset{\sim}{A}_1 =$	0,1	0,3	0,2	0	0,7	0,6	0,2	1	,
$\underset{\sim}{A}_2 =$	0,1	0	0	0	0,7	0,6	0,8	0,9	,
$\underset{\sim}{A}_3 =$	1	0	0,3	0	0,2	0	0	0,6	,
$\underset{\sim}{A}_4 =$	1	0,1	0,2	0,6	0,6	0	0	0,6	,
$\underset{\sim}{A}_5 =$	0,3	0,5	1	0,1	1	0,2	0,4	1	,
$\underset{\sim}{A}_6 =$	0,2	0,5	0	0,6	0,1	0	0,4	0,4	,
$\underset{\sim}{A}_7 =$	0,1	0,2	0,2	1	0,4	0,1	0,3	0,4	.

Choose these messages using their relative generalized Hamming distance.

(1) By taking the min–max transitive closure of the dissemblance relation,

(2) By not taking this transitive closure and considering ordinary min–addition. Answer the same questions using the relative euclidean distance between the messages.

II29. The messages $(\underset{\sim}{A}_1, \underset{\sim}{A}_2, \ldots, \underset{\sim}{A}_7)$ of Exercise II28 are transformed into messages $(\underset{\sim}{B}_1, \underset{\sim}{B}_2, \ldots, \underset{\sim}{B}_5)$ by the following relation:

$\underset{\sim}{\mathcal{R}}$	y_1	y_2	y_3	y_4	y_5
x_1	1	0,9	0	0,7	0,1
x_2	0,5	0,1	0,6	0,8	0,5
x_3	0,4	0,9	1	0,8	0,6
x_4	0,7	0,4	0,9	1	0
x_5	1	0,8	0,6	0,2	0,6
x_6	0,6	0,2	1	0,3	1
x_7	0,2	0,8	0	1	0
x_8	0,3	1	0	0	0,2

,

Choose the five messages $(\underset{\sim}{B}_1, \underset{\sim}{B}_2, \ldots, \underset{\sim}{B}_5)$ as was done for $(\underset{\sim}{A}_1, \underset{\sim}{A}_2, \ldots, \underset{\sim}{A}_8)$ in Exercise II28.

II30. Consider the ten fuzzy graphs below, taken as messages. Choose these messages as was done in Exercise II28.

$\underset{\sim}{G}_1$

	y_1	y_2	y_3
x_1	0,6	0	1
x_2	0,3	1	0,4
x_3	0,3	0,2	0,6

$\underset{\sim}{G}_2$

	y_1	y_2	y_3
x_1	0	0,1	0,7
x_2	1	0,3	0,2
x_3	1	0,8	0,5

$\underset{\sim}{G}_3$

	y_1	y_2	y_3
x_1	0,6	0,1	0,7
x_2	1	0,3	0,2
x_3	0,6	0,7	0,5

$\underset{\sim}{G}_4$

	y_1	y_2	y_3
x_1	0,5	0,1	0,9
x_2	0,3	1	0,4
x_3	0,2	0,1	0,5

$\underset{\sim}{G}_5$

	y_1	y_2	y_3
x_1	0,6	0,1	0,8
x_2	0,7	0,7	0,3
x_3	0,4	0,5	0,6

$\underset{\sim}{G}_6$

	y_1	y_2	y_3
x_1	0,1	0,3	0,7
x_2	0,5	0,3	0,2
x_3	0,9	1	0,6

$\underset{\sim}{G}_7$

	y_1	y_2	y_3
x_1	0,3	0,1	0,7
x_2	0,3	0,3	0,2
x_3	1	0,8	0,6

$\underset{\sim}{G}_8$

	y_1	y_2	y_3
x_1	0	0,1	0,7
x_2	0,9	0,3	0,2
x_3	1	0,7	0,5

$\underset{\sim}{G}_9$

	y_1	y_2	y_3
x_1	0	0,1	0,8
x_2	0,8	0,3	0,3
x_3	1	0,7	0,5

$\underset{\sim}{G}_{10}$

	y_1	y_2	y_3
x_1	0,5	0,2	0,7
x_2	0,7	0,7	0,4
x_3	0,4	0,5	0,6

II31. Carry out Exercise II28 once again using min–algebraic sum ($a \hat{+} b = a + b - ab$) transitive closure.

CHAPTER III

FUZZY LOGIC

31. INTRODUCTION

To associate the word *fuzzy* with the word *logic* is shocking. Logic, in the ordinary sense of the word, is a conceptualization of the mechanisms of thought, one that may never be fuzzy, but always rigorous and formal. Mathematicians researching these mechanisms of thought have noted, however, that it is not a matter of having, in fact, one unique logic (for example, boolean logic), but having as many logics as one wishes; it all depends on the axiomatization chosen. Of course, once the axioms have been selected, all the propositions that are built upon them must be linked together rigorously according to the stated rules of this axiomatization, without contradiction.

Boolean logic is the logic associated with the boolean theory of sets; fuzzy logic is associated in the same manner with the theory of fuzzy subsets. As we shall see in Chapter V, there is no unique theory of fuzzy subsets; one may construct as many as one wishes.

Human thought—superposition of intuition and rigor that, on the one hand, regards globally or in parallel (necessarily fuzzy) and on the other reasons logically and sequentially (necessarily formal)—is a fuzzy mechanism. The laws of thought that we may want to be included in the programs of computers are obligatorily formal; laws of thought in the dialogue of man and man are fuzzy. Ought one not say that the theory of fuzzy subsets, in its generalized form (or the one emerging, as we shall see later, in the theory of categories), is well adapted to this human dialogue? When software constructed with respect to a fuzzy logic becomes operational and when fuzzy hardware becomes industrially possible, then man–machine communication will be much more convenient, rapid, and better adapted to the solution of problems.

Of course, in a work so modest as this, the author is limited to an introduction. But this domain will progress so much the more quickly when there are more engineers who are able to comprehend this imaginative new theory from the mathematicians. My deep intention in writing this book is to stimulate the imaginations of my readers so that they will go much further than what is modestly presented here.

32. CHARACTERISTIC FUNCTION OF A FUZZY SUBSET. FUZZY VARIABLES

Let $\mu_{\underset{\sim}{A}}(x)$ be the membership function of the element x in the fuzzy subset $\underset{\sim}{A}$. In Sections 2–8 we have defined the principal operations that may be realized in considering fuzzy subsets with the same reference set.

In the present chapter, we shall suppose that the membership set is always

$$M = [0,1]. \tag{32.1}$$

Previously, we have recalled how to carry out the operations of a binary boolean algebra with respect to the algebraic operations of ordinary subsets.

We shall use the following notation:

$$a = \mu_A(x)\ ,\ b = \mu_B(x)\ ,\ \text{etc.} \tag{32.2}$$

We know that in boolean binary algebras the variables, such as $a, b, \ldots$, may take only the values 0 or 1. The correspondence between the operations of set theory and those of boolean binary algebra is reviewed below:

	Subsets	Corresponding operations
(32.3)	$A \cap B$.	$a \,.\, b$.
(32.4)	$A \cup B$.	$a \dotplus b$.
(32.5)	$\bar{A} = \complement_E A$.	$\bar{a}$.
(32.6)	$A \oplus B = (\bar{A} \cap B) \cup (A \cap \bar{B})$.	$a \oplus b = \bar{a} \,.\, b \dotplus a \,.\, \bar{b}$.

The principal correspondences that we see in (32.3)–(32.6) constitute a didactic introduction to what follows, and will hold not only for boolean characteristic functions and membership functions with $M = \{0, 1\}$ but also for fuzzy functions with $M = [0, 1]$.

Let x be an element of the reference set E and let $\underset{\sim}{A}, \underset{\sim}{B}, \ldots$ be fuzzy subsets of this reference set. Put

$$\underset{\sim}{a} = \mu_{\underset{\sim}{A}}(x)\ ,\quad \underset{\sim}{b} = \mu_{\underset{\sim}{B}}(x)\ , \ldots, \underset{\sim}{a}, \underset{\sim}{b}, \ldots \in M = [0\ ,1]\ . \tag{32.7}$$

We shall define, with respect to what was given in Sections 2–8, the following operations for the quantities, $a, b, \ldots$:

$$\underset{\sim}{a} \wedge \underset{\sim}{b} = \text{MIN}\ (\underset{\sim}{a}, \underset{\sim}{b}), \tag{32.8}$$

$$\underset{\sim}{a} \vee \underset{\sim}{b} = \text{MAX}\ (\underset{\sim}{a}, \underset{\sim}{b}), \tag{32.9}$$

$$\overline{\underset{\sim}{a}} = 1 - \underset{\sim}{a}, \tag{32.10}$$

(32.11) $$\underset{\sim}{a} \oplus \underset{\sim}{b} = (\overline{\underset{\sim}{a}} \wedge \underset{\sim}{b}) \vee (\underset{\sim}{a} \wedge \overline{\underset{\sim}{b}}).$$

Carrying forward what is defined in (7.18)–(7.32), we may write:

(32.12) $$\underset{\sim}{a} \wedge \underset{\sim}{b} = \underset{\sim}{b} \wedge \underset{\sim}{a},$$
(32.13) $$\underset{\sim}{a} \vee \underset{\sim}{b} = \underset{\sim}{b} \vee \underset{\sim}{a},$$
} commutativity

(32.14) $$(\underset{\sim}{a} \wedge \underset{\sim}{b}) \wedge \underset{\sim}{c} = \underset{\sim}{a} \wedge (\underset{\sim}{b} \wedge \underset{\sim}{c}),$$
(32.15) $$(\underset{\sim}{a} \vee \underset{\sim}{b}) \vee \underset{\sim}{c} = \underset{\sim}{a} \vee (\underset{\sim}{b} \vee \underset{\sim}{c}),$$
} associativity

(32.16) $$\underset{\sim}{a} \wedge \underset{\sim}{a} = \underset{\sim}{a},$$
(32.17) $$\underset{\sim}{a} \vee \underset{\sim}{a} = \underset{\sim}{a},$$
} idempotence

(32.18) $$\underset{\sim}{a} \wedge (\underset{\sim}{b} \vee \underset{\sim}{c}) = (\underset{\sim}{a} \wedge \underset{\sim}{b}) \vee (\underset{\sim}{a} \wedge \underset{\sim}{c}),$$
(32.19) $$\underset{\sim}{a} \vee (\underset{\sim}{b} \wedge \underset{\sim}{c}) = (\underset{\sim}{a} \vee \underset{\sim}{b}) \wedge (\underset{\sim}{a} \vee \underset{\sim}{c}),$$
} distributivity

(32.20) $$\underset{\sim}{a} \wedge 0 = 0,$$

(32.21) $$\underset{\sim}{a} \vee 0 = \underset{\sim}{a},$$

(32.22) $$\underset{\sim}{a} \wedge 1 = \underset{\sim}{a},$$

(32.23) $$\underset{\sim}{a} \vee 1 = 1,$$

(32.24) $$(\overline{\overline{\underset{\sim}{a}}}) = \underset{\sim}{a},$$

(32.25) $$\overline{\underset{\sim}{a} \wedge \underset{\sim}{b}} = \overline{\underset{\sim}{a}} \vee \overline{\underset{\sim}{b}},$$
(32.26) $$\overline{\underset{\sim}{a} \vee \underset{\sim}{b}} = \overline{\underset{\sim}{a}} \wedge \overline{\underset{\sim}{b}},$$
} De Morgan's theorems generalized to the case where $\mathbf{M} = [0, 1]$.

The proofs of all these formulas are trivial except perhaps those of (32.18), (32.19), (32.25), and (32.26).

We shall prove (32.18). For this, we suppose that the quantities $\underset{\sim}{a}$, $\underset{\sim}{b}$, and $\underset{\sim}{c}$ have their values in the following three distinct total orders (it is not useful to consider six)†:

(32.27) $$1)\ 0 \leqslant \underset{\sim}{a} \leqslant \underset{\sim}{b} \leqslant \underset{\sim}{c} \leqslant 1,\quad 2)\ 0 \leqslant \underset{\sim}{b} \leqslant \underset{\sim}{c} \leqslant \underset{\sim}{a} \leqslant 1 \text{ and } 3)\ 0 \leqslant \underset{\sim}{c} \leqslant \underset{\sim}{a} \leqslant \underset{\sim}{b} \leqslant 1.$$

We have

(32.28) 1) $$\underset{\sim}{a} \wedge (\underset{\sim}{b} \vee \underset{\sim}{c}) = \text{MIN}[\underset{\sim}{a}, \text{MAX}(\underset{\sim}{b}, \underset{\sim}{c})] = \text{MIN}(\underset{\sim}{a}, \underset{\sim}{c}) = \underset{\sim}{a},$$

(32.29) $$(\underset{\sim}{a} \wedge \underset{\sim}{b}) \vee (\underset{\sim}{a} \wedge \underset{\sim}{c}) = \text{MAX}[\text{MIN}(\underset{\sim}{a}, \underset{\sim}{b}), \text{MIN}(\underset{\sim}{a}, \underset{\sim}{c})] = \text{MAX}(\underset{\sim}{a}, \underset{\sim}{a}) = \underset{\sim}{a}.$$

(32.30) 2) $$\underset{\sim}{a} \wedge (\underset{\sim}{b} \vee \underset{\sim}{c}) = \text{MIN}[\underset{\sim}{a}, \text{MAX}(\underset{\sim}{b}, \underset{\sim}{c})] = \text{MIN}(\underset{\sim}{a}, \underset{\sim}{c}) = \underset{\sim}{c},$$

(32.31) $$(\underset{\sim}{a} \wedge \underset{\sim}{b}) \vee (\underset{\sim}{a} \wedge \underset{\sim}{c}) = \text{MAX}[\text{MIN}(\underset{\sim}{a}, \underset{\sim}{b}), \text{MIN}(\underset{\sim}{a}, \underset{\sim}{c})] = \text{MAX}(\underset{\sim}{b}, \underset{\sim}{c}) = \underset{\sim}{c}.$$

†If at least two of these quantities are equal, then the proof is immediate and trivial.

$$(32.32) \quad 3) \quad \underset{\sim}{a} \wedge (\underset{\sim}{b} \vee \underset{\sim}{c}) = \text{MIN}[\underset{\sim}{a}, \text{MAX}(\underset{\sim}{b}, \underset{\sim}{c})] = \text{MIN}(\underset{\sim}{a}, \underset{\sim}{b}) = \underset{\sim}{a},$$

$$(32.33) \quad (\underset{\sim}{a} \wedge \underset{\sim}{b}) \vee (\underset{\sim}{a} \wedge \underset{\sim}{c}) = \text{MAX}[\text{MIN}(\underset{\sim}{a}, \underset{\sim}{b}), \text{MIN}(\underset{\sim}{a}, \underset{\sim}{c})] = \text{MAX}(\underset{\sim}{a}, \underset{\sim}{c}) = \underset{\sim}{a}.$$

One may prove (32.19) in the same manner.

We shall prove (32.25). Let $0 \leqslant \underset{\sim}{a} < \underset{\sim}{b} \leqslant 1$†;

$$(32.34) \quad \text{MAX}\,[(1 - \underset{\sim}{a}), (1 - \underset{\sim}{b})] = 1 - \underset{\sim}{a},$$

$$(32.35) \quad \text{MIN}\,[\underset{\sim}{a}, \underset{\sim}{b}] = \underset{\sim}{a},$$

$$(32.36) \quad \text{MAX}\,[1 - \underset{\sim}{a}), (1 - \underset{\sim}{b})] + \text{MIN}\,[\underset{\sim}{a}, \underset{\sim}{b}] = 1 - \underset{\sim}{a} + \underset{\sim}{a} = 1\,;$$

then

$$(32.37) \quad \text{MAX}\,[(1 - \underset{\sim}{a}), (1 - \underset{\sim}{b})] = 1 - \text{MIN}\,[\underset{\sim}{a}, \underset{\sim}{b}],$$

or

$$(32.38) \quad \overline{\underset{\sim}{a}} \vee \overline{\underset{\sim}{b}} = \overline{\underset{\sim}{a} \wedge \underset{\sim}{b}}.$$

Very important remark. Properties (32.12)–(32.26) are all the properties of a boolean binary algebra with the following two exceptions:

$$(32.39),\ (32.40) \quad a\,.\,\bar{a} = 0 \quad \text{and} \quad a \dotplus \bar{a} = 1\,,$$

for which the corresponding expressions are not satisfied:

$$(32.41) \quad \underset{\sim}{a} \wedge \overline{\underset{\sim}{a}} \neq 0\,, \quad \text{except for} \quad \underset{\sim}{a} = 0 \quad \text{or} \quad \underset{\sim}{a} = 1\,,$$

$$(32.42) \quad \underset{\sim}{a} \vee \overline{\underset{\sim}{a}} \neq 1\,, \quad \text{except for} \quad \underset{\sim}{a} = 0 \quad \text{or} \quad \underset{\sim}{a} = 1\,.$$

Because of this, the structure obtained over the variables $\underset{\sim}{a}, \underset{\sim}{b}, \ldots$ for the operations $\wedge$, $\vee$, and $^-$ may not be considered as constituting an algebra in the sense given to this word in modern mathematics. Let it be understood also that the word *algebra*, as many other words in the language of mathematics, is not employed by all in the same sense.

Fuzzy variables. Functions of fuzzy variables. The variables $\underset{\sim}{a}, \underset{\sim}{b}, \ldots \in [0, 1]$ will be called *fuzzy variables*‡ in the present theory. The functions constructed with the aid of these variables will be called functions of fuzzy variables under the following condition:

Let $\underset{\sim}{f}(\underset{\sim}{a}, \underset{\sim}{b}, \ldots)$ be a function of $\underset{\sim}{a}, \underset{\sim}{b}, \ldots$. In order that this function may be called a *function of fuzzy variables* it is necessary and sufficient that $\underset{\sim}{f}$ depend only on fuzzy variables and that

†If at least two of these quantities are equal, then the proof is immediate and trivial.

‡The word *fuzzy* is not best for characterizing such variables; it is employed here in parallel with the word *boolean* used when one speaks of boolean variables such that $a \in \{0, 1\}$.

(32.43) $$0 \leqslant \underset{\sim}{f} \leqslant 1 \ .$$

Theorem I. If $\underset{\sim}{f}(\underset{\sim}{a}, \underset{\sim}{b}, \ldots)$ contains only fuzzy variables and the operators $\wedge$, $\vee$, and $^-$, then (32.43) is always satisfied.

Proof. This is evident. Each of the operations $\wedge$, $\vee$, or $^-$ on the variables $\underset{\sim}{a}, \underset{\sim}{b}, \ldots \in [0, 1]$ cannot produce a result outside the limits 0 and 1. Submitting such a result to the operations $\wedge$, $\vee$, $^-$ with other results of the same nature cannot produce a result outside these limits.

Simplification of functions of fuzzy variables. Functions of fuzzy variables may not be, as are boolean binary functions, the objects of truth tables permitting an ordered analysis (in the order of binary numbers) of these functions. Nor may they be simplified easily, as are boolean functions, because of the absence of the two properties (32.39) and (32.40). Also because of this, one may not decompose these functions in a disjunctive canonical form (with the aid of the minterms) or in a conjunctive canonical form (with the aid of maxterms.†

Sometimes using only properties (32.12)–(32.26), a certain number of simplifications may usefully be carried out. We shall see several examples of simplifications:

(32.44)
$$\underset{\sim}{f}(\underset{\sim}{a}, \underset{\sim}{b}) = \underset{\sim}{a} \vee (\underset{\sim}{a} \wedge \underset{\sim}{b})$$
$$= \underset{\sim}{a} \wedge (1 \vee \underset{\sim}{b}) \ ,$$ after (32.18) and (32.22) since $\underset{\sim}{a} \wedge (1 \vee \underset{\sim}{b}) = (\underset{\sim}{a} \wedge 1) \vee (\underset{\sim}{a} \wedge \underset{\sim}{b}) = \underset{\sim}{a} \vee (\underset{\sim}{a} \wedge \underset{\sim}{b})$
$$= \underset{\sim}{a} \wedge 1 \ ,$$ after (32.23),
$$= \underset{\sim}{a} \ ,$$ after (32.22).

Also

(32.45) $\underset{\sim}{a} \vee (\underset{\sim}{a} \wedge \underset{\sim}{b}) = \underset{\sim}{a}$. This is called the property of absorption.

In a similar fashion one may show that

(32.46) $\underset{\sim}{a} \wedge (\underset{\sim}{a} \vee \underset{\sim}{b}) = \underset{\sim}{a}$. This is the dual form of the property of absorption.

We consider another example:

(32.47)
$$\underset{\sim}{f}(\underset{\sim}{a}, \underset{\sim}{b}, \underset{\sim}{c}) = (\underset{\sim}{a} \wedge \underset{\sim}{b} \wedge \underset{\sim}{\bar{c}}) \vee [\underset{\sim}{\bar{a}} \wedge (\underset{\sim}{\bar{b}} \vee \underset{\sim}{c})] \vee \underset{\sim}{\bar{a}} \vee (\underset{\sim}{b} \wedge \underset{\sim}{\bar{c}})$$
$$= \underbrace{(\underset{\sim}{a} \wedge \underset{\sim}{b} \wedge \underset{\sim}{\bar{c}})}_{(1)} \vee \underbrace{(\underset{\sim}{\bar{a}} \wedge \underset{\sim}{\bar{b}})}_{(2)} \vee \underbrace{(\underset{\sim}{\bar{a}} \wedge \underset{\sim}{c})}_{(3)} \vee \underbrace{\underset{\sim}{\bar{a}}}_{(4)} \vee \underbrace{(\underset{\sim}{b} \wedge \underset{\sim}{\bar{c}})}_{(5)}$$
$$= (b \wedge \bar{c}) \vee \bar{a} \ ,$$

It is unnecessary to recall the important role of parentheses.

†As we have mentioned several times, we suppose that the reader has made at least an elementary study of modern mathematics, and in particular of boolean algebra.

We know that the number of distinct boolean functions obtained with the aid of distinct variables is equal to $2^{(2^n)}$. In the case of n fuzzy variables, the number of fuzzy functions constructed in an arbitrary fashion with these n variables and the operations $\wedge$, $\vee$, and $^-$ is likewise finite; we shall prove this later.

Remark. Any $\vee$ operation may be replaced by a $\wedge$ operation and vice versa. In fact

$$\underset{\sim}{a} \wedge \underset{\sim}{b} = \text{MIN}\,(\underset{\sim}{a}, \underset{\sim}{b}) \tag{32.48}$$

$$= 1 - \text{MAX}\,(\overline{\underset{\sim}{a}}, \overline{\underset{\sim}{b}})$$

$$= \overline{\overline{\underset{\sim}{a}} \vee \overline{\underset{\sim}{b}}}\,.$$

This is another way of presenting (32.25). One may do the same for (32.26).

Thus it is sufficient to use either the operators $\wedge$ and $^-$ or the operators $\vee$ and $^-$ in order to represent any function of fuzzy variables involving the symbols $\wedge$, $\vee$, and $^-$, but the notation is then very cumbersome.

Recall that in a boolean algebra a single operator suffices to represent an arbitrary boolean function.

Consider the Sheffer operator

$$a \mid b = \overline{a \,.\, b} \tag{32.49}$$

$$= \overline{a} \dotplus \overline{b}\,,$$

because

$$a \dotplus b = \overline{a} \mid \overline{b} = (a \mid a) \mid (b \mid b)\,, \tag{32.50}$$

$$a \,.\, b = \overline{a \mid b} = (a \mid b) \mid (a \mid b)\,, \tag{32.51}$$

$$\overline{a} = a \mid a\,. \tag{32.52}$$

Consider the Peirce operator

$$a \downarrow b = \overline{a \dotplus b} \tag{32.53}$$

$$= \overline{a} \,.\, \overline{b}\,,$$

because

$$a \dotplus b = \overline{a \downarrow b} = (a \downarrow b) \downarrow (a \downarrow b)\,, \tag{32.54}$$

$$a \,.\, b = \overline{a} \downarrow \overline{b} = (a \downarrow a) \downarrow (b \downarrow b)\,, \tag{32.55}$$

$$\overline{a} = a \downarrow a\,. \tag{32.56}$$

One may pass from a boolean expression using the Peirce operator to an expression involving the Sheffer operator and vice versa:

$$a \downarrow b = \overline{a} \,.\, \overline{b} = \overline{\overline{a} \mid \overline{b}} = \overline{(a \mid a) \mid (b \mid b)} \tag{32.57}$$

$$= \big((a \mid a) \mid (b \mid b)\big) \mid \big((a \mid a) \mid (b \mid b)\big)\,,$$

(32.58) $$a \mid b = \bar{a} \dot{+} \bar{b} = \overline{\bar{a} \downarrow \bar{b}} = \overline{(a \downarrow b) \downarrow (a \downarrow b)}$$
$$= \Big((a \downarrow b) \downarrow (a \downarrow b)\Big) \downarrow \Big((a \downarrow b) \downarrow (a \downarrow b)\Big).$$

The difficulties in writing appear rapidly, so that one practically sets aside the use of such operators: but one may construct electronic circuits with a single technology and this may be useful in certain cases.

For the case of fuzzy variables, we define the operators†

(32.59)

Sheffer: $$\underset{\sim}{a} \mid \underset{\sim}{b} = \overline{\underset{\sim}{a} \wedge \underset{\sim}{b}} = \bar{\underset{\sim}{a}} \vee \bar{\underset{\sim}{b}},$$

(32.60)

Peirce: $$\underset{\sim}{a} \downarrow \underset{\sim}{b} = \overline{\underset{\sim}{a} \vee \underset{\sim}{b}} = \bar{\underset{\sim}{a}} \wedge \bar{\underset{\sim}{b}}.$$

Any function of fuzzy variables may be written with the aid of only one of these operators. For

(32.61) (1) $$\underset{\sim}{a} \vee \underset{\sim}{b} = \bar{\underset{\sim}{a}} \mid \bar{\underset{\sim}{b}} = (\underset{\sim}{a} \mid \underset{\sim}{a}) \mid (\underset{\sim}{b} \mid \underset{\sim}{b}),$$

(32.62) $$\underset{\sim}{a} \wedge \underset{\sim}{b} = \overline{\underset{\sim}{a} \mid \underset{\sim}{b}} = (\underset{\sim}{a} \mid \underset{\sim}{b}) \mid (\underset{\sim}{a} \mid \underset{\sim}{b}),$$

(32.63) $$\bar{\underset{\sim}{a}} = \underset{\sim}{a} \mid \underset{\sim}{a}.$$

(32.64) (2) $$\underset{\sim}{a} \vee \underset{\sim}{b} = \overline{\underset{\sim}{a} \downarrow \underset{\sim}{b}} = (\underset{\sim}{a} \downarrow \underset{\sim}{b}) \downarrow (\underset{\sim}{a} \downarrow \underset{\sim}{b}),$$

(32.65) $$\underset{\sim}{a} \wedge \underset{\sim}{b} = \bar{\underset{\sim}{a}} \downarrow \bar{\underset{\sim}{b}} = (\underset{\sim}{a} \downarrow \underset{\sim}{a}) \downarrow (\underset{\sim}{b} \downarrow \underset{\sim}{b}),$$

(32.66) $$\bar{\underset{\sim}{a}} = \underset{\sim}{a} \downarrow \underset{\sim}{a}.$$

And one may pass from Peirce to Sheffer and vice versa using formulas (32.57) and (32.58) above.

As an example we see how to write a function of fuzzy variables that is not too complicated using the Sheffer operator:

(32.67) $$\underset{\sim}{f}(\underset{\sim}{a}, \underset{\sim}{b}, \underset{\sim}{c}) = \bar{\underset{\sim}{a}} \wedge (\underset{\sim}{b} \vee \bar{\underset{\sim}{c}})$$
$$= (\underset{\sim}{a} \mid \underset{\sim}{a}) \wedge \Big((\underset{\sim}{b} \mid \underset{\sim}{b}) \mid (\bar{\underset{\sim}{c}} \mid \bar{\underset{\sim}{c}})\Big)$$
$$= (\underset{\sim}{a} \mid \underset{\sim}{a}) \wedge \left((\underset{\sim}{b} \mid \underset{\sim}{b}) \,\Big|\, \Big((\underset{\sim}{c} \mid \underset{\sim}{c}) \mid (\underset{\sim}{c} \mid \underset{\sim}{c})\Big)\right)$$

†We keep the same symbols as for boolean variables; if this presents some unforseen difficulty, it will be appropriate to modify these.

$$= \left((\underset{\sim}{a} \mid \underset{\sim}{a}) \,\middle|\, \left((\underset{\sim}{b} \mid \underset{\sim}{b}) \,\middle|\, \left((\underset{\sim}{c} \mid \underset{\sim}{c}) \mid (\underset{\sim}{c} \mid \underset{\sim}{c}) \right) \right) \right) \,\middle|\, \left((\underset{\sim}{a} \mid \underset{\sim}{a}) \,\middle|\, \left((\underset{\sim}{b} \mid \underset{\sim}{b}) \,\middle|\, \left((\underset{\sim}{c} \mid \underset{\sim}{c}) \mid (\underset{\sim}{c} \mid \underset{\sim}{c}) \right) \right) \right).$$

This is a horribly complicated expression to express a function so simple as $\overline{\underset{\sim}{a}} \wedge (\overline{\underset{\sim}{b}} \vee \overline{\underset{\sim}{c}})$.

Table of values of a function of fuzzy variables. In order to study boolean binary functions one may use what is called a truth table, where one assigns to the binary variables all possible values and obtains thanks to this the values of the function. Such a truth table would not make sense for functions of fuzzy variables, but one may construct tables of a different nature that play a similar role.

In order to study a function of one fuzzy variable $\underset{\sim}{a}$, we shall examine its value in the following two cases:

$$\underset{\sim}{a} \leqslant \overline{\underset{\sim}{a}}, \qquad \overline{\underset{\sim}{a}} \leqslant \underset{\sim}{a}. \tag{32.68}$$

In order to study a function of two variables $\underset{\sim}{a}$, and $\underset{\sim}{b}$, we shall examine its value in the following eight cases†.

$$\begin{aligned}
&\underset{\sim}{a} \leqslant \underset{\sim}{b} \leqslant \overline{\underset{\sim}{b}} \leqslant \overline{\underset{\sim}{a}},\\
&\underset{\sim}{a} \leqslant \overline{\underset{\sim}{b}} \leqslant \underset{\sim}{b} \leqslant \overline{\underset{\sim}{a}},\\
&\overline{\underset{\sim}{a}} \leqslant \underset{\sim}{b} \leqslant \overline{\underset{\sim}{b}} \leqslant \underset{\sim}{a},\\
&\overline{\underset{\sim}{a}} \leqslant \overline{\underset{\sim}{b}} \leqslant \underset{\sim}{b} \leqslant \underset{\sim}{a},\\
&\underset{\sim}{b} \leqslant \underset{\sim}{a} \leqslant \overline{\underset{\sim}{a}} \leqslant \overline{\underset{\sim}{b}},\\
&\underset{\sim}{b} \leqslant \overline{\underset{\sim}{a}} \leqslant \underset{\sim}{a} \leqslant \overline{\underset{\sim}{b}},\\
&\overline{\underset{\sim}{b}} \leqslant \underset{\sim}{a} \leqslant \overline{\underset{\sim}{a}} \leqslant \underset{\sim}{b},\\
&\overline{\underset{\sim}{b}} \leqslant \overline{\underset{\sim}{a}} \leqslant \underset{\sim}{a} \leqslant \underset{\sim}{b}.
\end{aligned} \tag{32.69}$$

In order to study a function of the three variables $\underset{\sim}{a}$, $\underset{\sim}{b}$, and $\underset{\sim}{c}$, one considers the 48 cases below, presented without the $\leqslant$ sign and without the symbol $\sim$ in order to save space:

$a\ b\ c\ \bar{c}\ \bar{b}\ \bar{a}$	$a\ c\ b\ \bar{b}\ \bar{c}\ \bar{a}$	$b\ a\ c\ \bar{c}\ \bar{a}\ \bar{b}$
$a\ b\ \bar{c}\ c\ \bar{b}\ \bar{a}$	$a\ c\ \bar{b}\ b\ \bar{c}\ \bar{a}$	$b\ a\ \bar{c}\ c\ \bar{a}\ \bar{b}$
$a\ \bar{b}\ c\ \bar{c}\ b\ \bar{a}$	$a\ \bar{c}\ b\ \bar{b}\ c\ \bar{a}$	$b\ \bar{a}\ c\ \bar{c}\ a\ \bar{b}$
$a\ \bar{b}\ \bar{c}\ c\ b\ \bar{a}$	$a\ \bar{c}\ \bar{b}\ b\ c\ \bar{a}$	$b\ \bar{a}\ \bar{c}\ c\ a\ \bar{b}$
$\bar{a}\ b\ c\ \bar{c}\ \bar{b}\ a$	$\bar{a}\ c\ b\ \bar{b}\ \bar{c}\ a$	$\bar{b}\ a\ c\ \bar{c}\ \bar{a}\ b$
$\bar{a}\ b\ \bar{c}\ c\ \bar{b}\ a$	$\bar{a}\ c\ \bar{b}\ b\ \bar{c}\ a$	$\bar{b}\ a\ \bar{c}\ c\ \bar{a}\ b$

†I suggest calling these enumeration procedures *antipalindromes* since one forms palindrome sequences and superimposes complementation on these in an antisymmetric fashion. (A palindrome is a word or phrase that is identical to itself when one reads it from the end to the beginning: "Able was I ere I saw Elba.")

(33.70)

$$
\begin{array}{lll}
\bar{a}\,\bar{b}\,c\,\bar{c}\,b\,a & \bar{a}\,\bar{c}\,b\,\bar{b}\,c\,a & \bar{b}\,\bar{a}\,c\,\bar{c}\,a\,b \\
\bar{a}\,\bar{b}\,\bar{c}\,c\,b\,a, & \bar{a}\,\bar{c}\,\bar{b}\,b\,c\,a, & \bar{b}\,\bar{a}\,\bar{c}\,c\,a\,b, \\
b\,c\,a\,\bar{a}\,\bar{c}\,\bar{b} & c\,a\,b\,\bar{b}\,\bar{a}\,\bar{c} & c\,b\,a\,\bar{a}\,\bar{b}\,\bar{c} \\
b\,c\,\bar{a}\,a\,\bar{c}\,\bar{b} & c\,a\,\bar{b}\,b\,\bar{a}\,\bar{c} & c\,b\,\bar{a}\,a\,\bar{b}\,\bar{c} \\
b\,\bar{c}\,a\,\bar{a}\,c\,\bar{b} & c\,\bar{a}\,b\,\bar{b}\,a\,\bar{c} & c\,\bar{b}\,a\,\bar{a}\,b\,\bar{c} \\
b\,\bar{c}\,\bar{a}\,a\,c\,\bar{b} & c\,\bar{a}\,\bar{b}\,b\,a\,\bar{c} & c\,\bar{b}\,\bar{a}\,a\,b\,\bar{c} \\
\bar{b}\,c\,a\,\bar{a}\,\bar{c}\,b & \bar{c}\,a\,b\,\bar{b}\,\bar{a}\,c & \bar{c}\,b\,a\,\bar{a}\,\bar{b}\,c \\
\bar{b}\,c\,\bar{a}\,a\,\bar{c}\,b & \bar{c}\,a\,\bar{b}\,b\,\bar{a}\,c & \bar{c}\,b\,\bar{a}\,a\,\bar{b}\,c \\
\bar{b}\,\bar{c}\,a\,\bar{a}\,c\,b & \bar{c}\,\bar{a}\,b\,\bar{b}\,a\,c & \bar{c}\,\bar{b}\,a\,\bar{a}\,b\,c \\
\bar{b}\,\bar{c}\,\bar{a}\,a\,c\,b, & \bar{c}\,\bar{a}\,\bar{b}\,b\,a\,c, & \bar{c}\,\bar{b}\,\bar{a}\,a\,b\,c.
\end{array}
$$

In order to study a function of n variables, one considers

(32.71) $$\mathrm{P}_n \,.\, 2^n \quad \text{where} \quad \mathrm{P}_n = n! = n(n-1)\ldots 3.2.1\,.$$

cases.

By examining (32.68)–(32.70) one may establish the effect of antisymmetry owing to

(32.72) $$\text{if} \quad \underset{\sim}{x} \leqslant \underset{\sim}{y} \quad \text{then} \quad \bar{\underset{\sim}{y}} \leqslant \bar{\underset{\sim}{x}}\,.$$

To enumerate all possible cases without omission or repetition, one uses a lexicographic procedure. Establish, for example, the correspondence

(32.73)
$$
\begin{array}{ll}
1 & a\,, \\
2 & \bar{a}\,, \\
3 & b\,, \\
4 & \bar{b}\,.
\end{array}
$$

then one has the correspondences

(32.74)
$$
\begin{array}{llll}
1\,3 & a\,b & \text{from which} & a\,b\,\bar{b}\,\bar{a}\,, \\
1\,4 & a\,\bar{b} & \text{from which} & a\,\bar{b}\,b\,\bar{a}\,, \\
2\,3 & \bar{a}\,b & \text{from which} & \bar{a}\,b\,\bar{b}\,a\,, \\
2\,4 & \bar{a}\,\bar{b} & \text{from which} & \bar{a}\,\bar{b}\,b\,a\,, \\
3\,1 & b\,a & \text{from which} & b\,a\,\bar{a}\,\bar{b}\,, \\
3\,2 & b\,\bar{a} & \text{from which} & b\,\bar{a}\,a\,\bar{b}\,, \\
4\,1 & \bar{b}\,a & \text{from which} & \bar{b}\,a\,\bar{a}\,b\,, \\
4\,2 & \bar{b}\,\bar{a} & \text{from which} & \bar{b}\,\bar{a}\,a\,b\,.
\end{array}
$$

Other procedures may easily be imagined.

We consider an example. Enumerate the values of the function

(32.75) $$\underset{\sim}{f}(\underset{\sim}{a}, \underset{\sim}{b}) = (\underset{\sim}{a} \wedge \underset{\sim}{\bar{a}}) \vee (\underset{\sim}{\bar{a}} \wedge \underset{\sim}{b} \wedge \underset{\sim}{\bar{b}})\ ;$$

the result is supplied in the table in Figure 32.1.

$\leqslant$	$\leqslant$	$\leqslant$		$\underset{\sim}{a} \wedge \underset{\sim}{\bar{a}}$	$\underset{\sim}{\bar{a}} \wedge \underset{\sim}{b} \wedge \underset{\sim}{\bar{b}}$	$(\underset{\sim}{a} \wedge \underset{\sim}{\bar{a}}) \vee (\underset{\sim}{\bar{a}} \wedge \underset{\sim}{b} \wedge \underset{\sim}{\bar{b}})$
$\underset{\sim}{a}$	$\underset{\sim}{b}$	$\underset{\sim}{\bar{b}}$	$\underset{\sim}{\bar{a}}$	$\underset{\sim}{a}$	$\underset{\sim}{b}$	$\underset{\sim}{b}$
$\underset{\sim}{a}$	$\underset{\sim}{\bar{b}}$	$\underset{\sim}{b}$	$\underset{\sim}{\bar{a}}$	$\underset{\sim}{a}$	$\underset{\sim}{\bar{b}}$	$\underset{\sim}{\bar{b}}$
$\underset{\sim}{\bar{a}}$	$\underset{\sim}{b}$	$\underset{\sim}{\bar{b}}$	$\underset{\sim}{a}$	$\underset{\sim}{\bar{a}}$	$\underset{\sim}{\bar{a}}$	$\underset{\sim}{\bar{a}}$
$\underset{\sim}{\bar{a}}$	$\underset{\sim}{\bar{b}}$	$\underset{\sim}{b}$	$\underset{\sim}{a}$	$\underset{\sim}{\bar{a}}$	$\underset{\sim}{\bar{a}}$	$\underset{\sim}{\bar{a}}$
$\underset{\sim}{b}$	$\underset{\sim}{a}$	$\underset{\sim}{\bar{a}}$	$\underset{\sim}{\bar{b}}$	$\underset{\sim}{a}$	$\underset{\sim}{b}$	$\underset{\sim}{a}$
$\underset{\sim}{b}$	$\underset{\sim}{\bar{a}}$	$\underset{\sim}{a}$	$\underset{\sim}{\bar{b}}$	$\underset{\sim}{\bar{a}}$	$\underset{\sim}{b}$	$\underset{\sim}{\bar{a}}$
$\underset{\sim}{\bar{b}}$	$\underset{\sim}{a}$	$\underset{\sim}{\bar{a}}$	$\underset{\sim}{b}$	$\underset{\sim}{a}$	$\underset{\sim}{\bar{b}}$	$\underset{\sim}{a}$
$\underset{\sim}{\bar{b}}$	$\underset{\sim}{\bar{a}}$	$\underset{\sim}{a}$	$\underset{\sim}{b}$	$\underset{\sim}{\bar{a}}$	$\underset{\sim}{\bar{b}}$	$\underset{\sim}{\bar{a}}$

Fig. 32.1

Equality of two functions of fuzzy variables. We shall say that two functions of fuzzy variables $\underset{\sim}{f}_1$ and $\underset{\sim}{f}_2$ are equal (one also says identical) if they produce the same table of values through enumeration of all possible cases.

Mixed operations. The variables $\underset{\sim}{a}, \underset{\sim}{b}, \ldots \in [0, 1]$ may be submitted to operations other than $\wedge$, $\vee$, and $^{-}$ in order to form what will be called *mixed functions of fuzzy variables*.

Among such operations we shall include

(32.76) Product $\underset{\sim}{a} \,.\, \underset{\sim}{b}\,,$ where one can easily verify

(32.77) $$\underset{\sim}{a} \in [0,1]\,,\ \underset{\sim}{b} \in [0,1] \ \Rightarrow\ \underset{\sim}{a} \,.\, \underset{\sim}{b} \in [0,1]\,.$$

(32.78) Sum $$\underset{\sim}{a} \,\hat{+}\, \underset{\sim}{b} = \underset{\sim}{a} + \underset{\sim}{b} - \underset{\sim}{a} \,.\, \underset{\sim}{b}\,,$$

where the above property still holds.

Thus, the function

(32.79) $$\underset{\sim}{f}(\underset{\sim}{a}, \underset{\sim}{b}, \underset{\sim}{c}) = (\underset{\sim}{a} \,\hat{+}\, \underset{\sim}{b}) \wedge (\underset{\sim}{b} \,\hat{+}\, \underset{\sim}{\bar{c}}) \wedge \underset{\sim}{a} \wedge \underset{\sim}{c}\,,$$

is a mixed function.

Important remark. With the aid of a table of enumeration one may define for n variables

$$\text{N} = (2n)^{(n!\,2^n)} \tag{32.80}$$

distinct functions†; thus

$$\begin{aligned} n &= 1 & \text{N} &= (2.1)^2 = 2^2 = 4\,, \\ n &= 2 & \text{N} &= (2.2)^{2.2^2} = 4^8 = 65.536\,, \\ n &= 3 & \text{N} &= (2.3)^{6.2^3} = 6^{48}\,, \\ n &= 4 & \text{N} &= (2.4)^{24.2^4} = 8^{384}\,, \ \cdots\cdot \text{ etc.} \end{aligned} \tag{32.81}$$

Among all these functions, a considerably smaller number are formed by the functions of fuzzy variables expressible with the aid of the operations $\wedge$ and $\vee$ on the variables $\underset{\sim}{a}, \underset{\sim}{b}, \ldots$ and $\underset{\sim}{a}, \underset{\sim}{b}, \ldots$.

Convention. Unless otherwise noted, we shall call an analytic function of fuzzy variables, designated by $\underset{\sim}{f}$, any function of the variables $\underset{\sim}{a}, \underset{\sim}{b}, \ldots$ that may be expressed using only the operations $\wedge$ and $\vee$; the variables may occur either in their direct form or as their l's complement, that is $\underset{\sim}{a}, \underset{\sim}{b}, \ldots$.

In order to simplify language, already rather cumbersome, analytic functions of fuzzy variables will be called functions of fuzzy variables when this will not introduce error or confusion.

33. POLYNOMIAL FORMS

Given the double distributivity expressed by (32.18) and (32.19), any function $\underset{\sim}{f}(\underset{\sim}{a}, \underset{\sim}{b}, \ldots)$ may be expressed in a polynomial form with respect to $\wedge$ or with respect to $\vee$.

To begin, we consider an example. Let

$$\underset{\sim}{f}(\underset{\sim}{a}, \underset{\sim}{b}, \underset{\sim}{c}) = (\bar{\underset{\sim}{a}} \wedge \bar{\underset{\sim}{b}}) \vee (\underset{\sim}{a} \wedge \underset{\sim}{b} \wedge \bar{\underset{\sim}{c}})\,. \tag{33.1}$$

This function is presented in a polynomial form with respect to $\vee$ (two monomials in $\wedge$ connected by $\vee$). We may transform this into a polynomial form with respect to $\wedge$ by using (32.19); it becomes

$$\underset{\sim}{f}(\underset{\sim}{a}, \underset{\sim}{b}, \underset{\sim}{c}) = (\bar{\underset{\sim}{a}} \vee \underset{\sim}{a}) \wedge (\bar{\underset{\sim}{a}} \vee \underset{\sim}{b}) \wedge (\bar{\underset{\sim}{a}} \vee \bar{\underset{\sim}{c}}) \wedge (\bar{\underset{\sim}{b}} \vee \underset{\sim}{a}) \wedge (\bar{\underset{\sim}{b}} \vee \underset{\sim}{b}) \wedge (\bar{\underset{\sim}{b}} \vee \bar{\underset{\sim}{c}}). \tag{33.2}$$

We see another example. Let

†This is easy to prove. For n variables, $a, b, \ldots, l$, one must have $n!$ permutations. But, in each permutation one must have a or $\bar{a}$, b or $\bar{b}, \ldots, l$ or $\bar{l}$; thus 2^n times more permutations; this gives $n! \cdot 2^n$ as the number of rows in a table such as that in Figure 32.1. Each row may take a value among these $2n$ variables and their respective complements; thus one may define $(2n)^{(n! \cdot 2^n)}$ distinct functions with such tables.

(33.3) $$\underset{\sim}{f}(\underset{\sim}{a}, \underset{\sim}{b}, \underset{\sim}{c}) = (\underset{\sim}{a} \vee \underset{\sim}{b}) \wedge \underset{\sim}{c} \wedge (\overline{\underset{\sim}{a}} \vee \underset{\sim}{b} \vee \underset{\sim}{c}) = (\underset{\sim}{a} \vee \underset{\sim}{b}) \wedge \underset{\sim}{c}$$

by absorption of the third term by the second. Developing this using (32.18), we have

(33.4) $$\underset{\sim}{f}(\underset{\sim}{a}, \underset{\sim}{b}, \underset{\sim}{c}) = (\underset{\sim}{a} \wedge \underset{\sim}{c}) \vee (\underset{\sim}{b} \wedge \underset{\sim}{c}),$$

which gives a polynomial form with respect to $\vee$; whereas $(\underset{\sim}{a} \vee \underset{\sim}{b}) \wedge \underset{\sim}{c}$ is the corresponding polynomial form with respect to $\wedge$.

In the case of boolean functions, in order to show that two functions f and f' are identical, it suffices to check that they lead to the same truth table or that their disjunctive or conjunctive canonical forms are respectively the same. Concerning functions of fuzzy variables, one may define a similar but less strong notion.

Maximal monomial. Let $\underset{\sim}{f}(\underset{\sim}{a}, \underset{\sim}{b}, \ldots)$ be expressed in a polynomial form with respect to $\wedge$. A monomial of this polynomial form will be said to be maximal (one also says *principal monomial*) if it is absorbed by no other monomial of this polynomial form (a corresponding definition is made for a monomial in a polynomial form with respect to $\vee$).

Reduced polynomial form. Any polynomial form with respect to $\vee$ that does not contain a maximal monomial in $\wedge$ will be said to be a *reduced polynomial form with respect to* $\vee$. A symmetric definition, by permuting $\vee$ and $\wedge$, will define a *reduced polynomial form with respect to* $\wedge$.

An analytic function $\underset{\sim}{f}(\underset{\sim}{a}, \underset{\sim}{b}, \ldots)$ may correspond to several reduced polynomial forms. We shall see an example. The two reduced polynomial forms

(33.5) $$\underset{\sim}{f}(\underset{\sim}{a}, \underset{\sim}{b}) = (\underset{\sim}{a} \wedge \overline{\underset{\sim}{a}}) \vee (\underset{\sim}{a} \wedge \underset{\sim}{b}) \vee (\underset{\sim}{a} \wedge \overline{\underset{\sim}{b}})$$

and

(33.6) $$\underset{\sim}{f}(\underset{\sim}{a}, \underset{\sim}{b}) = (\underset{\sim}{a} \wedge \underset{\sim}{b}) \vee (\underset{\sim}{a} \wedge \overline{\underset{\sim}{b}})$$

correspond to the same analytic function, as one may verify by antipalindrome enumeration, as has been done, for example, in Figure 32.1.

For any analytic function, there exists at least one reduced polynomial form with respect to $\vee$ and at least one reduced polynomial form with respect to $\wedge$. One may pass from one to the other by developing with respect to $\wedge$ (respectively, with respect to $\vee$) and effecting the absorption of nonmaximal monomials. In Appendix C of Volume II we shall treat these notions in more detail.

Example. The function

(33.7) $$\underset{\sim}{f}(\underset{\sim}{a}, \underset{\sim}{b}, \underset{\sim}{c}) = (\overline{\underset{\sim}{a}} \wedge \underset{\sim}{b} \wedge \overline{\underset{\sim}{c}}) \vee (\overline{\underset{\sim}{b}} \wedge \underset{\sim}{c})$$

is presented in a reduced polynomial form with respect to $\vee$.

Its reduced polynomial form with respect to $\wedge$ is

$$\underset{\sim}{f}(\underset{\sim}{a}, \underset{\sim}{b}, \underset{\sim}{c}) = (\underset{\sim}{\bar{a}} \vee \underset{\sim}{\bar{b}}) \wedge (\underset{\sim}{\bar{a}} \vee \underset{\sim}{c}) \wedge (\underset{\sim}{b} \vee \underset{\sim}{\bar{b}}) \wedge (\underset{\sim}{b} \vee \underset{\sim}{c}) \wedge (\underset{\sim}{c} \vee \underset{\sim}{\bar{c}}) \wedge (\underset{\sim}{\bar{b}} \vee \underset{\sim}{\bar{c}}) . \tag{33.8}$$

Identity of two functions of fuzzy variables. A sufficient condition for two functions of fuzzy variables to be identical is that one can bring them to the same reduced polynomial form in $\wedge$ (respectively, in $\vee$). A necessary and sufficient condition is that one obtains the same table of values for the functions.

Theorem. The number of distinct reduced polynomial forms in n variables is finite and is a superior bound for the number of distinct analytic functions of n fuzzy variables.

As may be seen in the enumeration that follows, these reduced polynomials forms are enumerated as the elements of a distributive free lattice with $2n$ generators† and are enumerated in the same fashion. Thus, for $n = 1$, there are 4 distinct forms; for $n = 2$, there are 166; for $n = 3$, there are 7,828,532; . . . ; but this number of distinct forms always remains finite because the number of elements of a distributive free lattice with $2n$ generators is always finite if n is finite.

The enumeration of all reduced forms of n fuzzy variables does not seem to be an easy problem.

For one variable, it is trivial. One has

$$\underset{\sim}{a} \ , \quad \underset{\sim}{\bar{a}} \ , \quad \underset{\sim}{a} \wedge \underset{\sim}{\bar{a}} \ , \quad \underset{\sim}{a} \vee \underset{\sim}{\bar{a}} \ , \tag{33.9}$$

that is, four reduced forms. Note well that $\underset{\sim}{a} \wedge \underset{\sim}{\bar{a}}$, for example, is to be distinguished from $\underset{\sim}{a}$ since

$$\underset{\sim}{a} \wedge \underset{\sim}{\bar{a}} = \underset{\sim}{a} \quad \text{if} \quad \underset{\sim}{a} \leqslant \underset{\sim}{\bar{a}} \quad \text{and} \quad \underset{\sim}{a} \wedge \underset{\sim}{\bar{a}} = \underset{\sim}{\bar{a}} \quad \text{if} \quad \underset{\sim}{\bar{a}} \leqslant \underset{\sim}{a} . \tag{33.10}$$

For two variables, it is already no longer simple, and is in fact very complicated.

We use, for example, reduced polynomial forms in $\vee$ (monomials in $\wedge$). We know that to each form in $\vee$ there corresponds a form in $\wedge$ and vice versa (because of the two theorems of De Morgan).

We then see the enumeration‡ of all possible distinct reduced polynomials forms in $\vee$ $\underset{\sim}{f}(\underset{\sim}{a}, \underset{\sim}{b})$:

(1) $\underset{\sim}{f}(\underset{\sim}{a}, \underset{\sim}{b})$ containing one monomial:

(33.11)

1	a (1)	$1 \wedge 2$	$a \wedge \bar{a}$ (5)	$1 \wedge 2 \wedge 3$	$a \wedge \bar{a} \wedge b$ (11)		
2	$\bar{a}$ (2)	$1 \wedge 3$	$a \wedge b$ (6)	$1 \wedge 2 \wedge 4$	$a \wedge \bar{a} \wedge \bar{b}$ (12)		
3	b (3)	$1 \wedge 4$	$a \wedge \bar{b}$ (7)	$1 \wedge 3 \wedge 4$	$a \wedge b \wedge \bar{b}$ (13)		
4	$\bar{b}$ (4)	$2 \wedge 3$	$\bar{a} \wedge b$ (8)	$2 \wedge 3 \wedge 4$	$\bar{a} \wedge b \wedge \bar{b}$ (14)		
	4	$2 \wedge 4$	$\bar{a} \wedge \bar{b}$ (9)	4			
		$3 \wedge 4$	$b \wedge \bar{b}$ (10)			$1 \wedge 2 \wedge 3 \wedge 4$	$a \wedge \bar{a} \wedge b \wedge \bar{b}$ (15)
			6				1

†See this concept, for example, in reference [3K, p. 287].

‡For this enumeration, we use as before a lexicographic procedure avoiding omission and redundancy, moreover bearing in mind that a monomial must always be maximal with respect to others of the same polynomial. On the other hand, in order to simplify the notation, we have not placed the fuzzy symbol ~.

There are then 4 + 6 + 4 + 1 = 15 reduced forms constructed using one monomial.

(2) $\underset{\sim}{f}(\underset{\sim}{a}, \underset{\sim}{b})$ containing two monomials (these two monomials must never be contractable by absorption to a single monomial form):

(33.12)

$1 \vee 2$	$a \vee \bar{a}$	(16)
$1 \vee 3$	$a \vee b$	(17)
$1 \vee 4$	$a \vee \bar{b}$	(18)
$2 \vee 3$	$\bar{a} \vee b$	(19)
$2 \vee 4$	$\bar{a} \vee \bar{b}$	(20)
$3 \vee 4$	$b \vee \bar{b}$	(21)

6

$1 \vee (2 \wedge 3)$	$a \vee (\bar{a} \wedge b)$	(22)
$1 \vee (2 \wedge 4)$	$a \vee (\bar{a} \wedge \bar{b})$	(23)
$1 \vee (3 \wedge 4)$	$a \vee (b \wedge \bar{b})$	(24)
$2 \vee (1 \wedge 3)$	$\bar{a} \vee (a \wedge b)$	(25)
$2 \vee (1 \wedge 4)$	$\bar{a} \vee (a \wedge \bar{b})$	(26)
$2 \vee (3 \wedge 4)$	$\bar{a} \vee (b \wedge \bar{b})$	(27)
$3 \vee (1 \wedge 2)$	$b \vee (a \wedge \bar{a})$	(28)
$3 \vee (1 \wedge 4)$	$b \vee (a \wedge \bar{b})$	(29)
$3 \vee (2 \wedge 4)$	$b \vee (\bar{a} \wedge \bar{b})$	(30)
$4 \vee (1 \wedge 2)$	$\bar{b} \vee (a \wedge \bar{a})$	(31)
$4 \vee (1 \wedge 3)$	$\bar{b} \vee (a \wedge b)$	(32)
$4 \vee (2 \wedge 3)$	$\bar{b} \vee (\bar{a} \wedge b)$	(33)

12

$1 \vee (2 \wedge 3 \wedge 4)$	$a \vee (\bar{a} \wedge b \wedge \bar{b})$	(34)
$2 \vee (1 \wedge 3 \wedge 4)$	$\bar{a} \vee (a \wedge b \wedge \bar{b})$	(35)
$3 \vee (1 \wedge 2 \wedge 4)$	$b \vee (a \wedge \bar{a} \wedge \bar{b})$	(36)
$4 \vee (1 \wedge 2 \wedge 3)$	$\bar{b} \vee (a \wedge \bar{a} \wedge b)$	(37)

4

$(1 \wedge 2) \vee (1 \wedge 3)$	$(a \wedge \bar{a}) \vee (a \wedge b)$	(38)
$(1 \wedge 2) \vee (1 \wedge 4)$	$(a \wedge \bar{a}) \vee (a \wedge \bar{b})$	(39)
$(1 \wedge 2) \vee (2 \wedge 3)$	$(a \wedge \bar{a}) \vee (\bar{a} \wedge b)$	(40)
$(1 \wedge 2) \vee (2 \wedge 4)$	$(a \wedge \bar{a}) \vee (\bar{a} \wedge \bar{b})$	(41)
$(1 \wedge 2) \vee (3 \wedge 4)$	$(a \wedge \bar{a}) \vee (b \wedge \bar{b})$	(42)
$(1 \wedge 3) \vee (1 \wedge 4)$	$(a \wedge b) \vee (a \wedge \bar{b})$	(43)
$(1 \wedge 3) \vee (2 \wedge 3)$	$(a \wedge b) \vee (\bar{a} \wedge b)$	(44)
$(1 \wedge 3) \vee (2 \wedge 4)$	$(a \wedge b) \vee (\bar{a} \wedge \bar{b})$	(45)
$(1 \wedge 3) \vee (3 \wedge 4)$	$(a \wedge b) \vee (b \wedge \bar{b})$	(46)
$(1 \wedge 4) \vee (2 \wedge 3)$	$(a \wedge \bar{b}) \vee (\bar{a} \wedge b)$	(47)
$(1 \wedge 4) \vee (2 \wedge 4)$	$(a \wedge \bar{b}) \vee (\bar{a} \wedge \bar{b})$	(48)
$(1 \wedge 4) \vee (3 \wedge 4)$	$(a \wedge \bar{b}) \vee (b \wedge \bar{b})$	(49)
$(2 \wedge 3) \vee (2 \wedge 4)$	$(\bar{a} \wedge b) \vee (\bar{a} \wedge \bar{b})$	(50)
$(2 \wedge 3) \vee (3 \wedge 4)$	$(\bar{a} \wedge b) \vee (b \wedge \bar{b})$	(51)
$(2 \wedge 4) \vee (3 \wedge 4)$	$(\bar{a} \wedge \bar{b}) \vee (b \wedge \bar{b})$	(52)

15

$(1 \wedge 2) \vee (1 \wedge 3 \wedge 4)$	$(a \wedge \bar{a}) \vee (a \wedge b \wedge \bar{b})$	(53)
$(1 \wedge 2) \vee (2 \wedge 3 \wedge 4)$	$(a \wedge \bar{a}) \vee (\bar{a} \wedge b \wedge \bar{b})$	(54)
$(1 \wedge 3) \vee (1 \wedge 2 \wedge 4)$	$(a \wedge b) \vee (a \wedge \bar{a} \wedge \bar{b})$	(55)
$(1 \wedge 3) \vee (2 \wedge 3 \wedge 4)$	$(a \wedge b) \vee (\bar{a} \wedge b \wedge \bar{b})$	(56)
$(1 \wedge 4) \vee (1 \wedge 2 \wedge 3)$	$(a \wedge \bar{b}) \vee (a \wedge \bar{a} \wedge b)$	(57)
$(1 \wedge 4) \vee (2 \wedge 3 \wedge 4)$	$(a \wedge \bar{b}) \vee (\bar{a} \wedge b \wedge \bar{b})$	(58)
$(2 \wedge 3) \vee (1 \wedge 2 \wedge 4)$	$(\bar{a} \wedge b) \vee (a \wedge \bar{a} \wedge \bar{b})$	(59)
$(2 \wedge 3) \vee (1 \wedge 3 \wedge 4)$	$(\bar{a} \wedge b) \vee (a \wedge b \wedge \bar{b})$	(60)
$(2 \wedge 4) \vee (1 \wedge 2 \wedge 3)$	$(\bar{a} \wedge \bar{b}) \vee (a \wedge \bar{a} \wedge b)$	(61)
$(2 \wedge 4) \vee (1 \wedge 3 \wedge 4)$	$(\bar{a} \wedge \bar{b}) \vee (a \wedge b \wedge \bar{b})$	(62)
$(3 \wedge 4) \vee (1 \wedge 2 \wedge 3)$	$(b \wedge \bar{b}) \vee (a \wedge \bar{a} \wedge b)$	(63)
$(3 \wedge 4) \vee (1 \wedge 2 \wedge 4)$	$(b \wedge \bar{b}) \vee (a \wedge \bar{a} \wedge \bar{b})$	(64)

12

$(1 \wedge 2 \wedge 3) \vee (1 \wedge 2 \wedge 4)$	$(a \wedge \bar{a} \wedge b) \vee (a \wedge \bar{a} \wedge \bar{b})$	(65)
$(1 \wedge 2 \wedge 3) \vee (1 \wedge 3 \wedge 4)$	$(a \wedge \bar{a} \wedge b) \vee (a \wedge b \wedge \bar{b})$	(66)
$(1 \wedge 2 \wedge 3) \vee (2 \wedge 3 \wedge 4)$	$(a \wedge \bar{a} \wedge b) \vee (\bar{a} \wedge b \wedge \bar{b})$	(67)
$(1 \wedge 2 \wedge 4) \vee (1 \wedge 3 \wedge 4)$	$(a \wedge \bar{a} \wedge \bar{b}) \vee (a \wedge b \wedge \bar{b})$	(68)
$(1 \wedge 2 \wedge 4) \vee (2 \wedge 3 \wedge 4)$	$(a \wedge \bar{a} \wedge \bar{b}) \vee (\bar{a} \wedge b \wedge \bar{b})$	(69)
$(1 \wedge 3 \wedge 4) \vee (2 \wedge 3 \wedge 4)$	$(a \wedge b \wedge \bar{b}) \vee (\bar{a} \wedge b \wedge \bar{b})$	(70)

6

There are then $6 + 12 + 4 + 15 + 12 + 6 = 55$ reduced forms constructed using two monomials.

(3) $\underset{\sim}{f}(\underset{\sim}{a}, \underset{\sim}{b})$ containing three monomials (subject to the preceding remark on contradiction:

$1 \vee 2 \vee 3$	$a \vee \bar{a} \vee b$ (71)	$1 \vee 2 \vee (3 \wedge 4)$	$a \vee \bar{a} \vee (b \wedge \bar{b})$ (75)
$1 \vee 2 \vee 4$	$a \vee \bar{a} \vee \bar{b}$ (72)	$1 \vee 3 \vee (2 \wedge 4)$	$a \vee b \vee (\bar{a} \wedge \bar{b})$ (76)
$1 \vee 3 \vee 4$	$a \vee b \vee \bar{b}$ (73)	$1 \vee 4 \vee (2 \wedge 3)$	$a \vee \bar{b} \vee (\bar{a} \wedge b)$ (77)
$2 \vee 3 \vee 4$	$\bar{a} \vee b \vee \bar{b}$ (74)	$2 \vee 3 \vee (1 \wedge 4)$	$\bar{a} \vee b \vee (a \wedge \bar{b})$ (78)
		$2 \vee 4 \vee (1 \wedge 3)$	$\bar{a} \vee \bar{b} \vee (a \wedge b)$ (79)
4		$3 \vee 4 \vee (1 \wedge 2)$	$b \vee \bar{b} \vee (a \wedge \bar{a})$ (80)
		6	

$1 \vee (2 \wedge 3) \vee (2 \wedge 4)$	$a \vee (\bar{a} \wedge b) \vee (\bar{a} \wedge \bar{b})$ (81)
$1 \vee (2 \wedge 3) \vee (3 \wedge 4)$	$a \vee (\bar{a} \wedge b) \vee (b \wedge \bar{b})$ (82)
$1 \vee (2 \wedge 4) \vee (3 \wedge 4)$	$a \vee (\bar{a} \wedge \bar{b}) \vee (b \wedge \bar{b})$ (83)
$2 \vee (1 \wedge 3) \vee (1 \wedge 4)$	$\bar{a} \vee (a \wedge b) \vee (a \wedge \bar{b})$ (84)
$2 \vee (1 \wedge 3) \vee (3 \wedge 4)$	$\bar{a} \vee (a \wedge b) \vee (b \wedge \bar{b})$ (85)
$2 \vee (1 \wedge 4) \vee (3 \wedge 4)$	$\bar{a} \vee (a \wedge \bar{b}) \vee (b \wedge \bar{b})$ (86)
$3 \vee (1 \wedge 2) \vee (1 \wedge 4)$	$b \vee (a \wedge \bar{a}) \vee (a \wedge \bar{b})$ (87)
$3 \vee (1 \wedge 2) \vee (2 \wedge 4)$	$b \vee (a \wedge \bar{a}) \vee (\bar{a} \wedge \bar{b})$ (88)
$3 \vee (1 \wedge 4) \vee (2 \wedge 4)$	$b \vee (a \wedge \bar{b}) \vee (\bar{a} \wedge \bar{b})$ (89)
$4 \vee (1 \wedge 2) \vee (1 \wedge 3)$	$\bar{b} \vee (a \wedge \bar{a}) \vee (a \wedge b)$ (90)
$4 \vee (1 \wedge 2) \vee (2 \wedge 3)$	$\bar{b} \vee (a \wedge \bar{a}) \vee (\bar{a} \wedge b)$ (91)
$4 \vee (1 \wedge 3) \vee (2 \wedge 3)$	$\bar{b} \vee (a \wedge b) \vee (\bar{a} \wedge b)$ (92)
12	

(33.13)

$(1 \wedge 2) \vee (1 \wedge 3) \vee (1 \wedge 4)$	$(a \wedge \bar{a}) \vee (a \wedge b) \vee (a \wedge \bar{b})$ (93)
$(1 \wedge 2) \vee (1 \wedge 3) \vee (2 \wedge 3)$	$(a \wedge \bar{a}) \vee (a \wedge b) \vee (\bar{a} \wedge b)$ (94)
$(1 \wedge 2) \vee (1 \wedge 3) \vee (2 \wedge 4)$	$(a \wedge \bar{a}) \vee (a \wedge b) \vee (\bar{a} \wedge \bar{b})$ (95)
$(1 \wedge 2) \vee (1 \wedge 3) \vee (3 \wedge 4)$	$(a \wedge \bar{a}) \vee (a \wedge b) \vee (b \wedge \bar{b})$ (96)
$(1 \wedge 2) \vee (1 \wedge 4) \vee (2 \wedge 3)$	$(a \wedge \bar{a}) \vee (a \wedge \bar{b}) \vee (\bar{a} \wedge b)$ (97)
$(1 \wedge 2) \vee (1 \wedge 4) \vee (2 \wedge 4)$	$(a \wedge \bar{a}) \vee (a \wedge \bar{b}) \vee (\bar{a} \wedge \bar{b})$ (98)
$(1 \wedge 2) \vee (1 \wedge 4) \vee (3 \wedge 4)$	$(a \wedge \bar{a}) \vee (a \wedge \bar{b}) \vee (b \wedge \bar{b})$ (99)
$(1 \wedge 2) \vee (2 \wedge 3) \vee (2 \wedge 4)$	$(a \wedge \bar{a}) \vee (\bar{a} \wedge b) \vee (\bar{a} \wedge \bar{b})$ (100)
$(1 \wedge 2) \vee (2 \wedge 3) \vee (3 \wedge 4)$	$(a \wedge \bar{a}) \vee (\bar{a} \wedge b) \vee (b \wedge \bar{b})$ (101)
$(1 \wedge 2) \vee (2 \wedge 4) \vee (3 \wedge 4)$	$(a \wedge \bar{a}) \vee (\bar{a} \wedge \bar{b}) \vee (b \wedge \bar{b})$ (102)
$(1 \wedge 3) \vee (1 \wedge 4) \vee (2 \wedge 3)$	$(a \wedge b) \vee (a \wedge \bar{b}) \vee (\bar{a} \wedge b)$ (103)
$(1 \wedge 3) \vee (1 \wedge 4) \vee (2 \wedge 4)$	$(a \wedge b) \vee (a \wedge \bar{b}) \vee (\bar{a} \wedge \bar{b})$ (104)
$(1 \wedge 3) \vee (1 \wedge 4) \vee (3 \wedge 4)$	$(a \wedge b) \vee (a \wedge \bar{b}) \vee (b \wedge \bar{b})$ (105)
$(1 \wedge 3) \vee (2 \wedge 3) \vee (2 \wedge 4)$	$(a \wedge b) \vee (\bar{a} \wedge b) \vee (\bar{a} \wedge \bar{b})$ (106)
$(1 \wedge 3) \vee (2 \wedge 3) \vee (3 \wedge 4)$	$(a \wedge b) \vee (\bar{a} \wedge b) \vee (b \wedge \bar{b})$ (107)
$(1 \wedge 3) \vee (2 \wedge 4) \vee (3 \wedge 4)$	$(a \wedge b) \vee (\bar{a} \wedge \bar{b}) \vee (b \wedge \bar{b})$ (108)
$(1 \wedge 4) \vee (2 \wedge 3) \vee (2 \wedge 4)$	$(a \wedge \bar{b}) \vee (\bar{a} \wedge b) \vee (\bar{a} \wedge \bar{b})$ (109)
$(1 \wedge 4) \vee (2 \wedge 3) \vee (3 \wedge 4)$	$(a \wedge \bar{b}) \vee (\bar{a} \wedge b) \vee (b \wedge \bar{b})$ (110)

$(1 \wedge 4) \vee (2 \wedge 4) \vee (3 \wedge 4)$ | $(a \wedge \bar{b}) \vee (\bar{a} \wedge \bar{b}) \vee (b \wedge \bar{b})$ (111)
$(2 \wedge 3) \vee (2 \wedge 4) \vee (3 \wedge 4)$ | $(\bar{a} \wedge b) \vee (\bar{a} \wedge \bar{b}) \vee (b \wedge \bar{b})$ (112)

20

$(1 \wedge 2) \vee (1 \wedge 3) \vee (2 \wedge 3 \wedge 4)$ | $(a \wedge \bar{a}) \vee (a \wedge b) \vee (\bar{a} \wedge b \wedge \bar{b})$ (113)
$(1 \wedge 2) \vee (1 \wedge 4) \vee (2 \wedge 3 \wedge 4)$ | $(a \wedge \bar{a}) \vee (a \wedge \bar{b}) \vee (\bar{a} \wedge b \wedge \bar{b})$ (114)
$(1 \wedge 2) \vee (2 \wedge 3) \vee (1 \wedge 3 \wedge 4)$ | $(a \wedge \bar{a}) \vee (\bar{a} \wedge b) \vee (a \wedge b \wedge \bar{b})$ (115)
$(1 \wedge 2) \vee (2 \wedge 4) \vee (1 \wedge 3 \wedge 4)$ | $(a \wedge \bar{a}) \vee (\bar{a} \wedge \bar{b}) \vee (a \wedge b \wedge \bar{b})$ (116)
$(1 \wedge 3) \vee (1 \wedge 4) \vee (2 \wedge 3 \wedge 4)$ | $(a \wedge b) \vee (a \wedge \bar{b}) \vee (\bar{a} \wedge b \wedge \bar{b})$ (117)
$(1 \wedge 3) \vee (2 \wedge 3) \vee (1 \wedge 2 \wedge 4)$ | $(a \wedge b) \vee (\bar{a} \wedge b) \vee (a \wedge \bar{a} \wedge \bar{b})$ (118)
$(1 \wedge 3) \vee (3 \wedge 4) \vee (1 \wedge 2 \wedge 4)$ | $(a \wedge b) \vee (b \wedge \bar{b}) \vee (a \wedge \bar{a} \wedge \bar{b})$ (119)
$(1 \wedge 4) \vee (2 \wedge 4) \vee (1 \wedge 2 \wedge 3)$ | $(a \wedge \bar{b}) \vee (\bar{a} \wedge \bar{b}) \vee (a \wedge \bar{a} \wedge b)$ (120)
$(1 \wedge 4) \vee (3 \wedge 4) \vee (1 \wedge 2 \wedge 3)$ | $(a \wedge \bar{b}) \vee (b \wedge \bar{b}) \vee (a \wedge \bar{a} \wedge b)$ (121)
$(2 \wedge 3) \vee (2 \wedge 4) \vee (1 \wedge 3 \wedge 4)$ | $(\bar{a} \wedge b) \vee (\bar{a} \wedge \bar{b}) \vee (a \wedge b \wedge \bar{b})$ (122)
$(2 \wedge 3) \vee (3 \wedge 4) \vee (1 \wedge 2 \wedge 4)$ | $(\bar{a} \wedge b) \vee (b \wedge \bar{b}) \vee (a \wedge \bar{a} \wedge \bar{b})$ (123)
$(2 \wedge 4) \vee (3 \wedge 4) \vee (1 \wedge 2 \wedge 3)$ | $(\bar{a} \wedge \bar{b}) \vee (b \wedge \bar{b}) \vee (a \wedge \bar{a} \wedge b)$ (124)

12

$(1 \wedge 2) \vee (1 \wedge 3 \wedge 4) \vee (2 \wedge 3 \wedge 4)$ | $(a \wedge \bar{a}) \vee (a \wedge b \wedge \bar{b}) \vee (\bar{a} \wedge b \wedge \bar{b})$ (125)
$(1 \wedge 3) \vee (1 \wedge 2 \wedge 4) \vee (2 \wedge 3 \wedge 4)$ | $(a \wedge b) \vee (a \wedge \bar{a} \wedge \bar{b}) \vee (\bar{a} \wedge b \wedge \bar{b})$ (126)
$(1 \wedge 4) \vee (1 \wedge 2 \wedge 3) \vee (2 \wedge 3 \wedge 4)$ | $(a \wedge \bar{b}) \vee (a \wedge \bar{a} \wedge b) \vee (\bar{a} \wedge b \wedge \bar{b})$ (127)
$(2 \wedge 3) \vee (1 \wedge 2 \wedge 4) \vee (1 \wedge 3 \wedge 4)$ | $(\bar{a} \wedge b) \vee (a \wedge \bar{a} \wedge \bar{b}) \vee (a \wedge b \wedge \bar{b})$ (128)
$(2 \wedge 4) \vee (1 \wedge 2 \wedge 3) \vee (1 \wedge 3 \wedge 4)$ | $(\bar{a} \wedge \bar{b}) \vee (a \wedge \bar{a} \wedge b) \vee (a \wedge b \wedge \bar{b})$ (129)
$(3 \wedge 4) \vee (1 \wedge 2 \wedge 3) \vee (1 \wedge 2 \wedge 4)$ | $(b \wedge \bar{b}) \vee (a \wedge \bar{a} \wedge b) \vee (a \wedge \bar{a} \wedge \bar{b})$ (130)

6

$(1 \wedge 2 \wedge 3) \vee (1 \wedge 2 \wedge 4) \vee (1 \wedge 3 \wedge 4)$ | $(a \wedge \bar{a} \wedge b) \vee (a \wedge \bar{a} \wedge \bar{b}) \vee (a \wedge b \wedge \bar{b})$ (131)
$(1 \wedge 2 \wedge 4) \vee (1 \wedge 3 \wedge 4) \vee (2 \wedge 3 \wedge 4)$ | $(a \wedge \bar{a} \wedge \bar{b}) \vee (a \wedge b \wedge \bar{b}) \vee (\bar{a} \wedge b \wedge \bar{b})$ (132)
$(1 \wedge 3 \wedge 4) \vee (2 \wedge 3 \wedge 4) \vee (1 \wedge 2 \wedge 3)$ | $(a \wedge b \wedge \bar{b}) \vee (\bar{a} \wedge b \wedge \bar{b}) \vee (a \wedge \bar{a} \wedge b)$ (133)
$(2 \wedge 3 \wedge 4) \vee (1 \wedge 2 \wedge 3) \vee (1 \wedge 2 \wedge 4)$ | $(\bar{a} \wedge b \wedge \bar{b}) \vee (a \wedge \bar{a} \wedge b) \vee (a \wedge \bar{a} \wedge \bar{b})$ (134)

4

There are then $4 + 6 + 12 + 20 + 12 + 6 + 4 = 64$ reduced forms constructed from three monomials.

(4) $\underset{\sim}{f}(\underset{\sim}{a}, \underset{\sim}{b})$ containing four monomials:

$1 \vee 2 \vee 3 \vee 4$ | $a \vee \bar{a} \vee b \vee \bar{b}$ (135)

1

$1 \vee (2 \wedge 3) \vee (2 \wedge 4) \vee (3 \wedge 4)$	$a \vee (\bar{a} \wedge b) \vee (\bar{a} \wedge \bar{b}) \vee (b \wedge \bar{b})$ (136)
$2 \vee (1 \wedge 3) \vee (1 \wedge 4) \vee (3 \wedge 4)$	$\bar{a} \vee (a \wedge b) \vee (a \wedge \bar{b}) \vee (b \wedge \bar{b})$ (137)
$3 \vee (1 \wedge 2) \vee (1 \wedge 4) \vee (2 \wedge 4)$	$b \vee (a \wedge \bar{a}) \vee (a \wedge \bar{b}) \vee (\bar{a} \wedge b)$ (138)
$4 \vee (1 \wedge 2) \vee (1 \wedge 3) \vee (2 \wedge 3)$	$\bar{b} \vee (a \wedge \bar{a}) \vee (a \wedge b) \vee (\bar{a} \wedge b)$ (139)

4

(33.14)

$(1 \wedge 2) \vee (1 \wedge 3) \vee (1 \wedge 4) \vee (2 \wedge 3)$	$(a \wedge \bar{a}) \vee (a \wedge b) \vee (a \wedge \bar{b}) \vee (\bar{a} \wedge b)$ (140)
$(1 \wedge 2) \vee (1 \wedge 3) \vee (1 \wedge 4) \vee (2 \wedge 4)$	$(a \wedge \bar{a}) \vee (a \wedge b) \vee (a \wedge \bar{b}) \vee (\bar{a} \wedge \bar{b})$ (141)
$(1 \wedge 2) \vee (1 \wedge 3) \vee (1 \wedge 4) \vee (3 \wedge 4)$	$(a \wedge \bar{a}) \vee (a \wedge b) \vee (a \wedge \bar{b}) \vee (b \wedge \bar{b})$ (142)
$(1 \wedge 2) \vee (1 \wedge 3) \vee (2 \wedge 3) \vee (2 \wedge 4)$	$(a \wedge \bar{a}) \vee (a \wedge b) \vee (\bar{a} \wedge b) \vee (\bar{a} \wedge \bar{b})$ (143)
$(1 \wedge 2) \vee (1 \wedge 3) \vee (2 \wedge 3) \vee (3 \wedge 4)$	$(a \wedge \bar{a}) \vee (a \wedge b) \vee (\bar{a} \wedge b) \vee (b \wedge \bar{b})$ (144)
$(1 \wedge 2) \vee (1 \wedge 3) \vee (2 \wedge 4) \vee (3 \wedge 4)$	$(a \wedge \bar{a}) \vee (a \wedge b) \vee (\bar{a} \wedge \bar{b}) \vee (b \wedge \bar{b})$ (145)
$(1 \wedge 2) \vee (1 \wedge 4) \vee (2 \wedge 3) \vee (2 \wedge 4)$	$(a \wedge \bar{a}) \vee (a \wedge \bar{b}) \vee (\bar{a} \wedge b) \vee (\bar{a} \wedge \bar{b})$ (146)
$(1 \wedge 2) \vee (1 \wedge 4) \vee (2 \wedge 3) \vee (3 \wedge 4)$	$(a \wedge \bar{a}) \vee (a \wedge \bar{b}) \vee (\bar{a} \wedge b) \vee (b \wedge \bar{b})$ (147)
$(1 \wedge 2) \vee (1 \wedge 4) \vee (2 \wedge 4) \vee (3 \wedge 4)$	$(a \wedge \bar{a}) \vee (a \wedge \bar{b}) \vee (\bar{a} \wedge \bar{b}) \vee (b \wedge \bar{b})$ (148)
$(1 \wedge 2) \vee (2 \wedge 3) \vee (2 \wedge 4) \vee (3 \wedge 4)$	$(a \wedge \bar{a}) \vee (\bar{a} \wedge b) \vee (a \wedge \bar{b}) \vee (b \wedge \bar{b})$ (149)
$(1 \wedge 3) \vee (1 \wedge 4) \vee (2 \wedge 3) \vee (2 \wedge 4)$	$(a \wedge b) \vee (a \wedge \bar{b}) \vee (\bar{a} \wedge b) \vee (\bar{a} \wedge \bar{b})$ (150)
$(1 \wedge 3) \vee (1 \wedge 4) \vee (2 \wedge 3) \vee (3 \wedge 4)$	$(a \wedge b) \vee (a \wedge \bar{b}) \vee (\bar{a} \wedge b) \vee (b \wedge \bar{b})$ (151)
$(1 \wedge 3) \vee (1 \wedge 4) \vee (2 \wedge 4) \vee (3 \wedge 4)$	$(a \wedge b) \vee (a \wedge \bar{b}) \vee (\bar{a} \wedge \bar{b}) \vee (b \wedge \bar{b})$ (152)
$(1 \wedge 3) \vee (2 \wedge 3) \vee (2 \wedge 4) \vee (3 \wedge 4)$	$(a \wedge b) \vee (\bar{a} \wedge b) \vee (\bar{a} \wedge \bar{b}) \vee (b \wedge \bar{b})$ (153)
$(1 \wedge 4) \vee (2 \wedge 3) \vee (2 \wedge 4) \vee (3 \wedge 4)$	$(a \wedge \bar{b}) \vee (\bar{a} \wedge b) \vee (\bar{a} \wedge \bar{b}) \vee (b \wedge \bar{b})$ (154)

15

$(1 \wedge 2) \vee (1 \wedge 3) \vee (1 \wedge 4) \vee (2 \wedge 3 \wedge 4)$	$(a \wedge \bar{a}) \vee (a \wedge b) \vee (a \wedge \bar{b}) \vee (\bar{a} \wedge b \wedge \bar{b})$ (155)
$(1 \wedge 2) \vee (2 \wedge 3) \vee (2 \wedge 4) \vee (1 \wedge 3 \wedge 4)$	$(a \wedge \bar{a}) \vee (\bar{a} \wedge b) \vee (\bar{a} \wedge \bar{b}) \vee (a \wedge b \wedge \bar{b})$ (156)
$(1 \wedge 3) \vee (2 \wedge 3) \vee (3 \wedge 4) \vee (1 \wedge 2 \wedge 4)$	$(a \wedge b) \vee (\bar{a} \wedge b) \vee (b \wedge \bar{b}) \vee (a \wedge \bar{a} \wedge \bar{b})$ (157)
$(1 \wedge 4) \vee (2 \wedge 4) \vee (3 \wedge 4) \vee (1 \wedge 2 \wedge 3)$	$(a \wedge \bar{b}) \vee (\bar{a} \wedge \bar{b}) \vee (b \wedge \bar{b}) \vee (a \wedge \bar{a} \wedge b)$ (158)

4

$(1 \wedge 2 \wedge 3) \vee (1 \wedge 2 \wedge 4) \vee (1 \wedge 3 \wedge 4) \vee (2 \wedge 3 \wedge 4)$	$(a \wedge \bar{a} \wedge b) \vee (a \wedge \bar{a} \wedge \bar{b}) \vee (a \wedge b \wedge \bar{b}) \vee (\bar{a} \wedge b \wedge \bar{b})$ (159)

1

Thus, there are 1 + 4 + 15 + 4 + 1 = 25 reduced forms constructed from four monomials.

(5) $\underset{\sim}{f}(\underset{\sim}{a}, \underset{\sim}{b})$ containing five monomials:

(33.15)

$(1 \wedge 2) \vee (1 \wedge 3) \vee (1 \wedge 4) \vee (2 \wedge 3) \vee (2 \wedge 4)$	$(a \wedge \bar{a}) \vee (a \wedge b) \vee (a \wedge \bar{b}) \vee (\bar{a} \wedge b) \vee (\bar{a} \wedge \bar{b})$ (160)
$(1 \wedge 2) \vee (1 \wedge 3) \vee (1 \wedge 4) \vee (2 \wedge 3) \vee (3 \wedge 4)$	$(a \wedge \bar{a}) \vee (a \wedge b) \vee (a \wedge \bar{b}) \vee (\bar{a} \wedge b) \vee (b \wedge \bar{b})$ (161)
$(1 \wedge 2) \vee (1 \wedge 3) \vee (1 \wedge 4) \vee (2 \wedge 4) \vee (3 \wedge 4)$	$(a \wedge \bar{a}) \vee (a \wedge b) \vee (a \wedge \bar{b}) \vee (\bar{a} \wedge \bar{b}) \vee (b \wedge \bar{b})$ (162)
$(1 \wedge 2) \vee (1 \wedge 3) \vee (2 \wedge 3) \vee (2 \wedge 4) \vee (3 \wedge 4)$	$(a \wedge \bar{a}) \vee (a \wedge b) \vee (\bar{a} \wedge b) \vee (\bar{a} \wedge \bar{b}) \vee (b \wedge \bar{b})$ (163)

$(1 \wedge 2) \vee (1 \wedge 4) \vee (2 \wedge 3) \vee (2 \wedge 4) \vee (3 \wedge 4) \mid (a \wedge \bar{a}) \vee (a \wedge \bar{b}) \vee (\bar{a} \wedge b) \vee (\bar{a} \wedge \bar{b}) \vee (b \wedge \bar{b})$ (164)
$(1 \wedge 3) \vee (1 \wedge 4) \vee (2 \wedge 3) \vee (2 \wedge 4) \vee (3 \wedge 4) \mid (a \wedge b) \vee (a \wedge \bar{b}) \vee (\bar{a} \wedge b) \vee (\bar{a} \wedge \bar{b}) \vee (b \wedge \bar{b})$ (165)

6

These are then six reduced forms constructed from five monomials.

(6) $\underset{\sim}{f}(\underset{\sim}{a}, \underset{\sim}{b})$ containing six monomials:

(33.16)

$(1 \wedge 2) \vee (1 \wedge 3) \vee (1 \wedge 4) \vee (2 \wedge 3) \vee (2 \wedge 4) \vee (3 \wedge 4) \mid$

1

$(a \wedge \bar{a}) \vee (a \wedge b) \vee (a \wedge \bar{b}) \vee (\bar{a} \wedge b) \vee (\bar{a} \wedge \bar{b}) \vee (b \wedge \bar{b})$.
(166)

Thus, there is one reduced form constructed from six monomials.

In total, one has

x	functions with x monomials
1	15
2	55
3	64
4	25
5	6
6	1
	166

reduced polynomial forms.

34. ANALYSIS OF A FUNCTION OF FUZZY VARIABLES. METHOD OF MARINOS

We decompose $\mathbf{M} = [0, 1]$ into m joined intervals, closed on the left and open on the right, except the last:

(34.1) $\mathbf{I}_1 = [\alpha_0 = 0, \alpha_1[\quad , \quad \mathbf{I}_2 = [\alpha_1, \alpha_2[, \ldots\ldots , \quad \mathbf{I}_m = [\alpha_{m-1}, \alpha_m = 1]$,

where

(34.2) $\mathbf{M} = ([\alpha_0 = 0, \alpha_1[) \cup ([\alpha_1 , \alpha_2[) \cup \ldots\ldots \cup ([\alpha_{m-1} , \alpha_m = 1])$.

We then seek conditions so that a function of n fuzzy variables

(34.3) $\underset{\sim}{f}(\underset{\sim}{a}_1, \underset{\sim}{a}_2, \ldots, \underset{\sim}{a}_n)$, $\underset{\sim}{a}_i \in [0,1]$, $i = 1, 2, \ldots, n$,

will belong to an interval $\mathbf{I}_k$.

Example 1. We begin with an example.

Let

$$\underset{\sim}{f}(\underset{\sim}{a}, \underset{\sim}{b}, \underset{\sim}{c}) = (\overline{\underset{\sim}{a}} \wedge \overline{\underset{\sim}{b}}) \vee (\underset{\sim}{a} \wedge \underset{\sim}{b} \wedge \overline{\underset{\sim}{c}}) . \tag{34.4}$$

What conditions will give one

$$\underset{\sim}{f}(\underset{\sim}{a}, \underset{\sim}{b}, \underset{\sim}{c}) \in \mathbf{I}_k , \tag{34.5}$$

that is,

$$\alpha_{k-1} \leqslant \underset{\sim}{f}(\underset{\sim}{a}, \underset{\sim}{b}, \underset{\sim}{c}) < \alpha_k . \tag{34.6}$$

We examine (34.4). The member on the right is formed of two terms; thus it is necessary to take the largest. We begin with a first hypothesis.

Hypothesis I:

$$\overline{\underset{\sim}{a}} \wedge \overline{\underset{\sim}{b}} \geqslant \underset{\sim}{a} \wedge \underset{\sim}{b} \wedge \overline{\underset{\sim}{c}} . \tag{34.7}$$

This implies:

$$\alpha_{k-1} \leqslant \overline{\underset{\sim}{a}} \wedge \overline{\underset{\sim}{b}} < \alpha_k , \tag{34.8}$$

that is, explicitly,

$$\alpha_{k-1} \leqslant \text{MIN}\ (\overline{\underset{\sim}{a}}, \overline{\underset{\sim}{b}}) < \alpha_k , \tag{34.9}$$

or again

$$\alpha_{k-1} \leqslant \text{MIN}\ (1 - \underset{\sim}{a},\ 1 - \underset{\sim}{b}) < \alpha_k . \tag{34.10}$$

Since one may not place $\underset{\sim}{a}$ and $\underset{\sim}{b}$ arbitrarily with respect to one another, it is necessary that

$$1 - \underset{\sim}{a} \geqslant \alpha_{k-1} \quad \text{and} \quad 1 - \underset{\sim}{b} \geqslant \alpha_{k-1} , \tag{34.11}$$

and

$$1 - \underset{\sim}{a} < \alpha_k \quad \text{or/and} \quad 1 - \underset{\sim}{b} < \alpha_k . \tag{34.12}$$

This may be rewritten:

$$\underset{\sim}{a} \leqslant 1 - \alpha_{k-1} \quad \text{and} \quad \underset{\sim}{b} \leqslant 1 - \alpha_{k-1} \tag{34.13}$$

and

$$\underset{\sim}{a} > 1 - \alpha_k \quad \text{or/and} \quad \underset{\sim}{b} > 1 - \alpha_k \tag{34.14\dagger}$$

Hypothesis II:

$$\overline{\underset{\sim}{a}} \wedge \overline{\underset{\sim}{b}} < \underset{\sim}{a} \wedge \underset{\sim}{b} \wedge \overline{\underset{\sim}{c}} \tag{34.15}$$

This implies

$$\alpha_{k-1} \leqslant \underset{\sim}{a} \wedge \underset{\sim}{b} \wedge \overline{\underset{\sim}{c}} < \alpha_k , \tag{34.16}$$

explicitly, that is,

$$\alpha_{k-1} \leqslant \text{MIN}\ (\underset{\sim}{a}, \underset{\sim}{b}, \overline{\underset{\sim}{c}}) < \alpha_k , \tag{34.17}$$

†One ought not be astonished by the *and* in (34.13) and the *or/and* in (34.14). For the inferior limit α_{k-1}, it is necessary that $1 - \underset{\sim}{a}$ and $1 - \underset{\sim}{b}$ both be greater than or equal to α_{k-1}. But for α_k, it suffices that only one of the quantities $1 - \underset{\sim}{a}$ or $1 - \underset{\sim}{b}$ be less than α_k.

or again

(34.18) $$\alpha_{k-1} \leqslant \text{MIN}\,(\underset{\sim}{a}, \underset{\sim}{b}, 1 - \underset{\sim}{c}) < \alpha_k\,.$$

Since we may not place $\underset{\sim}{a}$, $\underset{\sim}{b}$, and $\underset{\sim}{c}$ arbitrarily with respect to one another, first it is necessary that

(34.19) $$\underset{\sim}{a} \geqslant \alpha_{k-1} \quad \text{and} \quad \underset{\sim}{b} \geqslant \alpha_{k-1} \quad \text{and} \quad 1 - \underset{\sim}{c} \geqslant \alpha_{k-1}$$

and

(34.20) $$\underset{\sim}{a} < \alpha_k \quad \text{or/and} \quad \underset{\sim}{b} < \alpha_k \quad \text{or/and} \quad 1 - \underset{\sim}{c} < \alpha_k\,.$$

This may be rewritten

(34.20) $$\underset{\sim}{a} \geqslant \alpha_{k-1} \quad \text{and} \quad \underset{\sim}{b} \geqslant \alpha_{k-1} \quad \text{and} \quad \underset{\sim}{c} \leqslant 1 - \alpha_{k-1}$$

and

(34.21) $$\underset{\sim}{a} < \alpha_k \quad \text{or/and} \quad \underset{\sim}{b} < \alpha_k \quad \text{or/and} \quad \underset{\sim}{c} > 1 - \alpha_k\,.$$

Finally, these results may be regrouped in the following fashion:

Property $\mathcal{P}_1$

(34.22)

$$[(\underset{\sim}{a} \leqslant 1 - \alpha_{k-1}) \text{ and } (\underset{\sim}{b} \leqslant 1 - \alpha_{k-1})] \text{ or/and } [(\underset{\sim}{a} \geqslant \alpha_{k-1}) \text{ and } (\underset{\sim}{b} \geqslant \alpha_{k-1}) \text{ and } (\underset{\sim}{c} \leqslant 1 - \alpha_{k-1})].$$

Property $\mathcal{P}_2$

(34.23)†

$$[(\underset{\sim}{a} > 1 - \alpha_k) \text{ or/and } (\underset{\sim}{b} > 1 - \alpha_k)] \text{ and } [(\underset{\sim}{a} < \alpha_k) \text{ or/and } (\underset{\sim}{b} < \alpha_k) \text{ or/and } (\underset{\sim}{c} > 1 - \alpha_k)]$$

In order that (34.6) be satisfied, it is necessary and sufficient that properties $\mathcal{P}_1$ and $\mathcal{P}_2$ be satisfied.

As a sample of the calculation of $\underset{\sim}{f}(\underset{\sim}{a}, \underset{\sim}{b}, \underset{\sim}{c})$ for particular numerical values, we suppose that

(34.25) $$\underset{\sim}{a} = 0{,}55 \quad , \quad \underset{\sim}{b} = 0{,}57 \quad , \quad \underset{\sim}{c} = 0{,}80\,.$$

Then one has

(34.26)

$$\begin{aligned}\underset{\sim}{f}(\underset{\sim}{a}, \underset{\sim}{b}, \underset{\sim}{c}) &= \underset{\sim}{f}(0{,}55\,;\,0{,}57\,;\,0{,}80)\\ &= (\overline{\underset{\sim}{a}} \wedge \overline{\underset{\sim}{b}}) \vee (\underset{\sim}{a} \wedge \underset{\sim}{b} \wedge \overline{\underset{\sim}{c}}) \quad \text{where} \quad \underset{\sim}{a} = 0{,}55\,;\, \underset{\sim}{b} = 0{,}57\,;\, \underset{\sim}{c} = 0{,}80\,.\\ &= (0{,}45 \wedge 0{,}43) \vee (0{,}55 \wedge 0{,}57 \wedge 0{,}20)\\ &= 0{,}43 \vee 0{,}20\\ &= 0{,}43\,.\end{aligned}$$

We now consider a complete numerical example.

†There is no Equation (34.24).

Example 2. Let

(34.27) $$f(\underset{\sim}{a}, \underset{\sim}{b}, \underset{\sim}{c}) = (\underset{\sim}{a} \wedge \overline{\underset{\sim}{b}}) \vee (\overline{\underset{\sim}{a}} \wedge \underset{\sim}{c}) \vee \overline{\underset{\sim}{c}},$$

and suppose that $[0, 1]$ is divided into three intervals

$$[0\,,\,0{,}2[\,,\ [0{,}2\,,\,0{,}3[\,,\ [0{,}3\,,\,1]\,.$$

First we consider the interval $[0, 0{,}2[$.

(34.28) Hypothesis I: $\underset{\sim}{a} \wedge \overline{\underset{\sim}{b}} > \overline{\underset{\sim}{a}} \wedge \underset{\sim}{c}, \quad \underset{\sim}{a} \wedge \overline{\underset{\sim}{b}} > \overline{\underset{\sim}{c}}.$

(34.29) One then has $\quad 0 \leqslant \underset{\sim}{a} \wedge \overline{\underset{\sim}{b}} < 0{,}2\,,$

(34.30) So $\quad 0 \leqslant \text{MIN}\,(\underset{\sim}{a},\ 1 - \underset{\sim}{b}) < 0{,}2\,,$

and

(34.31) $\quad \underset{\sim}{a} \geqslant 0 \quad$ and $\quad \underset{\sim}{b} \leqslant 1$

(34.32) $\quad \underset{\sim}{a} < 0{,}2 \quad$ or/and $\quad \underset{\sim}{b} > 0{,}8\,.$

(34.33) Hypothesis II: $\overline{\underset{\sim}{a}} \wedge \underset{\sim}{c} > \underset{\sim}{a} \wedge \overline{\underset{\sim}{b}}, \quad \overline{\underset{\sim}{a}} \wedge \underset{\sim}{c} > \overline{\underset{\sim}{c}}.$

(34.34) One then has $\quad 0 \leqslant \overline{\underset{\sim}{a}} \wedge \underset{\sim}{c} < 0{,}2\,,$

(34.35) So $\quad 0 \leqslant \text{MIN}\,(1 - \underset{\sim}{a}, \underset{\sim}{c}) < 0{,}2\,,$

and

(34.36) $\quad \underset{\sim}{a} \leqslant 1 \quad$ and $\quad \underset{\sim}{c} \geqslant 0\,,$

(34.37) $\quad \underset{\sim}{a} > 0{,}8$ or/and $\underset{\sim}{c} < 0{,}2\,.$

(34.38) Hypothesis III: $\overline{\underset{\sim}{c}} > \underset{\sim}{a} \wedge \overline{\underset{\sim}{b}}, \quad \overline{\underset{\sim}{c}} > \overline{\underset{\sim}{a}} \wedge \underset{\sim}{c}.$

(34.39) One then has $\quad 0 \leqslant \overline{\underset{\sim}{c}} < 0{,}2\,,$

(34.40) So $\quad 0 \leqslant 1 - \underset{\sim}{c} < 0{,}2\,,$

(34.41) $\quad 0{,}8 < \underset{\sim}{c} \leqslant 1\,.$

Now we consider the interval $[0{,}2, 0{,}3[$.

(34.42) Hypothesis I: $\underset{\sim}{a} \wedge \overline{\underset{\sim}{b}} > \overline{\underset{\sim}{a}} \wedge \underset{\sim}{c}, \quad \underset{\sim}{a} \wedge \overline{\underset{\sim}{b}} > \overline{\underset{\sim}{c}}.$

(34.43) $\quad 0{,}2 \leqslant \underset{\sim}{a} \wedge \overline{\underset{\sim}{b}} < 0{,}3\,,$

and

(34.44) $\quad \underset{\sim}{a} \geqslant 0{,}2 \quad$ and $\quad \underset{\sim}{b} \leqslant 0{,}8\,,$

(34.45) $\quad \underset{\sim}{a} < 0{,}3$ or/and $\underset{\sim}{b} > 0{,}7\,.$

(34.46) Hypothesis II: $\overline{\underset{\sim}{a}} \wedge \underset{\sim}{c} > \underset{\sim}{a} \wedge \overline{\underset{\sim}{b}}\,, \quad \overline{\underset{\sim}{a}} \wedge \underset{\sim}{c} > \overline{\underset{\sim}{c}},$

(34.47) $\quad 0{,}2 \leqslant \overline{\underset{\sim}{a}} \wedge \underset{\sim}{c} < 0{,}3\,,$

and

(34.48) $\quad \underset{\sim}{a} \leqslant 0{,}8 \quad$ and $\quad \underset{\sim}{c} \geqslant 0{,}2\,.$

(34.49) $\quad \underset{\sim}{a} > 0{,}7 \quad$ and $\quad \underset{\sim}{c} < 0{,}3\,.$

(34.50) Hypothesis III: $\overline{\underset{\sim}{c}} > \underset{\sim}{a} \wedge \overline{\underset{\sim}{b}}, \quad \overline{\underset{\sim}{c}} > \overline{\underset{\sim}{a}} \wedge \underset{\sim}{c},$

(34.51) $$0{,}2 \leqslant \bar{\underset{\sim}{c}} < 0{,}3 .$$
and
(34.52) $$\underset{\sim}{c} \leqslant 0{,}8 \quad \text{and} \quad \underset{\sim}{c} > 0{,}7 .$$

Lastly, we consider the interval [0,3, 1].

(34.53) Hypothesis I: $$\underset{\sim}{a} \wedge \bar{\underset{\sim}{b}} > \bar{\underset{\sim}{a}} \wedge \underset{\sim}{c}, \quad \underset{\sim}{a} \wedge \bar{\underset{\sim}{b}} > \bar{\underset{\sim}{c}}$$

(34.54) $$0{,}3 \leqslant \underset{\sim}{a} \wedge \bar{\underset{\sim}{b}} \leqslant 1 ,$$

(34.55) $$\underset{\sim}{a} \geqslant 0{,}3 \quad \text{and} \quad \underset{\sim}{b} \leqslant 0{,}7 ,$$
and
(34.56) $$\underset{\sim}{a} \leqslant 1 \quad \text{or/and} \quad \underset{\sim}{b} \geqslant 0 .$$

(34.57) Hypothesis II: $$\bar{\underset{\sim}{a}} \wedge \underset{\sim}{c} > \underset{\sim}{a} \wedge \bar{\underset{\sim}{b}}, \quad \bar{\underset{\sim}{a}} \wedge \underset{\sim}{c} > \bar{\underset{\sim}{c}},$$

(34.58) $$0{,}3 \leqslant \bar{\underset{\sim}{a}} \wedge \underset{\sim}{c} \leqslant 1 ,$$

(34.59) $$\underset{\sim}{a} \leqslant 0{,}7 \quad \text{and} \quad \underset{\sim}{c} \geqslant 0{,}3 ,$$
and
(34.60) $$\underset{\sim}{a} \geqslant 0 \quad \text{or/and} \quad \underset{\sim}{c} \leqslant 1 ,$$

(34.61) Hypothesis III: $$\bar{\underset{\sim}{c}} > \underset{\sim}{a} \wedge \bar{\underset{\sim}{b}}, \ \underset{\sim}{c} > \bar{\underset{\sim}{a}} \wedge \underset{\sim}{c},$$

(34.62) $$0{,}3 \leqslant \bar{\underset{\sim}{c}} \leqslant 1 ,$$
and
(34.63) $$\underset{\sim}{c} \leqslant 0{,}7 \quad \text{and} \quad \underset{\sim}{c} \geqslant 0 .$$

Finally, the results of this example may be regrouped in the following fashion:

(34.64) a) $$0 \leqslant \underset{\sim}{f}(\underset{\sim}{a}, \underset{\sim}{b}, \underset{\sim}{c}) < 0{,}2 .$$

Property $\mathcal{P}_1^{(1)}$:

(34.65) $$[(\underset{\sim}{a} \geqslant 0) \text{ and } (\underset{\sim}{b} \leqslant 1)] \text{ or/and } [(\underset{\sim}{a} \leqslant 1) \text{ and } (\underset{\sim}{c} \geqslant 0)] \text{ or/and } (\underset{\sim}{c} \leqslant 1) .$$

Property $\mathcal{P}_2^{(1)}$:

(34.66)

$$[(\underset{\sim}{a} < 0{,}2) \text{ or/and } (\underset{\sim}{b} > 0{,}8)] \text{ and } [(\underset{\sim}{a} > 0{,}8) \text{ or/and } (\underset{\sim}{c} < 0{,}2)] \text{ and } (\underset{\sim}{c} > 0{,}8) .$$

If properties $\mathcal{P}_1^{(1)}$ (34.64) and $\mathcal{P}_2^{(1)}$ (34.66) are verified, then one has (34.64).

(34.67) b) $$0{,}2 \leqslant \underset{\sim}{f}(\underset{\sim}{a}, \underset{\sim}{b}, \underset{\sim}{c}) < 0{,}3 .$$

Property $\mathcal{P}_1^{(2)}$:

(34.68)

$$[(\underset{\sim}{a} \geqslant 0{,}2) \text{ and } (\underset{\sim}{b} \leqslant 0{,}8)] \text{ or/and } [(\underset{\sim}{a} \leqslant 0{,}8) \text{ and } (\underset{\sim}{c} \geqslant 0{,}2)] \text{ or/and } (\underset{\sim}{c} \leqslant 0{,}8) .$$

Property $\mathcal{P}_2^{(2)}$:

(34.69)

$$[(\underset{\sim}{a} < 0{,}3) \text{ or/and } (\underset{\sim}{b} > 0{,}7)] \text{ and } [(\underset{\sim}{a} > 0{,}7) \text{ or/and } (\underset{\sim}{c} < 0{,}3)] \text{ and } (\underset{\sim}{c} > 0{,}7) .$$

If properties $\mathcal{P}_1^{(2)}$ (34.67) and $\mathcal{P}_2^{(2)}$ (34.68) are satisfied, then one has (34.67).

(34.70) c) $$0{,}3 \leqslant f(\underset{\sim}{a}, \underset{\sim}{b}, \underset{\sim}{c}) \leqslant 1 .$$

Property $\mathcal{P}_1{}^{(3)}$:

(34.71)

$$[(\underset{\sim}{a} \geqslant 0{,}3) \text{ and } (\underset{\sim}{b} \leqslant 0{,}7)] \text{ or/and } [(\underset{\sim}{a} \leqslant 0{,}7) \text{ and } (\underset{\sim}{c} \geqslant 0{,}3)] \text{ or/and } (\underset{\sim}{c} \leqslant 0{,}7) .$$

Property $\mathcal{P}_2{}^{(3)}$:

(34.72) $$[(\underset{\sim}{a} \leqslant 1) \text{ or/and } (\underset{\sim}{b} \geqslant 0)] \text{ and } [(\underset{\sim}{a} \geqslant 0) \text{ or/and } (\underset{\sim}{c} \leqslant 1)] \text{ and} (\underset{\sim}{c} \geqslant 0) .$$

If properties $\mathcal{P}_1{}^{(3)}$ (34.71) and $\mathcal{P}_2{}^{(3)}$ are satisfied then one has (34.70).

Important remark. We examine $\mathcal{P}_1$ (34.22) and $\mathcal{P}_2$ (34.23). One may see that properties $\mathcal{P}_1$ and $\mathcal{P}_2$ are dual to one another if one mutually replaces

($<$) by ($\geqslant$), ($\leqslant$) by ($>$), ($>$) by ($\leqslant$), ($\geqslant$) by ($<$)
(and) by (and/or), (and/or) by (and).

The same holds for $\mathcal{P}_1^{(1)}$ (34.65) and $\mathcal{P}_2^{(1)}$ (34.66), $\mathcal{P}_1^{(2)}$ (34.68) and $\mathcal{P}_2^{(2)}$ (34.69), and $\mathcal{P}_1^{(3)}$ (34.71) and $\mathcal{P}_2^{(3)}$ (34.72) (for the last two, $>$ becomes $\geqslant$ and $<$ becomes $\leqslant$, since the last interval is closed both on the left and the right).

This property is not fortuitous; it is general for all reduced polynomial forms with respect to $\vee$ or with respect to $\wedge$.

As an example, moreover, we consider a polynomial form with respect to $\wedge$.

Example III. Let

(34.73) $$f(\underset{\sim}{a}, \underset{\sim}{b}, \underset{\sim}{c}) = (\underset{\sim}{a} \vee \underset{\sim}{b}) \wedge (\bar{\underset{\sim}{b}} \vee \underset{\sim}{c}) .$$

Under what conditions does one have

(34.74) $$\alpha_{k-1} \leqslant f(\underset{\sim}{a}, \underset{\sim}{b}, \underset{\sim}{c}) < \alpha_k .$$

Hypothesis I:

(34.75) $$\underset{\sim}{a} \vee \underset{\sim}{b} < \bar{\underset{\sim}{b}} \vee \underset{\sim}{c}$$

This implies

(34.76) $$\alpha_{k-1} \leqslant \underset{\sim}{a} \vee \underset{\sim}{b} < \alpha_k ,$$

or again

(34.77) $$\alpha_{k-1} \leqslant \text{MAX } (\underset{\sim}{a}, \underset{\sim}{b}) < \alpha_k .$$

Since we may not arbitrarily place $\underset{\sim}{a}$ and $\underset{\sim}{b}$ with respect to one another, it is necessary that

(34.78) $$\underset{\sim}{a} \geqslant \alpha_{k-1} \quad \text{or/and} \quad \underset{\sim}{b} \geqslant \alpha_{k-1} ,$$

and

(34.79) $$\underset{\sim}{a} < \alpha_k \quad \text{and} \quad \underset{\sim}{b} < \alpha_k .$$

Hypothesis II:

(34.80) $$\bar{\underset{\sim}{b}} \vee \underset{\sim}{c} > \underset{\sim}{a} \vee \underset{\sim}{b}.$$

This implies

(34.81) $$\alpha_{k-1} \leqslant \bar{\underset{\sim}{b}} \vee \underset{\sim}{c} < \alpha_k .$$

or again

(34.82) $$\alpha_{k-1} \leqslant \text{MAX}\,(1 - \underset{\sim}{b}, \underset{\sim}{c}) < \alpha_k .$$

Thus

(34.83) $$\underset{\sim}{b} \leqslant 1 - \alpha_{k-1} \quad \text{or/and} \quad \underset{\sim}{c} \geqslant \alpha_{k-1}$$

and

(34.84) $$\underset{\sim}{b} > 1 - \alpha_k \quad \text{and} \quad \underset{\sim}{c} < \alpha_k .$$

Regrouping the results obtained, we have:

Property $\mathcal{P}'_1$:

(34.85)

$$[(\underset{\sim}{a} \geqslant \alpha_{k-1}) \text{ or/and } (\underset{\sim}{b} \geqslant \alpha_{k-1})] \text{ and } [(\underset{\sim}{b} \leqslant 1 - \alpha_{k-1}) \text{ or/and } (\underset{\sim}{c} \geqslant \alpha_{k-1})].$$

Property $\mathcal{P}'_2$:

(34.86) $$[(\underset{\sim}{a} < \alpha_k) \text{ and } (\underset{\sim}{b} < \alpha_k)] \text{ or/and } [(\underset{\sim}{b} > 1 - \alpha_k) \text{ and } (\underset{\sim}{c} < \alpha_k)] .$$

In order that (34.74) be satisfied, it is necessary and sufficient that properties $\mathcal{P}'_2$ and $\mathcal{P}'_1$ be satisfied.

We note that the property of duality reappears, but *or/and* has taken the place of *and*, and vice versa.

35. LOGICAL STRUCTURE OF A FUNCTION OF FUZZY VARIABLES

Recall that the propositional algebra, in which appear the propositions

(35.1) "and" denoted by Δ

(35.2) "or/and" denoted by ∇

(35.3) "complement" denoted by $^{-}$

follows exactly the same rules as those of boolean algebras†:

†In a number of works treating boolean propositional logic one uses the symbols $\wedge$, $\vee$, and $\neg$ in the same way as the other symbols $\Rightarrow$, $\Leftrightarrow$, which we have not used in the present chapter.. Let it be given that $\wedge$ and $\vee$ are used for min and max until we use Δ and ∇ in their place. Later one must use Δ and ∇ for inferior limit and superior limit in ordered concepts. Let it be understood that these sections of Chapter V are quite distinct and removed from what is considered here; the risks of confusion are minimal. In any case, the reader is warned.

Δ is associated with $\cap$

∇ is associated with $\cup$

$-$ is associated with $^{-}$.

In order to present the logical structure of the relations (strict or nonstrict inequalities) that appear in a fuzzy logical function, and considering an interval $[\alpha_{k-1}, \alpha_k]$ we will use the following symbols.

Let $\underset{\sim}{f}(\underset{\sim}{a}, \underset{\sim}{b}, \ldots, \underset{\sim}{l})$ be a function of the fuzzy variables $\underset{\sim}{a}, \underset{\sim}{b}, \ldots, \underset{\sim}{l}$ and let $[\alpha_{k-1}, \alpha_k]$ be an interval. If $\underset{\sim}{x}$ and $\underset{\sim}{y}$ are some variables of $\underset{\sim}{f}$, we shall use the following symbols:

$$\mathcal{P}_{\underset{\sim}{x}} = (\underset{\sim}{x} \mid \underset{\sim}{x} \geqslant \alpha_{k-1}), \tag{35.4}$$

$$\mathcal{P}_{\underset{\sim}{\bar{x}}} = (\underset{\sim}{x} \mid \underset{\sim}{x} \leqslant 1 - \alpha_{k-1}), \tag{35.5}$$

$$\mathcal{P}'_{\underset{\sim}{x}} = (\underset{\sim}{x} \mid \underset{\sim}{x} < \alpha_k), \tag{35.6}$$

$$\mathcal{P}'_{\underset{\sim}{\bar{x}}} = (\underset{\sim}{x} \mid \underset{\sim}{x} > 1 - \alpha_k). \tag{35.7}$$

Suppose that $\underset{\sim}{f}(\underset{\sim}{a}, \underset{\sim}{b}, \ldots, \underset{\sim}{l})$ may be presented in a reduced polynomial form with respect to $\vee$. In order to obtain the logical structure in the interval $[\alpha_{k-1}, \alpha_k[$, one proceeds as follows:

(1) Any expression of the form $\underset{\sim}{x} \wedge \underset{\sim}{y}$ will be replaced by an expression $\mathcal{P}_{\underset{\sim}{x}} \Delta \mathcal{P}_{\underset{\sim}{y}}$. Thus, for example, an expression such as $\underset{\sim}{\bar{a}} \wedge \underset{\sim}{b} \wedge \underset{\sim}{\bar{c}}$ will be replaced by $\mathcal{P}_{\underset{\sim}{\bar{a}}} \Delta \mathcal{P}_{\underset{\sim}{b}} \Delta \mathcal{P}_{\underset{\sim}{\bar{c}}}$

(2) The monomials of $\underset{\sim}{f}$ joined by the symbol $\vee$ will be replaced by the monomials in $\mathcal{P}$ obtained in (1) and be joined by the symbol ∇. Thus, for example, $(\underset{\sim}{\bar{a}} \wedge \underset{\sim}{b} \wedge \underset{\sim}{\bar{c}}) \vee (\underset{\sim}{\bar{b}} \wedge \underset{\sim}{c})$ will be replaced by $(\mathcal{P}_{\underset{\sim}{\bar{a}}} \Delta \mathcal{P}_{\underset{\sim}{b}} \Delta \mathcal{P}_{\underset{\sim}{\bar{c}}}) \nabla (\mathcal{P}_{\underset{\sim}{\bar{b}}} \Delta \mathcal{P}_{\underset{\sim}{c}})$.

(3) Form the dual of the logical expression obtained in (2) by replacing $\mathcal{P}_x$ by $\mathcal{P}'_{\underset{\sim}{x}}$, $\mathcal{P}_{\underset{\sim}{\bar{x}}}$ by $\mathcal{P}'_{\underset{\sim}{\bar{x}}}$, Δ by ∇, ∇ by Δ. Thus, for example, $(\mathcal{P}_{\underset{\sim}{\bar{a}}} \Delta \mathcal{P}_{\underset{\sim}{b}} \Delta \mathcal{P}_{\underset{\sim}{\bar{c}}}) \nabla (\mathcal{P}_{\underset{\sim}{\bar{b}}} \Delta \mathcal{P}_{\underset{\sim}{c}})$ becomes $(\mathcal{P}'_{\underset{\sim}{\bar{a}}} \nabla \mathcal{P}'_{\underset{\sim}{b}} \nabla \mathcal{P}'_{\underset{\sim}{\bar{c}}}) \Delta (\mathcal{P}'_{\underset{\sim}{\bar{b}}} \nabla \mathcal{P}'_{\underset{\sim}{c}})$.

(4) Join the results obtained in (2) and (3) with the symbol Δ. This gives the logical expression relative to $\underset{\sim}{f}$ in the interval $[\alpha_{k-1}, \alpha_k[$. Thus, for example, concerning the example already examined in (1)–(3), one sees for

$$\underset{\sim}{f}(\underset{\sim}{a}, \underset{\sim}{b}, \underset{\sim}{c}) = (\underset{\sim}{\bar{a}} \wedge \underset{\sim}{b} \wedge \underset{\sim}{\bar{c}}) \vee (\underset{\sim}{\bar{b}} \wedge \underset{\sim}{c}), \tag{35.8}$$

the logical expression

$$\mathcal{P} = [(\mathcal{P}_{\underset{\sim}{\bar{a}}} \Delta \mathcal{P}_{\underset{\sim}{b}} \Delta \mathcal{P}_{\underset{\sim}{\bar{c}}}) \nabla (\mathcal{P}_{\underset{\sim}{\bar{b}}} \Delta \mathcal{P}_{\underset{\sim}{c}})] \Delta [(\mathcal{P}'_{\underset{\sim}{\bar{a}}} \nabla \mathcal{P}'_{\underset{\sim}{b}} \nabla \mathcal{P}'_{\underset{\sim}{\bar{c}}}) \Delta (\mathcal{P}'_{\underset{\sim}{\bar{b}}} \nabla \mathcal{P}'_{\underset{\sim}{c}})]. \tag{35.9}$$

If $\underset{\sim}{f}(\underset{\sim}{a}, \underset{\sim}{b}, \ldots, \underset{\sim}{l})$ is represented in a polynomial form with respect to $\wedge$, rules (1)–(4) above are modified as follows:

(1) Any expression of the form $\underset{\sim}{x} \vee \underset{\sim}{y}$ will be replaced by an expression $\mathcal{P}_{\underset{\sim}{x}} \nabla \mathcal{P}_{\underset{\sim}{y}}$.

(2) The monomials of $\underset{\sim}{f}$ joined by the symbol $\wedge$ will be replaced by the corresponding monomials in $\mathcal{P}$ joined by the symbol Δ.

(3) Take the dual of (2).

(4) Join the results of (2) and (3) with the symbol Δ

We consider an example. Let

$$f(\underset{\sim}{a}, \underset{\sim}{b}, \underset{\sim}{c}, \underset{\sim}{d}) = (\underset{\sim}{a} \vee \underset{\sim}{\bar{b}}) \wedge (\underset{\sim}{b} \vee \underset{\sim}{\bar{d}}) \wedge (\underset{\sim}{\bar{c}} \vee \underset{\sim}{d}) . \tag{35.10}$$

One sees

(35.11)

$$\mathcal{P} = [(\mathcal{P}_{a} \nabla \mathcal{P}_{\bar{b}}) \, \Delta \, (\mathcal{P}_{b} \nabla \mathcal{P}_{\bar{d}}) \, \Delta \, (\mathcal{P}_{\bar{c}} \nabla \mathcal{P}_{d})] \, \Delta \, [(\mathcal{P}'_{a} \, \Delta \, \mathcal{P}'_{\bar{b}}) \nabla (\mathcal{P}'_{b} \, \Delta \, \mathcal{P}'_{\bar{d}}) \nabla (\mathcal{P}'_{\bar{c}} \, \Delta \, \mathcal{P}'_{d})].$$

Suppose, in order to give a numerical illustration, that

$$[\alpha_{k-1} , \alpha_k [\; = [0{,}6 , 0{,}8[. \tag{35.12}$$

Then expression (35.11) may be written

(35.13)

$$\left[\begin{pmatrix} \underset{\sim}{a} \geqslant 0{,}6 \\ \text{or/and}\ \underset{\sim}{b} \leqslant 0{,}4 \end{pmatrix} \text{ and } \begin{pmatrix} \underset{\sim}{b} \geqslant 0{,}6 \\ \text{or/and}\ \underset{\sim}{d} \leqslant 0{,}4 \end{pmatrix} \text{ and } \begin{pmatrix} \underset{\sim}{c} \leqslant 0{,}4 \\ \text{or/and}\ \underset{\sim}{d} \geqslant 0{,}6 \end{pmatrix}\right]$$

$$\text{and } \left[\begin{pmatrix} \underset{\sim}{a} < 0{,}8 \\ \text{and}\ \underset{\sim}{b} > 0{,}2 \end{pmatrix} \text{ or/and } \begin{pmatrix} \underset{\sim}{b} < 0{,}8 \\ \text{and}\ \underset{\sim}{d} > 0{,}2 \end{pmatrix} \text{ or/and } \begin{pmatrix} \underset{\sim}{c} > 0{,}2 \\ \text{and}\ \underset{\sim}{d} < 0{,}8 \end{pmatrix}\right]$$

It is interesting to develop logical expressions in $\mathcal{P}_{x}$, $\mathcal{P}_{\bar{x}}$, $\mathcal{P}'_{x}$, and $\mathcal{P}'_{\bar{x}}$; these give sufficient conditions for each monomial in a development with respect to ∇. An example shows this easily. Reconsider (35.8),

$$f(\underset{\sim}{a}, \underset{\sim}{b}, \underset{\sim}{c}) = (\underset{\sim}{\bar{a}} \wedge \underset{\sim}{b} \wedge \underset{\sim}{\bar{c}}) \vee (\underset{\sim}{\bar{b}} \wedge \underset{\sim}{c}) , \tag{35.14}$$

and suppose that

$$[\alpha_{k-1} , \alpha_k [\; = [0{,}3 , 0{,}8[. \tag{35.15}$$

We have calculated $\mathcal{P}$ [see (35.9)]. We now go on to develop (35.9) in order to form a polynomial in Δ:

$$\mathcal{P} = [(\mathcal{P}_{\bar{a}} \, \Delta \, \mathcal{P}_{b} \, \Delta \, \mathcal{P}_{\bar{c}}) \nabla (\mathcal{P}_{\bar{b}} \, \Delta \, \mathcal{P}_{c})] \, \Delta \, [(\mathcal{P}'_{\bar{a}} \nabla \mathcal{P}'_{b} \nabla \mathcal{P}'_{\bar{c}}) \, \Delta \, (\mathcal{P}'_{\bar{b}} \nabla \mathcal{P}'_{c})] \tag{35.16}$$

In order to condense the writing, we represent $\mathcal{P}_{x} \, \Delta \, \mathcal{P}_{y}$ by $\mathcal{P}_{x} \mathcal{P}_{y}$.

$$\begin{aligned} \mathcal{P} &= (\mathcal{P}_{\bar{a}} \mathcal{P}_{b} \mathcal{P}_{\bar{c}} \nabla \mathcal{P}_{\bar{b}} \mathcal{P}_{c}) (\mathcal{P}'_{\bar{a}} \nabla \mathcal{P}'_{b} \nabla \mathcal{P}'_{\bar{c}}) (\mathcal{P}'_{\bar{b}} \nabla \mathcal{P}'_{c}) \\ &= (\mathcal{P}_{\bar{a}} \mathcal{P}_{b} \mathcal{P}_{\bar{c}} \nabla \mathcal{P}_{\bar{b}} \mathcal{P}_{c}) (\mathcal{P}'_{\bar{a}} \mathcal{P}'_{\bar{b}} \nabla \mathcal{P}'_{\bar{a}} \mathcal{P}'_{c} \nabla \mathcal{P}'_{b} \mathcal{P}'_{\bar{b}} \nabla \mathcal{P}'_{b} \mathcal{P}'_{c} \nabla \mathcal{P}'_{\bar{b}} \mathcal{P}'_{\bar{c}} \nabla \mathcal{P}'_{c} \mathcal{P}'_{\bar{c}}) \\ &= \mathcal{P}_{\bar{a}} \mathcal{P}'_{\bar{a}} \mathcal{P}_{b} \mathcal{P}'_{\bar{b}} \mathcal{P}_{\bar{c}} \nabla \mathcal{P}_{\bar{a}} \mathcal{P}'_{\bar{a}} \mathcal{P}_{b} \mathcal{P}_{\bar{c}} \mathcal{P}'_{c} \nabla \mathcal{P}_{\bar{a}} \mathcal{P}_{b} \mathcal{P}'_{b} \mathcal{P}'_{\bar{b}} \mathcal{P}_{\bar{c}} \nabla \mathcal{P}_{\bar{a}} \mathcal{P}_{b} \mathcal{P}'_{b} \mathcal{P}_{\bar{c}} \mathcal{P}'_{c} \\ &\quad \nabla \mathcal{P}_{\bar{a}} \mathcal{P}_{b} \mathcal{P}'_{\bar{b}} \mathcal{P}_{\bar{c}} \mathcal{P}'_{\bar{c}} \nabla \mathcal{P}_{\bar{a}} \mathcal{P}_{b} \mathcal{P}_{\bar{c}} \mathcal{P}'_{c} \mathcal{P}'_{\bar{c}} \nabla \mathcal{P}'_{\bar{a}} \mathcal{P}_{\bar{b}} \mathcal{P}'_{\bar{b}} \mathcal{P}_{c} \nabla \mathcal{P}'_{\bar{a}} \mathcal{P}_{\bar{b}} \mathcal{P}_{c} \mathcal{P}'_{c} \\ &\quad \nabla \mathcal{P}_{\bar{b}} \mathcal{P}'_{b} \mathcal{P}'_{\bar{b}} \mathcal{P}_{c} \nabla \mathcal{P}_{\bar{b}} \mathcal{P}'_{b} \mathcal{P}_{c} \mathcal{P}'_{c} \nabla \mathcal{P}_{\bar{b}} \mathcal{P}'_{\bar{b}} \mathcal{P}_{c} \mathcal{P}'_{\bar{c}} \nabla \mathcal{P}_{\bar{b}} \mathcal{P}_{c} \mathcal{P}'_{c} \mathcal{P}'_{\bar{c}} . \end{aligned} \tag{35.17}$$

Each of these monomials is sufficient so that we have

$$0{,}3 \leqslant (\underset{\sim}{\bar{a}} \wedge \underset{\sim}{b} \wedge \underset{\sim}{\bar{c}}) \vee (\underset{\sim}{\bar{b}} \wedge \underset{\sim}{c}) < 0{,}8 . \tag{35.18}$$

We verify this, for example, for

$$\mathscr{P}_{\underset{\sim}{\bar{b}}}\ \mathscr{P}'_{\underset{\sim}{b}}\ \mathscr{P}'_{\underset{\sim}{\bar{b}}}\ \mathscr{P}_{\underset{\sim}{c}} \qquad \text{(ninth monomial)} \tag{35.19}$$

Applying the notations (35.4)–(35.7), one has

$$\begin{aligned}
&\mathscr{P}_{\underset{\sim}{\bar{b}}} &&: \underset{\sim}{b} \leqslant (1 - 0{,}3) \qquad \text{thus} \quad \underset{\sim}{b} \leqslant 0{,}7 ,\\
&\mathscr{P}'_{\underset{\sim}{b}} &&: \underset{\sim}{b} < 0{,}8 ,\\
&\mathscr{P}'_{\underset{\sim}{\bar{b}}} &&: \underset{\sim}{b} > 0{,}2 ,\\
&\mathscr{P}_{\underset{\sim}{c}} &&: \underset{\sim}{c} \geqslant 0{,}3 .
\end{aligned} \tag{35.20}$$

And thus

$$0{,}2 < \underset{\sim}{b} \leqslant 0{,}7 \quad \text{and} \quad \underset{\sim}{c} \geqslant 0{,}3 . \tag{35.21}$$

This constitutes a sufficient condition for (35.18) to be verified.

It is equally interesting to carry out the dual development with respect to Δ. Taking again the same example (35.14) and (35.15), we develop into polynomials with respect to $\wedge$; (35.14) gives

$$f(\underset{\sim}{a}, \underset{\sim}{b}, \underset{\sim}{c}) = (\underset{\sim}{\bar{a}} \vee \underset{\sim}{\bar{b}}) \wedge (\underset{\sim}{\bar{a}} \vee \underset{\sim}{c}) \wedge (\underset{\sim}{b} \vee \underset{\sim}{\bar{b}}) \wedge (\underset{\sim}{b} \vee \underset{\sim}{c}) \wedge (\underset{\sim}{c} \vee \underset{\sim}{\bar{c}}) \wedge (\underset{\sim}{\bar{b}} \vee \underset{\sim}{\bar{c}}) . \tag{35.22}$$

One obtains, if we suppress the sign Δ,

$$\begin{aligned}
\mathscr{P} = {} & (\mathscr{P}_{\underset{\sim}{\bar{a}}} \nabla \mathscr{P}_{\underset{\sim}{\bar{b}}})\ (\mathscr{P}_{\underset{\sim}{\bar{a}}} \nabla \mathscr{P}_{\underset{\sim}{c}})\ (\mathscr{P}_{\underset{\sim}{b}} \nabla \mathscr{P}_{\underset{\sim}{\bar{b}}})\ (\mathscr{P}_{\underset{\sim}{b}} \nabla \mathscr{P}_{\underset{\sim}{c}})\ (\mathscr{P}_{\underset{\sim}{c}} \nabla \mathscr{P}_{\underset{\sim}{\bar{c}}})\ (\mathscr{P}_{\underset{\sim}{\bar{b}}} \nabla \mathscr{P}_{\underset{\sim}{\bar{c}}})\\
& \times (\mathscr{P}'_{\underset{\sim}{\bar{a}}} \mathscr{P}'_{\underset{\sim}{\bar{b}}} \nabla \mathscr{P}'_{\underset{\sim}{\bar{a}}} \mathscr{P}'_{\underset{\sim}{c}} \nabla \mathscr{P}'_{\underset{\sim}{b}} \mathscr{P}'_{\underset{\sim}{\bar{b}}} \nabla \mathscr{P}'_{\underset{\sim}{b}} \mathscr{P}'_{\underset{\sim}{c}} \nabla \mathscr{P}'_{\underset{\sim}{c}} \mathscr{P}'_{\underset{\sim}{\bar{c}}} \nabla \mathscr{P}'_{\underset{\sim}{\bar{b}}} \mathscr{P}'_{\underset{\sim}{\bar{c}}}) .
\end{aligned} \tag{35.23}$$

Note that if one carries out the development with respect to ∇ one recovers (35.17) with the exception of the reduction of the nonmaximal monomials.†

†The notion of maximal monomial in these developments in Δ or in Δ is the same as that given in Section 33, but adapted here to logical functions such as (35.17) or (35.23). And, due to the properties $\mathscr{P}$ met in boolean logic, one may extend the reduction further by considering that it is always false that one have $\mathscr{P} \ \Delta\ \overline{\mathscr{P}}$ and always true that $\mathscr{P} \nabla \overline{\mathscr{P}}$ but we have no need of this here since the negation of $\mathscr{P}$ never appears.

TABLE OF PRINCIPAL FUNCTION OF TWO FUZZY VARIABLES AND OF THEIR LOGICAL STRUCTURE FOR AN INTERVAL $[\alpha_{k-1}, \alpha_k[$

	$f(\underset{\sim}{a}, \underset{\sim}{b})$	Polynomial form with respect to ∇	Polynomial form with respect to ∇
(35.24)	$\underset{\sim}{a} \wedge \underset{\sim}{b}$	$(\mathcal{P}_{a} \Delta \mathcal{P}'_{a} \Delta \mathcal{P}_{b}) \nabla (\mathcal{P}_{a} \Delta \mathcal{P}_{b} \Delta \mathcal{P}'_{b})$	$(\mathcal{P}_{a}) \Delta (\mathcal{P}_{b}) \Delta (\mathcal{P}'_{a} \nabla \mathcal{P}'_{b})$
(35.25)	$\underset{\sim}{\bar{a}} \wedge \underset{\sim}{b}$	$(\mathcal{P}_{\bar{a}} \Delta \mathcal{P}'_{\bar{a}} \Delta \mathcal{P}_{b}) \nabla (\mathcal{P}_{\bar{a}} \Delta \mathcal{P}_{b} \Delta \mathcal{P}'_{b})$	$(\mathcal{P}_{\bar{a}}) \Delta (\mathcal{P}_{b}) \Delta (\mathcal{P}'_{\bar{a}} \nabla \mathcal{P}'_{b})$
(36.26)	$\underset{\sim}{\bar{a}} \wedge \underset{\sim}{\bar{b}}$	$(\mathcal{P}_{\bar{a}} \Delta \mathcal{P}'_{\bar{a}} \Delta \mathcal{P}_{\bar{b}}) \nabla (\mathcal{P}_{\bar{a}} \Delta \mathcal{P}_{\bar{b}} \Delta \mathcal{P}'_{\bar{b}})$	$(\mathcal{P}_{\bar{a}}) \Delta (\mathcal{P}_{\bar{b}}) \Delta (\mathcal{P}'_{\bar{a}} \nabla \mathcal{P}'_{\bar{b}})$
(35.27)	$\underset{\sim}{a} \vee \underset{\sim}{b}$	$(\mathcal{P}_{a} \Delta \mathcal{P}'_{a} \Delta \mathcal{P}'_{b}) \nabla (\mathcal{P}'_{a} \Delta \mathcal{P}_{b} \Delta \mathcal{P}'_{b})$	$(\mathcal{P}'_{a}) \Delta (\mathcal{P}'_{b}) \Delta (\mathcal{P}_{a} \nabla \mathcal{P}_{b})$
(35.28)	$\underset{\sim}{\bar{a}} \vee \underset{\sim}{b}$	$(\mathcal{P}_{\bar{a}} \Delta \mathcal{P}'_{\bar{a}} \Delta \mathcal{P}'_{b}) \nabla (\mathcal{P}_{\bar{a}} \Delta \mathcal{P}_{b} \Delta \mathcal{P}'_{b})$	$(\mathcal{P}'_{\bar{a}}) \Delta (\mathcal{P}'_{b}) \Delta (\mathcal{P}_{\bar{a}} \nabla \mathcal{P}_{b})$
(35.29)	$\underset{\sim}{\bar{a}} \vee \underset{\sim}{\bar{b}}$	$(\mathcal{P}_{\bar{a}} \Delta \mathcal{P}'_{\bar{a}} \Delta \mathcal{P}'_{\bar{b}}) \nabla (\mathcal{P}_{\bar{a}} \Delta \mathcal{P}_{\bar{b}} \Delta \mathcal{P}'_{\bar{b}})$	$(\mathcal{P}'_{\bar{a}}) \Delta (\mathcal{P}'_{\bar{b}}) \Delta (\mathcal{P}_{\bar{a}} \nabla \mathcal{P}_{\bar{b}})$
(35.30)	$(\underset{\sim}{a} \wedge \underset{\sim}{\bar{b}}) \vee (\underset{\sim}{\bar{a}} \wedge \underset{\sim}{b})$	$(\mathcal{P}_{a} \Delta \mathcal{P}'_{a} \Delta \mathcal{P}'_{\bar{a}} \Delta \mathcal{P}_{\bar{b}}) \nabla (\mathcal{P}_{a} \Delta \mathcal{P}'_{a} \Delta \mathcal{P}'_{b} \Delta \mathcal{P}_{\bar{b}})$ $\nabla (\mathcal{P}_{a} \Delta \mathcal{P}'_{\bar{a}} \Delta \mathcal{P}_{\bar{b}} \Delta \mathcal{P}'_{\bar{b}}) \nabla (\mathcal{P}_{a} \Delta \mathcal{P}'_{b} \Delta \mathcal{P}_{\bar{b}} \Delta \mathcal{P}'_{\bar{b}})$ $\nabla (\mathcal{P}'_{a} \Delta \mathcal{P}_{\bar{a}} \Delta \mathcal{P}'_{\bar{a}} \Delta \mathcal{P}_{b}) \nabla (\mathcal{P}'_{a} \Delta \mathcal{P}_{\bar{a}} \Delta \mathcal{P}_{b} \Delta \mathcal{P}'_{b})$ $\nabla (\mathcal{P}_{\bar{a}} \Delta \mathcal{P}'_{\bar{a}} \Delta \mathcal{P}_{b} \Delta \mathcal{P}'_{\bar{b}}) \nabla (\mathcal{P}_{\bar{a}} \Delta \mathcal{P}_{b} \Delta \mathcal{P}'_{b} \Delta \mathcal{P}'_{\bar{b}})$	$(\mathcal{P}_{a} \nabla \mathcal{P}_{\bar{a}}) \Delta (\mathcal{P}_{a} \nabla \mathcal{P}_{b}) \Delta (\mathcal{P}_{\bar{a}} \nabla \mathcal{P}_{\bar{b}})$ $\Delta (\mathcal{P}_{b} \nabla \mathcal{P}_{\bar{b}}) \Delta (\mathcal{P}'_{a} \nabla \mathcal{P}'_{\bar{b}})$ $\Delta (\mathcal{P}'_{\bar{a}} \nabla \mathcal{P}'_{b})$
(35.31)	$(a \vee \bar{b}) \wedge (\bar{a} \vee b)$	$(\mathcal{P}_{a} \Delta \mathcal{P}'_{a} \Delta \mathcal{P}_{b} \Delta \mathcal{P}'_{\bar{b}}) \nabla (\mathcal{P}_{a} \Delta \mathcal{P}'_{\bar{a}} \Delta \mathcal{P}_{b} \Delta \mathcal{P}'_{b})$ $\Delta (\mathcal{P}_{a} \Delta \mathcal{P}_{\bar{a}} \Delta \mathcal{P}'_{a} \Delta \mathcal{P}'_{\bar{b}}) \nabla (\mathcal{P}_{a} \Delta \mathcal{P}_{\bar{a}} \Delta \mathcal{P}'_{\bar{a}} \Delta \mathcal{P}'_{b})$ $\Delta (\mathcal{P}'_{a} \Delta \mathcal{P}_{b} \Delta \mathcal{P}_{\bar{b}} \Delta \mathcal{P}'_{\bar{b}}) \nabla (\mathcal{P}'_{\bar{a}} \Delta \mathcal{P}_{b} \Delta \mathcal{P}_{\bar{b}} \Delta \mathcal{P}'_{b})$ $\nabla (\mathcal{P}_{\bar{a}} \Delta \mathcal{P}'_{a} \Delta \mathcal{P}_{\bar{b}} \Delta \mathcal{P}'_{\bar{b}}) \nabla (\mathcal{P}_{\bar{a}} \Delta \mathcal{P}'_{\bar{a}} \Delta \mathcal{P}_{\bar{b}} \Delta \mathcal{P}'_{b})$	$(\mathcal{P}_{a} \nabla \mathcal{P}_{\bar{b}}) \Delta (\mathcal{P}_{b} \nabla \mathcal{P}_{\bar{a}}) \Delta (\mathcal{P}'_{a} \nabla \mathcal{P}'_{b})$ $\Delta (\mathcal{P}'_{a} \nabla \mathcal{P}'_{\bar{a}}) \Delta (\mathcal{P}'_{b} \nabla \mathcal{P}'_{\bar{b}}) \Delta (\mathcal{P}'_{\bar{a}} \nabla \mathcal{P}'_{\bar{b}})$.

These expressions may all be simplified if one knows the position of $\alpha_{k-1} + \alpha_k$ with respect to 1.

Important remark. If a fuzzy variable $\underset{\sim}{a}$ takes its values in an interval

(35.32) $$\mathcal{D}_{\underset{\sim}{a}} = [a_1, a_2[\subset [0,1],$$

then the variable $\underset{\sim}{\bar{a}} = 1 - \underset{\sim}{a}$ takes its values in the interval

(35.33) $$\mathcal{D}_{\underset{\sim}{\bar{a}}} =]1 - a_2, 1 - a_1] \subset [0,1].$$

If $\underset{\sim}{a}$ takes its values in the interval

(35.34) $$\overline{\mathcal{D}}_{\underset{\sim}{a}} = [0, a_1[\cup [a_2, 1],$$

then $\underset{\sim}{\bar{a}}$ takes its values in the interval

(35.35) $$\overline{\mathcal{D}}_{\underset{\sim}{\bar{a}}} = [0, 1 - a_2] \cup]1 - a_1, 1].$$

36. COMPOSITION OF INTERVALS

Let

(36.1) $$\underset{\sim}{a} \in \mathcal{D}_{\underset{\sim}{a}} = [a_1, a_2[$$

and

(36.2) $$\underset{\sim}{b} \in \mathcal{D}_{\underset{\sim}{b}} = [b_1 , b_2[.$$

Then to which interval $\mathcal{D}_{\underset{\sim}{a} \wedge \underset{\sim}{b}}$ does $\underset{\sim}{a} \wedge \underset{\sim}{b}$ belong? It is easy to see that

(36.3) $$\underset{\sim}{a} \wedge \underset{\sim}{b} \in \mathcal{D}_{\underset{\sim}{a} \wedge \underset{\sim}{b}} = [a_1 \wedge b_1 , a_2 \wedge b_2[.$$

In the same manner, one may see that

(36.4) $$\underset{\sim}{a} \vee \underset{\sim}{b} \in \mathcal{D}_{\underset{\sim}{a} \vee \underset{\sim}{b}} = [a_1 \vee b_1 , a_2 \vee b_2[.$$

Example. Let

(36.5) $$\mathcal{D}_{\underset{\sim}{a}} = [0{,}6 , 0{,}9[\text{ and } \mathcal{D}_{\underset{\sim}{b}} = [0{,}2 , 0{,}7[$$

One sees

(36.6) $$\mathcal{D}_{\underset{\sim}{a} \wedge \underset{\sim}{b}} = [0{,}6 \wedge 0{,}2 , 0{,}9 \wedge 0{,}7[$$
$$= [0{,}2 , 0{,}7[,$$

(36.7) $$\mathcal{D}_{\underset{\sim}{a} \vee \underset{\sim}{b}} = [0{,}6 \vee 0{,}2 , 0{,}9 \vee 0{,}7[$$
$$= [0{,}6 , 0{,}9[.$$

Because of properties (32.14) and (32.15), one may see by extension that if

(36.8) $$\underset{\sim}{a} \in \mathcal{D}_{\underset{\sim}{a}} = [a_1 , a_2[, \quad \underset{\sim}{b} \in \mathcal{D}_{\underset{\sim}{b}} = [b_1 , b_2[, \quad \underset{\sim}{c} \in \mathcal{D}_{\underset{\sim}{c}} = [c_1 , c_2[,$$

then :

(36.9) $$\underset{\sim}{a} \wedge \underset{\sim}{b} \wedge \underset{\sim}{c} \in \mathcal{D}_{\underset{\sim}{a} \wedge \underset{\sim}{b} \wedge \underset{\sim}{c}} = [a_1 \wedge b_1 \wedge c_1 , \quad a_2 \wedge b_2 \wedge c_2[,$$

and

(36.10) $$\underset{\sim}{a} \vee \underset{\sim}{b} \vee \underset{\sim}{c} \in \mathcal{D}_{\underset{\sim}{a} \vee \underset{\sim}{b} \vee \underset{\sim}{c}} = [a_1 \vee b_1 \vee c_1 , \quad a_2 \vee b_2 \vee c_2[.$$

It is also interesting to examine the case of fuzzy variables that take their values in the complementary interval.

If

(36.11) $$\overline{\mathcal{D}}_{\underset{\sim}{a}} = [0 , a_1 [\cup [a_2 , 1] ,$$

and

(36.12) $$\overline{\mathcal{D}}_{\underset{\sim}{b}} = [0 , b_1 [\cup [b_2 , 1] ,$$

one may see the following results

(36.13) $$\underset{\sim}{f}(\underset{\sim}{a} , \underset{\sim}{b}) = \underset{\sim}{a} \wedge \underset{\sim}{b}, \quad \underset{\sim}{a} \in \mathcal{D}_{\underset{\sim}{a}}, \quad \underset{\sim}{b} \in \mathcal{D}_{\underset{\sim}{b}} ,$$

(36.14) then $$\mathcal{D}_{\underset{\sim}{a} \wedge \underset{\sim}{b}} = [a_1 \wedge b_1 , a_2 \wedge b_2[.$$

(36.15) $$\underset{\sim}{f}(\underset{\sim}{a} , \underset{\sim}{b}) = \underset{\sim}{a} \wedge \underset{\sim}{b} , \quad \underset{\sim}{a} \in \overline{\mathcal{D}}_{\underset{\sim}{a}} , \quad \underset{\sim}{b} \in \mathcal{D}_{\underset{\sim}{b}} ,$$

then

(36.16) $$\mathcal{D}_{\underset{\sim}{a} \wedge \underset{\sim}{b}} = [0\,, a_1 \wedge b_2\,[\cup [a_2 \wedge b_1\,, b_2[\,.$$

We have formed below the table for the eight fundamental cases for $\mathcal{D}_{\underset{\sim}{a}} = [a_1, a_2[$ and $\mathcal{D}_{\underset{\sim}{b}} = [b_1, b_2[$:

	$\underset{\sim}{f}(\underset{\sim}{a}\,,\underset{\sim}{b})$	Domain of $\underset{\sim}{a}$	Domain of $\underset{\sim}{b}$	Domain of $\underset{\sim}{f}(\underset{\sim}{a}\,,\underset{\sim}{b})$
(36.17)	$\underset{\sim}{a} \wedge \underset{\sim}{b}$	$\mathcal{D}_{\underset{\sim}{a}}$	$\mathcal{D}_{\underset{\sim}{b}}$	$[a_1 \wedge b_1\,, a_2 \wedge b_2[\,.$
(36.18)		$\overline{\mathcal{D}}_{\underset{\sim}{a}}$	$\mathcal{D}_{\underset{\sim}{b}}$	$[0\,, a_1 \wedge b_2\,[\cup [a_2 \wedge b_1\,, b_2[\,.$
(36.19)		$\mathcal{D}_{\underset{\sim}{a}}$	$\overline{\mathcal{D}}_{\underset{\sim}{b}}$	$[0\,, b_1 \wedge a_2\,[\cup [b_2 \wedge a_1\,, a_2[\,.$
(36.20)		$\overline{\mathcal{D}}_{\underset{\sim}{a}}$	$\overline{\mathcal{D}}_{\underset{\sim}{b}}$	$[0\,, a_1 \vee b_1\,[\cup [a_2 \wedge b_2\,, 1]$ (1).
(36.21)	$\underset{\sim}{a} \vee \underset{\sim}{b}$	$\mathcal{D}_{\underset{\sim}{a}}$	$\mathcal{D}_{\underset{\sim}{b}}$	$[a_1 \vee b_1\,, a_2 \vee b_2[\,.$
(36.22)		$\overline{\mathcal{D}}_{\underset{\sim}{a}}$	$\mathcal{D}_{\underset{\sim}{b}}$	$[b_1\,, a_1 \vee b_2\,[\cup [a_2 \vee b_1\,, 1]\,.$
(36.23)		$\mathcal{D}_{\underset{\sim}{a}}$	$\overline{\mathcal{D}}_{\underset{\sim}{b}}$	$[a_1\,, b_1 \vee a_2\,[\cup [b_2 \vee a_1\,, 1]\,.$
(36.24)†		$\overline{\mathcal{D}}_{\underset{\sim}{a}}$	$\overline{\mathcal{D}}_{\underset{\sim}{b}}$	$[0\,, a_1 \vee b_1\,[\cup [a_2 \wedge b_2\,, 1]$ (1).

There is, of course, no reason to confuse $\overline{\mathcal{D}}_{\underset{\sim}{a}}$ with $\mathcal{D}_{\overline{\underset{\sim}{a}}}$.

(36.25) $$\overline{\mathcal{D}}_{\underset{\sim}{a}} = [0\,, a_1\,[\cup [a_2\,, 1]\,,$$

(36.26) $$\mathcal{D}_{\overline{\underset{\sim}{a}}} =]1 - a_2\,, 1 - a_1]\,,$$

and finally

(36.27) $$\overline{\mathcal{D}}_{\overline{\underset{\sim}{a}}} = [0\,, 1 - a_2] \cup]\,1 - a_1\,, 1]\,.$$

With the aid of the expressions given in (36.17)–(36.24) one may form the domains of more complicated functions $\underset{\sim}{f}(\underset{\sim}{a}, \underset{\sim}{b}, \ldots)$.

Example. Determine the domain of

(36.28) $$\underset{\sim}{f}(\underset{\sim}{a}\,,\underset{\sim}{b}) = \overline{\underset{\sim}{a}} \wedge \underset{\sim}{b}\,;$$

knowing that

(36.29) $$\mathcal{D}_{\underset{\sim}{a}} = [a_1\,, a_2[\quad \text{and} \quad \mathcal{D}_{\underset{\sim}{b}} = [b_1\,, b_2[\,.$$

From (36.29) and (36.26) we extract

(36.30) $$\mathcal{D}_{\overline{\underset{\sim}{a}}} =]1 - a_2\,, 1 - a_1]$$

†Thus, (36.20) and (36.24) give the same interval.

Using (36.17), where we have replaced a_1 by $1 - a_2$ and a_2 by $1 - a_1$, there develops, taking account of the orientation of the square brackets:

$$
\begin{aligned}
(36.31)\quad \mathcal{D}_{\bar{\underset{\sim}{a}} \wedge \underset{\sim}{b}} &=]\,(1-a_2)\wedge b_1\,,\,(1-a_1)\wedge b_2\,], && 1-a_2 \leqslant b_1 \text{ and } 1-a_1 \leqslant b_2\,,\\
&= [\,(1-a_2)\wedge b_1\,,\,(1-a_1)\wedge b_2\,], && 1-a_2 > b_1 \text{ and } 1-a_1 \leqslant b_2\,,\\
&=]\,(1-a_2)\wedge b_1\,,\,(1-a_1)\wedge b_2\,[\,, && 1-a_2 \leqslant b_1 \text{ and } 1-a_1 > b_2\,,\\
&= [\,(1-a_2)\wedge b_1\,,\,(1-a_1)\wedge b_2\,[\,, && 1-a_2 > b_1 \text{ and } 1-a_1 > b_2\,,
\end{aligned}
$$

and thus

$$
\begin{aligned}
(36.32)\quad \mathcal{D}_{\bar{\underset{\sim}{a}} \wedge \underset{\sim}{b}} &=]1-a_2\,,\,1-a_1\,], && \text{if } 1-a_2 \leqslant b_1 \text{ and } 1-a_1 \leqslant b_2,\\
&= [b_1\,,\,1-a_1\,], && \text{if } 1-a_2 > b_1 \text{ and } 1-a_1 \leqslant b_2,\\
&=]1-a_2\,,\,b_2\,[\,, && \text{if } 1-a_2 \leqslant b_1 \text{ and } 1-a_1 > b_2,\\
&= [b_1\,,\,b_2\,[\,, && \text{if } 1-a_2 > b_1 \text{ and } 1-a_1 > b_2.
\end{aligned}
$$

Case of a discrete membership function. Suppose that the interval [0, 1] has been decomposed into 10 equal parts, thus determining 11 discrete values:

$$(36.33)\qquad \mathrm{M} = \{0\,,\,0{,}1\,,\,0{,}2\,,\,0{,}3\,,\,0{,}4\,,\,0{,}5\,,\,0{,}6\,,\,0{,}7\,,\,0{,}8\,,\,0{,}9\,,\,1\}$$

It is then convenient to establish tables for the principal functions to be considered; we have represented these in Figures 36.1–36.8. In the theory of functions of fuzzy variables, these tables play a role similar to that played by truth tables in the study of functions of boolean variables; but instead of assigning two values to each variable and the boolean laws, we shall have here dozens of values from 0 to 1 and the laws defined in Section 32 in (32.12)–(32.26).

In the tables given in Figures 36.1–36.12 decimal fractions such as 0,1, 0,2, 0,3, . . . , 0,9 have been represented by .1, .2, .3, . . . , .9 in order to save space.

$\underset{\sim}{a} \wedge \underset{\sim}{b}$ ($\underset{\sim}{a}$ \ $\underset{\sim}{b}$)	0	.1	.2	.3	.4	.5	.6	.7	.8	.9	1
0	0	0	0	0	0	0	0	0	0	0	0
.1	0	.1	.1	.1	.1	.1	.1	.1	.1	.1	.1
.2	0	.1	.2	.2	.2	.2	.2	.2	.2	.2	.2
.3	0	.1	.2	.3	.3	.3	.3	.3	.3	.3	.3
.4	0	.1	.2	.3	.4	.4	.4	.4	.4	.4	.4
.5	0	.1	.2	.3	.4	.5	.5	.5	.5	.5	.5
.6	0	.1	.2	.3	.4	.5	.6	.6	.6	.6	.6
.7	0	.1	.2	.3	.4	.5	.6	.7	.7	.7	.7
.8	0	.1	.2	.3	.4	.5	.6	.7	.8	.8	.8
.9	0	.1	.2	.3	.4	.5	.6	.7	.8	.9	.9
1	0	.1	.2	.3	.4	.5	.6	.7	.8	.9	1

Fig. 36.1

$\bar{\underset{\sim}{a}} \wedge \underset{\sim}{b}$ ($\underset{\sim}{a}$ \ $\underset{\sim}{b}$)	0	.1	.2	.3	.4	.5	.6	.7	.8	.9	1
0	0	.1	.2	.3	.4	.5	.6	.7	.8	.9	1
.1	0	.1	.2	.3	.4	.5	.6	.7	.8	.9	.9
.2	0	.1	.2	.3	.4	.5	.6	.7	.8	.8	.8
.3	0	.1	.2	.3	.4	.5	.6	.7	.7	.7	.7
.4	0	.1	.2	.3	.4	.5	.6	.6	.6	.6	.6
.5	0	.1	.2	.3	.4	.5	.5	.5	.5	.5	.5
.6	0	.1	.2	.3	.4	.4	.4	.4	.4	.4	.4
.7	0	.1	.2	.3	.3	.3	.3	.3	.3	.3	.3
.8	0	.1	.2	.2	.2	.2	.2	.2	.2	.2	.2
.9	0	.1	.1	.1	.1	.1	.1	.1	.1	.1	.1
1	0	0	0	0	0	0	0	0	0	0	0

Fig. 36.2

$\bar{a} \wedge \bar{b}$ (a \ b)	0	.1	.2	.3	.4	.5	.6	.7	.8	.9	1
0	1	.9	.8	.7	.6	.5	.4	.3	.2	.1	0
.1	.9	.9	.8	.7	.6	.5	.4	.3	.2	.1	0
.2	.8	.8	.8	.7	.6	.5	.4	.3	.2	.1	0
.3	.7	.7	.7	.7	.6	.5	.4	.3	.2	.1	0
.4	.6	.6	.6	.6	.6	.5	.4	.3	.2	.1	0
.5	.5	.5	.5	.5	.5	.5	.4	.3	.2	.1	0
.6	.4	.4	.4	.4	.4	.4	.4	.3	.2	.1	0
.7	.3	.3	.3	.3	.3	.3	.3	.3	.2	.1	0
.8	.2	.2	.2	.2	.2	.2	.2	.2	.2	.1	0
.9	.1	.1	.1	.1	.1	.1	.1	.1	.1	.1	0
1	0	0	0	0	0	0	0	0	0	0	0

Fig. 36.3

$a \vee b$ (a \ b)	0	.1	.2	.3	.4	.5	.6	.7	.8	.9	1
0	0	.1	.2	.3	.4	.5	.6	.7	.8	.9	1
.1	.1	.1	.2	.3	.4	.5	.6	.7	.8	.9	1
.2	.2	.2	.2	.3	.4	.5	.6	.7	.8	.9	1
.3	.3	.3	.3	.3	.4	.5	.6	.7	.8	.9	1
.4	.4	.4	.4	.4	.4	.5	.6	.7	.8	.9	1
.5	.5	.5	.5	.5	.5	.5	.6	.7	.8	.9	1
.6	.6	.6	.6	.6	.6	.6	.6	.7	.8	.9	1
.7	.7	.7	.7	.7	.7	.7	.7	.7	.8	.9	1
.8	.8	.8	.8	.8	.8	.8	.8	.8	.8	.9	1
.9	.9	.9	.9	.9	.9	.9	.9	.9	.9	.9	1
1	1	1	1	1	1	1	1	1	1	1	1

Fig. 36.4

$\bar{a} \vee b$ (a \ b)	0	.1	.2	.3	.4	.5	.6	.7	.8	.9	1
0	1	1	1	1	1	1	1	1	1	1	1
.1	.9	.9	.9	.9	.9	.9	.9	.9	.9	.9	1
.2	.8	.8	.8	.8	.8	.8	.8	.8	.8	.9	1
.3	.7	.7	.7	.7	.7	.7	.7	.7	.8	.9	1
.4	.6	.6	.6	.6	.6	.6	.6	.7	.8	.9	1
.5	.5	.5	.5	.5	.5	.5	.6	.7	.8	.9	1
.6	.4	.4	.4	.4	.4	.5	.6	.7	.8	.9	1
.7	.3	.3	.3	.3	.4	.5	.6	.7	.8	.9	1
.8	.2	.2	.2	.3	.4	.5	.6	.7	.8	.9	1
.9	.1	.1	.2	.3	.4	.5	.6	.7	.8	.9	1
1	0	.1	.2	.3	.4	.5	.6	.7	.8	.9	.1

Fig. 36.5

$\bar{a} \vee \bar{b}$ (a \ b)	0	.1	.2	.3	.4	.5	.6	.7	.8	.9	1
0	1	1	1	1	1	1	1	1	1	1	1
.1	1	.9	.9	.9	.9	.9	.9	.9	.9	.9	.9
.2	1	.9	.8	.8	.8	.8	.8	.8	.8	.8	.8
.3	1	.9	.8	.7	.7	.7	.7	.7	.7	.7	.7
.4	1	.9	.8	.7	.6	.6	.6	.6	.6	.6	.6
.5	1	.9	.8	.7	.6	.5	.5	.5	.5	.5	.5
.6	1	.9	.8	.7	.6	.5	.4	.4	.4	.4	.4
.7	1	.9	.8	.7	.6	.5	.4	.3	.3	.3	.3
.8	1	.9	.8	.7	.6	.5	.4	.3	.2	.2	.2
.9	1	.9	.8	.7	.6	.5	.4	.3	.2	.1	.1
1	1	.9	.8	.7	.6	.5	.4	.3	.2	.1	0

Fig. 36.6

$(a \wedge \bar{b}) \vee (\bar{a} \wedge b)$ (a \ b)	0	.1	.2	.3	.4	.5	.6	.7	.8	.9	1
0	0	.1	.2	.3	.4	.5	.6	.7	.8	.9	1
.1	.1	.1	.2	.3	.4	.5	.6	.7	.8	.9	.9
.2	.2	.2	.2	.3	.4	.5	.6	.7	.8	.8	.8
.3	.3	.3	.3	.3	.4	.5	.6	.7	.7	.7	.7
.4	.4	.4	.4	.4	.4	.5	.6	.6	.6	.6	.6
.5	.5	.5	.5	.5	.5	.5	.5	.5	.5	.5	.5
.6	.6	.6	.6	.6	.6	.5	.4	.4	.4	.4	.4
.7	.7	.7	.7	.7	.6	.5	.4	.3	.3	.3	.3
.8	.8	.8	.8	.7	.6	.5	.4	.3	.2	.2	.2
.9	.9	.9	.8	.7	.6	.5	.4	.3	.2	.1	.1
1	1	.9	.8	.7	.6	.5	.4	.3	.2	.1	0

Fig. 36.7

$(a \vee \bar{b}) \wedge (\bar{a} \vee b)$ (a \ b)	0	.1	.2	.3	.4	.5	.6	.7	.8	.9	1
0	1	.9	.8	.7	.6	.5	.4	.3	.2	.1	0
.1	.9	.9	.8	.7	.6	.5	.4	.3	.2	.1	.1
.2	.8	.8	.8	.7	.6	.5	.4	.3	.2	.2	.2
.3	.7	.7	.7	.7	.6	.5	.4	.3	.3	.3	.3
.4	.6	.6	.6	.6	.6	.5	.4	.4	.4	.4	.4
.5	.5	.5	.5	.5	.5	.5	.5	.5	.5	.5	.5
.6	.4	.4	.4	.4	.4	.5	.6	.6	.6	.6	.6
.7	.3	.3	.3	.3	.4	.5	.6	.7	.7	.7	.7
.8	.2	.2	.2	.3	.4	.5	.6	.7	.8	.8	.8
.9	.1	.1	.2	.3	.4	.5	.6	.7	.8	.9	.9
1	0	.1	.2	.3	.4	.5	.6	.7	.8	.9	1

Fig. 36.8

We see how to employ these tables.

Example I. Let

(36.34) $$\underset{\sim}{f}(\underset{\sim}{a}, \underset{\sim}{b}) = \overline{\underset{\sim}{a}} \wedge \underset{\sim}{b},$$

(36.35) with† $$\underset{\sim}{a} \in \{0,2 \; ; 0,3 \; ; 0,4 \; ; 0,5\},$$

(36.36) $$\underset{\sim}{b} \in \{0 \; ; 0,1\} \cup \{0,7 \; ; 0,8, \; 0,9\}.$$

We have reproduced Figure 36.2 in Figure 36.9. Examination of the shaded domains shows that

(36.37) $$\overline{\underset{\sim}{a}} \wedge \underset{\sim}{b} \in \{0 \; ; 0,1\} \cup \{0.5 \; , 0,6 \; , 0,7 \; , 0,8\}.$$

$\overline{\underset{\sim}{a}} \wedge \underset{\sim}{b}$	.0	.1	.2	.3	.4	.5	.6	.7	.8	.9	1
0	0	.1	.2	.3	.4	.5	.6	.7	.8	.9	1
.1	0	.1	.2	.3	.4	.5	.6	.7	.8	.9	.9
.2	0	.1	.2	.3	.4	.5	.6	.7	.8	.8	.8
.3	0	.1	.2	.3	.4	.5	.6	.7	.7	.7	.7
.4	0	.1	.2	.3	.4	.5	.6	.6	.6	.6	.6
.5	0	.1	.2	.3	.4	.5	.5	.5	.5	.5	.5
.6	0	1	.2	.3	.4	.4	.4	.4	.4	.4	.4
.7	0	1	.2	.3	.3	.3	.3	.3	.3	.3	.3
.8	0	.1	.2	.2	.2	.2	.2	.2	.2	.2	.2
.9	0	.1	.1	.1	.1	.1	.1	.1	.1	.1	.1
1	0	0	0	0	0	0	0	0	0	0	0

Fig. 36.9

Example II. Let

(36.38) $$\underset{\sim}{f}(\underset{\sim}{a}, \underset{\sim}{b}, \underset{\sim}{c}) = (\underset{\sim}{a} \wedge \overline{\underset{\sim}{b}} \wedge \underset{\sim}{c}) \vee \overline{\underset{\sim}{c}},$$

(36.39) where $$\underset{\sim}{a} \in \{0,3 \; ; 0,4 \; ; 0,5\},$$

(36.40) $$\underset{\sim}{b} \in \{0,1 \; ; 0,2\} \cup \{0,6\},$$

(36.41) $$\underset{\sim}{c} \in \{0 \; ; 0,1\} \cup \{0,7 \; ; 0,8 \; ; 0,9 \; ; 1\}.$$

First put

(36.42) $$\underset{\sim}{d} = \underset{\sim}{a} \wedge \overline{\underset{\sim}{b}}$$

and calculate the domain of $\underset{\sim}{d}$ with the aid of the table in Figure 36.10 (which is the transpose of that in Figure 36.2). One finds

(36.43) $$\underset{\sim}{d} = \underset{\sim}{a} \wedge \overline{\underset{\sim}{b}} \in \{0,3 \; ; 0,4 \; ; 0,5\}.$$

†Of course, this may be written $\{0, 0,1; 0,7; 0,8; 0,9\}$, but we have separated the nonconsecutive elements so that one may more easily see the nonconsecutive parts.

$\underset{\sim}{d} = \underset{\sim}{a} \wedge \overline{\underset{\sim}{b}}$ ($\underset{\sim}{a}$ \ $\underset{\sim}{b}$)	0	1	.2	.3	.4	.5	.6	.7	.8	.9	1
0	0	0	0	0	0	0	0	0	0	0	0
.1	.1	.1	.1	.1	.1	.1	.1	.1	.1	.1	0
.2	.2	.2	.2	.2	.2	.2	.2	.2	.2	.1	0
.3	.3	.3	.3	.3	.3	.3	.3	.3	.2	.1	0
.4	.4	.4	.4	.4	.4	.4	.4	.3	.2	.1	0
.5	.5	.5	.5	.5	.5	.5	.4	.3	.2	.1	0
.6	.6	.6	.6	.6	.6	.5	.4	.3	.2	.1	0
.7	.7	.7	.7	.7	.6	.5	.4	.3	.2	.1	0
.8	.8	.8	.8	.7	.6	.5	.4	.3	.2	.1	0
.9	.9	.9	.8	.7	.6	.5	.4	.3	.2	.1	0
1	1	9	.8	.7	.6	.5	.4	.3	.2	.1	0

Fig. 36.10

$\underset{\sim}{e} = \underset{\sim}{d} \wedge \underset{\sim}{c}$ ($\underset{\sim}{d}$ \ $\underset{\sim}{c}$)	0	1	.2	.3	.4	.5	.6	.7	.8	.9	1
0	0	0	0	0	0	0	0	0	0	0	0
.1	0	.1	.1	.1	.1	.1	.1	.1	.1	.1	.1
.2	0	.1	.2	.2	.2	.2	.2	.2	.2	.2	.2
.3	0	.1	.2	.3	.3	.3	.3	.3	.3	.3	.3
.4	0	.1	.2	.3	.4	.4	.4	.4	.4	.4	.4
.5	0	.1	.2	.3	.4	.5	.5	.5	.5	.5	.5
.6	0	.1	.2	.3	.4	.5	.6	.6	.6	.6	.6
.7	0	.1	.2	.3	.4	.5	.6	.7	.7	.7	.7
.8	0	.1	.2	.3	.4	.5	.6	.7	.8	.8	.8
.9	0	.1	.2	.3	.4	.5	.6	.7	.8	.9	.9
1	0	.1	.2	.3	.4	.5	.6	.7	.8	.9	1

Fig. 36.11

We now calculate the domain of

(36.44) $$\underset{\sim}{e} = \underset{\sim}{d} \wedge \underset{\sim}{c} = \underset{\sim}{a} \wedge \overline{\underset{\sim}{b}} \wedge \underset{\sim}{c}.$$

with the aid of the table in Figure 36.11 (which is nothing but that of Figure 36.1). One finds

(36.45) $$\underset{\sim}{e} = \underset{\sim}{d} \wedge \underset{\sim}{c} = \underset{\sim}{a} \wedge \overline{\underset{\sim}{b}} \wedge \underset{\sim}{c} \in \{0\ ;\ 0{,}1\} \cup \{0{,}3\ ;\ 0{,}4\ ;\ 0{,}5\}.$$

Finally, we calculate the domain of

(36.46) $$\underset{\sim}{f} = \underset{\sim}{e} \vee \overline{\underset{\sim}{c}} = (\underset{\sim}{d} \wedge \underset{\sim}{c}) \vee \overline{\underset{\sim}{c}} = (\underset{\sim}{a} \wedge \overline{\underset{\sim}{b}} \wedge \underset{\sim}{c}) \vee \overline{\underset{\sim}{c}},$$

with the aid of the table in Figure 36.12 (which is the transpose of that in Figure 36.5).

$\underset{\sim}{f} = \underset{\sim}{e} \vee \overline{\underset{\sim}{c}}$	0	.1	.2	.3	.4	.5	.6	.7	.8	.9	1
0	1	.9	.8	.7	.6	.5	.4	.3	.2	.1	0
.1	1	.9	.8	.7	.6	.5	.4	.3	.2	.1	.1
.2	1	.9	.8	.7	.6	.5	.4	.3	.2	.2	.2
.3	1	.9	.8	.7	.6	.5	.4	.3	.3	.3	.3
.4	1	.9	.8	.7	.6	.5	.4	.4	.4	.4	.4
.5	1	.9	.8	.7	.6	.5	.5	.5	.5	.5	.5
.6	1	.9	.8	.7	.6	.6	.6	.6	.6	.6	.6
.7	1	.9	.8	.7	.7	.7	.7	.7	.7	.7	.7
.8	1	.9	.8	.8	.8	.8	.8	.8	.8	.8	.8
.9	1	.9	.9	.9	.9	.9	.9	.9	.9	.9	.9
1	1	1	1	1	1	1	1	1	1	1	1

Fig. 36.12

One finds

$$f(a, b, c) \in \{0\ ;\ 0{,}1\ ;\ 0{,}2\ ;\ 0{,}3\ ;\ 0{,}4\ ;\ 0{,}5\} \cup \{0{,}9\ ;\ 1\}. \tag{36.47}$$

37. SYNTHESIS OF A FUNCTION OF FUZZY VARIABLES

We now examine the following problem. Being given variables a and b (we begin with two variables), realize a function $f(a, b)$ that takes its values in the interval $[\alpha_{k-1}, \alpha_k[$.

Considering what has been presented in (35.24)–(35.31), we may see that the solution is not unique. We may choose, for example, to realize $f(a, b)$ taking its values in the interval $[\alpha_{k-1}, \alpha_k[$ with the aid of a function of the type $a \wedge b$ (35.24).

For this, it is necessary to satisfy a polynomial form with respect to ∇ or with respect to Δ and that concerning the condition of type $\mathcal{R}$. We shall choose those pertaining to Δ (one might have taken many others, but these appear to be a little more complicated):

$$(\mathcal{R}_a)\ \Delta\ (\mathcal{R}_b)\ \Delta\ (\mathcal{R}'_a\ \nabla\ \mathcal{R}'_b), \tag{37.1}$$

that is, with reference to the conventions (35.4)–(35.7)

$$\left\{\begin{array}{l} \quad a \geqslant \alpha_{k-1} \\ \text{and} \quad b \geqslant \alpha_{k-1} \end{array}\right\} \quad \text{and} \quad \left\{\begin{array}{l} \quad a < \alpha_k \\ \text{or/and} \quad b < \alpha_k \end{array}\right\}. \tag{37.2}$$

One may realize a solution with any other function; for example, with the aid of $\bar{a} \wedge \bar{b}$ (35.25), for which we see

$$(\mathcal{R}_{\bar{a}})\ \Delta\ (\mathcal{R}_b)\ \Delta\ (\mathcal{R}'_{\bar{a}}\ \nabla\ \mathcal{R}'_b), \tag{37.3}$$

thus

$$\left\{\begin{array}{l} \quad a \leqslant 1 - \alpha_{k-1} \\ \text{and} \quad b \geqslant \alpha_{k-1} \end{array}\right\} \quad \text{and} \quad \left\{\begin{array}{l} \quad a > 1 - \alpha_k \\ \text{or/and} \quad b < \alpha_k \end{array}\right\}. \tag{37.4}$$

Returning to (37.1) and (37.2), suppose now that the inferior and superior limits for the variables a and b may be given by the following values:

$$\left\{\begin{array}{l} \quad a \geqslant w_1, \\ \text{and} \quad b \geqslant w_2 \end{array}\right\} \quad \text{and} \quad \left\{\begin{array}{l} \quad a < w_3, \\ \text{or/and} \quad b < w_4 \end{array}\right\}. \tag{37.5}$$

We may now introduce *coefficients of adjustment* λ_{ij}, also called multipliers in some technologies,

$$\lambda_{11} w_1 = \alpha_{k-1}, \quad \lambda_{12} w_2 = \alpha_{k-1}, \quad \lambda_{21} w_3 = \alpha_k \quad \text{and} \quad \lambda_{22} w_4 = \alpha_k, \tag{37.6}$$

that is,

(37.7) $$\lambda_{11} = \frac{\alpha_{k-1}}{w_1}, \quad \lambda_{12} = \frac{\alpha_{k-1}}{w_2}, \quad \lambda_{21} = \frac{\alpha_k}{w_3}, \quad \lambda_{22} = \frac{\alpha_k}{w_4},$$

In order to realize technologically a function $\underset{\sim}{f}(\underset{\sim}{a}, \underset{\sim}{b})$ that takes its values in the interval $[\alpha_{k-1}, \alpha_k[$ when the two given variables $\underset{\sim}{a}$ and $\underset{\sim}{b}$ are respectively located in the intervals $[w_1, w_2[$ and $[w_3, w_4[$, one may construct a scheme of elements such as that shown in Figure 37.1.

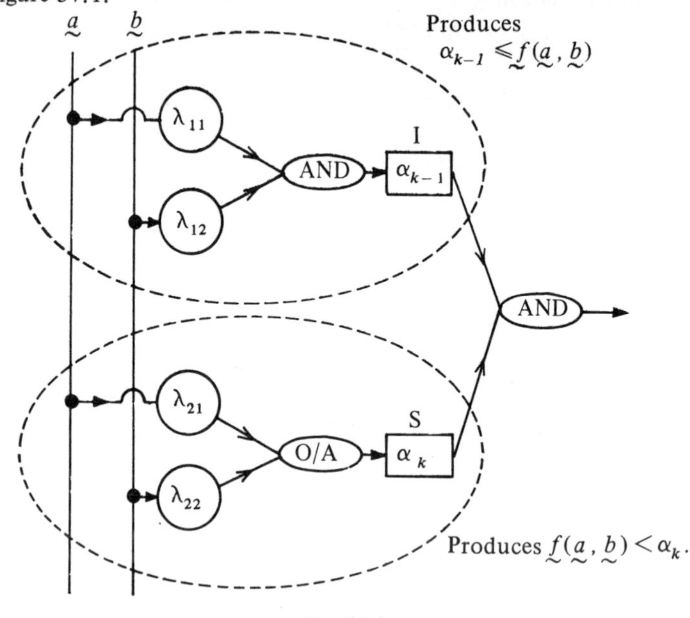

Fig. 37.1

For all schemes of elements of this type, we shall use the following symbols:

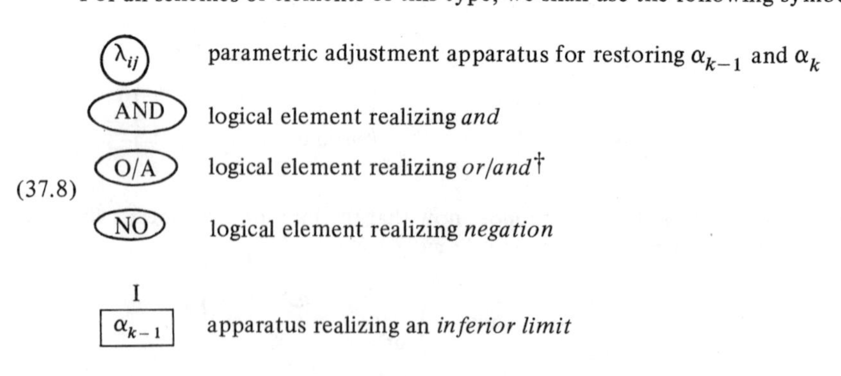

(37.8)

- λ_{ij}: parametric adjustment apparatus for restoring α_{k-1} and α_k
- AND: logical element realizing *and*
- O/A: logical element realizing *or/and*†
- NO: logical element realizing *negation*
- I α_{k-1}: apparatus realizing an *inferior limit*
- S α_k: apparatus realizing a *superior limit*

†The author prefers to use O/A for or/and rather than OR, which is confusing; computer scientists are inclined toward OR.

Example II. We shall realize a synthesis given

$$\alpha_{k-1} \leqslant \underset{\sim}{f}(\underset{\sim}{a}, \underset{\sim}{b}) < \alpha_k \tag{37.9}$$

using the function

$$\underset{\sim}{f}(\underset{\sim}{a}, \underset{\sim}{b}) = (\underset{\sim}{a} \wedge \overline{\underset{\sim}{b}}) \vee (\overline{\underset{\sim}{a}} \wedge \underset{\sim}{b}). \tag{37.10}$$

Referring to the rule given in Section 35, we see†

$$\mathcal{P} = [(\mathcal{P}_{\underset{\sim}{a}} \,\Delta\, \mathcal{P}_{\overline{\underset{\sim}{b}}}) \,\nabla\, (\mathcal{P}_{\overline{\underset{\sim}{a}}} \,\Delta\, \mathcal{P}_{\underset{\sim}{b}})] \,\Delta\, [(\mathcal{P}'_{\underset{\sim}{a}} \,\nabla\, \mathcal{P}'_{\overline{\underset{\sim}{b}}}) \,\Delta\, (\mathcal{P}'_{\overline{\underset{\sim}{a}}} \,\nabla\, \mathcal{P}'_{\underset{\sim}{b}})] \tag{37.11}$$

which may be written

$$\begin{pmatrix} \underset{\sim}{a} \geqslant \alpha_{k-1} \\ \text{and}\quad \underset{\sim}{b} \leqslant 1-\alpha_{k-1} \end{pmatrix} \text{ or/and } \begin{pmatrix} \underset{\sim}{a} \leqslant 1-\alpha_{k-1} \\ \text{and}\quad \underset{\sim}{b} \geqslant \alpha_{k-1} \end{pmatrix}$$
$$\text{and } \begin{pmatrix} \underset{\sim}{a} < \alpha_k \\ \text{or/and}\quad \underset{\sim}{b} > 1-\alpha_k \end{pmatrix} \text{ and } \begin{pmatrix} \underset{\sim}{a} > 1-\alpha_k \\ \text{or/and}\quad \underset{\sim}{b} < \alpha_k \end{pmatrix}. \tag{37.12}$$

If the limits are the following

$$\begin{pmatrix} \underset{\sim}{a} \geqslant w_1 \\ \text{and}\quad \underset{\sim}{b} \leqslant w_2 \end{pmatrix} \text{ or/and } \begin{pmatrix} \underset{\sim}{a} \leqslant w_3 \\ \text{and}\quad \underset{\sim}{b} \geqslant w_4 \end{pmatrix}$$
$$\text{and } \begin{pmatrix} \underset{\sim}{a} < w_5 \\ \text{or/and}\quad \underset{\sim}{b} > w_6 \end{pmatrix} \text{ and } \begin{pmatrix} \underset{\sim}{a} > w_7 \\ \text{or/and}\quad \underset{\sim}{b} < w_8 \end{pmatrix}. \tag{37.13}$$

one may see

(37.14)

$$\lambda_{11} = \frac{\alpha_{k-1}}{w_1}, \quad \lambda_{12} = \frac{1-\alpha_{k-1}}{w_2}, \quad \lambda_{13} = \frac{1-\alpha_{k-1}}{w_3}, \quad \lambda_{14} = \frac{\alpha_{k-1}}{w_4},$$

$$\lambda_{21} = \frac{\alpha_k}{w_5}, \quad \lambda_{22} = \frac{1-\alpha_k}{w_6}, \quad \lambda_{23} = \frac{1-\alpha_k}{w_7}, \quad \lambda_{24} = \frac{\alpha_k}{w_8}$$

One thus obtains the scheme of elements given in Figure 37.2.

Example III. Realize

$$\alpha_{k-1} \leqslant \underset{\sim}{f}(\underset{\sim}{a}, \underset{\sim}{b}, \underset{\sim}{c}, \underset{\sim}{d}) < \alpha_k, \tag{37.15}$$

where

$$\underset{\sim}{f}(\underset{\sim}{a}, \underset{\sim}{b}, \underset{\sim}{c}, \underset{\sim}{d}) = (\underset{\sim}{a} \vee \underset{\sim}{b} \vee \overline{\underset{\sim}{c}}) \wedge (\overline{\underset{\sim}{a}} \vee \underset{\sim}{c} \vee \underset{\sim}{d}). \tag{37.16}$$

Using the rule given in Section 35, we put

†One may also use the reduced expressions (35.30).

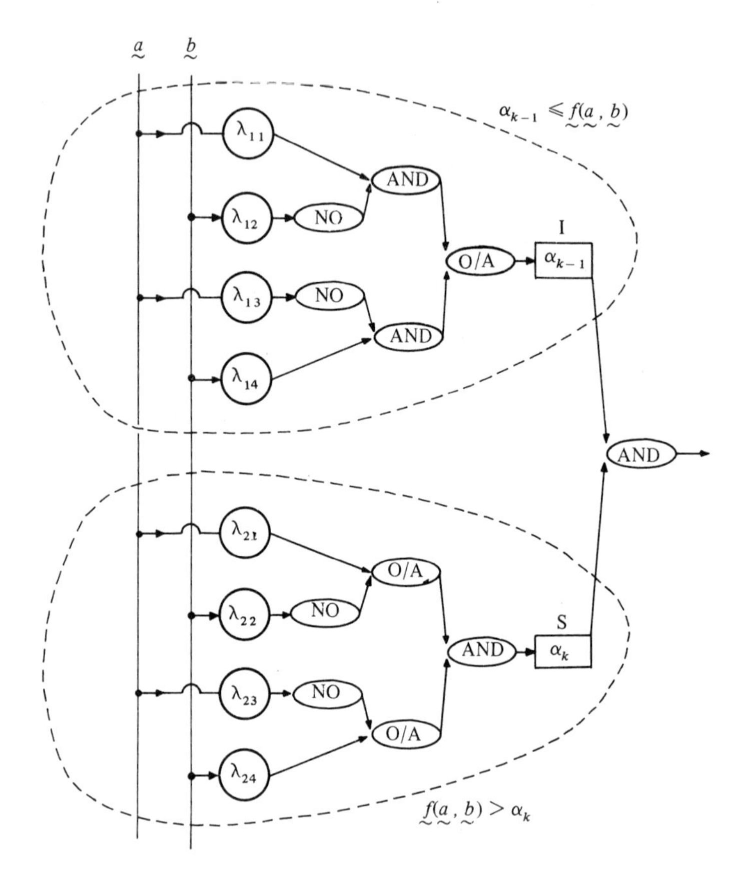

Fig. 37.2

(37.17)

$$\mathcal{P} = [(\mathcal{P}_{a} \nabla \mathcal{P}_{b} \nabla \mathcal{P}_{\bar{c}}) \Delta (\mathcal{P}_{\bar{a}} \nabla \mathcal{P}_{c} \nabla \mathcal{P}_{d})] \Delta [(\mathcal{P}'_{a} \Delta \mathcal{P}'_{b} \Delta \mathcal{P}'_{\bar{c}}) \nabla (\mathcal{P}'_{\bar{a}} \Delta \mathcal{P}'_{c} \Delta \mathcal{P}'_{d})] ,$$

This may be rewritten

$$\begin{pmatrix} & a \geqslant \alpha_{k-1} , \\ \text{or/and} & b \geqslant \alpha_{k-1} , \\ \text{or/and} & c \leqslant 1 - \alpha_{k-1} \end{pmatrix} \quad \text{and} \quad \begin{pmatrix} & a \leqslant 1 - \alpha_{k-1} \\ \text{or/and} & c \geqslant \alpha_{k-1} \\ \text{or/and} & d \geqslant \alpha_{k-1} \end{pmatrix}$$

(37.18) and

$$\begin{pmatrix} & a < \alpha_k \\ \text{and} & b < \alpha_k \\ \text{and} & c > 1 - \alpha_k \end{pmatrix} \quad \text{or/and} \quad \begin{pmatrix} & a > 1 - \alpha_k \\ \text{and} & c < \alpha_k \\ \text{and} & d < \alpha_k \end{pmatrix} .$$

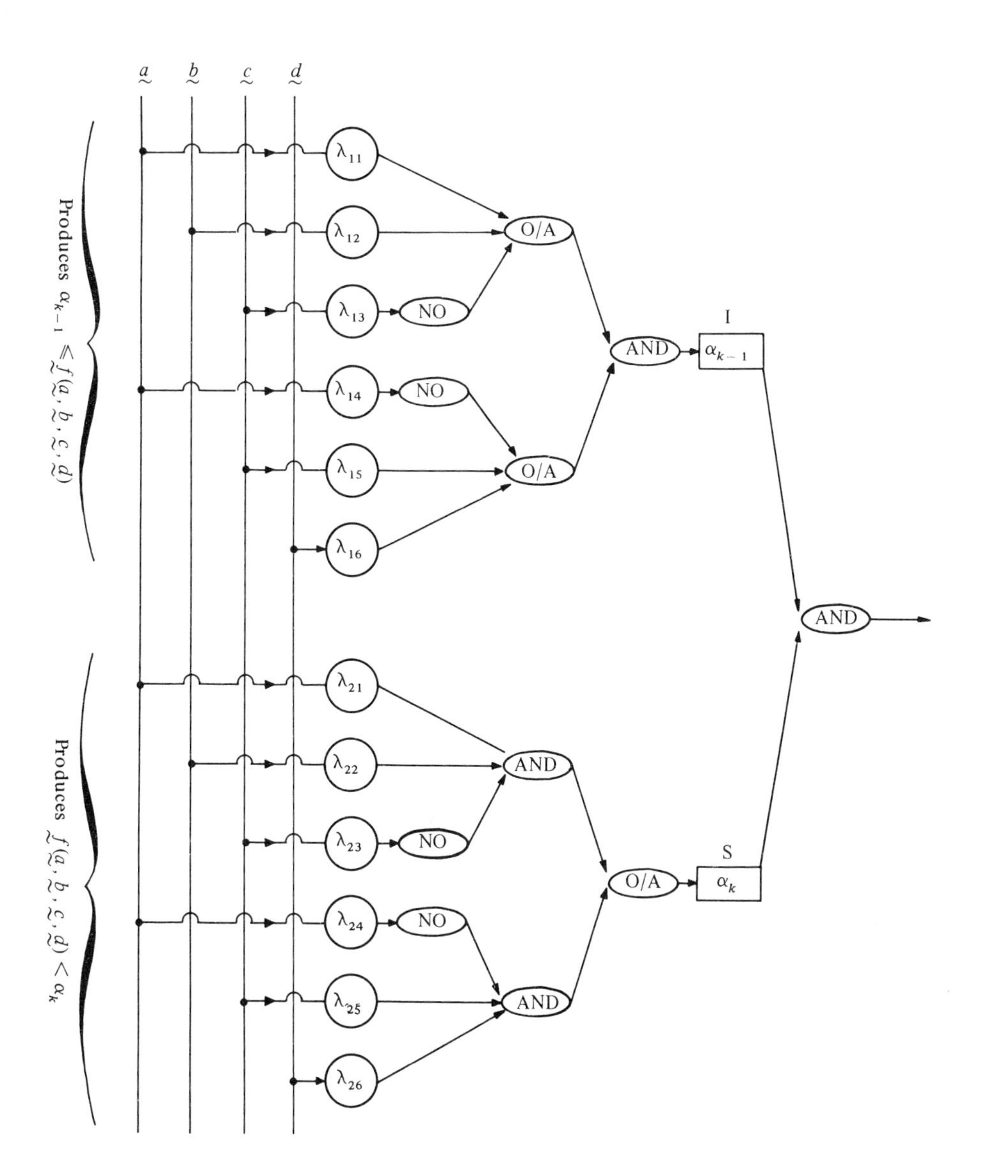

a
b
c
d
Produces $\alpha_{k-1} \leq f(a, b, c, d)$
Produces $f(a, b, c, d) < \alpha_k$
λ_{11}
λ_{12}
λ_{13}
λ_{14}
λ_{15}
λ_{16}
λ_{21}
λ_{22}
λ_{23}
λ_{24}
λ_{25}
λ_{26}
NO
NO
NO
NO
O/A
O/A
AND
AND
AND
O/A
I
α_{k-1}
S
α_k
AND

Fig. 37.3

Thus, for the limits

$$\begin{pmatrix} & \underset{\sim}{a} \geqslant w_1 , \\ \text{or/and} & \underset{\sim}{b} \geqslant w_2 , \\ \text{or/and} & \underset{\sim}{c} \leqslant w_3 \end{pmatrix} \quad \text{and} \quad \begin{pmatrix} & \underset{\sim}{a} \leqslant w_4 , \\ \text{or/and} & \underset{\sim}{c} \geqslant w_5 , \\ \text{or/and} & \underset{\sim}{d} \geqslant w_6 , \end{pmatrix}$$

(37.19) and

$$\begin{pmatrix} & \underset{\sim}{a} < w_7 , \\ \text{and} & \underset{\sim}{b} < w_8 , \\ \text{and} & \underset{\sim}{c} > w_9 . \end{pmatrix} \quad \text{or/and} \quad \begin{pmatrix} & \underset{\sim}{a} > w_{10} , \\ \text{and} & \underset{\sim}{c} < w_{11} , \\ \text{and} & \underset{\sim}{d} < w_{12} . \end{pmatrix}$$

and the λ_{ij}

$$\lambda_{11} = \frac{\alpha_{k-1}}{w_1} , \quad \lambda_{12} = \frac{\alpha_{k-1}}{w_2} , \quad \lambda_{13} = \frac{1-\alpha_{k-1}}{w_3} , \quad \lambda_{14} = \frac{1-\alpha_{k-1}}{w_4} ,$$

(37.20) $$\lambda_{15} = \frac{\alpha_{k-1}}{w_5} , \quad \lambda_{16} = \frac{\alpha_{k-1}}{w_6} , \quad \lambda_{21} = \frac{\alpha_k}{w_7} , \quad \lambda_{22} = \frac{\alpha_k}{w_8} , \quad \lambda_{23} = \frac{1-\alpha_k}{w_9} ,$$

$$\lambda_{24} = \frac{1-\alpha_k}{w_{10}} , \quad \lambda_{25} = \frac{\alpha_k}{w_{11}} , \quad \lambda_{26} = \frac{\alpha_k}{w_{12}} .$$

Figures 37.3 gives the realization.

Circuits of the type presented in Figures 37.1 – 37.3 will be called *primal–dual.*

(37.21) All circuits of primal type give rise to $\alpha_{k-1} \leqslant \underset{\sim}{f}(\underset{\sim}{a}, \underset{\sim}{b}, \ldots)$.

(37.22) All circuits of dual type give rise to $\underset{\sim}{f}(\underset{\sim}{a}, \underset{\sim}{b}, \ldots) < \alpha_k$.

It is not necessary to realize a primal–dual circuit in order to obtain

(37.23) $$\alpha_{k-1} \leqslant \underset{\sim}{f}(\underset{\sim}{a}, \underset{\sim}{b}, \ldots) < \alpha_k ,$$

one may also operate with respect to a monomial of reduced form with respect to ∇. We consider two examples.

Example III. In Figure 37.1 and with the aid of (37.1), (37.2), (37.5)–(37.7), we have seen how to present

(37.24) $$\alpha_{k-1} \leqslant \underset{\sim}{a} \wedge \underset{\sim}{b} < \alpha_k ,$$

with the aid of a primal–dual circuit obtained by the polynomial form with respect to Δ (35.24).

On the other hand, we now use a polynomial form with respect to Δ (35.24). For this, some monomials of

$$(37.25) \qquad \mathcal{P} = (\mathcal{P}_{\underset{\sim}{a}} \,\Delta\, \mathcal{P}'_{\underset{\sim}{a}} \,\Delta\, \mathcal{P}_{\underset{\sim}{b}}) \,\nabla\, (\mathcal{P}_{\underset{\sim}{a}} \,\Delta\, \mathcal{P}_{\underset{\sim}{b}} \,\Delta\, \mathcal{P}'_{\underset{\sim}{b}})$$

suffice. The symbol ∇ indicates to us that a single one of the conditions, that is a monomial, is sufficient. We take, for example, the first

$$(37.26) \qquad \mathcal{P}_{\underset{\sim}{a}} \,\Delta\, \mathcal{P}'_{\underset{\sim}{a}} \,\Delta\, \mathcal{P}_{\underset{\sim}{b}} ,$$

which gives

(37.27)

$$(a \geqslant \alpha_{k-1}) \text{ and } (a < \alpha_k) \text{ and } (b \geqslant \alpha_{k-1})$$

or

(37.28)

$$\begin{pmatrix} a \geqslant \alpha_{k-1} \\ \text{and } b \geqslant \alpha_{k-1} \end{pmatrix} \text{ and } (a < \alpha_k) .$$

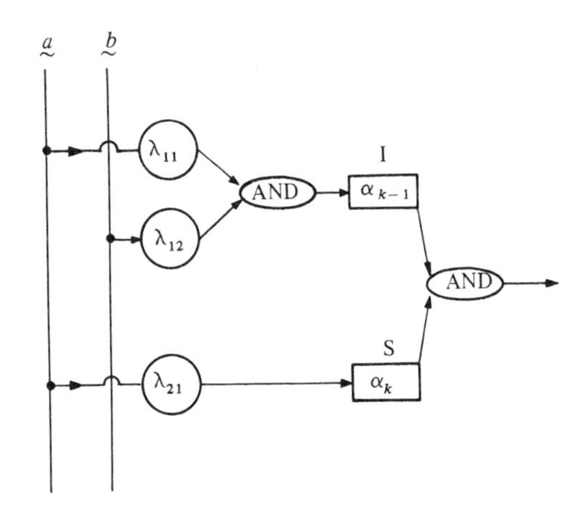

Fig. 37.4

Figure 37.4 gives the technological synthesis.

Remark. One ought not be astonished that a circuit such as that in Figure 37.4 may give the same results as a circuit such as that in Figure 37.1, that is,

$$(37.29) \qquad \alpha_{k-1} \leqslant \underset{\sim}{a} \wedge \underset{\sim}{b} < \alpha_k ;$$

since for the function $f(\underset{\sim}{a}, \underset{\sim}{b}) = \underset{\sim}{a} \wedge \underset{\sim}{b}$,

$$(37.30) \qquad (\alpha_{k-1} \leqslant \underset{\sim}{a} \wedge \underset{\sim}{b} < \alpha_k) \quad \Leftrightarrow \quad (\alpha_{k-1} \leqslant \underset{\sim}{a} \wedge \underset{\sim}{b} \quad \text{and} \quad \underset{\sim}{a} < \alpha_k) ,$$

which shows, moreover, the development with respect to ∇ [of (35.24)].

Example IV. We reconsider Example II treated in (37.9)–(37.14). This time, instead of taking the polynomial form with respect to Δ of (35.30) or carrying out the primal development, we shall use one of the terms of the development in ∇ of (35.30), for example, the sixth monomial:

$$(37.31) \qquad \mathcal{P}'_{\underset{\sim}{a}} \,\Delta\, \mathcal{P}_{\bar{\underset{\sim}{a}}} \,\Delta\, \mathcal{P}_{\underset{\sim}{b}} \,\nabla\, \mathcal{P}'_{\underset{\sim}{b}} .$$

The corresponding conditions are

$$(37.32) \qquad \begin{pmatrix} \underset{\sim}{a} \leqslant 1 - \alpha_{k-1} \\ \text{and } \; \underset{\sim}{b} \geqslant \alpha_{k-1} \end{pmatrix} \quad \text{and} \quad \begin{pmatrix} \underset{\sim}{a} < \alpha_k , \\ \text{and } \; \underset{\sim}{b} < \alpha_k . \end{pmatrix} .$$

From this one obtains the circuit shown in Figure 37.5, obviously simpler than that of Figure 37.2.

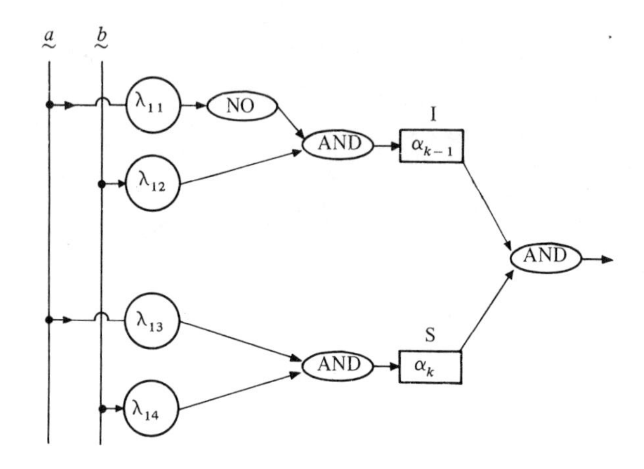

Fig. 37.5

Remark. Given that any function $f(a, b, \ldots)$ may be the object of a development of $\mathcal{P}$ in a polynomial form in ∇ in which each monomial contains only terms $\mathcal{P}_x$ or/and $\mathcal{P}_{\bar{x}}$ or/and $\mathcal{P}'_x$ or/and $\mathcal{P}'_{\bar{x}}$ joined by Δ one may always realize a circuit containing only **AND** and **NO**. But according to De Morgan's theorem, which permits us to write

(37.33) $$\overline{\mathcal{P} \,\Delta\, \mathcal{P}'} = \overline{\mathcal{P}} \,\nabla\, \overline{\mathcal{P}'} ,$$

(37.34) $$\overline{\mathcal{P} \,\nabla\, \mathcal{P}'} = \overline{\mathcal{P}} \,\Delta\, \overline{\mathcal{P}'} ,$$

by using conditions of the type $\overline{\mathcal{P}}$, one may carry out a development identical to that which gives a polynomial form in ∇, where we have replaced the $\mathcal{P}$ by $\overline{\mathcal{P}}$, the ∇ by Δ, and the Δ by ∇. Thus, one may use only the technological operators **O/A** and **NO.**

In fact, one may use extremely varied combinations of operators, as is the general custom of builders of hardware, Likewise, one may use a single operator, either that of Scheffer or that of Peirce, that is,

(37.35) $$\mathcal{P}_1 \mid \mathcal{P}_2 = \overline{\mathcal{P}}_1 \,\nabla\, \overline{\mathcal{P}}_2$$

or

(35.36) $$\mathcal{P}_1 \downarrow \mathcal{P}_2 = \overline{\mathcal{P}}_1 \,\Delta\, \overline{\mathcal{P}}_2 ,$$

but this would often be technologically inconvenient.

Mixed circuits. Calling primal the conditions that give

$$\alpha_{k-1} \leqslant \underset{\sim}{f}(\underset{\sim}{a}, \underset{\sim}{b}, \ldots) \tag{37.37}$$

and dual the conditions that give

$$\underset{\sim}{f}(\underset{\sim}{a}, \underset{\sim}{b}, \ldots) < \alpha_k , \tag{37.38}$$

one may have just as well a mixed circuit giving

$$\alpha_{k-1} \leqslant \underset{\sim}{f}_1(\underset{\sim}{a}, \underset{\sim}{b}, \ldots) \tag{37.39}$$

and

$$\underset{\sim}{f}_2(\underset{\sim}{a}, \underset{\sim}{b}, \ldots) < \alpha_k \tag{37.40}$$

It is sufficient to assemble using a technological operator **AND** a primal circuit for (37.39) with a dual circuit for (37.40). We consider an example.

Example. We shall realize

$$\alpha_{k-1} \leqslant \underset{\sim}{f}_1(\underset{\sim}{a}, \underset{\sim}{b}) = \overline{\underset{\sim}{a}} \wedge \underset{\sim}{b} \tag{37.41}$$

and

$$\underset{\sim}{f}_2(\underset{\sim}{a}, \underset{\sim}{b}, \underset{\sim}{c}) = (\underset{\sim}{a} \wedge \underset{\sim}{b}) \vee (\overline{\underset{\sim}{b}} \wedge \underset{\sim}{c}) < \alpha_k . \tag{37.42}$$

For $\underset{\sim}{f}_1$, one has the primal conditions

$$\mathcal{P}_{\overline{\underset{\sim}{a}}} \,\Delta\, \mathcal{P}_{\underset{\sim}{b}} , \tag{37.43}$$

that is,

$$\begin{pmatrix} \underset{\sim}{a} \leqslant 1 - \alpha_{k-1} , \\ \text{and } \underset{\sim}{b} \geqslant \alpha_{k-1} . \end{pmatrix} . \tag{37.44}$$

For $\underset{\sim}{f}_2$, one has the dual conditions

$$(\mathcal{P}'_{\underset{\sim}{a}} \,\nabla\, \mathcal{P}'_{\underset{\sim}{b}}) \,\Delta\, (\mathcal{P}'_{\overline{\underset{\sim}{b}}} \,\nabla\, \mathcal{P}'_{\underset{\sim}{c}}) , \tag{37.45}$$

that is,

$$\begin{pmatrix} \underset{\sim}{a} < \alpha_k \\ \text{or/and } \underset{\sim}{b} < \alpha_k \end{pmatrix} \text{ and } \begin{pmatrix} \underset{\sim}{b} > 1 - \alpha_k \\ \text{or/and } \underset{\sim}{c} < \alpha_k \end{pmatrix} . \tag{37.46}$$

Grouping (37.44) and (37.46) with the conjunction *and*, we have

$$\begin{pmatrix} \underset{\sim}{a} \leqslant 1 - \alpha_{k-1} \\ \text{and } \underset{\sim}{b} \geqslant \alpha_{k-1} \end{pmatrix} . \tag{37.47}$$

(37.47)

$$\text{and} \quad \left(\begin{array}{r} \underset{\sim}{a} < \alpha_k \\ \text{or/and} \quad \underset{\sim}{b} < \alpha_k \end{array} \right) \quad \text{and} \quad \left(\begin{array}{r} \underset{\sim}{b} > 1 - \alpha_k \\ \text{or/and} \quad \underset{\sim}{c} < \alpha_k \end{array} \right) .$$

We finally obtain the synthesis of the circuit of Figure 37.6.

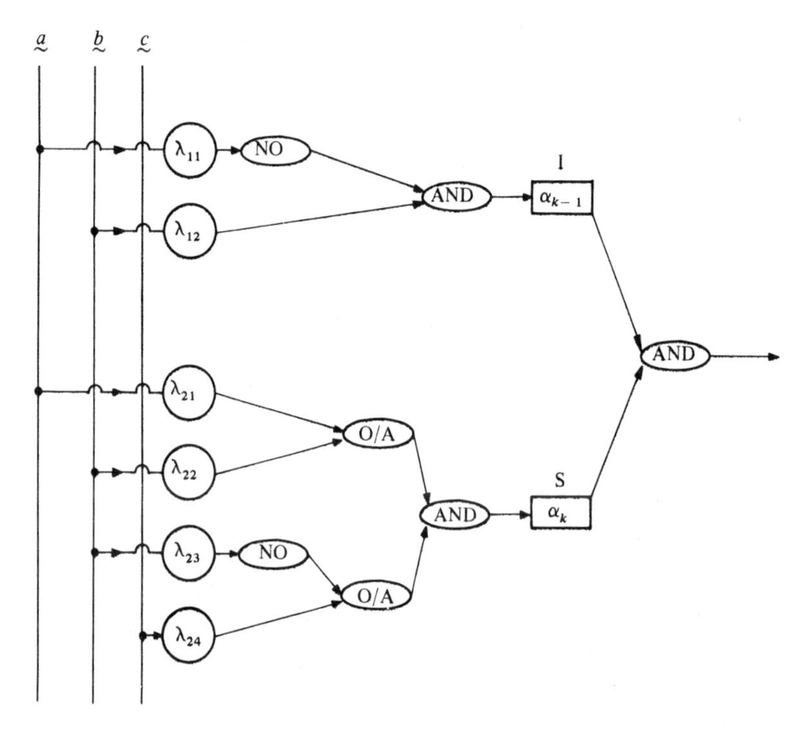

Fig. 37.6

Thus, for the circuit of Figure 37.6, one is assured that he has, at the same time

(37.48) $$\alpha_{k-1} \leqslant \overline{\underset{\sim}{a}} \wedge \underset{\sim}{b} \text{ and } (\underset{\sim}{a} \wedge \underset{\sim}{b}) \vee (\overline{\underset{\sim}{b}} \wedge \underset{\sim}{c}) < \alpha_k ,$$

through a suitable adjustment of the coefficients λ_{ij}.

All the considerations of the present section may be the objects of various extensions; it is likely that a number of readers will be interested in these.

Remark. Those who have knowledge of electronics are aware that the establishment of such a fuzzy technology is neither easy nor economical to realize (stable multipliers, fine control of potentials, etc.). But this is a path for research.

38. NETWORKS OF FUZZY ELEMENTS

It is interesting to use a network representation, as has been done in the theory of chains of contacts (see reference [2F]), in the theory of reliability of networks (see reference [7K]), and in many other theories where such associations may appear, in series or/and in parallel.

Fuzzy element of a network. To any fuzzy variable $\underset{\sim}{a} \in [0,\ 1]$, we shall associate an element $\underset{\sim}{a}$, designated by the same symbol. We shall then construct a network possessing elements such as $\underset{\sim}{a}$.

To the function $\underset{\sim}{a} \wedge \underset{\sim}{b}$, we shall associate a network such as that in Figure 38.1. To the function $\underset{\sim}{a} \vee \underset{\sim}{b}$, we shall associate a network such as that in Figure 38.2. The first will be called a series network; the second will be called a parallel network. In such networks we shall define an entry E and an exit S. The result of the operation carried out among the elements of the network is called the *flow of the network.*

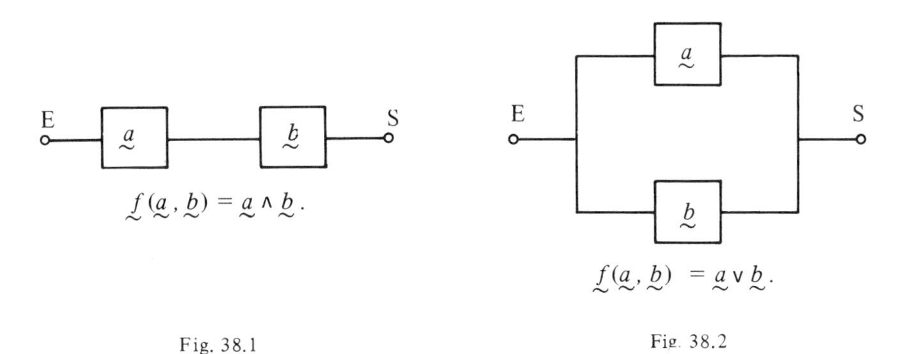

Fig. 38.1 Fig. 38.2

Thus, in the case of Figure 38.1, if $\underset{\sim}{a} = 0{,}7$ and $\underset{\sim}{b} = 0{,}4$, then the flow that goes from E to S is equal to 0,4. It would be equal to 0,7 in the network of Figure 38.2.

Theorem I. To any analytic function of fuzzy variables $\underset{\sim}{f}(\underset{\sim}{a}, \underset{\sim}{b}, \ldots)$ one may associate a network of fuzzy elements, where the series position is associated with the operation $\vee$ and the parallel position is associated with the operation $\wedge$.

Proof. We have already seen that to any function such as $\underset{\sim}{f}(\underset{\sim}{a}, \underset{\sim}{b}, \ldots)$ that is analytic, one may associate, by definition, a reduced polynomial form with respect to $\wedge$ or with respect to $\vee$. One may then associate to each of these a network..

Example. Thus, to the following function, presented in a reduced polynomial form with respect to $\vee$,

$$\underset{\sim}{f}(\underset{\sim}{a}, \underset{\sim}{b}, \underset{\sim}{c}) = (\underset{\sim}{a} \wedge \bar{\underset{\sim}{b}} \wedge \underset{\sim}{c}) \vee (\underset{\sim}{a} \wedge \underset{\sim}{b}) \vee (\bar{\underset{\sim}{a}} \wedge \underset{\sim}{b} \wedge \underset{\sim}{c}) \tag{38.1}$$

one will associate the network shown in Figure 38.3.

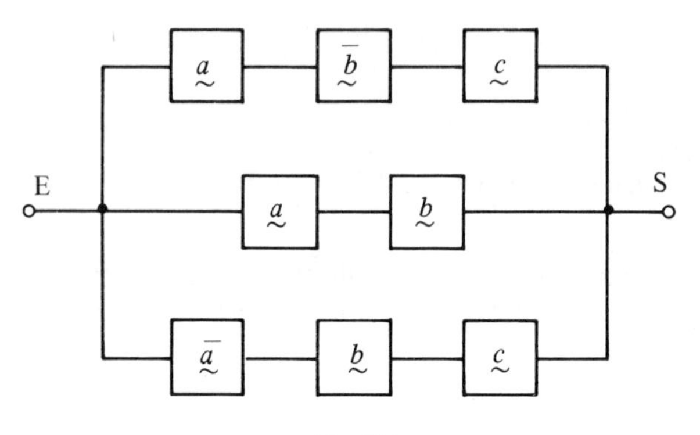

Fig. 38.3

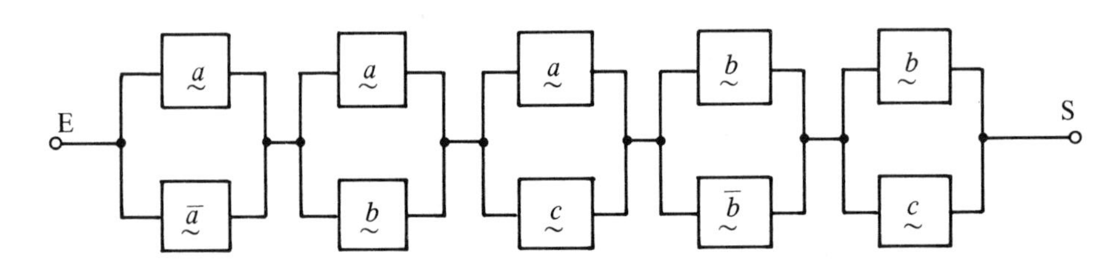

Fig. 38.4

The network corresponding to the same function but presented in a reduced polynomial form with respect to $\wedge$

$$(38.2)\qquad \underset{\sim}{f}(\underset{\sim}{a}, \underset{\sim}{b}, \underset{\sim}{c}) = (\underset{\sim}{a} \vee \overline{\underset{\sim}{a}}) \wedge (\underset{\sim}{a} \vee \underset{\sim}{b}) \wedge (\underset{\sim}{a} \vee \underset{\sim}{c}) \wedge (\underset{\sim}{b} \vee \overline{\underset{\sim}{b}}) \wedge (\underset{\sim}{b} \vee \underset{\sim}{c})$$

is represented in Figure 38.4.

Routes. A sequence of elements meeting to join E to S will be called a route or path†. Thus, in Figure 38.3,

$$(38.3)\qquad (\underset{\sim}{a}, \overline{\underset{\sim}{b}}, \underset{\sim}{c}) \qquad \text{is a route}$$

A route is *simple* if it does not contain the same element $\underset{\sim}{x}$ more than one time nor the element $\overline{\underset{\sim}{x}}$ more than one time.

Thus, in Figure 38.4,

$$(38.4)\qquad (\underset{\sim}{a}, \underset{\sim}{a}, \underset{\sim}{c}, \overline{\underset{\sim}{b}}, \underset{\sim}{b}) \qquad \text{is a route,}$$

†The word *path* has already been employed in another sense in Section 18; we gladly prefer to use the word *route.*

(38.5) $$(\underset{\sim}{a}, \underset{\sim}{c}, \underset{\sim}{\bar{b}}, \underset{\sim}{b}) \quad \text{is a simple route,}$$

A route is considered with respect to the operation $\wedge$, which is associative and commutative; the order in which the elements are presented in the sequence is therefore immaterial.

Maximal simple route. Let **I** be the ordinary set of simple routes of a network; then any simple route that does not contain another route of **I** is a maximal simple route. It is clear that a maximal simple route contains a maximum of $2n$ elements if n is the number of fuzzy variables of $\underset{\sim}{f}$.

Fundamental property. By putting all maximal simple routes in parallel, one constructs a network equivalent to a reduced polynomial with respect to $\underset{\sim}{f}$.

This property is evident, having been given the ways in which one constructs polynomial forms on the one hand and on the other series–parallel networks (series grouped in parallel) with respect to maximal simple routes.

Example I. We consider the network of Figure 38.5, which corresponds to the function

(38.6) $$\underset{\sim}{f}(\underset{\sim}{a}, \underset{\sim}{b}, \underset{\sim}{c}) = [\underset{\sim}{a} \wedge [(\underset{\sim}{b} \wedge \underset{\sim}{c}) \vee (\underset{\sim}{a} \wedge \underset{\sim}{\bar{c}})]] \vee \underset{\sim}{\bar{c}}.$$

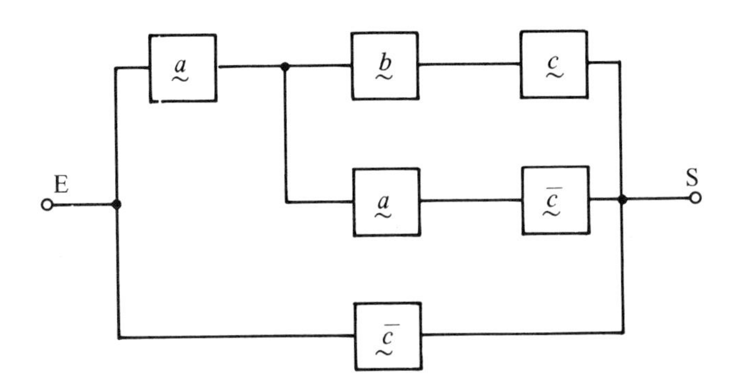

Fig. 38.5

The set of routes is

(38.7) $$\{(\underset{\sim}{a}, \underset{\sim}{b}, \underset{\sim}{c}),\ (\underset{\sim}{a}, \underset{\sim}{a}, \underset{\sim}{\bar{c}}),\ (\underset{\sim}{\bar{c}})\}.$$

The set of simple routes is

(38.8) $$\{(\underset{\sim}{a}, \underset{\sim}{b}, \underset{\sim}{c}),\ (\underset{\sim}{a}, \underset{\sim}{\bar{c}}),\ (\underset{\sim}{\bar{c}})\}$$

The set of maximal simple routes is

(38.9) $$\{(\underset{\sim}{a}, \underset{\sim}{b}, \underset{\sim}{c}),\ (\underset{\sim}{\bar{c}})\}.$$

This corresponds to the reduced polynomial form in $\vee$:

(38.10)
$$\underset{\sim}{f}(\underset{\sim}{a}\,,\underset{\sim}{b}\,,\underset{\sim}{c}) = (\underset{\sim}{a} \wedge \underset{\sim}{b} \wedge \underset{\sim}{c}) \vee (\underset{\sim}{\bar{c}})\,,$$

and to the simplified network in Figure 38.6.

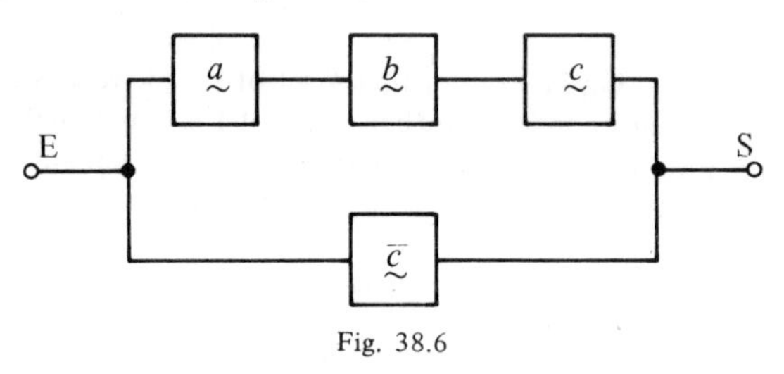

Fig. 38.6

Example II. We now see a more complicated case (Figure 38.7). The set of routes is†

(38.11)

$$\{(\underset{\sim}{a},\underset{\sim}{b},\underset{\sim}{\bar{c}})\,,\ (\underset{\sim}{a},\underset{\sim}{c},\underset{\sim}{a})\,,\ (\underset{\sim}{a},\underset{\sim}{c},\underset{\sim}{c},\underset{\sim}{\bar{a}},\underset{\sim}{\bar{c}})\,,\ (\underset{\sim}{a},\underset{\sim}{b},\underset{\sim}{\bar{a}},\underset{\sim}{c},\underset{\sim}{a})\,,\ (\underset{\sim}{a},\underset{\sim}{b},\underset{\sim}{\bar{a}},\underset{\sim}{c},\underset{\sim}{c},\underset{\sim}{b},\underset{\sim}{\bar{c}})\,,$$

$$(\underset{\sim}{b},\underset{\sim}{c},\underset{\sim}{a})\,,\ (\underset{\sim}{b},\underset{\sim}{\bar{a}},\underset{\sim}{\bar{c}})\,,\ (\underset{\sim}{b},\underset{\sim}{\bar{a}},\underset{\sim}{b},\underset{\sim}{c},\underset{\sim}{a})\,,\ (\underset{\sim}{b},\underset{\sim}{c},\underset{\sim}{c},\underset{\sim}{b},\underset{\sim}{\bar{c}})\,,\ (\underset{\sim}{b},\underset{\sim}{c},\underset{\sim}{c},\underset{\sim}{b},\underset{\sim}{\bar{a}},\underset{\sim}{c},\underset{\sim}{a})\}\,.$$

The set of simple routes is

(38.12)

$$\{(\underset{\sim}{a},\underset{\sim}{b},\underset{\sim}{\bar{c}})\,,\ (\underset{\sim}{a},\underset{\sim}{c})\,,\ (\underset{\sim}{a},\underset{\sim}{\bar{a}},\underset{\sim}{c},\underset{\sim}{\bar{c}})\,,\ (\underset{\sim}{a},\underset{\sim}{\bar{a}},\underset{\sim}{b},\underset{\sim}{c})\,,\ (\underset{\sim}{a},\underset{\sim}{\bar{a}},\underset{\sim}{b},\underset{\sim}{c},\underset{\sim}{\bar{c}})\,,\ (\underset{\sim}{a},\underset{\sim}{b},\underset{\sim}{c})\,,$$

$$(\underset{\sim}{\bar{a}},\underset{\sim}{b},\underset{\sim}{\bar{c}})\,,\ (\underset{\sim}{a},\underset{\sim}{\bar{a}},\underset{\sim}{b},\underset{\sim}{c})\,,\ (\underset{\sim}{b},\underset{\sim}{c},\underset{\sim}{\bar{c}})\,,\ (\underset{\sim}{a},\underset{\sim}{\bar{a}},\underset{\sim}{b},\underset{\sim}{c})\}\,.$$

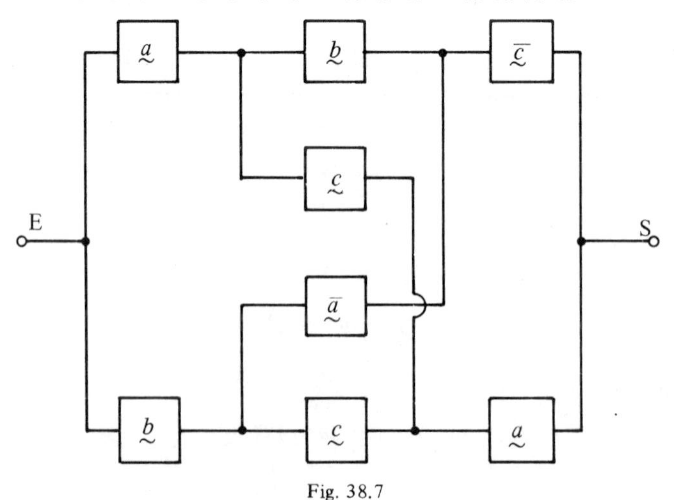

Fig. 38.7

†In fact, the number of routes will be infinite if one allows that these paths form circuits.

The set of maximal simple routes is

(38.13) $\{(\underset{\sim}{a}, \underset{\sim}{b}, \underset{\sim}{\bar{c}}) \; , \; (\underset{\sim}{\bar{a}}, \underset{\sim}{b}, \underset{\sim}{\bar{c}}) \; , \; (\underset{\sim}{b}, \underset{\sim}{c}, \underset{\sim}{\bar{c}}) \; , \; (\underset{\sim}{a}, \underset{\sim}{c})\}$.

This corresponds to the reduced polynomial form in

(38.14) $\underset{\sim}{f}(\underset{\sim}{a}, \underset{\sim}{b}, \underset{\sim}{c}) = (\underset{\sim}{a} \wedge \underset{\sim}{b} \wedge \underset{\sim}{\bar{c}}) \vee (\underset{\sim}{\bar{a}} \wedge \underset{\sim}{b} \wedge \underset{\sim}{\bar{c}}) \vee (\underset{\sim}{b} \wedge \underset{\sim}{c} \wedge \underset{\sim}{\bar{c}}) \vee (\underset{\sim}{a} \wedge \underset{\sim}{c})$,

and the series–parallel network of Figure 38.8.

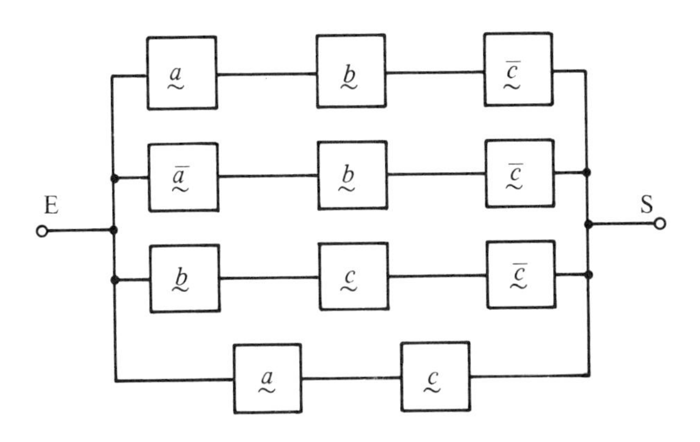

Fig. 38.8

Plane networks. If, in a network, there does not exist a connection between two elements intersecting another connection when the network between E and S is drawn in a plane, one says that the network is *realizable in a plane* or is *planar*. In the contrary case one says that the network is nonplanar. Thus, the network of Figure 38.5 is planar and that of Figure 38.7 is not.

We note the following property: Networks corresponding to polynomial forms in $\wedge$ or in $\vee$ are planar. In fact, to any polynomial form in $\wedge$ there corresponds a parallel–series network that is planar; and likewise to any polynomial form in $\vee$ there corresponds a series–parallel network that is planar (see, for example, Figures 38.1 and 38.2).

Dual of planar network. Let R be a planar network. Being planar, one may there define the faces $\alpha, \beta, \ldots$ bounded by the connections and the elements (see Figure 38.9). In each of these faces one may place an intersection point of the connections; and one may do the same for the face defined above ES and for the face defined below ES.

By applying the rule: join by a connection each element with the intersection point of any face to which it is adjacent, one forms a new network R', which will be called the dual of R.

In Figure 38.9 we have represented the network R' dual to R in dashed lines. Then, in Figure 38.10, R is represented directly.

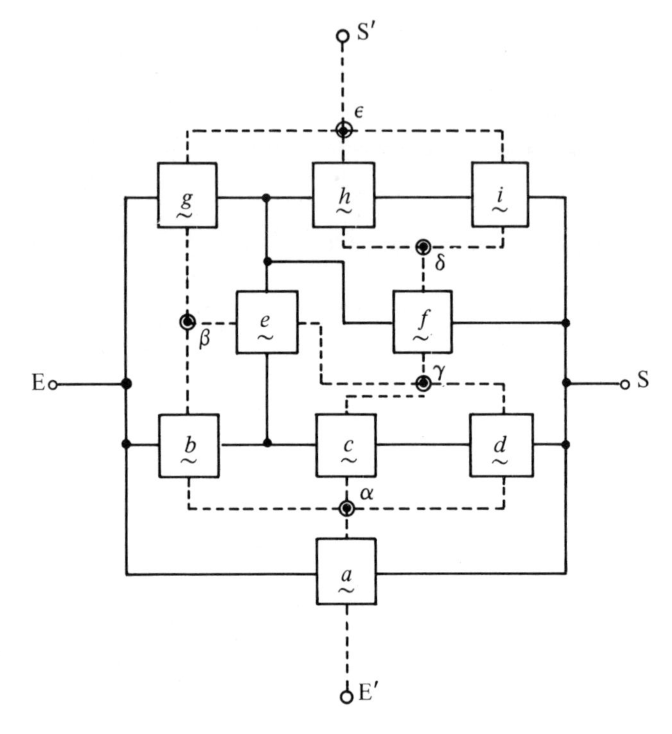

Fig. 38.9

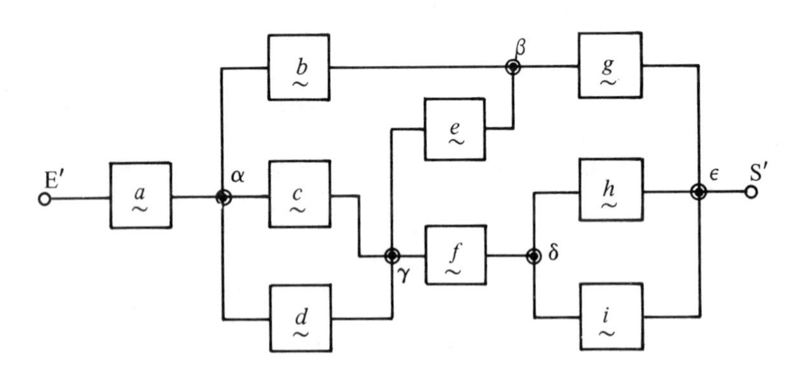

Fig. 38.10

A network and its dual have the easy to prove property:

(38.15) $$(R')' = R,$$

that is, the dual of the dual of a network is the network.

Method of antiroutes. Consider a planar† network R and its dual R' The routes corresponding to R are called antiroutes of R.

The maximal simple routes of R' will give the maximal simple antiroutes of R and the latter will give the polynomial form in R', of the function $\underset{\sim}{f}(\underset{\sim}{a}, \underset{\sim}{b}, \underset{\sim}{c})$ represented by the network R. With this polynomial form in R will be associated a parallel–series network equivalent to the given network.

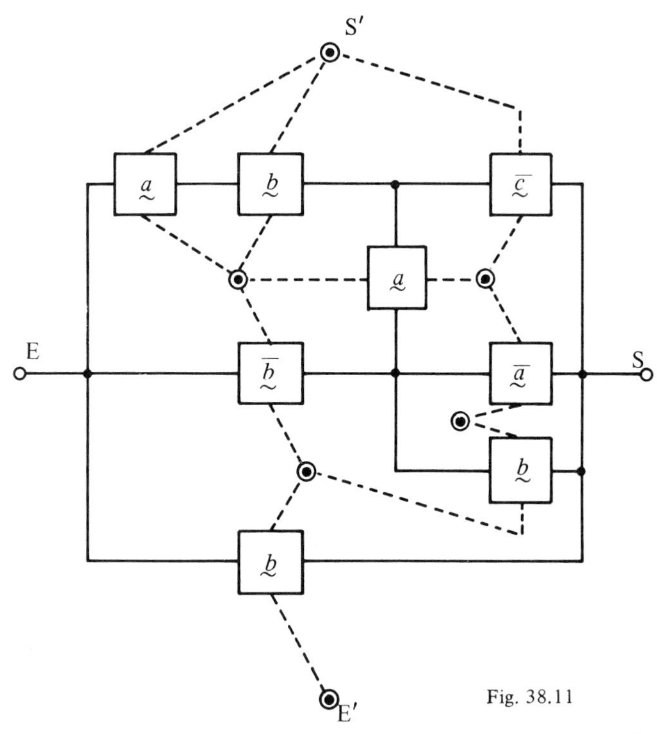

Fig. 38.11

Example. Consider the network R of Figure 38.11 whose dual R' is reproduced in Figure 38.12.

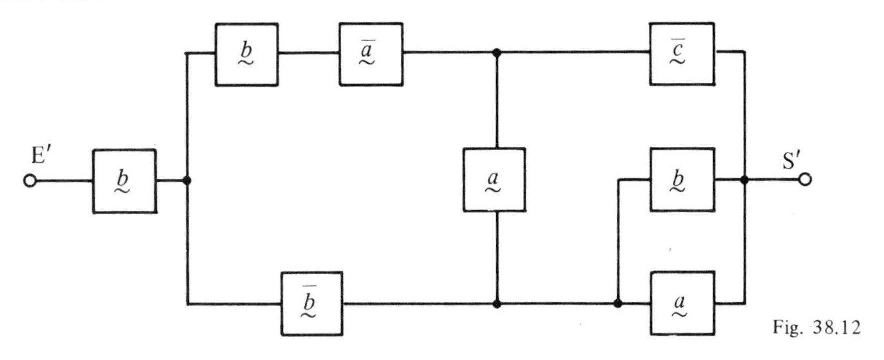

Fig. 38.12

†One may imagine the application of a similar method to the case of nonplanar circuits, but at the price of very great difficulties that would deflect our interest.

The routes of R' are the antiroutes of R. The set of routes of R', (that is, that of the antiroutes of R) is

(38.16) $$\{(\underset{\sim}{b}, \underset{\sim}{b}, \underset{\sim}{\bar{a}}, \underset{\sim}{\bar{c}}) , (\underset{\sim}{b}, \underset{\sim}{b}, \underset{\sim}{\bar{a}}, \underset{\sim}{a}, \underset{\sim}{b}) , (\underset{\sim}{b}, \underset{\sim}{b}, \underset{\sim}{\bar{a}}, \underset{\sim}{a}, \underset{\sim}{a}) , (\underset{\sim}{b}, \underset{\sim}{\bar{b}}, \underset{\sim}{a}, \underset{\sim}{\bar{c}}) , (\underset{\sim}{b}, \underset{\sim}{\bar{b}}, \underset{\sim}{b}) , (\underset{\sim}{b}, \underset{\sim}{\bar{b}}, \underset{\sim}{a})\}$$

The set† of simple antiroutes is

(38.17)

$$\{(\underset{\sim}{b}, \underset{\sim}{\bar{a}}, \underset{\sim}{\bar{c}}) , (\underset{\sim}{b}, \underset{\sim}{\bar{a}}, \underset{\sim}{a})_1 , (\underset{\sim}{b}, \underset{\sim}{\bar{a}}, \underset{\sim}{a})_2 , (\underset{\sim}{b}, \underset{\sim}{\bar{b}}, \underset{\sim}{a}, \underset{\sim}{\bar{c}}) , (\underset{\sim}{b}, \underset{\sim}{\bar{b}}) , (\underset{\sim}{b}, \underset{\sim}{\bar{b}}, \underset{\sim}{a})\}$$

and that of the minimal simple antiroutes reduces to

(38.18) $$\{(\underset{\sim}{\bar{a}}, \underset{\sim}{b}, \underset{\sim}{\bar{c}}) , (\underset{\sim}{a}, \underset{\sim}{\bar{a}}, \underset{\sim}{b}) , (\underset{\sim}{b}, \underset{\sim}{\bar{b}})\}.$$

Thus, the reduced form in $\wedge$ corresponding to the parallel–series network of Figure 38.11 is

(38.19) $$\underset{\sim}{f}(\underset{\sim}{a}, \underset{\sim}{b}, \underset{\sim}{c}) = (\underset{\sim}{\bar{a}} \vee \underset{\sim}{b} \vee \underset{\sim}{\bar{c}}) \wedge (\underset{\sim}{a} \vee \underset{\sim}{\bar{a}} \vee \underset{\sim}{b}) \wedge (\underset{\sim}{b} \vee \underset{\sim}{\bar{b}})$$

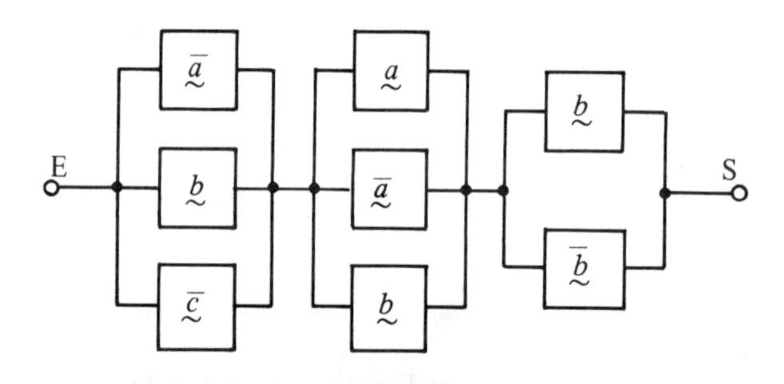

Fig. 38.13

By the method of routes one may find the polynomial form in $\vee$:

(38.20) $$\underset{\sim}{f}(\underset{\sim}{a}, \underset{\sim}{b}, \underset{\sim}{c}) = (\underset{\sim}{a} \wedge \underset{\sim}{\bar{b}} \wedge \underset{\sim}{\bar{c}}) \vee (\underset{\sim}{\bar{a}} \wedge \underset{\sim}{b}) \vee (\underset{\sim}{b}) ,$$

whose corresponding series–parallel network is shown in Figure 38.14. By carrying out the appropriate developments one may verify that (38.19) and (38.20) indeed represent the same function.

†The route may sometimes present itself with the same symbols, as $(b, a, a)_1$ and $(b, a, a)_2$; but they are distinct routes. The distinction disappears when one passes to maximal simple routes.

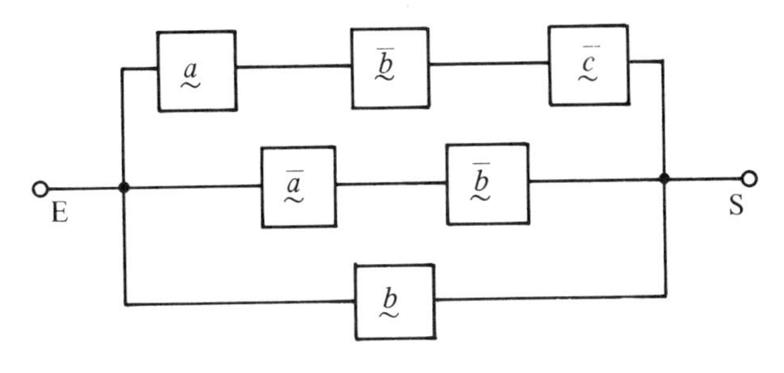

Fig. 38.14

Remark. We know that† any network of circuits may be duplicated with various technologies (diodes, bridges, transistors, integrated circuits, etc.). Thus, with the appropriate correspondences, what has been treated theoretically here may be duplicated in the case of more diverse technologies permitting the realization of fuzzy logics. But these fuzzy logics are actually very costly one fears (it is necessary to control the potentials finely, which is not the case for binary technologies). But the present remarks and restrictions certainly ought not to remain valid for long.

39. FUZZY PROPOSITIONS AND THEIR FUNCTIONAL REPRESENTATION

Fuzzy logic does not rest on truth tables as does formal logic, but upon operations realized on fuzzy subsets.

We begin with a comparative example based on the tale, "Little Red Riding Hood."‡ We consider two formal propositions for which one must verify a posteriori (after reading the story) whether they are true or false:

$\mathcal{P}_1$: the wolf is dressed in the guise of a grandmother

$\mathcal{P}_2$: the wolf has eaten the little girl.

The proposition $\mathcal{P}_1 \Delta \mathcal{P}_2$ will mean*: "the wolf is dressed as a grandmother and has eaten the little girl." In order that this be true, it is proper that the two statements be true; if only one or neither is true, this would not be coherent with the story of Little Red Riding Hood. Thus we have the truth table

†See, for example, reference [2F].

‡Pardon this rather naive example, which constitutes a very elementary didactic explication. Fuzzy logic will be considered as an application in the following volume.

*On the subject of the use of the symbols Δ and ∇, we refer to the footnote on p. 214 at the beginning of Section 35.

$\mathcal{P}_1$	$\mathcal{P}_2$	$\mathcal{P}_1 \Delta \mathcal{P}_2$
true	true	true
true	false	false
false	true	false
false	false	false

Fig. 39.1

But we now present the two logical statements in another fashion. There exists a set of animals

(39.1) $$\mathbf{E} = \{\text{cat, dog, wolf, fox, goat, rat, rabbit}\}.$$

Consider $\mathbf{A} \subset \mathbf{E}$, the formal subset of animals apt to dress as a grandmother:

(39.2) $$\mathbf{A} = \{(\text{cat} \mid 0{,}1), (\text{dog} \mid 1), (\text{wolf} \mid 1), (\text{fox} \mid 0{,}5), (\text{goat} \mid 1), (\text{rat} \mid 0), (\text{rabbit} \mid 0)\}.$$

that is,

(39.3) $$\mathbf{A} = \{\text{wolf}\}.$$

Consider $\mathbf{B} \subset \mathbf{E}$, the formal subset of animals likely to eat the little girl:

(39.4) $$\mathbf{B} = \{(\text{cat} \mid 0), (\text{dog} \mid 0), (\text{wolf} \mid 1), (\text{fox} \mid 0), (\text{goat} \mid 0), (\text{rat} \mid 0), (\text{rabbit} \mid 0)\}.$$

that is,

(39.5) $$\mathbf{B} \cap \{\text{wolf}\}.$$

The formal subset of animals prone to dressing as a grandmother and eating little girls is

(39.6) $$\mathbf{A} \cap \mathbf{B} = \{\text{wolf}\}.$$

Thus, through such a procedure we have verified that the wolf is indeed the cunning and cruel animal described in this celebrated story.

We consider now two statements from the fuzzy tale of Little Red Riding Hood. There exists a set of animals:

(39.7) $$\mathbf{E} = \{\text{cat, dog, wolf, fox, goat, rat, rabbit}\}$$

Consider $\mathbf{A} \subset \mathbf{E}$, the fuzzy subset of animals apt to dress as a grandmother:

(39.8) $$\mathbf{A} = \{(\text{cat} \mid 0{,}1), (\text{dog} \mid 1), (\text{wolf} \mid 1), (\text{fox} \mid 0{,}5), (\text{goat} \mid 1), (\text{rat} \mid 0), (\text{rabbit} \mid 0)\}.$$

Consider $\mathbf{B} \subset \mathbf{E}$, the fuzzy subset of animals likely to eat a little girl:

(39.9) $$\mathbf{B} = \{(\text{cat} \mid 0{,}1), (\text{dog} \mid 0{,}4), (\text{wolf} \mid 1), (\text{fox} \mid 0{,}7), (\text{goat} \mid 0), (\text{rat} \mid 0), (\text{rabbit} \mid 0)\}.$$

Then, the fuzzy subset of animals apt to dress as a grandmother and eat a little girl will be

(39.10) $\mathbf{A} \cap \mathbf{B} = \{(\text{cat} \mid 0{,}1), (\text{dog} \mid 0{,}4), (\text{wolf} \mid 1), (\text{fox} \mid 0{,}5), (\text{goat} \mid 0), (\text{rat} \mid 0), (\text{rabbit} \mid 0)\}.$

The tale may refer to the wolf, but also to a fox, a dog, or a cat.

The statements of the fuzzy logic, as the statements of the formal logic, are associated explicitly or implicitly to set theory, fuzzy for the former and formal for the latter.

With the operations $\cap$, $\cup$, and $-$ (intersection, union, and complementation) one associates in the formal logic the connectives Δ, ∇, and $\urcorner$ (conjunction *and*, disjunction *or/and*, negation *not*).

Passage to the fuzzy connectives Δ, ∇, and $\urcorner$ of the corresponding fuzzy logic does not present any difficulties since we have already defined the corresponding set operations in Section 5.

But it is necessary to give special attention to the other connectives:

implication

metaimplication

logical equivalence

We now go on to review these questions, first in formal logic then in fuzzy logic.

Consider two formal propositions $\mathcal{P}$ and $\mathcal{Q}$. The compound proposition "$\mathcal{P}$ implies $\mathcal{Q}$," denoted $\mathcal{P} > \mathcal{Q}$, corrcsponds to the truth table in Figure 39.2.

$\mathcal{P}$	$\mathcal{Q}$	$\mathcal{P} > \mathcal{Q}$
false	false	true
false	true	true
true	false	false
true	true	true

Fig. 39.2

To this compound proposition corresponds, for the subset **A** associated with $\mathcal{P}$ and the subset **B** associated with $\mathcal{Q}$, the set operation $\overline{\mathbf{A}} \cup \mathbf{B}$.

Now we consider the compound proposition "$\mathcal{P}$ metaimplies $\mathcal{Q}$," denoted $\mathcal{P} \Rightarrow \mathcal{Q}$. To this metaimplication one gives the following sense: when $\mathcal{P}$ is true, $\mathcal{Q}$ is always true (the syllogism rule is happily recovered here), but one may affirm nothing when $\mathcal{P}$ is false, $\mathcal{Q}$ may be just as well be true as false. Thus, a statement like "if the sea is made of sweet cider, I will change myself into a siren" is correct, the sea being, alas, evil to drink and certainly not made of sweet cider. The connection $\Rightarrow$ therefore reduces to: if $\mathcal{P} \Rightarrow \mathcal{Q}$, it is necessary that $\mathcal{P}$ be true only when $\mathcal{Q}$ is also.

One must guard therefore against confusing $\mathcal{P} > \mathcal{Q}$ and $\mathcal{P} \Rightarrow \mathcal{Q}$. The first is an operation of logic

(39.11) $\mathcal{P} > \mathcal{Q} = \overline{\mathcal{P}} \nabla \mathcal{Q}$ (in one notation)

$= (\urcorner \mathcal{P}) \nabla (\mathcal{Q})$ (in another notation)

The second is a metalogical operation that may not be brought to (39.11). But the habit has been taken of calling metaimplication implication and thus confusing the two. The compound proposition $\mathcal{P} > \mathcal{Q}$ does not introduce a cause and effect relation, nor a proof of $\mathcal{Q}$ with respect to $\mathcal{P}$, contrary to that which holds for $\mathcal{P} \Rightarrow \mathcal{Q}$.

One may present the false paradox introduced by $\mathcal{P} > \mathcal{Q}$ in the following terms: since the propositions $\mathcal{P}$ and $\mathcal{Q}$ have not been analyzed, since they occur only through their contents, since the only given accessible is the logical value of each, $\mathcal{P} > \mathcal{Q}$ may not introduce a relation of cause and effect. But if one knows a priori that $\mathcal{P}$ is true and that $\mathcal{P} > \mathcal{Q}$ is true, then one may conclude that $\mathcal{Q}$ is true.

We present an example cited in reference [3K]. Let $\mathcal{P}$ and $\mathcal{Q}$ be the following propositions, which we shall examine considering the table of Figure 39.2:

$\mathcal{P}$: Napoléon died at Saint-Hélène (true)
$\mathcal{Q}$: Vercingétorix wore a moustache (one is not sure)
$\mathcal{P} > \mathcal{Q}$ is true if $\mathcal{Q}$ is true
$\mathcal{P}$: Two and two are five (false)
$\mathcal{Q}$: 12 is a prime number (false)
$\mathcal{P} > \mathcal{Q}$ is true
$\mathcal{P}$: The moon is made of gruyère cheese (false)
$\mathcal{Q}$: 17 is prime (true)
$\mathcal{P} > \mathcal{Q}$ is true
$\mathcal{P}$: 17 is prime (true)
$\mathcal{Q}$: 16 is prime (false)
$\mathcal{P} > \mathcal{Q}$ is false

Logical equivalence is less ambiguous. This will be defined by the truth table in Figure 39.3.

$\mathcal{P}$	$\mathcal{Q}$	$\mathcal{P} \equiv \mathcal{Q}$
false	false	true
false	true	false
true	false	false
true	true	true

Fig. 39.3

Like implication, logical equivalence does not bring the contents of the two propositions into a causal relationship.

To this compound proposition corresponds, for the subset **A** associated with $\mathcal{P}$ and the subset **B** associated with $\mathcal{Q}$, the set operation $(\bar{\mathbf{A}} \cup \mathbf{B}) \cap (\mathbf{A} \cup \bar{\mathbf{B}})$.

Metaequivalence carries the same name usually: $\mathcal{P}$ is equivalent to $\mathcal{Q}$, denoted $\mathcal{P} \Rightarrow \mathcal{Q}$, that is, $\mathcal{P}$ metaimplies $\mathcal{Q}$ and $\mathcal{Q}$ metaimplies $\mathcal{P}$, because the symmetry leads to a truth table identical to that of equivalence $\mathcal{P} \equiv \mathcal{Q}$. This is why one may confuse these without ambiguity.

Fuzzy propositions of the types fuzzy implication and fuzzy equivalence will be defined respectively with reference to the operations $\overline{\underset{\sim}{\mathbf{A}}} \cup \underset{\sim}{\mathbf{B}}$ and $(\underset{\sim}{\mathbf{A}} \cup \overline{\underset{\sim}{\mathbf{B}}}) \cap (\overline{\underset{\sim}{\mathbf{A}}} \cup \underset{\sim}{\mathbf{B}})$. We insist on the fact that one must, as for intersection, union, and negation, pass through the reference set and the associated membership set.

In order to define a metaimplication in fuzzy logic we shall use the notion of a binary relation. A correspondence such as that represented in Figures 39.4 and 39.5 gives an example where $x_i \in \mathbf{E}_1, y_j \in \mathbf{E}_2$. One sees evident here:

(39.12)

$$\begin{array}{ll} \text{if } x = x_1 & \text{then } y = y_2, \\ \text{if } x = x_2 & \text{then } y = y_6, \\ \text{if } x = x_3 & \text{then } y = y_1, \\ \quad \cdots\cdot & \quad \cdots\cdot \\ \text{if } x = x_7 & \text{then } y = y_3. \end{array}$$

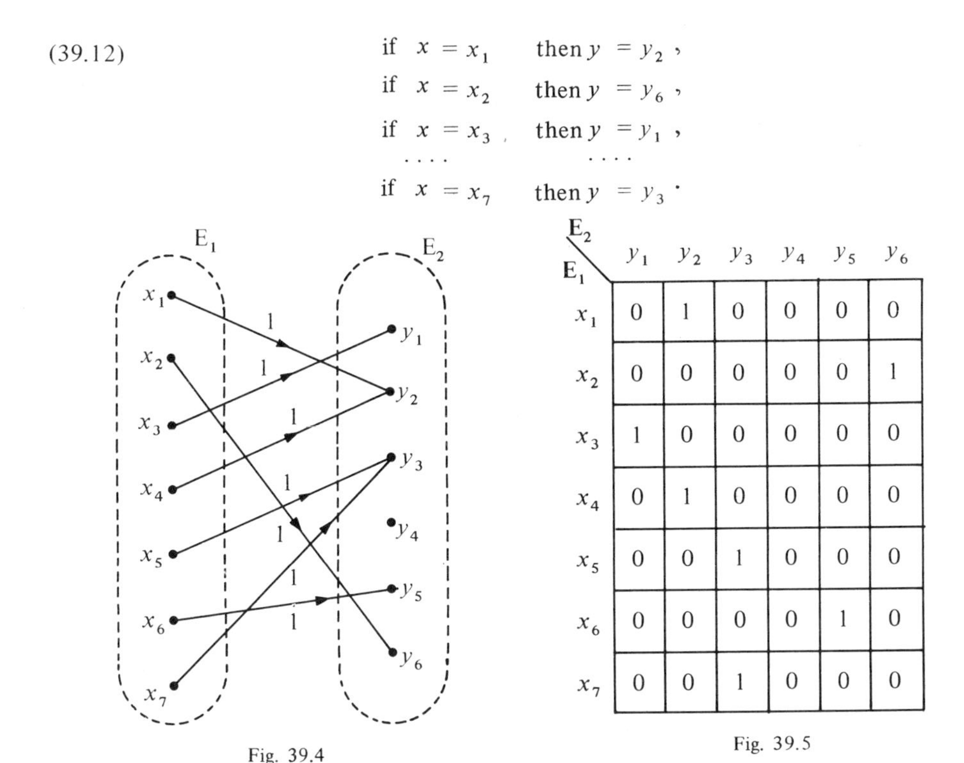

Fig. 39.4

E_2 \ E_1	y_1	y_2	y_3	y_4	y_5	y_6
x_1	0	1	0	0	0	0
x_2	0	0	0	0	0	1
x_3	1	0	0	0	0	0
x_4	0	1	0	0	0	0
x_5	0	0	1	0	0	0
x_6	0	0	0	0	1	0
x_7	0	0	1	0	0	0

Fig. 39.5

Figure 39.6, on the other hand, will correspond to an element of $\mathbf{E}_1$ a fuzzy subset of $\mathbf{E}_2$:

(39.12)

$$\begin{array}{lll} \text{if } x = x_1 & \text{then } \underset{\sim}{\mathbf{B}} & = \{(y_1 \mid 0{,}8), (y_2 \mid 1), (y_3 \mid 0{,}3), (y_4 \mid 1), (y_5 \mid 0{,}9), (y_6 \mid 0{,}9)\}, \\ \text{if } x = x_2 & \text{then } \underset{\sim}{\mathbf{B}} & = \{(y_1 \mid 0{,}2), (y_2 \mid 0{,}9), (y_3 \mid 1), (y_4 \mid 0), (y_5 \mid 0{,}6), (y_6 \mid 1)\}, \\ \text{if } x = x_3 & \text{then } \underset{\sim}{\mathbf{B}} & = \{(y_1 \mid 0{,}3), (y_2 \mid 0{,}8), (y_3 \mid 0{,}9), (y_4 \mid 1), (y_5 \mid 0{,}8), (y_6 \mid 0)\}, \\ \quad \cdots\cdot & \quad \cdots\cdot & \\ \text{if } x = x_7 & \text{then } \underset{\sim}{\mathbf{B}} & = \{(y_1 \mid 0{,}1), (y_2 \mid 1), (y_3 \mid 0), (y_4 \mid 0{,}9), (y_5 \mid 0{,}3), (y_6 \mid 1)\}. \end{array}$$

But in Section 15 we have defined the possibility of a correspondence between fuzzy subsets where $\underset{\sim}{\mathbf{A}} \subset \mathbf{E}_1$ and $\underset{\sim}{\mathbf{B}} \subset \mathbf{E}_2$; this was done with the aid of the notion of a conditioned fuzzy subset. The relation giving the fuzzy subset $\underset{\sim}{\mathbf{B}}$ corresponding to the fuzzy subset $\underset{\sim}{\mathbf{A}}$ is then

$E_1 \backslash E_2$	y_1	y_2	y_3	y_4	y_5	y_6
x_1	0,8	1	0,3	1	0,9	0,9
x_2	0,2	0,9	1	0	0,6	1
x_3	0,3	0,8	0,9	1	0,8	0
x_4	0,5	0	1	1	0,8	0,9
x_5	1	0,2	0,9	0,6	0	0,5
x_6	0,6	0,8	1	1	0,8	1
x_7	0,1	1	0	0,9	0,3	1

Fig. 39.6

(39.13) $$(\mu_{\underset{\sim}{B}}(y) = \underset{x \in E_1}{\text{MAX}} \; \text{MIN} \; (\mu_{\underset{\sim}{B}}(y \,\|\, x) \, , \, \mu_{\underset{\sim}{A}}(x)) \, .$$

We have given an example in Section 15 [see (15.3)–(15.11)] ; here we take up another example using the fuzzy relation of Figure 39.6.

Suppose that

(39.14) $$\underset{\sim}{A} = \{(x_1 \,|\, 0{,}2) \, , (x_2 \,|\, 0{,}3) \, , (x_3 \,|\, 0{,}5) \, , (x_4 \,|\, 1) \, , (x_5 \,|\, 0) \, , (x_6 \,|\, 0) \, , (x_7 \,|\, 0{,}8)\}$$

One sees successively

(39.15)

$$\mu_{\underset{\sim}{B}}(y_1) = \text{MAX} \, [\text{MIN} \, (0{,}8 \; ; \; 0{,}2) \, , \text{MIN} \, (0{,}2 \; ; \; 0{,}3) \, , \text{MIN} \, (0{,}3 \; ; \; 0{,}5) \, , \text{MIN} \, (0{,}5 \; ; \; 1) \, ,$$
$$\text{MIN} \, (1 \; ; \; 0) \, , \text{MIN} \, (0{,}6 \; ; \; 0) \, , \text{MIN}(0{,}1 \; ; \; 0{,}8)]$$
$$= \text{MAX} \, [0{,}2 \; ; \; 0{,}2 \; ; \; 0{,}3 \; ; \; 0{,}5 \; ; \; 0 \; ; \; 0 \; ; \; 0{,}1] = 0{,}5 \, ,$$

(39.16)

$$\mu_{\underset{\sim}{B}}(y_2) = \text{MAX} \, [\text{MIN} \, (1 \; ; \; 0{,}2) \, , \text{MIN} \, (0{,}9 \; ; \; 0{,}3) \; ; \; \text{MIN} \, (0{,}8 \; ; \; 0{,}5) \, , \text{MIN} \, (0 \; ; \; 1) \, ,$$
$$\text{MIN} \, (0{,}2 \; ; \; 0) \, , \text{MIN} \, (0{,}8 \; ; \; 0) \, , \text{MIN} \, (1 \; ; \; 0{,}8)$$
$$= \text{MAX} \, [0{,}2 \; ; \; 0{,}3 \; ; \; 0{,}5 \; ; \; 0; 0; 0; 0{,}8] = 0{,}8;$$

and in the same manner

(39.17)

$$\mu_{\underset{\sim}{B}}(y_3) = 1 \quad , \quad \mu_{\underset{\sim}{B}}(y_4) = 1 \quad , \quad \mu_{\underset{\sim}{B}}(y_5) = 0{,}8 \quad , \quad \mu_{\underset{\sim}{B}}(y_6) = 0{,}9 \, .$$

The calculations have been presented in Figure 39.7, where the operation * corresponds to max–min.

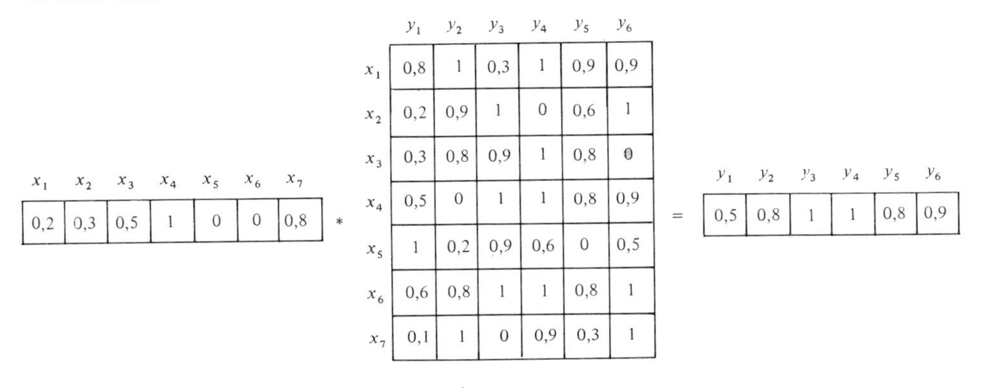

x_1	x_2	x_3	x_4	x_5	x_6	x_7
0,2	0,3	0,5	1	0	0	0,8

*

	y_1	y_2	y_3	y_4	y_5	y_6
x_1	0,8	1	0,3	1	0,9	0,9
x_2	0,2	0,9	1	0	0,6	1
x_3	0,3	0,8	0,9	1	0,8	0
x_4	0,5	0	1	1	0,8	0,9
x_5	1	0,2	0,9	0,6	0	0,5
x_6	0,6	0,8	1	1	0,8	1
x_7	0,1	1	0	0,9	0,3	1

=

y_1	y_2	y_3	y_4	y_5	y_6
0,5	0,8	1	1	0,8	0,9

Fig. 39.7

One therefore has: if

(39.18)

$$\underset{\sim}{A} = \{(x_1 | 0{,}2), (x_2 | 0{,}3), (x_3 | 0{,}5), (x_4 | 1), (x_5 | 0), (x_6 | 0), (x_7 | 0{,}8)\},$$

(39.19) then

$$\underset{\sim}{B} = \{(y_1 | 0{,}5), (y_2 | 0{,}8), (y_3 | 1), (y_4 | 1), (y_5 | 0{,}8), y_6 | 0{,}9)\}.$$

In this fashion we show that considering an if–then proposition corresponds well to what is used in formal relations.

Let

(39.20) $A = \{(x_1 | 0), (x_2 | 0), (x_3 | 0), (x_4 | 1), (x_5 | 0), (x_6 | 0), (x_7 | 0)\}$

that is

(39.21) $$A = \{x_4\}.$$

Referring to the correspondence given in Figure 39.5 and using (39.13) again, one finds

(39.22) $B = \{(y_1 | 0), (y_2 | 1), (y_3 | 0), (y_4 | 0), (y_5 | 0), (y_6 | 0)\}$,

that is,

(39.23) $$B = \{y_2\}$$

which may be stated

(39.24) $$\text{if} \quad A = \{x_4\}, \quad \text{then} \quad B = \{y_2\}$$

or again

(39.25) $$\text{if} \quad x = x_4, \quad \text{then} \quad y = y_2 .$$

One indeed recovers the if–then proposition such as that defined in (39.12).

We review all the propositions stated thus far:

fuzzy conjunction (fuzzy *and*): defined by $\underset{\sim}{\mathbf{A}} \cap \underset{\sim}{\mathbf{B}}$,

fuzzy disjunction (fuzzy *or*): defined by $\underset{\sim}{\mathbf{A}} \cup \underset{\sim}{\mathbf{B}}$,

fuzzy negation (fuzzy *not*): defined by $\overline{\underset{\sim}{\mathbf{A}}}$,

fuzzy implication: defined by $\overline{\underset{\sim}{\mathbf{A}}} \cup \underset{\sim}{\mathbf{B}}$,

fuzzy equivalence: defined by $(\underset{\sim}{\mathbf{A}} \cup \overline{\underset{\sim}{\mathbf{B}}}) \cap (\overline{\underset{\sim}{\mathbf{A}}} \cup \underset{\sim}{\mathbf{B}})$,

fuzzy *if–then*: defined by $\mu_{\underset{\sim}{\mathbf{B}}}(y) = \underset{x}{\mathrm{MAX}}\ \mathrm{MIN}\ (\mu_{\underset{\sim}{\mathbf{B}}}(y \parallel x), \mu_{\underset{\sim}{\mathbf{A}}}(x))$

(fuzzy metaimplication)

This last was not a fuzzy logic, but rather a fuzzy metalogic, proposition,

In Volume II, devoted to applications of the theory of fuzzy subsets, various sections will reconsider these notions in detail and will give a number of developments.

40. THE THEORY OF FUZZY SUBSETS AND THE THEORY OF PROBABILITY

Many persons, without too much thought, state: Why be interested in the theory of fuzzy subsets? The theory of probability serves very well for all that. There are, in fact, several common aspects between the two theories; but these theories relate to some considerations that it is appropriate to distinguish. We proceed first to review the basics of the theory of probability and then examine that which joins and that which separates these theories.

Axiomatics of the theory of probability.

(1) *Case of a finite reference set.* Let **E** be a finite reference set, $\mathcal{P}(\mathbf{E})$ its finite power set, and $\boldsymbol{\Delta}$ a subset of $\mathcal{P}(\mathbf{E})$ necessarily containing **E**. The subset $\boldsymbol{\Delta}$ will be called a family and one will say that this family is probabilizable if the following two conditions are satisfied:

(40.1) a) $\forall\, \mathrm{A} \in \boldsymbol{\Delta} : \overline{\mathrm{A}} \in \boldsymbol{\Delta}$,

(40.2) b) $\forall\, \mathrm{A} \in \boldsymbol{\Delta}$ and $\forall\, \mathrm{B} \in \boldsymbol{\Delta} :\quad \mathrm{A} \cup \mathrm{B} \in \boldsymbol{\Delta}$.

For example, let

(40.3) $$\mathbf{E} = \{a, b, c, d\}.$$

and

(40.4) $$\boldsymbol{\Delta} = \{\phi, \{b\}, \{c\}, \{b, c\}, \{a, d\}, \{a, b, d\}, \{a, c, d\}, \mathbf{E}\}.$$

The family $\boldsymbol{\Delta}$ is probabilizable. One may easily verify that, for all the elements of (40.4), conditions (40.1) and (40.2) are satisfied.

Properties (40.1) and (40.2) imply several others as the reader may easily prove†:

(40.5) *c*) $\phi \in \mathbf{\Delta}$.

(40.6) *d*) $\forall$ **A** and $\forall$ **B** : $\mathbf{A} \cap \mathbf{B} \in \mathbf{\Delta}$.

e) $\forall$ **A** and $\forall$ **B** : $\mathbf{A} - \mathbf{B} = \mathbf{A} \cap \overline{\mathbf{B}} \in \mathbf{\Delta}$.

A probabilizable family $\mathbf{\Delta}$ constitutes a ring for the operations $\oplus$ (disjunctive sum) and $\cap$ (intersection). Indeed, one verifies:

f) $\forall\, \mathbf{A}, \mathbf{B}, \mathbf{C} \in \Delta$

(40.7) $(\mathbf{A} \oplus \mathbf{B}) \oplus \mathbf{C} = \mathbf{A} \oplus (\mathbf{B} \oplus \mathbf{C})$, associativity for $\oplus$;

(40.8) $\mathbf{A} \oplus \phi = \phi \oplus \mathbf{A} = \mathbf{A}$, ϕ is the identity for $\oplus$;

(40.9) $\mathbf{A} \oplus \mathbf{A} = \phi$, every **A** has an inverse (it is its own inverse);

(40.10) $\mathbf{A} \oplus \mathbf{B} = \mathbf{B} \oplus \mathbf{A}$, commutativity.

Thus we have a commutative group with respect to the operation $\oplus$. On the other hand,

(40.11) $(\mathbf{A} \cap \mathbf{B}) \cap \mathbf{C} = \mathbf{A} \cap (\mathbf{B} \cap \mathbf{C})$, associativity for $\cap$;

and finally

(40.12) $(\mathbf{A} \oplus \mathbf{B}) \cap \mathbf{C} = (\mathbf{A} \cap \mathbf{C}) \oplus (\mathbf{B} \cap \mathbf{C})$, distributivity on the left and right

(40.13) $\mathbf{C} \cap (\mathbf{A} \oplus \mathbf{B}) = (\mathbf{C} \cap \mathbf{A}) \oplus (\mathbf{C} \cap \mathbf{B})$, with respect to $\oplus$.

Thus $(\Delta, \oplus, \cap)$ makes up a ring structure.

Finally, any family Δ forms a distributive and complemented lattice, that is, a boolean lattice, in which the order relation is inclusion. Thus for the family Δ given by (40.4), one obtains the boolean lattice represented in Figure 40.1.

A subset $\mathbf{F} \subset \mathcal{P}(\mathbf{E})$ is called a *probability base over* **E** if with respect to **F**, using complementation and union [(40.1) and (40.2)], one may arrive at a probabilistic family $\Delta \subset \mathcal{P}(\mathbf{E})$. One also says that **F** is a generator of Δ; such a *generator* is not in general unique.

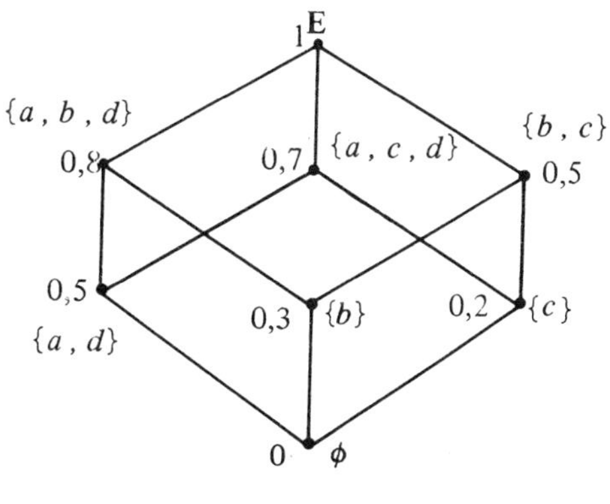

Fig. 40.1

†All of this is very classical. But, if necessary, the reader is referred to, for example, [8K].

For example, referring to (40.4) and Figure 40.1, one easily sees that

$$\mathbf{F} = \{\{a, d\}, \{b\}, \{c\}\}, \tag{40.14}$$

is a generator of (40.4).

(2) *Case of an infinite reference set (denumerable or not).* In this case $\mathcal{P}(\mathbf{E})$ is not denumerable; let Δ be a subset of $\mathcal{P}(\mathbf{E})$ necessarily containing $\mathbf{E}$. One will say that the family Δ is probabilizable if:

(40.15) g) $\forall\, \mathrm{A} \in \boldsymbol{\Delta} \;:\; \overline{\mathrm{A}} \in \boldsymbol{\Delta}$,

h) for any denumerable sequence, $\mathrm{A}_1, \mathrm{A}_2, \ldots, \mathrm{A}_n, \ldots$:

(40.16) $\mathrm{A}_1, \mathrm{A}_2, \ldots, \mathrm{A}_n, \ldots \in \boldsymbol{\Delta} \Rightarrow \mathrm{A}_1 \cup \mathrm{A}_2 \cup \ldots \cup \mathrm{A}_n \cup \ldots \in \boldsymbol{\Delta}$.

Equation (40.16) is purely and simply the extension of (40.2) to the case of reference sets denumerable or not.

Probability. Theoretical definition. Given a probabilizable family $\Delta \subset \mathcal{P}(\mathbf{E})$, a *probability* is a mapping of Δ into $\mathbf{R}^+$ having the following properties where the value taken by $\mathbf{X}$ in $\mathbf{R}^+$ is written $pr(\mathbf{X})$:

(40.17) *i*) $\forall\, \mathrm{A} \in \boldsymbol{\Delta} \;:\; pr(\mathrm{A}) \geqslant 0$,

(40.18) *j*) $\forall \mathrm{A} \in \boldsymbol{\Delta}$ and $\forall\, \mathrm{B} \in \boldsymbol{\Delta}$: $\mathrm{A} \cap \mathrm{B} = \phi \;\Rightarrow\; pr(\mathrm{A} \cup \mathrm{B}) = pr(\mathrm{A}) + pr(\mathrm{B})$,

(40.19) *k*) $pr(\mathrm{E}) = 1$.

Axioms (40.1), (40.2), (40.17)–(40.19), or (40.15)–(40.19) then allow one to attribute to each element of a family $\Delta \subset \mathcal{P}(\mathbf{E})$ a nonnegative number less than or equal to 1.

With respect to the five axioms† (a), (b), (i), (j), and (k) it is easy to prove a certain number of properties:

$$pr(\phi) = 0, \tag{40.20}$$

$$pr(\overline{\mathrm{A}}) = 1 + pr(\mathrm{A}), \tag{40.21}$$

$$pr(\mathrm{A}) + pr(\mathrm{B}) = pr(\mathrm{A} \cup \mathrm{B}) + pr(\mathrm{A} \cap \mathrm{B}), \tag{40.22}$$

$$\mathrm{B} \subset \mathrm{A} \;\Rightarrow\; pr(\mathrm{B}) \leqslant pr(\mathrm{A}). \tag{40.23}$$

Returning to the notion of a fuzzy subset, we insist on the following important point: "it is not sufficient to associate with a subset a number $p \in [0, 1]$ such that p is a probability; it is necessary that the subset and p satisfy the five fundamental axioms mentioned above."

Difference between the probability concept for fuzzy subsets and for ordinary subsets. We consider a very simple finite example. How does one proceed in the theory of fuzzy subsets?

†The axioms are sometimes called the axioms of Borel and Kolmogorov.

Let

(40.24) $$\mathrm{E} = \{a, b, c, d\}.$$

One defines a fuzzy subset by assigning to each element a value of the membership function; for example,

(40.25) $$\underset{\sim}{\mathrm{A}} = \{(a \mid 0{,}3), \ (b \mid 0{,}7), (c \mid 0), (d \mid 1)\}.$$

In probability theory one assigns the numbers $p \in [0, 1]$ to the ordinary subsets constituting a probabilizable family. Thus, letting $\mathbf{\Delta}$ be given by (40.4), one might have, for example,

(40.26)
$$pr(\phi) = 0, \ pr(\{b\}) = 0{,}3, \ pr(\{c\}) = 0{,}2, \ pr(\{a, d\}) = 0{,}5, \ pr(\{a, b, d\}) = 0{,}8,$$

$$pr(\{a, c, d\}) = 0{,}7, \ pr(\{b, c\}) = 0{,}5, \ pr(\mathrm{E}) = 1.$$

All these probabilities evidently satisfy (40.17)–(40.23).

As one sees here the two considerations are quite distinct, and one may conceive (and this is useful) of the assignment of probabilities to fuzzy subsets by taking each fuzzy subset belonging to a reference set formed by elements that are fuzzy subsets of another reference set. For example, assign a probability to $\underset{\sim}{\mathrm{A}}$ given by (40.25) and write

(40.27) $$pr(\underset{\sim}{\mathrm{A}}) = 0{,}6.$$

One may imagine a probability theory of fuzzy events. But, one must evidently distinguish between the two theories, that of fuzzy subsets and that of the probabilization of ordinary subsets.

The theory of fuzzy subsets is related to the theory of a vector lattice,† and probability theory to the theory of boolean lattice.

41. THE THEORY OF FUZZY SUBSETS AND THE THEORY OF FUNCTIONS OF STRUCTURE

Some interesting connections may be seen to hold between the theory of fuzzy variables, such as we have defined in Section 32 and the following sections, and the theory of functions of structure, such as one considers in the study of reliability of systems. First we proceed to recall the basics of the theory of functions of structure.‡

Functions of structure. We shall consider variables $a, b, \ldots \in \{0, 1\}$. For these binary variables, we shall use exclusively the operations*:

†Which, as we shall go on to see in Chapter V, is connected to a still more general concept, that of an ordered structure, or that of a preordered structure.

‡The notion of a function of structure that we review here plays an essential role in the theory of reliability. See [7K].

*In the concept of function of structure, one does not take into account complementation: $\bar{a} = 1 - a$.

(41.1) $a \, . \, b \, ,$ ordinary multiplication

(41.2) $a \hat{+} b = a + b - a \, . \, b \, ,$ where + is ordinary addition and – ordinary subtraction.

With respect to these operations we shall define functions of these variables, functions in which one introduces only the operations $\hat{+}$ and $.$

We examine first general properties relative to the variables $a, b, \ldots \in \{0,1\}$ and to the operations . and $\hat{+}$:

(41.3) $a \, . \, b = b \, . \, a \, ,$

commutativity

(41.4) $a \hat{+} b = b \hat{+} a \, .$

(41.5) $a \, . \, (b \, . \, c) = (a \, . \, b) \, . \, c \, ,$

associativity

(41.6) $a \hat{+} (b \hat{+} c) = (a \hat{+} b) \hat{+} c \, .$

(41.7) $a \, . \, a = a \, ,$

idempotence

(41.8) $a \hat{+} a = a \, .$

(41.9) $a \, . \, (b \hat{+} c) = a \, . \, b \hat{+} a \, . \, c \, ,$

distributivity

(41.10) $a \hat{+} (b \, . \, c) = (a \hat{+} b) \, . \, (a \hat{+} c) \, .$

(41.11) $a \, . \, 0 = 0 \, ,$

(41.12) $a \hat{+} 0 = a \, ,$

(41.13) $a \, . \, 1 = a \, ,$

(41.14) $a \hat{+} 1 = 1 \, ,$

A function of structure of the variables $a, b, \ldots$ will be designated by

(41.15) $$\varphi(a, b, \ldots) \, .$$

Thus

(41.16) $$\varphi(a, b, c) = a \hat{+} ab \hat{+} bc \, ,$$

is a function of structure.

Recall two properties that permit simplification of functions of structure by absorption:

(41.17) $$a(a \hat{+} b) = a \, ,$$

(41.18) $$a \hat{+} ab = a \, ,$$

which are consequences of (41.3) and (41.14).

Any function $\varphi(a, b, \ldots)$ may be expressed in a polynomial form with respect to $\cdot$ or with respect to $\hat{+}$ by using a notion of maximal monomials.

Thus

$$\varphi(a, b, c, d) = a \hat{+} bc \hat{+} bd \tag{41.19}$$

is formed of three maximal monomials. On the other hand,

$$\varphi(a, b, c, d) = a \hat{+} b \hat{+} bd \hat{+} cd \tag{41.20}$$

may be reduced to

$$\varphi(a, b, c, d) = a \hat{+} b \hat{+} cd. \tag{41.21}$$

A polynomial form that contains only maximal monomials will be said to be reduced or canonical.

Two functions of structure will be said to be equal or identical when they amount to the same polynomial form in the product . or in the sum $\hat{+}$. Of course, any canonical form in . may be transformed into a canonical form in $\hat{+}$ and vice versa.

To any function of structure one may associate a network representation, where one corresponds a series arrangement with the operation . and a parallel arrangement with the operation $\hat{+}$.

Example. Consider the function of structure corresponding to the network of Figure 41.1:

$$\varphi(a, b, c, d) = ab \hat{+} acd \hat{+} b \hat{+} abd. \tag{41.22}$$

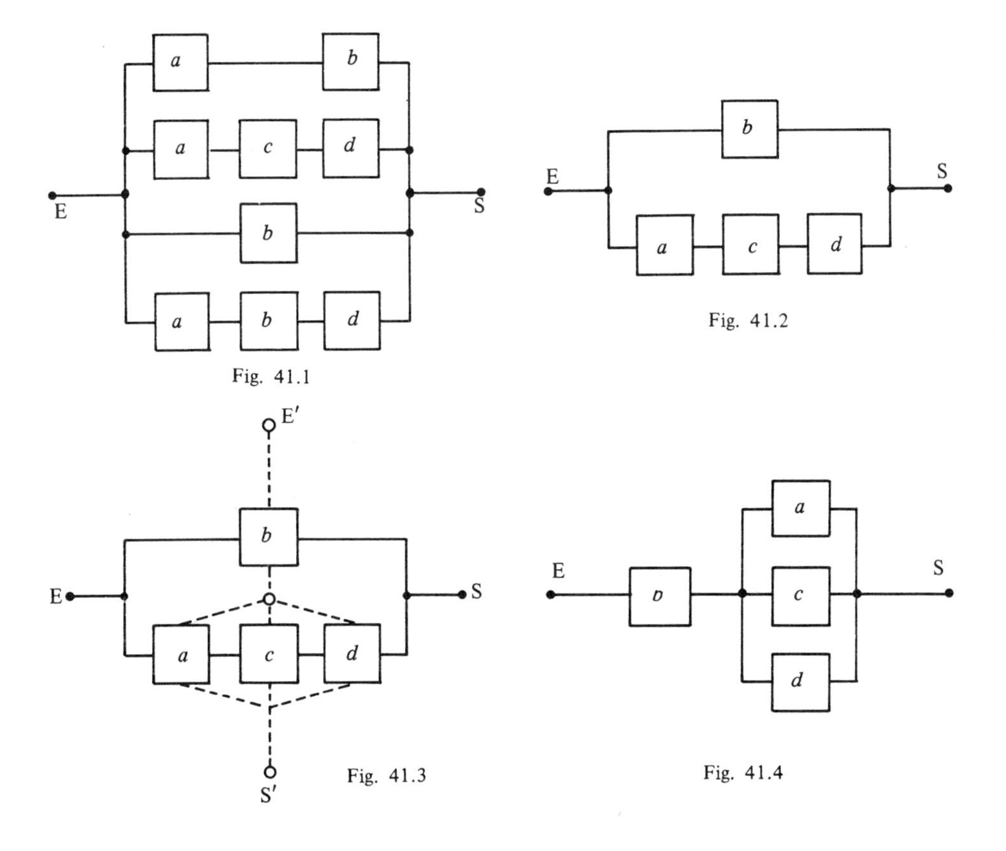

Fig. 41.1

Fig. 41.2

Fig. 41.3

Fig. 41.4

A first reduction gives

(41.23) $$\varphi(a, b, c, d) = ab \mathbin{\hat{+}} acd \mathbin{\hat{+}} b,$$

(41.24) since $$b \mathbin{\hat{+}} abd = b$$

A second reduction gives

(41.25) $$\varphi(a, b, c, d) = b \mathbin{\hat{+}} acd,$$

(41.26) since $$ab \mathbin{\hat{+}} b = b$$

(one could have carried out this second reduction at the same time as the first).

Thus, the canonical form is

(41.27) $$\varphi_1(a, b, c, d) = b \mathbin{\hat{+}} acd,$$

whose network is represented in Figure 41.2.

By taking the dual network in Figure 41.3 one obtains the dual canonical form through consideration of the three routes going from E to S in Figure 41.4:

(41.28) $$\varphi_1'(a, b, c, d) = ba \mathbin{\hat{+}} bc \mathbin{\hat{+}} bd,$$

then, by replacing . by $\hat{+}$, reciprocally,

(41.29) $$\varphi_2(a, b, c, d) = (a \mathbin{\hat{+}} b) . (b \mathbin{\hat{+}} c) . (b \mathbin{\hat{+}} d) .$$

The network corresponding to this second canonical form is presented in Figure 41.5.

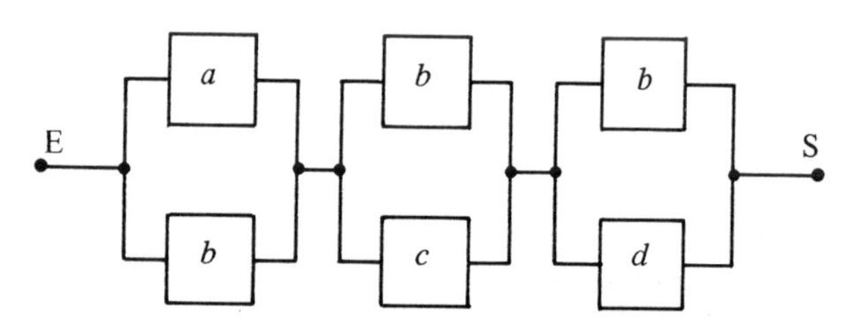

Fig. 41.5

Transition to probabilities. In order to avoid too abstract a presentation, we consider a concrete example.

Suppose that the variables $a, b, \ldots \in \{0, 1\}$ represent the states of the components A, B, . . . ∈ **E**, where **E** is the set of these components, a set that we shall designate by the words *complex equipment, system*, or simply *equipment*. One considers then that

(41.30)† X is functional if $x = 1$
X is not functional if $x = 0$,

where $x \in \{0, 1\}$ is the binary variable associated with the component X. In this case, $\varphi(a, b, \ldots)$ represents a binary function $\in \{0,1\}$ such that

† There is no Eq. (41.31)

(41.32) E is functional if $\varphi = 1$

E is not functional if $\varphi = 0$,

and $(a, b, \ldots)$ expresses the dependence of **E** on its components.

Let p_x be the probability that component X is functional and h_E the probability that equipment **E** is functional. We propose to calculate h_E as a function of $p_a, p_b, \ldots$.

In order to show how to carry out the corresponding calculations, we must recall the two idempotence formulas (41.7) and (41.8):

$$a \, . \, a = a , \tag{41.33}$$

$$a \hat{+} a = a . \tag{41.34}$$

But, if the calculation is carried out relative to ordinary addition, one would see

$$a + a = 2a ; \tag{41.35}$$

it being understood that this sum no longer is an element of $\{0, 1\}$.

In probability theory applied in the theory of reliability of systems considered here, we may write that, if p_x is the probability that X is functional, then $1 - p_x$ is the probability that it is not. Consider an equipment $\mathbf{E} = \{A, B\}$; one then sees, considering the probability of **E** functioning,

$$h_E = p_a . p_b \qquad \text{corresponding to} \quad \varphi(a, b) = ab , \tag{41.36}$$

$$1 - h_E = (1 - p_a).(1 - p_b) \qquad \text{corresponding to} \quad \varphi(a, b) = a \hat{+} b ; \tag{41.37}$$

this last expression may be written

$$h_E = 1 - (1 - p_a)(1 - p_b) \text{ corresponding to } \varphi(a, b) = a + b - ab \tag{41.38}$$

$$= p_a + p_b - p_a p_b .$$

We see that there exists an isomorphism† between the functions h_E and the functions φ. But, whereas one has the distributive property between $\hat{+}$ and . for $a, b, \ldots \in \{0,1\}$, distributivity no longer holds for these operations concerning $p_a, p_b, \ldots \in [0, 1]$. But distributivity is recovered if one considers the ordinary operations + and $\cdot$. In order to pass from φ functions to the functions h_E, one would reduce the operators $\hat{+}$ in φ into operators +, which leads also to use of $-$; then one may pass from φ to h_E by replacing the x's with p_x and by not neglecting to apply the idempotence property (41.7) whenever necessary.

Thus, we propose to establish the probability of functioning (also called the reliability) of a system whose function of structure is expressed by (41.22), that is a system **E** such that

E is functional	if	A and B are functional
E is functional	or/and	A, C, and D are functional

†For readers who are not familiar with this notion, we have represented a very detailed review in Section 57.

or/and B is functional

or/and A, B, and D are functional

One sees

(41.39) $\varphi(a, b, c, d) = ab \mathbin{\hat{+}} acd \mathbin{\hat{+}} b \mathbin{\hat{+}} abd = b \mathbin{\hat{+}} acd$

[see (41.27)] or again

(41.40) $$\varphi(a, b, c, d) = b + acd - abcd .$$

Whence

(41.41) $$h_E(p_a, p_b, p_c, p_d) = p_b + p_a p_c p_d - p_a p_b p_c p_d .$$

With respect to (41.29) one would have

(41.42)
$$\begin{aligned}\varphi(a, b, c, d) &= (a \mathbin{\hat{+}} b) . (b \mathbin{\hat{+}} c) . (b \mathbin{\hat{+}} d)\\ &= (ab \mathbin{\hat{+}} ac \mathbin{\hat{+}} b^2 \mathbin{\hat{+}} bc) . (b \mathbin{\hat{+}} d)\\ &= (ac \mathbin{\hat{+}} b) . (b \mathbin{\hat{+}} d)\\ &= abc \mathbin{\hat{+}} acd \mathbin{\hat{+}} b^2 \mathbin{\hat{+}} bd\\ &= b \mathbin{\hat{+}} acd .\end{aligned}$$

The general rule is well known among those interested in the reliability of systems: (1) Express φ with the aid of the operators +, −, and . . (2) Eliminate powers (using idempotence). (3) Replace the x's by p_x.

The operations $\hat{+}$ ***and . on fuzzy variables.*** We now consider the variables $\underset{\sim}{a}, \underset{\sim}{b}, \ldots \in [0, 1]$ and the following three operations:

(41.43) $\underset{\sim}{a} \cdot \underset{\sim}{b}$ ordinary multiplication;

(41.44) $\underset{\sim}{a} \mathbin{\hat{+}} \underset{\sim}{b} = \underset{\sim}{a} + \underset{\sim}{b} - \underset{\sim}{a} . \underset{\sim}{b}$ où (+) where + is ordinary addition and − is ordinary subtraction;

(41.45) $\bar{\underset{\sim}{a}} = 1 - \underset{\sim}{a}$ the complement of $\underset{\sim}{a}$

The following properties are then easy to verify:

(41.46) $\underset{\sim}{a} . \underset{\sim}{b} = \underset{\sim}{b} . \underset{\sim}{a}$

(41.47) $\underset{\sim}{a} \mathbin{\hat{+}} \underset{\sim}{b} = \underset{\sim}{b} \mathbin{\hat{+}} \underset{\sim}{a}$ } commutativity

(41.48) $\underset{\sim}{a} . (\underset{\sim}{b} . \underset{\sim}{c}) = (\underset{\sim}{a} . \underset{\sim}{b}) . \underset{\sim}{c}$

(41.49) $\underset{\sim}{a} \mathbin{\hat{+}} (\underset{\sim}{b} \mathbin{\hat{+}} \underset{\sim}{c}) = (\underset{\sim}{a} \mathbin{\hat{+}} \underset{\sim}{b}) \mathbin{\hat{+}} \underset{\sim}{c}$ } associativity

(41.50) $\underset{\sim}{a} . \underset{\sim}{a} \leqslant a ,$

(41.51) $\underset{\sim}{a} \mathbin{\hat{+}} \underset{\sim}{a} \geqslant a ,$

(41.52)† $$\underset{\sim}{a} \,.\, (\underset{\sim}{b} \,\hat{+}\, \underset{\sim}{c}) \leqslant \underset{\sim}{a} \,.\, \underset{\sim}{b} \,\hat{+}\, \underset{\sim}{a}\,\underset{\sim}{c} \,,$$

(41.53)‡ $$\underset{\sim}{a} \,\hat{+}\, \underset{\sim}{b}\,\underset{\sim}{c} \geqslant (\underset{\sim}{a} \,\hat{+}\, \underset{\sim}{b}) \,.\, (\underset{\sim}{a} \,\hat{+}\, \underset{\sim}{c}) \,,$$

(41.54) $$\underset{\sim}{a} \,.\, 0 = 0 \,,$$

(41.55) $$\underset{\sim}{a} \,\hat{+}\, 0 = \underset{\sim}{a} \,,$$

(41.56) $$\underset{\sim}{a} \,.\, 1 = \underset{\sim}{a} \,,$$

(41.57) $$\underset{\sim}{a} \,\hat{+}\, 1 = 1 \,,$$

(41.58) $$\overline{(\overline{\underset{\sim}{a}})} = \underset{\sim}{a} \,,$$

(41.59) $$\overline{\underset{\sim}{a} \,.\, \underset{\sim}{b}} = \overline{\underset{\sim}{a}} \,\hat{+}\, \overline{\underset{\sim}{b}} \,,$$

(41.60) $$\overline{\underset{\sim}{a} \,\hat{+}\, \underset{\sim}{b}} = \overline{\underset{\sim}{a}} \,.\, \overline{\underset{\sim}{b}} \,.$$

Theorems of De Morgan generalized for the operations $\hat{+}$, ., and $-$ on fuzzy variables.

Thus, the relations of idempotence [see (41.50) and (41.51)] and of distributivity [see (41.52) and (41.53)] are not satisfied.

One may sometimes consider the variables $\underset{\sim}{a}$, $\underset{\sim}{b}$, . . . as probabilities and allow that

$\underset{\sim}{a} \,.\, \underset{\sim}{b}$ represents the probability of occurrence of the formal independent events **A** and **B**,‡‡

$\underset{\sim}{a} \,\hat{+}\, \underset{\sim}{b}$ represents the probability of occurrence of the formal independent events **A** or/and **B**.

† One has
$$\underset{\sim}{a} \geqslant \underset{\sim}{a}^2 \,,$$
$$\underset{\sim}{a}\,\underset{\sim}{b}\,\underset{\sim}{c} \geqslant \underset{\sim}{a}^2\,\underset{\sim}{b}\,\underset{\sim}{c} \,,$$
$$-\underset{\sim}{a}\,\underset{\sim}{b}\,\underset{\sim}{c} \leqslant -\underset{\sim}{a}^2\,\underset{\sim}{b}\,\underset{\sim}{c} .$$
Adding $\underset{\sim}{a}\,\underset{\sim}{b} + \underset{\sim}{a}\,\underset{\sim}{c}$ to the two sides:
$$\underset{\sim}{a}\,\underset{\sim}{b} + \underset{\sim}{a}\,\underset{\sim}{c} - \underset{\sim}{a}\,\underset{\sim}{b}\,\underset{\sim}{c} \leqslant \underset{\sim}{a}\,\underset{\sim}{b} + \underset{\sim}{a}\,\underset{\sim}{c} - \underset{\sim}{a}^2\,\underset{\sim}{b}\,\underset{\sim}{c} \,,$$
thus
$$\underset{\sim}{a}\,(\underset{\sim}{b} \,\hat{+}\, \underset{\sim}{c}) \leqslant (\underset{\sim}{a} \,.\, \underset{\sim}{b}) \,\hat{+}\, (\underset{\sim}{a} \,.\, \underset{\sim}{c}) \,.$$

‡ One has
$$(\underset{\sim}{a}^2 - \underset{\sim}{a})\,(1 - \underset{\sim}{c})\,(1 - \underset{\sim}{b}) \leqslant 0 \,,$$
$$(\underset{\sim}{a}^2 - \underset{\sim}{a})\,(1 - \underset{\sim}{b} - \underset{\sim}{c} + \underset{\sim}{b}\,\underset{\sim}{c}) \leqslant 0 \,,$$
$$\underset{\sim}{a}^2 - \underset{\sim}{a} + \underset{\sim}{a}\,\underset{\sim}{c} - \underset{\sim}{a}^2\underset{\sim}{c} + \underset{\sim}{a}\,\underset{\sim}{b} - \underset{\sim}{a}^2\underset{\sim}{b} - \underset{\sim}{a}\,\underset{\sim}{b}\,\underset{\sim}{c} + \underset{\sim}{a}^2\underset{\sim}{b}\,\underset{\sim}{c} \leqslant 0 \,,$$
$$\underset{\sim}{a}^2 + \underset{\sim}{a}\,\underset{\sim}{c} - \underset{\sim}{a}^2\underset{\sim}{c} + \underset{\sim}{a}\,\underset{\sim}{b} - \underset{\sim}{a}^2\underset{\sim}{b} - \underset{\sim}{a}\,\underset{\sim}{b}\,\underset{\sim}{c} + \underset{\sim}{a}^2\underset{\sim}{b}\,\underset{\sim}{c} \leqslant \underset{\sim}{a} \,.$$
Adding $\underset{\sim}{b}\,\underset{\sim}{c} - \underset{\sim}{a}\,\underset{\sim}{b}\,\underset{\sim}{c}$ to the two sides:
$$\underset{\sim}{a}^2 + \underset{\sim}{a}\,\underset{\sim}{c} - \underset{\sim}{a}^2\underset{\sim}{c} + \underset{\sim}{a}\,\underset{\sim}{b} - \underset{\sim}{a}^2\underset{\sim}{b} - \underset{\sim}{a}\,\underset{\sim}{b}\,\underset{\sim}{c} + \underset{\sim}{a}^2\underset{\sim}{b}\,\underset{\sim}{c} + \underset{\sim}{b}\,\underset{\sim}{c} - \underset{\sim}{a}\,\underset{\sim}{b}\,\underset{\sim}{c} \leqslant \underset{\sim}{a} + \underset{\sim}{b}\,\underset{\sim}{c} - \underset{\sim}{a}\,\underset{\sim}{b}\,\underset{\sim}{c} .$$
$$(\underset{\sim}{a} + \underset{\sim}{b} - \underset{\sim}{a}\,\underset{\sim}{b}) \,.\, (\underset{\sim}{a} + \underset{\sim}{c} - \underset{\sim}{a}\,\underset{\sim}{c}) \leqslant \underset{\sim}{a} + \underset{\sim}{b}\,\underset{\sim}{c} - \underset{\sim}{a}\,\underset{\sim}{b}\,\underset{\sim}{c} \,,$$
thus:
$$(\underset{\sim}{a} \,\hat{+}\, \underset{\sim}{b}) \,.\, (\underset{\sim}{a} \,\hat{+}\, \underset{\sim}{c}) \leqslant \underset{\sim}{a} \,\hat{+}\, \underset{\sim}{b}\,\underset{\sim}{c} .$$

‡‡ One would have
$$a = pr(\mathbf{A}) \,,\; b = pr(\mathbf{B})$$
$$pr(\mathbf{A} \cup \mathbf{B}) = pr(\mathbf{A}) + pr(\mathbf{B}) - pr(\mathbf{A} \cap \mathbf{B})$$
with
$$pr(\mathbf{A} \cap \mathbf{B}) = pr(\mathbf{A}) \,.\, pr(\mathbf{B}) \,.$$
We have used the word formal attached to the notion of an event so that there will be no confusion with fuzzy subsets.

Thus, the operations defined in (41.43)–(41.50) may be applied to the calculation of probabilities.

At the same time that some have a tendency to confuse the theory of fuzzy subsets and probability theory, others will tend to include the notions of a function of fuzzy variables for the operations $\hat{+}$, ., $^-$ and of a function of structure for the operations $\hat{+}$ and . within the same theory. It is not only because one uses $\hat{+}$ and . in both theories that there is confusion.

The first theory concerns variables $\underset{\sim}{a}, \underset{\sim}{b}, \ldots \in [0, 1]$ for which one defines a complementation.† The second concerns variables $a, b, \ldots \in \{0,1\}$ for which one introduces no notion of complementation.

On the other hand, considering fuzzy variables $\underset{\sim}{a}, \underset{\sim}{b}, \ldots \in [0, 1]$ and the operations $\hat{+}$, ., and $^-$, one may take these as probabilities; whence there is a correspondence between the two theories if one considers only these three operations. But, if one considers fuzzy variables, $\underset{\sim}{a}, \underset{\sim}{b}, \ldots \in [0, 1]$ and the operations ($\vee$),($\wedge$), and $^-$, that is, max, min, and complement, the correspondence no longer exists.

Sometimes we may see a very interesting connection.

Notion of a performance function. In certain problems concerning functioning of systems one does not consider only whether the equipment fucntions or does not function, but that it operates at a certain performance level. For example:

functions perfectly

functions very satisfactorily

functions fairly well

functions rather poorly

does not function

Suppose, then, that to each component X_i of a system or set

$$E = \{X_1, X_2, \ldots, X_n\},$$

we attribute a fuzzy variable

$$\underset{\sim}{x}_i = \mu_{\underset{\sim}{A}}(X_i) \in [0,1],$$

where $\underset{\sim}{A}$ represents a state of the system **E** when one considers it functional with respect to each of its components. $\underset{\sim}{A}$ is indeed, in this case, a fuzzy subset.

If one admits that a level of the system is given by a function

$$\psi(\underset{\sim}{x}_1, \underset{\sim}{x}_2, \ldots, \underset{\sim}{x}_n) = \underset{\sim}{x}_1 \wedge \underset{\sim}{x}_2 \wedge \ldots \wedge \underset{\sim}{x}_n,$$

when functioning corresponds to a series network, and

$$\psi(\underset{\sim}{x}_1, \underset{\sim}{x}_2, \ldots, \underset{\sim}{x}_n) = \underset{\sim}{x}_1 \vee \underset{\sim}{x}_2 \vee \ldots \vee \underset{\sim}{x}_n,$$

when functioning corresponds to a parallel network, one recaptures various considerations from the theory of fuzzy subsets; but one must not consider complementation—this has no

†Or more precisely a pseudocomplementation of the type (5.17).

direct significance in such a theory of level of performance of a system.

The properties of the variables x_i, $i = 1, 2, \ldots, n$, are those that have been given in (32.12)–(32.23) with properties (32.24)–(32.26) no longer being significant.

We note that the properties of absorption

$$\underset{\sim}{a} \vee (\underset{\sim}{a} \wedge \underset{\sim}{b}) = \underset{\sim}{a},$$

$$\underset{\sim}{a} \wedge (\underset{\sim}{a} \vee \underset{\sim}{b}) = \underset{\sim}{a},$$

remain valid, which permits the simplification of formulas and the definition of monomials (in $\wedge$ or in $\wedge$) that may be maximal and of reduced polynomial forms.

Functions like ψ will be called *performance functions* We consider an example.

Example. See Figure 41.6. The function of structure is easily established

$$\psi(\underset{\sim}{a}, \underset{\sim}{b}, \underset{\sim}{c}) = [(\underset{\sim}{a} \wedge \underset{\sim}{b} \wedge \underset{\sim}{c}) \vee (\underset{\sim}{a} \wedge \underset{\sim}{c}) \vee \underset{\sim}{b}] \wedge \underset{\sim}{a}.$$

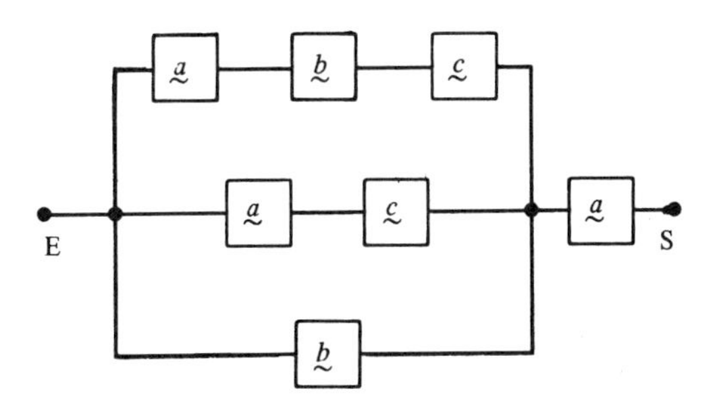

Fig. 41.6

By applying the rule of absorption to the polynomial within the square brackets, one has

$$\psi(\underset{\sim}{a}, \underset{\sim}{b}, \underset{\sim}{c}) = [(\underset{\sim}{a} \wedge \underset{\sim}{c}) \vee \underset{\sim}{b}] \wedge \underset{\sim}{a},$$

and by the properties of distributivity and idempotence

$$\psi(\underset{\sim}{a}, \underset{\sim}{b}, \underset{\sim}{c}) = (\underset{\sim}{a} \wedge \underset{\sim}{b}) \vee (\underset{\sim}{a} \wedge \underset{\sim}{c}).$$

One may study the function ψ with the aid of a table of values as we have done in Section 32 for the function ψ, but without introducing complements. One has the following cases:

one variable $\underset{\sim}{a}$: $\quad \underset{\sim}{a}$.

two variables $\underset{\sim}{a}$ and $\underset{\sim}{b}$: $\quad \underset{\sim}{a} \leqslant \underset{\sim}{b}$,

$\underset{\sim}{b} \leqslant \underset{\sim}{a}$.

three variables $\underset{\sim}{a}$, $\underset{\sim}{b}$, and $\underset{\sim}{c}$:

$$\underset{\sim}{a} \leqslant \underset{\sim}{b} \leqslant \underset{\sim}{c},$$
$$\underset{\sim}{a} \leqslant \underset{\sim}{c} \leqslant \underset{\sim}{b},$$
$$\underset{\sim}{b} \leqslant \underset{\sim}{a} \leqslant \underset{\sim}{c},$$
$$\underset{\sim}{b} \leqslant \underset{\sim}{c} \leqslant \underset{\sim}{a},$$
$$\underset{\sim}{c} \leqslant \underset{\sim}{a} \leqslant \underset{\sim}{b},$$
$$\underset{\sim}{c} \leqslant \underset{\sim}{b} \leqslant \underset{\sim}{a},$$

etc., for four, five, six, . . . variables.

If ψ is a performance function of n variables, the table will have $n!$ rows; each row may take n values; there are then

$$N = n^{n!}$$

distinct functions. Among these $n^{n!}$ functions, a small number may be presented in canonical† form (in $\wedge$ or in $\vee$) and thus are representable by a reliability network.

Taking up again the example of Figure 51.6, we have Figure 41.7.

$\leqslant$	$\leqslant$		$\underset{\sim}{a} \wedge \underset{\sim}{b}$	$\underset{\sim}{a} \wedge \underset{\sim}{c}$	$(\underset{\sim}{a} \wedge \underset{\sim}{b}) \vee (\underset{\sim}{a} \wedge \underset{\sim}{c})$
$\underset{\sim}{a}$	$\underset{\sim}{b}$	$\underset{\sim}{c}$	$\underset{\sim}{a}$	$\underset{\sim}{a}$	$\underset{\sim}{a}$
$\underset{\sim}{a}$	$\underset{\sim}{c}$	$\underset{\sim}{b}$	$\underset{\sim}{a}$	$\underset{\sim}{a}$	$\underset{\sim}{a}$
$\underset{\sim}{b}$	$\underset{\sim}{a}$	$\underset{\sim}{c}$	$\underset{\sim}{b}$	$\underset{\sim}{a}$	$\underset{\sim}{a}$
$\underset{\sim}{b}$	$\underset{\sim}{c}$	$\underset{\sim}{a}$	$\underset{\sim}{b}$	$\underset{\sim}{c}$	$\underset{\sim}{c}$
$\underset{\sim}{c}$	$\underset{\sim}{a}$	$\underset{\sim}{b}$	$\underset{\sim}{a}$	$\underset{\sim}{c}$	$\underset{\sim}{a}$
$\underset{\sim}{c}$	$\underset{\sim}{b}$	$\underset{\sim}{a}$	$\underset{\sim}{b}$	$\underset{\sim}{c}$	$\underset{\sim}{b}$

Fig. 41.7

Monotone property. If $\mathbf{E} = \{X_1, X_2, \ldots, X_n\}$, then let $\underset{\sim}{\mathbf{A}} \subset \mathbf{E}$ and $\underset{\sim}{\mathbf{B}} \subset \mathbf{E}$. Put

(41.61) $$a_i = \mu_{\underset{\sim}{A}}(X_i), \; i = 1, 2, \ldots, n,$$

(41.62) $$b_i = \mu_{\underset{\sim}{B}}(X_i), \; i = 1, 2, \ldots, n;$$

one has

(41.63) $$\underset{\sim}{B} \supset \underset{\sim}{A} \quad \Rightarrow \quad \psi(\underset{\sim}{b}_1, \underset{\sim}{b}_2, \ldots, \underset{\sim}{b}_n) \geqslant \psi(\underset{\sim}{a}_1, \underset{\sim}{a}_2, \ldots, \underset{\sim}{a}_n).$$

This property generalizes the well-known property of monotonicity of functions of structure where the level values are 0 or 1 (functioning or not). Other properties of these reliability networks in 0 or 1 may be generalized. Any function $\cup$ such that (41.63) is true will be called a monotone function; then the following three properties are equivalent:

†Here we can use the adjective *canonical* because we do not use complementation for variables.

ψ U is a monotone performance function;

ψ U is analytic, that is, may be written in a canonical form in $\wedge$ (or in $\vee$);

there exist a series–parallel network R_{sp} and a parallel–series network R_{sp} that are dual to one another.

One may draw other more or less trivial conclusions. Let R be a network. By including another network R'' in parallel with R one cannot increase its performance level.

It is appropriate to make an important remark on the subject of reliability on the one hand and performance level on the other. These are quite distinct notions.

We consider a system **E** made up of two components A and B. Suppose that $\underset{\sim}{a}$ is the performance level of level of A and $\underset{\sim}{b}$ that of B; $\underset{\sim}{a}, \underset{\sim}{b} \in [0, 1]$. Suppose also that the network of these two components is arranged in parallel. One then has

$$\psi(a, b) = \underset{\sim}{a} \vee \underset{\sim}{b}. \tag{41.64}$$

Finally, suppose that $\underset{\sim}{a} = \underset{\sim}{b}$; then

$$\psi(\underset{\sim}{a}, \underset{\sim}{b}) = \psi(\underset{\sim}{a}) = \psi(\underset{\sim}{b}) = \underset{\sim}{a}. \tag{41.65}$$

Thus, such a redundance does not modify the performance level.

Now consider the reliability of the same system **E**. If a and b are the state variables, $a, b \in \{0, 1\}$, one has

$$\varphi(a, b) = a \mathbin{\hat{+}} b = a + b - ab. \tag{41.66}$$

And from there

$$h(p_a, p_b) = p_a + p_b - p_a p_b. \tag{41.67}$$

Suppose that $p_a = p_b$; then one has

$$h(p_a, p_b) = 2p_a - p_a^2 \geqslant p_a \tag{41.68}$$

and

$$h(p_a, p_b) > p_a \quad \text{if} \quad p_a \neq 0 \quad \text{or} \quad p_a \neq 1. \tag{41.69}$$

Thus redundancy improves the reliability but not the level of performance.

The two concepts "performance level" and "reliability" then should not be confused. The first is associated with the theory of fuzzy subsets, the second with probability theory.

Thus, if each of two components is operating moderately well, their association in parallel gives moderate operation, but the reliability will be better. This shows well, from an example, all the difference existing between the two concepts.

What may be asserted concerning both performance and reliability is monotonicity. By including a network R' in parallel with a network R, one diminishes neither the performance nor the reliability. By including a network R' in series with a network R, one increases neither the performance nor the reliability.

We note that the theory of performance functions that we have seen outlined above may be generalized by considering the variables $\underset{\sim}{a}, \underset{\sim}{b}, \ldots$, that give the performance levels, as taking their values not in $[0, 1]$, but in a more general fashion in an arbitrary ordered set, as we shall see in Chapter V, where the theory of Zadeh is the object of an important extension.

The notion of performance may also be the object of different definitions that arise in a theory of taxonomy.†

42. EXERCISES

III1. Simplify the following functions of fuzzy variables:

α) $\underset{\sim}{f}(\underset{\sim}{a}, \underset{\sim}{b}) = (\underset{\sim}{a} \wedge \underset{\sim}{b}) \vee (\underset{\sim}{a} \wedge \underset{\sim}{b} \wedge \overline{\underset{\sim}{b}}) \vee (\overline{\underset{\sim}{a}} \wedge \underset{\sim}{a} \wedge \underset{\sim}{b})$,

β) $\underset{\sim}{f}(\underset{\sim}{a}, \underset{\sim}{b}) = (\underset{\sim}{a} \vee \overline{\underset{\sim}{a}} \vee \underset{\sim}{b} \vee \overline{\underset{\sim}{b}}) \wedge (\underset{\sim}{a} \vee \underset{\sim}{b} \vee \overline{\underset{\sim}{b}}) \wedge (\overline{\underset{\sim}{a}} \vee \underset{\sim}{b} \vee \overline{\underset{\sim}{b}})$.

γ) $\underset{\sim}{f}(\underset{\sim}{a}, \underset{\sim}{b}, \underset{\sim}{c}) = (\underset{\sim}{a} \vee \underset{\sim}{b} \vee \overline{\underset{\sim}{c}}) \wedge (\underset{\sim}{a} \vee \underset{\sim}{c}) \wedge (\underset{\sim}{a} \vee \overline{\underset{\sim}{c}}) \wedge \underset{\sim}{b}$.

δ) $\underset{\sim}{f}(\underset{\sim}{a}, \underset{\sim}{b}, \underset{\sim}{c}) = [[(\underset{\sim}{a} \wedge \underset{\sim}{b}) \vee (\underset{\sim}{a} \wedge \underset{\sim}{c})] \wedge (\underset{\sim}{b} \vee \underset{\sim}{c})] \vee \underset{\sim}{b}$.

III2. Express the functions of Exercise III1 explicitly with the aid of a table of values, as was done in example (32.75).

III3. Express the following functions in a reduced polynomial form with respect to $\wedge$:

α) $\underset{\sim}{f}(\underset{\sim}{a}, \underset{\sim}{b}) = [[(\underset{\sim}{a} \wedge \underset{\sim}{b}) \vee \overline{\underset{\sim}{a}}] \wedge [(\underset{\sim}{a} \vee \overline{\underset{\sim}{b}}) \wedge \underset{\sim}{a}]] \vee (\underset{\sim}{a} \wedge \underset{\sim}{b})$,

β) $\underset{\sim}{f}(\underset{\sim}{a}, \underset{\sim}{b}) = [[(\underset{\sim}{a} \wedge \overline{\underset{\sim}{b}}) \wedge \underset{\sim}{b}] \vee \overline{\underset{\sim}{b}}] \wedge (\underset{\sim}{a} \vee \underset{\sim}{b})$,

γ) $\underset{\sim}{f}(\underset{\sim}{a}, \underset{\sim}{b}, \underset{\sim}{c}) = (\underset{\sim}{a} \wedge \overline{\underset{\sim}{b}} \wedge \underset{\sim}{c}) \vee [(\underset{\sim}{a} \wedge \underset{\sim}{b}) \vee \overline{\underset{\sim}{c}}]$,

δ) $\underset{\sim}{f}(\underset{\sim}{a}, \underset{\sim}{b}, \underset{\sim}{c}) = (\underset{\sim}{a} \wedge \overline{\underset{\sim}{b}}) \vee (\overline{\underset{\sim}{a}} \wedge \underset{\sim}{c}) \vee (\overline{\underset{\sim}{a}} \wedge \underset{\sim}{c} \wedge \overline{\underset{\sim}{c}})$.

III4. The same question as III3, but this time giving a reduced polynomial form with respect to $\vee$ for each function.

III5. For each of the following functions, carry out the analysis according to the method of Marinos (see Section 34):

α) $\underset{\sim}{f}(\underset{\sim}{a}, \underset{\sim}{b}) = (\underset{\sim}{a} \wedge \overline{\underset{\sim}{b}}) \vee (\overline{\underset{\sim}{a}} \wedge \underset{\sim}{b})$,

β) $\underset{\sim}{f}(\underset{\sim}{a}, \underset{\sim}{b}, \underset{\sim}{c}) = (\underset{\sim}{a} \wedge \underset{\sim}{c}) \vee (\overline{\underset{\sim}{a}} \wedge \underset{\sim}{b}) \vee (\underset{\sim}{b} \wedge \underset{\sim}{c})$,

γ) $\underset{\sim}{f}(\underset{\sim}{a}, \underset{\sim}{b}, \underset{\sim}{c}) = [(\underset{\sim}{a} \wedge \underset{\sim}{b}) \vee \underset{\sim}{c}] \wedge \overline{\underset{\sim}{a}}$.

III6. Give the logical structures associated with the functions of Exercise III5 [as has been done for (35.8) and (35.9)]. The results should be expressed in developments with respect to ∇.

†Taxonomy is the science of the laws of classification, of the order that one may subjectively give in one's thoughts to abstract or concrete objects.

III7. Let

$$\underset{\sim}{a} \in \mathcal{D}_{\underset{\sim}{a}} = [a_1 \, , a_2[\, ,$$

$$\underset{\sim}{b} \in \mathcal{D}_{\underset{\sim}{b}} = [b_1 \, , b_2[\, ,$$

$$\underset{\sim}{c} \in \mathcal{D}_{\underset{\sim}{c}} = [c_1 \, , c_2[\, ,$$

What are the respective intervals in the interior of which the following functions take their values

α) $\underset{\sim}{f}(\underset{\sim}{a}, \underset{\sim}{b}) = \underset{\sim}{a} \wedge \overline{\underset{\sim}{b}}$,

β) $\underset{\sim}{f}(\underset{\sim}{a}, \underset{\sim}{b}, \underset{\sim}{c}) = (\underset{\sim}{a} \wedge \underset{\sim}{b}) \vee \underset{\sim}{c}$,

γ) $\underset{\sim}{f}(\underset{\sim}{a}, \underset{\sim}{b}) = \overline{\underset{\sim}{a}} \vee \overline{\underset{\sim}{b}}$.

Answer the question again, but with $\underset{\sim}{b} \in \overline{\mathcal{D}}_{\underset{\sim}{b}}$.

III8. Suppose that $\underset{\sim}{a}$, $\underset{\sim}{b}$, and $\underset{\sim}{f}(\underset{\sim}{a}, \underset{\sim}{b})$ take their values in $\{0; 0{,}1; 0{,}2; \ldots ; 0{,}9; 1\}$. Given the following functions at particular values for $\underset{\sim}{a}$ and $\underset{\sim}{b}$, give the respective domains of the functions $\underset{\sim}{f}$ [as has been done for Examples I and II of (36.34)–(36.43)].

α) $\underset{\sim}{f}(\underset{\sim}{a}, \underset{\sim}{b}) = \underset{\sim}{a} \vee (\overline{\underset{\sim}{a}} \wedge \underset{\sim}{b})$, $\quad \underset{\sim}{a} \in \{0{,}1 \; ; 0{,}2 \; ; 0{,}3 \}$,
$\quad \underset{\sim}{b} \in \{0{,}3 \} \cup \{0{,}9 \}$.

β) $\underset{\sim}{f}(\underset{\sim}{a}, \underset{\sim}{b}) = \underset{\sim}{a} \wedge \overline{\underset{\sim}{a}} \wedge \underset{\sim}{b}$, $\quad \underset{\sim}{a} \in \{0 \; ; 0{,}1\} \cup \{0{,}7 \; ; 0{,}8\}$,
$\quad \underset{\sim}{b} \in \{0{,}2 \; ; 0{,}3 \; ; 0{,}4 \}$.

III9. Give the synthesis schemes (as indicated in Section 37) for the function $\underset{\sim}{f}(\underset{\sim}{a}, \underset{\sim}{b}) = \underset{\sim}{a} \wedge \overline{\underset{\sim}{b}}$ by choosing (1) a development with respect to Δ; (2) a development with respect to $\widetilde{\nabla}$.

III10. For each of the networks of fuzzy elements shown below, determine the maximal simple routes, then the associated reduced polynomial form in ∨, and finally the network associated with this form.

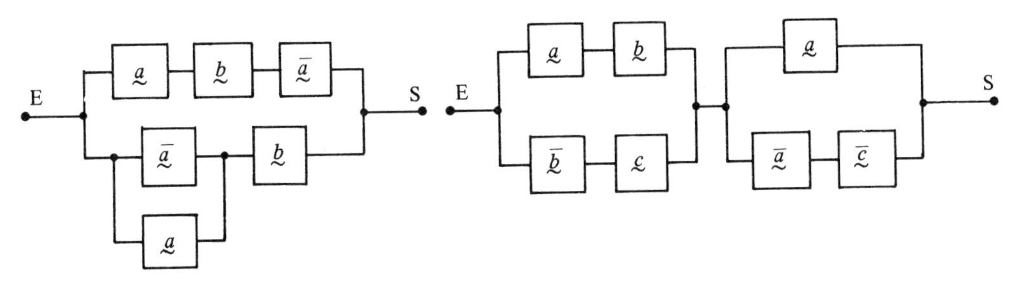

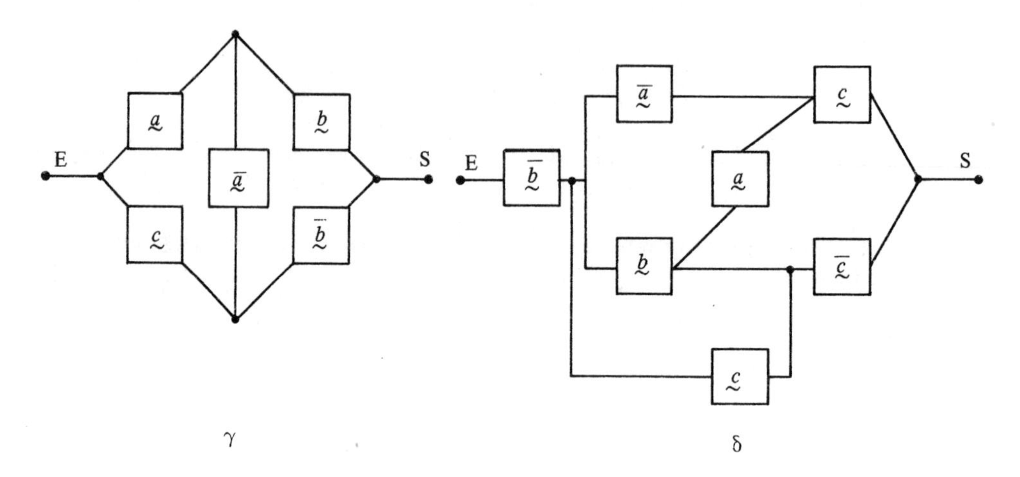

γ δ

III11. Consider the same question as Exercise III10 but taking the method of antiroutes; give the associated reduced polynomial form in $\wedge$ and the network associated with this form.

CHAPTER IV

THE LAWS OF FUZZY COMPOSITION

43. INTRODUCTION

Before studying various extensions concerning fuzzy subsets, which will be the important subject of Chapter V, we propose now that the reader concern himself with the laws of fuzzy composition.

Among these laws, the more common and useful, any mathematician would conjecture immediately, are those that form monoids, that is, those that are associative and have an identity. We shall show, moreover, that the structure of a group is not appropriate for the principal operations considered in the theory of fuzzy subsets—the notion of symmetry of a fuzzy subset is not found for the operators of this theory.

One knows what the actual importance of the theory of monoids or semigroups is in all that concerns information theory, codes, command of systems, etc. Our intention is to develop some corresponding useful notions in Volume II, which must follow what we have presented in this work.

What is presented in Chapter IV, which the reader should now go on to consider, is a first introduction to some important developments that will become more and more involving for him. The extension of ordinary concepts, is this not what interests the mathematician and the engineer?

44. REVIEW OF THE NOTION OF A LAW OF COMPOSITION

We review, in the interests of pedagogy, a certain number of very classical concepts from the theory of ordinary sets.

Law of internal composition. A *law of internal*† *composition on a set* **E** is a mapping from **E** $\times$ **E** into **E**. In other words, to each ordered pair $(x, y) \in \mathbf{E} \times \mathbf{E}$, one corresponds one and only one element $z \in \mathbf{E}$.

In practice, one represents this law by a symbol that, placed between x and y, serves to designate the element corresponding to the ordered pair (x, y). One often uses the sign $*$. Thus,

†Often the word *composition* is omitted, and one says "internal law" instead of "law of internal composition."

(44.1) $$x * y = z.$$

For particular types of laws one uses appropriate and common symbols like $+$, $\cdot$, $\times$, $\dot{+}$, , etc.

Often it is convenient to designate the mapping of $\mathbf{E} \times \mathbf{E}$ into $\mathbf{E}$ by a notation related to the elements of $\mathbf{E}$:

(44.2) $$(x, y) \rightsquigarrow z \quad , \quad x, y, z \in \mathbf{E}.$$

Law of external composition. Let $x \in \mathbf{E}_1, y \in \mathbf{E}_2$, and $z \in \mathbf{E}_3$. A mapping of $\mathbf{E}_1 \times \mathbf{E}_2$ into $\mathbf{E}_3$ is called a law of external composition. In other words, to each ordered pair (x, y) one corresponds an element $z \in \mathbf{E}_3$, and only one such element.

If and only if $\mathbf{E}_1 = \mathbf{E}_2 = \mathbf{E}_3$ is the law of composition internal.

Examples

(1) Let $\mathbf{E}_1 = \mathbf{E}_2 = \mathbf{R}$ (set of real numbers); if the law is that of ordinary addition $+$, this law is internal since the addition of a real with a real always gives a real; indeed one has $\mathbf{E}_3 = \mathbf{R}$.

(2) Let $\mathcal{P}(\mathrm{E})$ be the ordinary set of subsets of a set (the power set); then the operations intersection, union, difference, disjunctive sum, for example, define internal laws.

(3) If $\mathbf{E}_1 = \mathbf{E}_2 = \mathbf{R}^+$ (set of nonnegative reals) and if the law is that of the difference $x - y = z$, $x, y \in \mathbf{R}^+$, one obtains an external law, since it is possible that $z \notin \mathbf{R}^+$.

(4) If $\mathbf{E}_1 = \mathbf{E}_2 =$ set of free vectors in a plane and if one there defines $\times$ as the vector product (cross product) of the two vectors, one has a law of external composition.

Groupoid. An ordered pair formed by a set $\mathbf{E}$ and an internal law of composition $*$ defined on this set is called a groupoid. This is denoted $(\mathbf{E}, *)$.

Examples

(1) The law of composition presented in Figure 44.1 gives a groupoid.

E \ E	A	B	C	D	E
A	B	A	D	D	C
B	C	B	B	A	E
C	A	A	A	B	C
D	C	A	B	B	C
E	E	C	A	A	D

Fig. 44.1

(2) Examples (1) and (2) given for the notion of an internal law of composition above give groupoids.

(3) The greatest common divisor and least common multiple of positive integers define internal laws defined throughout the set $\mathbf{N}_0$ of positive integers. If $*_1$ indicates the greatest common divisor and $*_2$ the least common multiple, then $(\mathbf{N}_0, *_1)$ and $(\mathbf{N}_0, *_2)$ are groupoids.

45. LAW OF FUZZY INTERNAL COMPOSITION. FUZZY GROUPOID

The notions that we have proceeded to examine may be extended to fuzzy subsets in the following manner.

Let $\mathbf{E}$ be a reference set and $\underset{\sim}{\mathbf{A}} \subset \mathbf{E}$. Designate as we have done in Section 6, the set of fuzzy subsets of $\mathbf{E}$ by $\underset{\sim}{\mathcal{P}}(\mathbf{E})$. One may then write $\underset{\sim}{\mathbf{A}} \in \underset{\sim}{\mathcal{P}}(\mathbf{E})$. We have seen that if n = card $\mathbf{E}$ and m = card $\mathbf{M}$ are finite, then $\underset{\sim}{\mathcal{P}}(\mathbf{E})$ is finite.

One may then define a lawof internal composition on $\underset{\sim}{\mathcal{P}}(\mathbf{E})$,that is, a mapping from $\underset{\sim}{\mathcal{P}}(\mathbf{E}) \times \underset{\sim}{\mathcal{P}}(\mathbf{E})$ into $\underset{\sim}{\mathcal{P}}(\mathbf{E})$. In other words, to each ordered pair $(\underset{\sim}{\mathbf{A}}, \underset{\sim}{\mathbf{B}})$, where $\underset{\sim}{\mathbf{A}} \subset \mathbf{E}, \underset{\sim}{\mathbf{B}} \subset \mathbf{E}$, one corresponds a fuzzy subset $\underset{\sim}{\mathbf{C}} \subset \mathbf{E}$ and only one. If m and n are finite, one describes with these conditions a finite groupoid, and an infinite groupoid if m or/and n are not finite.

The laws of internal composition and the groupoids thus defined will be called *laws of fuzzy internal composition* or fuzzy internal laws and fuzzy groupoids.

We consider several examples.

Example I. Let

$$\mathbf{E} = \{A, B\} \quad \text{and} \quad \mathbf{M} = \{0, \tfrac{1}{2}, 1\}. \tag{45.1, 45.2}$$

One has, with reference to Figure 6.2,

$$\underset{\sim}{\mathcal{P}}(\mathbf{E}) = \{\{(A \mid 0), (B \mid 0)\}, \{(A \mid 0), (B \mid \tfrac{1}{2})\}, \{(A \mid \tfrac{1}{2}), (B \mid 0)\}, \{(A \mid \tfrac{1}{2}), (B \mid \tfrac{1}{2})\}, \{(A \mid 0), (B \mid 1\}, \{(A \mid 1), (B \mid 0)\}, \{(A \mid \tfrac{1}{2}), (B \mid 1)\}, \{(A \mid 1), (B \mid \tfrac{1}{2})\}, \{(A \mid 1), (B \mid 1)\}\}. \tag{45.3}$$

Designate, in order to simplify writing, for $\underset{\sim}{\mathbf{X}} \subset \mathbf{E}$,

$$\{(A \mid \mu_{\underset{\sim}{X}}(A)), (B \mid \mu_{\underset{\sim}{X}}(B))\} \tag{45.4}$$

by

$$(\mu_{\underset{\sim}{X}}(A), \mu_{\underset{\sim}{X}}(B)) \tag{45.5}$$

Thus $\{(A \mid \frac{1}{2}), (B \mid 0)\}$ will be designated by $(\frac{1}{2}, 0)$. With this notation, the table of Figure 45.1 represents a fuzzy groupoid.

$\mathscr{P}(\mathrm{E})$ \ $\mathscr{P}(\mathrm{E})$	$(0, 0)$	$(0, \frac{1}{2})$	$(\frac{1}{2}, 0)$	$(\frac{1}{2}, \frac{1}{2})$	$(0, 1)$	$(1, 0)$	$(\frac{1}{2}, 1)$	$(1, \frac{1}{2})$	$(1, 1)$
$(0, 0)$	$(0, 1)$	$(\frac{1}{2}, 0)$	$(\frac{1}{2}, \frac{1}{2})$	$(1, \frac{1}{2})$	$(0, 1)$	$(0, 1)$	$(1, \frac{1}{2})$	$(1, 1)$	$(1, \frac{1}{2})$
$(0, \frac{1}{2})$	$(1, 0)$	$(0, \frac{1}{2})$	$(\frac{1}{2}, 0)$	$(\frac{1}{2}, 1)$	$(0, 0)$	$(1, 0)$	$(0, 1)$	$(\frac{1}{2}, \frac{1}{2})$	$(\frac{1}{2}, 1)$
$(\frac{1}{2}, 0)$	$(0, 0)$	$(1, \frac{1}{2})$	$(0, \frac{1}{2})$	$(0, 1)$	$(1, \frac{1}{2})$	$(1, 1)$	$(1, 1)$	$(1, 1)$	$(\frac{1}{2}, \frac{1}{2})$
$(\frac{1}{2}, \frac{1}{2})$	$(1, 1)$	$(0, 0)$	$(1, 1)$	$(\frac{1}{2}, 1)$	$(1, 0)$	$(1, 0)$	$(\frac{1}{2}, \frac{1}{2})$	$(0, \frac{1}{2})$	$(0, 0)$
$(0, 1)$	$(\frac{1}{2}, \frac{1}{2})$	$(0, \frac{1}{2})$	$(1, \frac{1}{2})$	$(\frac{1}{2}, 0)$	$(\frac{1}{2}, 0)$	$(1, 0)$	$(\frac{1}{2}, 0)$	$(1, \frac{1}{2})$	$(\frac{1}{2}, 1)$
$(1, 0)$	$(\frac{1}{2}, 0)$	$(0, 0)$	$(\frac{1}{2}, \frac{1}{2})$	$(\frac{1}{2}, \frac{1}{2})$	$(0, 1)$	$(1, 0)$	$(0, 1)$	$(1, 1)$	$(1, \frac{1}{2})$
$(\frac{1}{2}, 1)$	$(0, 0)$	$(\frac{1}{2}, \frac{1}{2})$	$(0, 1)$	$(0, \frac{1}{2})$	$(1, 0)$	$(1, 0)$	$(\frac{1}{2}, 1)$	$(1, \frac{1}{2})$	$(1, 1)$
$(1, \frac{1}{2})$	$(0, 0)$	$(\frac{1}{2}, 1)$	$(0, \frac{1}{2})$	$(0, 0)$	$(1, \frac{1}{2})$	$(\frac{1}{2}, 1)$	$(0, 0)$	$(0, \frac{1}{2})$	$(\frac{1}{2}, 0)$
$(1, 1)$	$(\frac{1}{2}, \frac{1}{2})$	$(0, 1)$	$(0, 1)$	$(1, 1)$	$(1, 0)$	$(1, 0)$	$(\frac{1}{2}, 1)$	$(0, \frac{1}{2})$	$(1, \frac{1}{2})$

Fig. 45.1

Example II. If the operation $*$ being considered is intersection $\cap$ and if $\underset{\sim}{\mathbf{A}} \subset \mathbf{E}$ and $\underset{\sim}{\mathbf{B}} \subset \mathbf{E}$, one may form a groupoid with the fuzzy subsets $\underset{\sim}{\mathbf{A}} \cap \underset{\sim}{\mathbf{B}}$. And the same holds for the operations $\cup$, $\oplus$, defined in Section 5.

Construction of a fuzzy groupoid. It suffices to be given a reference set **E**, finite or not, to deduce $\mathscr{P}(\mathbf{E})$ explicitly or not, and to define a law $*$ that corresponds to each ordered pair of fuzzy subsets $(\underset{\sim}{\mathbf{A}}, \underset{\sim}{\mathbf{B}})$ one and only one fuzzy subset $\underset{\sim}{\mathbf{C}}$ $(\underset{\sim}{\mathbf{A}}, \underset{\sim}{\mathbf{B}}, \underset{\sim}{\mathbf{C}} \subset \mathbf{E})$.

We consider several examples.

Example I. Consider (45.1) and (45.2) again with the law

$$\underset{\sim}{\mathbf{A}} * \underset{\sim}{\mathbf{B}} = \underset{\sim}{\mathbf{A}} \cap \underset{\sim}{\mathbf{B}}, \tag{45.6}$$

that is,

$$\mu_{\underset{\sim}{\mathbf{A}} \cap \underset{\sim}{\mathbf{B}}}(x) = \mathrm{MIN}(\mu_{\underset{\sim}{\mathbf{A}}}(x)\ ,\ \mu_{\underset{\sim}{\mathbf{B}}}(x)) = \mu_{\underset{\sim}{\mathbf{A}}}(x) \wedge \mu_{\underset{\sim}{\mathbf{B}}}(x)\,. \tag{45.7}$$

One has thus constructed a groupoid; it is represented in Figure 45.2.

$\underset{\sim}{\mathfrak{L}}(E)$ \ $\underset{\sim}{\mathfrak{L}}(E)$	$(0, 0)$	$(0, \frac{1}{2})$	$(\frac{1}{2}, 0)$	$(\frac{1}{2}, \frac{1}{2})$	$(0, 1)$	$(1, 0)$	$(\frac{1}{2}, 1)$	$(1, \frac{1}{2})$	$(1, 1)$
$(0, 0)$	$(0, 0)$	$(0, 0)$	$(0, 0)$	$(0, 0)$	$(0, 0)$	$(0, 0)$	$(0, 0)$	$(0, 0)$	$(0, 0)$
$(0, \frac{1}{2})$	$(0, 0)$	$(0, \frac{1}{2})$	$(0, 0)$	$(0, \frac{1}{2})$	$(0, \frac{1}{2})$	$(0, 0)$	$(0, \frac{1}{2})$	$(0, \frac{1}{2})$	$(0, \frac{1}{2})$
$(\frac{1}{2}, 0)$	$(0, 0)$	$(0, 0)$	$(\frac{1}{2}, 0)$	$(\frac{1}{2}, 0)$	$(0, 0)$	$(\frac{1}{2}, 0)$	$(\frac{1}{2}, 0)$	$(\frac{1}{2}, 0)$	$(\frac{1}{2}, 0)$
$(\frac{1}{2}, \frac{1}{2})$	$(0, 0)$	$(0, \frac{1}{2})$	$(\frac{1}{2}, 0)$	$(\frac{1}{2}, \frac{1}{2})$	$(0, \frac{1}{2})$	$(\frac{1}{2}, 0)$	$(\frac{1}{2}, \frac{1}{2})$	$(\frac{1}{2}, \frac{1}{2})$	$(\frac{1}{2}, \frac{1}{2})$
$(0, 1)$	$(0, 0)$	$(0, \frac{1}{2})$	$(0, 0)$	$(0, \frac{1}{2})$	$(0, 1)$	$(0, 0)$	$(0, 1)$	$(0, \frac{1}{2})$	$(0, 1)$
$(1, 0)$	$(0, 0)$	$(0, 0)$	$(\frac{1}{2}, 0)$	$(\frac{1}{2}, 0)$	$(0, 0)$	$(1, 0)$	$(\frac{1}{2}, 0)$	$(1, 0)$	$(1, 0)$
$(\frac{1}{2}, 1)$	$(0, 0)$	$(0, \frac{1}{2})$	$(\frac{1}{2}, 0)$	$(\frac{1}{2}, \frac{1}{2})$	$(0, 1)$	$(\frac{1}{2}, 0)$	$(\frac{1}{2}, 1)$	$(\frac{1}{2}, \frac{1}{2})$	$(\frac{1}{2}, 1)$
$(1, \frac{1}{2})$	$(0, 0)$	$(0, \frac{1}{2})$	$(\frac{1}{2}, 0)$	$(\frac{1}{2}, \frac{1}{2})$	$(0, \frac{1}{2})$	$(1, 0)$	$(\frac{1}{2}, \frac{1}{2})$	$(1, \frac{1}{2})$	$(1, \frac{1}{2})$
$(1, 1)$	$(0, 0)$	$(0, \frac{1}{2})$	$(\frac{1}{2}, 0)$	$(\frac{1}{2}, \frac{1}{2})$	$(0, 1)$	$(1, 0)$	$(\frac{1}{2}, 1)$	$(1, \frac{1}{2})$	$(1, 1)$

Fig. 45.2

Example II. We attempt to define the "fuzzy positive integers." We begin by defining a fuzzy number $\underset{\sim}{1}$ with a membership function $\mu_{\underset{\sim}{1}}(n)$, arbitrary but such that

$$\sum_{n=0}^{\infty} \mu_{\underset{\sim}{1}}(n) = 1, \quad n = 0, 1, 2, 3, \ldots . \tag{45.8}$$

For example,

$$\underset{\sim}{1} = \{(0 \mid 0{,}1), \ (1 \mid 0{,}8), \ (2 \mid 0{,}1), \ \ldots \ (N > 2 \mid 0)\} \tag{45.9}$$

We form $\underset{\sim}{2}$ in the following fashion:

$$\begin{aligned}
\mu_{\underset{\sim}{2}}(0) &= \mu_{\underset{\sim}{1}}(0) \,.\, \mu_{\underset{\sim}{1}}(0) = (0{,}1) \,.\, (0{,}1) = 0{,}01 , \\
\mu_{\underset{\sim}{2}}(1) &= \mu_{\underset{\sim}{1}}(0) \,.\, \mu_{\underset{\sim}{1}}(1) + \mu_{\underset{\sim}{1}}(1) \,.\, \mu_{\underset{\sim}{1}}(0) = (0{,}1) \,.\, (0{,}8) + (0{,}8) \,.\, (0{,}1) = 0{,}16 , \\
\mu_{\underset{\sim}{2}}(2) &= \mu_{\underset{\sim}{1}}(0) \,.\, \mu_{\underset{\sim}{1}}(2) + \mu_{\underset{\sim}{1}}(1) \,.\, \mu_{\underset{\sim}{1}}(1) + \mu_{\underset{\sim}{1}}(2) \,.\, \mu_{\underset{\sim}{1}}(0) = (0{,}1) \,.\, (0{,}1) + \\
&\qquad + (0{,}8) \,.\, (0{,}8) + (0{,}1) \,.\, (0{,}1) = 0{,}66 , \\
\mu_{\underset{\sim}{2}}(3) &= \mu_{\underset{\sim}{1}}(1) \,.\, \mu_{\underset{\sim}{1}}(2) + \mu_{\underset{\sim}{1}}(2) \,.\, \mu_{\underset{\sim}{1}}(1) = (0{,}8) \,.\, (0{,}1) + (0{,}1) \,.\, (0{,}8) = 0{,}16 , \\
\mu_{\underset{\sim}{2}}(4) &= \mu_{\underset{\sim}{1}}(2) \,.\, \mu_{\underset{\sim}{1}}(2) = (0{,}1) \,.\, (0{,}1) = 0{,}01 , \\
\mu_{\underset{\sim}{2}}(N > 4) &= 0 .
\end{aligned} \tag{45.10}$$

Thus,

$$(45.11) \quad \underset{\sim}{2} = \{(0 \mid 0{,}01)\,,\,(1 \mid 0{,}16)\,,\,(2 \mid 0{,}66)\,,\,(3 \mid 0{,}16)\,,\,(4 \mid 0{,}01)\,,\ldots,\,(N > 4 \mid 0)\}$$

One then finishes $\underset{\sim}{3}$ by employing the formula that generalizes (45.10):

$$(45.12) \qquad \mu_{\underset{\sim}{A} * \underset{\sim}{B}}(N) = \sum_{r=0}^{N} \mu_{\underset{\sim}{A}}(r) \,.\, \mu_{\underset{\sim}{B}}(N - r) = \sum_{r=0}^{N} \mu_{\underset{\sim}{B}}(r) \,.\, \mu_{\underset{\sim}{A}}(N - r) \,.$$

One should recognize the convolution product used in probability theory and in transformations of linear functions.

For $\underset{\sim}{3}$, one sees

$$(45.13) \qquad \mu_{\underset{\sim}{3}}(N) = \mu_{\underset{\sim}{2} * \underset{\sim}{1}}(N) = \sum_{r=0}^{N} \mu_{\underset{\sim}{2}}(r) \,.\, \mu_{\underset{\sim}{1}}(N - r)\,, \qquad N \leqslant 6\,.$$

Thus

$$(45.14) \quad \underset{\sim}{3} = \{(0 \mid 0{,}001)\,,\,(1 \mid 0{,}024)\,,\,(2 \mid 0{,}195)\,,\,(3 \mid 0{,}560),\,(4 \mid 0{,}195)\,,\,(5 \mid 0{,}024)\,,\,(6 \mid 0{,}001)\,,\ldots,(N > 6 \mid 0)\}.$$

And thus it goes. Note that the fuzzy character is accentuated as the fuzzy integers increase.

We shall later concern ourselves with certain particular properties of groupoids. We note already that, for the groupoid we have constructed, one has the following properties:

$$(45.15) \qquad (\underset{\sim}{A} * \underset{\sim}{B}) * \underset{\sim}{C} = \underset{\sim}{A} * (\underset{\sim}{B} * \underset{\sim}{C})\,, \quad \text{associativity.}$$

$$(45.16) \qquad \underset{\sim}{A} * \underset{\sim}{B} = \underset{\sim}{B} * \underset{\sim}{A}\,, \quad \text{commutativity.}$$

One must choose $\mu_{\underset{\sim}{1}}(n)$ such that

$$(45.17) \qquad \sum_{n}^{\infty} \mu_{\underset{\sim}{1}}(n) = 1\,,$$

this due to the use of the convolution product (45.12).

Example III. This time we again take a membership function that may be considered as a law of probability. We consider two fuzzy subsets $\underset{\sim}{A} \subset \mathbf{R}$ and $\underset{\sim}{B} \subset \mathbf{R}$ with which we produce other fuzzy subsets (thus, we consider $\underset{\sim}{A}$ and $\underset{\sim}{B}$ as generators of an infinity of other fuzzy subsets). Let

$$(45.18) \qquad \mu_{\underset{\sim}{A}}(x) = \frac{1}{\sqrt{2\pi\sigma_1^2}}\, e^{-\frac{(x-a)^2}{2\sigma_1^2}}\,, \quad a\,,\,\sigma_1 \in \mathbf{R}^+\,,$$

$$(45.19) \qquad \mu_{\underset{\sim}{B}}(x) = \frac{1}{\sqrt{2\pi\sigma_2^2}}\, e^{-\frac{(x-b)^2}{2\sigma_2^2}} \quad b\,,\,\sigma_2 \in \mathbf{R}^+\,.$$

One then considers the composition product

$$(45.20)\quad \mu_{\underset{\sim}{A} * \underset{\sim}{B}}(x) = \int_{-\infty}^{\infty} \mu_{\underset{\sim}{A}}(t) \,.\, \mu_{\underset{\sim}{B}}(x-t) \,.\, dt = \int_{-\infty}^{\infty} \mu_{\underset{\sim}{B}}(t) \,.\, \mu_{\underset{\sim}{A}}(x-t) \,.\, dt$$

$$= \frac{1}{\sqrt{2\pi(\sigma_1^2 + \sigma_2^2)}}\, e^{-\frac{(x-a-b)^2}{2(\sigma_1^2+\sigma_2^2)}}.$$

This permits one to define the fuzzy number $\underset{\sim}{A} * \underset{\sim}{B}$.

In the same manner one generates other fuzzy numbers:

$$(45.21)\qquad \underset{\sim}{A} * \underset{\sim}{A}\,,\quad \underset{\sim}{B} * \underset{\sim}{B}\,,\quad \underset{\sim}{A} * \underset{\sim}{A} * \underset{\sim}{A}\,,\quad \underset{\sim}{A} * \underset{\sim}{A} * \underset{\sim}{B}\,,\quad \ldots,\quad \underset{\sim}{A}^r * \underset{\sim}{B}^s, \ldots$$

where the superscripts indicate that there are $r - 1$ compositions of $\underset{\sim}{A}$ and $s - 1$ compositions of $\underset{\sim}{B}$.

With the two fuzzy numbers $\underset{\sim}{A}$ and $\underset{\sim}{B}$ one then generates

$$(45.22)\qquad \underset{\sim}{A}\,, \underset{\sim}{B}\,,\quad \underset{\sim}{A} * \underset{\sim}{A}\,,\quad \underset{\sim}{A} * \underset{\sim}{B}\,,\quad \underset{\sim}{B} * \underset{\sim}{B}\,,\quad \ldots,\quad \underset{\sim}{A}^r * \underset{\sim}{B}^s, \ldots;$$

and the set

$$(45.23)\qquad Q = \{\underset{\sim}{A}\,,\ \underset{\sim}{B}\,,\ \underset{\sim}{A} * \underset{\sim}{A}\,,\ \underset{\sim}{A} * \underset{\sim}{B}\,,\ \underset{\sim}{B} * \underset{\sim}{B}\,,\ \ldots,\ \underset{\sim}{A}^r * \underset{\sim}{B}^s, \ldots\}$$

has the structure of a groupoid, which is moreover associative and commutative, since these properties are present in (45.20).

46. PRINCIPAL PROPERTIES CONCERNING FUZZY GROUPOIDS

Let $*$ be a law of internal composition of a fuzzy groupoid; we define several properties. The groupoid will be desingated by $(\underset{\sim}{\mathcal{P}}(E), *)$.

Commutativity. If, for all ordered pairs $(\underset{\sim}{A}, \underset{\sim}{B}) \in \underset{\sim}{\mathcal{P}}(E) \times \underset{\sim}{\mathcal{P}}(E)$ one has

$$(46.1)\qquad \underset{\sim}{A} * \underset{\sim}{B} = \underset{\sim}{B} * \underset{\sim}{A}\,,$$

one says that the law of internal composition is commutative; one also says that the groupoid is commutative. Thus, for example, the groupoid of Figure 45.2 is commutative, whereas that of Figure 45.1 is not. For example, in Figure 45.2 one may verify

$$(46.2)\qquad \{(A \mid \tfrac{1}{2}), (B \mid 1)\} \wedge \{(A \mid 1), (B \mid 0)\} = \{(A \mid \tfrac{1}{2}), (B \mid 0)\}\,,$$

$$(46.3)\qquad \{(A \mid 1), (B \mid 0)\} \wedge \{(A \mid \tfrac{1}{2}), (B \mid 1)\} = \{(A \mid \tfrac{1}{2}), (B \mid 0)\}\,.$$

Being given the definition of the law $*$ for fuzzy subsets, one may thence conclude that if

$$(46.4)\qquad \mu_{\underset{\sim}{A} * \underset{\sim}{B}}(x) = \mu_{\underset{\sim}{A}}(x) \odot \mu_{\underset{\sim}{B}}(x)\,,$$

commutativity for $\odot$ implies commutativity for $*$, and vice versa. An obvious example is given by (45.6) and (45.7).

Associativity. If

$$\forall \underset{\sim}{A}, \underset{\sim}{B}, \underset{\sim}{C} \subset E: \quad (\underset{\sim}{A} * \underset{\sim}{B}) * \underset{\sim}{C} = \underset{\sim}{A} * (\underset{\sim}{B} * \underset{\sim}{C}), \tag{46.5}$$

one says that the law is associative; one also says that the groupoid is associative.

Thus, the groupoid of Figure 45.2 is associative, whereas that of Figure 45.1 is not. Thus, in Figure 45.2 one may verify, using the abbreviated notation,

$$((\tfrac{1}{2}, \tfrac{1}{2}) \wedge (1, 0)) \wedge (\tfrac{1}{2}, 1) = (\tfrac{1}{2}, 0) \wedge (\tfrac{1}{2}, 1) = (\tfrac{1}{2}, 0), \tag{46.6}$$

$$(\tfrac{1}{2}, \tfrac{1}{2}) \wedge ((1, 0) \wedge (\tfrac{1}{2}, 1)) = (\tfrac{1}{2}, \tfrac{1}{2}) \wedge (\tfrac{1}{2}, 0) = (\tfrac{1}{2}, 0). \tag{46.7}$$

Being given the definition of the law $*$ for the fuzzy subsets, one may thence conclude that if

$$(\mu_{\underset{\sim}{A}}(x) \odot \mu_{\underset{\sim}{B}}(x)) \odot \mu_{\underset{\sim}{C}}(x) = \mu_{\underset{\sim}{A}}(x) \odot (\mu_{\underset{\sim}{B}}(x) \odot \mu_{\underset{\sim}{C}}(x)), \tag{46.8}$$

associativity for $\odot$ implies associativity for $*$, and vice versa.

Identity element. In the theory of ordinary sets, one defines for a law $*$ being considered a particular element $e \in \mathbf{E}$, if it exists, such that

$$\forall a \in \mathbf{E} : e * a = a. \tag{46.9}$$

This element is called a left identity. In the same manner, an element $e' \in \mathbf{E}$, if it exists, such that

$$\forall a \in \mathbf{E} : a * e' = a, \tag{46.10}$$

is called a right identity.

An element that is both a left identity and a right identity is called an identity.

When an identity element exists, it is always unique. If fact, if there existed another such element ϵ, one would have

$$\epsilon * e = e * \epsilon = \epsilon = e. \tag{46.11}$$

In a fuzzy groupoid one may define an identity in the same manner. We show first with an example that this is possible, and then pass to a general definition. We take the example of Figure 45.2. It is evident that (1, 1) is a left identity and a right identity, that is, an identity. In fact, $\forall x \in \{0, \frac{1}{2}, 1\}$ and $\forall y \in \{0, \frac{1}{2}, 1\}$,

$$(1,1) \wedge (x, y) = (x, y) \wedge (1,1) = (x, y), \tag{46.12}$$

We shall say that a fuzzy groupoid possesses a left identity $\underset{\sim}{U}$ for the law $*$ if

$$\forall \underset{\sim}{A} \subset \mathbf{E} : \quad \underset{\sim}{U} * \underset{\sim}{A} = \underset{\sim}{A}. \tag{46.13}$$

and possesses a right identity $\underset{\sim}{U}'$ for this law if

$$\forall \underset{\sim}{A} \subset \mathbf{E} : \quad \underset{\sim}{A} * \underset{\sim}{U}' = \underset{\sim}{A}. \tag{46.14}$$

and possesses a unique identity $\underset{\sim}{\mathbf{U}}$ if

$$\forall \underset{\sim}{\mathbf{A}} \subset \mathbf{E}: \quad \underset{\sim}{\mathbf{U}} * \underset{\sim}{\mathbf{A}} = \underset{\sim}{\mathbf{A}} * \underset{\sim}{\mathbf{U}} = \underset{\sim}{\mathbf{A}}. \tag{46.15}$$

We have seen with the example of Figure 45.2 the case of a fuzzy groupoid that possesses an identity. We now see the case of another that does not possess one. In the example treated in (45.8)–(45.16) one has the situation of a groupoid that does not possess an identity. With respect to $\underset{\sim}{1}$, one may not generate a fuzzy subset that has property (46.15), nor likewise one of the properties (46.13) or (46.14).

Inverse elements. Recall what is intended by an inverse element in the theory of ordinary sets.

We consider a law for which there exists an identity e. Then let there be two elements a and $\bar{a} \in \mathbf{E}$. If

$$\bar{a} * a = e, \tag{46.16}$$

one says that $\bar{a}$ is the left inverse of a. In the same manner, if

$$a * \bar{a}' = e, \tag{46.17}$$

one says that $\bar{a}'$ is the right inverse of a. Finally, if $\bar{a}' = a$, then

$$\bar{a} * a = a * \bar{a} = e, \tag{46.18}$$

and one says that $\bar{a}$ is the inverse of a.

In a fuzzy groupoid one may attempt to define an inverse for each element.

Again we take the example of Figure 45.2. We have seen that there exists an identity, which is (1, 1). It is clear that there is only one element that composed with itself is able to give (1, 1); this is (1, 1).

For any others such that $(a, b) \prec (1, 1)$ and $(a', b') \prec (1, 1)$, one has

$$(a, b) \wedge (a', b') \prec (1,1). \tag{46.19}$$

Thus the groupoid of Figure 45.2 does not have the property of possessing an inverse for each of its elements.

More generally, if the law $*$ is $\cup$ or $\cap$, one cannot have an inverse. In the case of $\cup$, there is an identity that is defined by: $\forall x \in \mathbf{E}: \mu_{\underset{\sim}{A}}(x) = 0$; and in the case of $\cap$, there is an identity defined by: $\forall x \in \mathbf{E}, \mu_{\underset{\sim}{A}}(x) = 1$. But for neither of these cases can one define an inverse, no matter what the fuzzy subset. One knows that

$$(\text{condition } \forall x \in \mathbf{E}: \quad \mu_{\underset{\sim}{A}}(x) = 0) \quad \Leftrightarrow \quad (\underset{\sim}{\mathbf{A}} = \phi), \tag{46.20}$$

$$(\text{condition } \forall x \in \mathbf{E}: \quad \mu_{\underset{\sim}{A}}(x) = 1) \quad \Leftrightarrow \quad (\underset{\sim}{\mathbf{A}} = \mathbf{E}). \tag{46.21}$$

But, if ϕ is the identity for $\cup$ and $\mathbf{E}$ the identity for $\cap$, these do not allow one to define inverses; any element such as $\underset{\sim}{\mathbf{B}}$ may not give

$$\underset{\sim}{\mathbf{A}} \cup \underset{\sim}{\mathbf{B}} = \phi \quad , \text{ unless } \underset{\sim}{\mathbf{B}} = \phi \text{ and } \underset{\sim}{\mathbf{A}} = \phi; \tag{46.22}$$

$$\underset{\sim}{\mathbf{A}} \cap \underset{\sim}{\mathbf{B}} = \mathbf{E} \quad , \text{ unless } \underset{\sim}{\mathbf{B}} = \mathbf{E} \text{ and } \underset{\sim}{\mathbf{A}} = \mathbf{E}. \tag{46.23}$$

One may verify in the same way that the laws†

$$\underset{\sim}{A} \oplus \underset{\sim}{B} = (\overline{\underset{\sim}{A}} \cap \underset{\sim}{B}) \cup (\underset{\sim}{A} \cap \overline{\underset{\sim}{B}}) \tag{46.24}$$

$$\underset{\sim}{A} \,\overline{\oplus}\, \underset{\sim}{B} = (\overline{\underset{\sim}{A}} \cup \underset{\sim}{B}) \cap (\underset{\sim}{A} \cup \overline{\underset{\sim}{B}}) \tag{46.25}$$

do not permit one to find this property of the presence of an inverse.

One may verify that this property also fails to hold for

$$\underset{\sim}{A} * \underset{\sim}{B} \quad \text{defined by} \quad \mu_{\underset{\sim}{A} * \underset{\sim}{B}}(x) = \mu_{\underset{\sim}{A}}(x) \cdot \mu_{\underset{\sim}{B}}(x) \tag{46.26}$$

$$\underset{\sim}{A} * \underset{\sim}{B} \quad \text{defined by} \quad \mu_{\underset{\sim}{A} * \underset{\sim}{B}}(x) = \int_{-\infty}^{\infty} \mu_{\underset{\sim}{A}}(t) \cdot \mu_{\underset{\sim}{B}}(x - t) \cdot dt \tag{46.27}$$

Distributivity. Let $*$ and $*'$ represent two laws of internal composition on the same set **E**. If one has

$$\forall \underset{\sim}{A}, \underset{\sim}{B}, \underset{\sim}{C} \subset E : \quad \underset{\sim}{A} * (\underset{\sim}{B} *' \underset{\sim}{C}) = (\underset{\sim}{A} * \underset{\sim}{B}) *' (\underset{\sim}{A} * \underset{\sim}{C}), \tag{46.28}$$

one says that the law $*$ is left distributive with respect to the law $*'$.

Likewise, if one has

$$\forall \underset{\sim}{A}, \underset{\sim}{B}, \underset{\sim}{C} \subset E : \quad (\underset{\sim}{A} *' \underset{\sim}{B}) * \underset{\sim}{C} = (\underset{\sim}{A} *' \underset{\sim}{C}) * (\underset{\sim}{B} *' \underset{\sim}{C}), \tag{46.29}$$

one says that the law $*$ is right distributive with respect to the law $*'$.

A law $*$ left and right distributive with respect to another law $*'$ is said to be distributive with respect to $*'$. One may then write

$$(\underset{\sim}{A} *' \underset{\sim}{B}) * (\underset{\sim}{C} *' \underset{\sim}{D}) = (\underset{\sim}{A} * \underset{\sim}{C}) *' (\underset{\sim}{A} * \underset{\sim}{D}) *' (\underset{\sim}{B} * \underset{\sim}{C}) *' (\underset{\sim}{B} * \underset{\sim}{D}). \tag{46.30}$$

One may verify, for example, that the law $\cap$ is distributive with respect to $\cup$, and vice versa, whereas this is not the case for the law $\oplus$:

$$\underset{\sim}{A} \oplus \underset{\sim}{B} = (\underset{\sim}{A} \cap \overline{\underset{\sim}{B}}) \cup (\overline{\underset{\sim}{A}} \cap \underset{\sim}{B}), \tag{46.31}$$

with $\cap$ or $\cup$.

Ordinary subset of a closed fuzzy subset with respect to a law of composition. Let‡ $\Delta \subset \mathcal{P}(E)$, where $\mathcal{P}(E)$ is provided with the law $*$. If for every ordered pair $(\underset{\sim}{A}, \underset{\sim}{B}) \in \Delta \times \Delta$, one has

†For $\underset{\sim}{A} \oplus \underset{\sim}{B}$ the identity is ϕ. If one puts $a = \mu_{\underset{\sim}{A}}(x)$ and $b = \mu_{\underset{\sim}{B}}(x)$, with $0 < a < b < 1$,

$$a * b = (a \wedge \bar{b}) \vee (\bar{a} \wedge b),$$

one may never have $a * b = 0$; thus, there exists no number b to associate with a as its inverse. Likewise for $\underset{\sim}{A} \,\overline{\oplus}\, \underset{\sim}{B}$.

‡Recall that the fuzzy subsets of **E** form a set written $\mathcal{P}(E)$; it then amounts to the same thing to say either **E** or $\mathcal{P}(E)$ is provided with the law $*$ if one considers that $*$ applies to the fuzzy subsets of **E**.

(46.32) $$\underset{\sim}{A} * \underset{\sim}{B} \in \boldsymbol{\Delta},$$

one says that Δ is closed with respect to $*$.

Consider, for example, the groupoid already presented in Figure 45.2. One may verify that

(46.33) $$\boldsymbol{\Delta}_1 = \{(0,0)\ ,\ (0\ ,\ \tfrac{1}{2})\ ,\ (\tfrac{1}{2}\ ,\ 0)\} \text{ is closed,}$$

(46.34) $$\boldsymbol{\Delta}_2 = \{(\tfrac{1}{2}\ ,\ 1)\ ,\ (1\ ,\ \tfrac{1}{2})\} \text{ is not closed.}$$

In Figure 46.1 we have represented the same groupoid as that in Figure 45.2 but with the law $\cup$:

(46.35) $$\boldsymbol{\Delta}_1 = \{(0,0)\ ,\ (0\ ,\ \tfrac{1}{2})\ ,\ (\tfrac{1}{2}\ ,\ 0)\} \text{ is not closed,}$$

(46.36) $$\boldsymbol{\Delta}_2 = \{(\tfrac{1}{2}\ ,\ 1)\ ,\ (1\ ,\ \tfrac{1}{2})\} \text{ is not closed,}$$

(46.37) $$\boldsymbol{\Delta}_3 = \{(\tfrac{1}{2}\ ,\ 1)\ ,\ (1\ ,\ \tfrac{1}{2})\ ,\ (1,1)\} \text{ is closed.}$$

$\underset{\sim}{\mathcal{P}}(\mathbf{E})$ \\ $\underset{\sim}{\mathcal{P}}(\mathbf{E})$	$(0, 0)$	$(0, \frac{1}{2})$	$(\frac{1}{2}, 0)$	$(\frac{1}{2}, \frac{1}{2})$	$(0, 1)$	$(1, 0)$	$(\frac{1}{2}, 1)$	$(1, \frac{1}{2})$	$(1, 1)$
$(0, 0)$	$(0, 0)$	$(0, \frac{1}{2})$	$(\frac{1}{2}, 0)$	$(\frac{1}{2}, \frac{1}{2})$	$(0, 1)$	$(1, 0)$	$(\frac{1}{2}, 1)$	$(1, \frac{1}{2})$	$(1, 1)$
$(0, \frac{1}{2})$	$(0, \frac{1}{2})$	$(0, \frac{1}{2})$	$(\frac{1}{2}, \frac{1}{2})$	$(\frac{1}{2}, \frac{1}{2})$	$(0, 1)$	$(1, \frac{1}{2})$	$(\frac{1}{2}, 1)$	$(1, \frac{1}{2})$	$(1, 1)$
$(\frac{1}{2}, 0)$	$(\frac{1}{2}, 0)$	$(\frac{1}{2}, \frac{1}{2})$	$(\frac{1}{2}, 0)$	$(\frac{1}{2}, \frac{1}{2})$	$(\frac{1}{2}, 1)$	$(1, 0)$	$(\frac{1}{2}, 1)$	$(1, \frac{1}{2})$	$(1, 1)$
$(\frac{1}{2}, \frac{1}{2})$	$(\frac{1}{2}, \frac{1}{2})$	$(\frac{1}{2}, \frac{1}{2})$	$(\frac{1}{2}, \frac{1}{2})$	$(\frac{1}{2}, \frac{1}{2})$	$(\frac{1}{2}, 1)$	$(1, \frac{1}{2})$	$(\frac{1}{2}, 1)$	$(1, \frac{1}{2})$	$(1, 1)$
$(0, 1)$	$(0, 1)$	$(0, 1)$	$(\frac{1}{2}, 1)$	$(\frac{1}{2}, 1)$	$(0, 1)$	$(1, 1)$	$(\frac{1}{2}, 1)$	$(1, 1)$	$(1, 1)$
$(1, 0)$	$(1, 0)$	$(1, \frac{1}{2})$	$(1, 0)$	$(1, \frac{1}{2})$	$(1, 1)$	$(1, 0)$	$(1, 1)$	$(1, \frac{1}{2})$	$(1, 1)$
$(\frac{1}{2}, 1)$	$(\frac{1}{2}, 1)$	$(\frac{1}{2}, 1)$	$(\frac{1}{2}, 1)$	$(\frac{1}{2}, 1)$	$(\frac{1}{2}, 1)$	$(1, 1)$	$(\frac{1}{2}, 1)$	$(1, 1)$	$(1, 1)$
$(1, \frac{1}{2})$	$(1, \frac{1}{2})$	$(1, \frac{1}{2})$	$(1, \frac{1}{2})$	$(1, \frac{1}{2})$	$(1, 1)$	$(1, \frac{1}{2})$	$(1, 1)$	$(1, \frac{1}{2})$	$(1, 1)$
$(1, 1)$	$(1, 1)$	$(1, 1)$	$(1, 1)$	$(1, 1)$	$(1, 1)$	$(1, 1)$	$(1, 1)$	$(1, 1)$	$(1, 1)$

Fig. 46.1

It is interesting to show how to obtain closed subsets for the examples of Figures 45.2 and 46.1 with the aid of a Hasse diagram of the vector lattice representing $\underset{\sim}{\mathcal{P}}(\mathbf{E})$. See Figure 46.2.

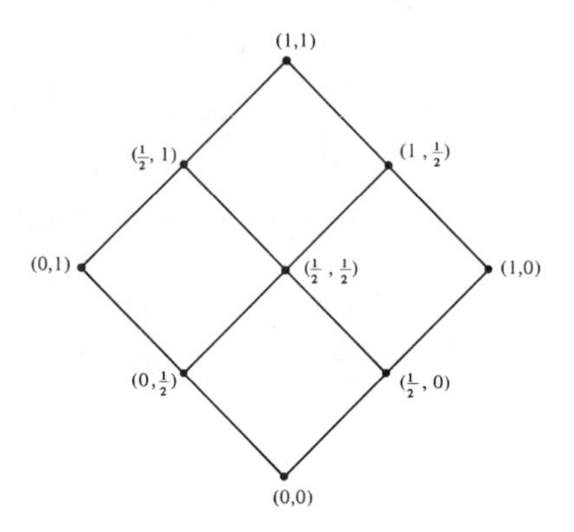

Fig. 46.2

The rule is as follows. For the operation $\cap$, any subset of $\underset{\sim}{\mathcal{P}}(\mathbf{E})$, in order to be closed, must contain the inferior limit of any pair $(\underset{\sim}{\mathbf{A}}, \underset{\sim}{\mathbf{B}})$, $\underset{\sim}{\mathbf{A}}, \underset{\sim}{\mathbf{B}} \in \Delta$. Thus, $\{(0, 0), (0, \frac{1}{2}), (\frac{1}{2}, 0), (1, 0)\}$ is closed for $\cap$. One may see this in Figure 46.2. On the other hand, $\{(0, \frac{1}{2}), (\frac{1}{2}, 0), (\frac{1}{2}, 1), (1, \frac{1}{2})\}$ is not closed for $\cap$. One applies the same rule for the operation $\cup$, but considers superior limits. Thus, $\{(0, 0), (0, \frac{1}{2}), (\frac{1}{2}, 0), (1, 0)\}$ is not closed for $\cup$, whereas $\{(0, \frac{1}{2}), (1, 0), (1, \frac{1}{2}), (1, 1)\}$ is closed with respect to $\cup$.

This property is general for any $\underset{\sim}{\mathcal{P}}(\mathbf{E})$, whatever $\mathbf{E}$ may be since, as we have seen, $\underset{\sim}{\mathcal{P}}(\mathbf{E})$ always forms a vector lattice for the inclusion relation (see Section 6) $\underset{\sim}{\mathbf{A}} \subset \underset{\sim}{\mathbf{B}}$ [that is, $\mu_{\underset{\sim}{\mathbf{A}}}(x) \leqslant \mu_{\underset{\sim}{\mathbf{B}}}(x)$] to which one may always refer $\underset{\sim}{\mathbf{A}} \cap \underset{\sim}{\mathbf{B}}$ and $\underset{\sim}{\mathbf{A}} \cup \underset{\sim}{\mathbf{B}}$.

Subgroupoids. Any subset $\Delta \subset \underset{\sim}{\mathcal{P}}(\mathbf{E})$ closed for a law $*$ will be called a *subgroupoid* of $(\mathbf{E}, *)$ and be denoted $(\Delta \subset \mathbf{E}, *)$ or $(\Delta, *)$ if confusion is not feared.

47. FUZZY MONOIDS

Any fuzzy groupoid that is associative and has an identity will be called a *fuzzy monoid.* We note that some authors do not require the obligatory presence of an identity is this definition, but we shall impose this requirement for all that is to be considered here.

If a monoid possesses in addition the property of commutativity, one calls it a *commutative monoid.*

All of the following fuzzy groupoids, defined by their membership functions and the internal law specified and indicated below, are monoids that are, moreover, commutative:

(47.1) $\quad (\underset{\sim}{\mathcal{P}}(\mathbf{E}), \cap)$ where $\mu_{\underset{\sim}{\mathbf{A}} \cap \underset{\sim}{\mathbf{B}}}(x) = \mu_{\underset{\sim}{\mathbf{A}}}(x) \wedge \mu_{\underset{\sim}{\mathbf{B}}}(x)$, $\underset{\sim}{\mathbf{A}}, \underset{\sim}{\mathbf{B}} \subset \mathbf{E}$.

Associativity is evident. The identity is the reference set $\mathbf{E}$.

(47.2) $\quad (\underset{\sim}{\mathcal{P}}(\mathbf{E}), \cup)$ where $\mu_{\underset{\sim}{\mathbf{A}} \cap \underset{\sim}{\mathbf{B}}}(x) = \mu_{\underset{\sim}{\mathbf{A}}}(x) \vee \mu_{\underset{\sim}{\mathbf{B}}}(x)$, $\underset{\sim}{\mathbf{A}}, \underset{\sim}{\mathbf{B}} \subset \mathbf{E}$.

Associativity is evident. The identity is ϕ.

(47.3) $(\mathcal{P}(\mathrm{E}), \cdot)$ where $\mu_{\underset{\sim}{\mathrm{A}} \cdot \underset{\sim}{\mathrm{B}}}(x) = \mu_{\underset{\sim}{\mathrm{A}}}(x) \cdot \mu_{\underset{\sim}{\mathrm{B}}}(x)$, $\underset{\sim}{\mathrm{A}}, \underset{\sim}{\mathrm{B}} \subset \mathrm{E}$.

This is associative, with identity **E**.

(47.4) $(\mathcal{P}(\mathrm{E}), \hat{+})$ where $\mu_{\underset{\sim}{\mathrm{A}} \hat{+} \underset{\sim}{\mathrm{B}}}(x) = \mu_{\underset{\sim}{\mathrm{A}}}(x) + \mu_{\underset{\sim}{\mathrm{B}}}(x) - \mu_{\underset{\sim}{\mathrm{A}}}(x) \cdot \mu_{\underset{\sim}{\mathrm{B}}}(x)$, $\underset{\sim}{\mathrm{A}}, \underset{\sim}{\mathrm{B}} \subset \mathrm{E}$.

Associative, with identity ϕ.†

(47.5) $(\mathcal{P}(\mathrm{E}), \oplus)$ where $\mu_{\underset{\sim}{\mathrm{A}} \oplus \underset{\sim}{\mathrm{B}}}(x) = [\mu_{\underset{\sim}{\mathrm{A}}}(x) \wedge (1 - \mu_{\underset{\sim}{\mathrm{B}}}(x))] \vee [\mu_{\underset{\sim}{\mathrm{B}}}(x) \wedge (1 - \mu_{\underset{\sim}{\mathrm{A}}}(x))]$, $\underset{\sim}{\mathrm{A}}, \underset{\sim}{\mathrm{B}} \subset \mathrm{E}$.

Associative, with identity ϕ.

A fuzzy monoid will be denoted $(\mathbf{E}, *)$ or, preferably, $(\mathcal{P}(\mathbf{E}), *)$.

We shall see several fuzzy groupoids that are not monoids:

Example I. Let $\underset{\sim}{\mathrm{A}} * \underset{\sim}{\mathrm{B}}$ be such that

(47.6) $$\mu_{\underset{\sim}{\mathrm{A}} * \underset{\sim}{\mathrm{B}}}(x) = |\mu_{\underset{\sim}{\mathrm{A}}}(x) - \mu_{\underset{\sim}{\mathrm{B}}}(x)|.$$

Put

(47.7) $$a = \mu_{\underset{\sim}{\mathrm{A}}}(x) \quad , \quad b = \mu_{\underset{\sim}{\mathrm{B}}}(x) \quad , \quad c = \mu_{\underset{\sim}{\mathrm{C}}}(x),$$

and denote

(47.8) $$a \odot b = |a - b|.$$

It is easy to show that

(47.9) $$(a \odot b) \odot c \neq a \odot (b \odot c).$$

that is,

(47.10) $$||a - b| - c| \neq |a - |b - c||.$$

For example, if

(47.11) $$a = 0{,}3 \quad , \quad b = 0{,}5 \quad , \quad c = 0{,}9 .$$

one has

(47.12) $$\begin{aligned} ||a - b| - c| &= ||0{,}3 - 0{,}5| - 0{,}9| \\ &= |0{,}2 - 0{,}9| = 0{,}7 . \end{aligned}$$

(47.13) $$\begin{aligned} |a - |b - c|| &= |0{,}3 - |0{,}5 - 0{,}9|| \\ &= |0{,}3 - 0{,}4| = 0{,}1 . \end{aligned}$$

This commutative groupoid is not a monoid since it is not associative.

Example II. With the notation of (47.7), put

(47.14) $$a \odot b = a + kb - ab \ , \quad k \in [0, 1].$$

†Equations (47.3) and (47.4) will give internal laws subject to the condition that $\mathbf{M} = [0, 1]$ or $\mathbf{M} = \{0, 1\}$. One would not have internal laws, for example, for $\mathbf{M} = \{0, \frac{1}{2}, 1\}$ since $(\frac{1}{2})(\frac{1}{2}) = \frac{1}{4} \notin \mathbf{M}$.

One has

$$(a \odot b) \odot c = (a + kb - ab) + kc - (a + kb - ab)\,c \qquad (47.15)$$
$$= a + kb + kc - ab - ac - kbc + abc\,,$$

$$a \odot (b \odot c) = a + k\,(b + kc - bc) - a\,(b + kc - bc) \qquad (47.16)$$
$$= a + kb + k^2 c - ab - kac - kbc + abc,$$

$$(a \odot b) \odot c - a \odot (b \odot c) = kc - k^2 c - ac + kac \qquad (47.17)$$
$$= c\,(1 - k)\,(k - a)\,.$$

Thus, associativity does not hold, unless $k = 1$.

Fuzzy submonoid. Let $(\underset{\sim}{\mathscr{P}}(\mathbf{E}), *)$ be a fuzzy monoid and $\Delta \subset \underset{\sim}{\mathscr{P}}(\mathbf{E})$ be closed for the law $*$, then Δ will be called a *fuzzy submonoid of* $(\underset{\sim}{\mathscr{P}}(\mathbf{E}), *)$ and will be designated by $(\Delta, *)$.

Example. Consider the monoid $(\underset{\sim}{\mathscr{P}}(\mathbf{E}), \cup)$ represented in Figure 46.1 .

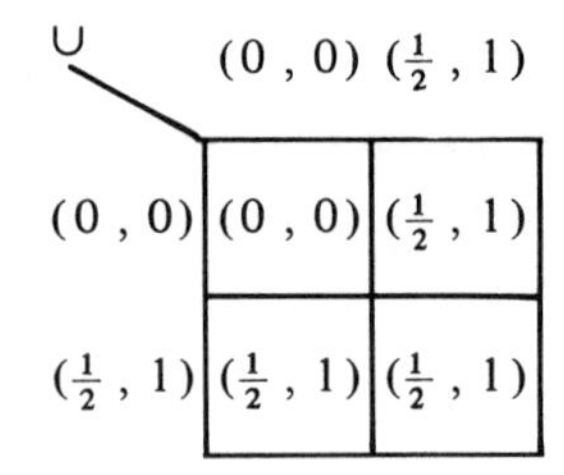

$\cup$	$(0, 0)$	$(\frac{1}{2}, 1)$
$(0, 0)$	$(0, 0)$	$(\frac{1}{2}, 1)$
$(\frac{1}{2}, 1)$	$(\frac{1}{2}, 1)$	$(\frac{1}{2}, 1)$

Fig. 47.1

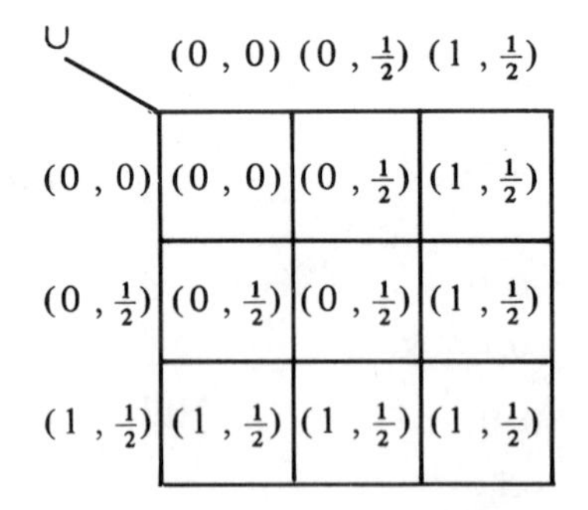

$\cup$	$(0, 0)$	$(0, \frac{1}{2})$	$(1, \frac{1}{2})$
$(0, 0)$	$(0, 0)$	$(0, \frac{1}{2})$	$(1, \frac{1}{2})$
$(0, \frac{1}{2})$	$(0, \frac{1}{2})$	$(0, \frac{1}{2})$	$(1, \frac{1}{2})$
$(1, \frac{1}{2})$	$(1, \frac{1}{2})$	$(1, \frac{1}{2})$	$(1, \frac{1}{2})$

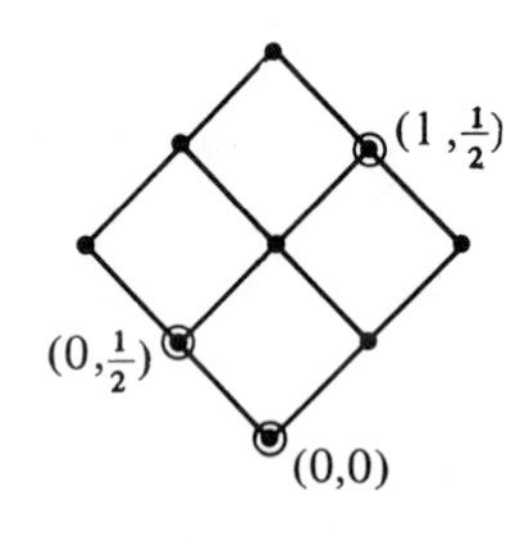

Fig. 47.2

Figures 47.1–47.3 represent submonoids of this monoid:

$$\mathbf{\Delta} = \{(0, 0), (\tfrac{1}{2}, 1)\} \qquad (47.18)$$

$$\mathbf{\Delta}' = \{(0, 0), (0, \tfrac{1}{2}), (1, \tfrac{1}{2})\} \qquad (47.19)$$

$$\mathbf{\Delta}'' = \{(0, 0), (0, \tfrac{1}{2}), (\tfrac{1}{2}, 0), (\tfrac{1}{2}, \tfrac{1}{2}), (\tfrac{1}{2}, 1)\}. \qquad (47.20)$$

There are several others of these which the reader might enumerate as an exercise.

Of course, all these monoids must include the identity (0, 0) [see (47.2)].

$\cup$	$(0, 0)$	$(0, \frac{1}{2})$	$(\frac{1}{2}, 0)$	$(\frac{1}{2}, \frac{1}{2})$	$(\frac{1}{2}, 1)$
$(0, 0)$	$(0, 0)$	$(0, \frac{1}{2})$	$(\frac{1}{2}, 0)$	$(\frac{1}{2}, \frac{1}{2})$	$(\frac{1}{2}, 1)$
$(0, \frac{1}{2})$	$(0, \frac{1}{2})$	$(0, \frac{1}{2})$	$(\frac{1}{2}, \frac{1}{2})$	$(\frac{1}{2}, \frac{1}{2})$	$(\frac{1}{2}, 1)$
$(\frac{1}{2}, 0)$	$(\frac{1}{2}, 0)$	$(\frac{1}{2}, \frac{1}{2})$	$(\frac{1}{2}, 0)$	$(\frac{1}{2}, \frac{1}{2})$	$(\frac{1}{2}, 1)$
$(\frac{1}{2}, \frac{1}{2})$	$(\frac{1}{2}, \frac{1}{2})$	$(\frac{1}{2}, \frac{1}{2})$	$(\frac{1}{2}, \frac{1}{2})$	$(\frac{1}{2}, \frac{1}{2})$	$(\frac{1}{2}, 1)$
$(\frac{1}{2}, 1)$	$(\frac{1}{2}, 1)$	$(\frac{1}{2}, 1)$	$(\frac{1}{2}, 1)$	$(\frac{1}{2}, 1)$	$(\frac{1}{2}, 1)$

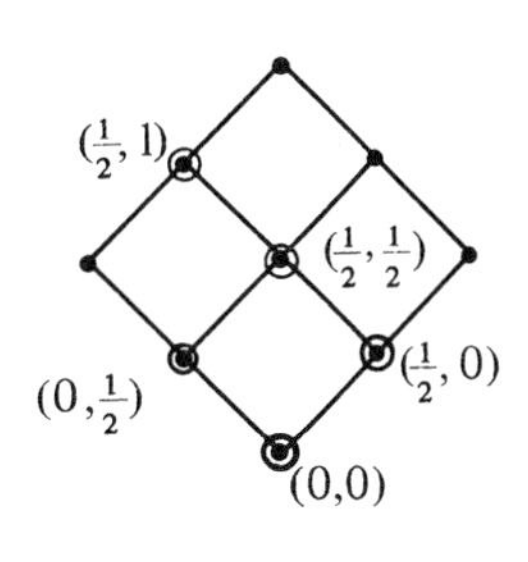

Fig. 47.3

Theorem. If $(\Delta, *)$ and $(\Delta', *)$ are submonoids of $(\underset{\sim}{\mathcal{P}}(\mathbf{E}), *)$, then

$$(\Delta \cap \Delta', *)$$

is a submonoid of $(\underset{\sim}{\mathcal{P}}(\mathbf{E}), *)$.

Proof. That intersection preserves associativity and the identity is evident. We show then that $\Delta \cap \Delta'$ remains closed for $*$.

Let $(\underset{\sim}{\mathbf{A}}, \underset{\sim}{\mathbf{B}}) \in \Delta \cap \Delta'$. Then $\underset{\sim}{\mathbf{A}} * \underset{\sim}{\mathbf{B}}$ belongs to Δ by hypothesis; it also belongs to Δ' by hypothesis (otherwise Δ or/and Δ' would not be closed for $*$); then $\underset{\sim}{\mathbf{A}} * \underset{\sim}{\mathbf{B}}$ belongs to $\Delta \cap \Delta'$ and $\Delta \cap \Delta'$ is closed with respect to $*$.

It is not the same for union $\cup$, which does not always preserve the closure property.

Fuzzy groups. One may ask the following interesting question: Do there exist actual groups that are fuzzy (not ordinary) if one considers the operations $\cap, \cup, \cdot, \hat{+}, \oplus$.

One knows that a group is a monoid such that each element possesses one and only one inverse.

In Chapter V, we shall show that a necessary condition for $(\underset{\sim}{\mathcal{P}}(\mathbf{E}), *)$ to have a group structure is that $\mathbf{M} = [0, 1]$ also have a group structure for an operation corresponding to $*$. We shall see that in any case $\mathbf{M} = [0, 1]$ may be endowed with a group structure for an operation $\circ$ to be defined.

$\mathbf{M} = [0, 1]$ is a vector lattice that may be reduced to a single chain forming a total order. We consider the operations $\wedge$ (min), $\vee$ (max), . (product), $\hat{+}$ (algebraic sum), $\oplus$ (disjunctive sum). For each of these operations, one has the associative property and there exists an identity, which is, depending on the case, 0 or 1; but it is easy to prove, almost in the same way, that for each of these operations, there does not exist an inverse for each element. We show this for $\wedge$. Consider a pair

$$(a\,,\,b) \in \mathbf{M} \times \mathbf{M}\,,$$

$\mathbf{M} = [0, 1]$ and such that $0 < a < b < 1$. The identity of $\wedge$ is 1. Does there exist then an a or a b such that

$$a \wedge b = 1\,, \tag{47.21}$$

This is impossible:

$$a \wedge b = a < 1\,. \tag{47.22}$$

On the other hand, if one takes $\mathbf{M} = \{0, 1\}$, one finds that a group is possible.

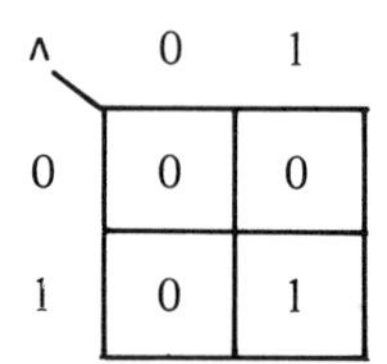

∧	0	1
0	0	0
1	0	1

This is not a group.
The identity is 1, but:

$0 \wedge 0 = 0\,,$
$0 \wedge 1 = 0\,,$
$1 \wedge 0 = 0\,,$
$1 \wedge 1 = 1\,,$

0 does not have an inverse.

∨	0	1
0	0	1
1	1	1

This is not a group.
The identity is 0, but:

$0 \vee 0 = 0\,,$
$0 \vee 1 = 1\,,$
$1 \vee 0 = 1\,,$
$1 \vee 1 = 1\,,$

1 does not have an inverse.

⊕	0	1
0	0	1
1	1	0

This is a group.
The identity is 0.
The inverse of 0 is 0.
The inverse of 1 is 1.

$\overline{\oplus}$	0	1
0	1	0
1	0	1

This is a group.
The identity is 1.
The inverse of 0 is 0.
The inverse of 1 is 1.

Fig. 47.6

Thus, we show in Figure 47.6 that one does not obtain a group for $\wedge$ or $\vee$ (one thus does not obtain a group any longer for $\cdot$ and $\hat{+}$, which in the boolean case give equivalent operations). On the contrary, one does obtain a group if one takes the operation $\oplus$. One also obtains a group if one considers the operation $\overline{\oplus}$ (inverse disjunctive sum). We note that the two groups $\oplus$ and $\overline{\oplus}$ are isomorphic† by permiting 0 and 1; a single group may represent the two.

It follows from this that if one considers any one of the operations $\cap$, $\cup$, $\cdot$, $\hat{+}$, $\oplus$, and $\mathbf{M} = [0, 1]$, one may not give $(\mathcal{P}(\mathbf{E}), *)$ a group structure.

If one takes $\mathbf{M} = \{0, 1\}$, it is only with $\oplus$ (or what amounts to the same thing, with $\overline{\oplus}$) that one may form a group. We consider as an example the ordinary group formed thus with

(47.23) $$\mathbf{E} = \{x, x_2, x_3\}.$$

$\oplus$	000	001	010	011	100	101	110	111
000	000	001	010	011	100	101	110	111
001	001	000	011	010	101	100	111	110
010	010	011	000	001	110	111	100	101
011	011	010	001	000	111	110	101	100
100	100	101	110	111	000	001	010	011
101	101	100	111	110	001	000	011	010
110	110	111	100	101	010	011	000	001
111	111	110	101	100	011	010	001	000

Fig. 47.7

If one puts

(47.24) $$abc = \{(x_1 \mid a), (x_2 \mid b), (x_3 \mid c)\}$$

in order to simplify writing, with

(47.25) $$a, b, c \in \{0, 1\},$$

one obtains the group represented in Figure 47.7. The identity is 000 and each element abc has itself for an inverse. This group $(\mathcal{P}(\mathbf{E}), \oplus)$ has been represented in Figure 47.8 by replacing the binary numbers abc by their corresponding decimals. One correctly notices certain properties (subgroups, latin squares, etc.). These properties are very general for these groups constructed with $\oplus$.

We shall return in Chapter V to all that corresponds to structures or configurations of a membership set **M**, which we shall generalize by examining other totally ordered configurations for **M**.

†For those in need of an explication of the theory of categories of sets, we shall present or review in Section 57 what is intended by the word *morphism*. The reader who does not know or has forgotten what an isomorphism is will see this further on.

	0	1	2	3	4	5	6	7
0	0	1	2	3	4	5	6	7
1	1	0	3	2	5	4	7	6
2	2	3	0	1	6	7	4	5
3	3	2	1	0	7	6	5	4
4	4	5	6	7	0	1	2	3
5	5	4	7	6	1	0	3	2
6	6	7	4	5	2	3	0	1
7	7	6	5	4	3	2	1	0

Fig. 47.8

48. FUZZY EXTERNAL COMPOSITION

Let $\mathbf{E}_1$ and $\mathbf{E}_2$ be two sets. If to each ordered pair $(\underset{\sim}{\mathbf{A}}_1, \underset{\sim}{\mathbf{A}}_2)$, $\underset{\sim}{\mathbf{A}}_1 \subset \mathbf{E}_1$, $\underset{\sim}{\mathbf{A}}_2 \subset \mathbf{E}_2$, one may correspond one and only one $\underset{\sim}{\mathbf{A}}_3 \subset \mathbf{E}_3$, one has a law of fuzzy external composition if $\mathbf{E}_3 \neq \mathbf{E}_1$ or/and $\mathbf{E}_3 \neq \mathbf{E}_2$. If $\mathbf{E}_3 = \mathbf{E}_1 = \mathbf{E}_2$, the law is internal.

We shall consider several examples of laws of fuzzy external composition.

Example I. We see first a purely discrete example. Let

(48.1)
(48.2) $\mathbf{E}_1 = \{A, B, C\}$, $\mathbf{M}_1 = \{0, \frac{1}{4}, \frac{1}{2}, \frac{3}{4}, 1\}$, card $\mathbf{E}_1 = 3$, card $\mathbf{M}_1 = 5$,

(48.3)
(48.4) $\mathbf{E}_2 = \{a, b, c, d\}$, $\mathbf{M}_2 = \{0, \frac{1}{2}, 1\}$, card $\mathbf{E}_2 = 4$, card $\mathbf{M}_2 = 3$,

(48.5)
(48.6) $\mathbf{E}_3 = \{\alpha, \beta\}$, $\mathbf{M}_3 = \{0, \frac{1}{3}, \frac{2}{3}, 1\}$, card $\mathbf{E}_3 = 2$, card $\mathbf{M}_3 = 4$,

Let $\underset{\sim}{\mathbf{A}}_1 \subset \mathbf{E}_1$ and $\underset{\sim}{\mathbf{A}}_2 \subset \mathbf{E}_2$; to each ordered pair such as $(\underset{\sim}{\mathbf{A}}_1, \underset{\sim}{\mathbf{A}}_2)$ we correspond one and only one $\underset{\sim}{\mathbf{A}}_3 \subset \mathbf{E}_3$ by means of a table. Thus, let

(48.7) $\underset{\sim}{\mathbf{A}}_1 = \{(A \mid \frac{1}{4}), (B \mid \frac{1}{2}), (C \mid 1)\}$ denoted $(\frac{1}{4}, \frac{1}{2}, 1)$,

(48.8) $\underset{\sim}{\mathbf{A}}_2 = \{(a \mid 0), (b \mid \frac{1}{2}), (c \mid 0), (d \mid 1)\}$ denoted $(0, \frac{1}{2}, 0, 1)$;

We suppose that the table corresponds to these two fuzzy subsets

(48.9) $\underset{\sim}{\mathbf{A}}_3 = \{(\alpha \mid \frac{1}{3}), (\beta \mid 1)\}$ denoted $(\frac{1}{3}, 1)$.

The table will possess $5^3 \times 3^4 = 125 \times 81$ cases; we do not present this, but give a small extract in Figure 48.1.

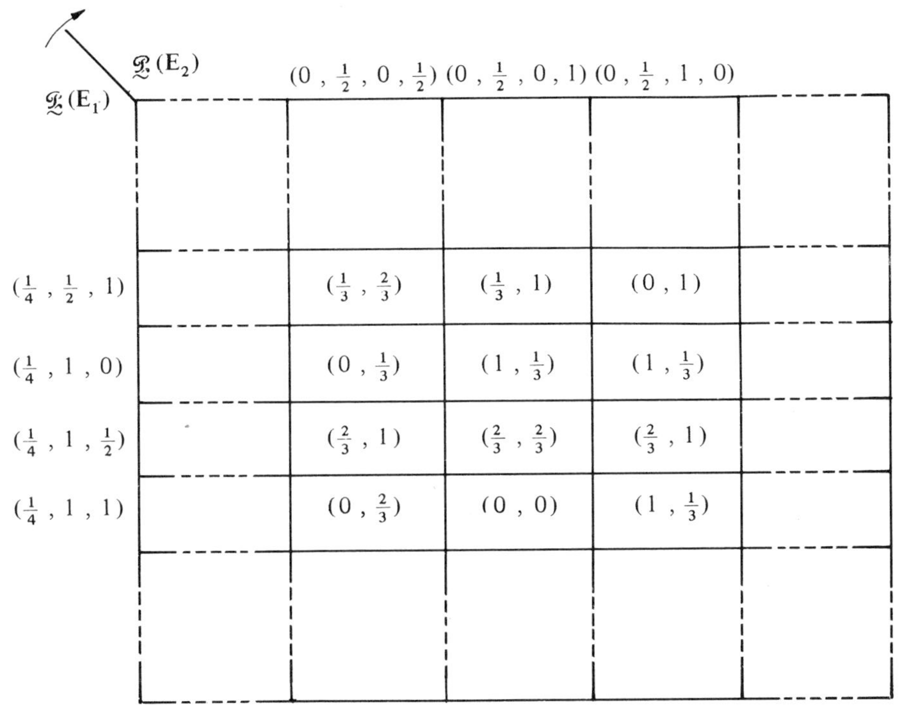

Fig. 48.1

Example II. We take the same example as above, but with the law

$$\mu_{\underset{\sim}{A}_3}(\alpha) = \bigwedge_x \bigwedge_y [\mu_{\underset{\sim}{A}_1}(x) \vee \mu_{\underset{\sim}{A}_2}(y)], \tag{48.10}$$

$$\mu_{\underset{\sim}{A}_3}(\beta) = \bigvee_x \bigvee_y [\mu_{\underset{\sim}{A}_1}(x) \wedge \mu_{\underset{\sim}{A}_2}(y)], \tag{48.11}$$

one obtains another composition table from which one goes on to calculate an element of $\underset{\sim}{\mathscr{P}}(E_1) \times \underset{\sim}{\mathscr{P}}(E_2)$. Let $\underset{\sim}{A}_1$ be given by (48.7) and $\underset{\sim}{A}_2$ be given by (48.8). One has

$$\begin{aligned}
(48.12)\quad \mu_{\underset{\sim}{A}_3}(\alpha) &= \bigwedge_x [\bigwedge_y (\tfrac{1}{4} \vee 0, \tfrac{1}{4} \vee \tfrac{1}{2}, \tfrac{1}{4} \vee 0, \tfrac{1}{4} \vee 1),\\
&\qquad \bigwedge_y (\tfrac{1}{2} \vee 0, \tfrac{1}{2} \vee \tfrac{1}{2}, \tfrac{1}{2} \vee 0, \tfrac{1}{2} \vee 1),\\
&\qquad \bigwedge_y (1 \vee 0, 1 \vee \tfrac{1}{2}, 1 \vee 0, 1 \vee 1)]\\
&= \bigwedge_x [\bigwedge_y (\tfrac{1}{4}, \tfrac{1}{2}, \tfrac{1}{4}, 1), \bigwedge_y (\tfrac{1}{2}, \tfrac{1}{2}, \tfrac{1}{2}, 1), \bigwedge_y (1, 1, 1, 1)]\\
&= \bigwedge_x (\tfrac{1}{4}, \tfrac{1}{2}, 1) = \tfrac{1}{4}.
\end{aligned}$$

(48.13) $$\mu_{\underset{\sim}{A}_3}(\beta) = \bigvee_x [\bigvee_y (\tfrac{1}{4} \wedge 0\,,\ \tfrac{1}{4} \wedge \tfrac{1}{2}\,,\ \tfrac{1}{4} \wedge 0\,,\ \tfrac{1}{4} \wedge 1),$$
$$\bigvee_y (\tfrac{1}{2} \wedge 0\,,\ \tfrac{1}{2} \wedge \tfrac{1}{2}\,,\ \tfrac{1}{2} \wedge 0\,,\ \tfrac{1}{2} \wedge 1),$$
$$\bigvee_y (1 \wedge 0\,,\ 1 \wedge \tfrac{1}{2}\,,\ 1 \wedge 0\,,\ 1 \wedge 1)]$$
$$= \bigvee_x [\bigvee_y (0\,,\ \tfrac{1}{4}\,,\ 0\,,\ \tfrac{1}{4})\,,\ \bigvee_y (0\,,\ \tfrac{1}{2}\,,\ 0\,,\ \tfrac{1}{2})\,,\ \bigvee_y (0\,,\ \tfrac{1}{2}\,,\ 0\,,\ 1)]$$
$$= \bigvee_x (\tfrac{1}{4}\,,\ \tfrac{1}{2}\,,\ 1) = 1\,.$$

Thus

(48.14) $$\mu_{\underset{\sim}{A}_3}(\alpha) = \tfrac{1}{4} \quad \text{and} \quad \mu_{\underset{\sim}{A}_3}(\beta) = 1\,.$$

To

(48.15) $$\underset{\sim}{A}_1 = (A \mid \tfrac{1}{4})\,,\ (B \mid \tfrac{1}{2})\,,\ (C \mid 1)$$

and

(48.16) $$\underset{\sim}{A}_2 = (a \mid 0)\,,\ (b \mid \tfrac{1}{2})\,,\ (c \mid 0)\,,\ (d \mid 1)\,,$$

one corresponds

(48.16′) $$\underset{\sim}{A}_3 = (\alpha \mid \tfrac{1}{4})\,,\ (\beta \mid 1)\,.$$

Remark. In the general case, let

$\mathbf{M}_1$ be associated with $\mathbf{E}_1$;
$\mathbf{M}_2$ be associated with $\mathbf{E}_2$;
$\mathbf{M}_3$ be associated with $\mathbf{E}_3$.

If $\underset{\sim}{\mathcal{P}}(E_3)$ is formed with respect to $\underset{\sim}{\mathcal{P}}(E_1)$ and $\underset{\sim}{\mathcal{P}}(E_2)$ by a law $*$ corresponding to

(48.17) $$\mu_{\underset{\sim}{A}_3}(x\,,y) = \mu_{\underset{\sim}{A}_1}(x) \odot \mu_{\underset{\sim}{A}_2}(y)\,,$$

$\mathbf{M}_3$ will be deduced from $\mathbf{M}_1$ and $\mathbf{M}_2$ by considering the formula of composition (48.17). Thus, in the example of (48.10) and (48.11), it is evident that

(48.18) $$\mathbf{M}_3 = \mathbf{M}_1 \cup \mathbf{M}_2 = \mathbf{M}_1 = \{0\,,\ \tfrac{1}{4}\,,\ \tfrac{1}{2}\,,\ \tfrac{3}{4}\,,\ 1\}.$$

Of course, (48.18) may not be a general formula.

In Section 36 we have shown how to compose intervals for the operations $\wedge$ and $\vee$. Analagous procedures may be employed for more general cases.

Example III. We shall construct a fuzzy graph whose vertices are fuzzy subsets; this will define a law of external composition.

Let

(48.19) $$\underset{\sim}{A} \subset E\,,\ \underset{\sim}{B} \subset E\,,$$

To any ordered pair $(\underset{\sim}{A}, \underset{\sim}{B}) \in \underset{\sim}{\mathcal{P}}(E) \times \underset{\sim}{\mathcal{P}}(E)$ one will correspond an element denoted

(48.20) $$\underset{\sim}{A} * \underset{\sim}{B} = c(\underset{\sim}{A}\,,\underset{\sim}{B})$$

The element c takes its values in a set $\mathbf{Q}$ defined by the operation $*$.

Suppose, for example, that

(48.21)
(48.22)
$$E = \{a, b\} \quad \text{and} \quad M = \{0, \tfrac{1}{2}, 1\}.$$

Suppose also that

$$c(\underset{\sim}{A}, \underset{\sim}{B}) = [\mu_{\underset{\sim}{A}}(a) \wedge \mu_{\underset{\sim}{B}}(a)] \vee [\mu_{\underset{\sim}{A}}(b) \wedge \mu_{\underset{\sim}{B}}(b)]. \tag{48.23}$$

With such a function, c takes its values in

$$Q = M = \{0, \tfrac{1}{2}, 1\}. \tag{48.24}$$

One obtains the fuzzy graph given in Figure 48.2.

*	$(0, 0)$	$(0, \frac{1}{2})$	$(0, 1)$	$(\frac{1}{2}, 0)$	$(\frac{1}{2}, \frac{1}{2})$	$(\frac{1}{2}, 1)$	$(1, 0)$	$(1, \frac{1}{2})$	$(1, 1)$
$(0, 0)$	0	0	0	0	0	0	0	0	0
$(0, \frac{1}{2})$	0	$\frac{1}{2}$	$\frac{1}{2}$	0	$\frac{1}{2}$	$\frac{1}{2}$	0	$\frac{1}{2}$	$\frac{1}{2}$
$(0, 1)$	0	$\frac{1}{2}$	1	0	$\frac{1}{2}$	1	0	$\frac{1}{2}$	1
$(\frac{1}{2}, 0)$	0	0	0	$\frac{1}{2}$	$\frac{1}{2}$	$\frac{1}{2}$	$\frac{1}{2}$	$\frac{1}{2}$	$\frac{1}{2}$
$(\frac{1}{2}, \frac{1}{2})$	0	$\frac{1}{2}$	$\frac{1}{2}$	$\frac{1}{2}$	$\frac{1}{2}$	$\frac{1}{2}$	$\frac{1}{2}$	$\frac{1}{2}$	$\frac{1}{2}$
$(\frac{1}{2}, 1)$	0	$\frac{1}{2}$	1	$\frac{1}{2}$	$\frac{1}{2}$	1	$\frac{1}{2}$	$\frac{1}{2}$	1
$(1, 0)$	0	0	0	$\frac{1}{2}$	$\frac{1}{2}$	$\frac{1}{2}$	1	1	1
$(1, \frac{1}{2})$	0	$\frac{1}{2}$	$\frac{1}{2}$	$\frac{1}{2}$	$\frac{1}{2}$	$\frac{1}{2}$	1	1	1
$(1, 1)$	0	$\frac{1}{2}$	1	$\frac{1}{2}$	$\frac{1}{2}$	1	1	1	1

Fig. 48.2

In this fashion one may construct fuzzy graphs that possess particular properties due to their construction. This is a conception of fuzzy graphs for which the elements or vertices are fuzzy subsets of the same reference set.

It is a matter here of an extension that may have concrete applications, for example, if $*$ corresponds to an evaluation of distance.

Example IV. Recall Example III and suppose now that $c(\underset{\sim}{\mathbf{A}}, \underset{\sim}{\mathbf{B}})$ is the relative generalized Hamming distance given by

$$\delta(\underset{\sim}{\mathbf{A}}, \underset{\sim}{\mathbf{B}}) = \frac{1}{2}\left(|\mu_{\underset{\sim}{\mathbf{A}}}(a) - \mu_{\underset{\sim}{\mathbf{B}}}(a)| + |\mu_{\underset{\sim}{\mathbf{A}}}(b) - \mu_{\underset{\sim}{\mathbf{B}}}(b)|\right). \tag{48.25}$$

This indeed gives a law of external composition. See Figure 48.3.

*	$(0,0)$	$(0,\frac{1}{2})$	$(0,1)$	$(\frac{1}{2},0)$	$(\frac{1}{2},\frac{1}{2})$	$(\frac{1}{2},1)$	$(1,0)$	$(1,\frac{1}{2})$	$(1,1)$
$(0,0)$	0	$\frac{1}{4}$	$\frac{1}{2}$	$\frac{1}{4}$	$\frac{1}{2}$	$\frac{3}{4}$	$\frac{1}{2}$	$\frac{3}{4}$	1
$(0,\frac{1}{2})$	$\frac{1}{4}$	0	$\frac{1}{4}$	$\frac{1}{2}$	$\frac{1}{4}$	$\frac{1}{2}$	$\frac{3}{4}$	$\frac{1}{2}$	$\frac{3}{4}$
$(0,1)$	$\frac{1}{2}$	$\frac{1}{4}$	0	$\frac{3}{4}$	$\frac{1}{2}$	$\frac{1}{4}$	1	$\frac{3}{4}$	$\frac{1}{2}$
$(\frac{1}{2},0)$	$\frac{1}{4}$	$\frac{1}{2}$	$\frac{3}{4}$	0	$\frac{1}{4}$	$\frac{1}{2}$	$\frac{1}{4}$	$\frac{1}{2}$	$\frac{3}{4}$
$(\frac{1}{2},\frac{1}{2})$	$\frac{1}{2}$	$\frac{1}{4}$	$\frac{1}{2}$	$\frac{1}{4}$	0	$\frac{1}{4}$	$\frac{1}{2}$	$\frac{1}{4}$	$\frac{1}{2}$
$(\frac{1}{2},1)$	$\frac{3}{4}$	$\frac{1}{2}$	$\frac{1}{4}$	$\frac{1}{2}$	$\frac{1}{4}$	0	$\frac{3}{4}$	$\frac{1}{2}$	$\frac{1}{4}$
$(1,0)$	$\frac{1}{2}$	$\frac{3}{4}$	1	$\frac{1}{4}$	$\frac{1}{2}$	$\frac{3}{4}$	0	$\frac{1}{4}$	$\frac{1}{2}$
$(1,\frac{1}{2})$	$\frac{3}{4}$	$\frac{1}{2}$	$\frac{3}{4}$	$\frac{1}{2}$	$\frac{1}{4}$	$\frac{1}{2}$	$\frac{1}{4}$	0	$\frac{1}{4}$
$(1,1)$	1	$\frac{3}{4}$	$\frac{1}{2}$	$\frac{3}{4}$	$\frac{1}{2}$	$\frac{1}{4}$	$\frac{1}{2}$	$\frac{1}{4}$	0

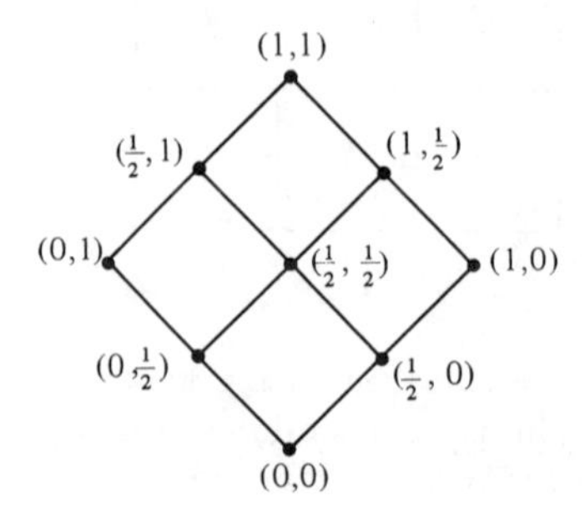

Fig. 48.3

Importance of the notion of a law of external composition of fuzzy subsets. This notion is important; it characterizes any system of evaluation of relations among fuzzy subsets of the same reference set, indeed of fuzzy subsets of different reference sets. The set in which $\mathcal{P}(E_1) \times \mathcal{P}(E_2)$ takes its values may be an ordinary set or more generally a set of fuzzy subsets or an ordinary power set (Figure 48.4).

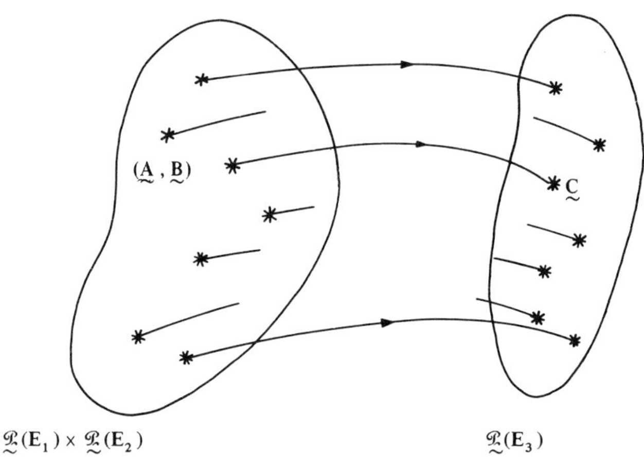

Fig. 48.4

The notion of distance between messages or fuzzy subsets of the same reference set gives an example (one of the more trivial examples) concerning this general notion.

We remark that the procedures for invention or ingenuity that one calls *biassociation*† are procedures essentially based on laws of external composition. One takes a concept **A**, which is an ordinary or fuzzy subset of a family of concepts $\mathbf{E}_1$, and another concept **B**, which is an ordinary or fuzzy subset of another (or eventually the same) family. The

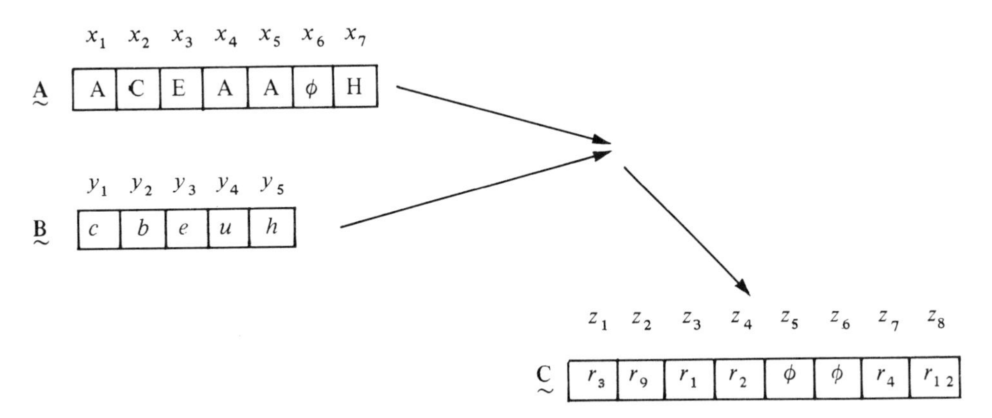

Fig. 48.5. The phenomenon of biassociation.

†See [6K].

biassociation of $\underset{\sim}{\mathbf{A}}$ and $\underset{\sim}{\mathbf{B}}$ is an external law $*$ that allows one to obtain a new concept $\underset{\sim}{\mathbf{C}}$, which is an ordinary or fuzzy subset of a third family (eventually the same as one of the preceding families) (Figure 48.5).

49. OPERATIONS ON FUZZY NUMBERS

We go on to consider various types of fuzzy numbers.

Exponential fuzzy integers. Consider a reference set

(49.1)
$$\mathbf{E} = \mathbf{R}^+ ,$$

and the fuzzy subset $\underset{\sim}{\mathbf{I}}_1$ such that

(49.2)
$$\mu_{\underset{\sim}{\mathbf{I}}_1}(x) = \lambda e^{-\lambda x}, \quad x \in \mathbf{R}^+ .$$

Then define $\underset{\sim}{\mathbf{I}}_2$ in the following fashion:

(49.3)
$$\begin{aligned} \mu_{\underset{\sim}{\mathbf{I}}_2}(x) &= \mu_{\underset{\sim}{\mathbf{I}}_1}(x) * \mu_{\underset{\sim}{\mathbf{I}}_1}(x) \\ &= \int_0^x \mu_{\underset{\sim}{\mathbf{I}}_1}(t) \cdot \mu_{\underset{\sim}{\mathbf{I}}_1}(x - t)\, dt \\ &= \int_0^x \lambda e^{-\lambda t} \cdot \lambda e^{-\lambda (x-t)}\, dt \\ &= \lambda^2 x e^{-\lambda x} . \end{aligned}$$

Next define $\underset{\sim}{\mathbf{I}}_3$ in the following manner:

(49.4)
$$\begin{aligned} \mu_{\underset{\sim}{\mathbf{I}}_3}(x) &= \mu_{\underset{\sim}{\mathbf{I}}_2}(x) * \mu_{\underset{\sim}{\mathbf{I}}_1}(x) = \mu_{\underset{\sim}{\mathbf{I}}_1}(x) * \mu_{\underset{\sim}{\mathbf{I}}_2}(x) \\ &= \int_0^x \lambda^2 t e^{-\lambda t} \cdot \lambda e^{-\lambda (x-t)}\, dt \\ &= \frac{\lambda^3 x^2 e^{-\lambda x}}{2} , \end{aligned}$$

and $\underset{\sim}{\mathbf{I}}_n$:

(49.5)
$$\begin{aligned} \mu_{\underset{\sim}{\mathbf{I}}_n}(x) &= \mu_{\underset{\sim}{\mathbf{I}}_{n-1}}(x) * \mu_{\underset{\sim}{\mathbf{I}}_1}(x) = \mu_{\underset{\sim}{\mathbf{I}}_1}(x) * \mu_{\underset{\sim}{\mathbf{I}}_{n-1}}(x) \\ &= \frac{\lambda^n x^{n-1} e^{-\lambda x}}{n\,!} . \end{aligned}$$

Note that

(49.6)
$$\begin{aligned} \underset{x}{\text{MAX}}\ \mu_{\underset{\sim}{\mathbf{I}}_n}(x) &= \underset{x}{\text{MAX}}\ \frac{\lambda^n x^{n-1} e^{-\lambda x}}{(n-1)\,!} \\ &= \lambda \frac{(n-1)^{(n-1)} e^{-(n-1)}}{(n-1)\,!} , \end{aligned}$$

for the value

(49.7)
$$x = \frac{n-1}{\lambda} .$$

Thus one may establish the values in Table 49.1.

Table 49.1

	$\underset{\sim}{I}_i$	$\mu_{\underset{\sim}{I}_i}(x)$	*Abscissa of the maximum*	*Ordinate of the maximum*
	$\underset{\sim}{I}_1$	$\lambda e^{-\lambda x}$	$x = 0$	λ
	$\underset{\sim}{I}_2$	$\lambda^2 x e^{-\lambda x}$	$x = \frac{1}{\lambda}$	λe^{-1}
(49.8)	$\underset{\sim}{I}_3$	$\frac{\lambda^3 x^2 e^{-\lambda x}}{2}$	$x = \frac{2}{\lambda}$	$\frac{\lambda 2^2 e^{-2}}{2}$
	.	.	.	.
	$\underset{\sim}{I}_n$	$\frac{\lambda^n x^{n-1} e^{-\lambda x}}{(n-1)!}$	$x = \frac{n-1}{\lambda}$	$\lambda \frac{(n-1)^{(n-1)} e^{-(n-1)}}{(n-1)!}$
	.	.	.	.

The fuzzy subsets

(49.9) $$\underset{\sim}{I}_1, \underset{\sim}{I}_2, \underset{\sim}{I}_3, \ldots, \underset{\sim}{I}_n, \ldots$$

will be called exponential fuzzy integers. $\underset{\sim}{I}_2$ will be called exponential fuzzy 1, $\underset{\sim}{I}_2$ will be called exponential fuzzy 2, etc.

The operation of composition defined by (49.3) is associative and commutative; thus, the set of fuzzy subsets $\underset{\sim}{I}_1$, $\underset{\sim}{I}_2$, $\underset{\sim}{I}_3$, . . . , $\underset{\sim}{I}_n$, . . . forms an associative and commutative groupoid.

This groupoid, moreover, has an identity, which we denote by $\underset{\sim}{I}_0$ and which is such that

(49.10) $$\mu_{\underset{\sim}{I}_0}(x) = \delta(x)$$

where $\delta(x)$ is the Dirac function such that

(40.11) $$\lim_{\epsilon \to 0} \int_0^{\epsilon} \delta(x)\, dx = 1 .$$

In fact,†

(49.12) $$\begin{aligned} \mu_{\underset{\sim}{I}_r}(x) * \mu_{\underset{\sim}{I}_0}(x) &= \mu_{\underset{\sim}{I}_0}(x) * \mu_{\underset{\sim}{I}_r}(x) \\ &= \int_0^t \mu_{\underset{\sim}{I}_r}(x) \cdot \delta(x - t)\, dt \\ &= \mu_{\underset{\sim}{I}_r}(x) \end{aligned}$$

Thus, this set of fuzzy subsets will be completed by $\underset{\sim}{I}_0$.

The monoid

(49.13) $$\underset{\sim}{I}_0, \underset{\sim}{I}_1, \underset{\sim}{I}_2, \ldots, \underset{\sim}{I}_n, \ldots$$

is isomorphic‡ to that of the natural numbers: :

†This is one of the properties of the Dirac function. This may be proved easily with the aid of operational calculus (symbolic calculus) or in the theory of Schwartz distributions.

‡If necessary, see Section 57.

(49.14) $$0\,,1\,,2\,,\ldots,n\,,\ldots.$$

On (49.8) we also remark that all of these exponential fuzzy integers have the abscissas of their maximums succeeding one another with intervals equal to $1/\lambda$.

Geometric fuzzy integers. Consider the reference set

(49.15) $$\mathbf{E} = \mathbf{N}$$

and the fuzzy subset $\underset{\sim}{\mathbf{J}}_1$ such that

(49.16) $$\mu_{\underset{\sim}{\mathbf{J}}_1}(x) = a(1-a)^{x-1}\,, \qquad a \in \mathbf{R}^+\,,\quad 0 < |a| < 1\,,\quad x = 1,2,3,\ldots.$$

We then define $\underset{\sim}{\mathbf{J}}_2$ in the following fashion:

(49.17)
$$\begin{aligned}
\mu_{\underset{\sim}{\mathbf{J}}_2}(x) &= \mu_{\underset{\sim}{\mathbf{J}}_1}(x) * \mu_{\underset{\sim}{\mathbf{J}}_1}(x)\\
&= \sum_{t=1}^{x-1} \mu_{\underset{\sim}{\mathbf{J}}_1}(t)\cdot \mu_{\underset{\sim}{\mathbf{J}}_1}(x-t)\\
&= \sum_{t=1}^{x-1} a(1-a)^{t-1}\cdot a(1-a)^{x-t-1}\\
&= a^2(1-a)^{x-2}\sum_{t=1}^{x-1} 1\\
&= (x-1)\,a^2(1-a)^{x-2}\,, \qquad x = 2,3,4,\ldots.
\end{aligned}$$

Then $\underset{\sim}{\mathbf{J}}_3$ is defined in the following way:

(49.18)
$$\begin{aligned}
\mu_{\underset{\sim}{\mathbf{J}}_3}(x) &= \mu_{\underset{\sim}{\mathbf{J}}_2}(x) * \mu_{\underset{\sim}{\mathbf{J}}_1}(x) = \mu_{\underset{\sim}{\mathbf{J}}_1}(x) * \mu_{\underset{\sim}{\mathbf{J}}_2}(x)\\
&= \sum_{t=2}^{x-1} \mu_{\underset{\sim}{\mathbf{J}}_2}(t)\cdot \mu_{\underset{\sim}{\mathbf{J}}_1}(x-t)\\
&= \sum_{t=2}^{x-1} (t-1)\,a^2(1-a)^{t-2}\cdot a(1-a)^{x-t-1}\\
&= a^3(1-a)^{x-3}\sum_{t=2}^{x-1}(t-1)\\
&= \frac{(x-2)(x-1)}{2}\,a^3(1-a)^{x-3} \qquad x = 3,4,5,\ldots.
\end{aligned}$$

More generally, in the same manner one obtains†

(49.19) $$\mu_{\underset{\sim}{\mathbf{J}}_r}(x) = C_{x-1}^{x-r}\cdot a^r\cdot(1-a)^{x-r}.$$

The abscissas of the maximums are $x = r, r+1, \ldots$ given in Table 49.2.

†This result is rather easy to obtain using the z transformation (which is equivalent to the Laplace transform but applied to N instead of R).

Table 49.2

	$\underset{\sim}{\mathbf{J}}_i$	$\mu_{\underset{\sim}{\mathbf{J}}_i}(x)$	Abscissa of the maximum
	$\underset{\sim}{\mathbf{J}}_1$	$a(1-a)^{x-1}$	$x = 1$
	$\underset{\sim}{\mathbf{J}}_2$	$(x-1)\,a^2(1-a)^{x-2}$	$\frac{1}{a} \leqslant x \leqslant 1 + \frac{1}{a}$
(49.20)	$\underset{\sim}{\mathbf{J}}_3$	$\frac{(x-2)(x-1)}{2}\,a^3(1-a)^{x-3}$	$\frac{2}{a} \leqslant x \leqslant 1 + \frac{2}{a}$
	. . .	. . .	. . .
	$\underset{\sim}{\mathbf{J}}_n$	$C_{x-1}^{x-n}\,a^n(1-a)^{x-n}$	$\frac{n-1}{a} \leqslant x \leqslant 1 + \frac{n-1}{a}$
	. . .	. . .	. . .

We remark that the abscissa x of a maximum (an abscissa that may not be unique) is not necessarily $x = n$ for $\underset{\sim}{\mathbf{J}}_n$; this depends on the value of a.

The fuzzy subsets

$$(49.21) \qquad \underset{\sim}{\mathbf{J}}_1, \underset{\sim}{\mathbf{J}}_2, \underset{\sim}{\mathbf{J}}_3, \ldots, \underset{\sim}{\mathbf{J}}_n, \ldots$$

will be called geometric fuzzy integers. $\underset{\sim}{\mathbf{J}}_1$ will be called geometric fuzzy 1, etc.

The set of fuzzy subsets (49.21) also forms a commutative monoid. This monoid possesses an identity that we designate by $\underset{\sim}{\mathbf{J}}_0$ and which is

$$(49.22) \qquad \begin{aligned} \mu_{\underset{\sim}{\mathbf{J}}_0}(x) &= 1, \quad x = 0 \\ &= 0, \quad x = 1, 2, 3, \ldots \end{aligned}$$

as one may verify by carrying out

$$(49.23) \qquad \mu_{\underset{\sim}{\mathbf{J}}_r}(x) * \mu_{\underset{\sim}{\mathbf{J}}_0}(x) = \mu_{\underset{\sim}{\mathbf{J}}_0}(x) * \mu_{\underset{\sim}{\mathbf{J}}_r}(x) = \mu_{\underset{\sim}{\mathbf{J}}_r}(x).$$

There is also an isomorphism between $\underset{\sim}{\mathbf{J}}_0, \underset{\sim}{\mathbf{J}}_1, \underset{\sim}{\mathbf{J}}_2, \ldots, \underset{\sim}{\mathbf{J}}_n, \ldots$ and the set $\mathbf{N}$ of natural numbers.

We remark again that according to (49.20), all these geometric fuzzy integers have the abscissas of their maximums succeeding one another by an interval that depends on a.

One may define by similar procedures other fuzzy natural numbers that one considers in probability laws, for example, in the binomial laws, the Poisson laws, the negative binomials, rectangular laws, normal, Euler (gamma), etc.

We shall limit ourselves below to the gaussian natural numbers (normal law).

Gaussian fuzzy integers. Consider a reference set

$$(49.24) \qquad E = R,$$

and the fuzzy subset $\underset{\sim}{\mathbf{K}}_1$ such that

$$(49.25) \qquad \mu_{\underset{\sim}{\mathbf{K}}_1}(x) = \frac{1}{\sqrt{2\pi\sigma_1^2}}\, e^{-\frac{(x-1)^2}{2\sigma_1^2}}.$$

Define $\underset{\sim}{\mathbf{K}}_2$ by

(49.26)
$$\mu_{\underset{\sim}{\mathbf{K}}_2}(x) = \mu_{\underset{\sim}{\mathbf{K}}_1}(x) * \mu_{\underset{\sim}{\mathbf{K}}_1}(x)$$
$$= \int_0^x \mu_{\underset{\sim}{\mathbf{K}}_1}(t) \cdot \mu_{\underset{\sim}{\mathbf{K}}_1}(x - t)\, dt$$
$$= \frac{1}{\sqrt{4\pi\sigma_1^2}} \cdot e^{-\frac{(x-2)^2}{4\sigma_1^2}}.$$

and continuing

(49.27)
$$\mu_{\underset{\sim}{\mathbf{K}}_r}(x) = \frac{1}{\sqrt{2\pi r \sigma_1^2}} \cdot e^{-\frac{(x-r)^2}{2r\sigma_1^2}}.$$

One may establish Table 49.3.

Table 49.3

$\underset{\sim}{\mathbf{K}}_i$	$\mu_{\underset{\sim}{\mathbf{K}}_i}(x)$	*Abscissa of the maximum*	*Ordinate of the maximum*
$\underset{\sim}{\mathbf{K}}_1$	$\frac{1}{\sqrt{2\pi\sigma_1^2}} \cdot e^{-\frac{(x-1)^2}{2\sigma_1^2}}$	1	$\frac{1}{\sqrt{2\pi\sigma_1^2}}$.
(49.28)			
$\underset{\sim}{\mathbf{K}}_2$	$\frac{1}{\sqrt{4\pi\sigma_1^2}} \cdot e^{-\frac{(x-2)^2}{4\sigma_1^2}}$	2	$\frac{1}{\sqrt{4\pi\sigma_1^2}}$.
$\underset{\sim}{\mathbf{K}}_3$	$\frac{1}{\sqrt{6\pi\sigma_1^2}} \cdot e^{-\frac{(x-3)^2}{6\sigma_1^2}}$	3	$\frac{1}{\sqrt{6\pi\sigma_1^2}}$.
⋮	⋮	⋮	⋮
$\underset{\sim}{\mathbf{K}}_r$	$\frac{1}{\sqrt{2\pi r\sigma_1^2}} \cdot e^{\frac{-(x-r)^2}{2r\sigma_1^2}}$	r	$\frac{1}{\sqrt{2\pi r\sigma_1^2}}$.
⋮	⋮	⋮	⋮

The fuzzy subsets

(49.29)
$$\underset{\sim}{\mathbf{K}}_1,\ \underset{\sim}{\mathbf{K}}_2,\ \underset{\sim}{\mathbf{K}}_3, \ldots, \underset{\sim}{\mathbf{K}}_r, \ldots$$

will be called gaussian fuzzy integers.

Indeed, here also one has a commutative monoid whose identity is $\underset{\sim}{\mathbf{K}}_0$, where

(49.30)
$$\mu_{\underset{\sim}{\mathbf{K}}_0}(x) = \delta(x),$$

with $\delta(x)$ such that

(49.31)
$$\lim_{\epsilon \to 0} \int_{x=-\epsilon}^{\epsilon} \delta(x)\, dx = 1,$$

(symmetric Dirac function).

Again one has an isomorphism with **N**, but this time the abscissas of the maximums of $\mu_{\underset{\sim}{K}_r}(x)$ are respectively equal to the values of the integer r being considered.

The gaussian fuzzy integers have the following important property: the dependence of the abscissa of the maximum (which is also the mean) on the variance is constant:

$$\frac{r}{r\sigma_1^2} = \frac{1}{\sigma_1^2} = C^{te}. \tag{49.32}$$

Thus, the greater a fuzzy number $\underset{\sim}{K}_r$ is (that is, the greater r is), the greater is its variance, that is, the greater is its fuzziness; but, with respect to r, the relative fuzziness is constant.

50. EXERCISES

IV1. Give the table representing the fuzzy groupoid such that

$$E = \{a, b\}, \quad M = \{0, \tfrac{1}{3}, \tfrac{2}{3}, 1\}$$

$$\mu_{\underset{\sim}{A} * \underset{\sim}{B}}(x) = \mu_{\underset{\sim}{A}}(x) \vee \mu_{\underset{\sim}{B}}(x).$$

IV2. We give the fuzzy groupoid expressed by the table below, giving the operation * relative to

$E = \{A, B\}$ and $M = \{0, 1\}$.

We have represented

$\{(A \mid \alpha), (B \mid \beta)\}$ by (α, β).

(1) Is this groupoid associative?

(2) Does it have an identity?

(3) If the answer is yes, what are the submonoids?

(4) For each ordered pair (α, β), does there exist an inverse, and if the answer is yes, what are the subgroups?

*	(0, 0)	(0, 1)	(1, 0)	(1, 1)
(0, 0)	(0, 0)	(0, 1)	(1, 0)	(1, 1)
(0, 1)	(0, 1)	(0, 0)	(1, 1)	(1, 0)
(1, 0)	(1, 0)	(1, 1)	(0, 1)	(0, 0)
(1, 1)	(1, 1)	(1, 0)	(0, 0)	(0, 1)

IV3. The table below represents a group for the operation *. Express this operation * with the aid of the symbols ∩ (intersection), ∪ (union), and ¯ (complementation) alone. We have represented

$\{(A \mid \alpha), (B \mid \beta)\}$ by (α, β).

For the same operation * give the group corresponding to

*	(0, 0)	(0, 1)	(1, 0)	(1, 1)
(0, 0)	(0, 0)	(0, 1)	(1, 0)	(1, 1)
(0, 1)	(0, 1)	(0, 0)	(1, 1)	(1, 0)
(1, 0)	(1, 0)	(1, 1)	(0, 0)	(0, 1)
(1, 1)	(1, 1)	(1, 0)	(0, 1)	(0, 0)

$$E = \{A, B, C\} \quad \text{instead of} \quad E = \{A, B\}.$$

IV4. Consider the fuzzy groupoids $(\mathcal{P}(E), *)$ where the membership function of **E** is $\mathbf{M} = [0, 1]$.

(1) $\mu_{A*B}(x) = (1 - \mu_A(x)) \wedge (1 - \mu_B(x))$.

(2) $\mu_{A*B}(x) = (\mu_A(x) \cdot \mu_B(x)) \wedge [(1 - \mu_A(x)) \cdot (1 - \mu_B(x))]$

(3) $\mu_{A*B}(x) = [(1 - \mu_A(x)) \cdot \mu_B(x)] \hat{+} [(1 - \mu_B(x)) \cdot \mu_A(x)]$

where $\hat{+}$ is such that $a \hat{+} b = a + b - a \cdot b$. Which are the fuzzy groupoids:

(a) that are commutative?

(b) that are associative?

(c) that possess an identity and if it exists, what is it?

(d) that are such that each fuzzy subset has an inverse?

IV5. Determine the following laws of external composition where

$$E = \{a, b\}, \quad M = \{0, \tfrac{1}{2}, 1\}.$$

$$(A, B) \in \mathcal{P}(E) \times \mathcal{P}(E), \quad x, y \in E,$$

and give the tables of these laws.

(a) $c(A, B) = \frac{1}{2}\sqrt{(\mu_A(a) - \mu_B(a))^2 + (\mu_A(b) - \mu_B(b))^2}$.

(b) $C \subset E' = \{X, Y\}, \quad C = A * B,$

$\mu_C(X) = \mu_A(x) \wedge \mu_B(y),$

$\mu_C(Y) = \mu_A(x) \vee \mu_B(y).$

CHAPTER V

GENERALIZATION OF THE NOTION OF FUZZY SUBSET

51. INTRODUCTION

We shall be concerned here with a very interesting generalization of the work of Zadeh carried out by Goguen [G3]. The fundamental idea is the following. In the theory of fuzzy subsets developed by Zadeh, the elements of a reference set **E** take their values in the set **M** = [0, 1]; Goguen proposes that one allow these elements to take their values in a set **L** having a more general structure. This extension is very rich, but introduces several complications about which we shall warn the reader.

In order to begin this study it is appropriate to review certain topics for readers who have forgotten or have never studied certain notions of lattice theory. We shall carry out this review since we have not forgotten that this book is intended for users who have not necessarily mastered modern mathematics in all its abstractions.

For this extension, Goguen has also used the works of Birkhoff [2B]; these are still the authority in lattice theory.†

Concerning terminology, "fuzzy subset" will be taken in the sense of Zadeh (**M** = [0, 1]); "**L**-fuzzy subset" will be taken in the sense of Goguen (**M** is generalized by **L**, which often later will be a lattice). The letter **L** comes from the word lattice.

52. OPERATIONS ON ORDINARY SETS

In Chapter I we reviewed how to carry out some operations on ordinary subsets of a reference set; we now go on to review three important operations—this time no longer concerning subsets of the same reference set, but concerning sets themselves, distinct or not.

Product of two sets. In fact, we have already spoken implicitly of this in Section 11, in order to introduce the notion of a graph. We return in detail to the notion of the product of two sets.

†See also [1F].

Let $\mathbf{E}_1$ and $\mathbf{E}_2$ be two sets and let $x \in \mathbf{E}_1$ and $y \in \mathbf{E}_2$. The set of ordered pairs (x, y) will be called the product of $\mathbf{E}_1$ and $\mathbf{E}_2$ or the product set formed by $\mathbf{E}_1$ with $\mathbf{E}_2$. This product set will be denoted $\mathbf{E}_1 \times \mathbf{E}_2$.

One has

(52.1) $$\mathbf{E}_1 \times \mathbf{E}_2 \neq \mathbf{E}_2 \times \mathbf{E}_1, \quad \text{unless} \quad \mathbf{E}_2 = \mathbf{E}_1 \quad \text{(noncommutativity).}$$

(52.2) $$(\mathbf{E}_1 \times \mathbf{E}_2) \times \mathbf{E}_3 = \mathbf{E}_1 \times (\mathbf{E}_2 \times \mathbf{E}_3) \quad \text{(associativity).}$$

Example†

(52.3)
(52.4) $$\mathbf{E}_1 = \{A, B\} \quad , \quad \mathbf{E}_2 = \{\alpha, \beta, \gamma\} .$$

(52.5) $$\mathbf{E}_1 \times \mathbf{E}_2 = \{(A, \alpha), (A, \beta), (A, \gamma), (B, \alpha), (B, \beta), (B, \gamma)\} ,$$

(52.6) $$\mathbf{E}_2 \times \mathbf{E}_1 = \{(\alpha, A), (\alpha, B), (\beta, A), (\beta, B), (\gamma, A), (\gamma, B)\} .$$

(52.7) If $$\mathbf{E}_3 = \{a, b\} ,$$

then :

(52.8) $$\begin{aligned}(\mathbf{E}_1 \times \mathbf{E}_2) \times \mathbf{E}_3 = \{&(A, \alpha, a), (A, \alpha, b), (A, \beta, a), (A, \beta, b) ,\\ &(A, \gamma, a), (A, \gamma, b), (B, \alpha, a), (B, \alpha, b) ,\\ &(B, \beta, a), (B, \beta, b), (B, \gamma, a), (B, \gamma, b)\} .\end{aligned}$$

One may likewise develop the right-hand side of (52.2).

For n sets $\mathbf{E}_1, \mathbf{E}_2, \ldots, \mathbf{E}_n$, one may define

(52.9) $$\mathbf{E}_1 \times \mathbf{E}_2 \times \cdots \times \mathbf{E}_n .$$

Order mattering, one may define, if all these sets are distinct

$$n\,! = n\,(n-1) \quad \ldots 2 \,.\, 1$$

distinct product sets.

Disjunctive sum of two sets. This may not be defined as we have done for subsets of the same reference set [see (5.34)] since we have not specified what one is to call the complement of a set (otherwise it would be a matter of a subset and of its complement referred to their reference set).

We therefore define $\mathbf{E}_1 \oplus \mathbf{E}_2$ as follows:

(52.10) $\mathbf{E}_1 \oplus \mathbf{E}_2$ is the set formed by the elements of $\mathbf{E}_1$ and of $\mathbf{E}_2$ with the exception of those that belong at the same time to $\mathbf{E}_1$ and to $\mathbf{E}_2$.

Example. Consider (52.3) and (52.4) again; then

†In a didactic spirit we generally take examples concerning very simple finite sets, but unless specified otherwise, the results apply to arbitrary sets, finite or not.

$$E_1 \oplus E_2 = \{A, B, \alpha, \beta, \gamma\}. \tag{52.11}$$

In this example E_1 and E_2 have no common element.

Two properties of (52.10) are

$$E_1 \oplus E_2 = E_2 \oplus E_1 \quad \text{(commutativity)}, \tag{52.12}$$

$$(E_1 \oplus E_2) \oplus E_3 = E_1 \oplus (E_2 \oplus E_3) \quad \text{(associativity)}. \tag{52.13}$$

A distributive property holds between the product and the disjunctive sum operations:

$$E_1 \times (E_2 \oplus E_3) = (E_1 \times E_2) \oplus (E_1 \times E_3), \tag{52.14}$$

$$(E_1 \oplus E_2) \times E_3 = (E_1 \times E_3) \oplus (E_2 \times E_3), \tag{52.15}$$

(distributivity on the left and on the right for $\oplus$).

We shall see an example.

Example. Consider (52.3), (52.4), and (52.7). Then

$$E_2 \oplus E_3 = \{\alpha, \beta, \gamma, a, b\}, \tag{52.16}$$

$$E_1 \times (E_2 \oplus E_3) = \{(A, \alpha), (A, \beta), (A, \gamma), (A, a), (A, b) \\ (B, \alpha), (B, \beta), (B, \gamma), (B, a), (B, b)\}, \tag{52.17}$$

$$E_1 \times E_2 = \{(A, \alpha), (A, \beta), (A, \gamma), (B, \alpha), (B, \beta), (B, \gamma)\}, \tag{52.18}$$

$$E_1 \times E_3 = \{(A, a), (A, b), (B, a), (B, b)\}, \tag{52.19}$$

$$(E_1 \times E_2) \oplus (E_1 \times E_3) = \{(A, \alpha), (A, \beta), (A, \gamma), (B, \alpha), (B, \beta), \\ (B, \gamma), (A, a), (A, b), (B, a), (B, b)\}. \tag{52.20}$$

One may verify easily that equality holds between (52.17) and (52.20).

Note that neither left nor right distributivity holds for the product. Again consider (52.3), (52.4), and (52.5)†:

$$E_1 \oplus (E_2 \times E_3) \neq (E_1 \oplus E_2) \times (E_1 \oplus E_3) \tag{52.21}$$

$$E_2 \times E_3 = \{(\alpha, a), (\alpha, b), (\beta, a), (\beta, b), (\gamma, a), (\gamma, b)\} \tag{52.22}$$

$$E_1 \oplus (E_2 \times E_3) = \{(A, B, (\alpha, a), (\alpha, b), (\beta, a), (\beta, b), (\gamma, a), (\gamma, b)\}, \tag{52.23}$$

$$E_1 \oplus E_2 = \{A, B, \alpha, \beta, \gamma\}, \tag{52.24}$$

$$E_1 \oplus E_3 = \{A, B, a, b\}, \tag{52.25}$$

$$(E_1 \oplus E_2) \times (E_1 \oplus E_3) = \{(A, A), (A, B), (A, a), (A, b), (B, A), (B, B), \\ (B, a), (B, b), (\alpha, A), (\alpha, B), (\alpha, a), (\alpha, b), \\ (\beta, A), (\beta, B), (\beta, a), (\beta, b), (\gamma, A), (\gamma, B), \\ (\gamma, a), (\gamma, b)\}. \tag{52.26}$$

†As for ordinary multiplication and addition of numbers:

$a(b + c) = ab + ac$: true

$a + bc = (a + b)(a + c)$: false

Set of mappings of $\mathbf{E}_1$ ***into*** $\mathbf{E}_2$. The set of functional mappings† of $\mathbf{E}_1$ into $\mathbf{E}_2$ will be denoted

$$\mathbf{E}_2^{\mathbf{E}_1} \qquad \text{(as a power)}$$

We may see an example immediately by taking (52.3) and (52.4) and by using some sketches (see Figure 52.1).

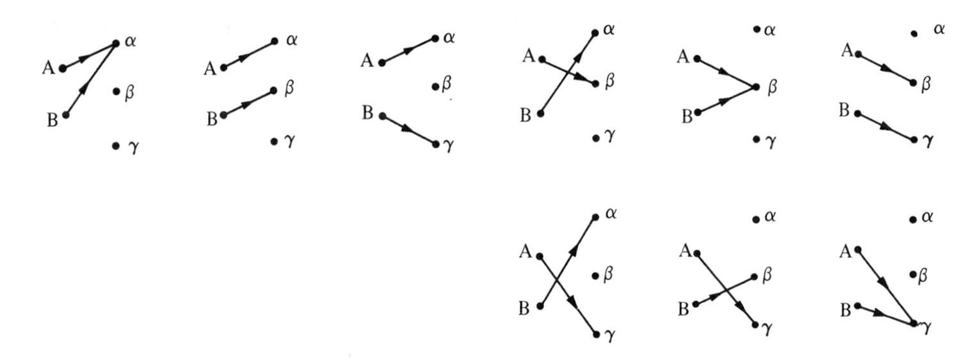

Fig. 52.1

One sees in this figure that if

(52.27)
(52.28)
$$\mathrm{E}_1 = \{\mathrm{A}\,,\mathrm{B}\} \quad \text{and} \quad \mathrm{E}_2 = \{\alpha\,,\beta\,,\gamma\}\,,$$

then

(52.29)
$$\begin{aligned}\mathrm{E}_2^{\mathrm{E}_1} = \{&\{(\mathrm{A}\,|\,\alpha)\,,(\mathrm{B}\,|\,\alpha)\},\{(\mathrm{A}\,|\,\alpha)\,,(\mathrm{B}\,|\,\beta)\},\{(\mathrm{A}\,|\,\alpha)\,,(\mathrm{B}\,|\,\gamma)\}\,,\\ &\{(\mathrm{A}\,|\,\beta)\,,(\mathrm{B}\,|\,\alpha)\},\{(\mathrm{A}\,|\,\beta)\,,(\mathrm{B}\,|\,\beta)\},\{(\mathrm{A}\,|\,\beta)\,,(\mathrm{B}\,|\,\gamma)\}\,,\\ &\{(\mathrm{A}\,|\,\gamma)\,,(\mathrm{B}\,|\,\alpha)\},\{(\mathrm{A}\,|\,\gamma)\,,(\mathrm{B}\,|\,\beta)\},\{(\mathrm{A}\,|\,\gamma)\,,(\mathrm{B}\,|\,\gamma)\}\}\,.\end{aligned}$$

The cardinality of $\mathrm{E}_2^{\mathrm{E}_1}$ is

(52.30)
$$\text{card}\,(\mathrm{E}_2^{\mathrm{E}_1}) = (\text{card}\ \mathrm{E}_2)^{(\text{card}\ \mathrm{E}_1)}\,.$$

(52.31) For our example, $\text{card}\,(\mathrm{E}_2^{\mathrm{E}_1}) = 3^2 = 9.$

(52.32) If E_1 or/and E_2 is infinite, $\text{card}(\mathrm{E}_2^{\mathrm{E}_1})$ is infinite.

Principal operations on sets. Recalling what follows from that specified above, if $\mathbf{E}_1$, $\mathbf{E}_2$, and $\mathbf{E}_3$ are sets, then

(52.33)
$$\mathrm{E}_1 \oplus \mathrm{E}_2 = \mathrm{E}_2 \oplus \mathrm{E}_1\ ,$$

†One also says single-valued mappings or simply functions.

(52.34) $$E_1 \oplus (E_2 \oplus E_3) = (E_1 \oplus E_2) \oplus E_3 \ ,$$

(52.35) $$E_1 \times (E_2 \times E_3) = (E_1 \times E_2) \times E_3 \ ,$$

(52.36) $$E_1 \times (E_2 \oplus E_3) = (E_1 \times E_2) \oplus (E_1 \times E_3) \ ,$$

(52.37) $$(E_1 \oplus E_2) \times E_3 = (E_1 \times E_3) \oplus (E_2 \times E_3) \ ,$$

(52.38) $$(E_1 \times E_2)^{E_3} = E_1^{E_3} \times E_2^{E_3} \ ,$$

(52.39)† $$(E_1^{E_2})^{E_3} = E_1^{E_2 \times E_3}$$

53. FUNDAMENTAL PROPERTIES OF THE SET OF MAPPINGS OF ONE SET INTO ANOTHER

The set of mappings of **E** into **L** is denoted

(53.1) $$\mathbf{L}^{\mathbf{E}}$$

We may then state the following fundamental property‡:

Any internal law $*$ defined on **L** induces a corresponding internal law $\circledast$ on $\mathbf{L}^{\mathbf{E}}$.

We immediately consider an example.

Example 1. Reconsider (52.29) and put

(53.2)
$$\begin{aligned}
A_1 &= \{(A|\alpha), (B|\alpha)\} , \\
A_2 &= \{(A|\alpha), (B|\beta)\} , \\
A_3 &= \{(A|\alpha), (B|\gamma)\} , \\
A_4 &= \{(A|\beta), (B|\alpha)\} , \\
A_5 &= \{(A|\beta), (B|\beta)\} , \\
A_6 &= \{(A|\beta), (B|\gamma)\} , \\
A_7 &= \{(A|\gamma), (B|\alpha)\} , \\
A_8 &= \{(A|\gamma), (B|\beta)\} , \\
A_9 &= \{(A|\gamma), (B|\gamma)\} .
\end{aligned}$$

†Pay attention to the parentheses:

$$(E_1^{E_2})^{E_3} \neq E_1^{(E_2^{E_3})}$$

One knows that exponentiation is not generally associative.

‡See a text on new mathematics, for example, [3K].

(53.3)
(53.4) with $$\mathbf{E} = \{A, B\} \quad \text{and} \quad \mathbf{L} = \{\alpha, \beta, \gamma\}.$$

Consider the internal law $*$ defined on $\mathbf{L}$

(53.5)

$*$	α	β	γ
α	β	β	α
β	γ	β	α
γ	β	α	γ

and see how this induces a law on $\mathbf{L}^{\mathbf{E}}$. For example,

(53.6)
$$\begin{aligned} A_2 \circledast A_6 &= \{(A|\alpha), (B|\beta)\} * \{(A|\beta), (B|\gamma)\} \\ &= \{(A|\alpha * \beta), (B|\beta * \gamma)\} \\ &= \{(A|\beta), (B|\alpha)\} \\ &= A_4 . \end{aligned}$$

One thus obtains a law $\circledast$ for $\mathbf{L}^{\mathbf{E}}$ as follows:

(53.7)

$\circledast$	A_1	A_2	A_3	A_4	A_5	A_6	A_7	A_8	A_9
A_1	A_5	A_5	A_4	A_5	A_5	A_4	A_2	A_2	A_1
A_2	A_6	A_5	A_4	A_6	A_5	A_4	A_3	A_2	A_1
A_3	A_5	A_4	A_6	A_5	A_4	A_6	A_2	A_1	A_3
A_4	A_8	A_8	A_7	A_5	A_5	A_4	A_2	A_2	A_1
A_5	A_9	A_8	A_7	A_6	A_5	A_4	A_3	A_2	A_1
A_6	A_8	A_7	A_9	A_5	A_4	A_6	A_2	A_1	A_3
A_7	A_5	A_5	A_4	A_2	A_2	A_1	A_8	A_8	A_7
A_8	A_6	A_5	A_4	A_3	A_2	A_1	A_9	A_8	A_7
A_9	A_5	A_4	A_6	A_2	A_1	A_3	A_8	A_7	A_9

Example 2. Consider a finite set

$$\mathrm{E} = \{x_1, x_2, \ldots, x_n\} \tag{53.8}$$

and

$$\mathrm{L} = \{0, 1\}; \tag{53.9}$$

then $\mathbf{L}^{\mathbf{E}}$ gives the power set (the set of ordinary subsets including ϕ).

The product operator $\cdot$ on **L** induces the operator $\cap$ on $\mathbf{L}^{\mathbf{E}}$; the sum operator $\dot{+}$ on **L** induces the operator $\cup$ on $\mathbf{L}^{\mathbf{E}}$; the complementation operator on **L** induces the complementation operator on $\mathbf{L}^{\mathbf{E}}$.

These conclusions remain valid for any **E** having the power of the natural numbers or of the continuum.

Example 3. Consider a finite set

$$\mathrm{E} = \{x_1, x_2, \ldots, x_n\} \tag{53.10}$$

and

$$\mathrm{L} = [0, 1]; \tag{53.11}$$

then $\mathbf{L}^{\mathbf{E}}$ gives the set of fuzzy subsets.

The operator $\wedge$ on **L** induces the operator $\cap$ on $\mathbf{L}^{\mathbf{E}}$; the operator $\vee$ on **L** induces $\cup$ on $\mathbf{L}^{\mathbf{E}}$; complementation on **L** induces complementation on $\mathbf{L}^{\mathbf{E}}$ $\cdot$ the product operator $\cdot$ on **L** induces $\cdot$ on $\mathbf{L}^{\mathbf{E}}$; the operator $\hat{+}$ on **L** induces $\oplus$ on $\mathbf{L}^{\mathbf{E}}$.

As we have noted, these conclusions remain valid when **E** has the power of the natural numbers or of the continuum.

Example 4. Let

$$\mathrm{E} = \{0, 1, 2, 3, 4, \ldots,\}, \tag{53.12}$$

$$\mathrm{L} = [0, 1], \tag{53.13}$$

then $\mathbf{L}^{\mathbf{E}}$ defines the infinite set of all the fuzzy integers. All the operations of Example 3 above also apply here.

Example 5. Let

$$\mathrm{E} = \mathbf{L}, \tag{53.14}$$

$$\mathrm{L} = [0, 1], \tag{53.15}$$

then $\mathbf{L}^{\mathbf{L}}$ represents the set of all the fuzzy numbers $\underset{\sim}{\mathbf{X}}$ such that $x \in \underset{\sim}{\mathbf{X}} \subset [0, 1]$, $\mu_{\underset{\sim}{X}}(x) \in [0, 1]$.

We return to the general case. It is easy to show that if $*$ is the law of **L** and $\mathbf{L}^{\mathbf{E}}$ the law for $\circledast$ induced by $*$, then one has the following formal implications:

$$* \text{ is associative} \Rightarrow \circledast \text{ is associative,} \tag{53.16}$$

$$* \text{ is commutative} \Rightarrow \circledast \text{ is commutative,} \tag{53.17}$$

$$* \text{ is idempotent} \Rightarrow \circledast \text{ is idempotent.} \tag{53.18}$$

We shall see, for example, how to prove (53.16). Let $\mathbf{E}_\alpha \in \mathbf{E}$ and let $l_r, l_s, l_t \in \mathbf{L}$. A set $\mathbf{A} \in \mathbf{L}^{\mathbf{E}}$ will have elements of the form $(\mathbf{E}_\alpha | l_r)$; likewise $\mathbf{B} \in \mathbf{L}^{\mathbf{E}}$ will have elements of the form $(\mathbf{E}_\alpha | l_s)$; and $\mathbf{C} \in \mathbf{L}^{\mathbf{E}}$ will have elements of the form $(\mathbf{E}_\alpha | l_t)$. Performing the operation $*$ on $\mathbf{L}$, one then obtains a set $\mathbf{D}$ whose elements are of the form $(\mathbf{E}_\alpha | l_r * l_s)$; this set $\mathbf{D} \in \mathbf{L}^{\mathbf{E}}$ will be denoted $\mathbf{A} \circledast \mathbf{B}$. With this notation we may write:

(53.19) $$(\mathrm{E}_\alpha | l_r * l_s) \in \mathrm{A} \circledast \mathrm{B} \ ,$$

(53.20) $$(\mathrm{E}_\alpha | (l_r * l_s) * l_t) \in (\mathrm{A} \circledast \mathrm{B}) \circledast \mathrm{C} \ ,$$

(53.21) $$(\mathrm{E}_\alpha | l_r * (l_s * l_t)) \in \mathrm{A} \circledast (\mathrm{B} \circledast \mathrm{C}) \ .$$

By comparing (53.20) and (53.21), one proves (53.16).

Possible structures of **L**. One may imagine that **L** has any structure that one may wish; for an operation $*$ one will obtain an operation $\circledast$ on $\mathbf{L}^{\mathbf{E}}$. One may imagine for **L** two associated operations $*_1$ and $*_2$; these will induce two operations $\circledast_1$ and $\circledast_2$ on $\mathbf{L}^{\mathbf{E}}$.

Thus, if there exists an operator structure (monoid, group, etc.) on **L**, one will have to check whether this structure is recovered in $\mathbf{L}^{\mathbf{E}}$, or what other structure holds in $\mathbf{L}^{\mathbf{E}}$.

In the theory of fuzzy subsets developed in the preceding chapters, **L** had the structure of a distributive vector lattice for the operations $\wedge$ and $\vee$; this lattice was the interval $[0, 1]$ of **R**; it induced a distributive vector lattice for $\cap$ and $\cup$ in $\mathbf{L}^{\mathbf{E}}$, forming the set of fuzzy subsets.

We may now carry out a very interesting generalization.

54. REVIEW OF SEVERAL FUNDAMENTAL STRUCTURES

Lattice.† Let **L** be an ordered set. Suppose that for any ordinary subset $\{X_i, X_j\}$ of **L**, there exists one and only one element of **L** constituting an inferior limit of $\{X_i, X_j\}$, and likewise there exists one and only one element of **L** constituting a superior limit of $\{X_i, X_j\}$; in this case, one says that **L** is a lattice or a reticular set.‡

We shall put

for the inferior limit of $\{X_i, X_j\}$: $X_i \,\Delta\, X_j$,
for the superior limit of $\{X_i, X_j\}$: $X_i \,\nabla\, X_j$.

The definition of a lattice may thus be written:

(54.1) $$(\forall X_i)(\forall X_j)(X_i \in \mathrm{L} \text{ and } X_j \in \mathrm{L}) :$$

†For more details on the notion of lattices, see, for example, references [1F, 1K, 3K].

‡Lattice or lattice work. A lattice is a long and slender piece of wood that one nails on the rafters of a roof. *Reticule* derives from the Latin *reticulum* signifying a small network or web.

(54.1) $$\exists\,!\; X_k = X_i \,\Delta\, X_j \quad \text{and} \quad X_k \in \mathbf{L}\ ,$$
$$\exists\,!\; X_l = X_i \,\nabla\, X_j \quad \text{and} \quad X_l \in \mathbf{L}\ ,$$

(the symbol $\exists !$ means "there exists one and only one").

The operations Δ and ∇ may also be considered as mappings of $\mathbf{L} \times \mathbf{L}$ into $\mathbf{L}$, which to any pair $\{X_i, X_j\}$, $X_i, X_j \in \mathbf{L}$, corresponds the element $X_i \,\Delta\, X_j$ on the one hand, and the element $X_i \,\Delta\, X_j$ on the other.

One may prove that, in a lattice, one has the four dual properties below. Let A, B, $C \in \mathbf{L}$:

(54.2) $$A \,\Delta\, B = B \,\Delta\, A$$
(54.3) $$A \,\nabla\, B = B \,\nabla\, A$$
commutativity,

(54.4) $$A \,\Delta\, (B \,\Delta\, C) = (A \,\Delta\, B) \,\Delta\, C$$
(54.5) $$A \,\nabla\, (B \,\nabla\, C) = (A \,\nabla\, B) \,\nabla\, C$$
associativity,

(54.6) $$A \,\Delta\, A = A$$
(54.7) $$A \,\nabla\, A = A$$
idempotence,

(54.8) $$A \,\Delta\, (A \,\nabla\, B) = A$$
(54.9) $$A \,\nabla\, (A \,\Delta\, B) = A$$
absorption.

Example. In Figure 54.1a we have presented an example of a lattice. In Figure 54.1b we have presented the diagram of maximal chains or the Hasse diagram.

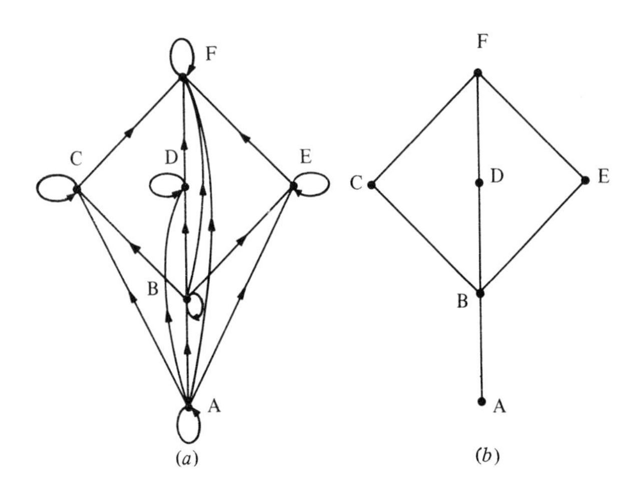

Fig. 54.1

We have :

(54.10) $$\mathbf{L} = \{A, B, C, D, E, F\}\ .$$

One may verify that:

$$
\begin{array}{lll}
 & A \,\Delta\, A = A\ , & A \,\nabla\, A = A\ , \\
 & A \,\Delta\, B = A\ , & A \,\nabla\, B = B\ , \\
 & A \,\Delta\, C = A\ , & A \,\nabla\, C = C\ , \\
 & \cdots\cdots\cdots & \cdots\cdots\cdots \\
(54.11) & C \,\Delta\, D = B\ , & C \,\nabla\, D = F\ , \\
 & \cdots\cdots\cdots & \cdots\cdots\cdots \\
 & E \,\Delta\, F = E\ , & E \,\nabla\, F = F\ , \\
 & F \,\Delta\, F = F\ , & F \,\nabla\, F = F\ .
\end{array}
$$

As an exercise, the reader should also verify properties (54.2)–(54.9).

Before passing to other explications, we recall for certain readers what one means by the notion of maximal chain. By examining the ordered set of Figure 54.1, one may write, using the notation $X_i \prec X_j$ if X_i precedes X_j:

$$
(54.12) \qquad \left\{ \begin{array}{l} A \prec B \prec C \prec F\ , \\ A \prec B \prec D \prec F\ , \\ A \prec B \prec E \prec F\ . \end{array} \right.
$$

Thus, this ordered set possesses three chains that qualify as maximal since each is not a part of another chain of the ordered set.

The nonoriented ordinary graph made up of the maximal chains is called the *diagram of maximal chains* or the Hasse diagram; it is represented in Figure 54.1b for our example.

Other examples. The totally ordered set $[a, b] \in \mathbf{R}$ is a lattice.

Let **L** be a totally ordered set, therefore a lattice, then $\mathbf{L} \times \mathbf{L}$ is also a lattice; likewise, $\mathbf{L}^r = \underbrace{\mathbf{L} \times \mathbf{L} \times \cdots \times \mathbf{L}}_{r \text{ times}}$, but these are partially ordered sets when $r > 1$.

If one looks back to Figures 6.1–6.6, one will see lattices presented by their Hasse diagrams.

Counterexamples. Figure 54.2 gives three counterexamples showing order relations that are not lattices. These are represented with the aid of Hasse diagrams. The diagram of Figure 54.2a does not represent a lattice. In fact, C is an inferior limit of {D, E}, but B is also; thus, there is at least one pair of elements that does not have a unique inferior limit. The diagram represented in Figure 54.2b does not represent a lattice since, for example, {D, E} does not have a superior limit. Likewise, Figure 54.2c does not represent a lattice. We shall give the maximal chains of these three order relations explicitly:

$$
\begin{array}{lll}
 & \text{Figure 54.2 a}: & A \prec B \prec D \prec F\ ,\ A \prec B \prec E \prec F\ ,\ A \prec C \prec D \prec F\ , \\
 & & A \prec C \prec E \prec F\ , \\
(54.13) & \text{Figure 54.2 b}: & A \prec B \prec D\ ,\ A \prec C \prec E \prec G\ ,\ A \prec C \prec F \prec G\ , \\
 & \text{Figure 54.2 c}: & A \prec C \prec E \prec G\ ,\ A \prec C \prec F \prec G\ ,\ B \prec D\ .
\end{array}
$$

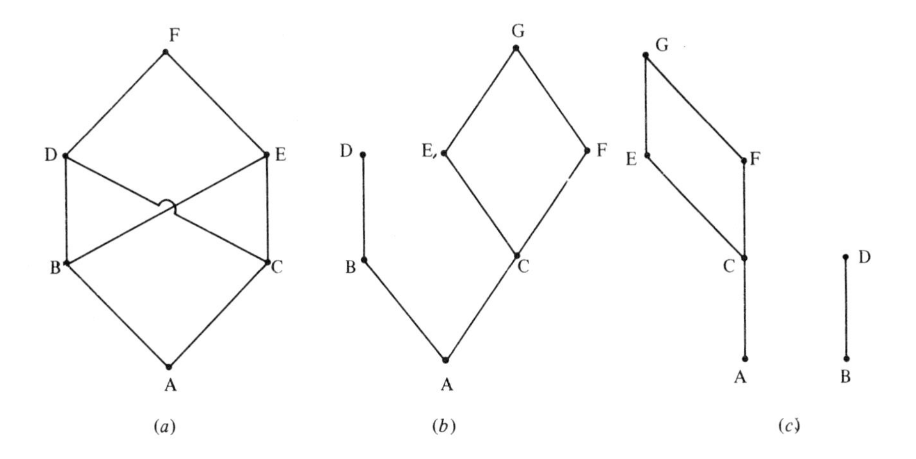

Fig. 54.2

We now review rapidly several important types of lattices.†

Modular lattice. One says that a lattice **L** is modular when, for three arbitrary elements $X_1, X_2, X_3 \in$ **L** one has

$$(54.14) \quad (X_1 \preccurlyeq X_3) \Rightarrow (X_1 \triangledown (X_2 \vartriangle X_3)) = ((X_1 \triangledown X_2) \vartriangle X_3) ,$$

where $\preccurlyeq$ is the symbol representing the order relation of the lattice.

Thus, the lattice of Figure 54.3 is modular; we verify this for A, B, and C. One has

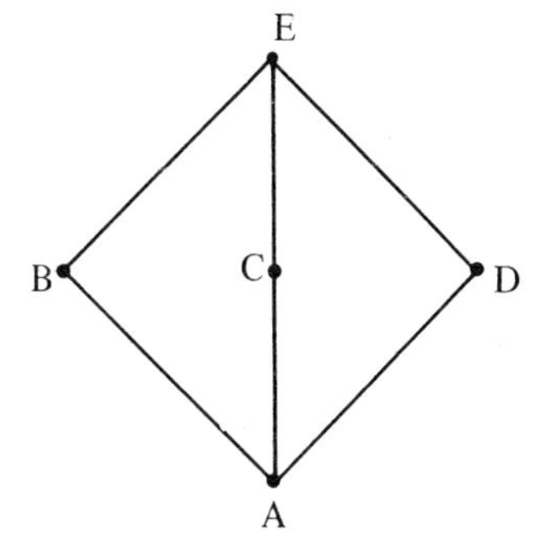

Fig. 54.3

$$(54.15) \qquad A \preccurlyeq C ,$$

and also, choosing B arbitrarily,

$$(54.16) \qquad A \triangledown (B \vartriangle C) = A \triangledown A = A ,$$

$$(54.17) \qquad (A \triangledown B) \vartriangle C = B \vartriangle C = A .$$

One may verify property (54.14) similarly for other triplets.

†We have need of these notions here, but we emphasize again that a reader who has not encountered lattices is referred to specialized texts recommended earlier.

Distributive lattice. One says that a lattice **L** is distributive when it satisfies the conditions

$$\forall X_1, X_2, X_3 \in \mathbf{L} :$$

$$(54.18) \quad \begin{aligned} X_1 \triangledown (X_2 \vartriangle X_3) &= (X_1 \triangledown X_2) \vartriangle (X_1 \triangledown X_3) , \\ X_1 \vartriangle (X_2 \triangledown X_3) &= (X_1 \vartriangle X_2) \triangledown (X_1 \vartriangle X_3) . \end{aligned}$$

One may verify that these conditions are indeed satisfied for the 20 triplets of the lattice of Figure 54.4. For example,

$$(54.19) \quad \begin{aligned} B \vartriangle (C \triangledown E) &= B \vartriangle E = B , \\ (B \vartriangle C) \triangledown (B \vartriangle E) &= A \triangledown B = B . \end{aligned}$$

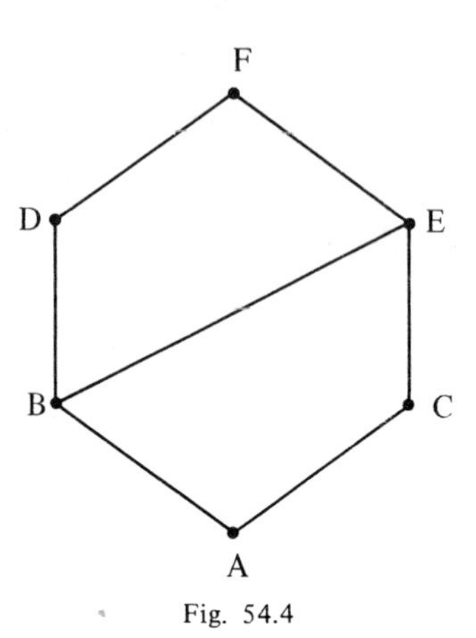

Fig. 54.4

One may show that any distributive lattice is modular.

Before proceeding to other types of lattices, recall what is meant by a sublattice. Consider a lattice **L** and let $\mathbf{A} \subset \mathbf{L}$, where **A** is ordered by the induced order. If $\forall X \in \mathbf{A}$, $\forall Y \in \mathbf{A}$, $X \vartriangle Y \in \mathbf{A}$ and $X \triangledown Y \in \mathbf{A}$, then **A** is a sublattice of **L**.

One may prove that any sublattice $\mathbf{L}'$ of a distributive lattice **L** is itself distributive.

Complemented lattice. We define first what is meant by a complement of an element of a lattice.

Suppose that a lattice **L** possesses an element, denoted 0, that is the inferior limit of the entire lattice **L**, and that it also possesses an element U that is the superior limit of this same lattice. Then, an element X_j is called a complement of X_i, both belonging to **L**, if

$$(54.20) \quad X_i \vartriangle X_j = 0 ,$$

$$(54.21) \quad X_i \triangledown X_j = U .$$

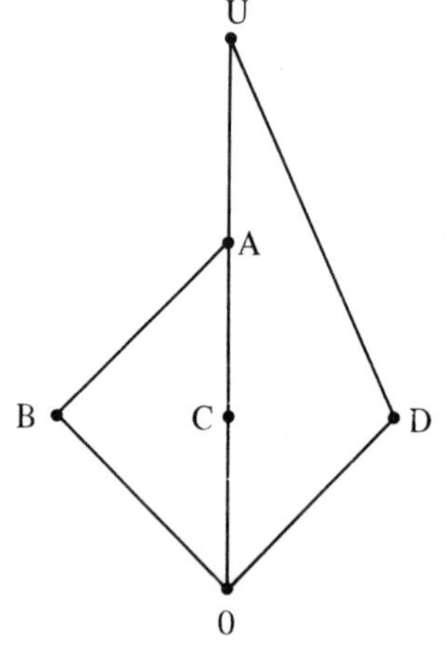

Fig. 54.5

A complement of X_i is denoted $\overline{X}_i$. This complement of X_i, when it exists, is not necessarily unique.

Thus, in the example of Figure 54.5, each element has a complement:

$$(54.22) \quad \overline{0} = U , \ \overline{D} = B \text{ or } C \text{ or } A , \ \overline{C} = D , \ \overline{B} = D , \ \overline{A} = D , \ \overline{U} = 0 .$$

One says that a lattice **L** is *complemented* when:

(1) It possesses a unique element $0 = \inf(\mathbf{L})$ and a unique element $U = \sup(\mathbf{L})$.

(2) Each $X_i \in \mathbf{L}$ possesses at least one complement in **L**.

As one may see, one may say that the lattice of Figure 54.5 is complemented.

One may show that in a distributive lattice, when the complement of an element X_i exists, it is always unique.

Distributive and complemented lattice or boolean lattice. A lattice that is at the same time distributive and complemented is called a boolean lattice.

The reader may verify that the lattice presented in Figure 54.6 is indeed boolean.

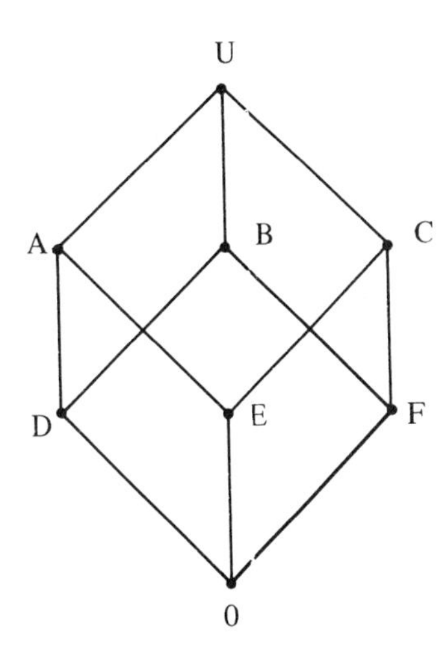

Fig. 54.6

We list several properties of boolean lattices:

(1) For each element, there exists one and only one complement.

(2) For each X_i, one has

$$\overline{(\overline{X}_i)} = X_i \ .$$

(3) The following relations hold:

$$\overline{X_i \, \Delta \, X_j} = \overline{X}_i \, \nabla \, \overline{X}_j$$

$$\overline{X_i \, \nabla \, X_j} = \overline{X}_i \, \Delta \, \overline{X}_j$$

(De Morgan's theorems).

(4) Any finite boolean lattice is isomorphic with a lattice of an ordinary power set of a set with respect to inclusion, and vice versa.

Vector lattice. Let **A**, **B**, . . . , **S** be n sets, each totally ordered by a relation $\prec$. The product set

$$\mathbf{A} \times \mathbf{B} \times \cdots \times \mathbf{S}$$

is ordered and forms a lattice called a *vector lattice* and the order relation is the dominance relation [(4.3) and (4.4)].

In Figure 54.7 we have represented a vector lattice formed by the product of the sets

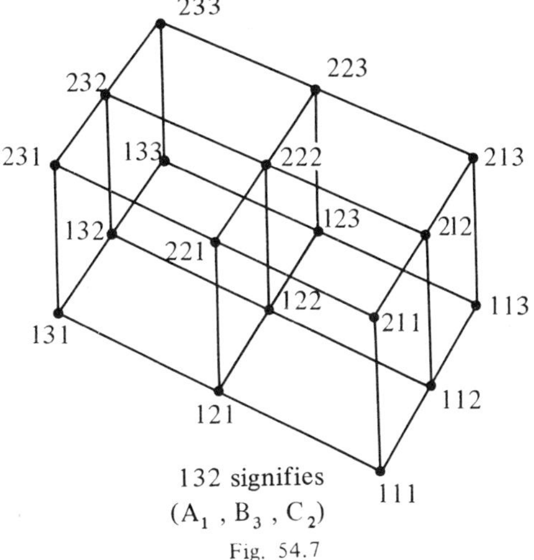

132 signifies
(A_1, B_3, C_2)

Fig. 54.7

$$\mathbf{A} = \{A_1, A_2\}\ ,$$
$$\mathbf{B} = \{B_1, B_2, B_3\}\ , \qquad (54.26)$$
$$\mathbf{C} = \{C_1, C_2, C_3\}\ .$$

We emphasize an important property: Any vector lattice is distributive but it is not complemented unless, evidently, the vector lattice is a boolean lattice.

Lexicographic vector lattice. This is a vector lattice that may be reduced to a total order, for example, that which is used in a dictionary (whence the name). One considers the following dominance relation. An n-tuple $(A_i, B_j, \ldots, S_t)$ will dominate an n-tuple $(A_{i'}, B_{j'}, \ldots, S_{t'})$ if the first r elements (arbitrarily beginning from the left) of the two n-tuples are equal, but the $(r+1)$th element of the first is greater (for the order relation in question) than the $(r+1)$th element of the second. One thus obtains a total order.

The two lattices presented in Figure 54.8 are lexicographic.

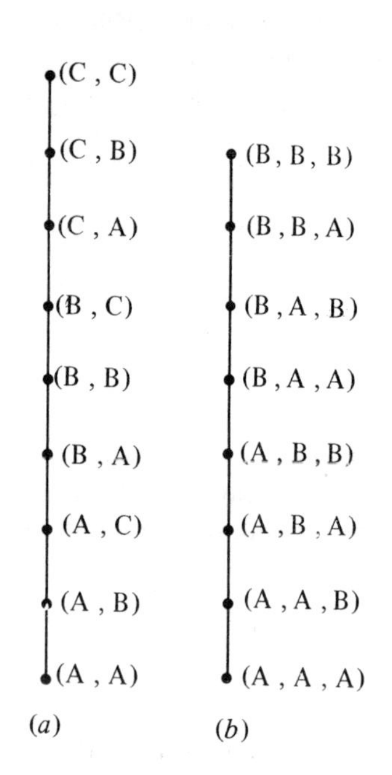

Fig. 54.8

Product of lattices. Let $\mathbf{L}_1$ and $\mathbf{L}_2$ be two lattices; then the product of these two sets gives a lattice. That is,

$$(54.27) \qquad (\mathbf{L}_1 \text{ is a lattice and } \mathbf{L}_2 \text{ is a lattice}) \Rightarrow (\mathbf{L}_1 \times \mathbf{L}_2 \text{ is a lattice}).$$

We consider an example. Let

$$(54.28) \qquad \mathbf{L}_1 = \{A, B, C, D, E, F\}\ ,$$

$$(54.29) \qquad \mathbf{L}_2 = \{\alpha, \beta, \gamma, \delta, \epsilon\}$$

(a) (b)

Fig. 54.9

each having, respectively, the lattice structure indicated in Figures 54.9a and 54.9b.

In the lattice $\mathbf{L}_1$ one notices the maximal chains

(54.30) $$F \prec E \prec B \prec A \ , \ F \prec E \prec C \prec A \ , \ F \prec E \prec D \prec A$$

and in lattice $\mathbf{L}_2$,

(54.31) $$\epsilon \prec \gamma \prec \beta \prec \alpha \ , \ \epsilon \prec \delta \prec \alpha \ .$$

Consider two ordered pairs $(X_i, Y_j), (X_i', Y_j') \in \mathbf{L}_1 \times \mathbf{L}_2$. If (X_i', Y_j') dominates (X_i, Y_j), one writes

(54.32) $$(X_{i'}, Y_{j'}) \succ (X_i, Y_j)$$

where $\succ$ represents here the order relation of dominance. Thus, $\mathbf{L}_1 \times \mathbf{L}_2$ is ordered and one may establish that relations (54.4)–(54.11) are verified for this structure; it is therefore a lattice.

In order to construct the lattice

(54.33) $$\mathbf{L} = \mathbf{L}_1 \times \mathbf{L}_2 ,$$

one examines all the ordered pairs (X_i, Y_j) with respect to one another in order to determine which are those that dominate the others; this will give the maximal chains of $\mathbf{L} = \mathbf{L}_1 \times \mathbf{L}_2$ and allow one to specify the product lattice.

The reader should involve himself in constructing the product lattice with respect to $\mathbf{L}_1$ and $\mathbf{L}_2$. Thus, by examining (54.30) and (54.31) he may write, for example,

(54.34) $$(F, \epsilon) \prec (F, \gamma) \prec (F, \beta) \prec (F, \alpha)$$

(54.35) $$(F, \epsilon) \prec (E, \epsilon) \prec \cdots \text{ etc} .$$

It is necessary to compare each pair with all the others; this is a rather long process. The operations may be simplified by considering the maximal chains (54.30) and (54.31), one with respect to the other.

An important particular case is that where $\mathbf{L}_1$ and $\mathbf{L}_2$ are totally ordered; then $\mathbf{L} = \mathbf{L}_1 \times \mathbf{L}_2$ forms a vector lattice.

An important general property must be emphasized. If $\mathbf{L}_1$ and $\mathbf{L}_2$ are distributive, then $\mathbf{L} = \mathbf{L}_1 \times \mathbf{L}_2$ is distributive.

Partially ordered set not forming a lattice. Of course, not all partially ordered sets form a lattice (as, for example, that represented in Figure 54.10). One sees, for example,

(54.36) $$H \nabla I \notin \mathbf{L} ,$$

(54.37) $$A \Delta B \notin \mathbf{L} .$$

When one has the property that for any ordinary subsets $\{X_i, X_j\}$ of $\mathbf{L}$, the superior limit of $\{X_i, X_j\}$ belongs to $\mathbf{L}$, one says then that $\mathbf{L}$ is a sup-semilattice.

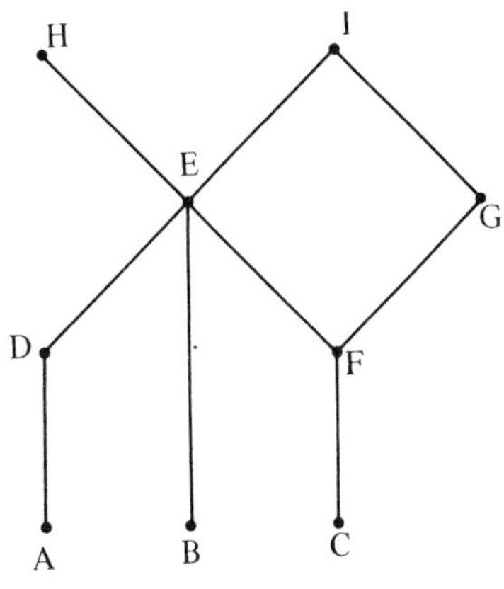

Fig. 54.10

In the parallel case using the inferior limit, one says that **L** is an inf-semilattice. A lattice, therefore, is at the same time an inf- and a sup-semilattice.

In Figure 54.11 we have presented some semilattices, and, as one may note, the ordered set represented in Figure 54.11 is neither an inf- nor a sup-semilattice.

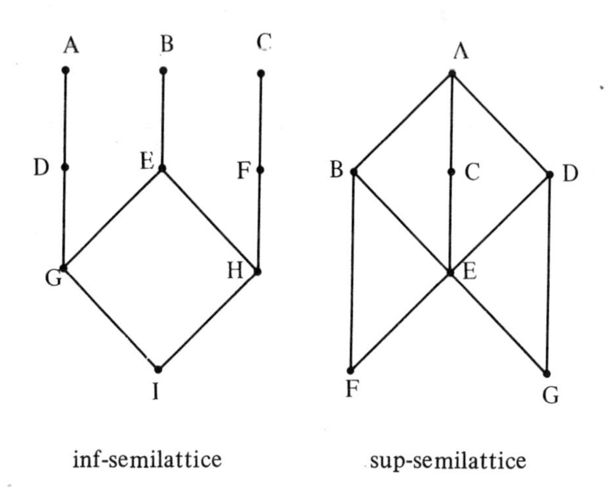

Fig. 54.11

Ordinal function of a partially ordered set. In Section 24 we have defined the notion of ordinal function for any graph without circuits. Given that a graph represents a partial order and concerns a finite set and is a graph without circuits, one may very usefully define its levels; this facilitates the analysis or synthesis of order relations. One may establish the ordinal function directly with reference to the Hasse diagram or diagram of maximal chains.

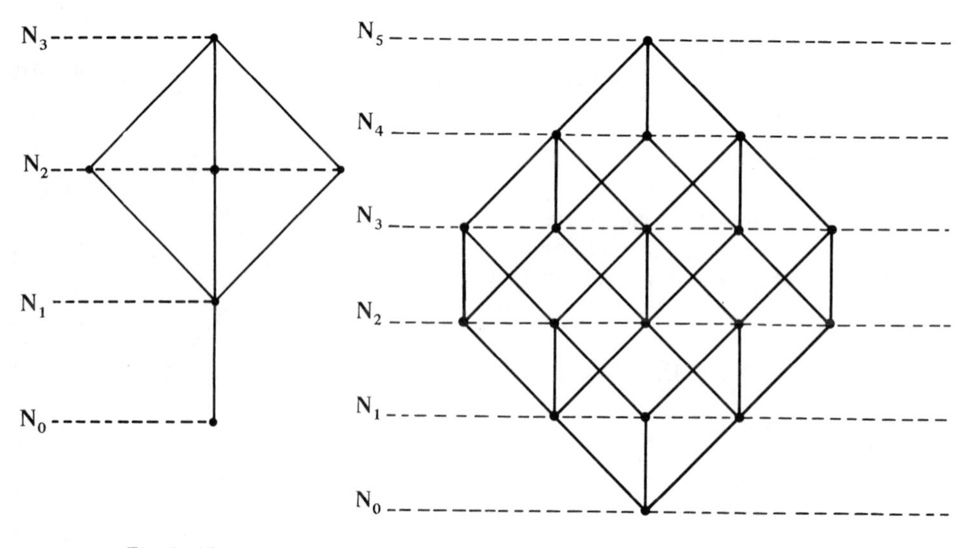

Fig. 54.12

Fig. 54.13

In Figures 54.12–54.14 one sees how to define levels for some order relations (lattice or not) with respect to their representations in Hasse diagrams.

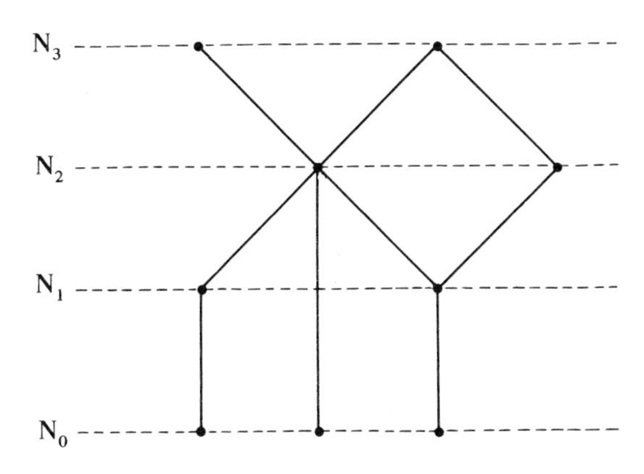

Fig. 54.14

Ring structure. Consider a set **F** provided with two internal laws $*$ and $\circ$ such that, for all $X_i, X_j, X_k \in \mathbf{F}$:

(54.38) (1) $(X_i * X_j) * X_k = X_i * (X_j * X_k)$, associativity for $*$.

(54.39) $X_i * e = e * X_i = X_i$, existence of an identity for $*$.

(54.40) $X_i * \overline{X}_i = \overline{X}_i * X_i = e$, existence of an inverse for all X_i.

(54.41) $X_i * X_j = X_j * X_i$, commutativity for $*$.

(54.42) (2) $(X_i \circ X_j) \circ X_k = X_i \circ (X_j \circ X_k)$, associativity for $\circ$.

(54.43) (3) $(X_i * X_j) \circ X_k = (X_i \circ X_k) * (X_j \circ X_k)$

(54.44) $X_k \circ (X_i * X_j) = (X_k \circ X_i) * (X_k \circ X_j)$

distributivity on the left and on the right with respect to $*$.

Then one says that **F** has a ring structure or that **F** is a ring. If the law $\circ$ is also commutative, one then says that the ring is commutative.

Figure 54.15 gives an example of a ring structure. A is the identity.

$*$	A	B	C	D
A	A	B	C	D
B	B	C	D	A
C	C	D	A	B
D	D	A	B	C

$\circ$	A	B	C	D
A	A	A	A	A
B	A	B	C	D
C	A	C	A	C
D	A	D	C	B

Fig. 54.15

Boolean lattice and boolean ring. A boolean lattice being a distributive and complemented lattice, one may verify that for all X_i, X_j, $X_k \in \mathbf{L}$:

(54.45) $$X_i \,\Delta\, X_j = X_j \,\Delta\, X_i \,,$$
(54.46) $$X_i \,\nabla\, X_j = X_j \,\nabla\, X_i \,,$$
commutativity.

(54.47) $$X_i \,\Delta\, (X_j \,\Delta\, X_k) = (X_i \,\Delta\, X_j) \,\Delta\, X_k \,,$$
(54.48) $$X_i \,\nabla\, (X_j \,\nabla\, X_k) = (X_i \,\nabla\, X_j) \,\nabla\, X_k \,,$$
associativity.

(54.49) $$X_i \,\Delta\, X_i = X_i \,,$$
(54.50) $$X_i \,\nabla\, X_i = X_i \,,$$
idempotence.

(54.51) $$X_i \,\Delta\, (X_j \,\nabla\, X_k) = (X_i \,\Delta\, X_j) \,\nabla\, (X_i \,\Delta\, X_k) \,,$$
(54.52) $$X_i \,\nabla\, (X_j \,\Delta\, X_k) = (X_i \,\nabla\, X_j) \,\Delta\, (X_i \,\nabla\, X_k) \,,$$
distributivity of ∇ with respect to Δ and vice versa.

(54.53) $$X_i \,\Delta\, \overline{X}_i = 0 \,,$$

(54.54) $$X_i \,\nabla\, \overline{X}_i = U \,.$$

(54.55) $$X_i \,\Delta\, 0 = 0 \,,$$

(54.56) $$X_i \,\nabla\, 0 = X_i \,.$$

(54.57) $$X_i \,\Delta\, U = X_i \,,$$

(54.58) $$X_i \,\nabla\, U = U \,.$$

(54.59) $$\overline{(\overline{X}_i)} = X_i$$

(54.60) $$\overline{X_i \,\Delta\, X_j} = \overline{X}_i \,\nabla\, \overline{X}_j \,,$$
(54.61) $$\overline{X_i \,\nabla\, X_j} = \overline{X}_i \,\Delta\, \overline{X}_j \,.$$
De Morgan's theorems.

Now we consider a set **E** and $\mathbf{L}^{\mathbf{E}}$, where **L** has a ring structure and where we define the operations ⊛ and ⊚ with respect to $*$ and $\circ$.

Then for any **A**, **B**, $\mathbf{C} \in \mathbf{L}^{\mathbf{E}}$, one may verify:

(54.62) (1) $$(\mathbf{A} \circledast \mathbf{B}) \circledast \mathbf{C} = \mathbf{A} \circledast (\mathbf{B} \circledast \mathbf{C}) \,,$$ associativity for ⊛.

(54.63) $$\mathbf{A} \circledast \phi = \phi \circledast \mathbf{A} = \mathbf{A} \,,$$ existence of an identity for ⊛.

(54.64) $$\mathbf{A} \circledast \mathbf{A} = \phi \,,$$ existence of an inverse, which is **A** itself.

(54.65) $$\mathbf{A} \circledast \mathbf{B} = \mathbf{B} \circledast \mathbf{A} \,,$$ commutativity for ⊛.

(54.66) (2) $$\mathbf{A} \circledcirc (\mathbf{B} \circledcirc \mathbf{C}) = (\mathbf{A} \circledcirc \mathbf{B}) \circledcirc \mathbf{C} \,,$$ associativity for ⊚.

(54.67) (3) $$(\mathbf{A} \circledast \mathbf{B}) \circledcirc \mathbf{C} = (\mathbf{A} \circledcirc \mathbf{C}) \circledast (\mathbf{B} \circledcirc \mathbf{C}) \,,$$
(54.68) $$\mathbf{C} \circledcirc (\mathbf{A} \circledast \mathbf{B}) = (\mathbf{C} \circledcirc \mathbf{A}) \circledast (\mathbf{C} \circledcirc \mathbf{B}) \,,$$
distributivity on the left and on the right.

Thus, the structure of $\mathbf{L}^{\mathbf{E}}$ is itself that of a ring, which one calls Boole's ring. One may show that in this ring, if $\mathbf{E} = \{0, 1\}$, the ⊛ corresponds to the disjunctive sum $\oplus$ and ⊚ corresponds to intersection $\cap$.

55. GENERALIZATION OF THE NOTION OF FUZZY SUBSET

First we consider a particular example.

First example. Suppose that

(55.1) $$\mathbf{L} = \{0, \alpha, \beta, 1\},$$

(55.2) $$\mathbf{E} = \{A, B\}.$$

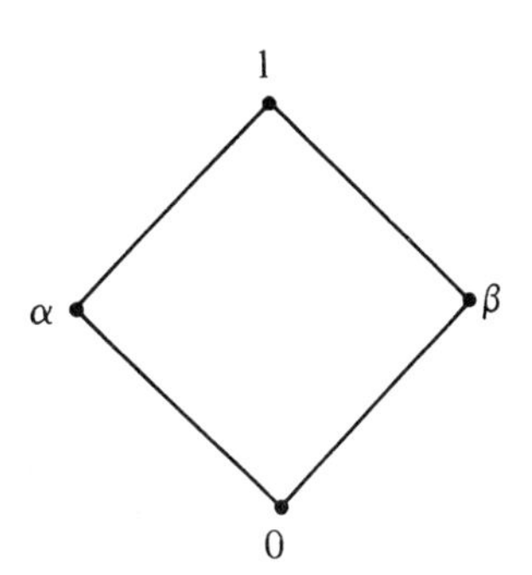

Fig. 55.1

Suppose also that **L** has the structure of a boolean lattice (that is, is distributive and complemented), like the one presented in Figure 55.1.

For the operations Δ and ∇, one may see the results below for **L**:

(55.3)

Δ	0	α	β	1
0	0	0	0	0
α	0	α	0	α
β	0	0	β	β
1	0	α	β	1

(55.4)

∇	0	α	β	1
0	0	α	β	1
α	α	α	1	1
β	β	1	β	1
1	1	1	1	1

We examine the properties of $\mathbf{L}^{\mathbf{E}}$. Put

(55.5) $$x_1, x_2, x_3, y_1, y_2, y_3 = \{0, \alpha, \beta, 1\},$$

(55.6) $$\mathbf{X}_1 = \{(A \mid x_1), (B \mid y_1)\},$$

(55.7) $$\mathbf{X}_2 = \{(A \mid x_2), (B \mid y_2)\},$$

(55.8) $$\mathbf{X}_3 = \{(A \mid x_3), (B \mid y_3)\}.$$

Put

(55.9) $$\mathbf{X}_1 \sqcap \mathbf{X}_2 = \{(A \mid x_1 \Delta x_2), (B \mid y_1 \Delta y_2)\},$$

(55.10) $$\mathbf{X}_1 \sqcup \mathbf{X}_2 = \{(A \mid x_1 \nabla x_2), (B \mid y_1 \nabla y_2)\}.$$

Since **L** has the structure of a boolean lattice, for Δ and ∇ one has the following properties:

associativity

commutativity

idempotence

absorption

distributivity of Δ with respect to ∇ and vice versa

existence of a unique complement for each element of **L**.

We then examine the properties of $\mathbf{L}^{\mathbf{E}}$. It is easy to see that $\sqcap$ is associative since Δ is. Likewise $\sqcup$ because of ∇. Also entirely easy to prove are commutativity, idempotence, and absorption.

We prove distributivity and complementation:

$$\begin{aligned}(55.11)\quad \mathbf{X}_1 \sqcup (\mathbf{X}_2 \sqcap \mathbf{X}_3) &= \{(\mathrm{A}\,|\,x_1 \nabla (x_2 \Delta x_3)), (\mathrm{B}\,|\,y_1 \nabla (y_2 \Delta y_3))\} \\ &= \{(\mathrm{A}\,|\,(x_1 \nabla x_2) \Delta (x_1 \nabla x_3)), (\mathrm{B}\,|\,(y_1 \nabla y_2) \Delta (y_1 \nabla y_3))\} \\ &= (\mathbf{X}_1 \sqcup \mathbf{X}_2) \sqcap (\mathbf{X}_1 \sqcup \mathbf{X}_3)\ .\end{aligned}$$

Put

$$(55.12)\quad \overline{\mathbf{X}}_1 = \{(\mathrm{A}\,|\,\overline{x}_1), (\mathrm{B}\,|\,\overline{y}_1)\}\ .$$

(55.13) We verify $\mathbf{X}_1 \sqcap \overline{\mathbf{X}}_1 = \{(\mathrm{A}\,|\,x_1 \Delta \overline{x}_1), (\mathrm{B}\,|\,y_1 \Delta \overline{y}_1)\} = \{(\mathrm{A}\,|\,0), (\mathrm{B}\,|\,0)\}$,

(55.14) Likewise, $\mathbf{X}_1 \sqcup \overline{\mathbf{X}}_1 = \{(\mathrm{A}\,|\,1), (\mathrm{B}\,|\,1)\}$.

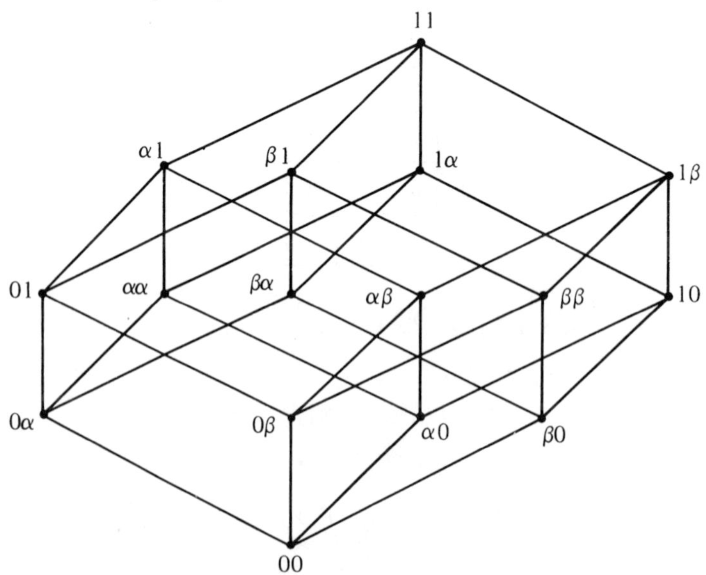

In this figure xy signifies $\{(\mathrm{A}\,|\,x), (\mathrm{B}\,|\,y)\}$.

Fig. 55.2

Thus, $\mathbf{L}^{\mathrm{E}}$ possesses the structure of a boolean lattice, just as **L**. This structure has been presented in Figure 55.2.

Another example. Again let :

$$\mathbf{L} = \{0, \alpha, \beta, 1\} \text{ where } 0 < \alpha < \beta < 1 \quad \text{and} \quad \mathbf{E} = \{A, B\} . \tag{55.15, 55.16}$$

But this time the structure of **L** is that of a lattice having a unique chain (Figure 55.3). This lattice is distributive but not complemented (that is, rather, a vector lattice). For the operations Δ and ∇, one has the results:

Fig. 55.3

(55.17)

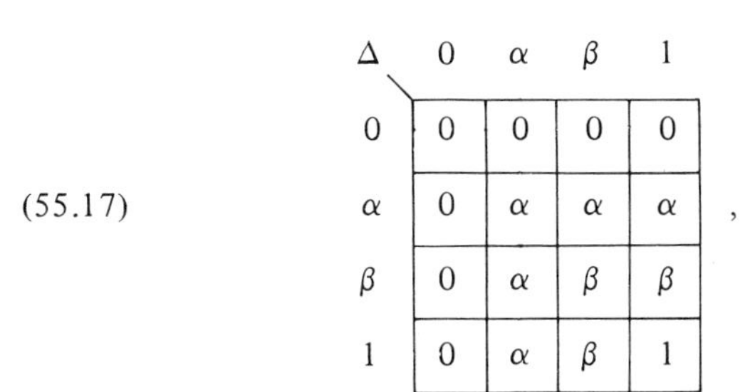

Δ	0	α	β	1
0	0	0	0	0
α	0	α	α	α
β	0	α	β	β
1	0	α	β	1

,

(55.18)

∇	0	α	β	1
0	0	α	β	1
α	α	α	β	1
β	β	β	β	1
1	1	1	1	1

.

L has the following properties:

associativity

commutativity

idempotence

absorption

distributivity with respect to Δ and with respect to ∇.

One may verify easily that $\mathbf{L}^{\mathrm{E}}$ possesses these same properties; it is also a vector lattice. It has been represented in Figure 55.4.

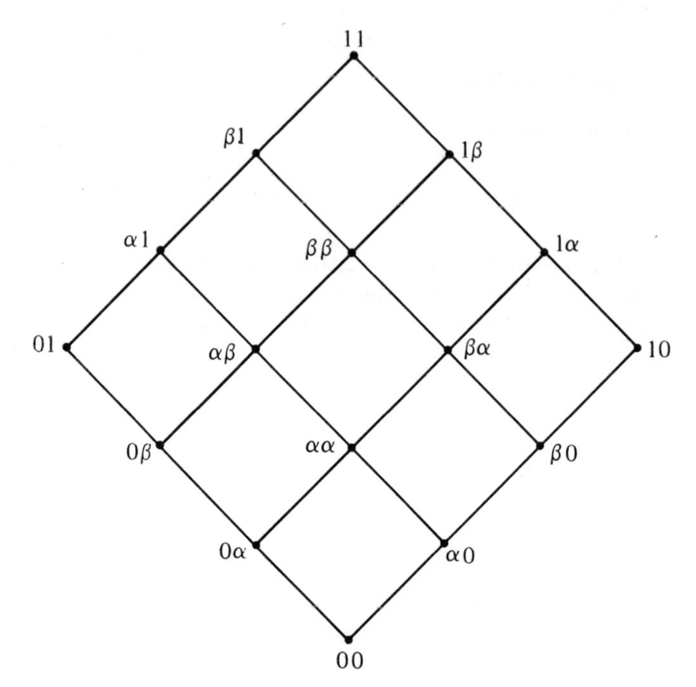

In this figure xy signifies
$\{(A \mid x), (B \mid y)\}$.

Fig. 55.4

Third example. Let

(55.19) $$\mathbf{L} = \{0, \alpha, \beta, \gamma\},$$

(55.20) $$\mathbf{E} = \{A, B\}.$$

The structure of **L**, being that of an inf-semilattice, is presented in Figure 55.5. Since it is an inf-semilattice, one may define only Δ; this gives

(55.21)

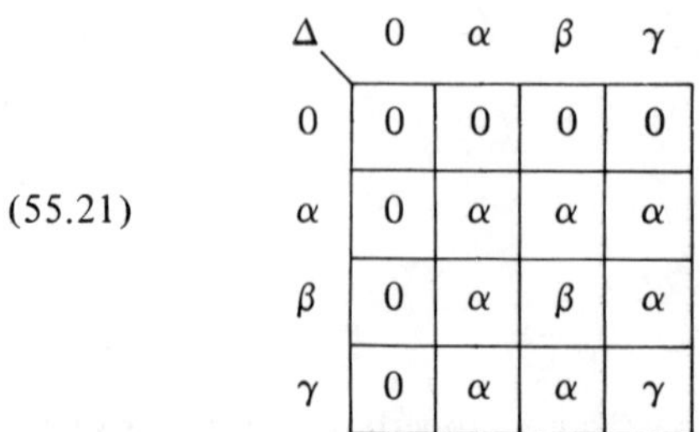

Δ	0	α	β	γ
0	0	0	0	0
α	0	α	α	α
β	0	α	β	α
γ	0	α	α	γ

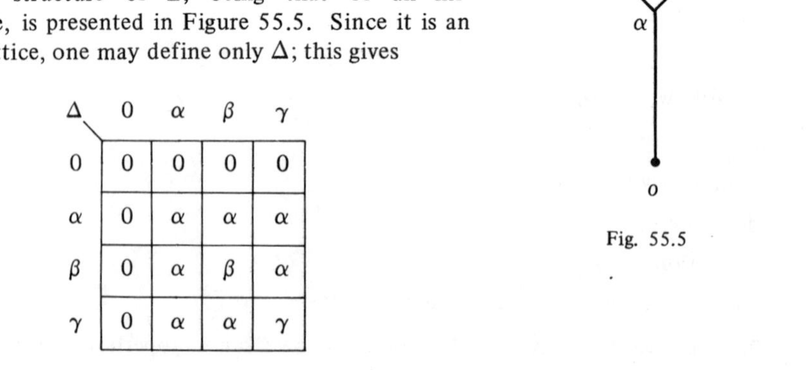

Fig. 55.5

For $\mathbf{L}^{\mathrm{E}}$, one obtains the following properties:

associativity for Δ

commutativity for Δ

idempotence for Δ

and one has the structure of an inf-semilattice (Figure 55.6).

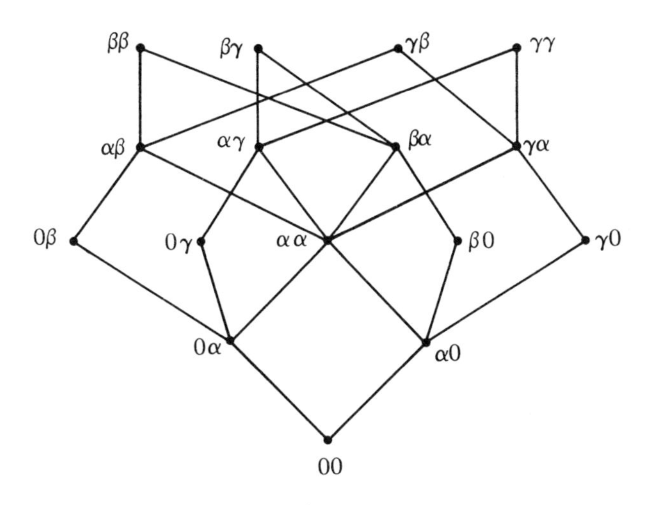

Fig. 55.6

Of course, by reconsidering the same example with the top of Figure 55.5 becoming the bottom and vice versa, one would obtain a sup-semilattice for $\mathbf{L}^E$.

Fourth example. Let

$$\mathbf{L} = \{\alpha , \beta , \gamma , \delta\} \tag{55.22}$$

$$\mathrm{E} = \{\mathrm{A} , \mathrm{B}\} \tag{55.23}$$

The structure of **L** is shown in Figure 55.7, which is not a semilattice. There no longer exists an inferior limit nor a superior limit for certain ordered pairs. One may sometimes define an $\mathbf{L}^E$ structure for a dominance relation; one then obtains the structure given in Figure 55.8.

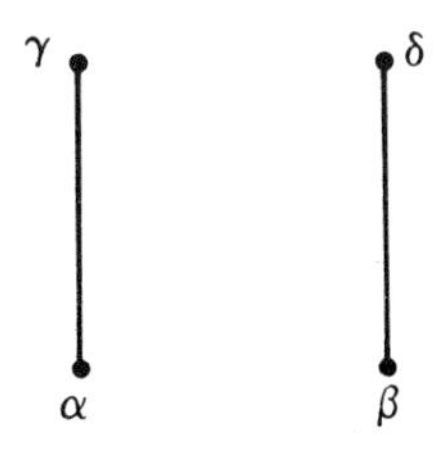

Fig. 55.7

Remark. One usually calls the graph representing a relation a configuration. Thus, a lattice may be considered as a configuration, similarly a semilattice, and finally therefore any graph representing a relation. But, as we have seen, a lattice is also a structure for the operations Δ and ∇, likewise a semilattice for one of the operations Δ or ∇. The graphs corresponding to the Hasse diagrams of Figures 55.7 and 55.8 are configurations, but these are no longer structures, at least for Δ or/and ∇.

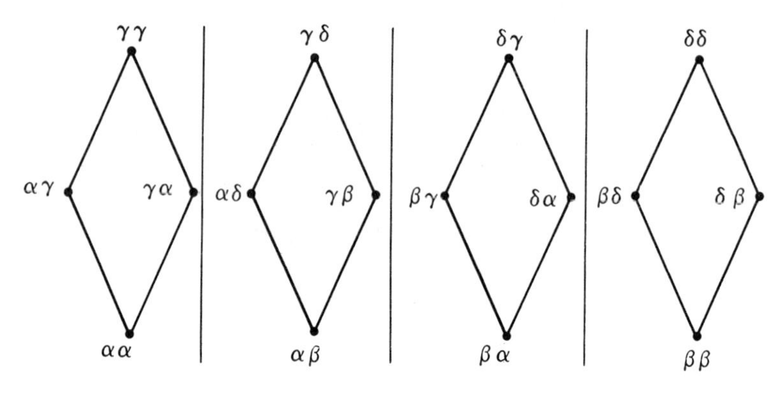

Fig. 55.8

Case where* L *has a preorder configuration. If **L** has an ordinary preorder configuration (that is, in the sense of the ordinary theory of graphs), we know that it is possible to find in this preordered set equivalence classes, and that these classes then form among themselves a (partial or total) order. This is how we shall proceed in the case where **L** has an ordinary preorder configuration.

We consider an example where

$$\mathbf{L} = \{\alpha, \beta, \gamma, \delta, \epsilon, \eta, \mu, \nu\} \tag{55.24}$$

$$\mathbf{E} = \{A, B\} \quad . \tag{55.25}$$

Suppose that **L** has a preorder structure like that indicated in the ordinary graph of Figure 55.9. In Figure 55.10 we have drawn the four preorder equivalence classes. In

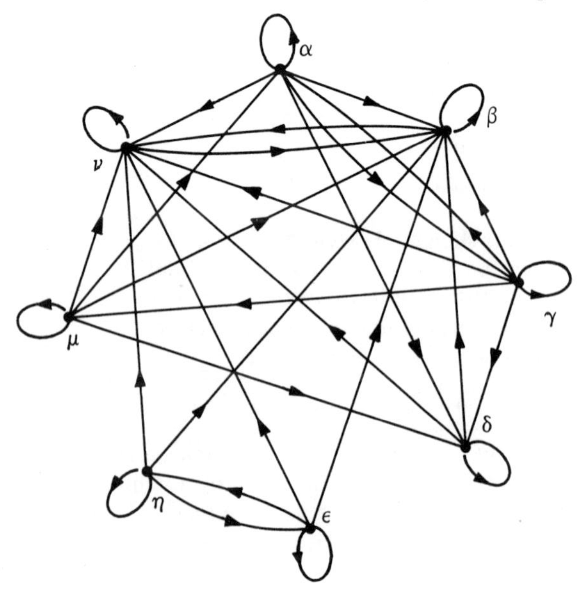

Fig. 55.9

Figure 55.11 we show the order of these classes, and in Figure 55.12, the maximal chains of this order.

We note that the classes form a sup-semilattice.

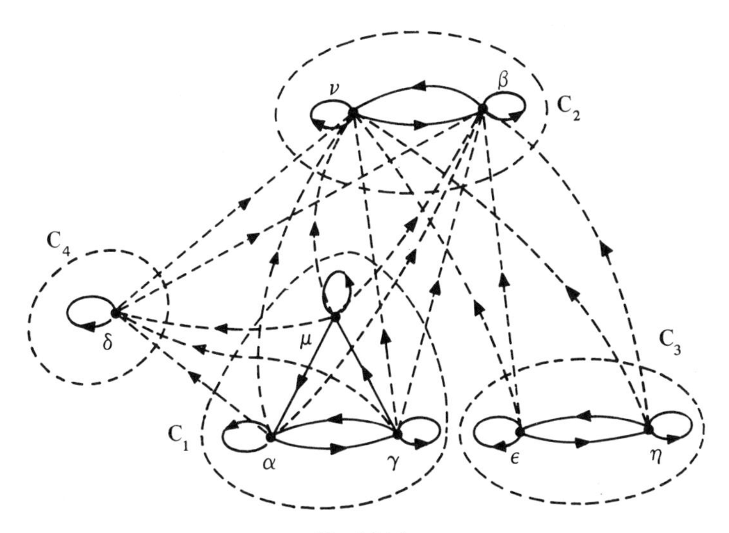

Fig. 55.10

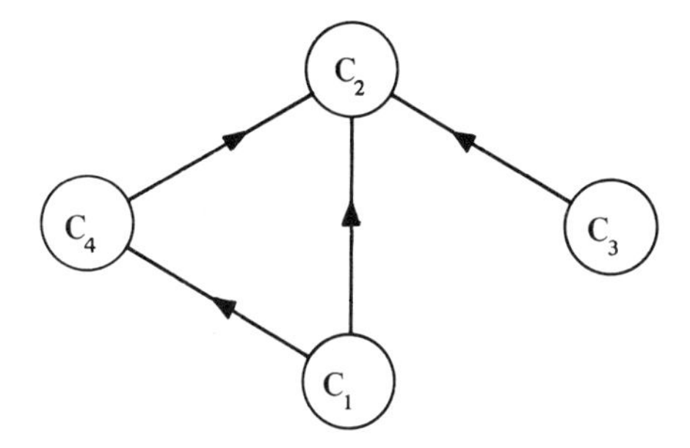

Fig. 55.11

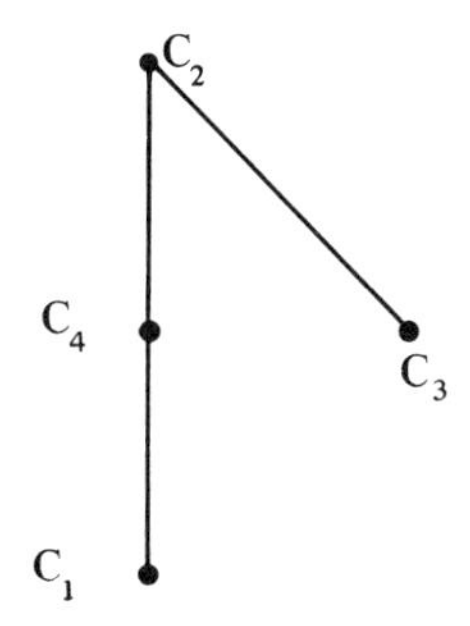

Fig. 55.12

For an immediate use, we note that

(55.26) $\nu \simeq \beta \quad , \quad \alpha \simeq \gamma \simeq \mu \quad , \quad \epsilon \simeq \eta \quad , \quad \delta \simeq \delta \quad .$

And, to simplify writing, we associate a representative to each class to represent the class:

(55.26′) $\alpha \in \mathbf{C}_1 \quad , \quad \beta \in \mathbf{C}_2 \quad , \quad \eta \in \mathbf{C}_3 \quad , \quad \delta \in \mathbf{C}_4 \quad .$

The sup-semilattice **L** may be expressed by the following relation:

(55.27)

$\triangledown$	α	β	η	δ
α	α	β	β	δ
β	β	β	β	β
η	β	β	η	β
δ	δ	β	β	δ

Figure 55.13 represents the sup-semilattice $\mathbf{L}^{\mathbf{E}}$, where xy is the representative of the class.

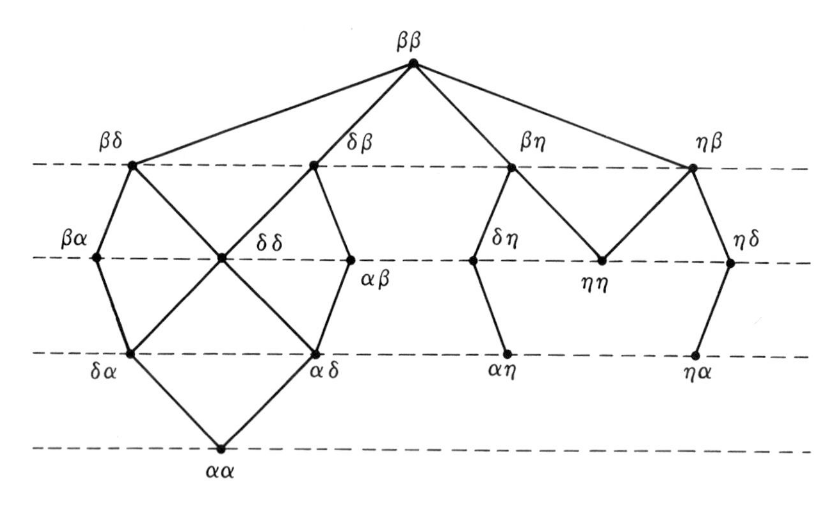

Fig. 55.13

There are 16 equivalence classes in this sup-semilattice. Thus, the class of $\alpha\beta$ is

(55.28) $\mathbf{C}_{\alpha\beta} = \{(\alpha , \beta) \ , \ (\alpha , \nu) \ , \ (\gamma , \beta) \ , \ (\gamma , \nu) \ , \ (\mu , \beta) \ , \ (\mu , \nu)\}$

And one may see that the 64 elements of $\mathbf{L}^{\mathbf{E}}$ are decomposed as follows:

(55.29)

$$
\begin{aligned}
&C_{\beta\beta} \quad \text{contains 4 elements,}\\
&C_{\beta\delta} \quad \text{contains 2 elements,}\\
&C_{\delta\beta} \quad \text{contains 2 elements,}\\
&C_{\beta\eta} \quad \text{contains 4 elements,}\\
&C_{\eta\beta} \quad \text{contains 4 elements,}\\
&C_{\beta\alpha} \quad \text{contains 6 elements,}\\
&C_{\delta\delta} \quad \text{contains 1 elements,}\\
&C_{\alpha\beta} \quad \text{contains 6 elements,}\\
&C_{\delta\eta} \quad \text{contains 2 elements,}\\
&C_{\eta\eta} \quad \text{contains 4 elements,}\\
&C_{\eta\delta} \quad \text{contains 2 elements,}\\
&C_{\delta\alpha} \quad \text{contains 3 elements,}\\
&C_{\alpha\delta} \quad \text{contains 3 elements,}\\
&C_{\alpha\eta} \quad \text{contains 6 elements,}\\
&C_{\eta\alpha} \quad \text{contains 6 elements,}\\
&C_{\alpha\alpha} \quad \text{contains 9 elements,}
\end{aligned}
$$

Thus, $\mathbf{L}^{\mathbf{E}}$ is an ordinary preorder containing 16 equivalence classes.

Case where* L *has a ring structure. Let

$$\mathbf{L} = \{e, \alpha, \beta, \gamma\},$$
$$\mathbf{E} = \{A, B\},$$

and

(55.30)

$*$	e	α	β	γ
e	e	α	β	γ
α	α	β	γ	e
β	β	γ	e	α
γ	γ	e	α	β

(55.31)

$\circ$	e	α	β	γ
e	e	e	e	e
α	e	α	β	γ
β	e	β	e	β
γ	e	γ	β	α

If one puts

$$\mathbf{A} = \{(\mathrm{A}\,|\,x_1)\ ,\ (\mathrm{B}\,|\,y_1)\}\ , \tag{55.32}$$

$$\mathbf{B} = \{(\mathrm{A}\,|\,x_2)\ ,\ (\mathrm{B}\,|\,y_2)\}\ , \tag{55.33}$$

where $x_1, x_2, y_1, y_2 \in \mathbf{L}$ and A, B $\in$ **E**, then

$$\mathbf{A} \circledast \mathbf{B} = \{(\mathrm{A}\,|\,x_1 * x_2)\ ,\ (\mathrm{B}\,|\,y_1 * y_2)\}\ , \tag{55.34}$$

$$\mathbf{A} \circledcirc \mathbf{B} = \{(\mathrm{A}\,|\,x_1 \circ x_2)\ ,\ (\mathrm{B}\,|\,y_1 \circ y_2)\}\ . \tag{55.35}$$

One obtains a ring structure for $\mathbf{L}^{\mathrm{E}}$; this is presented explicitly in (55.36) and (55.37), where we have used xy to denote $\{(\mathrm{A}\,|\,x), (\mathrm{B}\,|\,y)\}$.

(55.36)

⊛	ee	eα	eβ	eγ	αe	αα	αβ	αγ	βe	βα	ββ	βγ	γe	γα	γβ	γγ
ee	ee	eα	eβ	eγ	αe	αα	αβ	αγ	βe	βα	ββ	βγ	γe	γα	γβ	γγ
eα	eα	eβ	eγ	ee	αα	αβ	αγ	αe	βα	ββ	βγ	βe	γα	γβ	γγ	γe
eβ	eβ	eγ	ee	eα	αβ	αγ	αe	αα	ββ	βγ	βe	βα	γβ	γγ	γe	γα
eγ	eγ	ee	eα	eβ	αγ	αe	αα	αβ	βγ	βe	βα	ββ	γγ	γe	γα	γβ
αe	αe	αα	αβ	αγ	βe	βα	ββ	βγ	γe	γα	γβ	γγ	ee	eα	eβ	eγ
αα	αα	αβ	αγ	αe	βα	ββ	βγ	βe	γα	γβ	γγ	γe	eα	eβ	eγ	ee
αβ	αβ	αγ	αe	αα	ββ	βγ	βe	βα	γβ	γγ	γe	γα	eβ	eγ	ee	eα
αγ	αγ	αe	αα	αβ	βγ	βe	βα	ββ	γγ	γe	γα	γβ	eγ	ee	eα	eβ
βe	βe	βα	ββ	βγ	γe	γα	γβ	γγ	ee	eα	eβ	eγ	αe	αα	αβ	αγ
βα	βα	ββ	βγ	βe	γα	γβ	γγ	γe	eα	eβ	eγ	ee	αα	αβ	αγ	αe
ββ	ββ	βγ	βe	βα	γβ	γγ	γe	γα	eβ	eγ	ee	eα	αβ	αγ	αe	αα
βγ	βγ	βe	βα	ββ	γγ	γe	γα	γβ	eγ	ee	eα	eβ	αγ	αe	αα	αβ
γe	γe	γα	γβ	γγ	ee	eα	eβ	eγ	αe	αα	αβ	αγ	βe	βα	ββ	βγ
γα	γα	γβ	γγ	γe	eα	eβ	eγ	ee	αα	αβ	αγ	αe	βα	ββ	βγ	βe
γβ	γβ	γγ	γe	γα	eβ	eγ	ee	eα	αβ	αγ	αe	αα	ββ	βγ	βe	βα
γγ	γγ	γe	γα	γβ	eγ	ee	eα	eβ	αγ	αe	αα	αβ	βγ	βe	βα	ββ

(55.37)

$\odot$	ee	$e\alpha$	$e\beta$	$e\gamma$	αe	$\alpha\alpha$	$\alpha\beta$	$\alpha\gamma$	βe	$\beta\alpha$	$\beta\beta$	$\beta\gamma$	γe	$\gamma\alpha$	$\gamma\beta$	$\gamma\gamma$
ee	ee	ee	ee	ee	ee	ee	ee	ee	ee	ee	ee	ee	ee	ee	ee	ee
$e\alpha$	ee	$e\alpha$	$e\beta$	$e\gamma$	ee	$e\alpha$	$e\beta$	$e\gamma$	ee	$e\alpha$	$e\beta$	$e\gamma$	ee	$e\alpha$	$e\beta$	$e\gamma$
$e\beta$	ee	$e\beta$	ee	$e\beta$	ee	$e\beta$	ee	$e\beta$	ee	$e\beta$	ee	$e\beta$	ee	$e\beta$	ee	$e\beta$
$e\gamma$	ee	$e\gamma$	$e\beta$	$e\alpha$	ee	$e\gamma$	$e\beta$	$e\alpha$	ee	$e\gamma$	$e\beta$	$e\alpha$	ee	$e\gamma$	$e\beta$	$e\alpha$
αe	ee	ee	ee	ee	αe	αe	αe	αe	βe	βe	βe	βe	γe	γe	γe	γe
$\alpha\alpha$	ee	$e\alpha$	$e\beta$	$e\gamma$	αe	$\alpha\alpha$	$\alpha\beta$	$\alpha\gamma$	βe	$\beta\alpha$	$\beta\beta$	$\beta\gamma$	γe	$\gamma\alpha$	$\gamma\beta$	$\gamma\gamma$
$\alpha\beta$	ee	$e\beta$	ee	$e\beta$	αe	$\alpha\beta$	αe	$\alpha\beta$	βe	$\beta\beta$	βe	$\beta\beta$	γe	$\gamma\beta$	γe	$\gamma\beta$
$\alpha\gamma$	ee	$e\gamma$	$e\beta$	$e\alpha$	αe	$\alpha\gamma$	$\alpha\beta$	$\alpha\alpha$	βe	$\beta\gamma$	$\beta\beta$	$\beta\alpha$	γe	$\gamma\gamma$	$\gamma\beta$	$\gamma\alpha$
βe	ee	ee	ee	ee	βe	βe	βe	βe	ee	ee	ee	ee	βe	βe	βe	βe
$\beta\alpha$	ee	$e\alpha$	$e\beta$	$e\gamma$	βe	$\beta\alpha$	$\beta\beta$	$\beta\gamma$	ee	$e\alpha$	$e\beta$	$e\gamma$	βe	$\beta\alpha$	$\beta\beta$	$\beta\gamma$
$\beta\beta$	ee	$e\beta$	ee	$e\beta$	βe	$\beta\beta$	βe	$\beta\beta$	ee	$e\beta$	ee	$e\beta$	βe	$\beta\beta$	βe	$\beta\beta$
$\beta\gamma$	ee	$e\gamma$	$e\beta$	$e\alpha$	βe	$\beta\gamma$	$\beta\beta$	$\beta\alpha$	ee	$e\gamma$	$e\beta$	$e\alpha$	βe	$\beta\gamma$	$\beta\beta$	$\beta\alpha$
γe	ee	ee	ee	ee	γe	γe	γe	γe	βe	βe	βe	βe	αe	αe	αe	αe
$\gamma\alpha$	ee	$e\alpha$	$e\beta$	$e\gamma$	γe	$\gamma\alpha$	$\gamma\beta$	$\gamma\gamma$	βe	$\beta\alpha$	$\beta\beta$	$\beta\gamma$	αe	$\alpha\alpha$	$\alpha\beta$	$\alpha\gamma$
$\gamma\beta$	ee	$e\beta$	ee	$e\beta$	γe	$\gamma\beta$	γe	$\gamma\beta$	βe	$\beta\beta$	βe	$\beta\beta$	αe	$\alpha\beta$	αe	$\alpha\beta$
$\gamma\gamma$	ee	$e\gamma$	$e\beta$	$e\alpha$	γe	$\gamma\gamma$	$\gamma\beta$	$\gamma\alpha$	βe	$\beta\gamma$	$\beta\beta$	$\beta\alpha$	αe	$\alpha\gamma$	$\alpha\beta$	$\alpha\alpha$

Remark. It is interesting to compare the ring (55.30) and (55.31) with the boolean ring (55.3) and (55.4). We use the notation of (55.30) and (55.31) for (55.3) and (55.4) and compare them in Figure 55.14. On the left one has a ring [that of the example

*	e	α	β	γ
e	e	α	β	γ
α	α	β	γ	e
β	β	γ	e	α
γ	γ	e	α	β

$\circ$	e	α	β	γ
e	e	e	e	e
α	e	α	β	γ
β	e	β	e	β
γ	e	γ	β	α

Ring for addition and multiplication modulo 4

∇	e	α	β	γ
e	e	α	β	γ
α	α	α	γ	γ
β	β	γ	β	γ
γ	γ	γ	γ	γ

Δ	e	α	β	γ
e	e	e	e	e
α	e	α	e	α
β	e	e	β	β
γ	e	α	β	γ

Boole's ring

Fig. 55.14

(55.30) and (55.31)] that is a ring for addition and multiplication modulo 4 (to see this it is sufficient to take $c = 0$, $\alpha = 1$, $\beta = 2$, and $\gamma = 3$) and on the right, Boole's ring constructed on $\{e, \alpha, \beta, \gamma\}$.

This remark may be made more precise by recalling that Boole's ring is only one ring structure among many.

Transfer of properties. It is easy finally to prove after that which has been presented in the present section that:

(55.38) $\mathbf{L}$ is an ordinary preorder $\Rightarrow$ $\mathbf{L}$ is an ordinary preorder

(55.39) $\mathbf{L}$ is an ordinary order $\Rightarrow$ $\mathbf{L}$ is an ordinary order

(55.40) $\mathbf{L}$ is an inf-semilattice $\Rightarrow$ $\mathbf{L}$ is an inf-semilattice

(55.41) $\mathbf{L}$ is a sup-semilattice $\Rightarrow$ $\mathbf{L}$ is an sup-semilattice

(55.42) $\mathbf{L}$ is a lattice $\Rightarrow$ $\mathbf{L}$ is a lattice

(55.43) $\mathbf{L}$ is a ring $\Rightarrow$ $\mathbf{L}$ is a ring.

These properties come to be added to those that have already been specified in (53.16)–(53.18) by knowing:

Any law $*$ that is associative for $\mathbf{L}$ induces an associative law $\circledast$ for $\mathbf{L}^{\mathbf{E}}$; likewise if $*$ is commutative, $\circledast$ is commutative; similarly, if $*$ is idempotent, $\circledast$ is idempotent.

Other properties. Formulas (52.33)–(52.39) permit one to define other properties. Consider the case where

(55.44)† $$\mathbf{L} = \mathbf{L}_1 \times \mathbf{L}_2 \ .$$

One may easily prove several properties.

If $\mathbf{L}_1$ and $\mathbf{L}_2$ have ordinary preorder configurations, then $\mathbf{L}$ does also; similar results hold considering order, semilattice, lattice, or ring configurations.

One therefore also recovers these properties for associations among the product and exponentiation operations.

We consider an example.

Example. Let

(55.46) $$\mathbf{L}_1 = \{a, b, c\}$$

(55.47) $$\mathbf{L}_2 = \{\alpha, \beta\},$$

where $\mathbf{L}_1$ has the inf-semilattice structure represented in Figure 55.15 and $\mathbf{L}_2$ has that of a lattice (Figure 55.16).

†There is no Eq. (55.45).

Fig. 55.15 Fig. 55.16

The product

(55.48) $$\begin{aligned}\mathbf{L} &= \mathbf{L}_1 \times \mathbf{L}_2 \\ &= \{(a\,,\alpha)\,,(a\,,\beta)\,,(b\,,\alpha)\,,(b\,,\beta)\,,(c\,,\alpha)\,,(c\,,\beta)\}\end{aligned}$$

possesses an inf-semilattice structure (Figure 55.17).

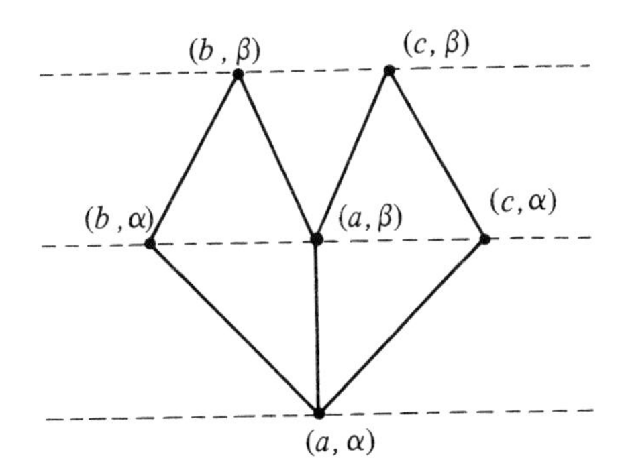

Fig. 55.17

Now let

(55.49) $$\mathbf{E} = \{A\,,B\,,C\,,D\}\,.$$

Then, the set $\mathbf{L}^{\mathbf{E}}$ with the following elements:

(55.50) $$(A\,|\,(x_1,y_1))\;,\;(B\,|\,(x_2,y_2))\;,\;(C\,|\,(x_3,y_3))\;,\;(D\,|\,(x_4,y_4))\,,$$

where $x_i = a\,,b\,,c$, $i = 1\,,2\,,3\,,4$, $y_j = \alpha\,,\beta$, $j = 1\,,2\,,3\,,4$,

is a set of $6^4 = 1296$ elements that has an inf-semilattice structure.

Another example: Higher order fuzzy subsets. Let

(55.51)
(55.52) $\mathbf{L}_1 = \{A\,,B\,,C\}$, $\mathbf{L}_2 = \{\alpha\,,\beta\}$, and $\mathbf{L}_3 = \{a\,,b\}$.
(55.53)

We investigate the representation of $\mathbf{L}_3^{\left(\mathbf{L}_2^{\mathbf{L}_1}\right)}$

First one has

(55.54)
$$\begin{aligned} L_2^{L_1} = \{&\{(A\,|\,\alpha), (B\,|\,\alpha), (G\,|\,\alpha)\}, \quad \{(A\,|\,\alpha), (B\,|\,\alpha), (C\,|\,\beta)\}, \\ &\{(A\,|\,\alpha), (B\,|\,\beta), (G\,|\,\alpha)\}, \quad \{(A\,|\,\alpha), (B\,|\,\beta), (C\,|\,\beta)\}, \\ &\{(A\,|\,\beta), (B\,|\,\alpha), (C\,|\,\alpha)\}, \quad \{(A\,|\,\beta), (B\,|\,\alpha), (C\,|\,\beta)\}, \\ &\{(A\,|\,\beta), (B\,|\,\beta), (C\,|\,\alpha)\}, \quad \{(A\,|\,\beta), (B\,|\,\beta), (C\,|\,\beta)\}\}. \end{aligned}$$

In order to simplify writing, put, respecting the arbitrary order given to the elements of $L_2^{L_1}$ above,

(55.55)
$$L_2^{L_1} = \{\underset{\sim}{A}_{\alpha\alpha\alpha}, \underset{\sim}{A}_{\alpha\alpha\beta}, \underset{\sim}{A}_{\alpha\beta\alpha}, \underset{\sim}{A}_{\alpha\beta\beta}, \underset{\sim}{A}_{\beta\alpha\alpha}, \underset{\sim}{A}_{\beta\alpha\beta}, \underset{\sim}{A}_{\beta\beta\alpha}, \underset{\sim}{A}_{\beta\beta\beta}\},$$

then one will see, for $L_3^{(L_2^{L_1})}$, $2^8 = 256$ elements, of which we give only one:

(55.56)
$$\underset{\approx}{A} = \{(\underset{\sim}{A}_{\alpha\alpha\alpha}\,|\,a), (\underset{\sim}{A}_{\alpha\alpha\beta}\,|\,a), (\underset{\sim}{A}_{\alpha\beta\alpha}\,|\,b), (\underset{\sim}{A}_{\alpha\beta\beta}\,|\,a), (\underset{\sim}{A}_{\beta\alpha\alpha}\,|\,b), (\underset{\sim}{A}_{\beta\alpha\beta}\,|\,a), (\underset{\sim}{A}_{\beta\beta\alpha}\,|\,b), (\underset{\sim}{A}_{\beta\beta\beta}\,|\,a)\}.$$

If $\mathbf{L}_1, \mathbf{L}_2, \mathbf{L}_3$ are lattices, then $L_3^{(L_2^{L_1})}$ is also a lattice.

We see appear here the notion of a fuzzy subset of order 2. And one may go still further to define fuzzy subsets of order n $(n > 2)$.

One should not confuse

(55.57)
$$L_3^{(L_2^{L_1})} \quad \text{with} \quad (L_3^{L_2})^{L_1},$$

this is a matter of another kind of fuzzification.

Relative generalized Hamming distance in the case where* L *is a lattice, and more generally an oriented graph. In order to define this notion, still more general than that which has been studied in Section 5, we first proceed to review what one calls the distance between two vertices in a nonoriented convex ordinary graph.†

The distance between two vertices of a nonoriented connected ordinary graph is the length of the shortest nonoriented chain (number of links of the shortest chain). We denote by $\mathscr{D}(X_i, X_j)$ the distance existing between X_i and X_j defined in this manner.

†Possibly one should consult a text on the theory of ordinary graphs: [1B, 1K, 2K, 2R]. Of course, one may define all sorts of other distances in a graph.

We verify that axioms (5.49)–(5.51) are indeed satisfied. Let **X** be the set of vertices of nonoriented graph under consideration; one must have

$$\forall \mathrm{X}_i, \mathrm{X}_j, \mathrm{X}_k \in \mathbf{X} :$$

(55.58) $$\mathscr{D}(\mathrm{X}_i, \mathrm{X}_j) \geqslant 0 \ ,$$

(55.59) $$\mathscr{D}(\mathrm{X}_i, \mathrm{X}_j) = \mathscr{D}(\mathrm{X}_j, \mathrm{X}_i) \ ,$$

(55.60) $$\mathscr{D}(\mathrm{X}_i, \mathrm{X}_k) \leqslant \mathscr{D}(\mathrm{X}_i, \mathrm{X}_j) + \mathscr{D}(\mathrm{X}_j, \mathrm{X}_k) \ .$$

And further

(55.61) $$\mathscr{D}(\mathrm{X}_i, \mathrm{X}_i) = 0 \ .$$

It is easy to check that these conditions are fulfilled for the notion of the distance between two vertices.

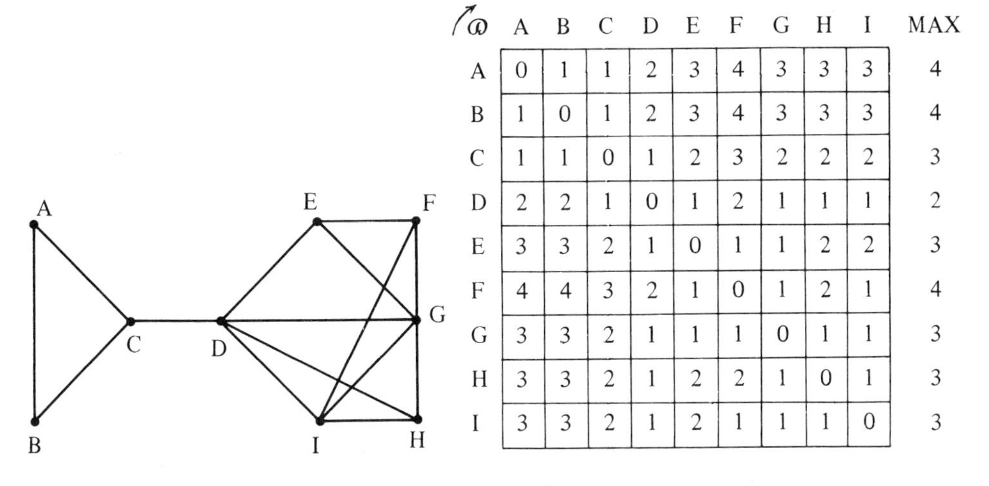

$\mathscr{D}$	A	B	C	D	E	F	G	H	I	MAX
A	0	1	1	2	3	4	3	3	3	4
B	1	0	1	2	3	4	3	3	3	4
C	1	1	0	1	2	3	2	2	2	3
D	2	2	1	0	1	2	1	1	1	2
E	3	3	2	1	0	1	1	2	2	3
F	4	4	3	2	1	0	1	2	1	4
G	3	3	2	1	1	1	0	1	1	3
H	3	3	2	1	2	2	1	0	1	3
I	3	3	2	1	2	1	1	1	0	3

Fig. 55.18

Fig. 55.19

In Figure 55.18 we have represented a nonoriented connected ordinary graph. Figure 55.19 gives the matrix of the distances $\mathscr{D}(\mathrm{X}_i, \mathrm{X}_j)$ in this graph.

We now consider the case where one has a set **E** whose fuzzy subsets take their values in **L**, knowing that **L** is an ordered set. We shall then construct the Hasse diagram of this ordered set, which will give a nonoriented ordinary graph; in this nonoriented ordinary graph we shall use the distance $\mathscr{D}(\mathrm{X}_i, \mathrm{X}_j)$ between the vertices, like that defined above.

Let

(55.62) $$\mathrm{E} = \{x_1, x_2, x_3, x_4, x_5, x_6\} \ ,$$

be a set whose membership function elements take their values in an ordered set **L** whose Hasse diagram has been represented in Figure 55.20. The matrix of the distances $\mathscr{D}(\mathrm{X}_i, \mathrm{X}_j)$ relative to this nonoriented ordinary graph is given in Figure 55.21.

Suppose now that we consider two fuzzy subsets of **E**, $\underset{\sim}{\mathbf{A}}$ and $\underset{\sim}{\mathbf{B}}$:

(55.63)

	x_1	x_2	x_3	x_4	x_5	x_6
$\underset{\sim}{A} =$	D	A	C	J	F	J

,

and

(55.64)

	x_1	x_2	x_3	x_4	x_5	x_6
$\underset{\sim}{B} =$	G	E	C	A	H	C

.

Fig. 55.20

$\mathcal{D}$	A	B	C	D	E	F	G	H	I	J	MAX
A	0	1	1	2	2	2	3	3	3	4	4
B	1	0	2	1	1	2	2	2	3	4	4
C	1	2	0	2	3	1	2	3	2	3	3
D	2	1	2	0	2	1	2	3	2	3	3
E	2	1	3	2	0	2	1	1	3	4	4
F	2	2	1	1	2	0	1	2	1	2	2
G	3	2	2	2	1	1	0	1	2	3	3
H	3	2	3	3	1	2	1	0	3	4	4
I	3	3	2	2	3	1	2	3	0	1	3
J	4	4	3	3	4	2	3	4	1	0	4

Fig. 55.21

We first evaluate the distances $\mathcal{D}(x_i, x_j)$ existing between the "values" or "vertices of the graph," these for each element $x_i \in \mathbf{E}$. These distances may be read off the matrix in Figure 55.21. One has

$$\begin{aligned}
&\mathcal{D}_{x_1}(D, G) = 2 \quad, \quad \mathcal{D}_{x_2}(A, E) = 2 \;, \\
(55.65) \qquad &\mathcal{D}_{x_3}(C, C) = 0 \quad, \quad \mathcal{D}_{x_4}(J, A) = 4 \;, \\
&\mathcal{D}_{x_5}(F, H) = 2 \quad, \quad \mathcal{D}_{x_6}(J, C) = 3 \;.
\end{aligned}$$

We may present these distances in a single row:

(55.66)

x_1	x_2	x_3	x_4	x_5	x_6
2	2	0	4	2	3

.

Now, recall what is meant by the diameter in a connected nonoriented ordinary graph: the largest of the shortest paths existing between any two vertices. Thus, for the graph in Figure 55.18, the diameter is equal to 4 (see the column on the right of the matrix in Figure 55.19, under the word max). For the graph in Figure 55.20, the diameter is again equal to 4. Then we divide each of the distances presented in (55.66) by 4 and obtain

(55.67)

x_1	x_2	x_3	x_4	x_5	x_6
0,5	0,5	0	1	0,5	0,75

In this manner we have returned to the case of numbers belonging to [0, 1].

Now the relative generalized Hamming distance† between $\underset{\sim}{\mathbf{A}}$ and $\underset{\sim}{\mathbf{B}}$ is defined by

(55.68)
$$\delta(\underset{\sim}{\mathbf{A}}, \underset{\sim}{\mathbf{B}}) = \frac{1}{6}[0{,}5 + 0{,}5 + 0 + 1 + 0{,}5 + 0{,}75]$$
$$= 0{,}54\ .$$

Such a conception of the relative generalized Hamming distance between two fuzzy subsets of the same reference set may be generalized still further in its turn if one allows that each element $x_i \in \mathbf{E}$, $i = 1, 2, \ldots, n$, may be evaluated according to a criterion that may or may not be peculiar to it. Whence we have the following general algorithm.

General algorithm. (1) Present each criterion in the form of a matrix of the shortest paths in a nonoriented ordinary graph.‡

(2) Consider two fuzzy subsets:

(55.69)

$\underset{\sim}{\mathbf{A}} =$

x_1	x_2	x_3		x_n
a_1	b_2	c_3	......	l_n

and

(55.70)

$\underset{\sim}{\mathbf{B}} =$

x_1	x_2	x_3		x_n
a'_1	b'_2	c'_3	......	l'_n

,

where a_1 and $a_1{}'$ are the evaluations in position for the criterion of x_1, to which will correspond a diameter λ_1. b_1 and $b_1{}'$ are the evaluations in position for the criterion of x_2, to which will correspond a diameter of λ_2. . . . l_n and $l_n{}'$ are the evaluations in position for the criterion of x_n, to which will correspond a diameter of λ_n.

(3) Calculate the distances $\mathcal{D}(a_1, a_1{}')$, $\mathcal{D}(b_2, b_2{}')$, . . . , $\mathcal{D}(l_m, l_n{}')$ and divide each distance by its diameter; thus,

(55.71)
$$\Delta(a_1, a_1') = \frac{\mathcal{D}(a_1, a_1')}{\lambda_1}, \Delta(b_2, b_2') = \frac{\mathcal{D}(b_2, b_2')}{\lambda_2}, \ldots, \Delta(l_n, l_n') = \frac{\mathcal{D}(l_n, l_n')}{\lambda_n}.$$

†Or a variant of the relative euclidean distance or even some other notion of distance.

‡Likewise one may, at the cost of certain restrictions, consider nonnegative and nonsymmetric matrices whose diagonals will be all 0's and such that their elements satisfy only condition (55.60) without having previously defined a graph.

(4) Form the relative generalized Hamming distance:

$$\delta(\underset{\sim}{A}, \underset{\sim}{B}) = \frac{1}{n}\,[\Delta(a_1, a_1') + \Delta(b_2, b_2') + \cdots + \Delta(l_n, l_n')] \,. \tag{55.72}$$

We shall consider a rather sophisticated example.

Example. See Figure 55.22. Let

(55.73) $\underset{\sim}{A} =$

x_1	x_2	x_3	x_4	x_5	x_6
B	γ	d	C	001	B
$\in \mathbf{L}_1$	$\in \mathbf{L}_3$	$\in \mathbf{L}_2$	$\in \mathbf{L}_1$	$\in \mathbf{L}_4$	$\in \mathbf{L}_1$

,

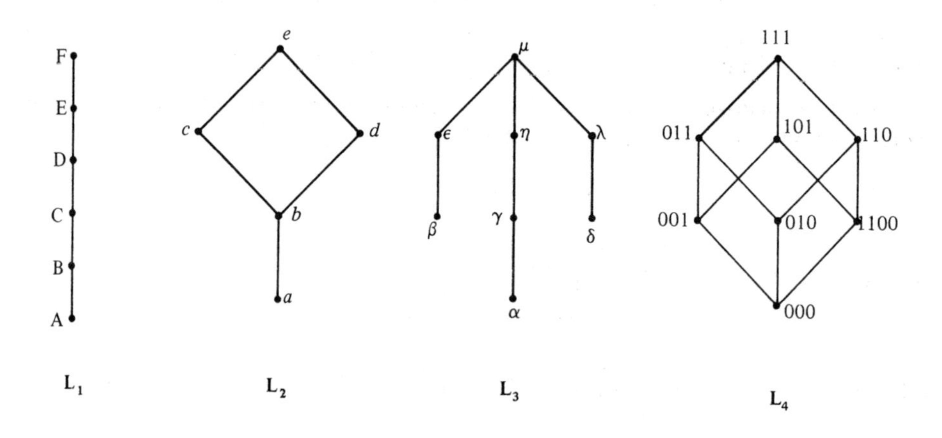

	A	B	C	D	E	F
A	0	1	2	3	4	5
B	1	0	1	2	3	4
C	2	1	0	1	2	3
D	3	2	1	0	1	2
E	4	3	2	1	0	1
F	5	4	3	2	1	0

	a	b	c	d	e
a	0	1	2	2	3
b	1	0	1	1	2
c	2	1	0	2	1
d	2	1	2	0	1
e	3	2	1	1	0

	α	β	γ	δ	ϵ	η	λ	μ
α	0	5	1	5	4	2	4	3
β	5	0	4	4	1	3	3	2
γ	1	4	0	4	3	1	3	2
δ	5	4	4	0	3	3	1	2
ϵ	4	1	3	3	0	2	2	1
η	2	3	1	3	2	0	2	1
λ	4	3	3	1	2	2	0	1
μ	3	2	2	2	1	1	1	0

	000	001	010	011	100	101	110	111
000	0	1	1	2	1	2	2	3
001	1	0	2	1	2	1	3	2
010	1	2	0	1	2	3	1	2
011	2	1	1	0	3	2	2	1
100	1	2	2	3	0	1	1	2
101	2	1	3	2	1	0	2	1
110	2	3	1	2	1	2	0	1
111	3	2	2	1	2	1	1	0

Fig. 55.22

(55.74)

$\underset{\sim}{B}$ =

x_1	x_2	x_3	x_4	x_5	x_6
D	λ	*a*	F	101	E
$\in \mathbf{L}_1$	$\in \mathbf{L}_3$	$\in \mathbf{L}_2$	$\in \mathbf{L}_1$	$\in \mathbf{L}_4$	$\in \mathbf{L}_1$

Then one has

(55.75) distances: $\mathcal{D}_i$:

x_1	x_2	x_3	x_4	x_5	x_6
2	3	2	3	1	3

(55.76) diameters: λ_i :

x_1	x_2	x_3	x_4	x_5	x_6
5	5	3	5	3	5

(55.77) relative distances: Δ_i :

x_1	x_2	x_3	x_4	x_5	x_6
2/5	3/5	2/3	3/5	1/3	3/5

and the relative generalized Hamming distance:

(55.78)
$$\delta(\underset{\sim}{A}, \underset{\sim}{B}) = \frac{1}{6}[2/5 + 3/5 + 2/3 + 3/5 + 1/3 + 3/5]$$
$$= 0{,}53 \ .$$

Heterogeneous fuzzy subset. Suppose that $\mu_{\underset{\sim}{A}}(x_i)$ takes its values in $\mathbf{L}_i$, $i = 1, 2, \ldots, n$. Then, the set of fuzzy subsets may be written

(55.79)†
$$\mathbf{L}_1^{\{x_1\}} \times \mathbf{L}_2^{\{x_2\}} \times \cdots \times \mathbf{L}_n^{\{x_n\}} \ ,$$

where the $\{x_i\}$, $i = 1, 2, \ldots, n$, are the singletons‡ of $\mathbf{E}$.

Any element of (55.79) will be called a heterogeneous fuzzy subset. Equations (55.73) and (55.74) constitute examples of heterogeneous fuzzy subsets of the same reference set.

If

(55.83)
$$\mathbf{L}_1 = \mathbf{L}_2 = \cdots = \mathbf{L}_n \ ,$$

then:

(55.84)
$$\mathbf{L}_1^{\{x_1\}} \times \mathbf{L}_2^{\{x_2\}} \times \cdots \times \mathbf{L}_n^{x_n} = \mathbf{L}^{\{x_1\} \cup \{x_2\} \cup \ldots \cup \{x_n\}}$$
$$= \mathbf{L}^{\mathbf{E}} \ .$$

And one recovers the concept of fuzzy subset that has been considered until now.

Remark. One may, evidently, go further in generalization by considering that certain $\mathbf{L}_i$ and $\mathbf{L}_j$ are not independent.

Thus, the ordinary subset $\{x_1, x_2\} \subset \{x_1, x_2, \ldots, x_n\}$ might take its values in $\mathbf{L}_{12}$ and no longer x_1 in $\mathbf{L}_1$ and x_2 in $\mathbf{L}_2$. This permits one to introduce a very interesting extension of the concept of fuzzy subset, which one may describe as interdependent

†There are no Eqs. (55.80)–(55.82).

‡A singleton is an ordinary subset having only one element. If x_i is this element, the singleton is denoted $\{x_i\}$ in the notation of ordinary set theory.

fuzzy subsets. By anticipating the treatment of this question in subsequent volumes of this work, the reader may imagine many sorts of interesting extensions.

56. OPERATIONS ON FUZZY SUBSETS WHEN L IS A LATTICE

We have seen that, by definition, in any lattice **L**, to any pair $\{\alpha, \beta\}$ one may correspond one and only element of **L** called the *inferior limit of* $\{\alpha, \beta\}$ and denoted $\alpha \,\Delta\, \beta$, and one may correspond one and only one element of **L** called the superior limit of $\{\alpha, \beta\}$ and denoted $\alpha \,\nabla\, \beta$. Thus the set of elements of **L** possesses two internal laws defined throughout, Δ and ∇.

One may then, for a lattice **L**, generalize all that has been developed in Section 3 and the following sections for membership sets **M**, which were limited there to totally ordered sets, which, as we know, are only particular cases of lattices.

We review, therefore, what has been presented at the beginning of this book, but this time replacing **M** by a lattice **L**.

Let **E** be a reference set and **L** be a lattice. One knows that the power set is $\mathbf{L}^{\mathbf{E}}$. Let $\alpha \in \mathbf{L}$. A fuzzy subset $\underset{\sim}{\mathbf{A}} \subset \mathbf{E}$, or equivalently $\underset{\sim}{\mathbf{A}} \in \mathbf{L}^{\mathbf{E}}$, will be such that to each $x \in \mathbf{E}$ one may associate an element $\alpha \in \mathbf{L}$; this element α will be denoted $\lambda_{\underset{\sim}{\mathbf{A}}}(x)$.

We shall see various extensions of the properties studied in Section 5; the properties will be examined on an example of a lattice (Figure 56.1) and on a set $\mathbf{E} = \{A, B, C\}$. We have chosen a simple example with a didactic purpose, but of course all this will remain valid for any reference set **E**, finite or not, and any lattice **L**, finite or not.

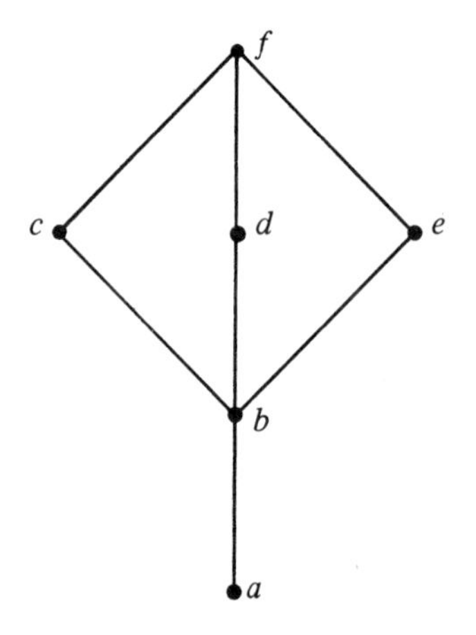

Fig. 56.1

Inclusion. Let $\preccurlyeq$ be the order relation of the lattice; one will write that $\underset{\sim}{\mathbf{A}}$ is included in $\underset{\sim}{\mathbf{B}}$ if

(56.1)
$$\forall x_i \in \mathbf{E} \quad : \quad \lambda_{\underset{\sim}{\mathbf{A}}}(x_i) \preccurlyeq \lambda_{\underset{\sim}{\mathbf{B}}}(x_i) \ ,$$

and this will be denoted

(56.2)
$$\underset{\sim}{\mathbf{A}} \subset \underset{\sim}{\mathbf{B}} \ .$$

One may also write

(56.3)
$$((\forall x_i \in \mathbf{E}) \quad : \quad \lambda_{\underset{\sim}{\mathbf{A}}}(x_i) \preccurlyeq \lambda_{\underset{\sim}{\mathbf{B}}}(x_i)) \quad \Leftrightarrow \quad (\underset{\sim}{\mathbf{A}} \subset \underset{\sim}{\mathbf{B}}) \ .$$

We see that two fuzzy subsets are comparable if: (1) The respective values taken by the membership function in the lattice **L** are comparable; (2) there exists a dominance relation between the two fuzzy subsets.

Example. See Figure 56.1. Let

$$\underset{\sim}{\mathbf{A}} = \{(\mathrm{A} \mid b), (\mathrm{B} \mid a), (\mathrm{C} \mid c)\}, \tag{56.4}$$

$$\underset{\sim}{\mathbf{B}} = \{(\mathrm{A} \mid d), (\mathrm{B} \mid e), (\mathrm{C} \mid c)\}; \tag{56.5}$$

$\underset{\sim}{\mathbf{A}}$ and $\underset{\sim}{\mathbf{B}}$ are indeed comparable since $b \preccurlyeq d$, $a \preccurlyeq e$, $c \preccurlyeq c$, and one has

$$\underset{\sim}{\mathbf{A}} \subset \underset{\sim}{\mathbf{B}}. \tag{56.6}$$

Again let

$$\underset{\sim}{\mathbf{C}} = \{(\mathrm{A} \mid f), (\mathrm{B} \mid b), (\mathrm{C} \mid d)\}. \tag{56.7}$$

One sees that $\underset{\sim}{\mathbf{C}}$ is not comparable with $\underset{\sim}{\mathbf{A}}$ since c and d, which occur for $\underset{\sim}{\mathbf{C}}$, are not comparable; neither is $\underset{\sim}{\mathbf{C}}$ comparable with $\underset{\sim}{\mathbf{B}}$.

Again let

$$\underset{\sim}{\mathbf{D}} = \{(\mathrm{A} \mid d), (\mathrm{B} \mid e), (\mathrm{C} \mid b)\}. \tag{56.8}$$

$\underset{\sim}{\mathbf{D}}$ is not comparable with $\underset{\sim}{\mathbf{A}}$; indeed, one has $b \preccurlyeq d$, $a \preccurlyeq e$, $c \succcurlyeq b$; but there does not exist a dominance between $\underset{\sim}{\mathbf{D}}$ and $\underset{\sim}{\mathbf{A}}$.

Equality. Two fuzzy subsets $\underset{\sim}{\mathbf{A}}$ and $\underset{\sim}{\mathbf{B}}$ are equal if and only if

$$\forall x_i \in \mathrm{E} \quad : \quad \lambda_{\underset{\sim}{\mathbf{A}}}(x_i) = \lambda_{\underset{\sim}{\mathbf{B}}}(x_i). \tag{56.9}$$

Equivalently one may write

$$(\forall x_i \in \mathrm{E} \quad : \quad \lambda_{\underset{\sim}{\mathbf{A}}}(x_i) = \lambda_{\underset{\sim}{\mathbf{B}}}(x_i)) \quad \Leftrightarrow \quad (\underset{\sim}{\mathbf{A}} = \underset{\sim}{\mathbf{B}}). \tag{56.10}$$

Complementation. It is necessary to linger here for a while. The complementation used by Zadeh when he considers a set $\mathbf{M} = [0, 1]$ is not that which is to be considered in lattice theory.

For the fuzzy subsets of Zadeh,

$$(\overline{\underset{\sim}{\mathbf{B}}} = \underset{\sim}{\mathbf{A}}) \quad \Leftrightarrow \quad (\forall x_i \in \mathrm{E} \ : \ \mu_{\underset{\sim}{\mathbf{B}}}(x_i) = 1 - \mu_{\underset{\sim}{\mathbf{A}}}(x_i)). \tag{56.11}$$

But, we have seen that not all lattices are complemented and that, in order that this property have sense, it is first necessary that (54.20) and (54.21) be satisfied; but it is necessary here that the complement also be unique. This will be the case if the lattice is distributive. Therefore we shall consider here distributive and complemented lattices, that is boolean lattices, in order to correspond to each element of $\mathbf{L}$ a unique complement, and as a consequence, to each element of $\mathbf{L}^{\mathrm{E}}$ a unique complement. And, since $\mathbf{L}$ is a boolean lattice, $\mathbf{L}^{\mathrm{E}}$ is also.

One may then write

$$(\overline{\underset{\sim}{\mathbf{B}}} = \underset{\sim}{\mathbf{A}}) \quad \Leftrightarrow \quad (\forall x_i \in \mathrm{E} \ : \ \mu_{\underset{\sim}{\mathbf{A}}}(x_i) \,\Delta\, \mu_{\underset{\sim}{\mathbf{B}}}(x_i) = 0 \quad \text{and} \quad \mu_{\underset{\sim}{\mathbf{A}}}(x_i) \,\nabla\, \mu_{\underset{\sim}{\mathbf{B}}}(x_i) = \mathrm{U}), \tag{56.12}$$

where 0 is the inferior limit of the boolean lattice $\mathbf{L}$ and U is the superior limit; this 0 and this U are not numbers here, but the extreme elements defined in (54.20) and (54.21).

The complementation used by Zadeh should, with regard to these considerations, be called pseudo-complementation. The only case where the two concepts coincide is when one considers $\mathbf{L} = \{0, 1\}$.

Remark. If the lattice **L** is a distributive and complemented lattice, then working with the membership functions of **E** is identical to working with probabilities. What shows how to generalize the present case also generalizes probability theory.

Intersection. One defines intersection

$$\underset{\sim}{A} \cap \underset{\sim}{B} \tag{56.13}$$

by the property

$$\forall x \in \mathbf{E} \quad : \quad \lambda_{\underset{\sim}{A} \cap \underset{\sim}{B}}(x) = \lambda_{\underset{\sim}{A}}(x) \,\Delta\, \lambda_{\underset{\sim}{B}}(x) \ . \tag{56.14}$$

We see that intersection may have sense only subject to the condition that the order relation that defines **L** be an inf-semilattice. Since by hypothesis we suppose here that **L** is a lattice, this is the case.

Example. See Figure 56.1. Consider (56.4) and (56.7); one has

$$\begin{aligned} \underset{\sim}{A} \cap \underset{\sim}{C} &= \{(A \mid b \,\Delta\, f), (B \mid a \,\Delta\, b), (C \mid c \,\Delta\, d)\} \\ &= \{(A \mid b), (B \mid a), (C \mid b)\} \ . \end{aligned} \tag{56.15}$$

Union. Onc will define the union

$$\underset{\sim}{A} \cup \underset{\sim}{B} \tag{56.16}$$

by the property

$$\forall x \in \mathbf{E} \quad : \quad \lambda_{\underset{\sim}{A} \cup \underset{\sim}{B}}(x) = \lambda_{\underset{\sim}{A}}(x) \,\nabla\, \lambda_{\underset{\sim}{B}}(x) \ . \tag{56.17}$$

We see that the union may have sense only subject to the condition that the order relation that defines **L** be a sup-semilattice. Since, by hypothesis, we suppose that **L** is a lattice, this is the case.

Example. See Figure 56.1. We consider (56.4) and (56.7); one has

$$\begin{aligned} \underset{\sim}{A} \cup \underset{\sim}{C} &= \{(A \mid b \,\nabla\, f), (B \mid a \,\nabla\, b), (C \mid c \,\nabla\, d)\} \\ &= \{(A \mid f), (B \mid b), (C \mid f)\} \ . \end{aligned} \tag{56.18}$$

Disjunctive sum. This will be defined only subject to the condition that **L** be a distributive and complemented lattice, that is, a boolean lattice; one will then have

$$\underset{\sim}{A} \oplus \underset{\sim}{B} = (\underset{\sim}{A} \cap \overline{\underset{\sim}{B}}) \cup (\overline{\underset{\sim}{A}} \cap \underset{\sim}{B}) \ . \tag{56.19}$$

Difference. Subject to the same restrictions as the disjunctive sum, one has

$$\underset{\sim}{A} - \underset{\sim}{B} = \underset{\sim}{A} \cap \overline{\underset{\sim}{B}} \ . \tag{56.20}$$

Properties of **L** *and those of* $\mathbf{L}^{\mathbf{E}}$. As we have seen, all the properties of $\mathbf{L}^{\mathbf{E}}$ for $\cap$, $\cup$, and complementation (when these exist) are deduced from those of **L** for Δ, ∇, and complementation (when these exist).

We note that the generalization of the theory of fuzzy subsets in the sense of Zadeh ($\mathbf{M} = [0, 1]$) is that where **L** is a vector lattice with

(56.21 $\quad \mathbf{L} = \mathbf{M}_1 \times \mathbf{M}_2 \times \cdots \times \mathbf{M}_n \quad$ with $\quad \mathbf{M}_i = [0\,,1]\,, i = 1\,, 2\,, \ldots\,, n$

In this case, if $\alpha_1 \in \mathbf{M}_1, \alpha_2 \in \mathbf{M}_2, \ldots, \alpha_n \in \mathbf{M}_n$ one will have

$$(56.22) \qquad \overline{(\alpha_1\,, \alpha_2\,, \ldots\,, \alpha_n)} = (1 - \alpha_1\,, 1 - \alpha_2\,, \ldots\,, 1 - \alpha_n)\;.$$

Another generalization concerns the case where one takes

$$(56.23) \qquad \mathbf{L} = \mathbf{M}_1 \times \mathbf{M}_2 \times \cdots \times \mathbf{M}_n$$

with $\mathbf{M}_i$, $i = 1, 2, \ldots, n$, having the configuration of a boolean lattice.

And, if one no longer introduces the notion of complementation (case of a boolean lattice) or of pseudo-complementation (in the sense of Zadeh), one may build a theory of fuzzy subsets for any kind of lattice.

Fuzzy variables. The notion of fuzzy variable defined in Section 32 may also be generalized to the case where **L** is defined by (56.21), that is, is a vector lattice. In this case, one may easily transpose the properties specified in (32.12)–(32.26).

Another conception of this may be taken; one may suppose that **L** is a boolean lattice where the complementation is that which is relative to the complemented lattice. One then recovers the properties of a boolean lattice, which are those of (32.12)–(32.26) by replacing $\wedge$ with Δ and $\vee$ with ∇ and by introducing, in addition, $a\Delta\bar{a} = 0$, and $a\nabla\bar{a} = 1$, where 0 and 1 will be respectively the inferior limit of the boolean lattice and the superior limit of this lattice.

57. REVIEW OF VARIOUS NOTIONS WITH A VIEW TOWARD INTRODUCING THE CONCEPT OF A CATEGORY

Category theory permits one to give a global idea of many of the results presented until this point in the present book. But, for those readers who have not studied this theory, it is appropriate not to begin without transition, without presenting or reviewing various intermediate definitions, which we now proceed to do in an essentially didactic spirit. The examples given generally concern finite reference sets, but the definitions apply to finite or infinite reference sets without reservation.

Correspondence. A correspondence Γ between one set $\mathbf{E}_1$ and a set $\mathbf{E}_2$ is defined if one is given an ordinary graph $\mathbf{G} \subset \mathbf{E}_1 \times \mathbf{E}_2$. One then says that **G** is the graph of Γ, $\mathbf{E}_1$ is the domain, and $\mathbf{E}_2$ is the range of Γ.

One designates by Γ^{-1} the inverse correspondence, where $\mathbf{E}_2$ is the domain and $\mathbf{E}_1$ the range.

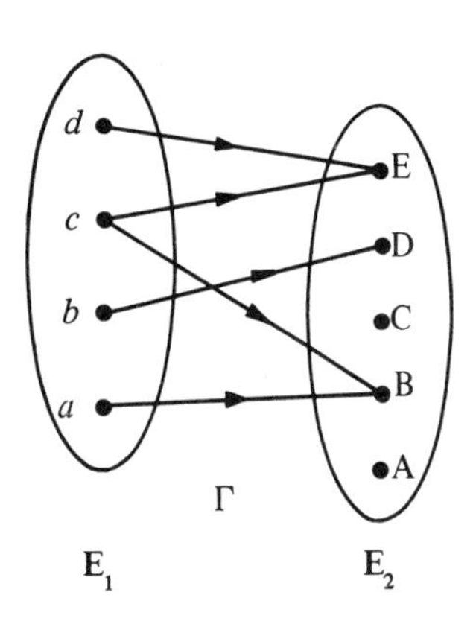

Fig. 57.1 Mapping.

Mapping. A correspondence such that to any $x \in \mathbf{E}_1$ there corresponds *at least* one $y \in \mathbf{E}_2$ is called a mapping of the set $\mathbf{E}_1$ into the set $\mathbf{E}_2$.

One then says that y is an *image* of x, and that x is a *variable* or *argument*.

Example. See Figure 57.1. Let

$$\mathbf{E}_1 = \{a, b, c, d\} \quad , \quad \mathbf{E}_2 = \{A, B, C, D, E\} . \tag{57.1, 57.2}$$

One has

$$\Gamma\{a\} = \{B\} \ , \ \Gamma\{b\} = \{D\} \ , \ \Gamma\{c\} = \{B, E\} \ , \ \Gamma\{d\} = \{E\} . \tag{57.3}$$

B is the image of a,

D is the image of b,

B and E are images of c,

E is the image of d.

Surjective mapping or surjection. If every $y \in \mathbf{E}_2$ is the image of at least one $x \in \mathbf{E}_1$, one says that one has a mapping of $\mathbf{E}_1$ onto $\mathbf{E}_2$; such a mapping is called a surjective mapping or a surjection.

Example. See Figure 57.2. Let

$$\mathbf{E}_1 = \{a, b, c, d, e, f\} \quad , \quad \mathbf{E}_2 = \{A, B, C, D\} . \tag{57.4, 57.5}$$

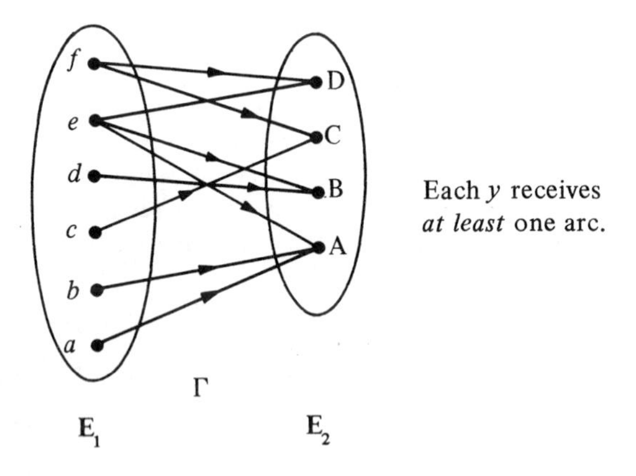

Fig. 57.2. Surjection

One has

$$\begin{aligned} &\Gamma\{a\} = \{A\} \ , \ \Gamma\{b\} = \{A\} \ , \ \Gamma\{c\} = \{C\} \ , \\ &\Gamma\{d\} = \{B\} \ , \ \Gamma\{e\} = \{A, B, D\} \ , \ \Gamma\{f\} = \{C, D\} . \end{aligned} \tag{57.6}$$

This is indeed a surjection since each $\Gamma^{-1}\{y\}$ is nonempty:

(57.7) $\Gamma^{-1}\{A\} = \{a, b, e\}$, $\Gamma^{-1}\{B\} = \{d, e\}$, $\Gamma^{-1}\{C\} = \{c, f\}$, $\Gamma^{-1}\{D\} = \{e, f\}$.

As one may see, $|\Gamma^{-1}\{y\}| \geqslant 1$ for all y characterizes a surjection.

Injective mapping or injection. If each $y \in \mathbf{E}_2$ is the image of only *one or no* $x \in \mathbf{E}_1$, one says that the mapping is an injective mapping of $\mathbf{E}_1$ *into* $\mathbf{E}_2$; such a mapping is called an injection.

Example. See Figure 57.3. Let

(57.8)
(57.9)
$$\mathbf{E}_1 = \{a, b, c, d\} \quad , \quad \mathbf{E}_2 = \{A, B, C, D, E, F\} \ .$$

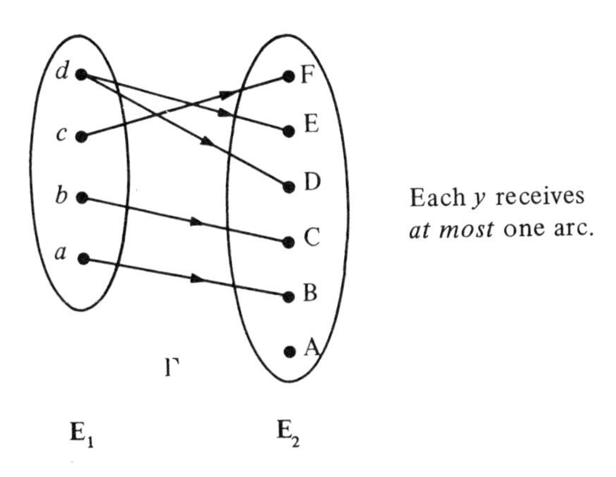

Fig. 57.3 Injection.

One has

(57.10) $\Gamma\{a\} = \{B\}$, $\Gamma\{b\} = \{C\}$, $\Gamma\{c\} = \{F\}$, $\Gamma\{d\} = \{D, E\}$

and

(57.11)
$$\Gamma^{-1}\{A\} = \phi \ , \ \Gamma^{-1}\{B\} = \{a\} \ , \ \Gamma^{-1}\{C\} = \{b\} \ , \ \Gamma^{-1}\{D\} = \{d\}$$
$$\Gamma^{-1}\{E\} = \{d\} \ , \ \Gamma^{-1}\{F\} = \{c\} \ .$$

As one may see, if we have for all $y \in \mathbf{E}_2$, $|\Gamma^{-1}\{y\}| \leqslant 1$, one indeed has an injection.

Bijective mapping or bijection. A mapping that is at the same time surjective and injective is called a bijective mapping or a bijection.

Example. See Figure 57.4. Let

(57.12)
(57.13)
$$\mathbf{E}_1 = \{a, b, c, d\} \quad , \quad \mathbf{E}_2 = \{A, B, C, D, E, F\} \ .$$

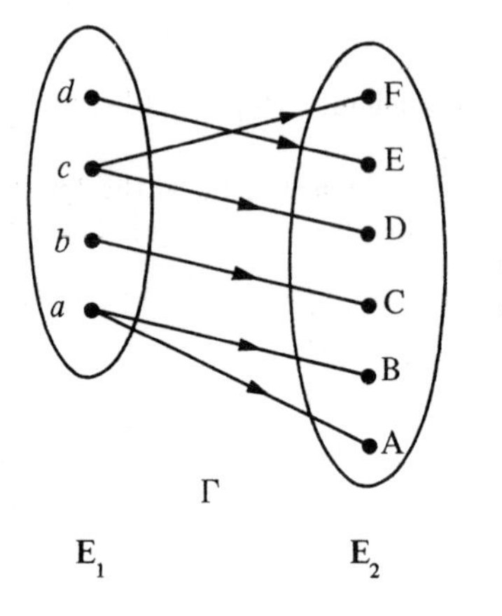

Fig. 57.4 Bijection.

One has

(57.14) $$\Gamma\{a\} = \{A, B\}\ ,\ \Gamma\{b\} = \{C\}\ ,\ \Gamma\{c\} = \{D, F\}\ ,\ \Gamma\{d\} = \{E\}\ .$$

and

(57.15) $$\begin{aligned}&\Gamma^{-1}\{A\} = \{a\}\ ,\ \Gamma^{-1}\{B\} = \{a\}\ ,\ \Gamma^{-1}\{C\} = \{b\}\ ,\ \Gamma^{-1}\{D\} = \{c\}\ ,\\ &\Gamma^{-1}\{E\} = \{d\}\ ,\ \Gamma^{-1}\{F\} = \{c\}\ .\end{aligned}$$

As one may see, if for all $y \in \mathbf{E}_2$, $|\Gamma^{-1}\{y\}| = 1$, then one has a bijection.

Function. A mapping that is such that

(57.16) $$\forall x \in \mathbf{E}_1\ ,\ |\Gamma\{x\}| = 1\ ,$$

is called a function.

In other words, a function of $\mathbf{E}_1$ into $\mathbf{E}_2$ is a mapping such that to each $x \in \mathbf{E}_1$ there corresponds *one and only one* $y \in \mathbf{E}_2$.

A function may be surjective, injective, or bijective.

Some examples are given in Figures 57.5–57.7.

One sees that if a function Γ is bijective, Γ^{-1} is also. In this case one has a one-to-one correspondence between $\mathbf{E}_1$ and $\mathbf{E}_2$.

Remark. Certain authors (and the present author has done so in several books for particular reasons) define a mapping as a function, that is, such that $\forall x \in \mathbf{E}_1, |\Gamma\{x\}| = 1$. In the present work we prefer

(57.17) $$\text{mapping} \quad : \quad \forall x \in \mathbf{E}_1 \ :\ |\Gamma\{x\}| \geqslant 1\ ,$$

(57.18) $$\text{function} \quad : \quad \forall x \in \mathbf{E}_1 \ :\ |\Gamma\{x\}| = 1\ .$$

Isotony between two ordered sets. Suppose that the set $\mathbf{E}_1$ is ordered by an order relation denoted $\preccurlyeq$, and likewise the set $\mathbf{E}_2$ by an order relation also denoted $\preccurlyeq$.† A mapping of

$\mathbf{E}_1$ into $\mathbf{E}_2$ will be said isotonic if it preserves the order, that is, if

$$(\forall x_i \in \mathbf{E}_1 \;,\; \forall x_j \in \mathbf{E}_1 \;:\; x_i \preccurlyeq x_j) \Rightarrow (\forall y_k \in \Gamma\{x_i\} \subset \mathbf{E}_2 \;,\; \forall y_l \in \Gamma\{x_j\} \subset \mathbf{E}_2 \;:\; y_k \preccurlyeq y_l) \;. \tag{57.19}$$

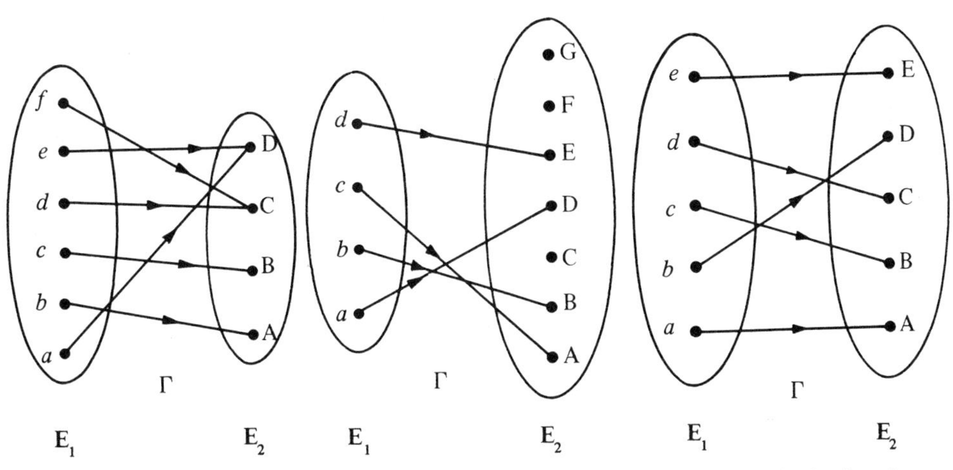

Fig. 57.5 Surjective function. Fig. 57.6 Injective function. Fig. 57.7 Bijective function.

In the case where the order is total, the isotony will be called an increasing monotony (the mapping is monotone increasing). We consider two examples.

Example 1. See Figure 57.8. Here we see two totally ordered sets $\mathbf{E}_1$ and $\mathbf{E}_2$ represented by their respective unique maximal chains. We see how the isotony defined by (57.19) is satisfied.

Consider, for example, b and c. One has $b \preccurlyeq c$. On the other hand, $\Gamma\{b\} = \{\mathrm{B}\}$ and $\Gamma\{c\} = \{\mathrm{B}, \mathrm{D}\}$; indeed one has $\mathrm{B} \preccurlyeq \mathrm{B}$ and $\mathrm{B} \preccurlyeq \mathrm{D}$. If one carries out the verification for other pairs of elements of $\mathbf{E}_1$, one will see that we indeed have an isotony (here monotone increasing).

Example 2. See Figure 57.9. This time $\mathbf{E}_1$ and $\mathbf{E}_2$ have order configurations that are not total orders. The maximal chains have been represented by their respective Hasse diagrams. As an exercise, we display all the pairs of elements constituting chains:

†Generally, the sets $\mathbf{E}_1$ and $\mathbf{E}_2$ may be distinct, and the order relations concerning them may also be different from one another; thus it would be appropriate to indicate one with a symbol different than the other, for example $\preccurlyeq$ and $\preccurlyeq'$ But we are in the habit of using the same symbol $\preccurlyeq$ whatever the order relation may be. This is also done, for example, where one uses the same symbol + for the integers and for the complex numbers—these are of course completely different concepts.

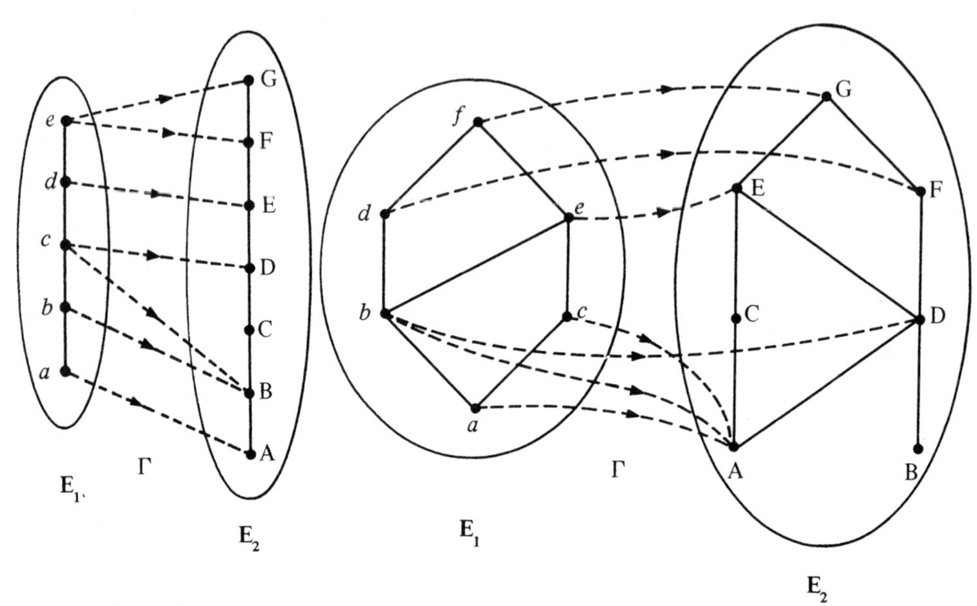

Fig. 57.8

Fig. 57.9

$$
\begin{array}{llll}
 & a \preccurlyeq b\ , \ \Gamma\{a\} = \{A\} & , \ \Gamma\{b\} = \{A\ , D\} & , \text{ one has: } A \preccurlyeq A\ , \ A \preccurlyeq D\ . \\
 & a \preccurlyeq c\ , \ \Gamma\{a\} = \{A\} & , \ \Gamma\{c\} = \{A\} & , \text{ one has: } A \preccurlyeq A\ . \\
 & b \preccurlyeq d\ , \ \Gamma\{b\} = \{A\ , D\} & , \ \Gamma\{d\} = \{F\} & , \text{ one has: } A \preccurlyeq F\ , \ D \preccurlyeq F\ . \\
(57.20) & b \preccurlyeq e\ , \ \Gamma\{b\} = \{A\ , D\} & , \ \Gamma\{e\} = \{E\} & , \text{ one has: } A \preccurlyeq E\ , \ D \preccurlyeq E\ . \\
 & c \preccurlyeq e\ , \ \Gamma\{c\} = \{A\} & , \ \Gamma\{e\} = \{E\} & , \text{ one has: } A \preccurlyeq E\ . \\
 & d \preccurlyeq f\ , \ \Gamma\{d\} = \{F\} & , \ \Gamma\{f\} = \{G\} & , \text{ one has: } F \preccurlyeq G\ . \\
 & e \preccurlyeq f\ , \ \Gamma\{e\} = \{E\} & , \ \Gamma\{f\} = \{G\} & , \text{ one has: } E \preccurlyeq G\ .
\end{array}
$$

We do not carry out the verification of isotony for the other ordered pairs (a, d), (a, f), etc., since isotony is evidently transitive and it therefore suffices to verify only for the elements of $\mathbf{E}_1$ that are adjacent to one another.

Antitony between two ordered sets. We consider property (57.19), but instead of writing $y_k \preccurlyeq y_l$, we write

$$
(57.21) \qquad y_l \preccurlyeq y_k \ ;
$$

in this the case, the mapping is said to be antitonic.

The reader should involve himself in finding an antitonic mapping by considering the ordered sets $\mathbf{E}_1$ and $\mathbf{E}_2$ in Figure 57.9.

Morphism between two ordered sets. A mapping Γ of an ordered set $\mathbf{E}_1$ into an ordered set $\mathbf{E}_2$ that is isotonic is a morphism.

Figures 57.8 and 57.9 represent morphisms.

Epimorphism between two ordered sets. This is a morphism in which the mapping Γ of $\mathbf{E}_1$ into $\mathbf{E}_2$ is surjective.

Monomorphism between two ordered sets. This is a morphism in which the mapping Γ of $\mathbf{E}_1$ into $\mathbf{E}_2$ is injective. If $\mathbf{E}_2{}' \subset \mathbf{E}_2$ is the subset of $\mathbf{E}_2$ whose elements each admit exactly one arc, it is also necessary that $\Gamma^{-1}\mathbf{E}_2{}'$ be isotonic with respect to $\mathbf{E}_2{}'$.

Isomorphism between two ordered sets. This is a morphism that is at the same time an epimorphism and a monomorphism; that is, such that the mapping Γ is bijective and Γ, thus also Γ^{-1}, are isotonic.

We go on to see several examples in order to better understand the content of these definitions.

Example 1. See Figure 57.10. First we verify that we have a morphism. For this, we examine all the ordered pairs (x, y), $x \in \mathbf{E}_1$, $y \in \mathbf{E}_2$, that compose the maximal chains:

$$\begin{aligned} &a \preccurlyeq b\ ,\ \Gamma\{a\} = \{A\}\ ,\ \Gamma\{b\} = \{B, C\}\text{, one has: } A \preccurlyeq B\ ,\ A \preccurlyeq C\ . \\ (57.22)\quad &b \preccurlyeq d\ ,\ \Gamma\{b\} = \{B, C\}\ ,\ \Gamma\{d\} = \{F\}\text{, one has: } B \preccurlyeq F\ ,\ C \preccurlyeq F\ . \\ &c \preccurlyeq d\ ,\ \Gamma\{c\} = \{E\}\ ,\ \Gamma\{d\} = \{F\}\text{, one has: } E \preccurlyeq F\ . \end{aligned}$$

One indeed has a morphism of $\mathbf{E}_1$ into $\mathbf{E}_2$. Is it an epimorphism? A priori no, since not all the $Y \in \mathbf{E}_2$ admit at least one arc (D admits no arc). Is it a monomorphism? For this, it is necessary to check whether the inverse mapping of $\mathbf{E}_2{}' = \{A, B, C, E, F\}$ into $\mathbf{E}_1$ is indeed isotonic.

One has

$$\begin{aligned} &A \preccurlyeq B\ ,\ \Gamma^{-1}\{A\} = \{a\}\ ,\ \Gamma^{-1}\{B\} = \{b\}\ \text{, one has : } a \preccurlyeq b\ . \\ (57.23)\quad &B \preccurlyeq C\ ,\ \Gamma^{-1}\{B\} = \{b\}\ ,\ \Gamma^{-1}\{C\} = \{b\}\ \text{, one has : } b \preccurlyeq b\ . \\ &C \preccurlyeq E\ ,\ \Gamma^{-1}\{C\} = \{b\}\ ,\ \Gamma^{-1}\{E\} = \{c\}\ \text{, one does not have : } b \preccurlyeq c\ . \end{aligned}$$

It is useless to go any further; this is not a monomorphism.

Example 2. See Figure 57.11. The ordered sets are the same as those of Figure 57.10, but the mapping is different. We check first whether this mapping is a morphism:

$$\begin{aligned} &a \preccurlyeq b\ ,\ \Gamma\{a\} = \{A, B\}\ ,\ \Gamma\{b\} = \{C, D\}\text{, one has: } A \preccurlyeq C\ ,\ A \preccurlyeq D\ , \\ &\qquad\qquad B \preccurlyeq C\ ,\ B \preccurlyeq D\ . \\ (57.24)\quad &b \preccurlyeq d\ ,\ \Gamma\{b\} = \{C, D\}\ ,\ \Gamma\{d\} = \{F\}\text{, one has: } C \preccurlyeq F, \\ &\qquad\qquad \text{but one does not have } D \preccurlyeq F \end{aligned}$$

Thus, this mapping is not a morphism.

Example 3. See Figure 57.12. The sets $\mathbf{E}_1$ and $\mathbf{E}_2$ are the same as those of Figures 57.10 and 57.11, but $\mathbf{E}_2$ is not ordered in the same fashion.

We check whether this mapping is a morphism:

$$
(57.25)\quad
\begin{array}{lll}
a \preccurlyeq b\ ,\ \Gamma\{a\} = \{\mathrm{A}, \mathrm{B}\}\ , & \Gamma\{b\} = \{\mathrm{B}, \mathrm{D}\}, \text{ one has: } & \mathrm{A} \preccurlyeq \mathrm{B}\ ,\ \mathrm{A} \preccurlyeq \mathrm{D}\ , \\
 & & \mathrm{B} \preccurlyeq \mathrm{B}\ ,\ \mathrm{B} \preccurlyeq \mathrm{D}\ . \\
b \preccurlyeq d\ ,\ \Gamma\{b\} = \{\mathrm{B}, \mathrm{D}\}\ , & \Gamma\{d\} = \{\mathrm{E}, \mathrm{F}\}, \text{ one has: } & \mathrm{B} \preccurlyeq \mathrm{E}\ ,\ \mathrm{B} \preccurlyeq \mathrm{F}\ , \\
 & & \mathrm{D} \preccurlyeq \mathrm{E}\ ,\ \mathrm{D} \preccurlyeq \mathrm{F}\ . \\
c \preccurlyeq d\ ,\ \Gamma\{c\} = \{\mathrm{C}\}\ , & \Gamma\{d\} = \{\mathrm{E}, \mathrm{F}\}, \text{ one has: } & \mathrm{C} \preccurlyeq \mathrm{E}\ ,\ \mathrm{C} \preccurlyeq \mathrm{F}\ .
\end{array}
$$

Thus, this is a morphism. Further it is an epimorphism since the mapping is surjective (each $\mathrm{Y} \in \mathbf{E}_2$ admits at least one arc). We see whether it is a monomorphism. A priori no, since the mapping is not injective; there is a $\mathrm{Y} \in \mathbf{E}_2$ that admits more than one arc.

Example 4. See Figure 57.13. Here we have some other sets. We see whether this mapping is a morphism:

$$
(57.26)\quad
\begin{array}{lll}
a \preccurlyeq b\ ,\ \Gamma\{a\} = \{\mathrm{A}\}\ , & \Gamma\{b\} = \{\mathrm{B}\}\ , & \text{one has: } \mathrm{A} \preccurlyeq \mathrm{B}\ . \\
b \preccurlyeq c\ ,\ \Gamma\{b\} = \{\mathrm{B}\}\ , & \Gamma\{c\} = \{\mathrm{B}\}\ , & \text{one has: } \mathrm{B} \preccurlyeq \mathrm{B}\ . \\
b \preccurlyeq d\ ,\ \Gamma\{b\} = \{\mathrm{B}\}\ , & \Gamma\{d\} = \{\mathrm{C}\}\ , & \text{one has: } \mathrm{B} \preccurlyeq \mathrm{C}\ . \\
d \preccurlyeq e\ ,\ \Gamma\{d\} = \{\mathrm{C}\}\ , & \Gamma\{e\} = \{\mathrm{C}\}\ , & \text{one has: } \mathrm{C} \preccurlyeq \mathrm{C}\ .
\end{array}
$$

It is indeed a morphism. It is also an epimorphism (each $\mathrm{Y} \in \mathbf{E}_2$ admits at least one arc), but it is not a monomorphism (there is at least one $\mathrm{Y} \in \mathbf{E}_2$ that admits more than one arc).

Example 5. See Figure 57.14. One has

$$
(57.27)\quad
\begin{array}{lll}
a \preccurlyeq b\ ,\ \Gamma\{a\} = \{\mathrm{A}\}\ , & \Gamma\{b\} = \{\mathrm{A}, \mathrm{B}\}, & \text{one has: } \mathrm{A} \preccurlyeq \mathrm{A}\ ,\ \mathrm{A} \preccurlyeq \mathrm{B}\ . \\
b \preccurlyeq c\ ,\ \Gamma\{b\} = \{\mathrm{A}, \mathrm{B}\}\ , & \Gamma\{c\} = \{\mathrm{C}\}\ , & \text{one has: } \mathrm{A} \preccurlyeq \mathrm{C}\ ,\ \mathrm{B} \preccurlyeq \mathrm{C}\ . \\
b \preccurlyeq d\ ,\ \Gamma\{b\} = \{\mathrm{A}, \mathrm{B}\}\ , & \Gamma\{d\} = \{\mathrm{C}\}\ , & \text{one has: } \mathrm{A} \preccurlyeq \mathrm{C}\ ,\ \mathrm{B} \preccurlyeq \mathrm{C}\ . \\
d \preccurlyeq e\ ,\ \Gamma\{d\} = \{\mathrm{C}\}\ , & \Gamma\{e\} = \{\mathrm{D}, \mathrm{E}\}, & \text{one has: } \mathrm{C} \preccurlyeq \mathrm{D}\ ,\ \mathrm{C} \preccurlyeq \mathrm{E}\ .
\end{array}
$$

This is indeed a morphism; it is also an epimorphism but not a morphism.

Example 6. See Figure 57.15. One has

$$
(57.28)\quad
\begin{array}{lll}
a \preccurlyeq b\ ,\ \Gamma\{a\} = \{\mathrm{A}\}\ , & \Gamma\{b\} = \{\mathrm{B}\}\ , & \text{one has: } \mathrm{A} \preccurlyeq \mathrm{B}\ . \\
b \preccurlyeq c\ ,\ \Gamma\{b\} = \{\mathrm{B}\}\ , & \Gamma\{c\} = \{\mathrm{D}\}\ , & \text{one has: } \mathrm{B} \preccurlyeq \mathrm{D}\ . \\
b \preccurlyeq d\ ,\ \Gamma\{b\} = \{\mathrm{B}\}\ , & \Gamma\{d\} = \{\mathrm{E}\}\ , & \text{one has: } \mathrm{B} \preccurlyeq \mathrm{E}\ .
\end{array}
$$

This is indeed a morphism but not an epimorphism (there is at least one $\mathrm{Y} \in \mathbf{E}_2$ that does not admit any arc). We check whether it is a monomorphism.

A first condition is satisfied: no $\mathrm{Y} \in \mathbf{E}_2$ admits more than one arc; but it is also necessary that $\Gamma^{-1}\mathbf{E}_2'$, where $\mathbf{E}_2' = \{\mathrm{A, B, D, E}\}$, be isotonic with respect to $\mathbf{E}_1$. We carry out this examination:

$$
(57.29)\quad
\begin{array}{lll}
\mathrm{A} \preccurlyeq \mathrm{B}\ ,\ \Gamma^{-1}\{\mathrm{A}\} = \{a\}\ , & \Gamma^{-1}\{\mathrm{B}\} = \{b\}\ , & \text{one has: } a \preccurlyeq b\ . \\
\mathrm{A} \preccurlyeq \mathrm{E}\ ,\ \Gamma^{-1}\{\mathrm{A}\} = \{a\}\ , & \Gamma^{-1}\{\mathrm{E}\} = \{d\}\ , & \text{one has: } a \preccurlyeq d\ . \\
\mathrm{B} \preccurlyeq \mathrm{E}\ ,\ \Gamma^{-1}\{\mathrm{B}\} = \{b\}\ , & \Gamma^{-1}\{\mathrm{E}\} = \{d\}\ , & \text{one has: } b \preccurlyeq d\ . \\
\mathrm{B} \preccurlyeq \mathrm{D}\ ,\ \Gamma^{-1}\{\mathrm{B}\} = \{b\}\ , & \Gamma^{-1}\{\mathrm{D}\} = \{c\}\ , & \text{one has: } b \preccurlyeq c\ .
\end{array}
$$

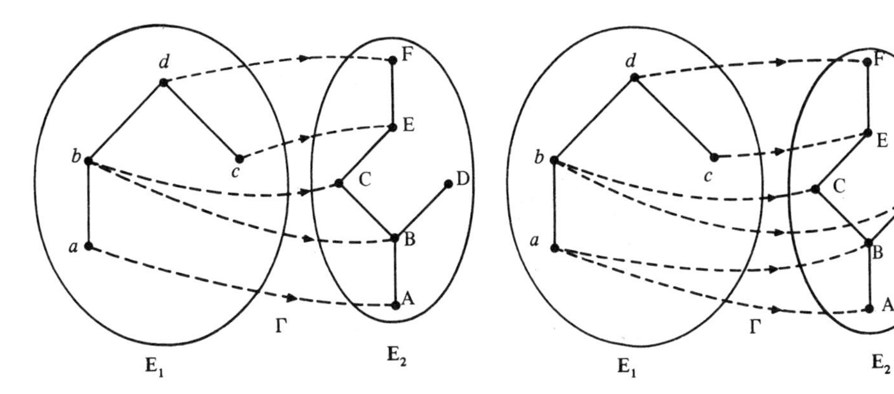

Fig. 57.10 A morphism that is neither an epimorphism nor a monomorphism.

Fig. 57.11 A mapping that is not a morphism.

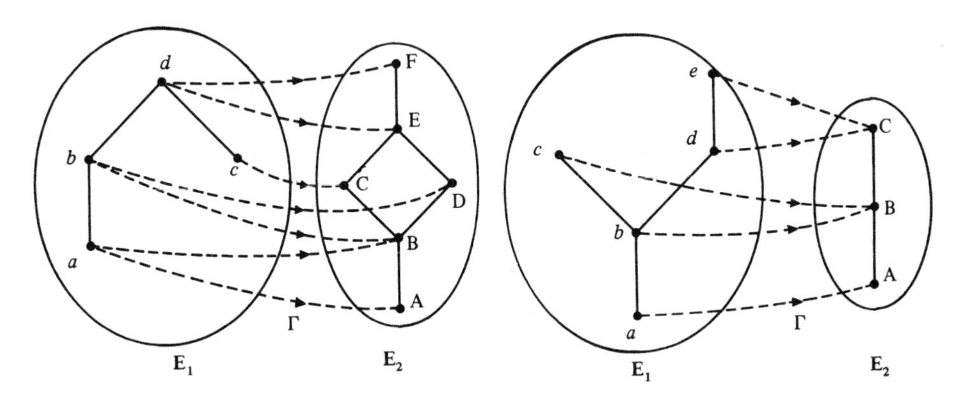

Fig. 57.12 Epimorphism.

Fig. 57.13 Another epimorphism.

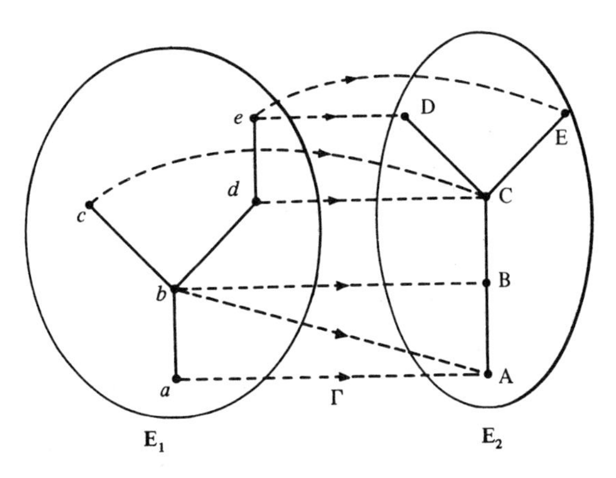

Fig. 57.14 Another epimorphism.

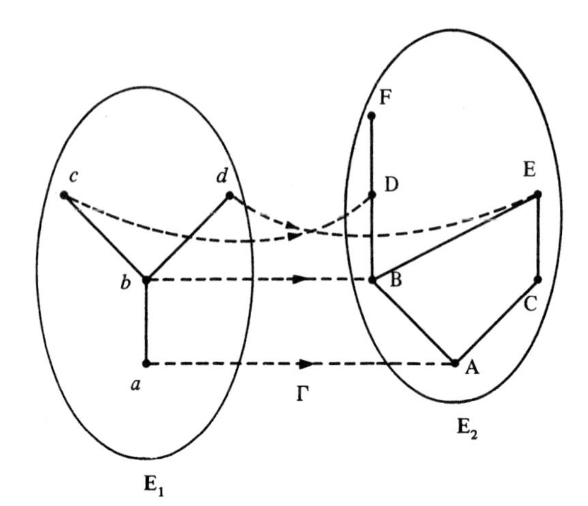

Fig. 57.15 Monomorphism.

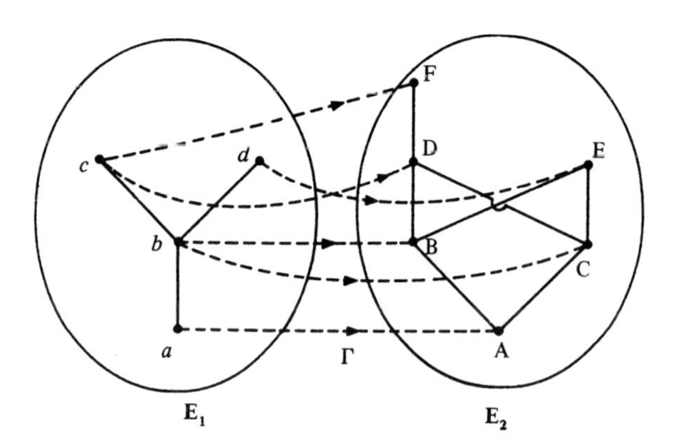

Fig. 57.16 Isomorphism.

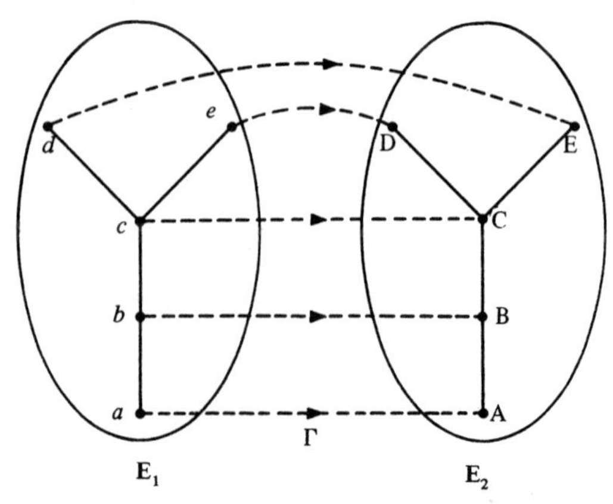

Fig. 57.17 Isomorphism.

Thus the mapping Γ^{-1} of $\mathbf{E}_2'$ into $\mathbf{E}_1$ is indeed isotonic. Therefore Γ is indeed a monomorphism.

Example 7. See Figure 57.16. One has

$$
\begin{array}{lllll}
 & a \preccurlyeq b , & \Gamma\{a\} = \{A\} , & \Gamma\{b\} = \{B , C\}, \text{ one has: } & A \preccurlyeq B , \ A \preccurlyeq C . \\
(57.30) & b \preccurlyeq c , & \Gamma\{b\} = \{B , C\} , & \Gamma\{c\} = \{D , F\}, \text{ one has: } & B \preccurlyeq D , \ B \preccurlyeq F , \\
 & & & & C \preccurlyeq D , \ C \preccurlyeq F . \\
 & b \preccurlyeq d , & \Gamma\{b\} = \{B , C\} , & \Gamma\{d\} = \{E\} , \text{ one has: } & B \preccurlyeq E , \ C \preccurlyeq E .
\end{array}
$$

Thus this is a morphism. Is it an epimorphism? Yes, since each $Y \in \mathbf{E}_2$ admits at least one arc (in fact, each Y admits one and only one).

We see whether it is a monomorphism. The first condition (each $Y \in \mathbf{E}_2$ admits at least one arc) is satisfied since each Y admits exactly one arc. We now examine the ordered pairs forming the maximal chains:

$$
\begin{array}{llllll}
 & A \preccurlyeq B , & \Gamma^{-1}\{A\} = \{a\} , & \Gamma^{-1}\{B\} = \{b\} , & \text{one has: } & a \preccurlyeq b . \\
 & A \preccurlyeq C , & \Gamma^{-1}\{A\} = \{a\} , & \Gamma^{-1}\{C\} = \{b\} , & \text{one has: } & a \preccurlyeq b . \\
 & B \preccurlyeq D , & \Gamma^{-1}\{B\} = \{b\} , & \Gamma^{-1}\{D\} = \{c\} , & \text{one has: } & b \preccurlyeq c . \\
(57.31) & B \preccurlyeq E , & \Gamma^{-1}\{B\} = \{b\} , & \Gamma^{-1}\{E\} = \{d\} , & \text{one has: } & b \preccurlyeq d . \\
 & C \preccurlyeq D , & \Gamma^{-1}\{C\} = \{b\} , & \Gamma^{-1}\{D\} = \{c\} , & \text{one has: } & b \preccurlyeq c . \\
 & C \preccurlyeq E , & \Gamma^{-1}\{C\} = \{b\} , & \Gamma^{-1}\{E\} = \{d\} , & \text{one has: } & b \preccurlyeq d . \\
 & D \preccurlyeq F , & \Gamma^{-1}\{D\} = \{c\} , & \Gamma^{-1}\{F\} = \{c\} , & \text{one has: } & c \preccurlyeq c .
\end{array}
$$

Thus, since Γ is an epimorphism and a monomorphism, it is an isomorphism. If one identifies B and C and identifies D and F, the isomorphism is apparent.

Example 8. See Figure 57.17. This is also an isomorphism, as the reader may verify. This may also be seen directly.

Endomorphism of an ordered set into itself. A morphism of an ordered set **E** into itself is called an endomorphism.

Automorphism of an ordered set into itself. An isomorphism of **E** onto itself is called an automorphism.

Duality between two ordered sets. Two ordered sets **E** and **E**$'$ are called dual with respect to one another if the inverse mappings Γ and Γ^{-1} are bijective and antitonic.

Example 9. See Figure 57.18. Endomorphism: We verify that the mapping is indeed a morphism of **E** into **E**:

$$
\begin{array}{llll}
a \preccurlyeq b , & \Gamma\{a\} = \{a , b\} , & \Gamma\{b\} = \{b , d\}, \text{ one has: } & a \preccurlyeq b , \ a \preccurlyeq d , \\
 & & & b \preccurlyeq b , \ b \preccurlyeq d .
\end{array}
$$

$$
(57.32) \quad
\begin{array}{llll}
b \preccurlyeq c \ , & \Gamma\{b\} = \{b\,, d\}\ , & \Gamma\{c\} = \{d\} & \text{one has} \quad b \preccurlyeq d \ , \ d \preccurlyeq d \ . \\
b \preccurlyeq d \ , & \Gamma\{b\} = \{b\,, d\}\ , & \Gamma\{d\} = \{d\} & \text{one has} \quad b \preccurlyeq d \ , \ d \preccurlyeq d \ . \\
c \preccurlyeq e \ , & \Gamma\{c\} = \{d\} & , \ \Gamma\{e\} = \{f\} & \text{one has} \quad d \preccurlyeq f \ . \\
c \preccurlyeq f \ , & \Gamma\{c\} = \{d\} & , \ \Gamma\{f\} = \{f\} & \text{one has} \quad d \preccurlyeq f \ . \\
d \preccurlyeq f \ , & \Gamma\{d\} = \{d\} & , \ \Gamma\{f\} = \{f\} & \text{one has} \quad d \preccurlyeq f \ . \\
f \preccurlyeq g \ , & \Gamma\{f\} = \{f\} & , \ \Gamma\{g\} = \{f\} & \text{one has} \quad f \preccurlyeq f \ .
\end{array}
$$

Thus, this mapping is indeed a morphism and is, therefore, an endomorphism of **E** into **E**.

Example 10. See Figure 57.19. Automorphism: One may easily verify that Γ is an isomorphism and also that Γ^{-1} is. This mapping is an automorphism of **E** onto itself.

Example 11. See Figure 57.20. Duality: This example concerns the duality between two ordered sets. We verify antitonicity:

$$
(57.33) \quad
\begin{array}{llll}
a \preccurlyeq b \ , & \Gamma\{a\} = \{a'\} & , \ \Gamma\{b\} = \{b'\} & , \text{one has:} \ b' \preccurlyeq a' \ . \\
b \preccurlyeq c \ , & \Gamma\{b\} = \{b'\} & , \ \Gamma\{c\} = \{c'\} & , \text{one has:} \ c' \preccurlyeq b' \ . \\
c \preccurlyeq d \ , & \Gamma\{c\} = \{c'\} & , \ \Gamma\{d\} = \{e'\} & , \text{one has:} \ e' \preccurlyeq c' \ . \\
c \preccurlyeq e \ , & \Gamma\{c\} = \{c'\} & , \ \Gamma\{e\} = \{d'\} & , \text{one has:} \ d' \preccurlyeq c' \ . \\
d \preccurlyeq f \ , & \Gamma\{d\} = \{e'\} & , \ \Gamma\{f\} = \{f'\} & , \text{one has:} \ f' \preccurlyeq e' \ . \\
e \preccurlyeq f \ , & \Gamma\{e\} = \{d'\} & , \ \Gamma\{f\} = \{f'\} & , \text{one has:} \ f' \preccurlyeq d' \ .
\end{array}
$$

It is verified. One may verify this for Γ^{-1} directly by symmetry. The mapping being bijective, the ordered sets **E** and **E**$'$ are dual.

Morphism of a structured set $\mathbf{E}_1$ into a structured set $\mathbf{E}_2$. We have reviewed the notion of morphism and other notions that follow for the case of ordered sets; we go on to take up this notion by considering this time two sets $\mathbf{E}_1$ and $\mathbf{E}_2$ each possessing a specified structure.

Let $\mathbf{E}_1$ and $\mathbf{E}_2$ be two sets for which one defines respectively laws of internal composition $*$ and $*'$, which are not necessarily defined throughout.

If†

$$
(57.34) \qquad \forall x\,, y \in \mathbf{E}_1 \quad : \quad \Gamma(x * y) = \Gamma x *' \Gamma y \quad (1)
$$

†We must write

$$
\Gamma\{(x * y)\} = \Gamma\{x\} *' \Gamma\{y\}
$$

but we have omitted the braces in order to simplify the notation. This is possible here since $\forall x : |\Gamma\, x\,| \leqslant 1$, because the mapping is functional.

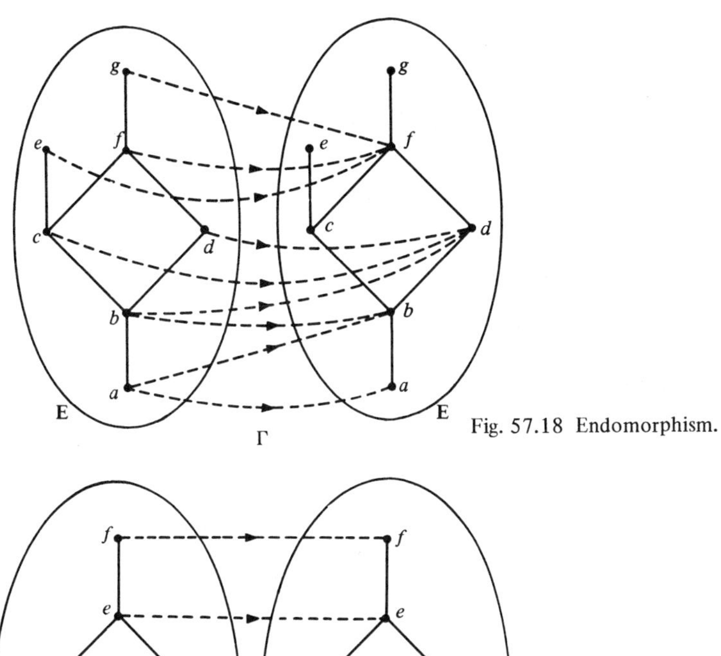

Fig. 57.18 Endomorphism.

Fig. 57.19 Automorphism.

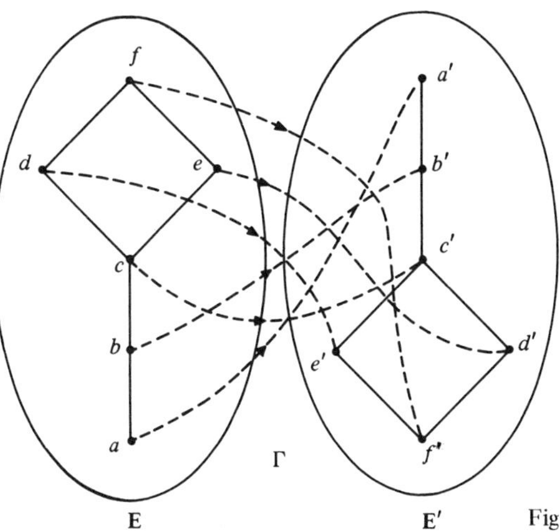

Fig. 57.20 Duality.

and if *the mapping* Γ *is functional, that is, is a function,* one says that one has realized a morphism of the structured set $\mathbf{E}_1$ into the structured set $\mathbf{E}_2$.

Note that in what is considered here, we restrict ourselves to a notion of morphism where the mappings are functional, that is, to any $x \in \mathbf{E}_1$ there corresponds one and only one $y \in \mathbf{E}_2$.

If the mapping Γ is injective, this morphism will be called a *monomorphism.*

If the mapping Γ is surjective, this morphism will be called an *epimorphism.*

If the mapping Γ is bijective, this morphism will be called an *isomorphism.*

When $\mathbf{E}_1 = \mathbf{E}_2$ and $* = *'$, this morphism will be called an endomorphism. In the case of an isomorphism, it will be called an automorphism.

As one may see, one recovers the definitions concerning ordered sets, but the condition of isotonicity is replaced here by condition (57.34) and the mapping is functional.

Example 1. Consider $\mathbf{E}_1$ with which one associates $*$ (Figure 57.21 on the left) and $\mathbf{E}_2$ with which one associates $*'$ (Figure 57.21 on the right). In the center of Figure 57.21 is defined a morphism of $\mathbf{E}_1$ into $\mathbf{E}_2$.

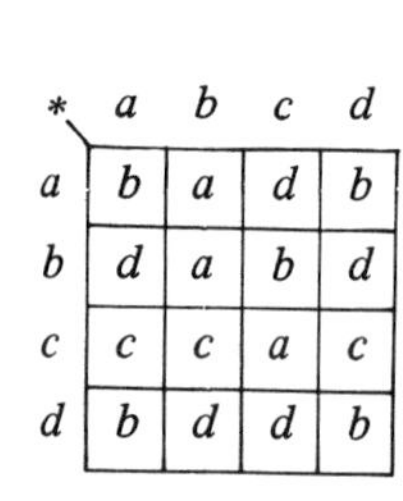

$*$	a	b	c	d
a	b	a	d	b
b	d	a	b	d
c	c	c	a	c
d	b	d	d	b

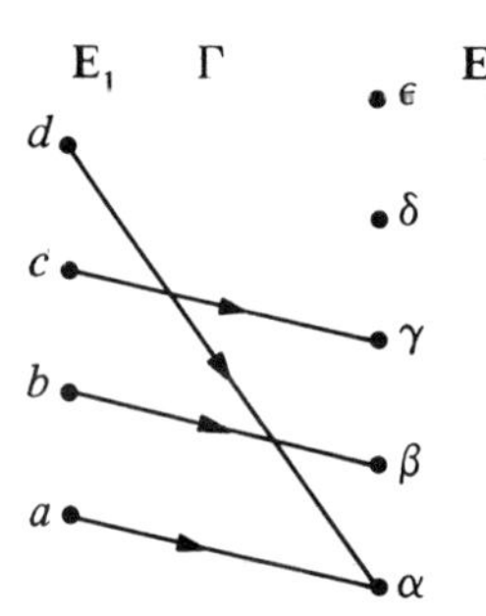

$*'$	α	β	γ	δ	ϵ
α	β	α	α	β	γ
β	α	α	β	β	α
γ	γ	γ	α	α	γ
δ	α	α	δ	α	γ
ϵ	δ	γ	ϵ	ϵ	γ

Fig. 57.21

As an exercise we verify that the mapping Γ is indeed a morphism of $\mathbf{E}_1$ into $\mathbf{E}_2$; one has

$$
\begin{array}{lll}
\Gamma(a * a) = \Gamma b = \beta & , & \Gamma a *' \Gamma a = \alpha *' \alpha = \beta \ , \\
\Gamma(a * b) = \Gamma a = \alpha & , & \Gamma a *' \Gamma b = \alpha *' \beta = \alpha \ , \\
\Gamma(a * c) = \Gamma d = \alpha & , & \Gamma a *' \Gamma c = \alpha *' \gamma = \alpha \ , \\
\Gamma(a * d) = \Gamma b = \beta & , & \Gamma a *' \Gamma d = \alpha *' \alpha = \beta \ , \\
\Gamma(b * a) = \Gamma d = \alpha & , & \Gamma b *' \Gamma a = \beta *' \alpha = \alpha \ , \\
\Gamma(b * b) = \Gamma a = \alpha & , & \Gamma b *' \Gamma b = \beta *' \beta = \alpha \ , \\
\Gamma(b * c) = \Gamma b = \beta & , & \Gamma b *' \Gamma c = \beta *' \gamma = \beta \ , \\
\Gamma(b * d) = \Gamma d = \alpha & , & \Gamma b *' \Gamma d = \beta *' \alpha = \alpha \ , \\
\Gamma(c * a) = \Gamma c = \gamma & , & \Gamma c *' \Gamma a = \gamma *' \alpha = \gamma \ , \\
\Gamma(c * b) = \Gamma c = \gamma & , & \Gamma c *' \Gamma b = \gamma *' \beta = \gamma \ ,
\end{array}
\tag{57.35}
$$

$$\Gamma(c * c) = \Gamma a = \alpha \quad , \quad \Gamma c *' \Gamma c = \gamma *' \gamma = \alpha \ ,$$
$$\Gamma(c * d) = \Gamma c = \gamma \quad , \quad \Gamma c *' \Gamma d = \gamma *' \alpha = \gamma \ ,$$
$$\Gamma(d * a) = \Gamma b = \beta \quad , \quad \Gamma d *' \Gamma a = \alpha *' \alpha = \beta \ ,$$
$$\Gamma(d * b) = \Gamma d = \alpha \quad , \quad \Gamma d *' \Gamma b = \alpha *' \beta = \alpha \ ,$$
$$\Gamma(d * c) = \Gamma d = \alpha \quad , \quad \Gamma d *' \Gamma c = \alpha *' \gamma = \alpha \ ,$$
$$\Gamma(d * d) = \Gamma b = \beta \quad , \quad \Gamma d *' \Gamma d = \alpha *' \alpha = \beta \ .$$

We have indeed verified for all ordered pairs $(x, y) \in \mathbf{E}_1 \times \mathbf{E}_1$, that (57.34) is satisfied. We remark that, as concerns the pairs $(u, v) \in \mathbf{E}_2 \times \mathbf{E}_2$, other than those formed with only α, β, and γ, one may place any element of $\mathbf{E}_2$ in these cases; these ordered pairs do not interfere in relation (57.35).

Example 2. Figure 57.22 represents a morphism of $\mathbf{E}_1$ into $\mathbf{E}_2$ in which we show that the relation associated with one of the sets need not necessarily be defined throughout; this is the case in $\mathbf{E}_2$ for the internal relation $*'$.

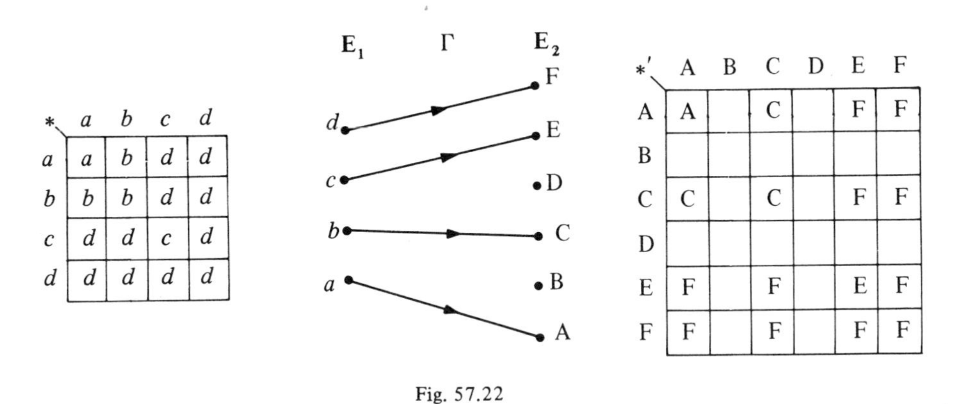

*	a	b	c	d
a	a	b	d	d
b	b	b	d	d
c	d	d	c	d
d	d	d	d	d

*'	A	B	C	D	E	F
A	A		C		F	F
B						
C	C		C		F	F
D						
E	F		F		E	F
F	F		F		F	F

Fig. 57.22

Example 3. Figure 57.23 represents a monomorphism of $\mathbf{E}_1$ into $\mathbf{E}_2$. It corresponds to that of Figure 57.15 with $* = \Delta$ and $*' = \Delta$.

*	a	b	c	d
a	a	a	a	a
b	a	b	b	b
c	a	b	c	b
d	a	b	b	d

$\mathbf{E}_1$ Γ $\mathbf{E}_2$

*'	A	B	C	D	E	F
A	A	A	A	A	A	A
B	A	B	A	B	B	B
C	A	A	C	A	C	A
D	A	B	A	D	B	D
E	A	B	C	B	E	B
F	A	B	A	D	B	F

Fig. 57.23

Example 4. Figure 57.24 represents an epimorphism of $\mathbf{E}_1$ into $\mathbf{E}_2$. It corresponds to that of Figure 57.13 with $* = \Delta$ and $*' = \Delta$.

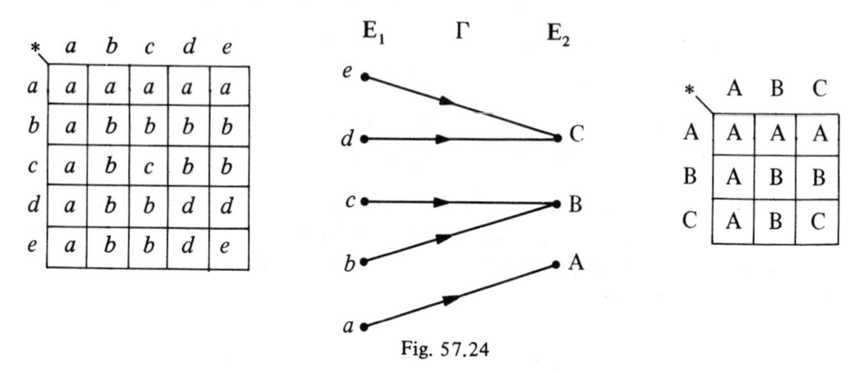

*	*a*	*b*	*c*	*d*	*e*
a	*a*	*a*	*a*	*a*	*a*
b	*a*	*b*	*b*	*b*	*b*
c	*a*	*b*	*c*	*b*	*b*
d	*a*	*b*	*b*	*d*	*d*
e	*a*	*b*	*b*	*d*	*e*

*	A	B	C
A	A	A	A
B	A	B	B
C	A	B	C

Fig. 57.24

Example 5. Figure 57.25 represents an isomorphism of $\mathbf{E}_1$ into $\mathbf{E}_2$. It corresponds to that of Figure 57.13 with $* = \Delta$ and $*' = \Delta$.

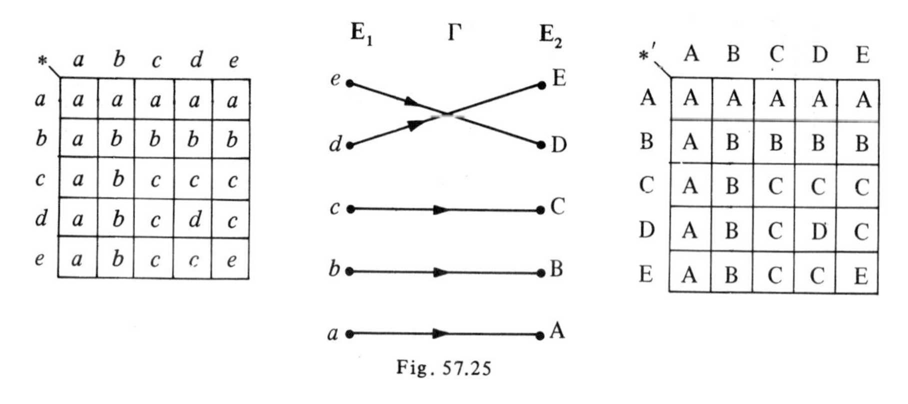

*	*a*	*b*	*c*	*d*	*e*
a	*a*	*a*	*a*	*a*	*a*
b	*a*	*b*	*b*	*b*	*b*
c	*a*	*b*	*c*	*c*	*c*
d	*a*	*b*	*c*	*d*	*c*
e	*a*	*b*	*c*	*c*	*e*

*'	A	B	C	D	E
A	A	A	A	A	A
B	A	B	B	B	B
C	A	B	C	C	C
D	A	B	C	D	C
E	A	B	C	C	E

Fig. 57.25

Example 6. Figure 57.26 represents an endomorphism of $\mathbf{E}$ into $\mathbf{E}$.

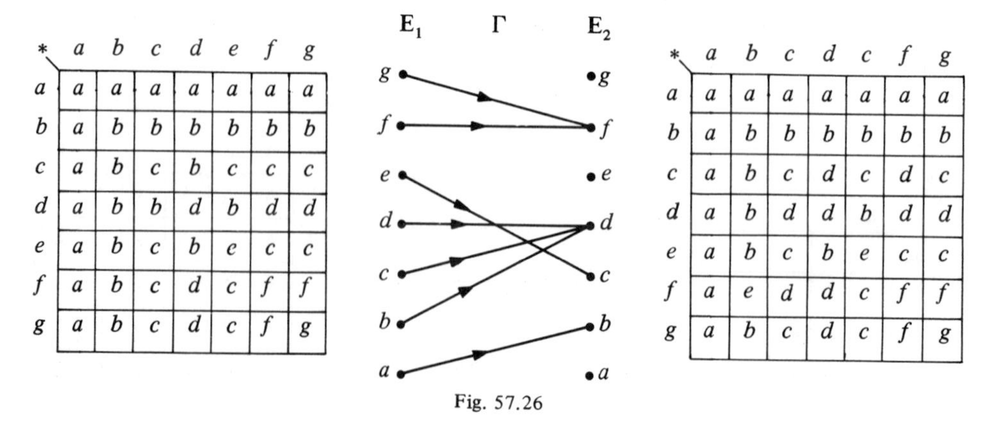

*	*a*	*b*	*c*	*d*	*e*	*f*	*g*
a	*a*	*a*	*a*	*a*	*a*	*a*	*a*
b	*a*	*b*	*b*	*b*	*b*	*b*	*b*
c	*a*	*b*	*c*	*b*	*c*	*c*	*c*
d	*a*	*b*	*b*	*d*	*b*	*d*	*d*
e	*a*	*b*	*c*	*b*	*e*	*c*	*c*
f	*a*	*b*	*c*	*d*	*c*	*f*	*f*
g	*a*	*b*	*c*	*d*	*c*	*f*	*g*

*	*a*	*b*	*c*	*d*	*c*	*f*	*g*
a	*a*	*a*	*a*	*a*	*a*	*a*	*a*
b	*a*	*b*	*b*	*b*	*b*	*b*	*b*
c	*a*	*b*	*c*	*d*	*c*	*d*	*c*
d	*a*	*b*	*d*	*d*	*b*	*d*	*d*
e	*a*	*b*	*c*	*b*	*e*	*c*	*c*
f	*a*	*e*	*d*	*d*	*c*	*f*	*f*
g	*a*	*b*	*c*	*d*	*c*	*f*	*g*

Fig. 57.26

Example 7. Figure 57.27 represents an automorphism of $\mathbf{E}$ onto $\mathbf{E}$. It corresponds to that of Figure 57.19 with $* = \Delta$.

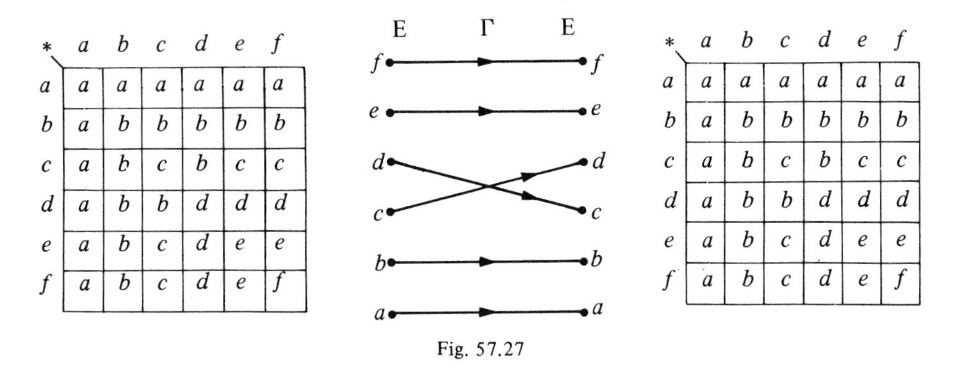

Fig. 57.27

Example 8. Figure 57.28 represents an automorphism of **E** onto **E** where one has on the left $*$, and on the right $*' \neq *$. In fact, referring to Figure 57.20, one sees that

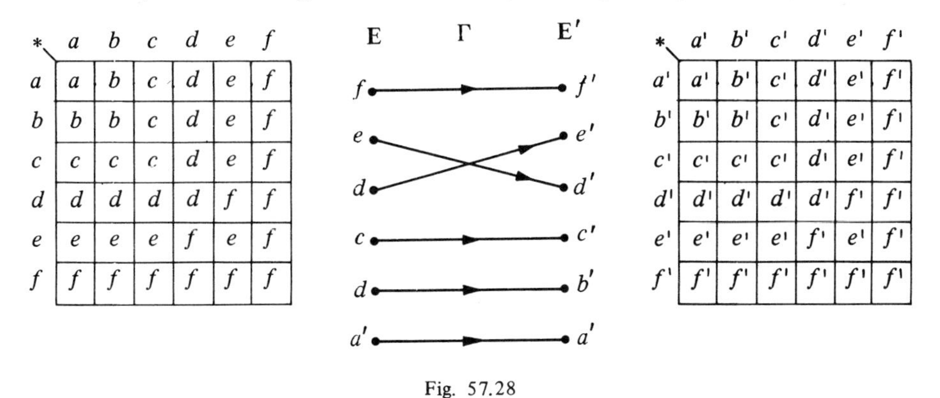

Fig. 57.28

$* = \nabla$ and $*' = \Delta$, where the morphism of **E** onto **E**$'$ constitutes a duality for the order relations represented in Figure 57.20.

We now consider some examples in which the $*$ are not structures corresponding to order relations.

Example 9. Figure 57.29 represents a monomorphism of the group (**E**, $*$) into the group (**E**$'$, $*'$).

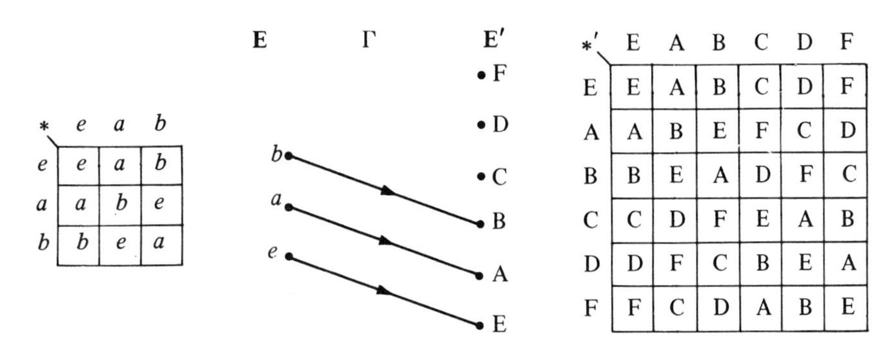

Fig. 57.29

Example 10. Figure 57.30 represents an epimorphism of the group $(\mathbf{E}, *)$ onto the group $(\mathbf{E}', *')$.

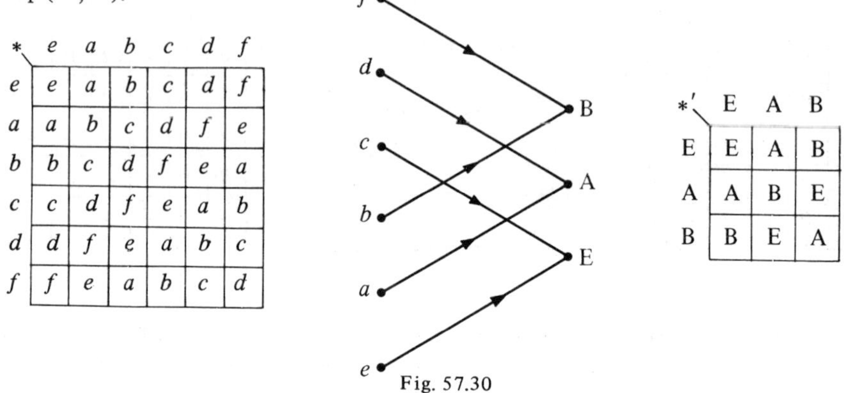

*	e	a	b	c	d	f
e	e	a	b	c	d	f
a	a	b	c	d	f	e
b	b	c	d	f	e	a
c	c	d	f	e	a	b
d	d	f	e	a	b	c
f	f	e	a	b	c	d

*'	E	A	B
E	E	A	B
A	A	B	E
B	B	E	A

Fig. 57.30

Example 11. Figure 57.31 represents an automorphism [an isomorphism of $(\mathbf{E}, *)$ into $(\mathbf{E}, *)$], but it is trivial. One may show that this depends on the fact that there exist only four finite groups of order 6 and that only one among these four permits a nontrivial automorphism [that of Figure 57.32, which is nonabelian (noncommutative)].

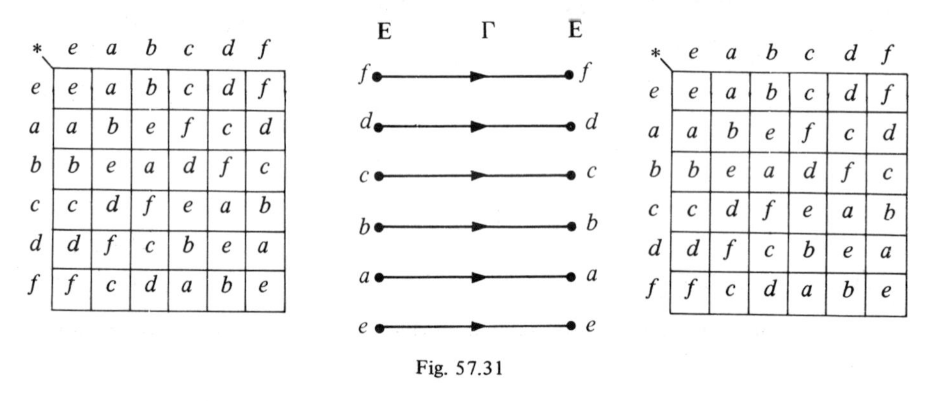

*	e	a	b	c	d	f
e	e	a	b	c	d	f
a	a	b	e	f	c	d
b	b	e	a	d	f	c
c	c	d	f	e	a	b
d	d	f	c	b	e	a
f	f	c	d	a	b	e

*	e	a	b	c	d	f
e	e	a	b	c	d	f
a	a	b	e	f	c	d
b	b	e	a	d	f	c
c	c	d	f	e	a	b
d	d	f	c	b	e	a
f	f	c	d	a	b	e

Fig. 57.31

Example 12. Figure 57.32 represents a nontrivial automorphism of $(\mathbf{E}, *)$ into $(\mathbf{E}, *)$. See the remark made in Example 11.

*	e	a	b	c	d	f
e	e	a	b	c	d	f
a	a	b	e	f	c	d
b	b	e	a	d	f	c
c	c	d	f	e	a	b
d	d	f	c	b	e	a
f	f	c	d	a	b	e

E Γ E

f, d, c, b, a, e → f, d, c, b, a, e

*	e	a	b	c	d	f
e	e	a	b	c	d	f
a	a	b	e	f	c	d
b	b	e	a	d	f	c
c	c	d	f	e	a	b
d	d	f	c	b	e	a
f	f	c	d	a	b	e

Fig. 57.32

Remark. All of the examples of the present section concern finite sets **E**, but all these considerations are applicable to arbitrary sets. As usual, we consider examples over finite sets with a didactic intent.

We shall later need to be able to compose morphisms with one another; for this we first proceed to review how to compose ordinary relations, then mappings.

Composition of two ordinary relations. Let **X**, **Y**, and **Z** be three sets; consider two ordinary graphs $\mathbf{G}_1 \subset \mathbf{X} \times \mathbf{Y}$ and $\mathbf{G}_2 \subset \mathbf{Y} \times \mathbf{Z}$ with which one associates the ordinary relations $\mathcal{R}_1$ and $\mathcal{R}_2$ (one may say also that $\mathcal{R}_1$ and $\mathcal{R}_2$ are correspondences associated with $\mathbf{G}_1$ and $\mathbf{G}_2$). Let $(x, y) \in \mathbf{G}_1$ and $(y, z) \in \mathbf{G}_2$; one will form an ordered pair (x, z) if and only if there exists a $y \in \mathbf{Y}$ such that $(x, y) \in \mathbf{G}_1$ and $(y, z) \in \mathbf{G}_2$. The set of ordered pairs (x, z) then forms the graph $\mathbf{G}_{1,2}$ composed of $\mathbf{G}_1$ and $\mathbf{G}_2$; it will be denoted

$$\mathbf{G}_{1,2} = \mathbf{G}_2 \circ \mathbf{G}_1 \tag{57.36}$$

and the corresponding relation will be denoted

$$\mathcal{R}_{1,2} = \mathcal{R}_2 \circ \mathcal{R}_1 \,. \tag{57.37}$$

We remark that this way of composing two ordinary relations is indeed that given in (13.10), which is, moreover, only a particular case of that given in (13.9). We have already given an example in Figure 13.3, and we present another in Figure 57.33.

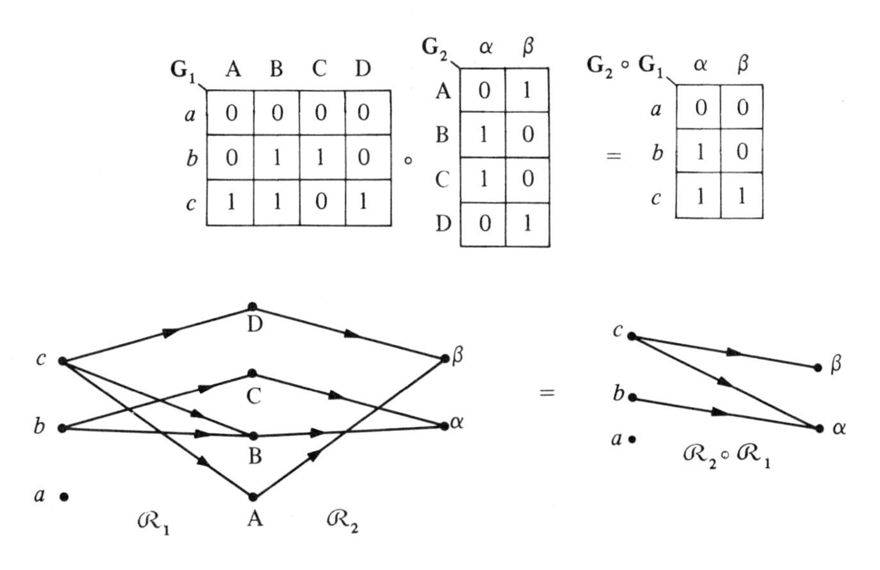

Fig. 57.33

Composition of mappings. Suppose that the law of composition of two ordinary relations is that which has been given in (57.36) and (57.37), and that the relations considered are *functional mappings or functions,* eventually surjections, injections, or bijections. One may then verify:

(57.38)

mapping	∘ mapping	= mapping,
mapping	∘ surjection	= mapping,
surjection	∘ mapping	= mapping,
mapping	∘ injection	= mapping,
injection	∘ mapping	= mapping,
mapping	∘ bijection	= mapping,
bijection	∘ mapping	= mapping,
surjection	∘ surjection	= surjection,
surjection	∘ injection	= mapping,
injection	∘ surjection	= mapping,
surjection	∘ bijection	= surjection,
bijection	∘ surjection	= surjection,
injection	∘ injection	= injection,
injection	∘ bijection	= injection,
bijection	∘ injection	= injection,
bijection	∘ bijection	= bijection.

In Figure 57.34 we have represented with the aid of a table the above properties, where the symbols are the initial letters of the respective words.

	B	I	S	M
B	B	I	S	M
I	I	I	M	M
S	S	M	S	M
M	M	M	M	M

Fig. 57.34

When the mappings Γ_1, Γ_2, and Γ_3 may be composed, the law of composition $\circ$ is associative:

(57.39) $$\Gamma_3 \circ (\Gamma_2 \circ \Gamma_1) = (\Gamma_3 \circ \Gamma_2) \circ \Gamma_1 .$$

For each mapping Γ, there exist a left identity and a right identity, which are the identity mappings $1_{\mathbf{X}}$ and $1_{\mathbf{Y}}$, where $\mathbf{Y} = \Gamma\mathbf{X}$, that is,

(57.40) $$\Gamma \circ 1_{\mathbf{X}} = 1_{\mathbf{Y}} \circ \Gamma = \Gamma .$$

If all the mappings being considered concern the same set $\mathbf{X}$ as domain and range, there exists then a unique identity 1 and the set of mappings Γ forms a monoid. If all the mappings considered are in addition bijective functions, then for each mapping there exists an inverse mapping and the set of these bijective functions forms a group.

Composition of morphisms.

Theorem I. Let

Γ_1 be a morphism of structured sets $\mathbf{X}$ into $\mathbf{Y}$;

Γ_2 a morphism of structured sets **Y** into **Z**;

then

$\Gamma_{1,2} = \Gamma_2 \circ \Gamma_1$ is a morphism of **X** into **Z**.

Proof. Let $*$ be the law associated with **X** and $*'$ that associated with **Y**. Since Γ_1 is a morphism of **X** into **Y**, one must have, according to (57.34),

(57.41) $\Gamma_1(\alpha * \beta) = \Gamma_1\alpha *' \Gamma_1\beta$, where $\alpha, \beta \in \mathbf{X}$, $\Gamma_1\alpha, \Gamma_1\beta \in \mathbf{Y}$.

Let $*''$ be the law associated with **Z**. Since Γ_2 is a morphism of **Y** into **Z**, one must have

(57.42) $\Gamma_2[\Gamma_1\alpha *' \Gamma_1\beta] = \Gamma_2(\Gamma_1\alpha) *'' \Gamma_2(\Gamma_1\beta)$ where $\Gamma_2(\Gamma_1\alpha), \Gamma_2(\Gamma_1\beta) \in \mathbf{Z}$.

Substituting for the member on the right of (57.41) in the member on the left of (57.42), one obtains

(57.43) $$\Gamma_2[\Gamma_1(\alpha * \beta)] = \Gamma_2(\Gamma_1\alpha) *'' \Gamma_2(\Gamma_1\beta).$$

(57.44) But $\Gamma_{1,2}x = \Gamma_2(\Gamma_1 x)$, by definition.

Then one indeed has

(57.45) $$\Gamma_{1,2}(\alpha * \beta) = \Gamma_{1,2}\alpha *'' \Gamma_{1,2}\beta .$$

We have considered the case where we have functional mappings Γ or functions between structured sets. The same proof will be taken up again for mappings, functional or not, between ordered sets. This forms another theorem.

Theorem II. Let

Γ_1 be a morphism of ordered sets **X** into **Y**;

Γ_2 be a morphism of ordered sets **Y** into **Z**;

then

$\Gamma_{1,2} = \Gamma_2 \circ \Gamma_1$ is a morphism of **X** into **Z**.

Proof. Since Γ_1 is a morphism of **X** into **Y**, one must have, according to the isotonic property (57.19),

(57.46) $(\forall x_i \in \mathbf{X}, \forall x_j \in \mathbf{X} : x_i \preccurlyeq x_j) \Rightarrow (\forall y_k \in \Gamma_1\{x_i\}, \forall y_l \in \Gamma_1\{x_j\} : y_k \preccurlyeq y_l)$.

Since Γ_2 is a morphism of **Y** into **Z**, one must have, again according to (57.19),

(57.47) $(\forall y_k \in \Gamma_1\{x_i\}, \forall y_l \in \Gamma_1\{x_j\} : y_k \preccurlyeq y_l) \Rightarrow (\forall z_m \in \Gamma_2\{y_k\}, \forall z_n \in \Gamma_2\{y_l\} : z_m \preccurlyeq z_n)$.

We note that the second term of (57.47) may be written

(57.48) $$(\forall z_m \in \Gamma_2(\Gamma_1\{x_i\}), \forall z_n \in \Gamma_2(\Gamma_1\{x_j\}) : z_m \preccurlyeq z_n) ,$$

or again

(57.49) $$(\forall z_m \in \Gamma_{1,2}\{x_i\} \quad , \forall z_n \in \Gamma_{1,2}\{x_j\} \quad : z_m \preccurlyeq z_n) .$$

By transitivity of implications, one may therefore write

(57.50) $(\forall x_i \in \mathbf{X}, \forall x_j \in \mathbf{X} : x_i \preccurlyeq x_j) \Rightarrow (\forall z_m \in \Gamma_{1,2}\{x_i\}, \forall z_n \in \Gamma_{1,2}\{x_j\} : z_m \preccurlyeq z_n)$.

This proves Theorem II.

Associativity in the laws of compositions of morphisms. If one considers the law of composition $\circ$ defined by (57.36) and (57.37) for ordinary relations and that one uses as a law of composition of morphisms one in the manner done in the immediately preceding considerations, then this law possesses the property of associativity. Thus if Γ_1, Γ_2, and Γ_3 are morphisms of **X** into **Y**, of **Y** into **Z**, and of **Z** into **V**, one will have

(57.51) $$\Gamma_3 \circ (\Gamma_2 \circ \Gamma_1) = (\Gamma_3 \circ \Gamma_2) \circ \Gamma_1 .$$

But, for other possible laws of composition, it would be appropriate to check this.

Throughout the present work, the laws of composition that we will consider for morphisms will all be associative.

58. THE CONCEPT OF CATEGORY

A category† **C** is a set of objects such that, for any ordered pair (X, Y), X ∈ **C**, Y ∈ **C**, there exists a set of morphisms Γ of X into Y having a specified property and denoted MOR(X, Y) [one allows the possibility that the set MOR(X, Y) may be empty]. These morphisms will be called **C**-morphisms and will have, by definition, the following properties:

(58.1) (1) $\mathrm{MOR}(X, Y) \cap \mathrm{MOR}(X', Y') = \phi$, if and only if $(X, Y) \neq (X', Y')$

(2) We suppose that a morphism Γ_1 of X into Y may be composed by a law of composition $\circ$ with a morphism Γ_2 of Y into Z (X ∈ **C**, Y ∈ **C**, Z ∈ **C**) to give a morphism $\Gamma_{1,2}$ of X into Z. The law $\circ$ must be specified. If $\Gamma_1 \in \mathrm{MOR}(X, Y)$, $\Gamma_2 \in \mathrm{MOR}(Y, Z)$, then $\Gamma_{1,2} = \Gamma_2 \circ \Gamma_1 \in \mathrm{MOR}(X, Z)$.

(3) We suppose that if $\Gamma_1 \in \mathrm{MOR}(X, Y)$, $\Gamma_2 \in \mathrm{MOR}(Y, Z)$, and $\Gamma_3 \in \mathrm{MOR}(Z, V)$. then

(58.2) $$\Gamma_3 \circ (\Gamma_2 \circ \Gamma_1) = (\Gamma_3 \circ \Gamma_2) \circ \Gamma_1 = \Gamma_3 \circ \Gamma_2 \circ \Gamma_1 ,$$

in other words, the law of composition must be associative.

(4) We suppose that for any X ∈ **C**, there exists a morphism that is an identity mapping of X onto itself. This morphism, denoted Γ° or 1, is such that for any

$$\Gamma \in \mathrm{MOR}(X, Y) \quad \text{and} \quad \Gamma' \in \mathrm{MOR}(Y, X) ,$$

one has

(58.3) $$\Gamma \circ \Gamma_X^\circ = \Gamma \qquad \text{and} \quad \Gamma_X^\circ \circ \Gamma' = \Gamma' .$$

†This is a restricted notion of this concept, but sufficient for the concrete applications considered here. For a deeper and more general study, see references [1E, 1M, 1R]. The notion of a category of classes is used by various authors instead of a category of sets.

The notion of a category is very widespread in mathematics. The following concepts are categories:

Groups with group morphisms: a group being designated by $G = (\mathbf{E}, *)$, where $\mathbf{E}$ is a set having the structure of a group and $*$ is the law of the group;

Sets and mappings between sets;

Lattices with morphisms of lattices; a lattice is designated by $\mathcal{C} = (\mathbf{E}, \preccurlyeq)$, where $\mathbf{E}$ is a set having the ordered configuration of a lattice and $\preccurlyeq$ is the order relation of the lattice; a lattice may also be defined by $\mathcal{C} = (\mathbf{E}, \Delta, \nabla)$, where Δ and ∇ are the laws corresponding respectively to the concepts of inferior limit and superior limit;

Semilattices $\mathcal{C} = (\mathbf{E}, \Delta)$ or $\mathcal{C} = (\mathbf{E}, \nabla)$;

Topological spaces with continuous mappings;

Measurable spaces with measurable transformations;

etc.

We proceed to consider several examples.

Example 1. This first example will be very elementary and will concern an easily enumerable case.

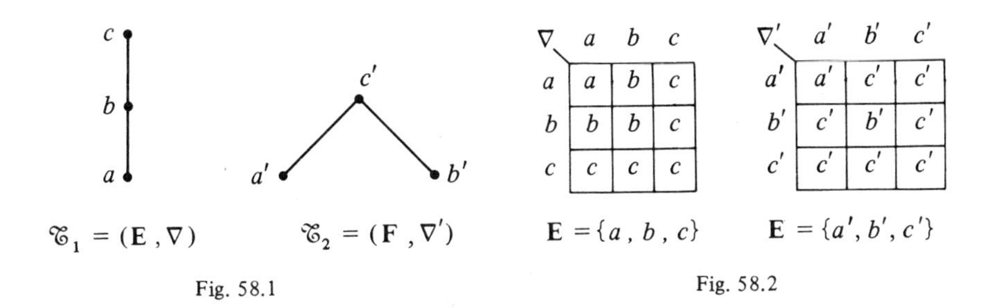

∇	a	b	c
a	a	b	c
b	b	b	c
c	c	c	c

∇'	a'	b'	c'
a'	a'	c'	c'
b'	c'	b'	c'
c'	c'	c'	c'

Fig. 58.1 Fig. 58.2

Let $\mathbf{E}$ and $\mathbf{F}$ be two ordered sets as indicated in Figure 58.1. These are two sup-semilattices $\mathcal{C}_1$ and $\mathcal{C}_2$ and will constitute the set of objects of the category $\mathbf{C}$ that we shall study.

These two sup-semilattices may also be considered as structured sets for the respective relations ∇ and ∇', that define the superior limits of the ordered pairs of $\mathbf{E}$ and of the ordered pairs of $\mathbf{F}$. See Figure 58.2.

The four sets $\mathrm{MOR}(\mathcal{C}_1, \mathcal{C}_1)$, $\mathrm{MOR}(\mathcal{C}_1, \mathcal{C}_2)$, $\mathrm{MOR}(\mathcal{C}_2, \mathcal{C}_1)$, and $\mathrm{MOR}(\mathcal{C}_2, \mathcal{C}_2)$ are given in Figure 58.3. The category formed by the two structured sets $\{\mathcal{C}_1, \mathcal{C}_2\}$ is such that

$$
\begin{aligned}
|\mathrm{MOR}(\mathcal{C}_1, \mathcal{C}_1)| &= 10 , \\
|\mathrm{MOR}(\mathcal{C}_1, \mathcal{C}_2)| &= 7 , \\
|\mathrm{MOR}(\mathcal{C}_2, \mathcal{C}_1)| &= 9 , \\
|\mathrm{MOR}(\mathcal{C}_2, \mathcal{C}_2)| &= 9 .
\end{aligned}
\tag{58.4}
$$

The identity mapping of $\mathcal{C}_1$ onto itself is Γ_{10} and the identity mapping of $\mathcal{C}_2$ onto itself is Γ_{35}. One has

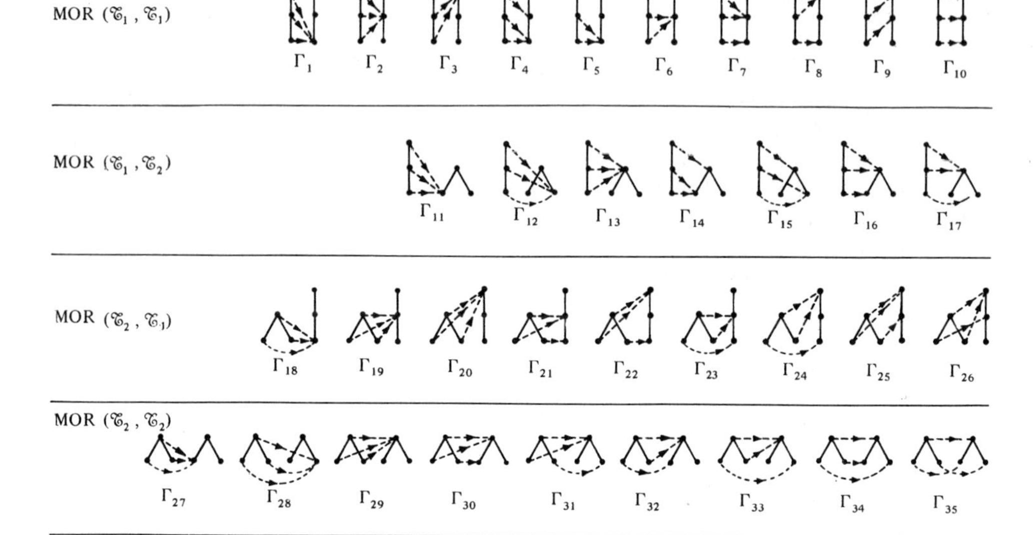

Fig. 58.3

$$
\begin{aligned}
&\Gamma_i \circ \Gamma_{10} = \Gamma_i \quad && i = 1, 2, \ldots, 17 \\
&\Gamma_{10} \circ \Gamma_i = \Gamma_i \quad && i = 1, 2, \ldots, 10, i = 18, 19, \ldots, 26 \\
&\Gamma_j \circ \Gamma_{35} = \Gamma_j \quad && j = 11, 12, \ldots, 17, j = 27, 28, \ldots, 35 \\
&\Gamma_{35} \circ \Gamma_j = \Gamma_j \quad && j = 18, 19, \ldots, 35 .
\end{aligned}
\tag{58.5}
$$

We remark that not all Γ_i may be composed with a Γ_j for the ordinary law of composition of relations; the chart in Figure 58.4 gives the results:

$$
\begin{aligned}
&\Gamma_{1\ \text{à}\ 10} \in \text{MOR}(\mathcal{C}_1, \mathcal{C}_1) , \\
&\Gamma_{11\ \text{à}\ 17} \in \text{MOR}(\mathcal{C}_1, \mathcal{C}_2) , \\
&\Gamma_{18\ \text{à}\ 26} \in \text{MOR}(\mathcal{C}_2, \mathcal{C}_1) , \\
&\Gamma_{27\ \text{à}\ 35} \in \text{MOR}(\mathcal{C}_2, \mathcal{C}_2) .
\end{aligned}
\tag{58.6}
$$

	Γ_{1-10}	Γ_{11-17}	Γ_{18-26}	Γ_{27-35}
Γ_{1-10}	Γ_{1-10}	Γ_{11-17}		
Γ_{11-17}			Γ_{1-10}	Γ_{11-17}
Γ_{18-26}	Γ_{18-26}	Γ_{27-35}		
Γ_{27-35}			Γ_{18-26}	Γ_{27-35}

Fig. 58.4

The category **C** gives all the **C**-morphisms existing between the two structured sets Γ_1 and Γ_2, between one another or between themselves. In other words, the category **C** gives all possible functions that may exist between the elements of an ordered pair (X, Y), X, Y $\in \{\mathscr{T}_1, \mathscr{T}_2\}$.

Example 2. We consider all finite groups having four or fewer elements. One knows that there exist:

1 group with 1 element;
1 group with 2 elements;
1 group with 3 elements;
2 groups with 4 elements.

These five groups,

$$\begin{aligned} G_1 &= (\{e\}, *_1) \ , \\ G_2 &= (\{e, a\}, *_2) \ , \\ G_3 &= (\{e, a, b\}, *_3) \ , \\ G_4 &= (\{e, a, b, c\}, *_4) \ , \\ G_5 &= (\{e, a, b, c\}, *_5) \ , \end{aligned} \tag{58.7}$$

have been represented in Figures 58.5–58.9. For each of these groups, *e* is the identity.

$*_1$	e
e	e

$G_1 = (\{e\}, *_1)$

Fig. 58.5

$*_2$	e	a
e	e	a
a	a	e

$G_2 = (\{e, a\}, *_2)$

Fig. 58.6

$*_3$	e	a	b
e	e	a	b
a	a	b	e
b	b	e	a

$G_3 = (\{e, a, b\}, *_3)$

fig. 58.7

$*_4$	e	a	b	c
e	e	a	b	c
a	a	e	c	b
b	b	c	e	a
c	c	b	a	e

$G_4 = (\{e, a, b, c\}, *_4)$

Fig. 58.8

$*_5$	e	a	b	c
e	e	a	b	c
a	a	e	c	b
b	b	c	a	e
c	c	b	e	a

$G_5 = (\{e, a, b, c\}, *_5)$

Fig. 58.9

We proceed to examine the category formed by the set of these five groups. The morphisms Γ considered here must be such that

$$\Gamma(x *_i y) = \Gamma x *_j \Gamma y \quad i, j = 1, 2, 3, 4, 5 \tag{58.8}$$

for a morphism Γ of G_i into G_j.

Enumerative listing of all the morphisms between finite groups of 4 or more elements.†

(G_1, G_1)

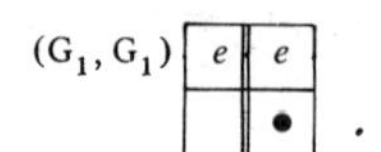

e	e
	•

.

(G_1, G_2)

e	e
	•

.

(G_1, G_3)

e	e
	•

.

(G_1, G_4)

e	e
	•

.

(G_1, G_5)

e	e
	•

.

(G_2, G_1)

a	e
e	e
	•

.

(G_2, G_2)

a	e	a
e	e	e
	•	•

.

(G_2, G_3)

a	e	a	b
e	e	e	e
	•	×	×

.

† • indicates a morphism.

(G_2, G_4)

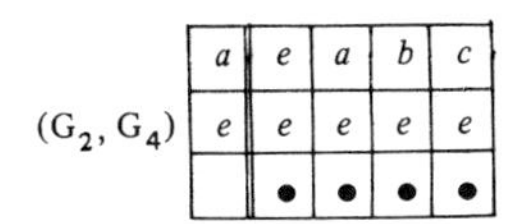

a	*e*	*a*	*b*	*c*
e	*e*	*e*	*e*	*e*
	●	●	●	●

.

(G_2, G_5)

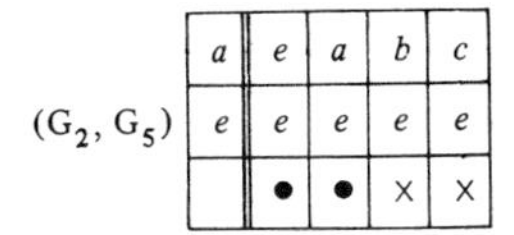

a	*e*	*a*	*b*	*c*
e	*e*	*e*	*e*	*e*
	●	●	×	×

.

(G_3, G_1)

b	*e*
a	*e*
e	*e*
	●

.

(G_3, G_2)

b	*e*	*a*	*e*	*a*
a	*e*	*e*	*a*	*a*
e	*e*	*e*	*e*	*e*
	●	×	×	×

.

(G_3, G_3)

b	*e*	*a*	*b*	*e*	*a*	*b*	*e*	*a*	*b*
a	*e*	*e*	*e*	*a*	*a*	*a*	*b*	*b*	*b*
e	*e*	*e*	*e*	*e*	*e*	*e*	*e*	*e*	*e*
	●	×	×	×	×	●	×	●	×

.

(G_3, G_4)

b	*e*	*a*	*b*	*c*	*e*	*a*	*b*	*c*	*e*	*a*	*b*	*c*	*e*	*a*	*b*	*c*
a	*e*	*e*	*e*	*e*	*a*	*a*	*a*	*a*	*b*	*b*	*b*	*b*	*c*	*c*	*c*	*c*
e	*e*	*e*	*e*	*e*	*e*	*e*	*e*	*e*	*e*	*e*	*e*	*e*	*e*	*e*	*e*	*e*
	●	×	×	×	×	×	×	×	×	×	×	×	×	×	×	×

.

(G_3, G_5)

b	*e*	*a*	*b*	*c*	*e*	*a*	*b*	*c*	*e*	*a*	*b*	*c*	*e*	*a*	*b*	*c*
a	*e*	*e*	*e*	*e*	*a*	*a*	*a*	*a*	*b*	*b*	*b*	*b*	*c*	*c*	*c*	*c*
e	*e*	*e*	*e*	*e*	*e*	*e*	*e*	*e*	*e*	*e*	*e*	*e*	*e*	*e*	*e*	*e*
	●	×	×	×	×	×	×	×	×	×	×	×	×	×	×	×

.

(G_4, G_1)

c	*e*
b	*e*
a	*e*
e	*e*
	●

.

(G_4, G_2)

c	*e*	*a*	*e*	*a*	*e*	*a*	*e*	*a*
b	*e*	*e*	*a*	*a*	*e*	*e*	*a*	*a*
a	*e*	*e*	*e*	*e*	*a*	*a*	*a*	*a*
e	*e*	*e*	*e*	*e*	*e*	*e*	*e*	*e*
	•	×	×	•	×	•	•	×

.

(G_4, G_3)

c	*e*	*a*	*b*	*e*	*a*	*b*	*e*	*a*	*b*	*e*	*a*	*b*	*e*	*a*	*b*	*e*	*a*	*b*	*e*	*a*	*b*	*e*	*a*	*b*	*e*	*a*	*b*
b	*e*	*e*	*e*	*a*	*a*	*a*	*b*	*b*	*b*	*e*	*e*	*e*	*a*	*a*	*a*	*b*	*b*	*b*	*e*	*e*	*e*	*a*	*a*	*a*	*b*	*b*	*b*
a	*e*	*e*	*e*	*e*	*e*	*e*	*e*	*e*	*e*	*a*	*a*	*a*	*a*	*a*	*a*	*a*	*a*	*a*	*b*	*b*	*b*	*b*	*b*	*b*	*b*	*b*	*b*
e	*e*	*e*	*e*	*e*	*e*	*e*	*e*	*e*	*e*	*e*	*e*	*e*	*e*	*e*	*e*	*e*	*e*	*e*	*e*	*e*	*e*	*e*	*e*	*e*	*e*	*e*	*e*
	•	×	×	×	×	×	×	×	×	×	×	×	×	×	×	×	×	×	×	×	×	×	×	×	×	×	×

.

(G_4, G_4)

c	*e*	*a*	*b*	*c*	*e*	*a*	*b*	*c*	*e*	*a*	*b*	*c*	*e*	*a*	*b*	*c*	*e*	*a*	*b*	*c*	*e*	*a*	*b*	*c*	*e*	*a*	*b*	*c*
b	*e*	*e*	*e*	*e*	*a*	*a*	*a*	*a*	*b*	*b*	*b*	*b*	*c*	*c*	*c*	*c*	*e*	*e*	*e*	*e*	*a*	*a*	*a*	*a*	*b*	*b*	*b*	*b*
a	*e*	*e*	*e*	*e*	*e*	*e*	*e*	*e*	*e*	*e*	*e*	*e*	*e*	*e*	*e*	*e*	*a*	*a*	*a*	*a*	*a*	*a*	*a*	*a*	*a*	*a*	*a*	*a*
e	*e*	*e*	*e*	*e*	*e*	*e*	*e*	*e*	*e*	*e*	*e*	*e*	*e*	*e*	*e*	*e*	*e*	*e*	*e*	*e*	*e*	*e*	*e*	*e*	*e*	*e*	*e*	*e*
	•	×	×	×	×	•	×	×	×	×	•	×	×	×	×	•	×	•	×	×	•	×	×	×	×	×	×	•

c	*e*	*a*	*b*	*c*	*e*	*a*	*b*	*c*	*e*	*a*	*b*	*c*	*e*	*a*	*b*	*c*	*e*	*a*	*b*	*c*	*e*	*a*	*b*	*c*	*e*	*a*	*b*	*c*
b	*c*	*c*	*c*	*c*	*e*	*e*	*e*	*e*	*a*	*a*	*a*	*a*	*b*	*b*	*b*	*b*	*c*	*c*	*c*	*c*	*e*	*e*	*e*	*e*	*a*	*a*	*a*	*a*
a	*a*	*a*	*a*	*a*	*b*	*b*	*b*	*b*	*b*	*b*	*b*	*b*	*b*	*b*	*b*	*b*	*b*	*b*	*b*	*b*	*c*	*c*	*c*	*c*	*c*	*c*	*c*	*c*
e	*e*	*e*	*e*	*e*	*e*	*e*	*e*	*e*	*e*	*e*	*e*	*e*	*e*	*e*	*e*	*e*	*e*	*e*	*e*	*e*	*e*	*e*	*e*	*e*	*e*	*e*	*e*	*e*
	×	×	•	×	×	×	•	×	×	×	×	•	•	×	×	×	×	•	×	×	×	×	×	•	×	×	•	×

c	*e*	*a*	*b*	*c*	*e*	*a*	*b*	*c*
b	*b*	*b*	*b*	*b*	*c*	*c*	*c*	*c*
a	*c*	*c*	*c*	*c*	*c*	*c*	*c*	*c*
e	*e*	*e*	*e*	*e*	*e*	*e*	*e*	*e*
	×	•	×	×	•	×	×	×

.

(G_4, G_5)

c	*e*	*a*	*b*	*c*	*e*	*a*	*b*	*c*	*e*	*a*	*b*	*c*	*e*	*a*	*b*	*c*	*e*	*a*	*b*	*c*	*e*	*a*	*b*	*c*	*e*	*a*	*b*	*c*
b	*e*	*e*	*e*	*e*	*a*	*a*	*a*	*a*	*b*	*b*	*b*	*b*	*c*	*c*	*c*	*c*	*e*	*e*	*e*	*e*	*a*	*a*	*a*	*a*	*b*	*b*	*b*	*b*
a	*e*	*e*	*e*	*e*	*e*	*e*	*e*	*e*	*e*	*e*	*e*	*e*	*e*	*e*	*e*	*e*	*a*	*a*	*a*	*a*	*a*	*a*	*a*	*a*	*a*	*a*	*a*	*a*
e	*e*	*e*	*e*	*e*	*e*	*e*	*e*	*e*	*e*	*e*	*e*	*e*	*e*	*e*	*e*	*e*	*e*	*e*	*e*	*e*	*e*	*e*	*e*	*e*	*e*	*e*	*e*	*e*
	•	×	×	×	×	•	×	×	×	×	×	×	×	×	×	×	×	•	×	×	•	×	×	×	×	×	×	×

c	*e*	*a*	*b*	*c*	*e*	*a*	*b*	*c*	*e*	*a*	*b*	*c*	*e*	*a*	*b*	*c*	*e*	*a*	*b*	*c*	*e*	*a*	*b*	*c*	*e*	*a*	*b*	*c*
b	*c*	*c*	*c*	*c*	*e*	*e*	*e*	*e*	*a*	*a*	*a*	*a*	*b*	*b*	*b*	*b*	*c*	*c*	*c*	*c*	*e*	*e*	*e*	*e*	*a*	*a*	*a*	*a*
a	*a*	*a*	*a*	*a*	*b*	*b*	*b*	*b*	*b*	*b*	*b*	*b*	*b*	*b*	*b*	*b*	*b*	*b*	*b*	*b*	*c*	*c*	*c*	*c*	*c*	*c*	*c*	*c*
e	*e*	*e*	*e*	*e*	*e*	*e*	*e*	*e*	*e*	*e*	*e*	*e*	*e*	*e*	*e*	*e*	*e*	*e*	*e*	*e*	*e*	*e*	*e*	*e*	*e*	*e*	*e*	*e*
	×	×	×	×	×	×	×	×	×	×	×	×	×	×	×	×	×	×	×	×	×	×	×	×	×	×	×	×

c	e	a	b	c	e	a	b	c
b	b	b	b	b	c	c	c	c
a	c	c	c	c	c	c	c	c
e	e	e	e	e	e	e	e	e
	×	×	×	×	×	×	×	×

.

(G_5, G_1)

c	e
b	e
a	e
e	e
	•

.

(G_5, G_2)

c	e	a	e	a	e	a	e	a
b	e	e	a	a	e	e	a	a
a	e	e	e	e	a	a	a	a
e	e	e	e	e	e	e	e	e
	•	×	×	•	×	×	×	×

.

(G_5, G_3)

c	e	a	b	e	a	b	e	a	b	e	a	b	e	a	b	e	a	b	e	a	b	e	a	b	e	a	b
b	e	e	e	a	a	a	b	b	b	e	e	e	a	a	a	b	b	b	e	e	e	a	a	a	b	b	b
a	e	e	e	e	e	e	e	e	e	a	a	a	a	a	a	a	a	a	b	b	b	b	b	b	b	b	b
e	e	e	e	e	e	e	e	e	e	e	e	e	e	e	e	e	e	e	e	e	e	e	e	e	e	e	e
	•	×	×	×	×	×	×	×	×	×	×	×	×	×	×	×	×	×	×	×	×	×	×	×	×	×	×

.

(G_5, G_4)

c	e	a	b	c	e	a	b	c	e	a	b	c	e	a	b	c	e	a	b	c	e	a	b	c	e	a	b	c
b	e	e	e	e	a	a	a	a	b	b	b	b	c	c	c	c	e	e	e	e	a	a	a	a	b	b	b	b
a	e	e	e	e	e	e	e	e	e	e	e	e	e	e	e	e	a	a	a	a	a	a	a	a	a	a	a	a
e	e	e	e	e	e	e	e	e	e	e	e	e	e	e	e	e	e	e	e	e	e	e	e	e	e	e	e	e
	•	×	×	×	×	•	×	×	×	×	•	×	×	×	×	•	×	×	×	×	×	×	×	×	×	×	×	×

c	e	a	b	c	e	a	b	c	e	a	b	c	e	a	b	c	e	a	b	c	e	a	b	c	e	a	b	c
b	c	c	c	c	e	e	e	e	a	a	a	a	b	b	b	b	c	c	c	c	e	e	e	e	a	a	a	a
a	a	a	a	a	b	b	b	b	b	b	b	b	b	b	b	b	b	b	b	b	c	c	c	c	c	c	c	c
e	e	e	e	e	e	e	e	e	e	e	e	e	e	e	e	e	e	e	e	e	e	e	e	e	e	e	e	e
	×	×	×	×	×	×	×	×	×	×	×	×	×	×	×	×	×	×	×	×	×	×	×	×	×	×	×	×

c	e	a	b	c	e	a	b	c
b	b	b	b	b	c	c	c	c
a	c	c	c	c	c	c	c	c
e	e	e	e	e	e	e	e	e
	×	×	×	×	×	×	×	×

.

(G_5, G_5)

c	e	a	b	c	e	a	b	c	e	a	b	c	e	a	b	c	e	a	b	c	e	a	b	c	e	a	b	c
b	e	e	e	e	a	a	a	a	b	b	b	b	c	c	c	c	e	e	e	e	a	a	a	a	b	b	b	b
a	e	e	e	e	e	e	e	e	e	e	e	e	e	e	e	e	a	a	a	a	a	a	a	a	a	a	a	a
e	e	e	e	e	e	e	e	e	e	e	e	e	e	e	e	e	e	e	e	e	e	e	e	e	e	e	e	e
	•	×	×	×	×	•	×	×	×	×	×	×	×	×	×	×	×	×	×	×	×	×	×	×	×	×	×	•

c	e	a	b	c	e	a	b	c	e	a	b	c	e	a	b	c	e	a	b	c	e	a	b	c	e	a	b	c
b	c	c	c	c	e	e	e	e	a	a	a	a	b	b	b	b	c	c	c	c	e	e	e	e	a	a	a	a
a	a	a	a	a	b	b	b	b	b	b	b	b	b	b	b	b	b	b	b	b	c	c	c	c	c	c	c	c
e	e	e	e	e	e	e	e	e	e	e	e	e	e	e	e	e	e	e	e	e	e	e	e	e	e	e	e	e
	×	×	×	×	×	×	×	×	×	×	×	×	×	×	×	×	×	×	×	×	×	×	×	×	×	×	×	×

c	e	a	b	c	e	a	b	c
b	b	b	b	b	c	c	c	c
a	c	c	c	c	c	c	c	c
e	e	e	e	e	e	e	e	e
	×	×	×	×	×	×	×	×

.

Fig. 58.10

In Figure 58.10 we have presented an enumerative lexicographic listing procedure with neither omission nor redundance.† The morphisms are indicated by •; the mappings that are not morphisms are indicated with ×.

This procedure of enumeration may be programmed without difficulty on a computer.

Figure 58.11 gives the category of groups having at most four elements. There are 59 morphisms in this category.

All possible subgroups of the groups having at most four elements may be obtained with respect to these morphisms. The law of composition of all these morphisms is an internal law; it is associative and possesses an identity as specified in (58.3).

†We have supposed in this lexicographic enumeration that the identity is unique for all the transformations. In order to program with an eye toward electronic calculation, test each mapping for $\Gamma(x * y) = \Gamma x *' \Gamma y$. This does not take very long. Also, one may considerably reduce the enumeration by introducing some constraints on the formation of subgroups.

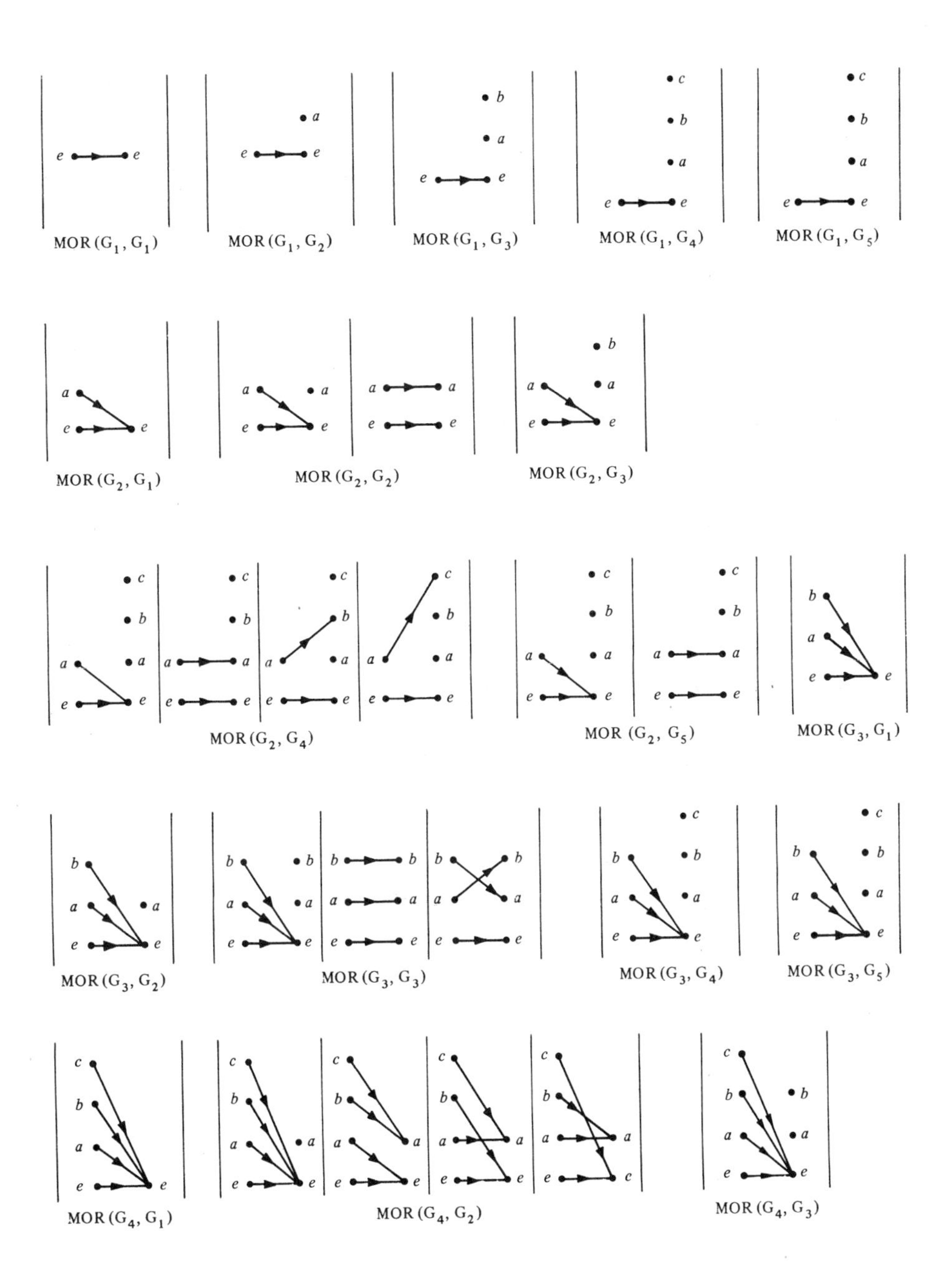
MOR (G_1, G_1)
MOR (G_1, G_2)
MOR (G_1, G_3)
MOR (G_1, G_4)
MOR (G_1, G_5)
MOR (G_2, G_1)
MOR (G_2, G_2)
MOR (G_2, G_3)
MOR (G_2, G_4)
MOR (G_2, G_5)
MOR (G_3, G_1)
MOR (G_3, G_2)
MOR (G_3, G_3)
MOR (G_3, G_4)
MOR (G_3, G_5)
MOR (G_4, G_1)
MOR (G_4, G_2)
MOR (G_4, G_3)

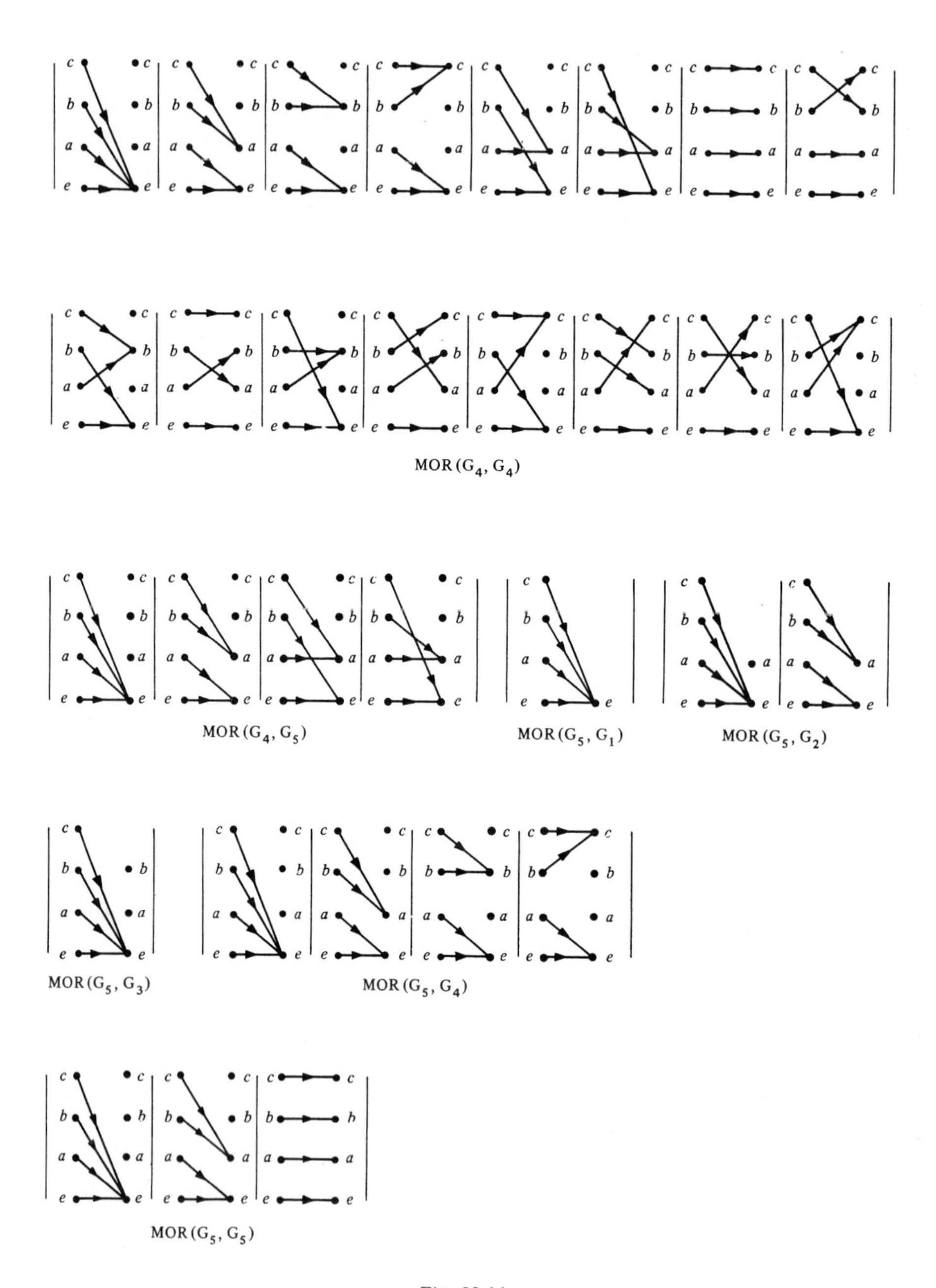

Fig. 58.11

Example 3. We consider the set of objects consisting of the intervals

(58.9) $\mathfrak{X} = (\mathrm{U}, *)$ where $\mathbf{U} = [\mathrm{X}, \mathrm{X}'] \subset [a, b]$ $a \in \mathbf{R}$, $b \in \mathbf{R}$,

that is, the set of closed continuous intervals in the continuum **R**.

U has the structure of a total order for the relation

(58.10) $[a, b] \preccurlyeq [a', b']$ if and only if' $a < a'$ or $a = a'$, $b \leqslant b'$.

and the operation ⊛ may be taken as an operation, not everywhere defined, obtained with respect to the order relation (58.9).

To any ordered pair $(\mathfrak{X}, \mathfrak{Y})$, where $\mathfrak{X}$ and $\mathfrak{Y}$ are defined by (58.9), one may correspond a set of morphisms f of $\mathfrak{X}$ into $\mathfrak{Y}$ having the property of all being functions. See Figures 58.12 and 58.13.

All these functions satisfy conditions (58.1)–(58.3).

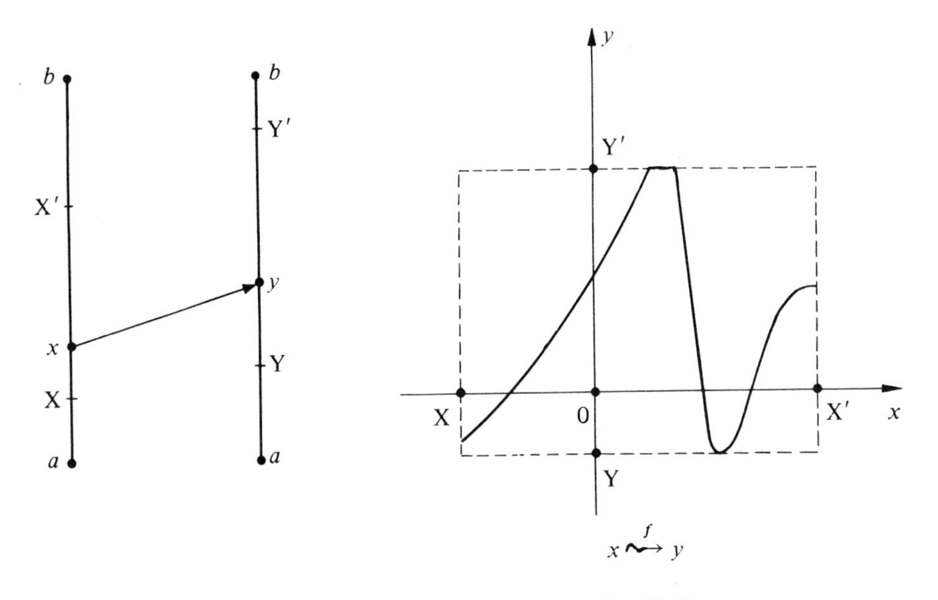

Fig. 58.12

Fig. 58.13

Then the set of objects $\mathfrak{X}$ constituted by all the closed intervals in $[a, b]$ forms a category **C** for the operation ⊛ corresponding to the induced order.

In the same manner one may define the category of domains of $\mathbf{R}^2$, of $\mathbf{R}^3$, etc.

58. FUZZY C-MORPHISMS

We begin with an example. We consider the two finite groups with four elements G_4 and G_5 represented in Figures 58.8 and 58.9. For convenience, we present these again here as Figures 59.1 and 59.2. In Figure 59.3 we have represented MOR(G_4, G_5), which has been calculated as an exercise in Example 2 of Section 58. By designating the morphisms of G_4 into G_5 by Γ_i, $i = 1, 2, 3, 4$, we have the reference set of morphisms of G_4 into G_5:

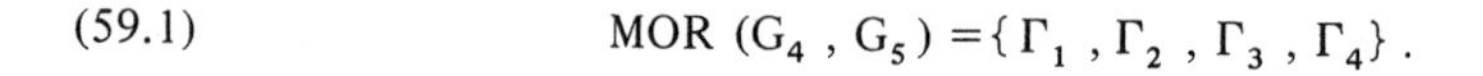

(59.1) $$\text{MOR}\,(G_4\,,\,G_5) = \{\Gamma_1\,,\Gamma_2\,,\Gamma_3\,,\Gamma_4\}\,.$$

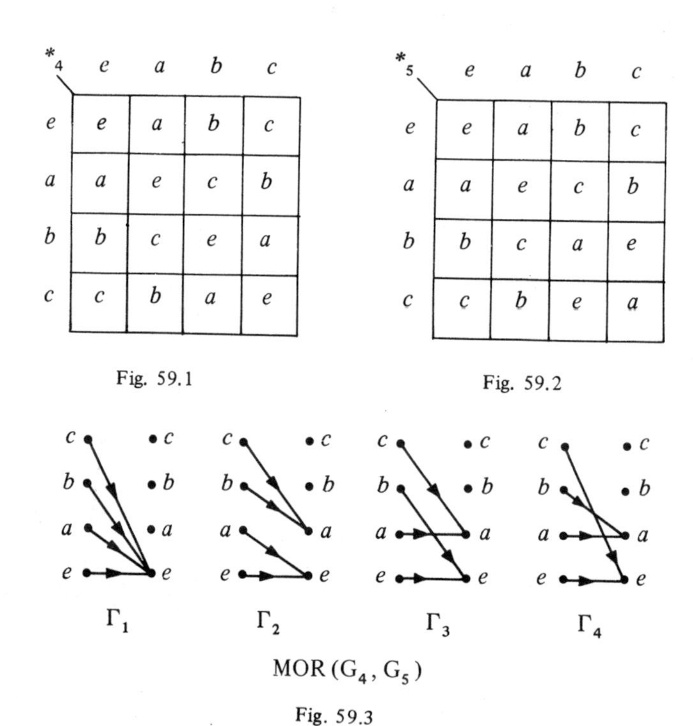

$*_4$	e	a	b	c
e	e	a	b	c
a	a	e	c	b
b	b	c	e	a
c	c	b	a	e

Fig. 59.1

$*_5$	e	a	b	c
e	e	a	b	c
a	a	e	c	b
b	b	c	a	e
c	c	b	e	a

Fig. 59.2

MOR (G_4, G_5)

Fig. 59.3

We consider first the fuzzy subsets of MOR(G_4, G_5) in the sense of Zadeh (that is, by taking $\mathbf{M} = [0, 1]$ as a membership set); we shall generalize this later. A fuzzy subset of MOR(G_4, G_5) will then be given by

(59.2) $$\underset{\sim}{\mathbf{\Gamma}} = \{(\Gamma_1 \mid \mu_\Phi(\Gamma_1)),\ (\Gamma_2 \mid \mu_\Phi(\Gamma_2)),\ (\Gamma_3 \mid \mu_\Phi(\Gamma_3)),\ (\Gamma_4 \mid \mu_\Phi(\Gamma_4))\}\,.$$

where

(59.3)
(59.4) $$\Phi = \text{MOR}\,(G_4\,,G_5)\,, \qquad \mu_\Phi(\Gamma_i) \in [0,1]\,, \qquad i = 1\,,2\,,3\,,4\,,$$

are respectively the reference set and the membership function with respect to which the fuzzy subset $\underset{\sim}{\mathbf{\Gamma}}$ is defined.

One will then say that $\underset{\sim}{\mathbf{\Gamma}}$ is a fuzzy morphism of G_4 into G_5.

It is advisable to submit $\underset{\sim}{\mathbf{\Gamma}}$ to the defined operations, for any fuzzy subset. One may then see how to generalize the notion of ordinary morphism to that of a fuzzy morphism.

Having presented this introductory example, we pass to a general definition.

Definition. Let **C** be a category and let X, Y $\in$ **C**; then an **L**-fuzzy **C**-morphism of X into Y will be a fuzzy **L**-subset of ordinary **C**-morphisms of X into Y. If we denote the set or reference set of morphisms of X into Y by Φ, i.e., $\Phi = \text{MOR}(X, Y)$, then an **L**-fuzzy **C**-morphism of X into Y will be an element of

$$\mathbf{L}^{\Phi} . \tag{59.5}$$

We shall denote this **L**-fuzzy **C**-morphism

$$\underset{\sim}{\mathbf{\Gamma}} \in \mathbf{L}^{\Phi} . \tag{59.6}$$

In the general case considered here, **L** is no longer only the interval [0, 1], but may be, as we have indicated in Section 55, an ordinary preorder, an ordinary order, an inf- or sup-semilattice, a lattice, a ring, or any structure satisfying (53.16)–(53.18).

The set of all fuzzy morphisms existing in a category **C** (a set that may or may not be finite, depending on the nature of **C**) defines what one calls the fuzzy category $\underset{\sim}{\mathbf{C}}$ associated with the category **C** with the reservation that, to each MOR(X, Y) one has associated and defined a structured set **L**.

We take up again the example presented at the beginning of the present section in order to examine what (59.5) represents.

Let

$$\Phi = \text{MOR}\,(G_4\,, G_5) \qquad \text{and} \qquad |\Phi| = 4 . \tag{59.7}$$

as represented in Figure 59.3, and

$$\mathbf{L} = \{0\,, \tfrac{1}{2}\,, 1\} \tag{59.8}$$

that is, here a total order with three numerical values $0, \frac{1}{2}$, and 1.

There will then exist $3^4 = 81$ fuzzy morphisms of G_4 into G_5; $\mathbf{L}^{\Phi}$ will represent the set of these 81 fuzzy morphisms. One of these 81 fuzzy morphisms will be, for example,

$$\underset{\sim}{\mathbf{\Gamma}} = \{(\Gamma_1 \mid 0)\,, (\Gamma_2 \mid \tfrac{1}{2})\,, (\Gamma_3 \mid 1)\,, (\Gamma_4 \mid \tfrac{1}{2})\} . \tag{59.9}$$

Composition of fuzzy morphisms.† Consider two fuzzy morphisms

$$\underset{\sim}{\mathbf{\Gamma}}_1 \subset \Phi \qquad \text{and} \qquad \underset{\sim}{\mathbf{\Gamma}}_2 \subset \Phi$$

where $\qquad \Phi = \text{MOR}\,(X\,, Y)\,, \qquad$ such that

$$\underset{\sim}{\mathbf{\Gamma}}_1 = \{\Gamma_i \in \Phi \mid l_{\Phi}(\Gamma_i) \in \mathbf{L}\}; \tag{59.10}$$

$$\underset{\sim}{\mathbf{\Gamma}}_2 = \{\Gamma_j \in \Phi \mid l_{\Phi}(\Gamma_j) \in \mathbf{L}\}; \tag{59.11}$$

†Where there will be no confusion on the nature of the category **C** and the structured set **L** associated with $\Phi = \text{MOR}(X, Y)$, we shall say simply morphism instead of **C**-fuzzy **L**-morphism.

Each Γ_i will be composed with each Γ_j according to the law of composition $\circ$ defined for the morphisms; this gives

$$\Gamma_{i,j} = \Gamma_j \circ \Gamma_i = \Gamma_k \in \Phi . \tag{59.12}$$

Then

$$\underset{\sim}{\Gamma}_{1,2} = \underset{\sim}{\Gamma}_2 \circ \underset{\sim}{\Gamma}_1 = \{\Gamma_k \mid l_\Phi(\Gamma_k) = \underset{i,j}{\nabla} (l_\Phi(\Gamma_i) \,\Delta\, l_\Phi(\Gamma_j))\} \tag{59.13}$$

will define the composition of $\underset{\sim}{\Gamma}_1$ with Γ_2.

The first example will render the significance of this complicated formula more evident.

Example 1. Reconsider Figures 59.1–59.3 and suppose that

$$\mathbf{L} = [0,1] . \tag{59.14}$$

Consider the reference set

$$\Phi = \text{MOR}\,(G_4 , G_5) = \{\Gamma_1 , \Gamma_2 , \Gamma_3 , \Gamma_4\} : \tag{59.15}$$

and let

$$\underset{\sim}{\Gamma}_1 = \{(\Gamma_1 \mid 0{,}5),\ (\Gamma_2 \mid 0{,}9),\ (\Gamma_3 \mid 0),\ (\Gamma_4 \mid 0{,}3)\} \tag{59.16}$$

and

$$\underset{\sim}{\Gamma}_2 = \{(\Gamma_1 \mid 0{,}8),\ (\Gamma_2 \mid 0{,}1),\ (\Gamma_3 \mid 0{,}5),\ (\Gamma_4 \mid 1)\} . \tag{59.17}$$

be two fuzzy morphisms.

The law of composition $\Gamma_i \circ \Gamma_j$ being the ordinary law of composition of two relations, as has been defined by (57.37), one may first determine the composition of each Γ_i with each Γ_j.

By examining Figure 59.3 one may easily see that we have

$$\begin{array}{ll}
\Gamma_{1,1} = \Gamma_1 \circ \Gamma_1 = \Gamma_1 , & \Gamma_{3,1} = \Gamma_1 \circ \Gamma_3 = \Gamma_1 , \\
\Gamma_{1,2} = \Gamma_2 \circ \Gamma_1 = \Gamma_1 , & \Gamma_{3,2} = \Gamma_2 \circ \Gamma_3 = \Gamma_1 , \\
\Gamma_{1,3} = \Gamma_3 \circ \Gamma_1 = \Gamma_1 , & \Gamma_{3,3} = \Gamma_3 \circ \Gamma_3 = \Gamma_2 , \\
\Gamma_{1,4} = \Gamma_4 \circ \Gamma_1 = \Gamma_1 , & \Gamma_{3,4} = \Gamma_4 \circ \Gamma_3 = \Gamma_3 , \\
\Gamma_{2,1} = \Gamma_1 \circ \Gamma_2 = \Gamma_1 , & \Gamma_{4,1} = \Gamma_1 \circ \Gamma_4 = \Gamma_1 , \\
\Gamma_{2,2} = \Gamma_2 \circ \Gamma_2 = \Gamma_1 , & \Gamma_{4,2} = \Gamma_2 \circ \Gamma_4 = \Gamma_1 , \\
\Gamma_{2,3} = \Gamma_3 \circ \Gamma_2 = \Gamma_2 , & \Gamma_{4,3} = \Gamma_3 \circ \Gamma_4 = \Gamma_4 , \\
\Gamma_{2,4} = \Gamma_4 \circ \Gamma_2 = \Gamma_2 , & \Gamma_{4,4} = \Gamma_4 \circ \Gamma_4 = \Gamma_4 .
\end{array} \tag{59.18}$$

In order to facilitate comprehension of the manner in which these compositions have been calculated, we give an example in Figure 59.4 concerning $\Gamma_{4,3} = \Gamma_3 \circ \Gamma_4 = \Gamma_4$.

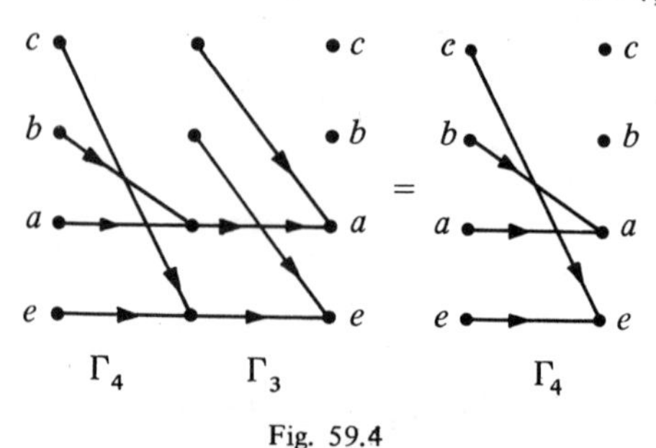

Fig. 59.4

Now we evaluate $\mu_\Phi(\Gamma_i) \wedge \mu_\Phi(\Gamma_j)$, for each ordered pair (Γ_i, Γ_j):

(59.19)
$$
\begin{array}{ll}
\Gamma_{1,1} = \Gamma_1 , & 0,5 \wedge 0,8 = 0,5 , \\
\Gamma_{1,2} = \Gamma_1 , & 0,5 \wedge 0,1 = 0,1 , \\
\Gamma_{1,3} = \Gamma_1 , & 0,5 \wedge 0,5 = 0,5 , \\
\Gamma_{1,4} = \Gamma_1 , & 0,5 \wedge 1 \;\; = 0,5 , \\
\Gamma_{2,1} = \Gamma_1 , & 0,9 \wedge 0,8 = 0,8 , \\
\Gamma_{2,2} = \Gamma_1 , & 0,9 \wedge 0,1 = 0,1 , \\
\Gamma_{2,3} = \Gamma_2 , & 0,9 \wedge 0,5 = 0,5 , \\
\Gamma_{2,4} = \Gamma_2 , & 0,9 \wedge 1 \;\; = 0,9 , \\
\Gamma_{3,1} = \Gamma_1 , & 0 \;\; \wedge 0,8 = 0 , \\
\Gamma_{3,2} = \Gamma_1 , & 0 \;\; \wedge 0,1 = 0 , \\
\Gamma_{3,3} = \Gamma_2 , & 0 \;\; \wedge 0,5 = 0 , \\
\Gamma_{3,4} = \Gamma_3 , & 0 \;\; \wedge 1 \;\; = 0 , \\
\Gamma_{4,1} = \Gamma_1 , & 0,3 \wedge 0,8 = 0,3 , \\
\Gamma_{4,2} = \Gamma_1 , & 0,3 \wedge 0,1 = 0,1 , \\
\Gamma_{4,3} = \Gamma_4 , & 0,3 \wedge 0,5 = 0,3 , \\
\Gamma_{4,4} = \Gamma_4 , & 0,3 \wedge 1 \;\; = 0,3 .
\end{array}
$$

Then for each Γ_k, $k = 1, 2, 3, 4$, we evaluate

$$\bigvee_{i,j} (\mu_\Phi(\Gamma_i) \wedge \mu_\Phi(\Gamma_j)) .$$

One has successively:

(59.20)
$$
\begin{array}{ll}
\text{For } \Gamma_1 : & \vee(0,5 \,;\, 0,1 \,;\, 0,5 \,;\, 0,5 \,;\, 0,8 \,;\, 0,1 \,;\, 0 \,;\, 0 \,;\, 0,3 \,;\, 0,1) = 0,8 , \\
\text{For } \Gamma_2 : & \vee(0,5 \,;\, 0,9 \,;\, 0) = 0,9 , \\
\text{For } \Gamma_3 : & \vee(0) = 0 , \\
\text{For } \Gamma_4 : & \vee(0,3 \,;\, 0,3) = 0,3 .
\end{array}
$$

And finally

$$(59.21) \quad \underset{\sim}{\mathbf{\Gamma}}_{1,2} = \underset{\sim}{\mathbf{\Gamma}}_2 \circ \underset{\sim}{\mathbf{\Gamma}}_1 = \{(\Gamma_1 \mid 0,8), (\Gamma_2 \mid 0,9), (\Gamma_3 \mid 0), (\Gamma_4 \mid 0,3)\} .$$

Remark. Of course, in order that $\underset{\sim}{\mathbf{\Gamma}}_1$ and $\underset{\sim}{\mathbf{\Gamma}}_2$ may be composed it is necessary that all the Γ_i may be composed with all the Γ_j and that one obtain at least once each $\Gamma_k \in \Phi$ in composing the Γ_i and the Γ_j.

Example 2. We show that formula (59.13) reduces to ordinary composition of two ordinary subsets of morphisms if $\mathbf{L} = \{0, 1\}$.

Again we use the example of Figures 59.1–59.3. Let

$$\boldsymbol{\Gamma}_1 = \{(\Gamma_1 \mid 0), (\Gamma_2 \mid 0), (\Gamma_3 \mid 1), (\Gamma_4 \mid 0)\} = \{\Gamma_3\} \tag{59.22}$$

and

$$\boldsymbol{\Gamma}_2 = \{(\Gamma_1 \mid 0), (\Gamma_2 \mid 0), (\Gamma_3 \mid 1), (\Gamma_4 \mid 1)\} = \{\Gamma_3, \Gamma_4\} \tag{59.23}$$

Application of the rules stated above gives

$$\boldsymbol{\Gamma}_{1,2} = \boldsymbol{\Gamma}_2 \circ \boldsymbol{\Gamma}_1 = \{(\Gamma_1 \mid 0), (\Gamma_2 \mid 1), (\Gamma_3 \mid 1), (\Gamma_4 \mid 0)\} = \{\Gamma_2, \Gamma_3\}. \tag{59.24}$$

We remark that if $\boldsymbol{\Gamma}_1$ and $\boldsymbol{\Gamma}_2$ are singletons (subsets containing only a single element), then the composition of $\boldsymbol{\Gamma}_1$ and $\boldsymbol{\Gamma}_2$ reduces to the ordinary composition of respective unique elements with their membership. Thus

$$\boldsymbol{\Gamma}_1 = \{(\Gamma_1 \mid 0), (\Gamma_2 \mid 0), (\Gamma_3 \mid 1), (\Gamma_4 \mid 0)\} \tag{59.25}$$

and

$$\boldsymbol{\Gamma}_2 = \{(\Gamma_1 \mid 0), (\Gamma_2 \mid 0), (\Gamma_3 \mid 0), (\Gamma_4 \mid 1)\} \tag{59.26}$$

gives

$$\boldsymbol{\Gamma}_{1,2} = \boldsymbol{\Gamma}_2 \circ \boldsymbol{\Gamma}_1 = \{(\Gamma_1 \mid 0), (\Gamma_2 \mid 0), (\Gamma_3 \mid 1), (\Gamma_4 \mid 0)\}, \tag{59.27}$$

which reduces to the composition of $\Gamma_{3,4} = \Gamma_4 \circ \Gamma_3 = \Gamma_3$.

Example 3. We take up Example 1 again where we have considered the two groups G_4 and G_5, but this time instead of taking the interval [0, 1] for **L**, we will take the finite distributive lattice indicated in Figure 59.5. In Figure 59.6 we present the tables for Δ and for ∇.

Consider two fuzzy morphisms of the reference set

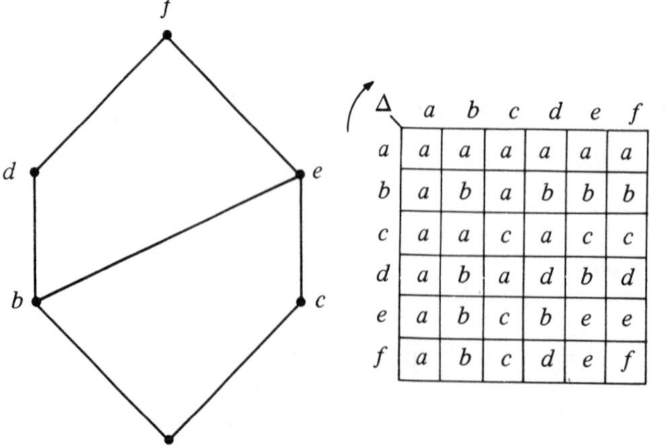

Fig. 59.5

Δ	a	b	c	d	e	f
a	a	a	a	a	a	a
b	a	b	a	b	b	b
c	a	a	c	a	c	c
d	a	b	a	d	b	d
e	a	b	c	b	e	e
f	a	b	c	d	e	f

∇	a	b	c	d	e	f
a	a	b	c	d	e	f
b	b	b	e	d	e	f
c	c	e	c	f	e	f
d	d	d	f	d	f	f
e	e	e	e	f	e	f
f	f	f	f	f	f	f

Fig. 59.6

(59.28) $$\Phi = \text{MOR}\,(G_4\,, G_5) = \{\Gamma_1\,, \Gamma_2\,, \Gamma_3\,, \Gamma_4\}\,,$$

(59.29) $$\underset{\sim}{\Gamma}_1 = \{(\Gamma_1 \mid b), \quad (\Gamma_2 \mid a), \quad (\Gamma_3 \mid f), \quad (\Gamma_4 \mid e)\}\,,$$

(59.30) $$\underset{\sim}{\Gamma}_2 = \{(\Gamma_1 \mid c), \quad (\Gamma_2 \mid d), \quad (\Gamma_3 \mid a), \quad (\Gamma_4 \mid d)\}\,,$$

Formulas (59.18) do not change, but those of (59.19) will evidently be different; we shall use the same notation however. For each ordered pair (Γ_i, Γ_j), we evaluate $(\Gamma_i\,, \Gamma_j) : l_\Phi(\Gamma_i)\ \Delta\ l_\Phi(\Gamma_j)$,

(59.31)
$$\begin{aligned}
\Gamma_{1,1} &= \Gamma_1 \circ \Gamma_1 = \Gamma_1\,, && b\ \Delta\ c = a\,,\\
\Gamma_{1,2} &= \Gamma_2 \circ \Gamma_1 = \Gamma_1\,, && b\ \Delta\ d = b\,,\\
\Gamma_{1,3} &= \Gamma_3 \circ \Gamma_1 = \Gamma_1\,, && b\ \Delta\ a = a\,,\\
\Gamma_{1,4} &= \Gamma_4 \circ \Gamma_1 = \Gamma_1\,, && b\ \Delta\ d = b\,,\\
\Gamma_{2,1} &= \Gamma_1 \circ \Gamma_2 = \Gamma_1\,, && a\ \Delta\ c = a\,,\\
\Gamma_{2,2} &= \Gamma_2 \circ \Gamma_2 = \Gamma_1\,, && a\ \Delta\ d = a\,,\\
\Gamma_{2,3} &= \Gamma_3 \circ \Gamma_2 = \Gamma_2\,, && a\ \Delta\ a = a\,,\\
\Gamma_{2,4} &= \Gamma_4 \circ \Gamma_2 = \Gamma_2\,, && a\ \Delta\ d = a\,,\\
\Gamma_{3,1} &= \Gamma_1 \circ \Gamma_3 = \Gamma_1\,, && f\ \Delta\ c = c\,,\\
\Gamma_{3,2} &= \Gamma_2 \circ \Gamma_3 = \Gamma_1\,, && f\ \Delta\ d = d\,,\\
\Gamma_{3,3} &= \Gamma_3 \circ \Gamma_3 = \Gamma_2\,, && f\ \Delta\ a = a\,,\\
\Gamma_{3,4} &= \Gamma_4 \circ \Gamma_3 = \Gamma_3\,, && f\ \Delta\ d = d\,,\\
\Gamma_{4,1} &= \Gamma_1 \circ \Gamma_4 = \Gamma_1\,, && e\ \Delta\ c = c\,,\\
\Gamma_{4,2} &= \Gamma_2 \circ \Gamma_4 = \Gamma_1\,, && e\ \Delta\ d = b\,,\\
\Gamma_{4,3} &= \Gamma_3 \circ \Gamma_4 = \Gamma_4\,, && e\ \Delta\ a = a\,,\\
\Gamma_{4,4} &= \Gamma_4 \circ \Gamma_4 = \Gamma_4\,, && e\ \Delta\ d = b\,,
\end{aligned}$$

Then, for each Γ_k, $k = 1, 2, 3, 4$, we will evaluate

$$\underset{i,j}{\nabla}\ (l_\Phi(\Gamma_i)\ \Delta\ l_\Phi(\Gamma_j))\,.$$

One proceeds successively. For

$$\Gamma_1 : (a \nabla b \nabla a \nabla b \nabla a \nabla a \nabla c \nabla d \nabla c \nabla b) = (a \nabla b \nabla c \nabla d)$$
$$= d \nabla c = f\,,$$

(59.32) $$\Gamma_2 : (a \nabla a \nabla a) = a$$

$$\Gamma_3 : d$$

$$\Gamma_4 : a \nabla b = b'$$

And finally :

(59.33) $$\underset{\sim}{\Gamma}_{1,2} = \underset{\sim}{\Gamma}_2 \circ \underset{\sim}{\Gamma}_1 = \{(\Gamma_1 \mid f), (\Gamma_2 \mid a), (\Gamma_3 \mid d), (\Gamma_4 \mid b\}$$

Remark. We may make the same remark as after (59.21)

Example 4. We consider Example 3 of Section 58 again. We considered the category **C** of all the intervals $[X, X'] \subset [a, b]$, $a \in \mathbf{R}$, $b \in \mathbf{R}$, where the morphisms were functions that correspond $[X, X'] \subset [a, b]$ to $[Y, Y'] \subset [a, b]$.

We then consider an ordered pair $(\mathfrak{X}, \mathcal{Y})$ where $\mathfrak{X} = (U, *)$, $U = [X, X'] \subset [a, b]$, and likewise $\mathcal{Y} = (V, *)$, $V = [Y, Y'] \subset [a, b]$. If f is a function of $\mathfrak{X}$ into $\mathcal{Y}$, one will write

(59.34) $$\mathfrak{X} \overset{}{\underset{f}{\leadsto}} \mathcal{Y}.$$

Then let Φ be the set of all these functions f (this is a set that is not finite; this does not matter). We then take a set **L** that may be a lattice, a ring, etc., as we have indicated in Section 54.

Then, the set $\mathbf{L}^\Phi$ will constitute the set of **L**-fuzzy **C**-morphisms:

(59.35) $$\underset{\sim}{F} = \{f_i \in \Phi \mid l_\Phi(f_i) \in \mathbf{L}\}.$$

And one may compose $\underset{\sim}{F}_1$ with $\underset{\sim}{F}_2$,

(59.36) $$\underset{\sim}{F}_1 = \{f_i \in \Phi \mid l_\Phi(f_i) \in \mathbf{L}\},$$

(59.37) $$\underset{\sim}{F}_2 = \{f_j \in \Phi \mid l_\Phi(f_j) \in \mathbf{L}\},$$

by using

(59.38) $$f_{i,j} = f_j \circ f_i = f_k \in \Phi.$$

And also

(59.39) $$\underset{\sim}{F}_{1,2} = \underset{\sim}{F}_2 \circ \underset{\sim}{F}_1 = \{f_k \mid l_\Phi(f_k) = \underset{i,j}{\nabla} (l_\Phi(f_i) \,\Delta\, l_\Phi(f_j))\}$$

will define the composition of $\underset{\sim}{F}_1$ with $\underset{\sim}{F}_2$.

Generalization of the composition of* L*-fuzzy* C*-morphisms. Formula (59.13) may be generalized. The law of composition of ordinary morphisms will be denoted $\circ$; thus $\Gamma_{i,j} = \Gamma_j \circ \Gamma_i$, $\Gamma_i, \Gamma_j \in \Phi = \mathrm{MOR}(X, Y)$; let $\square$ be the law of **L**, supposed to be associative:

(59.40) $$l_\Phi(\Gamma_{i,j}) = l_\Phi(\Gamma_j) \,\square\, l_\Phi(\Gamma_i)$$

We then replace (59.13) by

(59.41) $$\underset{\sim}{\Gamma}_{1,2} = \underset{\sim}{\Gamma}_2 \circ \underset{\sim}{\Gamma}_1 = \underset{i,j}{\cup} \{\Gamma_k = \Gamma_j \circ \Gamma_i \mid l_\Phi(\Gamma_k) = l_\Phi(\Gamma_j) \,\square\, l_\Phi(\Gamma_i)\},$$

the symbol $\cup_{ij}$ (union with respect to all the i and j giving k) has been introduced because different i and j may give the same k.

In this conception, the law of composition $\circ$ of ordinary morphisms may be different from the law of composition specified in (57.36), but since the morphisms considered here are **C**-morphisms, the law $\circ$ must be associative and possess an identity as indicated in (58.2) and (58.3).

Properties of **L**-*fuzzy* **C**-*morphisms.* We examine several properties of **L**-fuzzy **C**-morphisms.

Associativity. Let

(59.42)
$$\underset{\sim}{\mathbf{\Gamma}}_1, \underset{\sim}{\mathbf{\Gamma}}_2, \underset{\sim}{\mathbf{\Gamma}}_3 \in \mathbf{L}^\Phi,$$

then

(59.43)
$$\underset{\sim}{\mathbf{\Gamma}}_3 \circ (\underset{\sim}{\mathbf{\Gamma}}_2 \circ \underset{\sim}{\mathbf{\Gamma}}_1) = (\underset{\sim}{\mathbf{\Gamma}}_3 \circ \underset{\sim}{\mathbf{\Gamma}}_2) \circ \underset{\sim}{\mathbf{\Gamma}}_1 .$$

The proof is rather simple:

(59.44)
$$\underset{\sim}{\mathbf{\Gamma}}_2 \circ \underset{\sim}{\mathbf{\Gamma}}_1 = \bigcup_{i,j} \{\Gamma_k = \Gamma_j \circ \Gamma_i \mid l_\Phi(\Gamma_k) = l_\Phi(\Gamma_j) \,\square\, l_\Phi(\Gamma_i)\},$$

$$\underset{\sim}{\mathbf{\Gamma}}_3 \circ (\underset{\sim}{\mathbf{\Gamma}}_2 \circ \underset{\sim}{\mathbf{\Gamma}}_1) = \bigcup_{k,l} \{\Gamma_p = \Gamma_l \circ \Gamma_k \mid l_\Phi(\Gamma_p) = l_\Phi(\Gamma_l) \,\square\, l_\Phi(\Gamma_k)\}$$

$$= \bigcup_{k,l} [\bigcup_{i,j} \{\Gamma_p = \Gamma_l \circ \Gamma_k\,,\, \Gamma_k = \Gamma_j \circ \Gamma_i \mid l_\Phi(\Gamma_p) = l_\Phi(\Gamma_l) \,\square\, (l_\Phi(\Gamma_j) \,\square\, l_\Phi(\Gamma_i)\}] .$$

But $\Gamma_1 \circ (\Gamma_j \circ \Gamma_i) = (\Gamma_1 \circ \Gamma_j) \circ \Gamma_i$ since, by hypothesis, the law $\circ$ is associative (we are dealing with **C**-morphisms); and, on the other hand, **L** also having, by hypothesis, an associative law $\square$, one may write

(59.45)
$$\underset{\sim}{\mathbf{\Gamma}}_3 \circ (\underset{\sim}{\mathbf{\Gamma}}_2 \circ \underset{\sim}{\mathbf{\Gamma}}_1) = \bigcup_{i,j} [\bigcup_{kl} \{\Gamma_p = (\Gamma_l \circ \Gamma_j) \circ \Gamma_i \mid l_\Phi(\Gamma_p) = (l_\Phi(\Gamma_l) \,\square\, l_\Phi(\Gamma_j)) \,\square\, l_\Phi(\Gamma_i)\}]$$

$$= (\underset{\sim}{\mathbf{\Gamma}}_3 \circ \underset{\sim}{\mathbf{\Gamma}}_2) \circ \underset{\sim}{\mathbf{\Gamma}}_1 .$$

Existence of identities. There exists a right identity $\underset{\sim}{\Gamma}{}^\circ_{\mathrm{X}} \in \mathbf{L}^\Phi$ where $\Phi = \mathrm{MOR}(\mathrm{X}, \mathrm{Y})$ such that

(59.46)
$$\forall \underset{\sim}{\mathbf{\Gamma}} \in \Phi : \underset{\sim}{\mathbf{\Gamma}} \circ \underset{\sim}{\mathbf{\Gamma}}{}^\circ_{\mathrm{X}} = \underset{\sim}{\mathbf{\Gamma}}$$

and there exists a left identity $\underset{\sim}{\mathbf{\Gamma}}{}^\circ_{\mathrm{Y}} \in \mathbf{L}^\Phi$, such that

(59.47)
$$\forall \mathbf{\Gamma} \in \Phi : \underset{\sim}{\mathbf{\Gamma}}{}^\circ_{\mathrm{Y}} \circ \underset{\sim}{\mathbf{\Gamma}} = \underset{\sim}{\mathbf{\Gamma}} .$$

The identities are such that

$$\underset{\sim}{\Gamma}^{\circ}_{X} = \{\Gamma_i = \Gamma^{\circ}_{X} \mid l_{\Phi}(\Gamma_i) = 1\} , \tag{59.48}$$

$$\underset{\sim}{\Gamma}^{\circ}_{Y} = \{\Gamma_i = \Gamma^{\circ}_{Y} \mid l_{\Phi}(\Gamma_i) = 1\} . \tag{59.49}$$

One may write, in fact, to prove (59.46),

$$\begin{aligned} \underset{\sim}{\Gamma} \circ \underset{\sim}{\Gamma}^{\circ}_{X} &= \bigcup_j \{\Gamma_k = \Gamma_j \circ \Gamma^{\circ}_{X} \mid l_{\Phi}(\Gamma_k) = l_{\Phi}(\Gamma_j) \,\square\, l_{\Phi}(\Gamma^{\circ}_{X})\} \\ &= \bigcup_j \{\Gamma_k = \Gamma_j \mid l_{\Phi}(\Gamma_k) = l_{\Phi}(\Gamma_j)\} \\ &= \{\Gamma_j \mid l_{\Phi}(\Gamma_j)\} \\ &= \underset{\sim}{\Gamma} . \end{aligned} \tag{59.50}$$

Similarly one may show (59.47).

Fuzzy category. Since the $\underset{\sim}{\Gamma} \in \Phi$ are such that their composition is associative with a left identity and a right identity, as has been specified in (58.2) and (58.3), then the **L**-fuzzy **C**-morphisms permit one to define a fuzzy category of objects such as X.

Thus, any ordinary category is only a particular case of a fuzzy category. This fuzzy category will be denoted

$$\mathbf{L}^{\underset{\sim}{\mathbf{C}}} . \tag{59.51}$$

Ordinary subcategory. Recall the definition of the concept of a category at the beginning of Section 58.

Let **X** be a set of objects and Φ the set of morphisms. If $\mathbf{X}' \subset \mathbf{X}$ and $\Phi' \subset \Phi$ is that of the corresponding morphisms, then the category $\mathbf{C}'$ relative to $\mathbf{X}'$ will be called an ordinary subcategory of **C**; one may write

$$\mathbf{C}' \subset \mathbf{C} . \tag{59.52}$$

Fuzzy subcategory. If **X** is a set of objects and $\mathbf{L}^{\Phi}$ a set of fuzzy morphisms, and if $\mathbf{X}' \subset \mathbf{X}$ and $\Phi' \subset \Phi$, then the fuzzy category $\mathbf{L}^{\underset{\sim}{\mathbf{C}}'}$ associated with $\mathbf{X}'$ will be called a fuzzy subcategory.

60. EXERCISES

V1. Let

$$\mathrm{E}_1 = \{0 , 1 , 2 , 3\} \qquad \text{and} \qquad \mathrm{E}_2 = \{a , b , c\} ;$$

give

$$\mathrm{E}_1 \times \mathrm{E}_2 \qquad \text{and} \qquad \mathrm{E}_2 \times \mathrm{E}_1 .$$

V2. Let

$$\mathbf{E}_1 = \{0, 1, 2\}, \qquad \mathbf{E}_2 = \{4, 5\}, \qquad \mathbf{E}_3 = \{9, 10, 11\};$$

give

a) $\mathbf{E}_1 \oplus \mathbf{E}_2$ b) $\mathbf{E}_1 \times (\mathbf{E}_2 \oplus \mathbf{E}_3)$ c) $\mathbf{E}_1 \oplus (\mathbf{E}_2 \times \mathbf{E}_3)$.

V3. Let

$$\mathbf{E}_1 = \{0, \tfrac{1}{2}, 1\} \qquad \text{and} \qquad \mathbf{E}_2 = \{a, b\},$$

give

a) $\mathbf{E}_2^{\mathbf{E}1}$, b) $\mathbf{E}_1^{\mathbf{E}2}$.

V4. Let

$$\mathbf{E} = [a, b] \subset \mathbf{R},$$

$$\mathbf{L} = [0, 1],$$

describe

a) $\mathbf{L}^{\mathbf{E}}$ and b) $\mathbf{E}^{\mathbf{L}}$.

V5. Let $\mathbf{L} = \{0, 1, 2\}$ and let $*$ be the following law for $\mathbf{L}$ (sum modulo 3)

*	0	1	2
0	0	1	2
1	1	2	0
2	2	0	1

Give the law $\circledast$ of $\mathbf{L}^{\mathbf{E}}$, where

a) $\mathbf{E} = \{A, B\}$, b) $\mathbf{E} = \{A\}$, c) $\mathbf{E} = \{A, B, C\}$.

For each of the sets given in (a), (b), and (c), is the law $\circledast$ associative? commutative? idempotent?

V6. Among the eight lattices below, determine which are: (1) modular, (2) distributive, (3) complemented, (4) boolean.

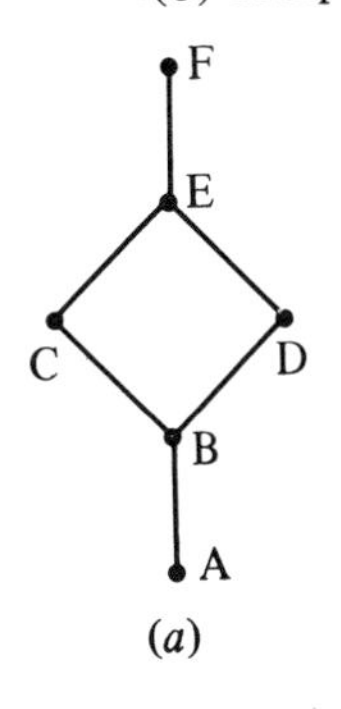

(*a*)

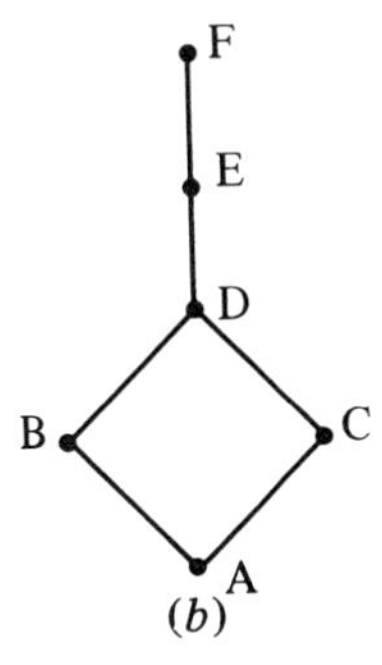

(*b*)

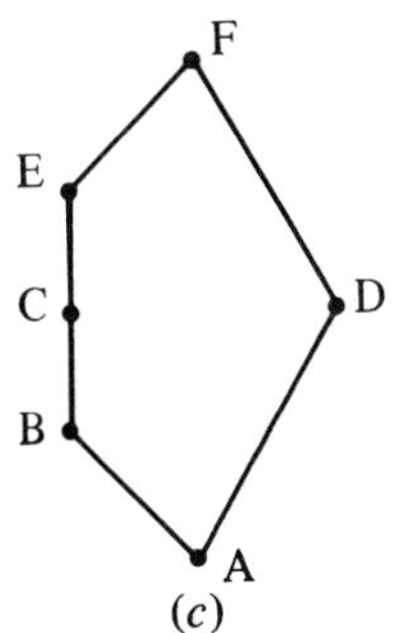

(*c*)

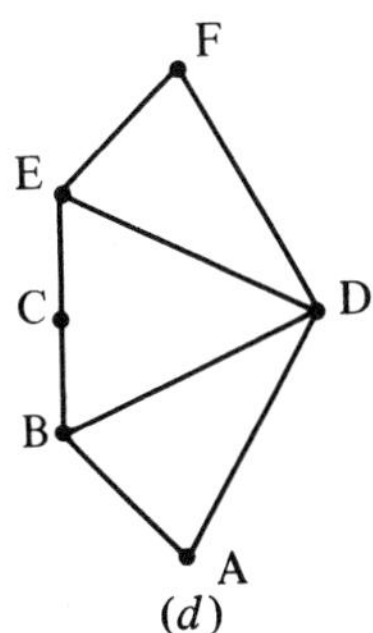

(*d*)

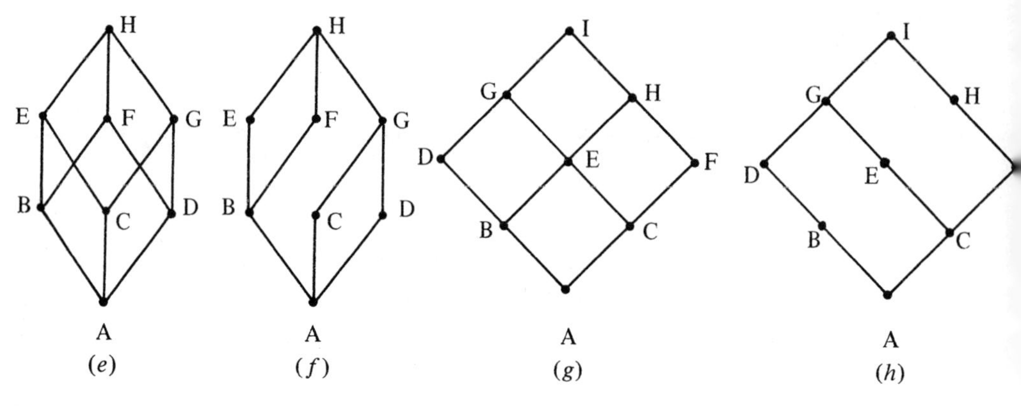

V7. Consider the lattices presented below:

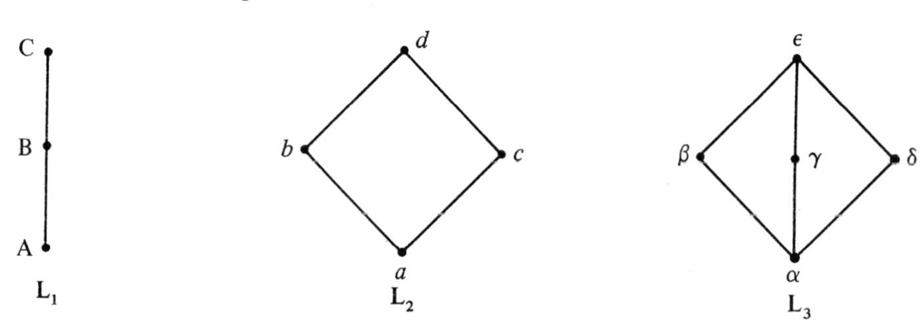

Give the following lattices by their Hasse diagrams:

a) $L_1 \times L_1$, b) $L_1 \times L_2$, c) $L_1 \times L_3$, d) $L_1 \times L_1 \times L_1$, e) $L_2 \times L_3$.

Among the five lattices obtained, which are: (a) distributive, (b) complemented, (c) boolean, (c) vectorial, (e) lexicographic?

V8. Consider the inf-semilattice shown here. Give the Hasse diagram of

$$L^E \quad \text{where} \quad E = \{A, B, C\}.$$

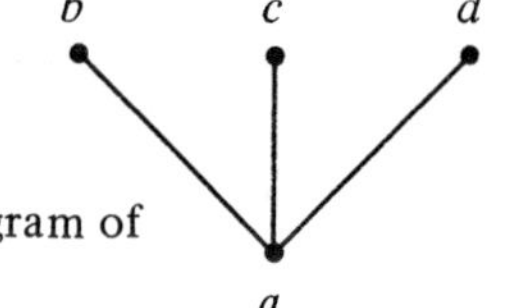

V9. Consider the three order relations expressed below by their respective Hasse diagrams:

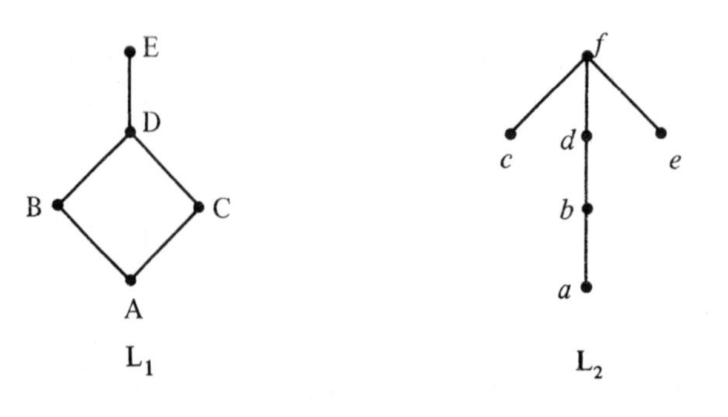

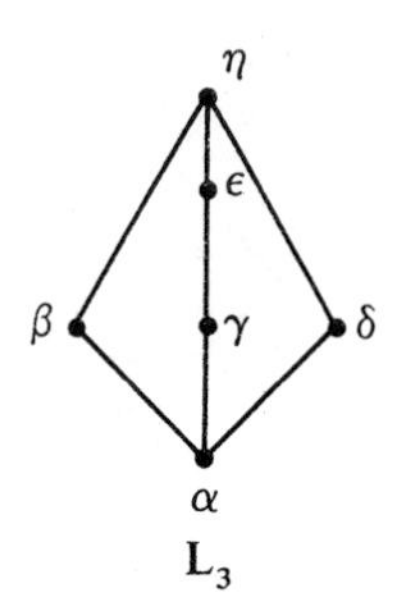

and a reference set

$$\mathbf{E} = \{x_1, x_2, x_3, x_4, x_5\}$$

where x_1, x_2, and x_3 are evaluated in $\mathbf{L}_1$;
x_3 is evaluated in $\mathbf{L}_2$;
x_4 is evaluated in $\mathbf{L}_3$.

Then consider the four fuzzy subsets of $\mathbf{E}$:

	x_1	x_2	x_3	x_4	x_5
$\underset{\sim}{X}_1 =$	A	D	c	α	B
$\underset{\sim}{X}_2 =$	A	E	b	ϵ	B
$\underset{\sim}{X}_3 =$	D	B	a	δ	A
$\underset{\sim}{X}_4 =$	E	A	e	ϵ	C

Give the matrix of the relative generalized Hamming distances corresponding to the ordered pairs $(\underset{\sim}{X}_i, \underset{\sim}{X}_j)$.

V10. Consider four sets $\mathbf{E}_i$ endowed respectively with laws $*_i$, $i = 1, 2, 3, 4$:

$*_1$	a
a	a

$*_2$	a	b
a	a	a
b	a	b

$*_3$	a	b	c
a	a	a	a
b	a	b	c
c	a	c	b

$*_4$	a	b	c	d
a	a	a	a	a
b	a	b	c	d
c	a	c	a	c
d	a	d	c	b

Let

$$H_1 = (\mathbf{E}_1, *_1), \quad H_2 = (\mathbf{E}_2, *_2), \quad H_3 = (\mathbf{E}_3, *_3), \quad H_4 = (\mathbf{E}_4, *_4),$$

give the morphisms

$$\text{MOR}(H_i, H_j), \qquad i,j = 1, 2, 3, 4.$$

Among these morphisms, list the epimorphisms, the monomorphisms, the isomorphisms, the endomorphisms, and the automorphisms.

V11. Consider the lattices shown below. Give the morphisms

$$\text{MOR}(\mathbf{L}_i, \mathbf{L}_j), \qquad i,j = 1, 2, 3.$$

and list the epimorphisms, the monomorphisms, etc.

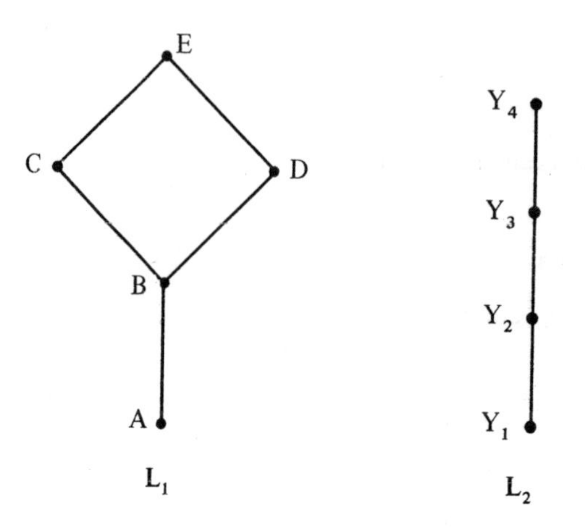

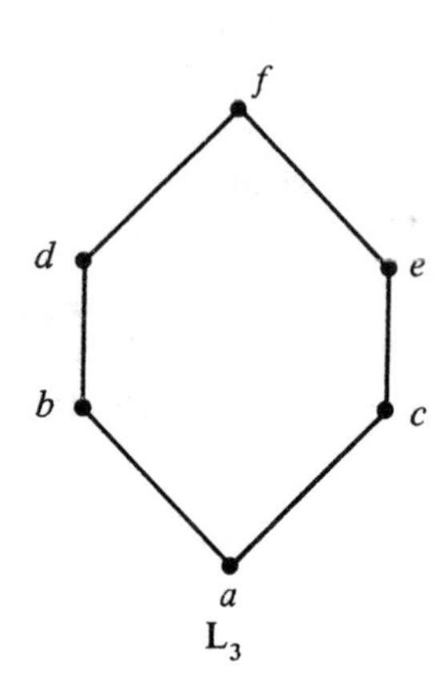

V12. Consider the reference set of morphisms

$$\Phi_{23} = \mathrm{MOR}(H_2, H_3)$$

wherc ($\mathrm{II}_2, *_2$) and ($H_3, *_3$) are as in Exercise V10. Let

$$\mathbf{L} = \{0, \tfrac{1}{2}, 1\}.$$

Give the **L**-fuzzy **C**-morphisms $\mathbf{L}^{\Phi_{2,3}}$.

V13. Consider the example given in Figures 59.1–59.3. For **L**, take the lattice shown here. Then consider two fuzzy morphisms of the reference set:

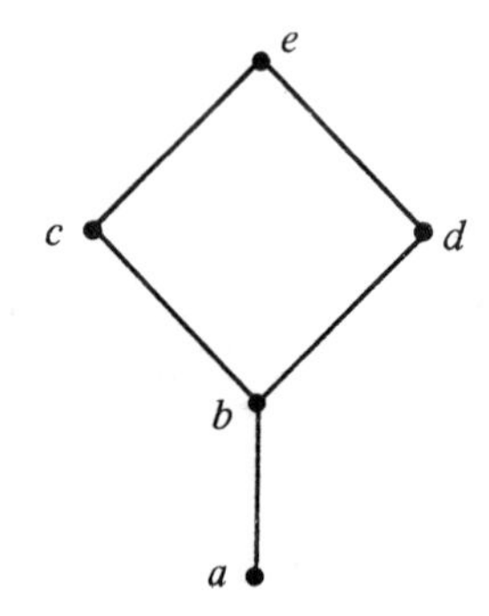

$$\Phi = \mathrm{MOR}\,(G_4, G_5) = \{\Gamma_1, \Gamma_2, \Gamma_3, \Gamma_4\},$$

$$\underset{\sim}{\mathbf{\Gamma}}_1 = \{(\Gamma_1 \mid a), (\Gamma_2 \mid c), (\Gamma_3 \mid e), (\Gamma_4 \mid a)\},$$

$$\underset{\sim}{\mathbf{\Gamma}}_2 = \{(\Gamma_1 \mid b), (\Gamma_2 \mid a), (\Gamma_3 \mid a), (\Gamma_4 \mid d)\}.$$

Give

a) $\underset{\sim}{\mathbf{\Gamma}}_1 \circ \underset{\sim}{\mathbf{\Gamma}}_2$ b) $\underset{\sim}{\mathbf{\Gamma}}_2 \circ \underset{\sim}{\mathbf{\Gamma}}_1$ and c) $\underset{\sim}{\mathbf{\Gamma}}_1^3 = \underset{\sim}{\mathbf{\Gamma}}_1 \circ \underset{\sim}{\mathbf{\Gamma}}_1 \circ \underset{\sim}{\mathbf{\Gamma}}_1$.

APPENDIX A

GENERAL PROOF PROCEDURE FOR OPERATIONS INVOLVING MAXIMUMS AND MINIMUMS

At various points in this book we have passed very rapidly over certain proofs concerning the operation "maximum of" or "minimum of." This is because such proofs are immediate consequences of properties defined in a lattice for superior or/and inferior limits.

Let Δ be the inferior limit and ∇ the superior limit of two elements (X_i, X_j) in a lattice. In a lattice one has the four dual properties [already reviewed in (54.2)–(54.9)]:

$$\left.\begin{aligned} &\text{(A.1)} & X_i \,\Delta\, X_j &= X_j \,\Delta\, X_i \,, \\ &\text{(A.2)} & X_i \,\nabla\, X_j &= X_j \,\nabla\, X_i \,. \end{aligned}\right\} \quad \text{commutativity}$$

$$\left.\begin{aligned} &\text{(A.3)} & X_i \,\Delta\, (X_j \,\Delta\, X_k) &= (X_i \,\Delta\, X_j) \,\Delta\, X_k \,, \\ &\text{(A.4)} & X_i \,\nabla\, (X_j \,\nabla\, X_k) &= (X_i \,\nabla\, X_j) \,\nabla\, X_k \,. \end{aligned}\right\} \quad \text{associativity}$$

$$\left.\begin{aligned} &\text{(A.5)} & X_i \,\Delta\, X_i &= X_i \,, \\ &\text{(A.6)} & X_i \,\nabla\, X_i &= X_i \,. \end{aligned}\right\} \quad \text{idempotence}$$

$$\left.\begin{aligned} &\text{(A.7)} & X_i \,\Delta\, (X_i \,\nabla\, X_j) &= X_i \,, \\ &\text{(A.8)} & X_i \,\nabla\, (X_i \,\Delta\, X_j) &= X_i \,. \end{aligned}\right\} \quad \text{absorption}$$

If, in addition, the lattice is distributive [see (54.18) and (54.19)]:

$$\left.\begin{aligned} &\text{(A.9)} & X_i \,\nabla\, (X_j \,\Delta\, X_k) &= (X_i \,\nabla\, X_j) \,\Delta\, (X_i \,\nabla\, X_k) \,, \\ &\text{(A.10)} & X_i \,\Delta\, (X_j \,\nabla\, X_k) &= (X_i \,\Delta\, X_j) \,\nabla\, (X_i \,\Delta\, X_k) \,. \end{aligned}\right\} \quad \text{distributivity}$$

And if it is complemented [see (54.20) and (54.21)]:

$$\left.\begin{aligned} &\text{(A.11)} & X_i \,\Delta\, \overline{X}_i &= 0 \,, \\ &\text{(A.12)} & X_i \,\nabla\, \overline{X}_i &= U \,. \end{aligned}\right\} \quad \text{complementation}$$

We shall consider several examples showing how to prove various formulas systematically.

Case where $\mathbf{L} = [0,1]$. Thus this case applies to all fuzzy subsets in the sense of Zadeh.

The totally ordered set $[0, 1]$ is a distributive but not complemented lattice. Thus all the properties (A.1)–(A.10) are satisfied, and one may denote Δ by $\wedge$ and ∇ by $\vee$; and one may call the inferior limit the minimum and likewise call the superior limit the maximum.

Suppose that one wants to prove (7.24), that is,

$$\underset{\sim}{\mathbf{A}} \cap (\underset{\sim}{\mathbf{B}} \cup \underset{\sim}{\mathbf{C}}) = (\underset{\sim}{\mathbf{A}} \cap \underset{\sim}{\mathbf{B}}) \cup (\underset{\sim}{\mathbf{A}} \cap \underset{\sim}{\mathbf{C}}). \tag{A.13}$$

This will be verified if $\forall\, x_i, x_j, x_k \in \mathbf{E}$:

$$\mu_{\underset{\sim}{\mathbf{A}}}(x_i) \wedge [\mu_{\underset{\sim}{\mathbf{B}}}(x_i) \vee \mu_{\underset{\sim}{\mathbf{C}}}(x_i)] = [\mu_{\underset{\sim}{\mathbf{A}}}(x_i) \wedge \mu_{\underset{\sim}{\mathbf{B}}}(x_i)] \vee [\mu_{\underset{\sim}{\mathbf{A}}}(x_i) \wedge \mu_{\underset{\sim}{\mathbf{C}}}(x_i)]. \tag{A.14}$$

Since $\mathbf{L}$ is a distributive lattice, this property is always satisfied.

We consider a more complicated case, that of proving (13.15), that is,

$$\underset{\sim}{\mathcal{R}} \circ (\underset{\sim}{\mathcal{Z}}_1 \cup \underset{\sim}{\mathcal{Z}}_2) = (\underset{\sim}{\mathcal{R}} \circ \underset{\sim}{\mathcal{Z}}_1) \cup (\underset{\sim}{\mathcal{R}} \circ \underset{\sim}{\mathcal{Z}}_2). \tag{A.15}$$

This will be verified if

$\forall x_i \in \mathbf{E}_1$, $\forall y_j \in \mathbf{E}_2$, $\forall z_k \in \mathbf{E}_3$ with the relations $x_i \underset{\sim}{\mathcal{R}}\, y_j$, $y_j \underset{\sim}{\mathcal{Z}}_1\, z_k$, $y_j \mathcal{Z}_2\, z_k$,

one has

$$\mu_{\underset{\sim}{\mathcal{R}} \circ (\underset{\sim}{\mathcal{Z}}_1 \cup \underset{\sim}{\mathcal{Z}}_2)}(x_i, z_k) = \mu_{\underset{\sim}{\mathcal{R}} \circ \underset{\sim}{\mathcal{Z}}_1}(x_i, z_k) \vee \mu_{\underset{\sim}{\mathcal{R}} \circ \underset{\sim}{\mathcal{Z}}_2}(x_i, z_k). \tag{A.16}$$

One may write

$$\begin{aligned} \mu_{\underset{\sim}{\mathcal{R}} \circ (\underset{\sim}{\mathcal{Z}}_1 \cup \underset{\sim}{\mathcal{Z}}_2)}(x_i, z_k) = {} & [\mu_{\underset{\sim}{\mathcal{R}}}(x_i, y_1) \wedge (\mu_{\underset{\sim}{\mathcal{Z}}_1}(y_1, z_k) \vee \mu_{\underset{\sim}{\mathcal{Z}}_2}(y_1, z_k))] \\ & \vee [\mu_{\underset{\sim}{\mathcal{R}}}(x_i, y_2) \wedge (\mu_{\underset{\sim}{\mathcal{Z}}_1}(y_2, z_k) \vee \mu_{\underset{\sim}{\mathcal{Z}}_2}(y_2, z_k))] \\ & \cdots \\ & \vee [\mu_{\underset{\sim}{\mathcal{R}}}(x_i, y_n) \wedge (\mu_{\underset{\sim}{\mathcal{Z}}_1}(y_n, z_k) \vee \mu_{\underset{\sim}{\mathcal{Z}}_2}(y_n, z_k))] \end{aligned} \tag{A.17}$$

$$\begin{aligned} \mu_{\underset{\sim}{\mathcal{R}} \circ \underset{\sim}{\mathcal{Z}}_1}(x_i, z_k) = {} & [\mu_{\underset{\sim}{\mathcal{R}}}(x_i, y_1) \wedge \mu_{\underset{\sim}{\mathcal{Z}}_1}(y_1, z_k)] \\ & \vee [\mu_{\underset{\sim}{\mathcal{R}}}(x_i, y_2) \wedge \mu_{\underset{\sim}{\mathcal{Z}}_1}(y_2, z_k)] \\ & \cdots \\ & \vee [\mu_{\underset{\sim}{\mathcal{R}}}(x_i, y_n) \wedge \mu_{\underset{\sim}{\mathcal{Z}}_1}(y_n, z_k)]. \end{aligned} \tag{A.18}$$

$$\begin{aligned} \mu_{\underset{\sim}{\mathcal{R}} \circ \underset{\sim}{\mathcal{Z}}_2}(x_i, z_k) = {} & [\mu_{\underset{\sim}{\mathcal{R}}}(x_i, y_1) \wedge \mu_{\underset{\sim}{\mathcal{Z}}_2}(y_1, z_k)] \\ & \vee [\mu_{\underset{\sim}{\mathcal{R}}}(x_i, y_2) \wedge \mu_{\underset{\sim}{\mathcal{Z}}_2}(y_2, z_k)] \\ & \vdots \\ & \vee [\mu_{\underset{\sim}{\mathcal{R}}}(x_i, y_n) \wedge \mu_{\underset{\sim}{\mathcal{Z}}_2}(y_n, z_k)]. \end{aligned} \tag{A.19}$$

In order to simplify writing, put

$$a_\alpha = \mu_{\underset{\sim}{\mathcal{R}}}(x_i, y_\alpha), \quad b_\beta = \mu_{\underset{\sim}{\mathcal{Q}}_1}(y_\beta, z_k), \quad c_\gamma = \mu_{\underset{\sim}{\mathcal{Q}}_2}(y_\gamma, z_k) \tag{A.20}$$

$\alpha, \beta, \gamma = 1, 2, \ldots n.$

Then (A.17)–(A.19) may be expressed

$$\mu_{\underset{\sim}{\mathcal{R}} \circ (\mathcal{Q}_1 \cup \mathcal{Q}_2)}(x_i, z_k) = [a_1 \wedge (b_1 \vee c_1)] \vee [a_2 \wedge (b_2 \vee c_2)] \vee \ldots \vee [a_n \wedge (b_n \vee c_n)], \tag{A.21}$$

$$\mu_{\underset{\sim}{\mathcal{R}} \circ \underset{\sim}{\mathcal{Q}}_1}(x_i, z_k) = (a_1 \wedge b_1) \vee (a_2 \wedge b_2) \vee \ldots \vee (a_n \wedge b_n), \tag{A.22}$$

$$\mu_{\underset{\sim}{\mathcal{R}} \circ \underset{\sim}{\mathcal{Q}}_2}(x_i, z_k) = (a_1 \wedge c_1) \vee (a_2 \wedge c_2) \vee \ldots \vee (a_n \wedge c_n), \tag{A.23}$$

One also has

$$\begin{aligned} \mu_{\underset{\sim}{\mathcal{R}} \circ \underset{\sim}{\mathcal{Q}}_1}(x_i, z_k) \vee \mu_{\underset{\sim}{\mathcal{R}} \circ \underset{\sim}{\mathcal{Q}}_2}(x_i, z_k) &= [(a_1 \wedge b_1) \vee (a_2 \wedge b_2) \vee \ldots \vee (a_n \wedge b_n)] \\ &\qquad \vee [(a_1 \wedge c_1) \vee (a_2 \wedge c_2) \vee \ldots \vee (a_n \wedge c_n)] \\ &= [(a_1 \wedge b_1) \vee (a_1 \wedge c_1)] \\ &\qquad \vee [(a_2 \wedge b_2) \vee (a_2 \wedge c_2)] \\ &\qquad \ldots \\ &\qquad \vee [(a_n \wedge b_n) \vee (a_n \wedge c_2)] \end{aligned} \tag{A.24}$$

because of the associativity for $\vee$.

By comparing (A.21) and (A.24) and remarking that because of distributivity

$$a_\alpha \wedge (b_\alpha \vee c_\alpha) = (a_\alpha \wedge b_\alpha) \vee (a_\alpha \wedge c_\alpha), \quad \alpha = 1, 2, \ldots, n; \tag{A.25}$$

one indeed has

$$\mu_{\underset{\sim}{\mathcal{R}} \circ (\underset{\sim}{\mathcal{Q}}_1 \cup \underset{\sim}{\mathcal{Q}}_2)}(x_i, z_k) = \mu_{\underset{\sim}{\mathcal{R}} \circ \underset{\sim}{\mathcal{Q}}_1}(x_i, z_k) \vee \mu_{\underset{\sim}{\mathcal{R}} \circ \underset{\sim}{\mathcal{Q}}_2}(x_i, z_k) = \mu_{(\underset{\sim}{\mathcal{R}} \circ \underset{\sim}{\mathcal{Q}}_1) \cup (\underset{\sim}{\mathcal{R}} \circ \underset{\sim}{\mathcal{Q}}_2)}(x_i, z_k). \tag{A.26}$$

This proves (A.13)

We examine another proof, that of (13.16). Here it is a matter of proving nondistributivity:

$$\underset{\sim}{\mathcal{R}} \circ (\underset{\sim}{\mathcal{Q}}_1 \cap \underset{\sim}{\mathcal{Q}}_2) \neq (\underset{\sim}{\mathcal{R}} \circ \underset{\sim}{\mathcal{Q}}_1) \cap (\underset{\sim}{\mathcal{R}} \circ \underset{\sim}{\mathcal{Q}}_2). \tag{A.27}$$

We shall use the same notations as for (A.13). Since we are proving that the distributive property is not true for some $\underset{\sim}{\mathbf{A}}$, $\underset{\sim}{\mathbf{B}}$, and $\underset{\sim}{\mathbf{C}}$, we restrict ourselves to a reference set where $\alpha, \beta, \gamma = 1, 2$ in (A.27). We have

$$\mu_{\underset{\sim}{\mathcal{R}} \circ (\underset{\sim}{\mathcal{Z}}_1 \cap \underset{\sim}{\mathcal{Z}}_2)}(x_i, z_k) = [a_1 \wedge (b_1 \wedge c_1)] \vee [a_2 \wedge (b_2 \wedge c_2)] \tag{A.28}$$

$$= (a_1 \wedge b_1 \wedge c_1) \vee (a_2 \wedge b_2 \wedge c_2)$$

$$\mu_{\underset{\sim}{\mathcal{R}} \circ \underset{\sim}{\mathcal{Z}}_1}(x_i, z_k) \wedge \mu_{\underset{\sim}{\mathcal{R}} \circ \underset{\sim}{\mathcal{Z}}_2}(x_i, z_k) = [(a_1 \wedge b_1) \vee (a_2 \wedge b_2)] \wedge [(a_1 \wedge c_1) \vee (a_2 \wedge c_2)]. \tag{A.29}$$

We expand (A.28) and (A.29) in order to show that they are different quantities:

$$\mu_{\underset{\sim}{\mathcal{R}} \circ (\underset{\sim}{\mathcal{Z}}_1 \cap \underset{\sim}{\mathcal{Z}}_2)}(x_i, z_k) = (a_1 \vee a_2) \wedge (a_1 \vee b_2) \wedge (a_1 \vee c_2) \wedge (a_2 \vee b_1) \wedge (b_1 \vee b_2)$$
$$\wedge (b_1 \vee c_2) \wedge (a_2 \vee c_1) \wedge (b_2 \vee c_1) \wedge (c_1 \vee c_2) \tag{A.30}$$

$$\mu_{\underset{\sim}{\mathcal{R}} \circ \underset{\sim}{\mathcal{Z}}_1}(x_i, z_k) \wedge \mu_{\underset{\sim}{\mathcal{R}} \circ \underset{\sim}{\mathcal{Z}}_2}(x_i, z_k) = (a_1 \vee a_2) \wedge (a_1 \vee b_2) \wedge (a_2 \vee b_1) \wedge (b_1 \vee b_2)$$
$$\wedge (a_1 \vee a_2) \wedge (a_1 \vee c_2) \wedge (a_2 \vee c_1) \wedge (c_1 \vee c_2) \tag{A31}$$

One may verify the inequality by elimination since

$$a_1 \vee a_2 \neq (b_1 \vee c_2) \wedge (b_2 \vee c_1). \tag{A.32}$$

APPENDIX B

DECOMPOSITION INTO MAXIMAL SIMILITUDE SUBRELATIONS

The problem of the decomposition of a resemblance relation into maximal similitude subrelations when the resemblance relation (or the corresponding notion of distance) does not allow one to obtain the similitude classes for distances less than or equal to a given distance is related to that of obtaining the maximal planar ordinary subgraphs of the associated ordinary graph. There are algorithms for this at our disposal; we cite two of them. The first is due to the engineer Y. Malgrange.†

First we develop this algorithm in a more general case and then return to the particular case that especially interests us.

Malgrange algorithm. Obtaining maximal complete submatrices or principal submatrices. Exposition of this algorithm requires a certain number of preliminary definitions.

In a matrix with binary elements (0 or 1), that is, in a boolean matrix associated with a graph (more strongly with a Berge graph), a *complete submatrix* is a submatrix all of whose elements without exception are 1. A principal submatrix, also called a maximal complete submatrix, is a complete submatrix that is not contained in any other complete submatrix. Thus, in Figure B.1 one finds seven principal submatrices in the matrix [M]. A covering of a boolean matrix is a set of complete submatrices covering all the coefficients of value 1 in this matrix.

[M] =

	a	*b*	*c*	*d*	*e*	*f*
A	0	1	0	1	1	1
B	1	0	1	0	0	0
C	0	1	0	1	1	0
D	0	1	1	0	0	1

†A very brilliant engineer, too soon missing. He has been one of my friends and colleagues at Machines Bull. Malgrange had established this algorithm in order to treat, on a computer, various covering problems of graph theory. See for example, "Graphs, Dynamic Programming and Games," by A. Kaufmann (Academic Press, New York, 1967).

Let **I** be the set of rows and **J** the set of columns of a boolean matrix. Each complete submatrix is defined by the ordered pair of ordinary subsets $(\mathbf{I}_p, \mathbf{I}_q)$ with $\mathbf{I}_p \subset \mathbf{I}$ and $\mathbf{J}_q \subset \mathbf{J}$. One may show that the operations $\dot\cup$ and $\dot\cap$ that, to two arbitrary complete submatrices of a boolean matrix [M], say

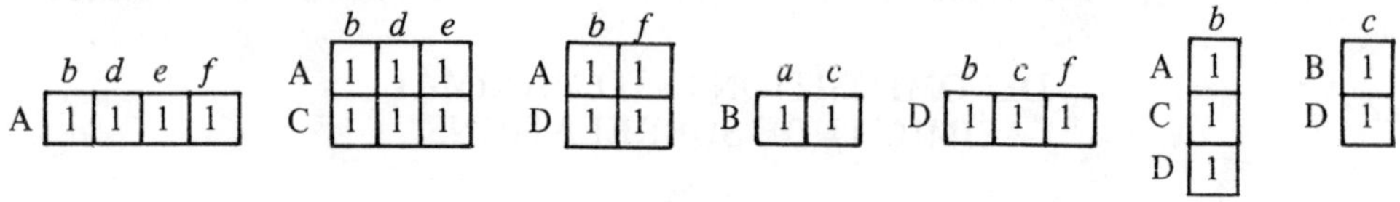

$$\begin{array}{c|cccc} & b & d & e & f \\ \hline A & 1 & 1 & 1 & 1 \end{array} \quad \begin{array}{c|ccc} & b & d & e \\ \hline A & 1 & 1 & 1 \\ C & 1 & 1 & 1 \end{array} \quad \begin{array}{c|cc} & b & f \\ \hline A & 1 & 1 \\ D & 1 & 1 \end{array} \quad \begin{array}{c|cc} & a & c \\ \hline B & 1 & 1 \end{array} \quad \begin{array}{c|ccc} & b & c & f \\ \hline D & 1 & 1 & 1 \end{array} \quad \begin{array}{c|c} & b \\ \hline A & 1 \\ C & 1 \\ D & 1 \end{array} \quad \begin{array}{c|c} & c \\ \hline B & 1 \\ D & 1 \end{array}$$

Fig. B.1

(B.1) $\qquad [\mathrm{M}_1]$ defined by $(\mathbf{I}_1, \mathbf{J}_1)$

(B.2) $\qquad [\mathrm{M}_2]$ defined by $(\mathbf{I}_2, \mathbf{J}_2)$

correspond the two submatrices

(B.3) $\qquad [\mathrm{M}_1] \dot\cup [\mathrm{M}_2] = [\mathrm{M}']$ defined by the ordered pair $(\mathbf{I}_1 \cup \mathbf{I}_2, \mathbf{J}_1 \cap \mathbf{J}_2)$,

(B.4) $\qquad [\mathrm{M}_1] \dot\cap [\mathrm{M}_2] = [\mathrm{M}'']$ defined by the ordered pair $(\mathbf{I}_1 \cap \mathbf{I}_2, \mathbf{J}_1 \cup \mathbf{J}_2)$,

are internal operations in the set **M** of complete submatrices of [M].

Alternate application of the rules below until exhaustion to the complete submatrices of a covering†

(B.5) $$\mathbf{C} = \{[\mathrm{M}_1], [\mathrm{M}_2], \ldots, [\mathrm{M}_p]\},$$

permits one to obtain the principal submatrices of the matrix [M] in a finite number of iterations.

First rule. Suppress all matrices $[\mathrm{M}_k]$ contained in another submatrix of the covering **C**.

Second rule. Add to **C** the submatrices obtained by the operations $\dot\cup$ and $\dot\cap$ defined above and applied to all pairs of matrices $[\mathrm{M}_k]$ and $[\mathrm{M}_l]$ kept in the covering (unless the complete submatrix is contained in a submatrix appearing in **C**, which avoids an infinite process).

We consider an example.

Example. We propose to calculate the principal submatrices of the boolean matrix of Figure B.1.

Phase 1. We choose a covering:

(B.6)
$$[\mathrm{M}_1] = \begin{array}{c|cccc} & b & d & e & f \\ \hline A & 1 & 1 & 1 & 1 \end{array}, \quad [\mathrm{M}_2] = \begin{array}{c|cc} & a & c \\ \hline B & 1 & 1 \end{array}, \quad [\mathrm{M}_3] = \begin{array}{c|ccc} & b & d & e \\ \hline C & 1 & 1 & 1 \end{array}, \quad [\mathrm{M}_4] = \begin{array}{c|ccc} & b & c & f \\ \hline D & 1 & 1 & 1 \end{array}.$$

†The symbol **C** has also been used to represent a category; we do not think that there will be any confusion between these two concepts represented by the same letter.

Phase 2 (Second rule)

We calculate the unions and intersections:

(B.7) $\quad \mathbf{I}_1 \cup \mathbf{I}_2 = \{A, B\} \quad , \quad \mathbf{J}_1 \cap \mathbf{J}_2 = \phi,$

(B.8) $\quad \mathbf{I}_1 \cup \mathbf{I}_3 = \{A, C\} \quad , \quad \mathbf{J}_1 \cap \mathbf{J}_3 = \{b, d, e\} \ ,$

from which a new submatrix:

(B.9) $[M_5] =$

	b	d	e
A	1	1	1
C	1	1	1

(B.10) $\quad \mathbf{I}_1 \cup \mathbf{I}_4 = \{A, D\} \quad , \quad \mathbf{J}_1 \cap \mathbf{J}_4 = \{b, f\} \ ,$

from which a new submatrix:

(B.11) $[M_6] =$

	b	f
A	1	1
D	1	1

(B.12) $\quad \mathbf{I}_2 \cup \mathbf{I}_3 = \{B, C\} \quad , \quad \mathbf{J}_2 \cap \mathbf{J}_3 = \phi$

(B.13) $\quad \mathbf{I}_2 \cup \mathbf{I}_4 = \{B, D\} \quad , \quad \mathbf{J}_2 \cap \mathbf{J}_4 = \{c\} \ ,$

from which a new submatrix:

(B.14) $[M_7] =$

	c
B	1
D	1

(B.15) $\quad \mathbf{I}_3 \cup \mathbf{I}_4 = \{C, D\} \quad , \quad \mathbf{J}_3 \cap \mathbf{J}_4 = \{b\} \ ,$

from which a new submatrix:

(B.16) $[M_8] =$

	b
C	1
D	1

Since all the intersections $\mathbf{I}_i \cap \mathbf{I}_j$, $\forall\ i, j$, are empty, it is useless to calculate the $\mathbf{J}_i \cup \mathbf{J}_j$.

Phase 3 (First rule)

The new covering is

(B.17) $$C' = \{[M_1], [M_2], [M_4], [M_5], [M_6], [M_7], [M_8]\}.$$

$[M_3]$ being contained in $[M_5]$ has not been conserved.

Phase 4 (Second rule)

In a didactic spirit we indicate all the details of the evaluations even though, evidently, certain of the submatrices have already been obtained or are empty:

(B.18) $I_1 \cup I_5 = \{A, C\}$, $J_1 \cap J_5 = \{b, d, e\}$, gives $[M_5]$,

(B.19) $I_1 \cup I_6 = \{A, D\}$, $J_1 \cap J_6 = \{b, f\}$, gives $[M_6]$,

(B.20) $I_1 \cup I_7 = \{A, B, D\}$, $J_1 \cap J_7 = \phi$,

(B.21) $I_1 \cup I_8 = \{A, C, D\}$, $J_1 \cap J_8 = \{b\}$,

This gives a new submatrix: :

(B.22) $[M_9] =$

	b
A	1
C	1
D	1

(B.23) $I_2 \cup I_5 = \{A, B, C\}$, $J_2 \cap J_5 = \phi$,

(B.24) $I_2 \cup I_6 = \{A, B, D\}$ $J_2 \cap J_6 = \phi$,

(B.25) $I_2 \cup I_7 = \{B, D\}$, $J_2 \cap J_7 = \{c\}$, gives $[M_7]$;

(B.26) $I_2 \cup I_8 = \{B, C, D\}$, $J_2 \cap J_8 = \phi$.

(B.27) $I_4 \cup I_5 = \{A, C, D\}$, $J_4 \cap J_5 = \{b\}$, gives $[M_9]$;

(B.28) $I_4 \cup I_6 = \{A, D\}$, $J_4 \cap J_6 = \{b, f\}$, gives $[M_6]$;

(B.29) $I_4 \cup I_7 = \{B, D\}$, $J_4 \cap J_7 = \{c\}$, gives $[M_7]$;

(B.30) $I_4 \cup I_8 = \{C, D\}$, $J_4 \cap J_8 = \{b\}$, included in $[M_9]$;

(B.31) $I_5 \cup I_6 = \{A, C, D\}$, $J_5 \cap J_6 = \{b\}$, gives $[M_9]$;

(B.32) $I_5 \cup I_7 = \{A, B, C, D\}$, $J_5 \cap J_7 = \phi$,

(B.33) $I_5 \cup I_8 = \{A, C, D\}$, $J_5 \cap J_8 = \{b\}$, gives $[M_9]$;

(B.34) $I_6 \cup I_7 = \{A, B, D\}$, $J_6 \cap J_7 = \phi$,

(B.35) $\mathbf{I}_6 \cup \mathbf{I}_8 = \{A, C, D\}$, $\mathbf{J}_6 \cap \mathbf{J}_8 = \{b\}$, gives $[M_9]$,

(B.36) $\mathbf{I}_7 \cup \mathbf{I}_8 = \{B, C, D\}$, $\mathbf{J}_7 \cap \mathbf{J}_8 = \phi$,

(B.37) $\mathbf{I}_1 \cap \mathbf{I}_5 = \{A\}$, $\mathbf{J}_1 \cup \mathbf{J}_5 = \{b, d, e, f\}$, gives $[M_1]$;

(B.38) $\mathbf{I}_1 \cap \mathbf{I}_6 = \{A\}$, $\mathbf{J}_1 \cup \mathbf{J}_6 = \{b, d, e, f\}$, gives $[M_1]$;

(B.39) $\mathbf{I}_1 \cap \mathbf{I}_7 = \phi$, $\mathbf{I}_1 \cap \mathbf{I}_8 = \phi$,

(B.40) $\mathbf{I}_2 \cap \mathbf{I}_5 = \phi$, $\mathbf{I}_2 \cap \mathbf{I}_6 = \phi$,

(B.41) $\mathbf{I}_2 \cap \mathbf{I}_7 = \{B\}$, $\mathbf{J}_2 \cup \mathbf{J}_7 = \{a, c\}$, gives $[M_2]$;

(B.42) $\mathbf{I}_2 \cap \mathbf{I}_8 = \phi$, $\mathbf{I}_4 \cap \mathbf{I}_5 = \phi$;

(B.43) $\mathbf{I}_4 \cap \mathbf{I}_6 = \{D\}$, $\mathbf{J}_4 \cup \mathbf{J}_6 = \{b, c, f\}$, gives $[M_4]$;

(B.44) $\mathbf{I}_4 \cap \mathbf{I}_7 = \{D\}$, $\mathbf{J}_4 \cup \mathbf{J}_7 = \{b, c, f\}$, gives $[M_4]$;

(B.45) $\mathbf{I}_4 \cap \mathbf{I}_8 = \{D\}$, $\mathbf{J}_4 \cup \mathbf{J}_8 = \{b, c, f\}$, gives $[M_4]$;

(B.46) $\mathbf{I}_5 \cap \mathbf{I}_6 = \{A\}$, $\mathbf{J}_5 \cup \mathbf{J}_6 = \{b, d, e, f\}$, gives $[M_1]$;

(B.47) $\mathbf{I}_5 \cap \mathbf{I}_7 = \phi$,

(B.48) $\mathbf{I}_5 \cap \mathbf{I}_8 = \{C\}$, $\mathbf{J}_5 \cup \mathbf{J}_8 = \{b, d, e\}$, included in $[M_5]$;

(B.49) $\mathbf{I}_6 \cap \mathbf{I}_7 = \{D\}$, $\mathbf{J}_6 \cup \mathbf{J}_7 = \{b, d, f\}$, gives $[M_4]$;

(B.50) $\mathbf{I}_6 \cap \mathbf{I}_8 = \{D\}$, $\mathbf{J}_6 \cup \mathbf{J}_8 = \{b, f\}$, included in $[M_4]$;

(B.51) $\mathbf{I}_7 \cap \mathbf{I}_8 = \{D\}$, $\mathbf{J}_7 \cup \mathbf{J}_8 = \{b, c\}$, included in $[M_4]$;

Phase 5 (First rule)

The new covering is

(B.52) $$C'' = \{[M_1], [M_2], [M_4], [M_5], [M_6], [M_7], [M_9]\},$$

$[M_8]$ has been eliminated since it is contained in $[M_9]$.

Phase 6 (Second rule)

Calculation of the unions and intersections shows that it is not possible to find complete submatrices that are not equal to some submatrices or contained in some submatrices of the preceding covering; thus this is a covering that gives the set of principal submatrices:

(B.53)
$$[M_1] = \begin{array}{c|c|c|c|c|} & b & d & e & f \\ \hline A & 1 & 1 & 1 & 1 \\ \hline \end{array}, \quad [M_2] = \begin{array}{c|c|c|} & a & c \\ \hline B & 1 & 1 \\ \hline \end{array}, \quad [M_4] = \begin{array}{c|c|c|c|} & b & c & f \\ \hline D & 1 & 1 & 1 \\ \hline \end{array}, \quad [M_5] = \begin{array}{c|c|c|c|} & b & d & e \\ \hline A & 1 & 1 & 1 \\ \hline C & 1 & 1 & 1 \\ \hline \end{array},$$

$$[M_6] = \begin{array}{c|c|c|} & b & f \\ \hline A & 1 & 1 \\ \hline D & 1 & 1 \\ \hline \end{array}, \quad [M_7] = \begin{array}{c|c|} & c \\ \hline B & 1 \\ \hline D & 1 \\ \hline \end{array}, \quad [M_9] = \begin{array}{c|c|} & b \\ \hline A & 1 \\ \hline C & 1 \\ \hline D & 1 \\ \hline \end{array},$$

The search for maximal similitude subrelations. We now go on to apply the Malgrange algorithm to the search for maximal similitude subrelations.

As an example we consider the symmetric and reflexive ordinary graph in Figure B.2a and we propose to determine in the associated boolean matrix (Figure B.2b) the principal submatrices forming a covering of this matrix; among these principal submatrices, those that are square will give the subrelations that we are seeking.

With the objective of beginning with the complete submatrices judged a priori as being rather close to those that we seek (a heuristic consideration, therefore not proven to be indispensable), we start by reordering the elements of the matrix in such a manner that one finds these elements in an order corresponding to the rows (then the columns) having the greatest number of 1's. This gives the matrix represented in Figure B.2c.

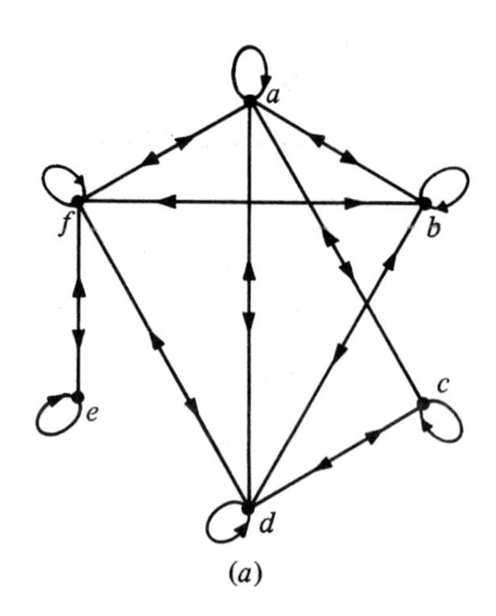

(*a*)

	a	*b*	*c*	*d*	*e*	*f*	
a	1	1	1	1	0	1	5
b	1	1	0	1	0	1	4
c	1	0	1	1	0	0	3
d	1	1	1	1	0	1	5
e	0	0	0	0	1	1	2
f	1	1	0	1	1	1	5

(*b*)

	a	*d*	*f*	*b*	*c*	*e*	
a	1	1	1	1	1	0	5
d	1	1	1	1	1	0	5
f	1	1	1	1	0	1	5
b	1	1	1	1	0	0	4
c	1	1	0	0	1	0	3
e	0	0	1	0	0	1	2

(*c*)

Fig. B.2

Phase 1. We take the following covering:

(B.54)

$[M_1] =$

	a	d	f	b
a	1	1	1	1
d	1	1	1	1
f	1	1	1	1
b	1	1	1	1

, $[M_2] =$

	a	d	c
c	1	1	1

, $[M_3] =$

	f	e
e	1	1

$[M_4] =$

	c
a	1
d	1
c	1

, $[M_5] =$

	e
f	1
e	1

.

Phase 2 (Second rule)

(B.55) $\quad \mathbf{I}_1 \cup \mathbf{I}_2 = \{a, d, f, b, c\}$, $\mathbf{J}_1 \cap \mathbf{J}_2 = \{a, d\}$,

from which a new submatrix:

(B.56) $[M_6] =$

	a	d
a	1	1
d	1	1
f	1	1
b	1	1
c	1	1

(B.57) $\quad \mathbf{I}_1 \cap \mathbf{I}_2 = \phi$,

(B.58) $\quad \mathbf{I}_1 \cup \mathbf{I}_3 = \{a, d, f, b, e\}$, $\mathbf{J}_1 \cap \mathbf{J}_3 = \{f\}$,

from which a new submatrix:

(B.59) $[M_7] =$

	f
a	1
d	1
f	1
b	1
e	1

.

$$(B.60)\quad \begin{aligned} &\mathbf{I}_1 \cap \mathbf{I}_3 = \phi\,, & & \\ &\mathbf{I}_1 \cup \mathbf{I}_4 = \{a, d, f, b, c\}\,, & &\mathbf{J}_1 \cap \mathbf{J}_4 = \phi\,, \\ &\mathbf{I}_1 \cap \mathbf{I}_4 = \{a, d\}\,, & &\mathbf{J}_1 \cup \mathbf{J}_4 = \{a, d, f, b, c\}\,, \end{aligned}$$

from which a new submatrix:

(B.61) $[\mathrm{M}_8] =$

	a	d	f	b	c
a	1	1	1	1	1
d	1	1	1	1	1

$$(B.62)\quad \begin{aligned} &\mathbf{I}_1 \cup \mathbf{I}_5 = \{a, d, f, b, e\}\,, & &\mathbf{J}_1 \cap \mathbf{J}_5 = \phi\,, \\ &\mathbf{I}_1 \cap \mathbf{I}_5 = \{f\}, & &\mathbf{J}_1 \cup \mathbf{J}_5 = \{a, d, f, b, e\}, \end{aligned}$$

from which a new submatrix:

(B.63) $[\mathrm{M}_9] =$

	a	d	f	b	e
f	1	1	1	1	1

$$(B.64)\quad \begin{aligned} &\mathbf{I}_2 \cup \mathbf{I}_3 = \{c, e\}, & &\mathbf{J}_2 \cap \mathbf{J}_3 = \phi\,, & & \\ &\mathbf{I}_2 \cap \mathbf{I}_3 = \phi\,, & & & & \\ &\mathbf{I}_2 \cup \mathbf{I}_4 = \{a, d, c\}, & &\mathbf{J}_2 \cap \mathbf{J}_4 = \{c\}, & &\text{gives } [\mathrm{M}_4]\,; \\ &\mathbf{I}_2 \cap \mathbf{I}_4 = \{c\}, & &\mathbf{J}_2 \cup \mathbf{J}_4 = \{a, d, c\}, & &\text{gives } [\mathrm{M}_2]\,, \\ &\mathbf{I}_2 \cup \mathbf{I}_5 = \{c, f, e\}, & &\mathbf{J}_2 \cap \mathbf{J}_5 = \phi\,, & & \\ &\mathbf{I}_2 \cap \mathbf{I}_5 = \phi\,, & & & & \\ &\mathbf{I}_3 \cup \mathbf{I}_4 = \{a, d, c, e\}, & &\mathbf{J}_3 \cap \mathbf{J}_4 = \phi\,, & & \\ &\mathbf{I}_3 \cap \mathbf{I}_4 = \phi\,, & & & & \\ &\mathbf{I}_3 \cup \mathbf{I}_5 = \{f, e\}, & &\mathbf{J}_3 \cap \mathbf{J}_5 = \{e\}, & &\text{gives } [\mathrm{M}_5]\,, \\ &\mathbf{I}_3 \cap \mathbf{I}_5 = \{e\}, & &\mathbf{J}_3 \cup \mathbf{J}_5 = \{f, e\}, & &\text{gives } [\mathrm{M}_3]\,, \\ &\mathbf{I}_4 \cup \mathbf{I}_5 = \{a, d, c, e, f\}, & &\mathbf{J}_4 \cap \mathbf{J}_5 = \phi\,, & & \\ &\mathbf{I}_4 \cap \mathbf{I}_5 = \phi\,. & & & & \end{aligned}$$

Phase 3 (Second rule). The new covering is†

$$(B.65)\quad C' = \{[\mathrm{M}_1], [\mathrm{M}_2], [\mathrm{M}_3], [\mathrm{M}_4], [\mathrm{M}_5], [\mathrm{M}_6], [\mathrm{M}_7], [\mathrm{M}_8], [\mathrm{M}_9]\}.$$

†Note that the algorithm may be simplified in the particular case of symmetric boolean matrices, but we have preferred for this didactic discussion to use the original algorithm.

Phase 4 (First rule)

(B.66)

$$
\begin{array}{lll}
\mathbf{I}_1 \cup \mathbf{I}_6 = \{a, d, f, b, c\}, & \mathbf{J}_1 \cap \mathbf{J}_6 = \{a, d\} & \text{gives } [\mathrm{M}_6], \\
\mathbf{I}_1 \cap \mathbf{I}_6 = \{a, d, f, b\}, & \mathbf{J}_1 \cup \mathbf{J}_6 = \{a, d, f, b\} & \text{gives } [\mathrm{M}_1], \\
\mathbf{I}_1 \cup \mathbf{I}_7 = \{a, d, f, b, e\}, & \mathbf{J}_1 \cap \mathbf{J}_7 = \{f\}, & \text{gives } [\mathrm{M}_7], \\
\mathbf{I}_1 \cap \mathbf{I}_7 = \{a, d, f, b\}, & \mathbf{J}_1 \cup \mathbf{J}_7 = \{a, d, f, b\}, & \text{gives } [\mathrm{M}_1], \\
\mathbf{I}_1 \cup \mathbf{I}_8 = \{a, d, f, b\}, & \mathbf{J}_1 \cap \mathbf{J}_8 = \{a, d, f, b\}, & \text{gives } [\mathrm{M}_1], \\
\mathbf{I}_1 \cap \mathbf{I}_8 = \{a, d\}, & \mathbf{J}_1 \cup \mathbf{J}_8 = \{a, d, f, b, c\}, & \text{gives } [\mathrm{M}_8], \\
\mathbf{I}_1 \cup \mathbf{I}_9 = \{a, d, f, b\}, & \mathbf{J}_1 \cap \mathbf{J}_9 = \{a, d, f, b\}, & \text{gives } [\mathrm{M}_1], \\
\mathbf{I}_1 \cap \mathbf{I}_9 = \{f\}, & \mathbf{J}_1 \cup \mathbf{J}_9 = \{a, d, f, b, e\}, & \text{gives } [\mathrm{M}_9], \\
\mathbf{I}_2 \cup \mathbf{I}_6 = \{a, d, f, b, c\}, & \mathbf{J}_2 \cap \mathbf{J}_6 = \{a, d\}, & \text{gives } [\mathrm{M}_6], \\
\mathbf{I}_2 \cap \mathbf{I}_6 = \{c\}, & \mathbf{J}_2 \cup \mathbf{J}_6 = \{a, d, c\}, & \text{gives } [\mathrm{M}_2] \\
\mathbf{I}_2 \cup \mathbf{I}_7 = \{a, d, f, b, e, c\}, & \mathbf{J}_2 \cap \mathbf{J}_7 = \phi, & \\
\mathbf{I}_2 \cap \mathbf{I}_7 = \phi, & & \\
\mathbf{I}_2 \cup \mathbf{I}_8 = \{a, d, c\}, & \mathbf{J}_2 \cap \mathbf{J}_8 = \{a, d, c\}, &
\end{array}
$$

giving a new submatrix:

(B.67) $[\mathrm{M}_{10}] =$

	a	d	c
a	1	1	1
d	1	1	1
c	1	1	1

(B.68)

$$
\begin{array}{lll}
\mathbf{I}_2 \cap \mathbf{I}_8 = \phi, & & \\
\mathbf{I}_2 \cup \mathbf{I}_9 = \{c, f\}, & \mathbf{J}_2 \cap \mathbf{J}_9 = \{a, d\}, & \text{included in } [\mathrm{M}_6]. \\
\mathbf{I}_2 \cap \mathbf{I}_9 = \phi, & & \\
\mathbf{I}_3 \cup \mathbf{I}_6 = \{a, d, f, b, c, e\}, & \mathbf{J}_3 \cap \mathbf{J}_6 = \phi, & \\
\mathbf{I}_3 \cap \mathbf{I}_6 = \phi, & & \\
\mathbf{I}_3 \cup \mathbf{I}_7 = \{a, d, f, b, e\}, & \mathbf{J}_3 \cap \mathbf{J}_7 = \{f\}, & \text{gives } [\mathrm{M}_7], \\
\mathbf{I}_3 \cap \mathbf{I}_7 = \{e\}, & \mathbf{J}_3 \cup \mathbf{J}_7 = \{f, e\}, & \text{gives } [\mathrm{M}_3], \\
\mathbf{I}_3 \cup \mathbf{I}_8 = \{a, d, e\}, & \mathbf{J}_3 \cap \mathbf{J}_8 = \{f\}, & \text{included in } [\mathrm{M}_7], \\
\mathbf{I}_3 \cap \mathbf{I}_8 = \phi, & & \\
\mathbf{I}_3 \cup \mathbf{I}_9 = \{f, e\}, & \mathbf{J}_3 \cap \mathbf{J}_9 = \{f, e\}, &
\end{array}
$$

giving a new submatrix:

(B.69)
$$[M_{11}] = \begin{array}{c|c|c|} & f & e \\ \hline f & 1 & 1 \\ \hline e & 1 & 1 \\ \hline \end{array} .$$

(B.70)

$I_3 \cap I_9 = \phi$,

$I_4 \cup I_6 = \{a, d, f, b, c\}$, $J_4 \cap J_6 = \phi$,

$I_4 \cap I_6 = \{a, d, c\}$, $J_4 \cup J_6 = \{a, d, c\}$, gives $[M_{10}]$,

$I_4 \cup I_7 = \{a, d, f, b, c, e\}$, $J_4 \cap J_7 = \phi$,

$I_4 \cap I_7 = \{a, d\}$, $J_4 \cup J_7 = \{f, c\}$ included in $[M_8]$,

$I_4 \cup I_8 = \{a, d, c\}$, $J_4 \cap J_8 = \{c\}$ gives $[M_4]$,

$I_4 \cap I_8 = \{a, d\}$, $J_4 \cup J_8 = \{a, d, f, b, c\}$ gives $[M_8]$,

$I_4 \cup I_9 = \{a, d, c, f\}$, $J_4 \cap J_9 = \phi$,

$I_4 \cap I_9 = \phi$,

$I_5 \cup I_6 = \{a, d, f, b, c . e\}$, $J_5 \cap J_6 = \phi$,

$I_5 \cap I_6 = \{f\}$, $J_5 \cup J_6 = \{a, d, e\}$, included in $[M_9]$,

$I_5 \cup I_7 = \{a, d, f, b, e\}$, $J_5 \cap J_7 = \phi$,

$I_5 \cap I_7 = \{f, e\}$, $J_5 \cup J_7 = \{f, e\}$, gives $[M_{11}]$

$I_5 \cup I_8 = \{a, d, f, e\}$, $J_5 \cap J_8 = \phi$,

$I_5 \cap I_8 = \phi$,

$I_5 \cup I_9 = \{f, e\}$, $J_5 \cap J_9 = \{e\}$, gives $[M_5]$

$I_5 \cap I_9 = \{f\}$, $J_5 \cup J_9 = \{a, d, f, b, e\}$, gives $[M_9]$,

$I_6 \cup I_7 = \{a, d, f, b, c, e\}$, $J_6 \cap J_7 = \phi$,

$I_6 \cap I_7 = \{a, d, f, b\}$, $J_6 \cup J_7 = \{a, d, f\}$, included in $[M_1]$,

$I_6 \cup I_8 = \{a, d, f, b, c\}$, $J_6 \cap J_8 = \{a, d\}$, gives $[M_6]$,

$I_6 \cap I_8 = \{a, d\}$, $J_6 \cup J_8 = \{a, d, f, b, c\}$, gives $[M_8]$,

$I_6 \cup I_9 = \{a, d, f, b, c\}$, $J_6 \cap J_9 = \{a, d\}$, gives $[M_6]$,

$I_6 \cap I_9 = \{f\}$, $J_6 \cup J_9 = \{a, d, f, b, e\}$, gives $[M_9]$,

$I_7 \cup I_8 = \{a, d, f, b, e\}$, $J_7 \cap J_8 = \{f\}$, gives $[M_7]$,

$I_7 \cap I_8 = \{a, d\}$, $J_7 \cup J_8 = \{a, d, f, b, c\}$, gives $[M_8]$,

$I_7 \cup I_9 = \{a, d, f, b, e\}$,	$J_7 \cap J_9 = \{f\}$,	gives $[M_7]$,
$I_7 \cap I_9 = \{f\}$,	$J_7 \cup J_9 = \{a, d, f, b, e\}$,	gives $[M_9]$,
$I_8 \cup I_9 = \{a, d, f\}$,	$J_8 \cap J_9 = \{a, d, f, b\}$,	included in $[M_1]$,
$I_8 \cap I_9 = \phi$.		

Phase 5 (Second rule). The new covering is

$$C'' = \{[M_1], [M_6], [M_7], [M_8], [M_9], [M_{10}], [M_{11}]\}. \tag{B.71}$$

We have eliminated $[M_2]$, $[M_3]$, $[M_4]$, $[M_5]$; these are included in other submatrices.

Phase 6 (First rule). The reader may verify that one cannot obtain new matrices. Then the covering that gives all the principal matrices is $\mathbf{C}''$ expressed by (B.71), namely:

(B.72)

$[M_1] =$

	a	d	f	b
a	1	1	1	1
d	1	1	1	1
f	1	1	1	1
b	1	1	1	1

, $[M_6] =$

	a	d
a	1	1
d	1	1
f	1	1
b	1	1
c	1	1

, $[M_7] =$

	f
a	1
d	1
f	1
b	1
e	1

,

$[M_8] =$

	a	d	f	b	c
a	1	1	1	1	1
d	1	1	1	1	1

, $[M_9] =$

	a	d	f	b	e
f	1	1	1	1	1

,

$[M_{10}] =$

	a	d	c
a	1	1	1
d	1	1	1
c	1	1	1

, $[M_{11}] =$

	f	e
f	1	1
e	1	1

.

One finds three square submatrices $[M_1]$, $[M_{10}]$, and $[M_{11}]$; these give the three subrelations that are nondisjoint. Figure B.3 represents these three nondisjoint subrelations.

We note that obtaining the matrices $[M_6] = [M_8]'$ and $[M_7] = [M_9]'$ is not useless; these give what one calls the couplings holding between the subrelations.

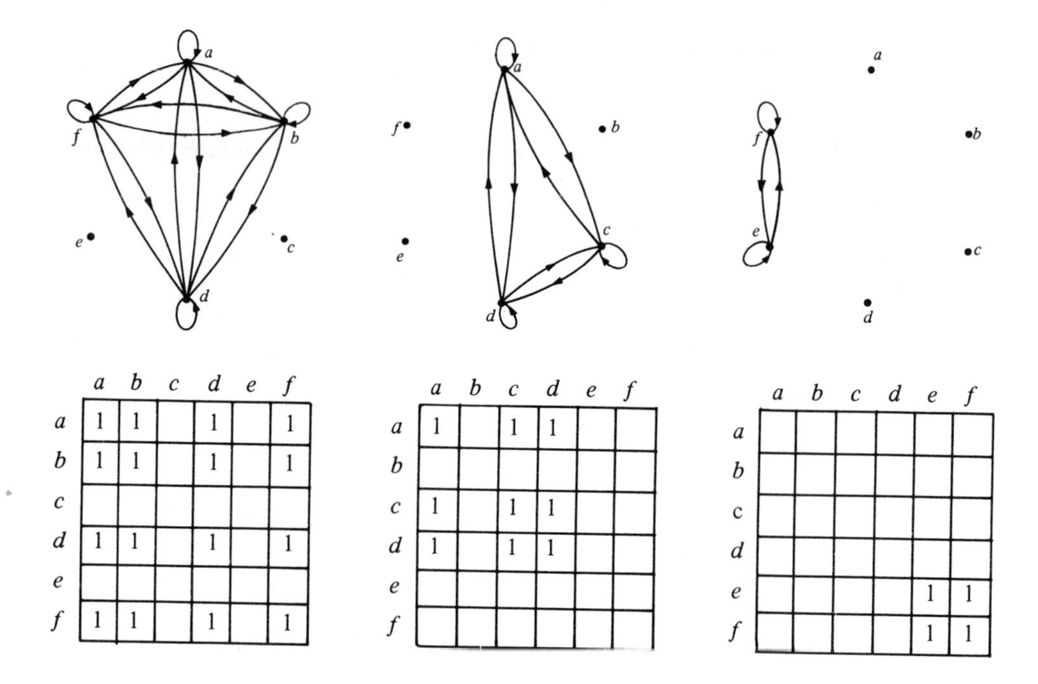

	a	b	c	d	e	f
a	1	1		1		1
b	1	1		1		1
c						
d	1	1		1		1
e						
f	1	1		1		1

	a	b	c	d	e	f
a	1		1	1		
b						
c	1		1	1		
d	1		1	1		
e						
f						

	a	b	c	d	e	f
a						
b						
c						
d						
e					1	1
f					1	1

Fig. B.3

Another, more rapid method.† This method is valid exclusively for symmetric square matrices, which is the case in which we are particularly interested here.

We consider the semimatrix above the principal diagonal (see, for example, Figure B.4)

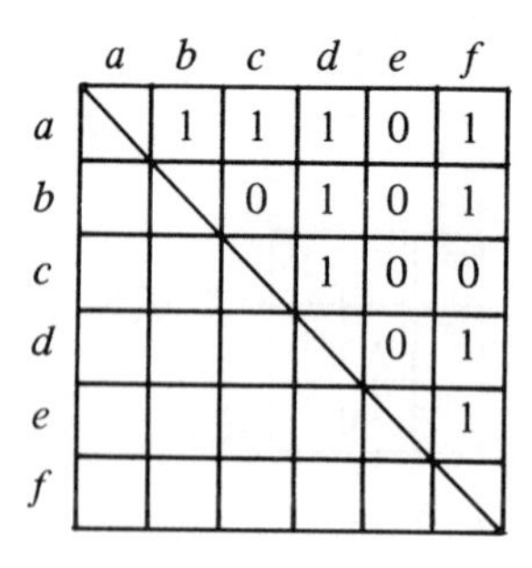

	a	b	c	d	e	f
a		1	1	1	0	1
b			0	1	0	1
c				1	0	0
d					0	1
e						1
f						

Fig. B.4

†This method has been proposed by M. Etienne Pichat, Professor at the Institut d'Informatique d'Enterprise, Conservatoir National des Arts et Metiers, Paris. Consult [IP] and see also Pichat, E., Algorithm for Finding the Maximal Elements of a Finite Universal Algebra. *Inform. Processing* 68 (1969).

Row by row, we consider the zeros. For each zero, by taking the elements as boolean variables, associate with the boolean sum sign $\dot{+}$ the index element of the row and the corresponding elements of the columns in which the zeros are found, which will be joined in a boolean products $\boldsymbol{.}$. If no zero is found in a row, we shall consider that the sum is equal to 1.

Form the product of the results and express the function obtained in minimal terms.† Then take the complement of each term. One thus obtains the maximal subrelations consitfuting a covering.

Thus, considering the example given in Figure B.2, whose half-matrix is presented in Figure B.4,

(B.73)

Line	a	: one forms	$a \dot{+} e$,
Line	b	: one forms	$b \dot{+} c\,e$,
Line	c	: one forms	$c \dot{+} e\,f$,
Line	d	: one forms	$d \dot{+} e$,
Line	e	: one forms	1 ,
Line	f	: one forms	1 ,

One then has

(B.74)

$$\begin{aligned}
S &= (a \dot{+} e) \,.\, (b \dot{+} c\,e) \,.\, (c \dot{+} e\,f) \,.\, (d \dot{+} e) \,.\, 1 \,.\, 1 \\
&= (a \dot{+} e)\cdot(b \dot{+} c\,e) \,.\, (c \dot{+} e\,f) \,.\, (d \dot{+} e) \\
&= (ab \dot{+} ace \dot{+} be \dot{+} ce)\,(c \dot{+} ef)\,(d \dot{+} e) \\
&= (abc \dot{+} abef \dot{+} ce \dot{+} cef \dot{+} bce \dot{+} bef)\,(d \dot{+} e) \\
&= (abc \dot{+} bef \dot{+} ce)\,(d \dot{+} e) \\
&= abcd \dot{+} abce \dot{+} bdef \dot{+} bef \dot{+} ced \dot{+} ce \\
&= abcd \dot{+} bef \dot{+} ce \,.
\end{aligned}$$

We calculate S′, where the respective terms are the complements of those of S. One has

(B.75)
$$S' = ef \dot{+} acd \dot{+} abdf$$

This gives the three subsets

(B.76)
$$\{e\,,f\}\ ,\ \{a\,,c\,,d\}\ ,\ \{a\,,b\,,d\,,f\},$$

for which one has the principal submatrices forming the covering.

Remark. One may be interested in the elements common to the nondisjoint relations two by two; one obtains these immediately by intersections:

†Following the rules of simplification of boolean expressions, but using the properties $x + x = x;\ x \cdot x = x;\ x + xy = x$.

$$\{e, f\} \cap \{a, c, d\} = \phi,$$

(B.77)
$$\{a, c, d\} \cap \{a, b, d, f\} = \{a, d\},$$

$$\{a, b, d, f\} \cap \{e, f\} = \{f\}.$$

The search for maximal similitude subrelations in a fuzzy preorder $\underset{\sim}{\mathcal{R}}$. Either one of the two preceding algorithms may be used to determine these subrelations. It suffices to consider the associated boolean matrix $\mathcal{R}$ such that

(B.78)
$$\mu_{\mathcal{R}}(x, y) = \mu_{\mathcal{R}}(y, x) = 1 \quad \text{if} \quad \mu_{\underset{\sim}{\mathcal{R}}}(x, y) = \mu_{\underset{\sim}{\mathcal{R}}}(y, x) > 0.$$

and

(B.79)
$$\mu_{\mathcal{R}}(x, y) = 0 \quad \text{and} \quad \mu_{\mathcal{R}}(y, x) = 0$$

$$\text{if} \quad \mu_{\underset{\sim}{\mathcal{R}}}(x, y) \neq \mu_{\underset{\sim}{\mathcal{R}}}(y, x)$$

$$\text{or} \quad \mu_{\underset{\sim}{\mathcal{R}}}(x, y) = \mu_{\underset{\sim}{\mathcal{R}}}(y, x) = 0$$

Example. Consider again the example given in Figure 21.3, which we have reproduced in Figure B.5. In this figure we have given the boolean matrix associated with $\underset{\sim}{\mathcal{R}}$ and designated by $\mathcal{R}$ conforming to (B.78) and (B.79).

$\underset{\sim}{\mathcal{R}}$	A	B	C	D
A	1	0,2	0,2	0,5
B	0,2	1	0,2	0,2
C	0,5	0,2	1	0,5
D	0,2	0,2	0,2	1

$\mathcal{R}$	A	B	C	D
A	1	1	0	0
B	1	1	1	1
C	0	1	1	0
D	0	1	0	1

Fig. B.5

We use the second method. One has

(B.80)
$$\begin{aligned} S &= (a \dotplus cd)(c \dotplus d) \\ &= ac \dotplus ad \dotplus cd \dotplus cd \\ &= ac \dotplus ad \dotplus cd. \end{aligned}$$

(B.81)
$$S' = bd \dotplus bc \dotplus ab.$$

Thus, there exist three maximal similitude subrelations in the preorder.These concern the ordinary subsets

(B.82)
$$\{b, d\}, \ \{b, c\}, \text{ and } \{a, b\}.$$

These are the subrelations given in Figure 21.3.

CONCLUSION

The goal of this first volume has been to realize, in a certain fashion, a "recycling" for readers used to boolean mathematics, where, to imitate Alfred de Musset, "it is necessary that a door be open or closed." Of course, at this stage the reader is left with his hunger for knowledge of applications. These will be the purpose of other volumes of this work. But, while waiting for the publication that follows, he will find in the bibliography of the present book listings of articles where these applications are mentioned and explicated. One suspects that they are numerous, and they will become more and more so. Human thought is sometimes global (nonsequential), sometimes logical (sequential), and more often a mixture of these modes. Machines for treating information are nearly all structured sequentially and for programming (except analog machines and certain hybrid devices); but throughout one now works for the realization of information systems whose internal and external units of all kinds carry out their tasks in parallel. It is necessary that we, in this domain, reinvent nature in which the marvelous secrets of genetic codes and memories, or neurobiological infrastructure, of processes of storing and operating at the level of neurons, axons, synapses, of dendrites, and of neurology, we are almost totally ignorant.

The path proposed by Zadeh is paved with difficulties, as are all paths opening in the direction of the most fertile generalizations; but the bolt has been sprung, the bolt imposed on analysis and modeling by boolean reasoning. It has been about 130 years since George Boole published in the face of general indifference his fundamental work on the laws of thought; it has been only in the past 20 years that boolean algebra has been used by engineers and introduced into all curricula, from kindergarten to universities. And now all is called into question. Not quite! Boolean algebra will always be useful, wherever precision in calculation and the use of predicates may be required; but it has received, thanks to Zadeh, an unlimited extension, having its sources, moreover, in set theory, which is Boole's theory presented in a neighboring fashion. Following Zadeh, Goguen has again opened other doors and in the leading areas that one penetrates through these doors, there are again new doors, for an exploration that is always larger in the marvelous and unending adventure of mathematics.

In this first volume, several topics that ought to have had their place are missing; I have discarded these with great regret, but otherwise would have had no room; for example,

convex fuzzy subsets,
fuzzy topology,

necessary developments concerning fuzzy propositional logic, fuzzy events, etc.

I shall endeavor to insert these questions in the volumes that follow.

I must repeat that the present work is elementary, intended only for persons who have in the scientific domain interests in applications. It is didactic, and the author has no pretense of progressing through the theoretical knowledge with respect to this work. But it ought to give rich and new ideas to engineers and to all those who have more and more need of mathematics in order to solve their concrete problems.

I would like to express a very sincere wish. I hope that my readers, interested and initiated by my modest work, will go further, very much further, and still further. The humanistic sciences are in need of a mathematics appropriate to our nature, to our fuzzy attitudes, to the nuances of our behavior, to our mensurations, to our multiple criteria. If this first volume is sufficiently stimulating, numerous articles concerning theoretical aspects or applications will be published by readers; concurrent books will see the light of day. All this will rapidly improve our methods with a view toward achieving humane sciences.

BIBLIOGRAPHY

A1 Asai, K., and S. Kitajima, Learning control of multimodal systems of fuzzy automata, *in* "Pattern Recognition and Machine Learning" (K.S.Fu, ed.). Plenum Press, New York, 1971.

A2 Arbib, M.A., Semi-ring languages. Notes, Dep. of Electrical Engineering, Stanford Univ., Stanford, California, 1970.

A3 Asai, K., and S. Kitajima, Optimising control using fuzzy automata, *Automatica–J. IFAC 8* (1972), 101–104.

A4 Asai, K., and S. Kitajima, A method for optimizing control of multimodal systems using fuzzy automata, *Information Sci.* **3** (1971), 343–353.

B1 Banaschewski, B., Injective hulls in the category of distributive lattices, *J. Reine Angew. Math.* **232** (1971), 102–109.

B2 Banaschewski, B., and G. Burns, Categorical characterization of the McNeville completion, *Arch. Math. (Basel)* **18** (1967), 369–377.

B3 Bellman, R.E., R. Kalaba, and L.A. Zadeh, Abstraction and pattern classification, *J. Math. Anal. Appl.* **13** (1967), 369–377.

B4 Bellman, R.E., and L.A. Zadeh, Decision making in a fuzzy environment, *Management Sci.* **17** (4) (December 1970).

B5 Boiscescu, V., Sur les algébres de Lukaciewicz, *in* "Logique, Automatique, Informatique." Acad. Rep. Soc. de Roumanie, Bucharest, 1971.

B6 Brown, J.G., Fuzzy sets and boolean lattices. Rep. 1957, Ballistic Res. Lab., Aberdeen, Maryland, January 1969.

B7 Brown, J.G., A note on fuzzy sets, *Information and Control* **18** (1971), 32–39.

B8 Black, M., Vagueness: An exercise in logical analysis, *Philos. Sci.* **17** (1970), 141–164.

B9 Bunge, M.C., Categories of set-valued functors. Ph.D. Thesis, Dep. of Mathematics, Univ. of California, 1966.

B10 Bellman, R.E., and M. Giertz, On the analytic formalism of the theory of fuzzy sets, *Information Sci.* **5** (1973), 149–156.

B11 Black, M., Reasoning with loose concepts, *Dialogue* **2** (1973), 1–12.

B12 Bezdek, J.C., Numerical taxonomy with fuzzy sets, *J. Math. Bio.* **1** (1974), 57–71.

C1 Chang, C.L., Fuzzy topological spaces, *J. Math. Anal. Appl.* **24** (1968), 182–190.

C2 Chang, S.K., Picture processing grammar and its applications, *Information Sci.* **3** (1971), 121–148.

C3 Chang, S.K., Fuzzy programs, theory and applications, *P.I.B. Proc. Comput. Automata* **21** (1971).

C4 Chang, S.K., Automated interpretation and editing of fuzzy line drawing, *in Proc. 1971 Spring Joint Computer Conf., AFIPS Proc.*, pp. 393–399, 1971.

C5 Chang, S.S.L., Fuzzy dynamic programming and the decision making process, *in Proc. Princeton Conf. Information Sci. Syst. 3rd*, pp. 200–203, 1969.

C6 Chang, S.S.L., and L.A. Zadeh, Fuzzy mapping and control, *IEEE Trans. Systems, Man, and Cybernetics* **SMC2** (1972), 30–34.

C7 Cignoli, R., Estudio algebraice de logicas polival entes; algebra de Moisil de orden *n* Ph.D. Thesis, Univ. Mac. de Sur-Bahia-Blanco, Argentina.

C8 Chang, S.K., On the execution of fuzzy programs using finite-state machines, *IEEE Trans. Computers* **C21** (1972), 231–253; *idem* [S4].

C9 Conche, B., and P. Courant, Applications des concepts flous aux problemes multi-criteres. Seminaire Bernard Roy, Univ. of Paris, Dauphine, 1972.

C10 Conche, B., J.P. Jouault, and P.M. Luan, Application des concepts flous a la programmation en langages quasi-naturel. Seminaires Bernard Roy, Univ. of Paris, Dauphine, 1973.

C11 Conche, B., Elements d'une méthode de classification par utilisation d'un automate flou. J.E.E.F.L.N., Univ. of Paris, Dauphine, 1973.

C12 Chang, C.C., Infinite-valued logic as a basis for set theory, *Proc. Internat. Congr. Logic, Methodol. Philos. Sci.* North Holland, Amsterdam, 1964.

C13 Chapin, E.W., An axiomatization of the set theory of Zadeh, *Notices Amer. Math. Soc.* **687-02-4** (1971), 753.

C14 Cohen, P.J., and R. Hirsch, Non-Cantorian set theory, *Scientific American* (December 1967), 101–116.

C15 Cools, M., and Peteau, M., STIM 5: Un programme de stimulation inventive utilisant la theorie des sous-ensembles flous. IMAGO discussion paper, Univ. Catholique Louvain, November 1973.

C16 Cools, M., La semantique dans les processus didactiques. IMAGO discussion paper, Univ. Catholique Louvain, November 1973.

C17 Chang, C.L., Fuzzy algebras; fuzzy functions and their applications to function approximation. Div. of Computer Res. and Technol., Nat. Inst. Health, Bethesda, Maryland, 1971.

D1 Diday, E., Optimisation en classification automatique et reconnaissance des formes, *Notes Scientifique IRIA,* **6** bull. 12, May–June 1972.

D2 DeLuca, A., and S. Termini, A definition of a non-probabilistic entropy in the setting of fuzzy set theory, *Information and Control* **20** (1972), 301–312.

D3 DeLuca, A., and S. Termini, Algorithmic aspects in the analysis of complex systems. Rep. LC/54/71AL, Lab. di Cibernetica del CNR, Naples, Italy, February 1971.

D4 Dubois, T., Une methode d'evaluation par les sous-ensembles flous appliquée a la simulation. IMAGO discussion paper, Univ. Catholique Louvain, January 1974.

D5 Dunn, J.C., A graph theoretic analysis of pattern classification via Tamura's fuzzy relations, *IEEE Trans. Systems, Man, and Cybernetics* **SMC4** (1974), 310–313.

F1 Floyd, R.W., Non-deterministic algorithms, *J. Assoc. Comput. Mach.* **14** (1967), 636–644.

F2 Fu, K.S., A critical review of learning control research, *in* "Pattern Recognition and Machine Learning" (K.S.Fu, ed.). Plenum Press, New York, 1971.

F3 Fee, W.G., and K.S. Fu, A formulation of fuzzy automata and its applications as a model of learning systems, *IEEE Trans. Systems, Simulation, Computing* **SSC5** (1969) 215–223.

F4 Fung, L., and K.S. Fu, The kth optimal policy algorithm for decision making in a fuzzy environment, *Proc. IRAC Symp. Identification System Parameter Estimation 3rd,* 1973.

F5 Fine, K., Vagueness, truth, and logic. Dep. of Logic, Univ. of Edinburgh, 1973.

G1 Georgescu, G., Les algebres de Lukaciewicz theta-valentes, *in* "Logique, Automatique, Informatique," pp. 99–176. Acad. Rep. Soc. de Roumaine, Bucharest, 1971.

G2 Gitman, I., Organization of data, a model and computational algorithm that uses the notion of fuzzy sets. Ph.D. Thesis, McGill University, Montreal, 1970.

G3 Goguen, J.A., *L*-Fuzzy sets, *J. Math. Anal. Appl.* **18** (1967), 145–174.

G4 Goguen, J.A., The logic of inexact concepts, *Synthese* **19** (1969), 325–373.

G5 Goguen, J.A., Some comments on applying mathematical systems theory. Dep. of Computer Sci., Univ. of California, Los Angeles, 1973.

G6 Goguen, J.A., Axioms, extensions, and applications for fuzzy sets: Languages and representations of concepts. Dep. of Computer Sci., Univ. of California, Los Angeles, 1973.

G7 Gitman, I., and M.D. Levine, An algorithm for detecting unimodal fuzzy sets and its application as a clustering technique, *IEEE Trans. Computers* **C19** (1970), 583–593.

G8 Goguen, J.A., Categories of fuzzy sets: Applications of non-Cantorian set theory. Ph.D. Thesis, Univ. of California, Berkeley, May 1968.

G9 Goguen, J.A., Categories of *L*-fuzzy sets, *Bull. Amer. Math. Soc.* **75** (1969), 622–624.

G10 Goguen, J.A., Mathematical representation of hierarchically organized systems, global systems dynamics, Karger, 1970.

G11 Goguen, J.A., Representing inexact concepts. Inst. for Computer Res. Quart. Rep. 20, Univ. of Chicago, February 1970.

G12 Goguen, J.A., Hierarchical inexact data structures in artificial intelligence, *Proc. Hawaii Internat. Conf. Systems Sci. 5th, Honolulu, 1972.*

G13 Goguen, J.A., The fuzzy Tychonoff theorem, *J. Math. Anal. Appl.* (1973).

G14 Goguen, J.A., J. Thatcher, E. Wagner, and J. Wright, A junction between computer science and category theory, basic concepts and examples II, universal constructions, *IBM Res. Rep.*, 1973.

G15 Goguen, J.A., Systems theory concepts in computer science, *Proc. Hawaii Internat. Conf. Systems Sci. 6th, Honolulu, 1973*, pp. 77–80.

G16 Gluss, B., Fuzzy multistage decision making, fuzzy states and terminal regulators and their relationship to non-fuzzy quadratic state and terminal regulators, *Internat. J. Control* **17** (1973), 177–192.

G17 Gentilhomme, Y., Les ensembles flous en linguistique, *Cahiers de Linguistique Theorique et Appliquée* **47** (1968).

G18 Goguen, J.A., Concept representation in natural and artificial languages: axioms, extensions, and applications for fuzzy sets, *Internat. J. Man–Machine Studies* **6** (1974), 513–561.

H1 Hirai, H., K. Asai, and S. Katajima, Fuzzy automaton and its applications to learning control systems. Memo number 10, Faculty of Engineering, Osaka City Univ., Japan, 1968.

H2 Henry-Labordere, A., and P. de Backer, Intelligence creatrice, automatization de processus d'association, *Revue METRA Internat.* **Rep. 69,** December 1972.

H3 Hintikka, J., J. Moravcsik, and P. Suppes (eds.), "Approach to Natural Languages." Reidel, Dordrecht, Netherlands, 1973.

J1 Jouault, J.P., and Pham Minh Luan, Application des concepts flous a la programmation en lagages quasi-naturels. Institut Informatique d'Enterprise, C.N.A.M. Paris, 1975.

J2 Jones, A., Towards the right solution, *Internat. J. Math. Educ. Sci. Tech.* **5** (1974), 337-357.

K1 Killing, R., Fuzzy planner. Tech. Rep. 168, Computer Sci. Dep., Univ. of Wisconsin, Madison, February 1973.

K2 Kitajima, S., and K. Isai, A method of learning control varying search domain by fuzzy automata, *Japan–US Seminar, Florida, October, 1973.*

K3 Kalmanson, D., Recherche cardio-vasculaire et théorie des ensembles flous, *La Nouvelle Presse Médicale* **41** (November 1973), 2757–2760.

K4 Kaufmann, A., M. Cools, and T. Dubois, Stimulation inventive dans un dialogue homme–machine utilisant la méthode des morphologies et la théorie des sous-ensembles flous. IMAGO discussion paper 6, Univ. Catholique Louvain, October 1973.

K5 Kandel, A., On minimization of fuzzy functions, *IEEE Trans. Computers* **C22** (1973),

K6 Kandel, A., Comment on an algorithm that generates fuzzy prime implicants by Lee and Chang, *Information and Control* **22** (1973),

K7 Knuth, D., Semantics of context-free languages, *Math. Systems Theory* **2** (1968), 127–145.

K8 Kluska-Nawarecka, S., E. Mysona-Burska, and E. Nawarecki, Algorithmes de commande pour certaines classes de problèmes opérationnels construits avec utilisation de la simulation numérique, Doc. Inst. de Fonderie, Cracovie, et Inst. d'Informatique, Académie des Mines et Métallurgie de Cracovie, Poland, January 1975.

K9 Kandel, A., and L. Yelowitz, Fuzzy chains, *IEEE Trans. Systems, Man, and Cybernetics* **SMC4** (September 1974),

K10 Kandel, A., On the analysis of fuzzy logic, *in Proc. Internat. Conf. Systems Sci. 6th, Honolulu, January 1973.*

K11 Kandel, A., Application of fuzzy logic in the detection of static hazards in combinatorial switching machines. Computer Sci. Rep. 122, New Mexico Inst. of Mining and Technology, Sorocco, New Mexico, April 1973.

K12 Kandel, A., Comments on the minimization of fuzzy functions, *IEEE Trans. Computers* **C22** (1973), 217.

K13 Kaufmann, A., "Introduction a la Théorie des sous-ensembles flous," Tome I: Eléments théoriques de base; Tome II: Langages, sémantique, et logique; Tome III: Classification et reconnaissance des formes, automates, machines de Turing, problèmes multi-critéres, systèmes. Masson, Paris, 1973–1975.

K14 Kaufmann, A., M. Cools, and T. Dubois, "Exercices avec Solutions sur la Theories des Sous-ensembles Flous." Masson, Paris, 1975.

L1 Lee, R.C.T., Fuzzy logic and the resolution principle, *in Proc. Joint Artificial Intelligence Conf. 2nd, London, 1971.*

L2 Lee, E.T., and L.A. Zadeh, Note on the fuzzy languages, *Information Sci.* **1** (1969), 421–454.

L3 Lukaciewicz, J., Logike trojwartoscieweg, *Ruch. Filosofiezne* **169** (1920).

L4 Lukaciewicz, J., and A. Tarski, Untersuchungen uber den Aussagenkalkul, *R. Soc. Lettres (Varsovie)* **XXIII** (1930), 30–50.

L5 Lakoff, G., Hedges: A study of meaning criteria and the logic of fuzzy concepts, *in Proc. Regional Meeting Chicago Linguistic Soc. 8th, Univ. of Chicago, April 1972.*

L6 Lientz, B.F., On time dependent fuzzy sets, *Information Sci.* **4** (1972), 367–376.

L7 Lombaerde, J., Mesures d'entropie en théorie des sous-ensembles flous. IMAGO discussion paper 12, Univ. Catholique Louvain, January 1974.

L8 Lee, E.T., and C.L. Chang, Some properties of fuzzy logic, *Information and Control* **19** (1971), 417–431.

L9 Lee, E.T., and L.A. Zadeh, Fuzzy languages and their acceptance by automata, *in Proc. Princeton Conf. Information Sci. Systems 4th,* p. 399, 1970.

L10 Lowen, R., Topologie générale, topologie floue, *C.R. Acad. Sci. Paris Ser. A* **278** (1974), 925.

L11 Lakoff, G., Linguistic and natural logic, *in* "Semantics of Natural Languages" (D. Davidson and G. Harman, eds.). Reidel, Dordrecht, Netherlands, 1971.

M1 Marinos, P.M., Fuzzy logic and its applications to switching systems, *IEEE Trans. Computers* **C18** (1969), 343–348.

M2 Marinos, P.N., Fuzzy logic. Tech. memo 66-2344-1, Bell Telephone Lab, Holmdel, New Jersey, August 1966.

M3 Mizumoto, N., K. Toyota, and K. Tanaka, Some considerations on fuzzy automata, *J. Computer Systems* **3** (1969), 409–422.

M4 Moisil, G.C., Recherches sur les logiques non-chrysipiennes, *Ann. Sci. Univ. Jassy* **24** (1940).

M5 Moisil, G.C., Notes sur les logiques non-chrysipiennes, *Ann. Sci. Univ. Jassy* **27** (1941).

M6 Moisil, G.C., "Incercari Vechi si Noi de Logica Neclasica." Stiintifica, Bucharest, 1965.

M7 Moisil, G.C., Logique modale, *Disquisitiones Mat. Phys.* **II1** (1942), 3–98.

M8 Moisil, G.C., Lukasiewiczian algebras. Centre de Calcul de l'Universite de Bucarest, October 1968.

M9 Moisil, G.C., Sur les logiques de Ludaciewicz a un nombre fini de variables, *Rev. Roumaine Math. Pures Appl.* **IX** (1964), 905.

M10 Mizumoto, M., J. Toyoka, and K. Tanaka, *N*-fold fuzzy grammars, *Information Sci.* **5** (1973) 25–43.

M11 McNaughton, R., A theorem about infinite-valued sentencial logic, *J. Symbolic Logic* **16** (1951), 1–13.

M12 Malvache, N., and P. Vidal, Application des systèmes flous a la modèlisation des phenomènes de prise de décision et d'appréhension des informations visuelles chez l'homme, A.T.P.–C.N.R.S. **1K05,** Paris, July 1974.

M13 Malvache, N., Analyse et identification des systèmes visuel et manuel en vision frontale et periphérique chez l'homme, Ph.D. Thesis, Lelle, April 1973.

M14 Malvache, N., G. Hilbred, and P. Vidal, Perception visuelle: champ de vision latérale, modèle de la fonction du regard. Rapport de synthèse, Contrat D.R.M.E., 71-251, Paris, 1973.

M15 Malvache, N., and D. Willayes, Représentation et minimisation de fonctions floue. Centre Univ. de Valenciennes, France, October 1974.

M16 Mizumoto, M., Fuzzy automata and fuzzy grammars, Ph.D. Thesis, Faculty of Engineering, Osaka, University, 1971.

N1 Nadiu, G.S., Sur la logique de Heytung, *in* "Logique, Automatique, Informatique," pp. 42–70. Acad. Rep. Soc. de Roumanie, Bucharest, 1971.

N2 Nazu, M., and N. Honda, Fuzzy events realized by finite probabilistic automata, *Information and Control* **12** (1968), 248–303.

N3 Nazu, M., and N. Honda, Mapping induced by PSGM-mapping and some recursively unsolvable problems of finite probabilistic automata, *Information and Control* **15** (1969), 250–273.

N4 Netto, A.B., Fuzzy classes, *Notices Amer. Math. Soc.* **68T-H28** (1968), 945.

N5 Negoita, C.V., On the application of the fuzzy set separation theorem for automatic classification in information retrieval systems, *Information Sci.* **5** (1973), 279–286.

N6 Negoita, C.V., and D.A. Falescu, Multimi Vagi si Applicatille Ler. Editura Tehnica, Bucharest, 1974.

O1 Oniga, T., Developpements de la logique trivalente, *Rev. Cybernetica Namur,* 1974.

P1 Post, E.L., Introduction to a general theory of elementary propositions, *Amer. J. Math.* **43** (1921), 163–185.

P2 Paz, A., Fuzzy star functions. Probabilistic automata and their applications in nonprobabilistic automata, *J. Computer Systems Sci.* **1** (1967), 371 – 390.

P3 Preparata, F.D., and R.T. Yeh, Continuous valued logic, *J. Computer Systems Sci.* **6** (1972), 397–418.

P4 Poston, T., Fuzzy geometry, Ph.D. Thesis, Univ. of Warwick, England, 1971.

P5 Prugovecki, E., Measurement in quantum mechanics as a stochastic process on spaces of fuzzy events, *Found. Phys.* **4** (1974).

P6 Prugovecki, E., Fuzzy sets in the theory of measurement of incompatible observables, *Found. Phys.* **4** (1974).

P7 Prugovecki, E., A postulational framework for theories of simultaneous measurements of several observables, *Found. Phys.* **3** (1973), No. 4.

R1 Ruspini, E.R., Numerical methods for fuzzy clustering, *Information Sci.* **2** (1970), 319–350.

R2 Ruspini, E.R., A new approach to clustering, *Information and Control* **15** (1969), 22–32.

R3 Rosenfeld, A., Fuzzy groups, *J. Math. Anal. Appl.* **35** (1971), 512–517.

R4 Russell, B., Vagueness, *Austral. J. Phil.* **1** (1923), 84 – 92.

R5 Reisinger, L., On fuzzy thesauri, *in Compstat 1974, Proc. Computational statist., Vienna, 1974.*

S1 Santos, E.S., Maximin automata, *Information and Control* **13** (1968), 363–367.

S2 Santos, E.S., Fuzzy algorithms, *Information and Control* **17** (1970), 326–339.

S3 Santos, E.S., and N.G. Wee, General formulation of sequential machines, *Information and Control* **12** (1970), 5–10.

S4 Shi-Kuo-Chang, On the execution of fuzzy algorithms using finite state machines, *IEEE Trans. Computers* **C21** (1972).

S5 Stone, M.H., Topological representation of distributive lattices and Brouwerian logics, *Cas. Math. Phys.* **67** (1937), 1–25.

S6 Santos, E.S., On reduction of maximin machines, *J. Math. Anal. Appl.* **37**(3) (1972).

S7 Santos, E.S., Max-product machines, *J. Math. Anal. Appl.* **37**(3) (1972).

S8 Santos, E.S., Maximin sequential-like machines and chains, *Math. Systems Theory* **3**(4) (1969).

S9 Siy, P., and C.S. Chen, Minimization of fuzzy functions, *IEEE Trans. Computers* **C21** (1972), 100–102.

S10 Sanchez, E., Equations de relations floues. Ph.D. Thesis, Marseille, 1974.

S11 Serfati, M., "Algebres de Boole avec une Introduction a la Théorie des Sous-Ensembles Floues." C.D.U., Paris, 1974.

S12 Sugeno, M., Fuzzy measures and fuzzy integrals, *Trans. Soc. Instruments and Control Engineers* (in Japanese) **9**(3) (1973).

S13 Sugeno, M., Constructing fuzzy measures and grading similarity of patterns by fuzzy integrals, *Trans. Soc. Instruments and Control Engineers* (in Japanese) **9**(3) (1973).

S14 Sugeno, M., and T. Terano, An approach to the identification of human characteristics by applying fuzzy integrals, in *Proc, IFAC Symp. Identification Sys. Parameter Estimation 3rd, 1973.*

S15 Sugeno, M., Tsukamoto, Y., and T. Terano, Subjective evaluation of fuzzy objects. Preprint of *IFAC Symp. Stochastic Control,* 1974.

S16 Sugeno, M., Theory of fuzzy integrals and its applications. Ph.D. Thesis, Tokyo Inst. of Technology, 1974.

S17 Siy, P., Fuzzy logic and handwritten character recognition. Ph.D. Thesis, Univ. of Akron, Ohio, 1973.

T1 Tamura, S., S. Higuchi, and K. Tanaka, Pattern classification based on fuzzy relations, *IEEE Trans. Systems, Man, and Cybernetics* **SMC1** (1971).

T2 Tsichritzis, D., Fuzzy properties and almost solvable problems. Tech. rep. 70, Dep. of Electrical Engineering, Princeton Univ., Princeton, New Jersey, 1968.

T3 Tsichritzis, D., Measures on countable sets. Tech. rep. 8, Univ. of Toronto, Dep. of Computer Science, Toronto, Canada, 1969.

T4 Tamura, S., and K. Tanaka, Learning of fuzzy languages, *IEEE Trans. Systems, Man, and Cybernetics* **SMC3** (1973), 98–102.

T5 Thomason, M.G., The effect of logic operations on fuzzy logic, *IEEE Trans. Systems, Man, and Cybernetics* **SMC4** (1974), 309–310.

V1 Vincke, P., Une application de la théorie des graphes flous. Univ. Libre de Bruxelles, 1973.

V2 Vincke, P., La théorie des Ensembles flous. Mémoire de licence en sciences mathématiques, Univ. Libre de Bruxelles, 1973.

V3 Van Velthoven, G., Application of fuzzy set theory to criminal investigation, *First European Congr. Operations Res., EURO 1, Bruxelles, January 1975.*

V4 Vayer, A., Ebauche d'une étude théorique des faits administratifs. Document Centres d'Entrainement aux Méthodes d'Education Active, December 1974.

W1 Wee, W.G., On generalizations of adaptive algorithms and applications of the fuzzy set concepts to pattern classification. Tech. rep. TREE 67-7, Dep. of Electrical Engineering, Purdue Univ., Lafayette, Indiana, June 1967.

W2 Wee, W.G., and K.S. Fu, A formulation of fuzzy automata and its application as a model of learning systems, *IEEE Trans. Systems Sci. and Cybernetics* **SSC5** (1969), 215–223.

W3 Watanable, S., Modified concepts of logic, probability, and information based on generalized continuous characteristic functions, *Information and Control* **15** (1969), 1–21.

W4 Wong, C.K., Covering properties of fuzzy topological spaces, IBM res. rep., Yorktown Heights, New York, 1971.

W5 Wong, C.K., Fuzzy topology: product and quotient theorems, *J. Math. Anal. Appl.* (1973).

W6 Wong, C.K., Fuzzy points and local properties of fuzzy topology. Rep. IVCDCS-R-73-561, Univ. of Illinois, Urbana, Champaign, 1973.

W7 Wong, G.A., and D.C. Shen, On the learning behavior of fuzzy automata. Paper presented at *Internat. Cong. Cybernetics Systems, Oxford, 1972.*

Y1 Yoeli, M., A note on a generalization of boolean matrix theory, *Amer. Math. Monthly* 38 (1961), 552–557.

Z1 Zadeh, L.A., Fuzzy sets, *Information and Control* 8 (1965), 338–353.

Z2 Zadeh, L.A., Fuzzy sets and systems, *Proc. Symp. Systems Theory, Brooklyn Polytech. Inst. Brooklyn, New York,* pp. 29–39, 1966.

Z3 Zadeh, L.A., Shadows of fuzzy sets, *Problems in Transmission of Information* **2** (1966), 37–44 (in Russian).

Z4 Zadeh, L.A., Fuzzy algorithms, *Information and Control* **12** (1968), 99–102.

Z5 Zadeh, L.A., Probability measure of fuzzy events, *J. Math. Anal. Appl.* **10** (1968), 421–427.

Z6 Zadeh, L.A., Biological applications of the theory of fuzzy sets and systems, *in* "Biocybernetics of the Central Nervous Systems" (L.C. Proctor, ed.), pp. 199–206, discussion by W. Kilmer, pp. 207–212. Little Brown, Boston, 1969.

Z7 Zadeh, L.A., Toward a theory of fuzzy systems. E.R.L. rep. 69.2, Electrical Res. Lab., Univ. of California, Berkeley, July 1969.

Z8 Zadeh, L.A., Toward fuzziness in computer systems, fuzzy algorithms and languages. Dep. of Electrical Engineering and Computer Sci., Univ. of California, Berkeley, 1969.

Z9 Zadeh, L.A., Quantitative fuzzy semantics, *Information Sci.* **3** (1970), 1–17.

Z10 Zadeh, L.A., Similarity relations and fuzzy ordering. E.R.L. memo M277, Electrical Res. Lab., Univ. of California, Berkeley, 1970.

Z11 Zadeh, L.A., Fuzzy languages and their relation to human and machine intelligence, *Proc. Conf. Man and Computer,* Institute de la Vie, Bordeaux, France, 1970; E.R.L. memo M302, Univ. of. California, Berkeley, August 1971.

Z12 Zadeh, L.A., Machine intelligence versus human intelligence, *Proc. Conf. Sci. and Society, Herceg Novi, Yugoslavia,* pp. 127–134, June 1969.

Z13 Zadeh, L.A., A system-theoretic view of behavior modification. Center for the Study of Democratic Institutions, Santa Monica, California, January 1972; E.R.L. memo M320, Univ. of California, Berkeley, January 1972.

Z14 Zadeh, L.A., On fuzzy algorithms. E.R.L. memo M325, Univ. of California, Berkeley, February 1972.

Z15 Zadeh, L.A., A rationale for fuzzy control, *J. Dynamical Systems Measurement and Control* 3 (1972), 3–4.

Z16 Zadeh, L.A., A fuzzy set theoretic interpretation of linguistic hedges. E.R.L. memo M335, Univ. of California, Berkeley, April 1972.

Z17 Zadeh, L.A., Outline of a new approach to the analysis of complex systems and decision processes. E.R.L. memo M342, Univ. of California, Berkeley, July 1972.

Z18 Zadeh, L.A., Linguistic cybernetics, *Proc. Internat. Symp. Systems Sci. and Cybernetics, Oxford Univ., 1972.*

Z19 Zadeh, L.A., The concept of linguistic variable and its application to approximate reasoning. E.R.L. memo M411, Univ. of California, Berkeley, October 1973.

Z20 Zadeh, L.A., Numerical versus linguistic variables, *Newspaper of the Circuits and Systems Society* 7 (1974), 3.

Z21 Zadeh, L.A., On the analysis of large-scale systems. E.R.L. memo M418, Univ. of California, Berkeley, January 1974; *Proc. Conf. Systems Theory and Environmental Problems, Bavarian Acad. Sci., Munich, 1973.*

Z22 Zadeh, L.A., Fuzzy logic and its applications to approximate reasoning, *Proc. IFIP Congr., August 1974.*

Z23 Zimmermann, H.J., and H. Gehring, Fuzzy information profiles for information selection, Congress book, Volume II, *Internat. Congr. AFCET 4th, Paris, 1975.*

Z24 Zadeh, L.A., A fuzzy algorithm approach to the definition of complex of imprecise concepts. E.R.L. memo M474, Univ. of California, Berkeley, 1974.

Z25 Zadeh, L., Fuzzy logic and approximate reasoning, E.R.L. memo M479, Univ. of California, Berkeley, 1974.

REFERENCES

1B Berge, C., "Théorie des Graphes et Ses Applications." Dunod, Paris.

2B Bruter, C.P., "Les Matroïdes–Nouvel Outil Mathématique." Dunod, Paris.

3B Busacker, R.G., and Saaty, T.L., "Finite Graphs and Networks–An Introduction with Applications". McGraw-Hill, New York.

1C Cullman, G., "Cours de Calcul Informationnel." Albin Michel, Paris.

1D DePalma, R., "Cours de Calcul Automatique." Albin Michel, Paris.

1E Ehresmann, C., "Catégories et Structures." Dunod, Paris.

1F Faure, R., and Heurgon, E., "Structures Ordonnées et Algèbres de Boole." Gauthier-Villars, Paris.

2F Faure, R., Kaufmann, A., and Denis-Papin, M., "Cours de Calcul Booléien Appliqué." Albin Michel, Paris.

1G Gross, M., and Lentin, A., "Notions sur les Grammaires Formelles." Gauthier-Villars, Paris.

1H Harary, F., Norman, R.Z., and Cartwright, D., "Structural Models–An Introduction to the Theory of Directed Graphs." Wiley, New York.

2H Hopcroft, J.E., and Ullman, J.D., "Formal Languages and Their Relations to Automata." Addison-Wesley, Reading, Massachusetts.

1K Kaufmann, A., "Introduction à la Combinatorique." Dunod, Paris.

2K Kaufmann, A., "Méthodes et Modèles de la Recherche Operationelle," Tome II. Dunod, Paris.

3K Kaufmann, A., and Precigout, M., "Cours de Mathématiques Nouvelles pour Recyclage des Ingénieurs." Dunod, Paris.

4K Kaufmann, A., and Denis-Papin, M., "Cours de Calcul Matriciel Appliqué." Albin Michel, Paris.

5K Kaufmann, A., "Des Points et des Flèches. . .La Théorie des Graphes." Science Poche, Dunod, Paris.

6K Kaufmann, A., Fustier, M., and Drevet, A., "L'Inventique, Nouvelles Méthodes de Créativité." E.M.E., Paris.

7K Kaufmann, A., "La Confiance Technique–Théorie Mathématique de la Fiabilité." Science Poche, Dunod, Paris.

8K Kaufmann, A., "Cours Moderne de Calcul de Probabilites." Albin Michel, Paris.

1M Mitchell, B., "Theory of Categories." Academic Press, New York.

1P Pichat, E., "Contribution à l'Algorithmique non Numérique dans les Ensembles Ordonnées." Thesis, Faculty of Sciences, Univ. of Grenoble, Grenoble, France.

1R Robert, P., "Algèbre–Introduction aux Catégories et Structures." Course at Faculty of Sciences, Univ. of Caen, Caen, France.

2R Roy, B., "Algèbre Moderne et Théorie des Graphes," Vols. I and II. Dunod, Paris.

1S Szasz, G., "Introduction to Lattice Theory." Academic Press, New York.

1V Vincke, A., "La Théorie des Ensembles Flous." Mémoire de License en Sciences Mathématiques, Free Univ. of Brussels, Brussels.

1Z Zadeh, L.A., *et al.*, "Fuzzy Sets and Their Applications to Cognitive and Decision Processes." Academic Press, New York, 1975.

INDEX

A

Algebra, boolean binary, 192
Algebraic product, 34
  of two relations, 53
Algorithm of Malgrange, 387
Antipalindrome enumeration, 198
Antiroutes, method of, 241
Antisymmetry, 105
  perfect, 108
Antitonic, 342
Automorphism, 347

B

Berge's graph, 43
Biassociation, 289
Bijection, 339
Bivalued correspondence, 340
Boolean binary algebra, 192
Boolean lattice, 309
Boolean product, 3
  of lattices, 310, 327
  of two sets, 297

C

Category, 358
  fuzzy, 378
Characteristic function, 192
Circuit in a fuzzy graph, 95
Closure
  max–min transitive, 87
  max–product transitive, 134
C-morphism, 361
  L-fuzzy, 371
Commutative ring, 313
Complementation, 10, 337
  pseudo, 335
Complemented lattice, 368
Complement of a relation, 55
Composition
  external, law of, 268, 269
  fuzzy, law of, 267
  internal, law of, 267, 269
  of *L*-fuzzy *C*-morphisms, 371
  max–min, 60
  max–product, 65
  max–star, 65.
  of morphisms, 356
  of two fuzzy relations, 60
  of two nearest ordinary relations, 68
Conditioned fuzzy subset, 71
Conjunction, fuzzy, 245
Container of a fuzzy relation, 49
Correspondence, 357
  bivalued, 340
  univalued, 340

D

Decomposition
  into maximal subrelations, 155
  for a perfect order relation, 160
  of a similitude relation,
   theorem of, 142
Decomposition tree, 150
DeMorgan, theorem of, 32, 33, 36, 196,
  309, 314
Diameter of a graph, 330
Difference, 12, 336
Disjunction, fuzzy, 245
Disjunctive sum, 12, 248, 336
  of two relations, 56
Dissimilitude, relation of
  min–max, 127
  ordinary min–addition, 155
Distance,
  euclidean, 19
    relative, 19
  Hamming, 15
    generalized, 18, 238
    generalized relative, 19
    relative, 16
  min–max
    in a resemblance relation, 132
    in a similitude relation, 130
  notion of, 15, 16
Distributive lattice, 308
Dominance relation, 27
Duality of ordinary sets, 347
Dual of a planar network, 239

E

Element in a fuzzy network, 235
Endomorphism, 347
Entropy, 25, 26
Equality, 10, 335
Equivalence
  fuzzy, 247
  logical, 245
Euclidean distance, 19
  relative, 19
Euclidean norm, 20

F

Form
 polynomial, 201
 reduced polynomial, 202
Function, 340
 characteristic, 1, 192
 common membership, 167 – 180
 of fuzzy variables, 194
 ordinal, of a fuzzy order relation, 124
 ordinal, of an ordinary finite graph, 118
 of performance, 260
 of structure, theory of, 253
Fuzzy binary relation, 28
Fuzzy category, 378
Fuzzy conjunction, 245
Fuzzy exponential integrers, 290
Fuzzy gaussian integers, 293
Fuzzy geometric integers, 292
Fuzzy graph, 41, 281
Fuzzy groupoid, 269
Fuzzy implication, 247
Fuzzy logic, 191, 243
Fuzzy membership, 6
Fuzzy monoid, 278
Fuzzy preorder relation, 98
Fuzzy relation, 46
Fuzzy subcategory, 348
Fuzzy subset, 4 – 6
Fuzzy total order relation, 112
Fuzzy variables, 192, 194, 337

G

Graph
 Berge's, 43
 fuzzy, 41, 281
 transitive, of distance, 145
Groupoid, 368
 fuzzy, 269

H

Hamming distance, 15
 generalized, 18, 328
 generalized relative, 19
 relative, 16
Heterogeneous fuzzy subset, 71

I

Implication, 245
 fuzzy, 247
Inclusion, 8, 334
Index of fuzziness, 217
Inf-semilattice, 312
Injection, 339
Intersection, 2, 336
 of two relations, 51
Isomorphism
 of ordered sets, 343
 of structured sets, 350
Isotonic, 340

L

Lattice, 304
 boolean, 309
 complemented, 308
 distributive, 308
  complemented, 309
 lexicographic, 310
 modular, 307
 product of, 310, 327
 vector, 309
Law
 of external composition, 268, 269
 of fuzzy composition, 267
 internal, 267, 269
Length, path, 94
Lexicographic lattice, 310
L-fuzzy *C*-morphism, 371
Logic, fuzzy, 191, 243

M

Malgrange's algorithm, 387
Mapping, 337
 bijective, 339
 injective, 339
 surjective, 338
Maximal monomial, 202
Maximal similitude subrelation, 387
Max–min composition, 60
Max–min transitivity, 79
Max–product composition, 65
Max–product transitivity, 97
Max–star composition, 60
Max–star transitivity, 97
Membership, fuzzy, 6
Metaimplication, 245
Method of antiroutes, 241
Method of Marinos, 208
Modular lattice, 307
Monoid, fuzzy, 278
Monomial, maximal, 202
Monomorphisms
 of ordered sets, 343
 of structured sets, 350
Monotonicity, 262, 341
Morphisms
 of ordered sets, 342
 of structured sets, 348

N

Nearest ordinary subset, 22
Negation, fuzzy, 245
Network
 of fuzzy elements, 235
 planar, 239
Nonstrict order relation, 113
Norm, euclidean, 20

O

Operator
 of Peirce, 196
 of Sheffer, 196
Ordinal function
 of fuzzy order relation, 129
 of ordinary finite graph, 118
Ordinary relation nearest to a fuzzy relation, 59
Ordinary subset, 5
 of level α, 27, 65

P

Path
 in a fuzzy graph, 94
 length of a, 94
 strongest, 4
Peirce operator, 196
Perfect antisymmetry, 108
Perfectly ordered similitude classes, 162
Perfect order relation, 113
Planar network, 239
Polynomial forms, 201
Preorder
 antireflexive, 100
 fuzzy reducible, 105
 relation of fuzzy, 98
Probability, theory of, 250
Product
 algebraic, 34
  of two relations, 53
 boolean, 3
 of lattices, 310, 327
 of two sets, 297
Projection of a fuzzy relation, 47
Pseudo-complementation, 335

R

Reduced polynomial forms, 202
Reflexitivity, 79
Relation,
 dominance, 7
 fuzzy, 46
 fuzzy binary, 78
 fuzzy partial order, 112
 fuzzy preorder, 98
 fuzzy total order, 112
 nonstrict order, 113
 ordinary, nearest to a fuzzy relation, 59
 perfect order, 113
 strict order, 113
Reliability, probability of, 257
Ring, commutative, 313
Route
 of a fuzzy network, 336
 maximal simple, 237

S

Semipreorder, fuzzy, 100
Set
 of fuzzy subsets, 30, 34
 of mappings, 300
 reticulated, 304
Sheffer operator, 196
Similitude, relation of, 101
Strict order relation, 113
Strongest path, 94
Structure of a ring, 237
Subcategory, fuzzy, 348
Subrelation, maximal similitude, 387
Subset
 conditioned fuzzy, 71
 fuzzy, 4 – 6
 heterogeneous fuzzy, 71
 nearest ordinary, 22
 of level α, 27, 65
 ordinary, 5
Sum
 algebraic, 34
  of two relations, 54
 boolean, 3
 disjunctive, 12, 248, 336
  of two relations, 56
Sup-semi-lattice, 34
Support of a fuzzy relation, 98
Surjection, 338
Symmetry, 79

T

Taxonomy, 264
Transitive graph of distance, 145
Transitivity
 max–min, 79
 max–product, 97
 max–star, 97
Tree of decomposition, 150

U

Union, 3, 11, 336
  of two relations, 50
Univalent correspondence, 340

V

Variables, fuzzy, 192, 194, 337
Vector lattice, 309